INDEX to RECORD and TAPE REVIEWS

A Classical Music Buying Guide

1975

—

Antoinette O. Maleady

Chulainn Press

San Anselmo, California • 1976

ML 156
.9
.M 28
1975

ISBN 0-917600-01-0
Library of Congress Catalog Card No. 72-3355
Manufactured in the United States of America
Copyright © 1976 by Antoinette O. Maleady

CHULAINN PRESS
Post Office Box 770
San Anselmo, California 94960

SSb

To Sarah

CONTENTS

INTRODUCTION

This Index brings together in one volume a listing of all classical music recordings reviewed in 1975 in the major reviewing media of the United States, England and Canada.

The Index has four sections. Section I is a straight listing by composer. Section II, "Music in Collections," lists records or tapes with several composers on one disc or tape. Section II main entries are arranged alphabetically by name of the manufacturer, and then serially by manufacturer's number within each main entry. The work of any composer in each of these collections appears in Section I under the name of that composer, with a reference to the Section II entry. Section III lists, alphabetically by title, anonymous works, with a reference to their location in Section I or Section II. Section IV is a Performer Index to recordings listed in Sections I and II, with reference by citation number to its location within the section. Each citation gives, if available, the disc label and number; variant labels and numbers; tape cassette, cartridge or reel numbers; quadraphonic disc or tape numbers; reissues; location of the reviews and the reviewers evaluation of the recording. The main entry for each recording is in upper case letters. Tapes or discs reviewed in 1975 that were also reviewed in 1974, 1973, 1972 and/or 1971 have all the reviews brought together in the 1975 Index.

A key is provided for understanding the entries of the four sections. The entries are fictitious entries to show the various possibilities of form.

Section I

MACONCHY, Elizabeth
1219 Ariadne, soprano and orchestra. WALTON: Songs: Daphne; Old Sir
[entry   Faulk; A song for the Lord Mayor's table; Through gilded
  no.]   trellises. Heather Harper, s [Heather Harper, soprano]; Paul
              Hamburger, pno [Paul Hamburger, piano]; ECO [English Chamber
              Orchestra]; Raymond Leppard [conductor]. Columbia M 30443
              [disc number] (2) [number of discs in set]. Tape (c) MT 30443
              [cassette number] (ct) MA 30443 [cartridge number] (r) L 30443
              [reel number] (Q) MQ 30443 [quadraphonic disc number] Tape (ct)
              MAQ 30443 [quadraphonic tape cartridge number]. (also CBS
              72941) [recording also available on CBS label]
                    ++Gr 9-74 p1025 [evaluation excellent; review in Gramo-
                            phone, September 9, 1974, page 1025]
                     +Gr 11-75 p667 tape [evaluation good, review in Gramo-
                            phone, November 1975, p667, page 667 for tape]
                     /MJ 5-73 p73 [evaluation fair, review in Music Journal,
                            May, 1973 page 73]
                     -NR 4-72 p4 [evaluation poor, review in New Records,
                            April, 1972, page 4]

Section II

LONDON

OS 36578 (also Decca SXL 3315) [label number. Also available on Decca
      label]
2395 Baroque flute sonatas. BLAVET: Sonata, flute, no. 2, F major.
[entry   GAULTIER: Suite, G minor. HANDEL: Sonata, flute, op. 1, no.
  no.]   5, G major. LOEILLET: Sonata, flute, F major. VINCI: Sonata,
       flute, D major. André Previn, flt [André Previn, flute];
       Raymond Leppard, hpd [Raymond Leppard, harpsichord]; Claude
       Viala, vlc [Claude Viala, violoncello]; AMF [Academy of St.
       Martin-in-the-Fields]; Raymond Leppard [conductor]
              +-NYT 4-30-75 pD32 [evaluation mixed, review in New
                    York Times, April 30, 1975, page D32]

Section III

Kyrie trope: Orbis factor. cf TALLIS: Missa salve intemerata. [Anony-
    mous work, appears in Section I under entry for Tallis]
Deo gracias Anglia, Agincourt carol. cf BASF 25 22286-1. [Anonymous
    work, appears in Section II under entry BASF 25 22286-1]

Section IV

Baker, Janet, mezzo-soprano (contralto) 231, 494, 939, 1376, 1776,
    2513 [performer, listed in reviews as mezzo-soprano or con-
    tralto. Citation numbers in Section I and II, where artist
    has performed]

ABBREVIATIONS

Periodicals Indexed

AR	American Recorder
ARG	American Record Guide
ARSC	Association for Recorded Sound Collections Journal
Audio	Audio
CJ	Choral Journal
CL	Clavier
Gr	Gramophone
GTR	Guitar Review
Ha	Harpers
HF	High Fidelity
HFN	Hi-Fi News & Record Review
HPD	Harpsichord
LJ	Library Journal/School Library Journal Previews
MJ	Music Journal
MQ	Musical Quarterly
MT	Musical Times
NR	New Records
NYT	New York Times
OC	Opera Canada
ON	Opera News
Op	Opera, London
PNM	Perspectives of New Music
RR	Records and Recording
SFC	San Francisco Examiner and Chronicle (This World)
SR	Saturday Review/World
St	Stereo Review
ST	Strad
STL	Sunday Times, London
Te	Tempo

Performers

Orchestral

AMF	Academy of St. Martin-in-the-Fields
BBO	Berlin Bach Orchestra
BeSO	Berlin Symphony Orchestra
BPhO	Berlin Philharmonic Orchestra
BPO	Boston "Pops" Orchestra
Brno PO	Brno State Philharmonic Orchestra
BRSO	Berlin Radio Symphony Orchestra
BSO	Boston Symphony Orchestra
CnSO	Cincinnati Symphony Orchestra
CO	Cleveland Orchestra
COA	Concertgebouw Orchestra, Amsterdam

ix

CPhO	Czech Philharmonic Orchestra
CSO	Chicago Symphony Orchestra
DBS	Deutsche Bach Solisten
ECO	English Chamber Orchestra
FK	Frankfurt Kantorei
HCO	Hungarian Chamber Orchestra
HRT	Hungarian Radio and Television
HSO	Hungarian State Symphony Orchestra
LAPO	Los Angeles Philharmonic Orchestra
LOL	Little Orchestra, London
LPO	London Philharmonic Orchestra
LSO	London Symphony Orchestra
MB	Munich Bach Orchestra
MPAC	Munich Pro Arte Chamber Orchestra
MPO	Moscow Philharmonic Orchestra
MRSO	Moscow Radio Symphony Orchestra
NPhO	New Philharmonia Orchestra
NSL	New Symphony, London
NWE	Netherlands Wind Ensemble
NYP	New York Philharmonic Orchestra
ORTF	O.R.T.F. Philharmonic Orchestra
OSCCP	Orchestre de la Société des Concerts du Conservatoire de Paris
OSR	l'Orchestre de la Suisse Romande
PCO	Prague Chamber Orchestra
PH	Philharmonia Hungarica
PhO	Philharmonia Orchestra, London
PO	Philadelphia Orchestra
PSO	Prague Symphony Orchestra
ROHO	Royal Opera House Orchestra, Covent Garden
RPO	Royal Philharmonic Orchestra
SBC	Stuttgart Bach Collegium
SCO	Stuttgart Chamber Orchestra
SDR	Stuttgart S.D.R. Symphony Orchestra
SPO	Stuttgart Philharmonic Orchestra
SSO	Sydney Symphony Orchestra
VCM	Vienna Concentus Musicus
VPM	Vienna Pro Musica
VPO	Vienna Philharmonic Orchestra
VSO	Vienna Symphony Orchestra
VSOO	Vienna State Opera Orchestra

Instrumental

acc	accordion	mand	mandolin
bal	balalaika	mar	marimba
bs	bass	ob	oboe
bsn	bassoon	ond	ondes martenot
c	celesta	org	organ
cimb	cimbalon	perc	percussion
cld	clavichord	pic	piccolo
clt	clarinet	pno	piano
cor	cornet	rec	recorder
dr	drums	sax	saxophone
Eh	English horn	sit	sitar
flt	flute	tpt	trumpet

Fr hn	French horn		trom	trombone
gtr	guitar		vib	vibraphone
harm	harmonica		v	viol
hn	horn		vla	viola
hp	harp		vlc	violoncello
hpd	harpsichord		vln	violin
lt	lute		z	zither

Vocal

bar	baritone		con	contralto
bs	bass		c-t	countertenor
ms	mezzo-soprano		t	tenor
s	soprano			

Qualitative Evaluation of Recordings

++	excellent or very good		/	fair
+	good		–	poor
+–	mixed		*	no evaluation

COMPOSERS

ACHRON, Joseph
    Hebrew dance. cf RCA ARM 4-0942/7.
    Hebrew dance, op. 35, no. 1. cf Discopaedia MB 1010.
    Hebrew lullaby, op. 33. cf Da Camera Magna SM 93399.
    Hebrew lullaby, op. 33. cf RCA ARM 4-0942/7.
    Hebrew melody, op. 33. cf Discopaedia MB 1010.
    Hebrew melody, op. 33 (2). cf RCA ARM 4-0942/7.
    Scher. cf Da Camera Magna SM 93399.
    Stimmung. cf RCA ARM 4-0942/7.
ADAM, Adolphe-Charles
1   Giselle. Monte Carlo Opera Orchestra; Richard Bonynge. London
      CSA 2226 (2). Tape (c) K 42226 (r) 2226.
          +SFC 6-17-73 p31   ++SFC 3-9-75 p27 tape
2   Giselle (arr. Busser). Si j'étais roi: Overture. OSCCP: Albert
      Wolff, Jean Martinon. Decca SPA 384. Tape (c) KCSP 15384.
      (Reissues from RCA SB 2018, SXL 2008)
          +Gr 4-75 p1871   +RR 5-75 p26
          +-HFN 7-75 p90 tape
    Giselle: Peasant pas de deux; Grand pas de deux; Finale. cf
      Decca SDDJ 393/5.
    Giselle: Peasant pas de deux, Act 1; Grand pas de deux and finale,
      Act 2. cf Decca DPA 515/6.
    Si j'étais roi: Overture. cf Giselle.
ADAM, Claus
3   Concerto, violoncello and orchestra. BARBER: Die Natali, op. 37.
      Stephen Kates, vlc; Louisville Orchestra; Jorge Mester. Louis-
      ville LS 745.
          +HF 12-75 p85   +SFC 8-17-75 p22
ADAMS
    The holy city. cf RCA ARL 1-0561.
ADDISON, John
4   Divertimento, op. 9. DODGSON: Sonata, brass. BENNETT: Commedia
      IV. GARDNER: Theme and variations, op. 7. Philip Jones Brass
      Ensemble. Argo ZRG 813.
          ++Gr 12-75 p1066   ++RR 12-75 p35
          ++HFN 12-75 p156
ADSON, John
    Ayres for cornetts and sagbutts. cf Klavier KS 536.
AGRELL, Johann
5   Sonata, harpsichord, no. 1, B flat major. Sonata, harpsichord,
      no. 5, D major. ROMAN: Sonata, harpsichord, C major. Sonata
      IX. Eva Nordenfelt, hpd. Orion ORS 74157.
          +Audio 12-75 p104   ++NR 11-74 p13
    Sonata, harpsichord, no. 5, D major. cf Sonata, harpsichord, no.
      1, B flat major.

AHRENDT, Karl
  Montage. cf Century/Advent OU 97 734.
AICH, Arnt von
6   Liederbuch: Songs and settings. Pöhlert Renaissance Ensemble;
       Werner Pöhlert. Musical Heritage Society MHS 1914.
         +HF 4-75 p94
ALAIN, Jehan
  Fantasie, no. 2. cf Orion ORS 74161.
  Variations on a theme by Jannequin, op. 78. cf BIS LP 7.
ALBENIZ, Isaac
  Bajo la palmera, op. 212. cf RCA ARL 1-0456.
  Cantos de España, op. 232, no. 5: Seguidillas. cf Decca SPA 372.
  España, op. 165, no. 2: Tango; no. 3: Malagueña. cf Supraphon
       50919.
  España, op. 165, no. 2: Tango. cf Argo ZRG 805.
  España, op. 165, no. 2: Tango (arr. Elman). cf Decca SPA 405.
  España, op. 165, no. 2: Tango. cf L'Oiseau-Lyre DSLO 7.
  España, op. 165, no. 5: Capriccio catalano. cf Pye GSGC 14154.
  Iberia: Evocación. cf RCA ARL 1-0456.
  Marcha militar. cf International Piano Library IPL 5005/6.
  Piezas características, no. 12: Torre bermeja. cf Nonesuch H 71233.
  Piezas características, no. 12: Torre bermeja. cf London STS
       15224.
  Piezas características, no. 12: Torre bermeja. cf RCA ARL 1-0485.
  Recuerdos de viaje, op. 71, no. 2: Leyenda. cf Nonesuch H 71233.
  Recuerdos de viaje, op. 71, no. 6: Rumores de la caleta. cf
       London STS 15224.
7   Suite española, no. 1: Granada; no. 3: Sevillanas. GRANADOS: Tona-
       dillas al estilo antiguo: La maja de Goya. Danza española,
       op. 37, no. 5: Andaluza. FALLA: El sombrero de tres picos:
       Danza del molinero. Homenaje a Debussy. TURINA: Fandanguillo,
       op. 36. Homenaje a Tárrega: Garrotin; Soleares. Ráfaga, op.
       53. Sevillana, op. 29. Sonata, guitar, op. 61: Andante.
       Konrad Ragossnig, gtr. Turnabout TV 34494S.
         +Gr 2-75 p1523   +-RR 2-75 p46
  Suites españolas, no. 1: Granada; no. 5: Asturias. cf London
       STS 15224.
  Suite española, no. 3: Sevillanas. cf Canon CNN 4983.
  Suite española, no. 3: Sevillanas. cf CBS 72950.
  Suite española, no. 3: Sevillanas (arr. Heifetz). cf Decca SPA
       405.
  Suite española, no. 3: Sevillanas. cf RCA ARM 4-0942/7.
  Suite española, no. 5: Asturias. cf Philips 6833 159.
d'ALBERT, Eugene
8   Concerto, piano, no. 2, op. 12, E major. REINECKE: Concerto,
       piano, no. 1, op. 72, F sharp minor. Michael Ponti, pno;
       Luxembourg Radio Orchestra; Pierre Cao. Candide CE 31078.
       (also Vox STGBY 681)
         +-Gr 6-75 p36   +NR 8-74 p4
         +HF 8-74 p83   /RR 5-75 p29
         /HFN 5-75 p123   +-St 10-74 p123
ALBERT, Heinrich
  Fantasia. cf Amberlee ACL 501X.
ALBINONI, Tommaso
9   Adagio, organ and string, G minor. MARCELLO: Concerto, 2 guitars
       (arr. Lagoya). VIVALDI: Concerto, 2 guitars, P 134, C major

(arr. Lagoya. Concerto, organ, A minor (trans. Bach).
Concerto, organ, D minor. Pierre Cochereau, org; Ida Presti,
Alexander Lagoya, gtr; Munich Pro Arte Chamber Orchestra; Kurt
Redel. Philips 6581 010.
+-Gr 2-75 p1504   +RR 2-75 p29
Adagio, organ and strings, G minor. cf HMV ASD 3017.
Adagio, organ and strings, G minor. cf HMV (Q) ASD 3131.
10   Concerto, oboe, op. 7, no. 3, B flat major. CIMAROSA: Concerto,
     oboe, C minor (arr. Benjamin). MARCELLO: Concerto, oboe, C
     minor. Evelyn Rothwell, ob; Pro Arte Orchestra; John Barbirolli.
     Pye GSGC 15011. Tape (c) ZCCCB 15011. (Reissue from GSGC 14023)
          *Gr 8-75 p335   +HFN 8-75 p89
          +Gr 12-75 p1121 tape   +RR 7-75 p21
11   Concerto, oboe, op. 7, no. 6, D major. HANDEL: Concerti, oboe, nos.
     8 and 9. MOZART: Concerto, oboe, K 314, C major. Quartet,
     oboe, K 370, F major. Andre Lardrot, ob; Boskovsky Quartet,
     VSOO; Felix Prohaska. Bach (Vanguard) Guild HM 40S.
          +Gr 4-75 p1980   +-RR 4-75 p28
          ++MJ 11-74 p48   +SFC 7-28-74 p34
     Concerto, oboe, op. 7, no. 6, D major: 1st movement. cf London
     SPC 21100.
12   Concerto, oboe, op. 9, no. 2, D minor. BACH, C.P.E.: Sonata, oboe
     and harpsichord, G minor. FIALA: Concerto, cor anglais, E flat
     major. HANDEL: Sonata, oboe, op. 1, no. 8, C minor. Edith
     Picht-Axenfeld, Leslie Pearson, hpd; Marçal Cervera, Rama
     Jucker, vlc; Ursula Holliger, hp; Heinz Holliger, cor anglais;
     ECO; Raymond Leppard. Philips 6833 097.
          +-Gr 4-75 p1825   ++RR 4-75 p20
          ++HFN 6-75 p102
     Concerto, oboe, op. 9, no. 2, D minor: Adagio, organ and strings.
     cf Pye GSGC 14153.
     Concerto, organ, B flat major. cf Saga 5374.
     Concerto, trumpet, B flat major. cf Erato STU 70871.
     Concerto, trumpet, B flat major. cf RCA CRL 2-7002.
13   Concerto, trumpet, C major. HERTEL: Concerto, trumpet, 2 oboes
     and 2 bassoons. TELEMANN: Concerto, trumpet, D major. Gerard
     Schwarz, tpt; Ronald Roseman, Susan Weiner, Virginia Brewer, ob;
     Donald MacCourt, William Scribner, bsn; Edward Brewer, hpd.
     Desto DC 6438. (also Peerless)
          +HFN 12-75 p148   +St 11-75 p140
14   Concerti, trumpet, F major, D minor. HANDEL: Sonata, trumpet, A
     major. MARTINI: Sonata al postcommunio. Toccata. VALENTINO:
     Sonata, trumpet, D minor. Claude Molenat, tpt; Georges Rabol,
     org; Willy Lockwood, double bs; Armand Cavallaro, perc.
     Vanguard SRV 319.
          ++NR 10-75 p5
     Sonata, trumpet and organ, D major. cf RCA CRL 2-7001.
     Sonata, trumpet and organ, F major. cf RCA CRL 2-7001.
ALBRECHTSBERGER, Johann
     Fugue on Ite Missa est, Alleluia. cf Hungaroton SLPX 11601/2 (2).
ALBRIGHT, William
     Take that. cf Opus One 22.
ALCOCK, Walter
15   Introduction and passacaglia. HOWELLS: Psalm prelude, Set 2.
          PEETERS: Chorale preludes, op. 68: Ach blieb mit Deiner Gnade;
     O Gott, du frommer Gott; Wachet auf. Suite modale, op. 43.

Mervyn J. Byers, org. President Gemini GMX 5022.
    +Gr 4-75 p1843   +RR 1-75 p51
  Voluntary IV, C minor. cf Wealden WS 160.
ALEXANDROV, Alexander
  Patriotic war; Soviet army song. cf HMV ASD 3116.
ALEXANDROV, Boris
  The guards greeted the spring in Berlin. cf HMV ASD 3116.
ALFONSO X, El Sabio (attrib.)
16 Las cantigas de Santa Maria: Prologue and cantigas 1, 7, 8, 10, 56,
    117, 158, 166, 189, 200, 221, 231, 318, 327, 340, 377. ANON.
    (13th and 14th C.): Benedicamus Verbum Patris; Laudemus virginem;
    Splendens sceptrigera; Stella spendens in monte. The Waverly
    Consort; Nicholas Kepros, narrator; Jan DeGaetani, ms; Constan-
    tine Cassolas, t; Kay Jaffee, rec, rauschpfeife, psaltery,
    organetto; Sally Longemann, shawm, rec, nun's fiddle; Judith
    Davidoff, medieval fiddles; Michael Jaffee, Moorish gtr,
    psalteries; Michael Jaffee. Vanguard VSD 71175.
        ++AR 2-75 p20   ++St 5-73 p85
        ++LJ 9-73 p49
  Rosa das rosas. cf CRD CRD 1019.
  Rosa das rosas. cf DG 2530 504.
  Santa Maria. cf DG 2530 504.
ALFORD, Harry L.
  Purple carnival. cf Columbia 33513.
ALFVEN, Hugo
  Dalecarlian rhapsody, no. 3, op. 48 (Swedish rhapsody). cf
    Symphony, no. 3, op. 23, E major.
17 Midsommarvaka, no. 1, op. 19 (Midsummer vigil). GRIEG: Peer Gynt,
    op. 46: Suite, no. 1. SIBELIUS: Finlandia, op. 26. En saga,
    op. 9. Legends, op. 22: Swan of Tuonela. Kuolema, op. 44:
    Valse triste. PO; Eugene Ormandy. Columbia MS 7674. Tape
    (c) 16-11-0184 (ct) 18-11-0184.
        +St 6-75 p49
18 The mountain king. The prodigal son. Royal Swedish Orchestra;
    Hugo Alfvén. Swedish Discofil SLT 33182.
        +St 6-75 p48
  The prodigal son. cf The mountain king.
  Skogen sover. cf RCA SER 5704/6.
  Songs: I long for you; Take my heart. cf RCA SER 5719.
19 Symphony, no. 3, op. 23, E major. Dalecarlian rhapsody, no. 3,
    op. 48 (Swedish rhapsody). Stockholm Concert Association Orch-
    estra; Stockholm Radio Orchestra; Hugo Alfvén. Odeon EO
    53-34620.
        +St 6-75 p49
20 Symphony, no. 4, op. 39, C minor. Sven Erik Vikström, t; Gunilla
    af Malmborg, s; Stockholm Philharmonic Orchestra; Nils
    Grevillius. Swedish Society Discofil SLT 33186.
        +-RR 2-75 p29
ALISON, Richard
  Dolorosa pavan. cf L'Oiseau-Lyre 12BB 203/6.
  ALL AMERICAN MUSIC CONCERT. cf Century/Advent OU 97 734.
ALLITSEN
  The Lord is my light. cf RCA ARL 1-0562.
d'ALMEIDA, Francisco Antonio
21 La spinalba (The crazy old man). Lydia Marimpietri, Romana Righetti,
    s; Laura Zannini, ms; Rena Garaziotti, con; Ugo Benelli, Fernando

Serafim, t; Otello Borgonovo, Teodoro Rovetto, bar; Gulbenkian
Foundation Chamber Orchestra; Gianfranco Rivoli. Philips
839 710/12 (3).
+OC 4-5-75 p48

ALONSO (16th century)
La tricotea. cf Telefunken SAWT 9620/1.

ALTHEN
Thou, blessed country. cf RCA SER 5719.

ALWYN, William
22 Fantasy waltzes. Sonata alla toccata. Sheila Randall, pno.
Lyrita RCS 16. (Reissue, 1960)
+HFN 12-75 p156   ++RR 7-75 p44
Sinfonietta, strings. cf Symphony, no. 2.
Sonata alla toccata. cf Fantasy waltzes.
23 Symphony, no. 2. Sinfonietta, strings. LPO; William Alwyn.
Lyrita SRCS 85.
+Gr 11-75 p801   ++RR 11-75 p36
+HFN 11-75 p147
24 Symphonies, nos. 4 and 5. LPO; William Alwyn. Lyrita SRCS 76.
+Gr 5-75 p1961   ++RR 5-75 p26
++HFN 5-75 p23

d'AMBROSIO, Alfredo
Canzonetta, op. 6. cf Discopaedia MB 1007.
Serenade. cf RCA ARM 4-0942/7.
AMERICA SINGS, THE GREAT SENTIMENTAL AGE--STEPHEN FOSTER TO CHARLES
IVES. cf Vox SVBX 5304.
AMERICANA FOR SOLO WINDS AND STRING ORCHESTRA. cf Era 1001.

AMIRKANIAN, Carles
Heavy aspirations. cf 1750 Arch 1752.
Just. cf 1750 Arch 1752.

AMMERBACH, Elias
Wer das Töcterlein haben will. cf Philips 6775 006.

ANCHIETTA, Juan de
Con amores, la mi madre. cf DG 2530 598.

ANDERSON, Beth
Torero piece. cf 1750 Arch 1752.

ANDERSON, Thomas Jefferson, Jr.
25 Squares. HAKIM: Visions of Ishwara. WILSON: Akwan. Richard
Bunger, pno; Baltimore Symphony Orchestra; Paul Freeman.
Columbia M 33434.
+Audio 12-75 p106   ++NR 9-75 p5
+HF 10-75 p65
Variations on a theme by M. B. Tolson. cf Nonesuch H 71302/3.

d'ANGLEBERT, Jean Henri
Chaconne, D major. cf Harmonia Mundi HMU 334.
Le tombeau de M. de Chambonnieres. cf Harmonia Mundi HMU 334.
ANNUS MIRABILIS, 1873. cf Rubini RS 300.
THE ANTIPHONAL ORGANS OF THE CATHEDRAL OF FREIBURG. cf Columbia
M 33514.

APPLETON, Jon
Apolliana. cf Works, selections (Folkways FTS 33437).
C.C.C.P. cf Works, selections (Folkways FTS 33437).
Chef d'oeuvre. cf Works, selections (Folkways FTS 33437).
Hommage to Orpheus. cf Works, selections (Folkways FTS 33437).
Ofa atu Tonga. cf Works, selections (Folkways FTS 33437).
Nevsehir. cf Works, selections (Folkways FTS 33437).

Sones de San Blas. cf Works, selections (Folkways FTS 33437).
Sounds and electronic tape. cf Works, selections (Folkways FTS
    33437).
Times square times ten. cf Works, selections (Folkways FTS 33437).
26 Works, selections: Apolliana. C.C.C.P. Chef d'oeuvre. Hommage
    to Orpheus. Ofa atu Tonga. Nevsehir. Sones de San Blas.
    Times square times ten. Sounds and electronic tape. Folkways
    FTS 33437.
        -NR 10-75 p11
ARBAN, Joseph Jean-Baptiste
    Carnival in Venice. cf London SPC 21100.
    Fantasie and variations on "The carnival of Venice". cf Nonesuch
    H 71298.
ARCADELT, Jacques
    Il ciel che rado. cf Hungaroton SLPX 11549.
    Si grand'è la pieta. cf Hungaroton SLPX 11549.
ARDITI, Luigi
    Parla. cf Pearl GEM 121/2.
ARENSKY, Anton
27 Concerto, piano, no. 2. PADEREWSKI: Fantaisie polonaise, op. 19.
    Felicja Blumenthal, piano; Innsbruck Symphony Orchestra; Robert
    Wagner. Everest SDBR 3376.
        +NR 5-75 p7
    Concerto, violin, A minor: Tempo di valse. cf RCA ARM 4-0942/7.
28 The fountain of Bakhchisarai, op. 46. TCHAIKOVSKY: Dmitri the
    imposter and Vassili Shuisky: Introduction and mazurka. Undine,
    excerpts. Irina Arkhipova, ms; Tamara Milashkina, s; Evgenii
    Rakov, t; Moscow Radio Symphony Orchestra and Chorus, Bolshoi
    Theatre Orchestra; Aleksandr Melik-Pashayev, Evgenii Akulov.
    Westminster WGS 8300.
        +-HF 10-75 p85   +-St 9-75 p106
    ARIAS FROM GREAT ITALIAN OPERAS. cf RCA ARL 1-0702.
    ARIAS FROM ITALIAN AND FRENCH OPERAS. cf BASF BC 21723.
ARISTAKESIAN, Emin
29 Sinfonietta, piano, xylophone and strings. GUBAJDULINA: Concor-
    danza, chamber ensemble. LEDENEV: Pieces, harp and string
    quartet, op. 16 (6). SHOSTAKOVITCH: Romances on words of
    Alexander Block, op. 127 (7). Quartet, strings, no. 2, op. 68.
    Brigita Sulcová, s; Radoslav Kvapil, pno; Petr Sprunk, xylophone;
    Libuse Váchalová, hp; Dvořák Piano Trio, Suk Quartet, Skvor
    Quartet, Musica Viva Pragensis, Musici de Prague; Zbyněk Vostřák,
    František Belfín. Panton 110 342/43.
        +RR 10-75 p89
ARKANGELSKII, Alexandr
    Blazhen razumevayei (Blessed are they that understand). cf Monitor
    MFS 757.
30 Blajen razoumnevai; Otche nach. CHRISTOV: Velikaia ekteniaa; Vo
    tzarstvii tvoem; Herouvimska; Tebe poem; Dostoino est. GRETCH-
    ANINOV: Verouiou; Slava...Edinorodnii. CHESNOKOV: Spassi, Boje,
    lioudi tvoi. Svetoslav Obretenov Bulgarian Choir; Georgui
    Robey. Harmonia Mundi HMB 101.
        ++Gr 11-75 p893
ARNE, Thomas
    Artaxerxes: Arias. cf London OS 26277.
    As you like it: Under the greenwood tree. cf BBC REB 191.
    As you like it: Under the greenwood tree. cf HMV SLS 5022.

31   Overtures, nos. 1-8. Academy of Ancient Music; Christopher Hogwood,
        cond and hpd. L'Oiseau-Lyre Florilegium DSLO 503.
              +Gr 11-74 p883   +RR 11-74 p17
             ++HF 8-75 p79   +SFC 11-30-75 p34
              +NR 6-75 p6
32   Sonatas, harpsichord, nos. 1-8. Christopher Hogwood, hpd.
        L'Oiseau-Lyre Florilegium DSLO 502.
              +Gr 11-74 p922   +RR 11-74 p17
              +HF 9-75 p81
     Sonata, harpsichord, no. 3, G major. cf Angel SB 3816.
     Sonata, harpsichord, no. 3, G major: Gigue. cf RCA LRL 1-5087.
     Songs: O ravishing delight; Where the bee sucks. cf HMV RLS 714.
ARNIM, Bettina von
     O schaudre nicht. cf DG Archive 2533 149.
ARNOLD
     Colonel Bogey march. cf BASF 292 2116-4 (2).
ARNOLD, Malcolm
33   Divertimento, flute, oboe and clarinet, op. 37. HINDEMITH:
        Septet, wind instruments. FRANCAIX: Quartet, flute, oboe,
        clarinet and bassoon. SPISAK: Sonatina, oboe, clarinet and
        bassoon. Alois Rybin, bs clt; Václav Junek, tpt; Czech Philhar-
        monic Wind Ensemble. Supraphon 50582.
              +Gr 11-74 p921   ++RR 1-75 p43
     A grand grand overture. cf HMV SLA 870.
     Quintet, brass. cf Crystal S 201.
     Shanties, wind quintet (3). cf Decca 396.
     United Nations, excerpts. cf HMV SLA 870.
ARRIAGA Y BALZOLA, Juan
34   Los esclavos felices: Overture. Symphony, D minor. ECO; Jesus
        Lopez Cobos. Pye NEL 2016.
              +Gr 6-75 p31   +RR 5-75 p26
35   Quartets, strings, nos. 1-3. Chilingirian Quartet. CRD CRD 1012/3
        (2).
              +Gr 6-75 p56   +RR 6-75 p57
              +HFN 9-75 p91E
     Symphony, D minor. cf Los esclavos felices: Overture.
     THE ART OF MARILYN HORNE. cf London OS 26277.
     THE ART OF THE RECORDER. cf HMV SLS 5022 (2).
     THE ART SONG IN AMERICA, vol. II. cf Duke University Press
        DWR 7306.
ARUTANIAN (Arutyanian), Aleksander
36   Concerto, trumpet. MOLTER: Concertos, trumpet, nos. 1 and 2.
        TELEMANN: Concerto, trumpet. David Hickman, tpt; Colorado
        Philharmonic Orchestra; Walter Charles. Clarino SLP 1005.
              +NR 10-74 p6   ++St 3-75 p110
ASCHER
     Alice, where art thou. cf Discopaedia MB 1006.
ASENCIO
     Dipso. cf RCA ARL 1-0864.
ASHFORTH, Alden
37   Byzantia (Two journeys after Yeats). Dennis Heath, t; James
        Bossert, org; Alden Ashforth, synthesizers. Orion ORS 74164.
             ++HF 10-75 p70   ++NR 10-75 p11
ASHLEY, Robert
     In Sara Mencken, Christ and Beethoven there were men and women.
        cf 1750 Arch 1752.

ATTAINGNANT, Pierre
    Basse danse La Brosse: Tripla, Tourdion; Basse danse La Gatta;
        Basse danse La Magdalena. cf DG Archive 2533 111.
    Content desir Basse danse. cf L'Oiseau-Lyre 12BB 203/6.
    Tourdion. cf L'Oiseau-Lyre SOL 329.
ATTERBERG, Kurt
38   Symphony, no. 2, op. 6, F major. Swedish Radio Symphony Orchestra;
        Stig Westerberg. Swedish Society Discofil SLT 33179.
            +HFN 11-75 p148   +-RR 2-75 p29
ATTEY, John
    Sweet was the song the Virgin sang. cf Turnabout TV 34569.
AUBER, Daniel
39   Le cheval de bronze: Overture. Fra Diavolo: Overture. BERLIOZ:
        Benevenuto Cellini, op. 23. BOIELDIEU: Calif of Bagdad: Over-
        ture. La dame blanche: Overture. HEROLD: Zampa: Overture.
        Paris Opera Orchestra; Paris Concert Orchestra; Pierre-Michel
        Le Conte. Audio Fidelity FCS 50060.
            /NR 10-75 p2
    Fra Diavolo: Agnese la Zitella. cf Rubini RS 300.
    Fra Diavolo: Overture. cf Le cheval de bronze: Overture.
40   Marco spada. LSO; Richard Bonynge. Decca SXL 6707. (also London
        6923)
            +Gr 7-75 p173   +RR 7-75 p21
            +HFN 7-75 p71
    AUGMENTED HISTORY OF THE VIOLIN ON RECORDS, 1920-1950. cf
        Thomas L. Clear TC 2580.
AULIN, Tor
    Aquarelles: Humoresque. cf Discopaedia MB 1008.
    AUSTRALIAN NATIONAL ANTHEM: Advance Australia fair. cf Decca
        SXL 6719.
AZZAIOLO, Filippo
    Quando le sera; Sentomi la formicula. cf L'Oiseau-Lyre 12BB 203/6.
BABBITT, Milton
    All set. cf Nonesuch H 71302/3.
41   Philomel, soprano, recorded soprano, synthesized sound. LERDAHL:
        Wake. Bethany Beardslee, s; BSO Chamber Players; David Epstein.
        DGG/Acoustic Research 0654 083.
            +HF 2-71 p97   +SFC 4-18-71 p31
            +MJ 2-75 p29   +SR 5-71 p94
            +MQ 7-72 p501   +St 5-71 p95
            *NYT 1-3-71 pD22
BABELL, William
    Concerto, 2 alto recorders, harpsichord and strings, no. 6, F major:
        Andante. cf Strobe SRCM 123.
BACARISSE, Salvador
    Concertino, guitar and orchestra, op. 72, A minor. cf HALFFTER:
        Concerto, guitar.
BACEWICZ, Grazyna
    Sonata, piano, no. 2. cf Avant AV 1012.
BACH, Carl Philipp Emanuel
42   Concerto, flute, W 22, D minor. Concerto, harpsichord, fortepiano
        and orchestra, W 47, E flat major. Ernst Groschel, pianoforte;
        Rudolph Zartner, hpd; Karl Leder, flt; Munich Pro Arte Chamber
        Orchestra; Kurt Redel. Pye GSGC 15007.
            +Gr 8-75 p315   +- RR 8-75 p28
            +-HFN 9-75 p92

43   Concerti, flute, W 166, A minor; W 169, G major. Jean-Pierre
        Rampal, flt; L'Oiseau-Lyre Orchestral Ensemble; Louis de Froment.
        L'Oiseau-Lyre OLS 183. (Reissue from OL 50121)
            +-Gr 7-75 p173   ++NR 10-75 p5
44   Concerti, harpsichord, Wq 14, Wq 43, E major and G major. BACH, J. C.
        (arr. Mozart): Concerto, harpsichord, no. 1, K 107, D major.
        Trevor Pinnock, hpd; English Concert Orchestra; Trevor Pinnock.
        CRD CRD 1011. Tape (c) CRD 4011.
            +Gr 6-75 p31   +HFN 9-75 p92
            +Gr 10-75 p721 tape   ++RR 6-75 p29
            +HFN 12-75 p173   +RR 11-75 p91 tape
45   Concerto, harpsichord, strings and continuo, Wq 23, D minor. Con-
        certo, oboe, strings and continuo, Wq 165, E flat major.
        BACH, J. C.: Sinfonia concertante, violin, violoncello and
        orchestra, A major. Sinfonia, double orchestra, op. 18, no. 1,
        E flat major. Gustav Leonhardt, hpd; Helmut Hucke, ob; Franz-
        josefMaier, vln; Angelica May, vlc; Collegium Aureum; Reinhard
        Peters. BASF BHM 23 29047/6. Tape (c) KBAC 2-7026.
            +Gr 5-73 p2041   ++RR 4-75 p76 tape
            +RR 6-73 p45
      Concerto, harpsichord, strings and continuo, Wq 23, D minor: 1st
        movement. cf BASF BAB 9005.
      Concerto, harpsichord Wq 46, F major: 1st movement. cf BASF BAB
        9005.
      Concerto, harpsichord,fortepiano and orchestra, W 47, E flat major.
        cf Concerto, flute, W 22, D minor.
46   Concerto, oboe, W 164, B flat major. Concerto, oboe, W 165,
        E flat major. BACH, J. S.: Cantata, no. 12, Weinen, Klagen,
        Sorgen, Zagen: Sinfonia. Cantata, no. 21, Ich hatte viel
        Bekümmernis: Sinfonia. Heinz Holliger, ob; ECO; Raymond
        Leppard. Philips 6500 830.
            ++NR 12-75 p7   ++St 12-75 p116
      Concerto, oboe, strings and continuo, Wq 165, E flat major. cf
        Concerto, harpsichord, strings and continuo, Wq 23, D minor.
      Concerto, oboe, W 165, E flat major. cf Concerto, oboe, W 164,
        B flat major.
47   Fantasia and fugue, C minor. Fugue, D minor. Prelude, D minor.
        Sonatas, organ (6). Xavier Darasse, org. Arion ARN 236006 (2).
            ++RR 11-75 p68
      Fugue, D minor. cf Fantasia and fugue, C minor.
      Prelude, D minor. cf Fantasia and fugue, C minor
      Sonata, flute and harpsichord, G major. cf Armstrong 721-4.
      Sonatas, flute and harsichord, Wq 134 and 125, G major and B flat
        major. cf Pandora PAN 103.
48   Sonata, harpsichord, no. 1, A minor. Sonata, harpsichord, no. 2,
        A flat major. DUPHLY: Pièces de clavecin (8). John Gibbons,
        hpd. Cambridge CRS 2530.
            ++NR 4-75 p11
      Sonata, harpsichord, no. 2, A flat major. cf Sonata, harpsichord,
        no. 1, A minor.
      Sonata, oboe and harpsichord, G minor. cf ALBINONI: Concerto,
        oboe, op. 9, no. 2, D minor.
      Sonatas, organ (6). cf Fantasia and fugue, C minor.
BACH, Johann Christian
49   Concerto, harpsichord, A major. HAYDN: Concerto, harpsichord,

op. 21, D major. George Malcolm, hpd; AMF; Neville Marriner.
  London STS 15172.
          ++NR 6-75 p6
  Concerto, harpsichord, no. 1, K 107, D major. cf BACH, C. P. E.:
    Concertos, harpsichord, Wq 14, Wq 43, E major and G major.
50 Concerti, keyboard, op. 7, nos. 1-3; op. 13, no. 4. Ingrid Haebler,
      fortepiano; Vienna Capella Academica; Eduard Melkus. Philips
      6500 846.
          ++Gr 7-75 p173   ++RR 6-75 p29
          ++HFN 10-75 p135
51 Concerti, keyboard, op. 7, nos. 4, 5; op. 13, nos. 2, 4, 5.
      Ingrid Haebler, fortepiano; Vienna Capella Academica; Eduard
      Melkus. Philips 6500 847.
          +Gr 4-75 p1801   ++RR 5-75 p27
          ++HFN 5-75 p123   ++STL 6-8-75 p36
  Concerto, keyboard, op. 13, no. 4. cf Concerti, keyboard, op. 7,
    nos. 1-3.
  Marches for First and Second Battalions of Guard Regiment in
    Hanover. cf Classics for Pleasure CFP 40230.
  Quartet, 2 flutes, viola and violoncello, D major. cf Quartet, 2
    flutes, violin and violoncello, C major.
52 Quartet, 2 flutes, violin and violoncello, C major. Quartet, 2
      flutes, viola and violoncello, D major. BACH, W. F.: Duet,
      flute, F major. BACH, W. F. E.: Trio, 2 flutes and viola,
      G major. Jean-Pierre Rampal, Eugenia Zukerman, flt; Pinchas
      Zukerman, vln, vla; Charles Tunnell, vlc. Columbia M 33310.
          ++HF 11-75 p96   ++NR 8-75 p8
  Quintet, F major. cf Quintets, op. 11, nos. 4 and 6, D major,
    E flat major.
53 Quintets, op. 11, nos. 4 and 6, D major, E flat major. Quintet,
      F major. Sonatas, op. 2, D major and G major. Collegium Pro
      Arte. L'Oiseau-Lyre OLS 182. (Reissue from OL 50046, 50015)
          +Gr 7-75 p199   +NR 12-75 p8
          +-HFN 6-75 p83   +RR 7-75 p38
  Sinfonia, double orchestra, op. 18, no. 1, E flat major. cf
    BACH, C. P. E.: Concerto, harpsichord, strings and continuo,
    Wq 23, D minor.
  Sinfonia concertante, violin, violoncello and orchestra, A major.
    cf BACH, C. P. E.: Concerto, harpsichord, strings and continuo,
    Wq 23, D minor.
  Sonatas, op. 2, D major and G major. cf Quintets, op. 11, nos.
    4 and 6, D major and E flat major.
54 Sonatas, flute, opp. 16 and 19. Ingrid Dingfelder, flt; Rita
      Koors-Myers, hpd and pno. Musical Heritage Society MHS 1974/5 (2).
          +-HF 9-75 p81   +MJ 10-75 p38
          +MJ 7-75 p34
  Sonata, harpsichord, op. 5, no. 2, D major. cf Angel SB 3816.
55 Sonatas, keyboard, op. 17, nos. 1-6. Brigitte Haudebourg, hpd.
      Arion ARN 38259.
          +HFN 11-75 p148   +RR 10-75 p69
56 Sonatas, keyboard, op. 17, nos. 1-6. Ingrid Haebler, fortepiano.
      Philips 6500 848.
          ++Gr 5-75 p1992   ++RR 6-75 p59
          ++HFN 8-75 p71D
57 Symphony, no. 3, op. 3, E flat major. BACH, J. S.: Brandenburg
      concerto, no. 2, S 1047, F major. HAYDN: Quartet, strings,

op. 3, no. 5, F major: Serenade. MOZART: Concerto, bassoon,
K 191, B flat major. Michael Chapman, bsn; AMF; Neville
Marriner. Philips 6833 122. (Reissues from 6700 045, 6707 013,
6706 020, 6580 066)
    ++Gr 2-75 p1504   +RR 1-75 p30

BACH, Johann Christoph Friedrich
58   Symphonies (7). Cologne Chamber Orchestra; Helmut Müller-Brühl.
    Nonesuch HB 73027 (2).
    +-HF 3-75 p75   +SR 1-25-75 p51
    +-HFN 7-75 p71   +St 4-75 p98
    ++NR 1-75 p3

BACH, Johann Lorenz
    Suite, G major: Minuet, Gavotte. cf Philips 6580 098.

BACH, Johann Sebastian
59   Adagio, S 968, G major. Allegro, S 1019, E minor. Sonatas,
    harpsichord, S 964-966. János Sebestyén, hpd. Hungaroton SLPX
    11613.
    +RR 1-75 p46   ++SFC 6-15-75 p24
    Allabreve, S 589, D major. cf Organ works (DG 2533 140).
    Allabreve, S 589, D major. cf Organ works (DG 2722 014).
    Allabreve, S 589, D major. cf Organ works (Telefunken BC 24104).
    Allabreve, S 589, D major. cf Organ works (Telefunken EK 6-35082).
    Allegro, S 1019, E minor. cf Adagio, S 968, G major.
60   Anna Magdalena notebook: Menuett, S 114-116, 118, 120-121; Untitled
    piece, S 131; Schaffs mit mir Gott, S 514; Bist du bei mir,
    S 508; Little preludes, S 924-930; S 933-942; S 999. Kornél
    Zempléni, pno. Hungaroton SLPX 11657.
    +NR 6-75 p14   +-RR 4-75 p41
    Anna Magdalena notebook, S 508: Little suite. cf CBS 77513.
    Applicatio, S 994, C major. cf Harpsichord works (DG Archive
    2722 015).
    Aria, S 587, F major. cf Organ works (Telefunken BC 25102 T/1-2).
    Aria variata alla maniera Italiana, S 989, A minor. cf Harpsichord
    works (DG 2722 020).
    Arioso (arr. Franko). cf Decca SPA 405.
    Ave Maria (arr. Gounod). cf Polydor 2489 519.
    Ave Maria (arr. Gounod). cf Saga 7029.
    Bist du bei mir (arr. Grace). cf Vista VPS 1021.
61   Bourrée, E minor. Prelude, S 999, C minor. Sarabande, E minor.
    SANZ: Suite española. SCARLATTI, D.: Sonata, guitar, E minor.
    WEISS: Suite, E major. Narciso Yepes, gtr. DG 2538 236. Tape
    (c) 3318 038.
    +-RR 7-75 p69 tape
62   Brandenburg concerti, nos. 1-6, S 1046-51. Collegium Aureum.
    BASF BMH 23 20331 (2). Tape (c) KBACC 3007/8.
    +-Gr 2-73 p1496   +-RR 5-73 p39
    +-HFN 2-73 p332   +RR 4-75 p76 tape
63   Brandenburg concerti, nos. 1-6, S 1046-51. Concerti, harpsichord,
    nos. 1 and 2, S 1052-3. Suites, orchestra, S 1066-9. George
    Malcolm, hpd; SCO; Karl Münchinger. Decca 14BB 213/7 (5).
    (Reissues from SXL 2125/7, 2300/1; 6101).
    +-Gr 10-75 p601   +-RR 10-75 p36
    +-HFN 9-75 p109
64   Brandenburg concerti, nos. 1-6, S 1046-51. Suites, orchestra,
    S 1067-8. Stuttgart Chamber Orchestra; Karl Münchinger. Decca

SPA 382/3 (2). Tape (c) KCSP 382/3. (Reissues from SXL
2125/7, 2300/1)
+-Gr 5-75 p1961   +-RR 5-75 p26
-HFN 5-75 p124   -RR 10-75 p96 tape

65   Brandenburg concerti, nos. 1-6, S 1046-51. BPhO; Herbert von
     Karajan. DG 139 005/6 (2).
     +-NR 5-75 p4

66   Brandenburg concerti, nos. 1-6, S 1046-51. Munich Bach Orchestra;
     Karl Richter. DG Archive 2708 013. Tape (c) 3370 002 (ct)
     3375 001.
     ++Gr 11-73 p1015 tape   +HF 11-75 p149 tape

67   Brandenburg concerti, nos. 1-6, S 1046-51. Suites, orchestra,
     S 1066-9. Munich Bach Orchestra; Karl Richter. DG Archive
     tape (c) 3376 003.
     +HF 11-75 p149 tape

68   Brandenburg concerti, nos. 1-6, S 1046-51 (original version). AMF;
     Neville Marriner. Philips 6500 186/7 (2). Tape (c) 7300 158/9.
     (also Philips 6700 045)
     +Gr 7-72 p479   +NR 3-74 p4
     +Gr 5-75 p2031 tape   +RR 9-72 p48
     ++HF 3-74 p80   +-RR 4-75 p76 tape
     +HFN 9-72 p1659   /SFC 12-16-73 p40
     +MJ 7-74 p30   +St 2-74 p113

69   Brandenburg concerti, nos. 1-6, S 1046-51. Würzburg Camerata
     Accademica; Hans Reinartz. Pye GGCD 301 (2).
     +-Gr 8-75 p315   +-RR 10-75 p37
     +-HFN 8-75 p92

70   Brandenburg concerti, nos. 1-6, S 1046-51. Jean-Pierre Rampal,
     Alain Marion, flt; Pierre Pierlot, Jacques Chambon, Claude
     Maisonneuve, ob; Maurice André, tpt; R. Tassin, hn; Paul Hongne,
     bsn; Gérard Jarry, vln; Anne-Marie Beckensteiner, hpd; Jean-
     François Paillard Chamber Orchestra; Jean-François Paillard.
     RCA CRL 2-5801 (2). Tape (ct) ARS 1-7004/5.
     +-HF 9-75 p81   +SFC 6-1-75 p21
     +NR 5-75 p4   ++St 6-75 p94

71   Brandenburg concerti, nos. 1-6, S 1046-51. Virtuosi of England;
     Arthur Davison. Vanguard SRV 313/4 (2). (also Classics for
     Pleasure CFP 40010/2)
     -Gr 10-72 p679   +NR 7-75 p3
     +HF 9-75 p81   +RR 10-72 p53
     +-HFN 9-72 p1659

72   Brandenburg concerti, nos. 1-6, S 1046-51. ECO; Johannes Somary.
     Vanguard VSD 71208/9 (2).
     ++SFC 11-30-75 p34

     Brandenburg concerti, nos. 1-3, S 1046-48, CCV 5007. cf Camden
     CCV 5000-12, 5014-24.

73   Brandenburg concerto, no. 2, S 1047, F major. HOFFMAN: Miniatures,
     solo trumpet (4). NERUDA: Concerto, trumpet. SCARLATTI, A.:
     Endimione and Cintia, aria. Stuart Steffen, s; David Hickman,
     tpt; Festival Chamber Orchestra; Legh Burns. Clarino SLP 1009.
     +NR 2-75 p4

     Brandenburg concerto, no. 2, S 1047, F major. cf BACH, J. C.:
     Symphony, no. 3, op. 3, E flat major.
     Brandenburg concerto, no. 2, S 1047, F major: 1st movement. cf
     Columbia CXM 32088.
     Brandenburg concerto, no. 3, S 1048, G major. cf Philips 6747 199.

Brandenburg concerto, no. 3, S 1048, G major: 1st movement. cf
   Sound Superb SPR 90049.
Brandenburg concerto, no. 3, S 1048, G major: Finale. cf CBS
   77513.
74 Brandenburg concerto, no. 5, S 1050, D major. Chorale preludes:
   Ich ruf zu dir, Herr Jesu Christ, S 639; Nun komm der Heiden
   Heiland, S 699; Wir glauben all an einen Gott, S 680. PO;
   Leopold Stokowski. CBS 30061.
      +-RR 12-75 p35
75 Cantatas, S 1, 4, 6, 11 (Ascension oratorio), 12, 44, 61, 64-65,
   67, 104, 111, 121, 124, 158, 171, 182. Christmas oratorio,
   S 248. Magnificat, S 243, D major. Edith Mathis, s; Anna
   Reynolds, ms; Peter Schreier, t; Dietrich Fischer-Dieskau, bar;
   Munich Bach Orchestras and Choirs; Karl Richter. DG Archive
   2722 018 (11). (Some reissues)
      +-HFN 12-75 p147
Cantata, no. 4, Christ lag in Todesbanden, Jesus Christus Gottes
   Sohn. cf Works, selections (Decca PFS 4278).
76 Cantata, no. 11, Lobet Gott in seinen Reichen. Cantata, no. 80,
   Ein feste Burg ist unser Gott. Felicity Palmer, s; Helen
   Watts, alto; Robert Tear, t; Michael Rippon, bs; Amor Artis
   Chorale; ECO; Johannes Somary. Vanguard VSD 71193. (Q) VSQ
   30040.
      +-St 6-75 p94
77 Cantata, no. 12, Weinen, Klagen, Sorgen, Zagen: Sinfonia. Suite,
   orchestra, S 1067, B minor. Suite, orchestra, S 1068, D major.
   OSR; Roger Reversy, ob; Ernest Ansermet. Decca ECS 754.
   (Reissue from SXL 6004)
      +-Gr 1-75 p1335
Cantata, no. 12, Weinen, Klagen, Sorgen, Zagen: Sinfonia. cf
   Works, selections (Supraphon 50871/2).
Cantata, no. 12, Weinen, Klagen, Sorgen, Zagen: Sinfonia. cf
   BACH, C. P. E.: Concerto, oboe, W 164, B flat major.
78 Cantatas, nos. 21, 34, 39, 51, 55-56, 60, 68, 76, 93, 106, 129,
   175, 189, 201, 211-212. Edith Mathis, Adele Stolte, Elizabeth
   Speiser, s; Anna Reynolds, Hertha Topper, Eva Fleischer,
   Ingeborg Springer, ms; Peter Schreier, Ernst Häfliger, Hans-
   Joachim Rotzsch, t; Dietrich Fischer-Dieskau, Kurt Moll, Theo
   Adam, Gunter Leib, Siegfried Lorenz, bs; Munich Bach Orchestra
   and Choir; Leipzig Gewandhaus Orchestra; Berlin Solisten-
   vereinignung; Berlin Chamber Orchestra; Kurt Thomas, Helmut
   Koch. DG Archive 2722 019 (11).
      +Gr 12-75 p1086   +-RR 12-75 p87
Cantata, no. 21, Ich hatte viel Bekümmernis: Sinfonia. cf BACH,
   C. P. E.: Concerto, oboe, W 164, B flat major.
79 Cantata, no. 23, Du wahrer Gott und Davids Sohn. Cantata, no.
   72, Alles nur nach Gottes Willen. Ingeborg Reichelt, s;
   Barbara Scherler, con; Friedrich Melzer, t; Bruce Abel, bs;
   Heinrich Schütz Choir; Würtemburg Chamber Orchestra; Fritz
   Werner. Erato STU 70775.
      +-Gr 5-75 p1999   +-RR 4-75 p57
Cantata, no. 29: Sinfonia. cf Organ works (Advent 5010).
80 Cantatas, nos. 31-34. Walter Gampert, treble; Kurt Equiluz, Marius
   van Altena, t; Rene Jacobs, c-t; Siegmund Nimsgern, Max van
   Egmond, bs; Vienna Boys'Choir; Chorus Viennensis; VCM; Leonhardt
   Consort; Gustav Leonhardt, Nikolaus Harnoncourt. Telefunken

SKW 9/1-2 (2).
    ++AR 2-75 p21   ++NR 12-74 p10
    ++Gr 12-74 p1193   ++RR 1-75 p53
    ++HF 7-75 p69   +STL 1-12-75 p36
    ++MJ 10-75 p45

81   Cantatas, nos. 35-38. Kurt Equiluz, t; Paul Esswood, c-t; Ruud
      van der Meer, bs; Herbert Tachezi, org; Vienna Boys' Choir;
      Chorus Viennensis; VCM; Nikolaus Harnoncourt. Telefunken SKW
      10/1-2 (2).
        ++AR 5-75 p51   +NR 2-75 p7
        ++Gr 12-74 p1194   ++RR 1-75 p53
        ++HF 7-75 p69   +STL 1-12-75 p36

82   Cantatas, nos. 39-42. René Jacobs, Paul Esswood, c-t; Marius von
      Altena, Kurt Equiluz, t; Max von Egmond, Ruud van der Meer, bs;
      Hanover Boys' Choir; Leonhardt Consort, Vienna Boys' Choir;
      Chorus Viennensis; VCM; Gustav Leonhardt, Hans Gillesberger,
      Nikolaus Harnoncourt. Telefunken EX 6-35269 (2). (also SKW 11)
        ++Gr 8-75 p351   ++NR 10-75 p6
        ++HFN 7-75 p72   +RR 9-75 p60

83   Cantatas, nos. 43-46. Soloists; VCM; Nikoluas Harnoncourt.
      Telefunken 6-35283 (2).
        ++NR 12-75 p10

84   Cantata, no. 51, Jauchzet Gott in allen Landen. MOZART: Exsultate
      Jubilate, K 165. Elisabeth Schwarzkopf, s; Harold Jackson,
      tpt; PO; Peter Gellhorn, Walter Susskind. Seraphin 60013.
        ++St 4-75 p67
    Cantata, no. 51, Jauchzet Gott, Jauchzet Gott in allen Landen. cf
      Philips 6833 105.
    Cantata, no. 72, Alles nur nach Gottes Willen. cf Cantata, no. 23,
      Du wahrer Gott und Davids Sohn.

85   Cantata, no. 78, Jesu, der du meine Seele. Cantata, no. 106,
      Gottes Zeit ist die allerbeste Zeit. Teresa Stich-Randall, s;
      Dagmar Hermann, con; Anton Dermota, t; Hans Braun, bs; Anton
      Heiller, org; Bach Orchestra and Choir; Felix Prohaska. Bach
      Guild HM 21. (Reissue from Top Rank 35/008)
        +Gr 11-75 p870   +-RR 10-75 p80
        +-HFN 10-75 p153
    Cantata, no. 79, Now thank we all our God. cf Cantata, no. 156:
      Arioso.
    Cantata, no. 80, Ein feste Burg ist unser Gott. cf Cantata, no.
      11, Lobet Gott in seinen Reichen.
    Cantata, no. 80, A mighty fortress is our God. cf CBS 77513.
    Cantata, no. 106, Gottes Zeit ist die allerbeste Zeit. cf Cantata,
      no. 78, Jesu, der du meine Seele.
    Cantata, no. 106: Sonatina, Magnificat, S 243, D major: Esurientes.
      cf HMV SLS 5022.
    Cantata, no. 110, Unser Mund sei voll Lachens. cf Magnificat,
      S 243, D major.

86   Cantata, no. 131, Aus der Tiefe rufe ich, Herr. HANDEL: Sing unto
      God, anthem. Wendy Eathorne, s; Paul Esswood, c-t; Neil Jenkins,
      t; John Noble, bar; London Bach Society Chorus; Steinitz Bach
      Players; Paul Steinitz. HMV CSD 3741. (also Nonesuch H 71294.
      Tape (r) NST 71294)
        +Gr 11-73 p962   ++HFN 12-73 p106
        ++HF 10-74 p87   +NR 8-74 p7
        +HF 7-75 p102 tape   +RR 11-73 p80

Cantata, no. 140, Sleepers, awake. cf Works, selections (Ember
    ECL 9007).
Cantata, no. 140, Sleepers, awake. cf CBS 77513.
Cantata, no. 140, Sleepers, awake. cf Cantata, no. 156: Arioso.
Cantata, no. 147, Jesu joy of man's desiring. cf Cantata, no. 156:
    Arioso.
Cantata, no. 147, Jesu, joy of man's desiring. cf Works, selections
    (Ember ECL 9007).
Cantata, S 147, Jesu, joy of man's desiring. cf Works, selections
    (HMV ASD 2971).
Cantata, S 147, Jesu, joy of man's desiring. cf CBS 77513.
Cantata, S 147, Jesu, joy of man's desiring. cf CRD CRD 1008.
Cantata, no. 147, Jesu, joy of man's desiring. cf Decca SDD 463.
Cantata, no. 147, Jesu, joy of man's desiring. cf Klavier KS 536.
Cantata, no. 147, Jesu, joy of man's desiring. cf Pye GH 589.
Cantata, no. 147, Jesu, joy of man's desiring. cf Pye GSGC 14153.
Cantata, no. 147, Jesu, joy of man's desiring. cf RCA CRL 2-7001.
Cantata, no. 147, Jesu, joy of man's desiring. cf University of
    Colorado, unnumbered.
Cantata, no. 152, Tritt auf die Glaubensbahn: Concerto. cf Works,
    selections (Supraphon 50871/2).
87   Cantata, no. 156: Arioso. Cantata, no. 140, Sleepers, awake.
    Cantata, no. 147, Jesu, joy of man's desiring. Cantata, no. 79,
    Now thank we all our God. Suite, orchestra, S 1068, D major:
    Air for the G string. Toccata and fugue, D minor. Fugue, S 578,
    G minor. Virgil Fox, org. RCA ARL 1-0476.
        -MJ 3-75 p33   *NR 6-74 p12
Cantata, no. 156: Arioso. cf Works, selections (RCA ARL 1-0880).
88   Cantata, no. 161,Komm, du süsse Todesstunde. Cantata, no. 165,
    Gott soll allein mein Herze haben. Julia Hamari, con, Joszef
    Réti, t; Gabor Lehotka, org; Zsuzsa Petris, hpd; Laszlo Som,
    double bs; Ferenc Liszt Academy Chamber Orchestra and Choir;
    Frigyes Sándor. Qualiton SLPX 1284.
        ++LJ 9-75 p37
Cantata, no. 165, Gott soll allein mein Herze haben. cf Cantata,
    no. 16, Komm, du süsse Todesstunde.
Cantata, no. 182, Himmelskönig, sei willkommen: Sonata. cf
    Works, selections (Supraphon 50871/2).
Cantata, no. 208, Sheep may safely graze. cf Works, selections
    (Ember ECL 9007).
Cantata, no. 208, Schafe können sicher weiden. cf HMV SLS 5022.
Cantata, no. 209, Non sa che sia dolore: Sinfonia. cf Works,
    selections (Supraphon 50871/2).
Canone perpetuo, S 1079, no. 9. cf Works, selections (Supraphon
    50871/2).
Canzona, S 588, D minor. cf Organ works (CBS 78256).
Canzona, S 588, D minor. cf Organ works (DG 2533 140).
Canzona, S 588, D minor. cf Organ works (DG 2722 014).
Canzona, S 588, D minor. cf Organ works (Pelca 40586).
Canzona, S 588, D minor. cf Organ works (Telefunken BC 24104).
Canzona, S 588, D minor. cf Organ works (Telefunken EK 6-35082).
Capriccio sopra la lontananze del suo fratello dilettissimo, S 992,
    B flat major. cf Harpsichord works (DG 2522 020).
89   Chaconne (Busoni). BEETHOVEN: Sonata, piano, no. 4, op. 7, E flat
    major. BRAHMS: Variations on a theme by Paganini, op. 35, A
    minor. Arturo Benedetti Michelangeli, pno. Rococo 2069.
        +-NR 2-75 p11

Chaconne, D minor. cf MOZART: Fantasia, K 475, C minor.

Chorales (Schubler) (6). cf Organ works (DG Archive 2722 016).

Chorale preludes of diverse kinds (18). cf Organ works (DG
Archive 2722 016).

90   Chorale preludes (Orgelbüchlein), S 599-644. Peter Hurford, org;
Alban Singers. Argo ZRG 776/78 (3).
      +Gr 10-74 p723   +RR 10-74 p83
      -HF 5-75 p69   ++STL 2-9-75 p37
      ++MT 7-75 p628

Chorale prelude: Gelobet seist du, Jesu Christ, S 604. cf Organ
works (CBS 78256).

91   Chorale prelude, O Mensch bewein dein Sünde grosse, S 622. Prelude
and fugue, S 533, E minor. BERTHIER: Short pieces (5).
DRISCHNER: Chorale prelude on 'O heiliger Geist, O heiliger
Gott'. Choralfantasia 'Lobe den Herren'. Nordische Fantasie.
Partita on 'Wilhemnus von Naussauen'. MICHEELSEN: Introduction
and chorale on 'Erhalt uns, Herr dei deinem Wort'. Toccata on
'Lobe den Herren'. Bryan Hesford, org. Wealden WS 125.
      +-HFN 9-75 p101   +-RR 9-75 p61

Chorale prelude: O Mensch bewein dein Sünde gross, S 622. cf
Organ works (CBS 78256).

Chorale prelude: Ich ruf zu dir, Herr Jesu Christ, S 639. cf
Organ works (CBS 78256).

Chorale prelude, Ich ruf zu dir, S 639. cf Wealden WS 110.

Chorale preludes: Ich ruf zu dir, Herr Jesu Christ, S 639; Nun
freut euch, lieben Christen mein, S 734; Nun komm der Heiden
Heiland, S 599. cf Works, selections (HMV ASD 2971).

Chorale preludes: Ich ruf zu dir, Herr Jesu Christ, S 639; Nun
komm der Heiden Heiland, S 699; Wir glauben all an einen Gott,
S 680. cf Brandenburg concerto, no. 5, S 1050, D major.

Chorale prelude: Wenn wir in höchsten Nöten sein, S 641. cf Organ
works (CBS 78256).

Chorale preludes, S 645-50. cf Decca 5BBA 1013-5.

Chorale prelude: Wachet auf, ruft uns die Stimme, S 645. cf
London STS 15160.

Chorale prelude: Meine Seele erhebt den Herren, S 648. cf Pelca
PSRK 41013/6.

92   Chorale preludes: Ach bleib bei uns, S 649. Allein Gott in der
Hoh sei Ehr, S 711, S 715, S 717. Christ lag in Todesbanden,
S 695, S 718. Christum wir sollen loben schon, S 696: Fughetta.
Gelobet seist du, Jesu Christ, S 722, S 722a. Gelobet seist du
Jesu Christ, S 697: Fughetta. Gottes Sohn ist kommen, S 724.
Gottes Sohn ist kommen, S 703: Fughetta. Herr Christ, der ein'ge
Gottes Sohn, S 698: Fughetta. In dich hab ich gehaffet, Herr,
S 712. In dulci jubilo, S 729, S 729a. Jesu, meine Freude,
S 713: Fantasia. Kommst du nun Jesu, S 650. Lob sei dem
almachtigen Gott, S 704: Fughetta. Lobt Gott ihr Christen
allzugleich, S 732, S 732a. Meine Seele erhebet den Herren,
S 648. Nun komm der Heiden Heiland, S 699: Fughetta. Vom
Himmel hoch, S 700, S 738. Vom Himmel hoch, S 769: Canonic
variations. Vom Himmel hoch, S 701: Fughetta. Wachet auf,
S 645. Wer nur den lieben, Gott, S 647. Wir Christenleut,
S 710. Wo soll ich fliehen hin, S 646, S 694. Fugue on the
Magnificat, S 733. Michael Chapuis, org. Telefunken BC 25099-1/
1-2 (2).
      +AR 5-75 p51   +-NR 9-74 p11

```
    +Gr 6-74 p71                    +RR 4-74 p59
    +HF 9-74 p85
```
93 Chorale preludes, S 651-S 688a. Michael Chapuis, org. Telefunken
    EK 6-35083 (2).
```
       ++Gr 10-75 p652              +RR 10-75 p69
       +-HFN 10-75 p136
```
    Chorale prelude: An Wasserflussen Babylon, S 653b. cf Organ works
       (Telefunken BC 25102 T/1-2)
    Chorale prelude: O Lamm Gottes unschuldig, S 656. cf Organ works
       (Telefunken BC 25102 T/1-2).
94 Chorale preludes: Nun komm, der Heiden Heiland, S 659-61. Sei
    gegrüsset, Jesu gütig, S 678. Fantasia and Fugue, S 537, C
    minor. William Tinker, org. Advent S 5013.
```
       -NR 12-75 p14
```
    Chorale prelude: Nun komm, der Heiden Heiland, S 659. cf RCA
    VH 020.
    Chorale prelude: Nun komm der Heiden Heiland, S 659. cf Organ
    works (CBS 78256).
95 Chorale preludes, S 669-671: Kyrie, Gott Vater in Ewigkeit;
    Christe, alle Welt Trost; Kyrie, Gott Heiliger Geist. KREBS:
    Trio, C minor. MONNIKENDAM: Toccata. SWEELINCK: Variations on
    'Mein Junges Leben Hatt Ein End'. WILLS: Prelude and fugue
    (Alkmaar). Robert Munns, org. Wealden WS 118.
```
       +RR 2-75 p47
```
    Chorale preludes: Sei gegrüsset Jesus gütig, D 678; Vom Himmel hoch
    variations, S 679. cf Organ works (DG Archive 2722 016).
    Chorale prelude: Wir glauben all an einen Gott, S 680. cf Works,
    selections (Decca PFS 4278).
    Chorale prelude: Vom Himmel hoch da komm ich her, S 700. cf
    Hungaroton SLPX 11548.
96 Chorale preludes: Jesu, meine Freude, S 713: Fantasia; Vom Himmel
    hoch, S 701: Fughetta. Motets, nos. 1-6, S 225-228, 230-231.
    Magnificat, S 243, D major: Vom Himmel hoch; Freut euch und
    jubiliert. David Lumsden, org; Louis Halsey Singers; Louis
    Halsey, David Lumsden. L'Oiseau-Lyre SOL 340/1 (2). (Some
    reissues from Unicorn UNS 252)
```
       +-Gr 7-75 p224               +RR 7-75 p55
       ++HFN 7-75 p73
```
97 Chorale preludes: Ach Gott und Herr, S 714. Erbarm dich mein, O
    Herre Gott, S 721. Valet will ich dir geben, S 735. Ein feste
    Burg ist unser Gott, S 720. Herr Gott, dich loben wir, S 725.
    Herr Jesu Christ, dich zu uns wend, S 709. Herr Jesu Christ,
    dich zu uns wend, S 726. Herzlich tut mich verlangen, S 727.
    Jesus, meine Zuversicht, S728. Liebster Jesu, wir sind hier,
    S 706; S 730/1. Nun freut euch, lieben Christen, g'mein, S 723.
    Christ der du bist der helle Tag, S 766. O Gott, du frommer
    Gott, S 767. Sei gegrüsset, Jesus gütig, S 768. Vater unser im
    Himmelreich, S 737. Valet will ich dir geben, S 736. Wer nur
    den lieben Gott lasst walten, S 691-690. Michael Chapuis, org.
    Telefunken BC 25103/1-2.
```
       +-NR 4-75 p10
```
98 Chorale preludes: Ach Gott und Herr, S 714; Christ, der du bist
    helle Tag, S 766: Partita; Erbarm dich mein, O Herre Gott, S 721;
    Ein feste Burg ist unser Gott, S 720; Herr Jesu Christ, dich zu
    uns wend, S 709/726; Herr Gott, dich loben wir, S 725; Herzlich
    tut mich verlangen, S 727; Jesus meine Zuversicht, S 728; Liebster

Jesu, wir sind hier, S 706/730/731; Nun freut euch, lieben
Christen g'mein, S 734; O Gott, du frommer Gott, S 767: Partita;
Sei gegrüsset Jesus gütig, S 768: Partite diverse; Valet will
ich dir geben, S 735/736; Vater unser im Himmelreich, S 737;
Wer nur den lieben Gott lässt walten, S 691/690. Helmut Walcha,
org. Telefunken EK 6-35081 (2).
    ++Gr 7-75 p213   +RR 9-75 p56
    +HFN 7-75 p72

Chorale preludes: Allein Gott in der Hoh sei Ehr, S 715; Nun freut
euch, lieben Christen g'mein, S 734; O Mensch, bewein den Sunde
gross, S 622. cf Organ works (Pelca 40586).

Chorale preludes: Ein feste Burg, S 720; Komm süsser Tod; Wachet
auf, S 645. cf Works, selections (RCA ARL 1-0880).

Chorale prelude: Herzlich tut, mich verlangen, S 727. cf Organ
works (CBS 78256).

Chorale prelude: In dulci jubilo, S 729. cf HMV Tape (c) TC MCS 13.

Chorale prelude: In dulci jubilo, S 729. cf Hungaroton SLPX 11548.

Chorale prelude: Ach Gott, vom Himmel sich darein, S 741. cf
Organ works (Telefunken BC 25102 T/1-2).

Christmas oratorio, S 248. cf Cantatas (DG Archive 2722 018).

Christmas oratorio, S 248: Flösst, mein Heiland. cf Philips 6833 105.

Christmas oratorio, S 248: Sinfonia. cf Decca SDD 411.

Christmas oratorio, S 248: Sinfonia. cf HMV ASD 3017.

99   Chromatic fantasia and fugue, S 903, D minor. English suite, no. 4,
S 809, F major. Partita, harpsichord, no. 5, S 829, G major.
Toccata, S 912, D major. András Schiff, pno. Hungaroton SLPX
11690.
    +NR 6-75 p14   +-RR 4-75 p42

100 Chromatic fantasia and fugue, S 903, D minor. Concerto, harpsichord,
S 972, D major. Duets, S 802-805. Concerto, harpsichord, S 971,
F major. János Sebestyén, hpd. Hungaroton SHLX 90042.
    +-RR 1-75 p45

Chromatic fantasy and fugue, S 903, D minor. cf Harpsichord works
(DG 2722 020).

Chromatic fantasia and fugue, S 903, D minor. cf Harpsichord works
(Supraphon 111 0750).

Clavierübung, Pt. 3. cf Organ works (DG Archive 2722 016).

Clavierübung: Duetto III. cf Angel S 36076.

101 Concerto, flute and strings, G minor (arr. from S 1056). Concerto,
oboe, violin and strings, D minor (arr. from S 1060). Concerto,
3 violins and strings, D major (arr. from S 1064). Carmel
Kaine, Ronald Thomas, Richard Studt, vln; Neil Black, ob;
William Bennett, flt; Christopher Hogwood, Nicholas Kramer, hpd;
AMF; Neville Marriner. Argo ZRG 820. Tape (c) KZRC 820.
    ++Gr 12-75 p1022   +RR 12-75 p36
    ++HFN 12-75 p147

102 Concerto, flute, violin and harpsichord, S 1044, A minor. Concerto,
harpsichord, no. 1, S 1052, D minor. Maria Teresa Garatti, hpd;
Severino Gazzelloni, flt; Salvatore Accardo, vln; I Musici.
Philips 6500 692. Tape (c) 7300 289.
    +-Gr 4-75 p1801   ++RR 7-75 p69 tape
    ++RR 3-75 p20

103 Concerto, harpsichord, S 971, F major (Italian). Partita, harpsichord, S 831, B minor. Prelude, fugue and allegro, S 998,
E flat major. Gustav Leonhardt, hpd. BASF BAC 3095.
    +Gr 10-75 p657   +RR 10-75 p69
    +HFN 9-75 p92

Concerto, harpsichord, S 971, F major.   cf Chromatic fantasia and
fugue, S 903, D minor.
Concerto, harpsichord, S 971, F major.   cf Harpsichord works
(DG 2722 020).
Concerto, harpsichord, S 971, F major.   cf Harpsichord works
(Supraphon 111 0750).
Concerto, harpsichord, S 971, F major.   cf CBS 73396.
Concerto, harpsichord, S 971, F major.   cf Pandora PAN 101.
Concerto, harpsichord, S 972, D major.   cf Chromatic fantasia and
fugue, S 903, D minor.
Concerti, harpsichord, nos. 1 and 2, S 1052-3.   cf Brandenburg
concerti, nos. 1-6, S 1046-51.
104   Concerto, harpsichord, nos. 1-7, S 1052-1058.   Zuzana Ruzickova,
hpd; Prague Chamber Soloists; Václav Neumann.   Supraphon
110 1451/3 (3).
        ++NR 9-75 p6
Concerto, harpsichord, no. 1, S 1052, D minor.   cf Concerto, flute,
violin and harpsichord, S 1044, A minor.
Concerto, harpsichord, no. 1, S 1052, D minor.   cf Concerto,
violin, S 1045, D major: Sinfonia.
Concerto, harpsichord, no. 1, S 1052, D minor: 2d movement.   cf
BASF BAB 9005.
105   Concerto, harpsichord, no. 2, S 1053, E major.   Concerto, harpsi-
chord, no. 4, S 1055, A major.   Concerto, harpsichord, no. 5,
S 1056, F minor.   Gustav Leonhardt, hpd; Leonhardt Consort.
Telefunken 6-41099.   Tape (c) CT 4-41099.
        +RR 9-75 p76 tape
106   Concerti, harpsichord, nos. 3-5, 7, S 1054-1056, S 1058.   Zoltán
Kocsis, pno; Ferenc Liszt Music Academy Orchestra; Albert
Simon.   Hungaroton SLPX 11711.
        /RR 4-75 p20
Concerto, harpsichord, no. 4, S 1055, A major.   cf Concerto, harp-
sichord, no. 2, S 1053, E major.
Concerto, harpsichord, no. 5, S 1056, F minor.   cf Concerto,
harpsichord, no. 2, S 1053, E major.
107   Concerto, 2 harpsichords, S 1060, C minor.   Concerto, 2 harpsichords,
S 1062, C minor.   Concerto, 3 harpsichords, S 1064, C major.
Isolde Ahlgrimm, Hans Pischner, Zuzana Ruzickova, hpd; Dresden
Philharmonic Orchestra; Kurt Redel.   Philips 6580 089.
        /Gr 6-75 p31   +RR 6-75 p26
        -HFN 6-75 p83
Concerto, 2 harpsichords, S 1060, C minor.   cf Classic Record
Library SQM 80-5731.
Concerto, 2 harpsichords, S 1062, C minor.   cf Concerto, 2 harpsi-
chords, S 1060, C minor.
108   Concerto, 3 harpsichords, S 1063, D minor.   Concerto, 2 harpsichords,
S 1064, C major.   Concerto, 4 harpsichords, S 1065, A minor.
Karl Richter, Hedwig Bilgram, Iwona Fütterer, Ulrike Schott,
hpd; Munich Bach Orchestra; Karl Richter.   DG Archive 2733 171.
Tape (c) 3310 129   (Reissue from 2722 009).
        +-Gr 4-75 p1801   ++RR 4-75 p20
        +HF 7-75 p69   ++SFC 11-30-75 p34
        ++NR 6-75 p6
Concerto, 3 harpsichords, S 1064, C major.   cf Concerto, 2 harpsi-
chords, S 1060, C minor.
Concerto, 3 harpsichords, S 1064, C major.   cf Concerto, 3 harpsi-
chords, S 1063, D minor.

Concerto, 4 harpsichords, S 1065, A minor. cf Concerto, 3 harpsi-
    chords, S 1063, D minor.
Concerto, oboe, violin and strings, D minor (arr. from S 1060). cf
    Concerto, flute and strings, G minor (arr. from S 1056).
Concerto, oboe, violin and strings, S 1060, D minor. cf Concerto,
    violin, S 1045, D major: Sinfonia.
109 Concerti, organ, nos. 1-6, S 592-7. Karl Richter, org. DG Archive
    2533 170. Tape (c) 3310 128.
        +-Gr 6-75 p67   ++NR 5-75 p15
        -HF 8-75 p79   +RR 6-75 p59
        +HFN 5-75 p124   ++SFC 10-5-75 p38
Concerti, organ, nos. 1-6, S 592-597. cf Organ works (Telefunken
    BC 25102 T/1-2).
110 Concerti, organ, no. 1, S 592, G major; no. 2, S 593, A minor;
    no. 3, S 594, C major; no. 5, S 596, D minor. Lionel Rogg,
    org. Seraphim S 60245.
        +NR 8-75 p14
Concerto, organ, no. 2, S 593, A minor. cf Organ works (Advent 5010).
Concerto, organ, no. 3, S 594, C major: Recitative. cf CBS
    Classics 61579.
Concerto, organ, no. 4, S 595, C major. cf Pelca PSRK 41013/6.
Concerto, organ, no. 6, S 597, E flat major (attrib.). cf Argo
    ZRG 783.
111 Concerti, violin and strings, S 1041-1043. Alice Harnoncourt, vln;
    VCM; Nikolaus Harnoncourt. Telefunken SAWT 6-41227. Tape (c)
    4-41227CX.
        +-HFN 7-75 p90 tape   +RR 7-75 p68 tape
112 Concerto, violin and strings, S 1041, A minor. Concerto, violin,
    S 1052, D minor (after Concerto, harpsichord). Concerto, violin
    and oboe, S 1060, C minor. Itzhak Perlman, vln; Neil Black, ob;
    ECO; Daniel Barenboim. Angel S 37076. (also HMV ASD 3076)
        +-Gr 6-75 p31   +-MT 8-75 p713
        +HF 6-75 p83   +NR 5-75 p7
        +HFN 6-75 p83   +RR 6-75 p26
113 Concerto, violin and strings, S 1041, A minor. Concerto, violin
    and strings, S 1042, E major. Concerto, 2 violins and strings,
    S 1043, D minor. Felix Ayo, Robert Michelucci, vln; I Musici.
    Philips 6580 021. Tape (c) 7317 105 (Reissues from SABL 142, 31)
        +-HFN 3-72 p497   +RR 10-75 p96 tape
114 Concerti, violin and strings, S 1041-2, A minor and E major.
    Concerto, 2 violins and strings, S 1043, D minor. Arthur
    Grumiaux, Koji Toyoda, vln; NPhO; Raymond Leppard, Edo de Waart.
    Philips Tape (c) 7300 304.
        +HFN 9-75 p110 tape   +RR 9-75 p75 tape
Concerto, violin and strings, S 1041, A minor. cf Concerto, violin
    and strings, S 1042, E major.
Concerto, violin and strings, S 1041, A minor. cf Philips 6747 199.
115 Concerto, violin and strings, S 1042, E major. Concerto, violin
    and strings, S 1056, G minor. Concerto, 2 violins and strings,
    S 1043, D minor. Itzhak Perlman, Pinchas Zukerman, vln; ECO;
    Daniel Barenboim. Angel S 36841. Tape (c) 4XS 36841 (ct) 8XS
    36841. (also HMV ASD 2783 Tape (c) TC ASD 2783)
        +-Gr 12-72 p1131   +NR 2-73 p6
        +-Gr 10-74 p771 tape   ++RR 1-75 p68 tape
        +- HF 5-73 p73   /+St 3-73 p114
        +HFN 1-73 p111

116   Concerto, violin and strings, S 1042, E major.   Concerto, violin
         and strings, S 1041, A minor.   Sonata, flute and harpsichord,
         S 1020, G minor.   Sonata, violin and figured bass, E minor.
         Manoug Parikian, vln; Alexander Molzalm, vlc; Herbert Hoffman,
         hpd; Severino Gazzelloni, flt; Bruno Canino, hpd; Baden Chamber
         Orchestra; Alexander Krannhals.   Audio Fidelity FCS 50057.
            -NR 4-75 p5
      Concerto, violin and strings, S 1042, E major.   cf Concerto,
         violin and strings, S 1041, A minor (Philips 6580 021).
117   Concerto, violin and strings, S 1045, D major: Sinfonia.   Concerto,
         oboe, violin and strings, S 1060, D minor.   Concerto, harpsichord,
         no. 1, S 1052, D minor.   Alice Harnoncourt, vln; Jurg Schaeft-
         lein, ob; Herbert Tachezi, hpd; VCM; Nikolaus Harnoncourt.
         Telefunken SAWT 9557. Tape (c) CX 4-41227.
            +Gr 9-70 p418   +SFC 5-9-71 p32
            +Gr 8-75 p376 tape   +St 5-72 p71
            +NR 1-71 p6
      Concerto, violin and strings, S 1052, D minor (after Concerto,
         harpsichord).   cf Concerto, violin, no. 1, S 1041, A minor
         (Angel 37076).
      Concerto, violin and strings, S 1056, G minor.   cf Concerto,
         violin and strings, S 1042, E major.
      Concerto, 2 violins and strings, S 1043, D minor.   cf Concerto,
         violin and strings, S 1041, A minor.
      Concerto, 2 violins and strings, S 1043, D minor.   cf Concerto,
         violin, S 1042, E major.
      Concerto, 2 violins and strings, S 1043, D minor.   cf Concerti,
         violin and strings, S 1041-2, A minor and E major.
      Concerto, 2 violins and strings, S 1043, D minor.   cf Pearl GEM 132.
      Concerto, 2 violins and strings, S 1043, D minor.   cf RCA ARM
         4-0942/7.
      Concerto, 2 violins and strings, S 1043, D minor.   cf RCA CRL 6-0720.
      Concerto, 3 violins and strings, D major (arr. from S 1064).   cf
         Concerto, flute and strings, G minor (arr. from S 1056).
118   Concerto, 3 violins and strings, D minor.   TELEMANN: Concerto, 3
         violins and strings, F major.   VIVALDI: Concerto, 3 violins,
         strings and harpsichord, F major.   Jean-Pierre Wallez, vln;
         Instrumental Ensemble of France.   Decca 7246.
            +HFN 8-75 p72   +RR 8-75 p44
119   Concerto, violin and oboe, S 1060, D minor.   HANDEL: Concerto, harp,
         op. 4, no. 6, B flat major.   VIVALDI: Concerto, 2 mandolins,
         P 133, G major.   Concerto, 4 violins and strings, no. 10, P 148,
         B minor.   Leo Driehuys, ob; Robert Michelucci, Walter Gallozzi,
         Anna Maria Cotogni, Luciano Vicari, vln; Maria Teresa Garatti,
         org, hpd; Enzo Altobelli, vlc; Ursula Holliger, hp; André Pépin,
         Jean-Claude Hermanjat, flt; Gino Vescovo, Tommaso Rota, mand;
         I Musici.   Philips 6580 067.   (Some reissues)
            +-Gr 4-75 p1819   +RR 3-75 p26
            +HFN 6-75 p102
      Concerto, violin and oboe, S 1060, C minor.   cf Concerto, violin,
         no. 1, S 1041, A minor.
      Contrapunctus.   cf Klavier KS 536.
      Contrapunctus IX.   cf Klavier KS 536.
      Contrapunctus IX.   cf Kendor KE 22174.
      Duets, S 802-805.   cf Chromatic fantasia and fugue, S 903, D minor.
      Duets, S 802-5.   cf Organ works (DG 2533 140).

English suites, nos. 1-6, S 806-811. cf Harpsichord works
  (DG 2722 020).
English suite, no. 3, S 808, G minor: Gavotte. cf Pye GSGC 14154.
English suite, no. 3, S 808, G minor: Sarabande, Gavotte, Musette.
  cf RCA ARM 4-0942/7.
English suite, no. 4, S 809, F major. cf Chromatic fantasia and
  fugue, S 903, D minor.
English suite, no. 6, S 811: Gavottes, nos. 1 and 2. cf RCA ARM
  4-0942/7.
Fantasia. cf Klavier KS 536.
120 Fantasia, S 562, C minor. Fantasia, S 572, G major. Prelude and
  fugue, S 552, E flat major. Prelude and fugue, S 538, D minor.
  Toccata and fugue, S 565, D minor. Helmut Walcha, org.
  DG Tape (c) 3318 018.
      ++Gr 1-75 p1402 tape   ++RR 1-75 p68 tape
Fantasias, S 562, C minor; S 572, G major. cf Organ works (DG
  2722 014).
121 Fantasia, S 562, C minor, excerpt. Preludes and fugues, S 531-3,
  535, 547-50. Toccata and fugue, S 540, F major. Michael
  Chapuis, org. Telefunken BC 25101/1-2.
      +AR 5-75 p51   +RR 9-74 p70
      ++Gr 9-74 p543
Fantasia, S 563, B minor. cf Wealden WS 160.
Fantasias, S 563, B minor; S 570, C major. cf Organ works
  (Telefunken BC 25098 T/1-2).
Fantasia, S 570, C major. cf Wealden WS 160.
Fantasia, S 572, G major. cf Fantasia, S 562, C minor.
Fantasia, S 572, G major. cf Organ works (Telefunken BC 24104).
Fantasia, S 572, G major. cf Organ works (Telefunken EK 6-35082).
Fantasia, S 573, C major. cf Organ works (Telefunken BC 25102 T/1-2).
Fantasia, S 906, C minor. cf Harpsichord works (DG 2722 020).
Fantasia, S 906, C minor. cf Harpsichord works (Supraphon 111 0750).
122 Fantasias and fugues, S 537, C minor; S 542, G minor. Preludes and
  fugues, S 534, F minor; S 536, A major; S 539, D minor; S 541,
  G major; S 543, A minor; S 545, C major; S 546, C minor. Toccata
  and fugue, S 538, D minor. Michael Chapuis, org. Telefunken
  BC 25100 (2).
      +-Gr 7-74 p229   +-NR 9-74 p11
      +HF 9-74 p83   +RR 6-74 p63
Fantasias and fugues, S 537, C minor; S 542, G minor. cf Organ
  works (DG 2722 014).
Fantasia and fugue, S 537, C minor. cf Chorale preludes (Advent
  S 5013).
Fantasia and fugue, S 537, C minor. cf Organ works (Telefunken
  BC 25102 T/1-2).
Fantasia and fugue, S 537, C minor. cf HMV HLM 7065.
Fantasia and fugue, S 542, G minor. cf Organ works (CBS 78256).
Fantasia and fugue, S 542, G minor. cf Pye TPLS 13066.
Fantasia and fugue, S 562, C minor, fragment. cf Organ works
  (Telefunken BC 25101 T/1-2).
Fantasia and fugue, S 904, A minor. cf Harpsichord works (DG
  Archive 2722 015).
Fantasia and fugue, S 944, A minor. cf Harpsichord works (DG
  2722 020).
123 Flute works: Sonatas, flute and harpsichord, S 1030-35. Sonata,
  flute (violin) and harpsichord, S 1020, G minor. Partita,

flute, S 1013, A minor. Sonata, flute, violin and harpsichord,
S 1038, G major. Sonata, 2 flutes and harpsichord, S 1039,
G major. Partita, flute and harpsichord, S 987, C minor. Jean-
Pierre Rampal, Alain Marion, flt; Robert Veyron-Lacroix, hpd;
Jordi Saval, vla da gamba; Robert Gendre, vln. RCA CRL 3-5820.
  ++HR 7-75 p69   ++SFC 6-1-75 p21
  ++NR 5-75 p9

124 French suites, nos. 1-4, S 812-15. Glenn Gould, pno. Columbia
  M 32347. Tape (c) MT 32347. (also CBS 73393)
   +Gr 4-75 p1991   +RR 4-75 p42
   ++HF 2-74 p79   ++SFC 12-16-73 p47
   +HF 5-74 p122 tape

French suites, nos. 1-6, S 812-17. cf Harpsichord works (DG
 2722 020).

Fuga IV. cf Crystal S 202.

Fugue, G minor. cf Works, selections (RCA ARL 1-0880).

Fugues, S 574, G minor; S 579, B minor. cf Organ works (Telefunken
 BC 24104).

Fugue, S 574, C minor; S 579, B minor. cf Organ works (Telefunken
 EK 6-35082).

Fugues, S 574, C minor; S 578, G minor; S 579, B minor; S 1080,
 D minor. cf Organ works (DG 2722 014).

Fugues, S 575, C minor; S 578, G minor. cf Organ works (Telefunken
 BC 25098 T/1-2).

Fugue, S 577, G major. cf Organ works (Telefunken BC 25102 T/1-2).

125 Fugue, S 578, G minor (Little). Pastorale, S 590, F major.
  Preludes and fugues, S 532, D major; S 535, G minor; S 536,
  A major; S 550, G major. Helmut Walcha, org. DG Archive
  2533 160.
   ++St 1-75 p106

Fugue, S 578, G minor. cf Cantata, no. 156: Arioso.

Fugue, S 578, G minor. cf Works, selections (Ember ECL 9007).

Fugue, S 578, G minor. cf Columbia CXM 32088.

126 Fugue, lute, S 1000, G minor. Prelude, S 999, C minor. Prelude,
  fugue and allegro, S 998, E flat major. Suites, lute (guitar),
  S 995-997, 1006a. John Williams, gtr. CBS 79203 (2).
   ++Gr 1-75 p652   +RR 11-75 p68
   ++HFN 11-75 p148   +-SFC 10-27-75 p24
   ++HFN 12-75 p173 tape ++St 12-75 p116
   +NR 10-75 p13   ++STL 10-5-75 p36

127 Fugue, lute, S 1000, G minor. Suite, lute, S 995, G minor. Suite,
  lute, S 1006a, E major. Narciso Yepes, gtr. DG 2530 461.
   +NR 3-75 p14

128 Fugue, lute, S 1000, G minor. Prelude, S 999, C minor. Prelude,
  fugue and allegro, S 998, E flat major. Suite, lute, S 995, G
  minor. Suite, lute, S 996, E minor. Suite, lute, S 997, C
  minor. Suite, lute, S 1006a, E major. Narciso Yepes, lt. DG
  Archive 2708 030 (2).
   +-Gr 4-74 p1875   ++SFC 1-19-75 p27
   +HF 5-74 p81   +St 5-74 p95
   +-RR 4-74 p60

Fugue, lute, S 1000, G minor. cf Lute works (Columbia M2 33510).

Fugue, lute, S 1000, G minor. cf Works, selections (Erato STU 70885).

Fugue on the Magnificat, S 733. cf Pelca PSRK 41013/6.

Gavottes, 1 and 2. cf Works, selections (Ember ECL 9007).

Der Geist hilft unsrer Schwachheit auf, S 226. cf Works, selections
 (Supraphon 50871/2).

Geistliches Lied, no. 51, S 487: Mein Jesu. cf Works, selections
(Decca PFS 4278).
129 Harpsichord works: Applicatio, S 994, C major. Fantasia and
fugue, S 904, A minor. Inventions, 2 part, S 772-6 (15). Inventions, 3 part, S 787-801 (15). Preludes, S 924-30, 939-42,
999 (12); S 933-8 (6). Minuets, nos. 1-3, S 841-843. Suites,
S 818a, S 819, A minor, E flat major. Variations, harpsichord,
S 998. Das wohltemperierte Klavier, S 846-893. Toccatas,
S 912-916. Ralph Kirkpatrick, Helmut Walcha, Karl Richter,
hpd. DG Archive 2722 015 (11).
　　+-Gr 6-75 p67　　　　　　　　+-HFN 6-75 p84
130 Harpsichord works: Aria variata alla maniera Italiana, S 989, A
minor. Capriccio sopra la lontananze del suo fratello dilettissimo, S 992, B flat major. Chromatic fantasy and fugue, S 903,
D minor. English suites, nos. 1-6, S 806-811. Fantasia, S 906,
C minor. French suites, nos. 1-6, S 812-817. Fantasia and
fugue, S 944, A minor. Concerto, S 971, F major. Partitas,
harpsichord, nos. 1-6, S 825-830. Partita, harpsichord, S 831,
B minor. Huguette Dreyfus, Ralph Kirkpatrick, hpd. DG
2722 020 (10). Reissues)
　　+RR 11-75 p67
131 Harpsichord works: Chromatic fantasia and fugue, S 903, D minor.
Fantasia, S 906, C minor. Concerto, harpsichord, S 971, F
major. Toccata and fugue, S 911, C minor. Zuzana Ruzickova,
hpd. Supraphon 111 0750. (Reissue from Erato ERA 9035)
　　+-RR 7-73 p66　　　　　　　　+-RR 12-75 p77
Instrumental movement, S 1040, F major. cf Works, selections
(Supraphon 50871/2).
Inventions, 2 part, S 772-6 (15). cf Harpsichord works (DG Archive
2722 015).
Invention,2 part, no. 1, S 772, C major. cf International Piano
Library IPL 5005/6.
Inventions, 3 part, S 787-801 (15). cf Harpsichord works (DG
Archive 2722 015).
Kleines harmonisches Labyrinth, S 591, C major. cf Organ works
(Telefunken BC 25102 T/1-2).
132 Die Kunst der Fuge (The art of the fugue), S 1080. AMF; Neville
Marriner. Philips 6747 172 (2).
　　+Gr 10-75 p601　　　　　　　　++NR 12-75 p4
　　++HFN 9-75 p92　　　　　　　　++RR 9-75 p26
Die Kunst der Fuge, S 1080. cf Organ works (DG Archive 2722 016).
133 The art of the fugue, nos. 1, 4-5, 7, 9, 11, 13-17, 19, S 1080.
Fritz Heitmann, org. Telefunken AJ 6-41905. (Reissues from
E 3777/8, LGM 65009L)
　　+Gr 11-75 p854
Leonore overture, no. 3, op. 72b. cf Concerto, piano, no. 4,
op. 58, G major.
134 Lute works: Fugue, S 1000, G minor. Prelude, S 999, C minor.
Prelude, flute and allegro, S 998, E major. Suite, S 996, E
minor. Suite, S 997, G minor. Suite, S 995, G minor. Suite,
S 1006a, E major. John Williams, gtr. Columbia M2 33510 (2).
(also CBS 79203 Tape (c) 40-79203)
　　++Gr 1-75 p652　　　　　　　　+RR 11-75 p68
　　++HFN 11-75 p148　　　　　　　+-SFC 10-27-75 p24
　　++HFN 12-75 p173 tape　　　　++St 12-75 p116
　　　+NR 10-75 p13　　　　　　　　++STL 10-5-75 p36

135 Magnificat, S 243, D major. Cantata, no. 110, Unser Mund sei voll
     Lachens. Theo Altmeyer, t; Siegmund Nimsgern, bs; Tölzer Boys'
     Choir; Collegium Aureum; Gerhard Schmidt-Gaden. BASF BAC 3067.
             +Gr 2-75 p1523   +RR 1-75 p54
     Magnificat, S 243, D major. cf Cantatas (DG Archive 2722 018).
     Magnificat, S 243, D major: Vom Himmel hoch; Freut euch und
       jubiliert. cf Chorale preludes: Jesu, meine Freude, S 713:
       Fantasia.
136 Mass, B minor (Missa 1733). Rotraud Hansmann, Emiko Ilyama, s;
     Helen Watts, con; Kurt Equiluz, t; Max van Egmond, bs; Vienna
     Boys' Choir; Vienna Chorus; VCM; Nikolaus Harnoncourt. Tele-
     funken Tape (c) 4-41135.
             +HFN 7-75 p90 tape   +RR 7-75 p69 tape
     Masses, S 232-236. cf Works, selections (DG Archive 2722 017).
137 Mass, S 232, B minor. Gundula Janowitz, s; Christa Ludwig, ms;
     Peter Schreier, t; Robert Kerns, bar; Karl Ridderbusch, bs;
     Vienna Singverein; BPhO; Herbert von Karajan. DG 2740 112 (3).
     Tape (c) 3371 012. (also DG 2709 049)
             -Gr 10-74 p737   ++RR 11-74 p85
             +-HF 1-75 p74   +RR 2-75 p74 tape
             +-NR 1-75 p11   ++St 1-75 p103
             -NYT 11-10-74 pD1
138 Mass, S 232, B minor. Yvonne Perrin, Wally Stämpfli, s; Magali
     Schwartz, ms; Claudine Perret, con; Olivier Dufour, t; Philippe
     Huttenlocher, bar; Niklaus Tuller, bs; Lausanne Instrumental
     Ensemble; Michel Corboz. Erato STU 70715/7 (3). (also RCA
     FVL 2-5715. Musical Heritage Society Tape (c) MHC 2104)
             +Gr 12-72 p1181   ++MJ 2-75 p39
             +HF 3-75 p75   +RR 12-72 p84
             +-HF 8-75 p116 tape   ++SFC 1-19-75 p27
139 Mass, S 232, B minor. Felicity Palmer, s; Helen Watts, alto;
     Robert Tear, t; Michael Rippon, bs; Amor Artis Chorale; ECO;
     Johannes Somary. Pye Vanguard VSD 71190/92. (Q) VSQ 30037/39.
             +-Audio 8-75 p80   +-NR 2-75 p7
             +-Gr 7-75 p223   +-RR 6-75 p77
             +HF 3-75 p75   +-SFC 1-19-75 p28
             +HFN 6-75 p84   +St 1-75 p103
             +MJ 2-75 p31
     Mass, S 232, B minor: Agnus Dei. cf Telefunken AJ 6-41867.
     Minuet. cf Klavier KS 536.
     Minuet, G major. cf Connoisseur Society (Q) CSQ'2066.
     Minuets, nos. 1-3, S 841-843. cf Harpsichord works (DG Archive
       2722 015).
     Minuets, nos. 1 and 2. cf Works, selections (Ember ECL 9007).
     Motets, nos. 1-6, S 225-228, 230-231. cf Chorale preludes: Jesu,
       meine Freude, S 713: Fantasia.
140 Motets, nos. 3, Jesu, meine Freude, S 227; no. 5, Komm, Jesu,
     komm, S 229; no. 6, Lobet den Herrn, alle Heiden, S 230.
     Cantata, no. 118, O Jesu Christ, mein's Lebens Licht. Aeolian
     Singers; Sebastian Forbes. London STS 15187.
             -HF 10-74 p87   +NR 12-74 p10
             -MJ 1-75 p48
     Motets: Fürchte dich nicht, S228; Der Geist hilft unser Schwachheit
       auf, S 226; Ich lasse dich nicht, S Anh 159; Jesu meine Freude,
       S 227; Komm, Jesu, komm, S 229; Lobet den Herrn, S 230; Sei
       Lobe und Preis, S 231; Singet dem Herrn, S 225. cf Works,
       Selections (DG Archive 2722 017).

Musette, D major. cf Connoisseur Society (Q) CSQ 2066.
Das Musikalisches Opfer, S 1079, C minor. cf Works, selections
(DG Archive 2722 013).

141 Organ works: Cantata, no. 29: Sinfonia. Concerto, no. 2, S 593,
     A minor. Preludes and fugues, S 532, D major; S 544, B minor.
     Toccata and fugue, S 565, D minor. Michael Murray, org.
     Advent 5010.
        +HF 7-75 p70   +NR 4-75 p10
142 Organ works: Canzona, S 588, D minor. Fantasia and fugue, S 542,
     G minor. Preludes and fugues, S 531, C major; S 532, D major;
     S 543, A minor. Toccata, adagio and fugue, S 546, C major.
     Chorale preludes: Gelobet seist du, Jesu Christ, S 604; Herzlich
     tut, mich verlangen, S 727; Ich ruf zu dir, Herr Jesu Christ,
     S 639; Nun komm der Heiden Heiland, S 659; O Mensch bewein
     dein Sunde gross, S 622; Wenn wir in höchsten Nöten sein, S 641.
     Albert Schweitzer, org. CBS 78256 (2). (Reissues from Philips
     ALB 3198, 3196, 3197)
        -Gr 1-75 p1365   -RR 1-75 p45
143 Organ works: Allabreve, S 589, D major. Canzona, S 588, D minor.
     Duets, S 802/5 (4). Trio sonata, no. 1, S 525, E flat major.
     Trio sonata, no. 6, S 530, G major. Helmut Walcha, org.
     DG 2533 140.
        +NR 8-75 p13
144 Organ works: Allabreve, S 589, D major. Canzona, S 588, D minor.
     Fantasias, S 562, C minor; S 572, G major. Fantasias and
     fugues, S 537, C minor; S 542, G minor. Fugues, S 574, C minor;
     S 578, G minor; S 579, B minor; S 1080, D minor. Passacaglia
     and fugue, S 582, C minor. Pastorale, S 590, F major. Preludes
     and fugues, S 531-536, S 539, S 541, S 543-552. Trio sonatas,
     S 525-530 (6). Toccata, adagio and fugue, S 564, C major.
     Toccatas and fugues, S 538, D minor; S 540, F major; S 565,
     D minor. Helmut Walcha, org. DG 2722 014 (8). (Reissue from
     2722 002/1)
        *Gr 8-75 p347   +-HFN 7-75 p72
145 Organ works: Chorales (Schubler) (6). Chorale preludes of diverse
     kinds (18). Chorale preludes: Sei gegrüsset Jesus gütig, D 678;
     Vom Himmel hoch variations, S 679. Clavierübung, Pt. 3. Die
     Kunst der Fuge, S 1080. Helmut Walcha, org. DG Archive
     2722 016 (8).
        ++RR 12-75 p77
146 Organ works: Canzona, S 588, D minor. Chorale preludes: Allein
     Gott in der Hoh sei Ehr, S 715; Nun freut euch, lieben Christen
     g'mein, S 734; O Mensch, bewein dein Sunde gross, S 622.
     Prelude and fugue, S 536, A major. Toccata, adagio and fugue,
     S 564, C major. Toccata and fugue, S 565, D minor. Konrad
     Voppel, org. Pelca PSR 40586.
        +NR 11-75 p15
147 Organ works: Allabreve, S 589, D major. Canzona, S 588, D minor.
     Fantasia, S 572, G major. Fugues, S 574, G minor; S 579, B
     minor. Passacaglia and fugue, S 582, C minor. Pastorale, S
     590, F major. Prelude and fugue, S 551, A minor. Toccatas (3).
     Michel Chapuis, org. Telefunken BC 24104 (2).
        +NR 7-75 p15   ++SFC 10-26-74 p24
148 Organ works: Fantasias, S 563, B minor; S 570, C major. Fugues,
     S 578, G minor; S 575, C minor. Preludes, S 569, A minor;
     S 568, G major. Trio, S 584, G minor. Sonatas, organ, nos.

1-6, S 525-530. Michel Chapuis, org. Telefunken BC
25098 T/1-2 (2).

+AR 5-75 p51	+MQ 10-74 p685
+Gr 4-74 p1876	+-NR 3-74 p14
+HF 9-74 p83	+RR 3-74 p51
++HFN 3-74 p22	++SFC 1-6-74 p32
++MJ 5-74 p52	+-St 6-74 p115

149 Organ works: Fantasia and fugue, S 562, C minor, fragment. Toccata
and fugue, S 540, F major. Preludes and fugues, S 544, B minor;
S 549, C minor; S 550, G major; S 533, E minor; S 531, C major;
S 535, G minor; S 532, D major; S 547, C major; S 548, E minor.
Michel Chapuis, org. Telefunken BC 25101 T/1-2 (2).

+AR 5-75 p51	+St 1-75 p106
+NR 11-74 p9	

150 Organ works: Aria, S 587, F major. Concerti, organ, nos. 1-6,
S 592-597. Chorale preludes: Ach Gott, vom Himmel sich darein,
S 741; An Wasserflussen Babylon, S 653b; O Lamm Gottes unschuldig,
S 656. Fantasia, S 573, C major. Fantasia and fugue, S 537,
C minor. Fugue, S 577, G major. Kleines harmonisches Labyrinth,
S 591, C major. Prelude, trio and fugue, S 545b, B major. Trio,
S 586, G major. Trio, S 1027a, G major. Michel Chapuis, org.
Telefunken BC 25102 T/1-2 (2).

+AR 5-75 p51	+-NR 2-75 p13
+Gr 2-75 p1515	+RR 2-75 p47

151 Organ works: Allabreve, S 589, D major. Canzona, S 588, D minor.
Fantasia, S 572, G major. Fugue, S 574, C minor; S 579, B
minor. Passacaglia and fugue, S 582, C minor. Pastorale, S
590, F major. Prelude and fugue, S 551, A minor. Toccata,
adagio and fugue, S 564, C major. Toccatas and fugues, S 565,
D minor; S 566, E major. Trio, S 583, D minor. Helmut Walcha,
org. Telefunken EK 6-35082 (2).

+Gr 7-75 p213	+RR 9-75 p56
++HFN 6-75 p84	

Partita, B flat major: Gigue. cf Angel S 36076.
Partita, B flat major: Gigue. cf International Piano Library IPL
5005/6.
Partita, B flat major: Menuets I and II. cf Angel S 36076.
Partita, flute, S 1013, A minor. cf Flute works (RCA CRL 3-5820).
Partita, flute, S 1013, A minor. cf Sonatas, flute, S 1030-5.
152 Partita, flute and harpsichord, S 987, C minor. CZERNY: Duo
concertante, flute and piano, op. 129. Alexander Murray, Murray
flute; Martha Goldstein, hpd and pno. Pandora PAN 102.
        -AR 2-75 p24
Partita, flute and harpsichord, S 987, C minor. cf Flute works
(RCA CRL 3-5820).
Partitas, harpsichord, nos. 1-6, S 825-830. cf Harpsichord works
(DG 2722 020).
Partita, harpsichord, no. 5, S 829, G major. cf Chromatic fantasia
and fugue, S 903, D minor.
Partita, harpsichord, S 831, B minor. cf Harpsichord works
(DG 2722 020).
Partita, harpsichord, S 831, B minor. cf Concerto, harpsichord,
S 971, F major.
Partita, violin, no. 2, S 1004, D minor. cf FRANCK: Sonata, violin
and piano, A major.
Partita, violin, no. 2, S 1004, D minor: Chaconne. cf Works,

selections (HMV ASD 2971).
Partita, violin, no. 2, S 1004, D minor: Chaconne. cf Works,
   selections (RCA ARL 1-0880).
Partita, violin, no. 2, S 1004, D minor: Chaconne. cf RCA ARL
   2-0512.
Partita, violin, no. 3, S 1006, E major: Gavotte, Preludio. cf
   Discopaedia MB 1003.
Partita, violin, no. 3, S 1006, E major: 2 minuets. cf Discopaedia
   MB 1010.
Partita, violin, no. 3, S 1006, E minor: Minuets 1 and 2. cf
   RCA ARM 4-0942/7.
Partita, violin, no. 3, S 1006, E major: Preludio. cf Works,
   selections (RCA ARL 1-0880).
Partita, violin, no. 3, S 1006, E major: Prelude, Loure, Gigue.
   cf CBS 76420.
Passacaglia and fugue, S 582, C minor. cf Organ works (DG 2722 014)
Passacaglia and fugue, S 582, C minor. cf Organ works (Telefunken
   BC 24104).
Passacaglia and fugue, S 582, C minor. cf Organ works (Telefunken
   EK 6-35082).
Passacaglia and fugue, S 582, C minor. cf Toccata and fugue,
   S 565, D minor.
Passacaglia and fugue, S 582, C minor. cf Works, selections (Decca
   PFS 4278).
Passacaglia and fugue, S 582, C minor. cf University of Iowa
   Press, unnumbered.
Pastorale, S 590, F major. cf Fugue, S 578, G minor
Pastorale, S 590, F major. cf Organ works (DG 2722 014).
Pastorale, S 590, F major. cf Organ works (Telefunken BC 24104).
Pastorale, S 590, F major. cf Organ works (Telefunken EK 6-35082).
Pastorale, S 590, F major: Aria. cf CBS Classics 61579.
Praeludium, E major. cf CBS 77513.
Prelude, C major. cf Works, selections (Ember ECL 9007).
Prelude, C minor. cf Connoisseur Society (Q) CSQ 2065.
Prelude, G minor. cf Columbia Special Products AP 12411.
Preludes, S 925, 929, 933, 941, 936. cf Pelca PSR 40598.
Preludes, S 568, G major; S 569, A minor. cf Organ works
   (Telefunken BC 25098 T/1-2).
Preludes, S 924-30, 929-42, 999 (12); S 933-38 (6). cf Harpsichord
   works (DG Archive 2722 015).
Prelude, S 929, B minor. cf Works, selections (HMV ASD 2971).
153 Prelude, S 999, C minor. Prelude, fugue and allegro, S 998, E flat
     major. Suite, lute, S 996, E minor. Suite, lute, S 997, C
     minor. Narciso Yepes, gtr. DG 2530 462.
        +NR 3-75 p14
Prelude, S 999, C minor. cf Bourrée, E minor.
Prelude, S 999, C minor. cf Fugue, S 1000, G minor.
Prelude, S 999, C minor. cf Fugue, lute, S 1000, G minor.
Prelude, S 999, C minor. cf Lute works (Columbia M2 33510).
Prelude, S 999, C minor. cf Works, selections (Erato STU 70885).
154 Prelude and fugue, B flat minor. MARTINU: Etudes and polkas.
     SCRIABIN: Etudes, op. 8, nos. 2, 8, 12. SCHUBERT: Impromptus,
     op. 142, nos. 3 and 4, D 935, B flat major and F minor. Martina
     Maixerova, pno. Panton 110 2329.
        +-HFN 10-75 p136
Prelude and fugue, F minor. cf Wealden WS 110.

Preludes and fugues, S 531, C major; S 532, D major; S 543, A
  minor. cf Organ works (CBS 78256).
Preludes and fugues, S 531-3, 535, 547-50. cf Fantasia, S 562,
  C minor, excerpt.
Preludes and fugues, S 531-536, S 539, S 541, S 543-552. cf Organ
  works (DG 2722 014).
Preludes and fugues, S 532-43. cf Toccata and fugue, S 565, D
  minor.
Prelude and fugue, S 532, D major. cf Organ works (Advent 5010).
Preludes and fugues, S 532, D major; S 535, G minor; S 536, A
  major; S 550, G major. cf Fugue, S 578, G minor.
Prelude and fugue, S533, E minor. cf Chorale prelude: O Mensch
  bewein dein Sunde grosse, S 622.
Preludes and fugues, S 534, F minor; S 536, A major; S 539, D
  minor; S 541, G major; S 543, A minor; S 545, C major; S 546,
  C minor. cf Fantasias and fugues, S 537, C minor; S 542, G
  minor.
Prelude and fugue, S 536, A major. cf Organ works (Pelca 40586).
Prelude and fugue, S 538, D minor. cf Fantasia, S 562, C minor.
Prelude and fugue, S 538, D minor. cf Wealden WS 139.
Prelude and fugue, S 543, A minor. cf Works, selections (HMV
  ASD 2971).
Prelude and fugue, S 544, B minor. cf Organ works (Advent 5010).
Preludes and fugues, S 544, B minor; S 549, C minor; S 550, G
  major; S 533, E minor; S 531, C major; S 535, G minor; S 532,
  D major; S 547, C major; S 548, E minor. cf Organ works
  (Telefunken BC 25101 T/1-2).
Prelude and fugue, S 546, C minor. cf HMV HLM 7065.
Prelude and fugue, S 546, C minor. cf Telefunken DX 6-35265.
Prelude and fugue, S 551, A minor. cf Organ works (Telefunken
  BC 24104).
Prelude and fugue, S 551, A minor. cf Organ works (Telefunken
  EK 6-35082).
Prelude and fugue, S 552, E flat major. cf Fantasia, S 562, C
  minor.
Prelude and fugue, S 552, E flat major. cf Abbey LPB 738.
Prelude, fugue and allegro, S 998, E flat major. cf Fugue, lute,
  S 1000, G minor (CBS 79203).
Prelude, fugue and allegro, S 998, E flat major. cf Fugue, lute,
  S 1000, G minor (DG Archive 2708 030).
Prelude, fugue and allegro, S 998, E flat major. cf Concerto,
  harpsichord, S 971, F major.
Prelude, fugue and allegro, S 998, E flat major. cf Lute works
  (Columbia M2 33510).
Prelude, fugue and allegro, S 998, E flat major. cf Prelude, S 999,
  C minor.
Prelude, fugue and allegro, S 998, E flat major. cf Works,
  selections (Erato STU 70885).
Prelude, fugue and allegro, S 998, E flat major. cf BASF BAB 9005.
Prelude, fugue and allegro, S 998, E flat major. cf CBS 72526.
Prelude, trio and fugue, S 545b, B major. cf Organ works
  (Telefunken BC 25102 T/1-2).
Quia respexit. cf Works, selections (Ember ECL 9007).
155 St. John Passion, S 245. Marianne Koehnlein-Goebel, Elly Ameling,
  s; Julia Hamari, con Dieter Ellenbeck, Wolfgang Isenhardt,
  Werner Hollweg, t; Walter Berry, Allan Ahrans, Manfred Ackermann,

Hermann Prey, bs; Stuttgart Hymnus Boys' Choir; Stuttgart
Chamber Orchestra; Karl Munchinger. Decca SET 590/2 (3).
　　+—Gr 4-75 p1844   +RR 4-75 p58
　St. John Passion, S 245. cf St. Matthew Passion, S 244.
　St. John Passion, S 245: Es ist vollbracht. cf Telefunken AJ
　　6-41867.
156　St. Luke Passion, S 246. Charlotte Lehmann, s; Elizabeth Kunstler,
　　con; Georg Jelden, t; Ulrich Schaible, bs; Balinger Kantorie;
　　Musica Tubingen Collegium Chamber Orchestra; Gerhard Rehm. Oryx
　　BACH 1123/5 (3).
　　　+—Gr 8-75 p351   +HFN 8-75 p71
157　St. Matthew Passion, S 244. St. John Passion, S 245. Irmgard
　　Seefried, Antonia Fahberg, Evelyn Lear, s; Hertha Töpper, con;
　　Ernst Häfliger, t; Dietrich Fischer-Dieskau, bar; Keith Engen,
　　Max Proebstl, Hermann Prey, bs; Munich Bach Orchestra and Choir;
　　Munich Boys' Choir; Karl Richter. DG Archive 2722 010 (7).
　　Tape (c) 3376 001/2. (Reissue from SAPM 198 009/12, SAPM
　　198 329/30)
　　　+Gr 11-74 p935   +—RR 12-74 p19
　　　+HF 11-75 p149 tape
158　St. Matthew Passion, S 244. Jo Vincent, s; Ilona Durigo, ms;
　　Karl Erb, Louis van Tulder, t; Willem Ravelli, Hermann Schey, bs;
　　Piet van Egmond, org; Johannes den Hertog, hpd; Amsterdam
　　Toonkunst Choir; Zanglust Boys' Choir; COA; Willem Mengelberg.
　　Philips 6747 168 (3). (Reissue from Columbia SL 179)
　　　+HF 8-75 p76   +НYT 8-10-75 pD14
　　　+NR 7-75 p9   +St 8-75 p94
159　St. Matthew Passion, S 244: Kommt, ihr Töchter, helft mir klagen;
　　Blute nur, du liebes Herz; Ich will bei meinem Jesu wachen;
　　O Mensch, bewein dein Sunde grosse; Erbarme dich, mein Gott;
　　Und von der sechsten Stunde; Wenn einmal soll scheiden; Und
　　siehe da, der Vorhang im Tempel zerris...Wahrlich, dieser ist
　　Gottes Sohn gewesen; Mache dich, mein Herze, rein; Wir setzen
　　uns mit Tränen nieder. Irmgard Seefried, s; Hertha Töpper, con;
　　Ernst Häfliger, t; Dietrich Fischer-Dieskau, bar; Kieth Engen,
　　bs; Munich Boys' Choir; Munich Bach Orchestra and Chorus; Karl
　　Richter. DG 2538 126. (Reissue from SAM 198 009/12)
　　　+Gr 5-75 p1999   +RR 5-75 p57
　　　+HFN 6-75 p84
160　St. Matthew Passion, S 244: Kommt, ihr Töchter, helft mir klagen;
　　Da versammleten sich die Hohenpriester; Aber am ersten Tage
　　der süssen Brot; O Schmerz...Was ist die Ursach; Ich will bei
　　meinem Jesu wachen...So schalfen unsre Sünden; Und er kam zu
　　seinen Jüngern; So ist mein Jesu nun gefangen...Lasst ihn,
　　haltet; Ach, nun ist mein Jesu hin...Wo ist denn dein Freund;
　　Auf das Fest aber hatte; Und da sie an die Stätte kamen; Sehet,
　　Jesus hat die Hand...Wohin; Und von der sechsten Stunden an;
　　Nun ist der Herr zur Ruh...Mein Jesu, gute Nacht; Wir setzen
　　uns mit Tränen nieder. Paul Esswood, Tom Sutcliffe, c-t;
　　Kurt Equiluz, t; Karl Ridderbusch, Max van Egmond, bs; Vienna
　　Boys' Choir; Regensburg Cathedral Boys' Choir; Cambridge,
　　King's College Chapel Choir; VCM; Nikolaus Harnoncourt, David
　　Willcocks. Telefunken SAWT 9606. (Reissue from SAWT 9572/5)
　　　+Gr 3-75 p1687   +—RR 3-75 p56
　　St. Matthew Passion, S 244: O Haupt, voll Blut und Wunden. cf
　　HMV TC MCS Tape (c) 14.

St. Matthew Passion, S 244: O sacred head surrounded. cf Pye GH 589.
Sarabande. cf Klavier KS 536.
Sarabande, E minor. cf Bourrée, E minor.
Sonata, solo flute, S 1013, A minor. cf Works, selections
    (DG Archive 2722 013).
Sonata, flute and harpsichord, G minor. cf Klavier KS 537.
Sonata, flute (violin) and harpsichord, S 1020, G minor. cf
    Flute works (RCA CRL 3-5820).
Sonata, flute and harpsichord, S 1020, G minor. cf Concerto,
    violin and strings, S 1042, E major.
Sonata, flute and harpsichord, S 1020, G minor. cf Works, selec-
    tions (DG Archive 2722 013).
161 Sonatas, flute and harpsichord, S 1030-5. Partita, flute, S 1013,
    A minor. Stephen Preston, flt; Trevor Pinnock, hpd; Jordi
    Savall, vla da gamba. CRC CRD 1014/5 (2).
        +Gr 8-75 p336   +STL 11-2-75 p38
        +RR 8-75 p54
Sonatas, flute and harpsichord, S 1030-35. cf Flute works
    (RCA CRL 3-5820).
Sonatas, flute and harpsichord, S 1030-35. cf Works, selections
    (DG Archive 2722 013).
162 Sonata, flute (oboe) and harpsichord, S 1030, G minor. COUPERIN, F.:
    Les goûts réunis: Concert, no. 9, E major. MARAIS: Les folies
    d'Espagne: Couplets. Heinz Holliger, ob, ob d'amore; Christiane
    Jaccottet, hpd; Marcal Cervera, vla da gamba. Philips 6500 618.
        ++Gr 6-75 p56   ++NR 5-75 p9
        ++HF 6-75 p107   ++RR 6-75 p59
        ++HFN 6-75 p101
Sonata, flute and harpsichord, S 1031, E flat major: Siciliano.
    cf Works, selections (Ember ECL 9007).
Sonata, flute and harpsichord, S 1031, E flat major: Siciliano.
    cf Works, selections(HMV ASD 2971).
Sonata, flute, violin and harpsichord, S 1038, G major. cf Flute
    works (RCA CRL 3-5820).
Sonata, 2 flutes and harpsichord, S 1039, G major. cf Flute
    works (RCA CRL 3-5820).
Sonata, 2 flutes and harpsichord, S 1039, G major. cf Works,
    selections (DG Archive 2722 013).
Sonatas, harpsichord, S 964-966. cf Adagio, S 968, G major.
Sonatas, organ, nos. 1-6, S 525-530. cf Organ works (Telefunken
    BC 25098 T/1-2).
Sonata, organ, no. 6, S 530, G major. cf Wealden WS 145.
Sonatas, viola da gamba, nos. 1-3, S 1027-1029. cf Works,
    selections (DG Archive 2722 013).
163 Sonatas, viola da gamba and harpsichord, nos. 1-3, S 1027-9.
    Leonard Rose, vlc; Glenn Gould, pno. Columbia M 32934. (also
    CBS 76373)
            -Gr 5-75 p1980   +-NR 11-74 p7
        +-HF 12-74 p97   +RR 5-75 p51
        +-HFN 5-75 p124   +-St 3-75 p96
164 Sonatas, viola da gamba and harpsichord, nos. 1-3, S 1027-29.
    TELEMANN: Sonata, viola da gamba and harpsichord, A minor.
    Eva Heinitz, vla da gamba; Malcolm Hamilton, hpd. Delos DELS
    15341.
        +-St 3-75 p96
165 Sonatas, viola and harpsichord, nos. 2, 4-6. Paul Hersch, vla;

Laurette Goldberg, hpd. 1750 ARCH S 1756.
+NR 12-75 p8
Sonatas, violin, no. 1, S 1001, G minor; no. 3, S 1005, C major.
cf RCA ARM 4-0942/7.
Sonata, violin and figured bass, E minor. cf Concerto, violin and
strings, S 1042, E major.
Sonatas, 2 violins and harpsichord, S 1036-1039, S 1079. cf
Works, selections (Supraphon 50871/2).
166 Sonatas and partitas, solo violin, S 1001-6. Nathan Milstein, vln.
DG 2709 047 (3).
++Gr 4-75 p1830   ++SFC 10-26-75 p24
++NR 12-75 p15   ++STL 5-4-75 p37
++RR 4-75 p47
167 Sonatas and partitas, solo violin, S 1001-6. Georges Enesco, vln.
Olympic 8117/3 (3).
+-HF 3-75 p76   +-NR 11-74 p11
168 Sonatas and partitas, solo violin, S 1001-6. Bretislav Novotný,
vln. Supraphon 111 1101/3 (3).
+HF 3-75 p76   +NR 2-75 p14
169 Sonatas and partitas, solo violin, S 1001-6. Paul Zukofsky, vln.
Vanguard VSD 71194/6 (3).
+-HF 3-75 p76   -SFC 1-19-75 p28
++MJ 2-75 p40   +SR 1-11-75 p51
+-NR 2-75 p14
170 Sonata and partita, solo violin, S 1003, A minor. DEBUSSY:
Sonata, violin and piano, no. 3, G minor. MOZART: Divertimento,
no. 15, D 287, B flat major. Joseph Szigeti, vln; Andor
Foldes, pno; Chamber Orchestra; Max Goberman. Rococo 2062.
(Reissues from various Columbia 78 rpm originals)
+HF 2-75 p87
171 Sonata and partita, solo violin, S 1004, D minor. Sonata and
partita, solo violin, S 1005, C major. Kyung-Wha Chung, vln.
Decca SXL 6721. Tape (c) KSXC 6721.
+-Gr 11-75 p859   +-RR 11-75 p67
++HFN 11-75 p149   +STL 11-2-75 p38
Sonata and partita, solo violin, S 1005, C major. cf Sonata and
partita, solo violin, S 1004, D minor
Songs: Ach, das nicht, S 439; Die bittre Leidenzeit, S 450; Brich
entzwei, S 444; Dir, Dir Jehova, S 452; Eins is not; S 453; Es
kostet viel, S 459; Gib dich zufrieden, S 521; Gott lebet noch,
S 461; Gott, wie gross, S 462; Die guldne Sonne, S 451; Ich
lass dich nicht, S 467; Ich steh an deiner Krippen, S 469; Ihr
Gestirn, S 476; Komm, susser Tod, S 478; Kommt, Seelen, S 479;
Kommt wieder, S 480; Der lieben Sonne Licht, S 446; Liebster
Herr Jesu, S 484; Mein Jesu, S 487; O Jesulein, S 493; So gehst
du nun, S 500; So gibst du nun, S 501; Steh ich bei meinem Gott,
S 503; Vergiss mein nicht, S 505; Wie wohl ist mir, S 517; Wo
ist mein Schaflein, S 507. cf Works, selections (DG 2722 017).
Song: Es kostet viel, S 459. cf Telefunken AJ 6-41867.
Suites, S 818a, S 819, A minor, E flat major. cf Harpsichord works
(DG Archive 2722 015).
Suites, lute (guitar), S 995-997, 1006a. cf Fugue, S 1000, G minor.
172 Suite, lute, S 995, G minor. CONRADI: Suite, lute, C major.
WEISS: Tombeau sur la mort de M. Comte de Logy. Eugen M. Dom-
bois, baroque lute. Philips Seon 6565 018.
+Gr 3-75 p1677   +RR 3-75 p39

Suite, lute, S 995, G minor. cf Lute works (Columbia M2 33510).
Suite, lute, S 995, G minor. cf Fugue, lute, S 1000, G minor.
Suite, lute, S 995, G minor. cf Fugue, lute, S 1000, G minor
  (DG Archive 2708 030).
Suite, lute, S 995, G minor. cf Philips 6833 159.
Suite, lute, S 995, G minor: Prelude, Presto. cf Works, selections
  (Erato STU 70885).
Suite, lute, S 996, E minor. cf Fugue, lute, S 1000, G minor.
Suite, lute, S 996, E minor. cf Lute works (Columbia M2 33510).
Suite, lute, S 996, E minor. cf Prelude, S 999, C minor.
Suite, lute, S 996, E minor. cf Works, selections (Erato STU 70885).
Suite, lute, S 996, E minor. cf Swedish Society SLT 33189.
Suite, lute, S 996, E minor: Prelude and bourrée. cf Amberlee ACL
  501X.
Suite, lute, S 997, C minor. cf Fugue, lute, S 1000, G minor.
Suite, lute, S 997, C minor. cf Lute works (Columbia M2 33510).
Suite, lute, S 997, C minor. cf Prelude, S 999, C minor.
Suite, lute, S 1006a, E major. cf Fugue, lute, S 1000, G minor.
Suite, lute, S 1006a, E major. cf Fugue, lute, S 1000, G minor
  (DG 2530 461).
Suite, lute, S 1006a, E major. cf Lute works (Columbia M2 33510).
173 Suites, orchestra, S 1066-1069. Ars Rediviva Orchestra; Milan
  Munclinger. Supraphon 110 1361/2 (2).
        +-HFN 8-75 p71   +RR 7-75 p22
        +NR 4-75 p5
Suites, orchestra, S 1066-1069. cf Brandenburg concerti, nos.
  1-6, S 1046-51 (Decca 14BB 213/7).
Suites, orchestra, S 1066-1069. cf Brandenburg concerti, nos.
  1-6, S 1046-51 (DG Archive Tape (c) 3376 003).
Suite, orchestra, S 1066, C major: Minuets, nos. 1, 2. cf
  Philips 6580 098.
174 Suite, orchestra, S 1067, B minor. Suite, orchestra, S 1068,
  D major. BPhO; Herbert von Karajan. DG 139 007. (Reissue)
        +NR 7-75 p3
Suite, orchestra, S 1067, B minor. cf Cantata, no. 12, Weinen,
  Klagen, Sorgen, Zagen: Sinfonia.
Suite, orchestra, S 1067, B minor: Minuet, Badinerie. cf RCA LRL
  1-5094.
Suite, orchestra, S 1067, B minor: Polonaise. cf Philips 6580 098.
Suite, orchestra, S 1067, B minor: Rondeau, Minuet, Badinerie.
  cf Decca SPA 394.
Suites, orchestra, S 1067-8. cf Brandenburg concerti, nos. 1-6,
  S 1046-51.
Suite, orchestra, S 1068, D major. cf Cantata, no. 12, Weinen,
  Klagen, Sorgen, Zagen.
Suite, orchestra, S 1068, D major. cf Suite, orchestra, S 1067,
  B minor.
Suite, orchestra, S 1068, D major. cf Decca SDD 463.
Suite, orchestra, S 1068, D major: Air. cf Works, selections
  (RCA ARL 1-0880).
Suite, orchestra, S 1068, D major: Air on the G string. cf Works,
  selections (Ember ECL 9007).
Suite, orchestra, S 1068, D major: Air on the G string. cf CBS
  77513.
Suite, orchestra, S 1068, D major: Air on the G string. cf Ember
  GVC 42.

Suite, orchestra, S 1068, D major: Air (arr. Rampal). cf Ember
  ECL 9040.
Suite, orchestra, S 1068, D major: Air. cf HMV ASD 3017.
Suite, orchestra, S 1068, D major: Air. cf Orion ORS 73114.
Suite, orchestra, S 1068, D major: Air on the G string. cf
  Pearl GEM 132.
Suite, orchestra, S 1068, D major: Air for the G string. cf
  Cantata, no. 156: Arioso.
Suite, orchestra, S 1068, D major: Air. cf Rococo 2072.
Suite, violin, no. 3, D major: Air. cf Discopaedia MB 1002.
Suite, violin, no. 3, D major: Air. cf Discopaedia MB 1003.
Suite, violin, no. 3, D major: Air. cf Discopaedia MB 1004.
Suites, solo violoncello, nos. 1-6, S 1007-1012. cf Works,
  selections (DG Archive 2722 013).
Suite, solo violoncello, no. 1, S 1007, G major: 3 movements.
  cf RCA ARL 1-0864.
Suite, solo violoncello, no. 3, S 1009, C major: Bourrées, nos.
  1, 2. cf HMV SEOM 19.
175 Suite, solo violoncello, no. 5, S 1011, C minor. BRITTEN: Suite,
  violoncello, no. 1, op. 72. Frans Helmerson, vlc. BIS LP 5.
        ++HFN 11-75 p148   ++RR 11-75 p68
Toccatas (3). cf Organ works (Telefunken BC 24104).
Toccatas, S 912-916. cf Harpsichord works (DG Archive 2722 015).
Toccata, S 912, D major. cf Chromatic fantasia and fugue, S
  903, D minor
Toccata, S 912, D major. cf Saga 5402.
176 Toccata, adagio and fugue, S 564, C major. Toccatas and fugues,
  S 538, D minor; S 540, F major; S 565, D minor. E. Power
  Biggs, org. Columbia M 32933. Tape (c) MT 32933 (ct) MA 32933
  (Q) MQ 32933 Tape (ct) MAQ 32933
        +Audio 8-75 p80   ++St 3-75 p102 Quad
        ++NR 2-75 p13
Toccata, adagio and fugue, S 564, C major. cf Organ works
  (CBS 78256).
Toccata, adagio and fugue, S 564, C major. cf Organ works
  (DG 2722 014).
Toccata, adagio and fugue, S 564, C major. cf Organ works (Pelca
  40586).
Toccata, adagio and fugue, S 564, C major. cf Organ works
  (Telefunken EK 6-35082).
Toccata, adagio and fugue, S 564, C major: Toccata. cf HMV HLM
  7065.
Toccata and fugue, D minor. cf Cantata, no. 156: Arioso.
Toccata and fugue, D minor. cf CBS 77513.
Toccata and fugue, D minor. cf Monitor MCS 2143.
Toccata and fugue, S 538, D minor. cf Fantasias and fugues,
  S 537, C minor; S 542, G minor.
Toccata and fugue, S 538, D minor. cf Polydor 2460 252.
Toccatas and fugues, S 538, D minor; S 540, F major; S 565, D
  minor. cf Organ works (DG 2722 014).
Toccatas and fugues, S 538, D minor; S 540, F major; S 565, D
  minor. cf Toccata, adagio and fugue, S 564, C major.
Toccata and fugue, S 540, F major. cf Fantasia, S 562, C minor,
  excerpt.
Toccata and fugue, S 540, F major. cf Organ works (Telefunken
  BC 25101).

177 Toccata and fugue, S 565, D minor. Passacaglia and fugue, organ,
     S 582, C minor. Preludes and fugues, S 532-543. Daniel
     Chorzempa, org. Philips 6500 214. Tape (c) 7300 108.
            ++Gr 9-74 p592 tape   ++NR 7-72 p13
            ++HF 7-72 p75   ++RR 1-75 p68 tape
            ++HFN 2-72 p302   ++SFC 5-21-72 p46
            ++MJ 9-72 p61
     Toccata and fugue, S 565, D minor. cf Works, selections (HMV ASD
        2971).
     Toccata and fugue, S 565, D minor. cf Fantasia, S 562, C minor.
     Toccata and fugue, S 565, D minor. cf Organ works (Advent 5010).
     Toccata and fugue, S 565, D minor. cf Organ works (Pelca 40586).
     Toccata and fugue, S 565, D minor. cf Works, selections (Decca
        PFS 4278).
     Toccata and fugue, S 565, D minor. cf University of Iowa Press,
        unnumbered.
     Toccata and fugue, S 565, D minor. cf Pye GH 589.
     Toccatas and fugues, S 565, D minor; S 566, E major. cf Organ
        works (Telefunken EK 6-35082).
     Toccata and fugue, S 910, F sharp minor. cf Toccata and fugue,
        S 916, G major.
     Toccata and fugue, S 911, C minor. cf Harpsichord works (Supra-
        phon 111 0750).
178 Toccata and fugue, S 916, G major. Toccata and fugue, S 910,
     F sharp minor. BOHM: Suite, no. 1, C minor. Suite, no. 7,
     F major. FISCHER: Suite, no. 9, D minor: Passacaglia. PACHELBEL:
     Aria sebaldina (Variations, F minor). Colin Tilney, hpd. Argo
     ZRG 780.
            +Gr 2-75 p1520   +RR 2-75 p48
            +NR 8-75 p14
     Trio, S 583, D minor. cf Organ works (Telefunken EK 6-35082).
     Trio, S 584, G minor. cf Organ works (Telefunken BC 25098 T/1-2).
     Trio, S 586, G major. cf Organ works (Telefunken BC 25102 T/1-2).
     Trio, S 1027a, G major. cf Organ works (Telefunken BC 25102 T/1-2).
     Trio, violin, oboe, bassoon and harpsichord, S 1040. cf Strobe
        SRCM 123.
     Trio sonatas, S 525-530. cf Organ works (DG 2722 014).
     Trio sonata, no. 1, S 525, E flat major. cf Organ works (DG 2533
        140).
     Trio sonata, no. 1, S 525, E flat major. cf Pelca PSRK 41013/6.
     Trio sonata, no. 6, S 530, G major. cf Organ works (DG 2533 140).
179 Variations, harpsichord, S 988 (Goldberg). David Sanger, hpd.
        Saga 5395.
            +-Gr 7-75 p213   +RR 6-75 p58
            +HFN 7-75 p72
180 Variations, harpsichord, S 988. Gustav Leonhardt, hpd. Telefunken
     SAWT 9474. Tape (c) 4-41198CX
            ++HFN 7-75 p90 tape   ++RR 7-75 p69 tape
     Variations, harpsichord, S 988. cf Harpsichord works (DG Archive
        2722 015).
     Das wohltemperierte Klavier, S 846-893. cf Harpsichord works
        (DG Archive 2722 015).
     The well-tempered clavier, Bk I: Prelude, S 853, E flat minor. cf
        Works, selections (Decca PFS 4278).
181 The well-tempered clavier, Bk II, S 870-893. Glenn Gould, pno.
     Columbia D3M 31525 (3). (also CBS 78277) (Reissues from singles).

```
            +Gr 4-75 p1830              +SFC 4-8-73 p29
            +RR 2-75 p48               ++St 3-73 p114
```
182 The well-tempered clavier, Bk II, S 870-893. Anthony Newman,
    hpd, clv, org. Columbia M2 32875 (2).
```
            +Audio 9-75 p69            ++NR 2-75 p12
            -HF 2-75 p87               ++SFC 10-27-74 p6
```
183 The well-tempered clavier, Bk II, S 870-893. Sviatoslav Richter,
    pno. Melodiya/Angel SRC 4120 (3).
```
            +-HF 2-75 p87              -NR 2-75 p14
```
    Das wohltemperierte Clavier, Bk II, S 870-893: Prelude, G major.
    cf London SPC 21100.
184 Works, selections: Chorale prelude, Wir glauben all an einen Gott,
    S 680. Cantata, no. 4, Christ lag in Todesbanden, Jesus
    Christus Gottes Sohn. Gesitliches Lied, no. 51, S 487: Mein
    Jesu. Passacaglia and fugue, S 582, C minor. The well-tempered
    clavier, Bk I: Prelude, S 853, E flat minor. Toccata and fugue,
    S 565, D minor. CPhO; Leopold Stokowski. Decca PFS 4278.
    (also London SPC 21096. Tape (c) M 51096 (ct) M 81096 (r)
    L 475096).
```
            +-Gr 5-74 p2025            -NR 8-74 p4
            +HF 11-74 p94             +RR 4-74 p34
            +MJ 2-75 p41              +SFC 4-28-74 p29
```
185 Works, selections: Das Musikalisches Opfer, S 1079, C minor.
    Sonata, solo flute, S 1013, A minor. Sonatas, flute and harpsi-
    chord, S 1020, G minor; S 1030-1035. Sonata, 2 flutes and
    harpsichord, S 1039, G major. Sonatas, viola da gamba, nos.
    1-3, S 1027-1029. Suites, solo violoncello, nos. 1-6, S 1007-
    1012. Aurèle Nicolet, Christiane Nicolet, flt; Karl Richter, C.
    Jacottet, Eduard Müller, Hedwig Bilgram, hpd; Johannes Fink,
    August Wenzinger, vla da gamba; Pierre Fournier, Fritz Kiskalt,
    vlc; Otto Büchner, Kurt Gunter, vln; Siegfried Meinecke, vla;
    Karl Richter, cond. DG Archive 2722 013 (7). (Some reissues).
```
            +Gr 5-75 p1980            +HFN 7-75 p71
```
186 Works, selections: Masses, S 232-S 236. Motets: Fürchte dich nicht,
    S 228; Der Geist hilft unser Schwachheit auf, S 226; Ich lasse
    dich nicht, S Anh 159; Jesu meine Freude, S 227; Komm, Jesu
    komm, S 229; Lobet den Herrn, S 230; Sei Lobe und Preis, S 231;
    Singet dem Herrn, S 225. Songs: Ach, das nicht, S 439; Die
    bittre Leidenzeit, S 450; Brich entzwei, S 444; Dir, Dir Jehova,
    S 452; Eins is not, S 453; Es kostet viel, S 459; Gib dich
    zufrieden, S 521; Gott lebet noch, S 461; Gott, wie gross, S 462;
    Die guldne Sonne, S 451; Ich lass dich nicht, S 467; Ich steh
    an deiner Krippen, S 469; Ihr Gestirn, S 476; Komm, susser Tod,
    S 478; Kommt, Seelen, S 479; Kommt wieder, S 480; Der lieben
    Sonne Licht, S 446; Liebster Herr Jesu, S 484; Mein Jesu, S 487;
    O Jesulein, S 493; So gehst du nun, S 500; So gibst du nun,
    S 501; Steh ich bei meinem Gott, S 503; Vergiss mein nicht, S 505;
    Wie wohl ist mir, S 517; Wo ist mein Schaflein, S 507. Maria
    Stader, Renate Krahmer, Elisabeth Speiser,s; Hertha Töpper,
    Annelies Burmeister, con; Ernest Häfliger, Peter Schreier, t;
    Dietrich Fischer-Dieskau, Keith Engen, Theo Adam, bs; Regens-
    burg Cathedral Choir; Vienna Capella Academica; Dresden Kreuz-
    chor; Munich Bach Orchestra and Choir; Dresden Philharmonic
    Orchestra; Hedwig Bilgram, positive organ; Karl Richter, Martin
    Flämig, Hans-Martin Schneidt. DG Archive 2722 017. (Some re-
    issues from 2710 001, 2708 031)
```
            +Gr 9-75 p499             +RR 8-75 p60
```

187 Works, selections: Cantata, no. 140: Sleepers, awake. Cantata,
       no. 147, Jesu, joy of man's desiring. Cantata, no. 208, Sheep
       may safely graze. Fugue, S 578, G minor. Gavottes, 1 and 2.
       Prelude, C major. Minuets, 1 and 2. Quia respexit. Sonata,
       flute and harpsichord, S 1031, E flat major: Siciliano. Suite,
       orchestra, S 1068, D major: Air on the G string. Laurindo
       Almeida, gtr. Ember ECL 9007. (also Orion ORS 7277)
              +-Gr 5-75 p2023   -RR 6-75 p58
              +-NR 7-73 p12
188 Works, selections: Fugue, S 1000, G minor. Prelude, S 999, C minor.
       Prelude, fugue and allegro, S 998, E flat major. Suite, lute,
       S 996, E minor. Suite, lute, S 995, G minor: Prelude, Presto.
       Turibio Santos, gtr. Erato STU 70885.
              +Gr 12-75 p1075   +RR 12-75 p77
189 Works, selections: Cantata, S 147, Jesu, joy of man's desiring
       (trans. Hess). Chorale preludes: Ich ruf zu dir, Herr Jesu
       Christ, S 639 (trans. Busoni); Nun freut euch, lieben Christen
       mein, S 734 (trans. Busoni); Nun komm der Heiden Heiland, S 599
       (trans. Busoni). Partita, violin, no. 2, S 1004: Chaconne,
       D minor (trans. Busoni). Prelude, S 929, B minor (trans.
       Siloti). Prelude and fugue, S 543, A minor (trans. Liszt).
       Sonata, flute and harpsichord, S 1031, E flat major: Siciliano
       (trans. Lustner). Toccata and fugue, S 565, D minor (trans.
       Busoni). Alexis Weissenberg, pno. HMV ASD 2971. (also Angel
       S 37088).
              +Gr 7-74 p226   -RR 7-74 p58
              ++NR 6-75 p14   +St 10-75 p105
190 Works, selections: Cantata, no. 156: Arioso. Chorale preludes:
       Ein feste Burg, S 720; Komm susser Tod; Wachet auf, S 645.
       Fugue, G minor. Partita, violin, no. 2, S 1004, D minor:
       Chaconne. Partita, violin, no. 3, S 1006, E major: Preludio.
       Suite, orchestra, S 1068, D major: Air. LSO; Leopold Stokowski.
       RCA ARL 1-0880. (Q) ARD 1-0880.
              +NR 7-75 p3   +SFC 6-15-75 p24
              +RR 9-75 p26
191 Works, selections: Cantata, no. 12, Weinen, Klagen, Sorgen, Zagen:
       Sinfonia. Cantata, no. 152, Tritt auf die Glaubensbahn: Con-
       certo. Cantata, no. 156, Ich steh mit einem Fuss in Grabe:
       Sinfonia. Cantata, no. 182, Himmelskönig, sei willkommen:
       Sonata. Cantata, no. 209, Non sa che sia dolore: Sinfonia.
       Canone perpetuo, S 1079, no. 9. Instrumental movement, S 1040,
       F major. Der Geist hilft unsrer Schwachheit auf, S 226. Son-
       atas, 2 violins and harpsichord, S 1036-1039, S 1079. Prague
       Ars Rediviva Ensemble; Milan Munclinger. Supraphon 50871/2 (2).
              +Gr 4-75 p1819   ++RR 3-75 p20
BACH, Wilhelm Friedemann
     Duet, flute, F major. cf BACH, J. C.: Quartet, 2 flutes, violin
       and violoncello, C major.
BACH, Wilhelm Friedrich Ernst
     Trio, 2 flutes and viola, G major. cf BACH, J. C.: Quartet, 2
       flutes, violin and violoncello, C major.
BACHRACH
     What's new pussycat. cf Columbia CXM 32088.
BACK
192 Sonata alla ricercare. GARRETA: Sonata, piano, C minor. José
       Ribera, pno. HMV 4E 061 35136.
              +-Gr 12-75 p1061

193 Octet, winds and strings. WELLESZ: Octet, winds and strings, op.
     67. Vienna Octet. London STS 15243.
          ++HF 3-75 p88   +NR 11-74 p6
BAGLEY, E.
     National emblem. cf Michigan University SM 0002.
BAINES, Francis
     Fanfare. cf HMV SLA 870.
     Hoffnung Festival overture, excerpts. cf HMV SLA 870.
     Introductory music. cf HMV SLA 870.
BAINTON, Edgar
     And I saw a new heaven. cf Argo ZRG 789.
BAIRSTOW, Edward
     Let all mortal flesh keep silence. cf HMV HQS 1350.
BAKALEINIKOFF
     Brahmsiana. cf Discopaedia MB 1007.
BAKER, David N.
194 Sonata, violoncello and piano. WHITE: Concerto, violin, F sharp
     minor (ed. Glass and Moore). Aaron Rosand, vln; Janos Starker,
     vlc; Alain Plaines, pno; LSO; Paul Freeman. Columbia M 33432.
          +HF 10-75 p65   ++NR 9-75 p9
BAKFARK, Balint
     Fantasias (4). cf DG Archive 2533 294.
     Fantasias, nos. 1, 8, 9, 10. cf Hungaroton SLPX 11549.
     Gagliarda. cf Hungaroton SLPX 11549.
     Non dite mal (galiarda). cf Hungaroton SLPX 11549.
     Schöner Deutscher Tanz. cf Hungaroton SLPX 11549.
BALASSA, Sándor
195 Cantata V, op. 21. Legend, op. 12. Requiem for Lajos Kossák,
     op. 15. Erika Sziklay, s; Sándor Palcsó, t; Endre Utö, bar;
     HRT Orchestra and Chorus; György Lehel, Ferenc Sapszon.
     Hungaroton SLPX 11681.
          +HFN 11-75 p149   ++RR 9-75 p67
          +NR 7-75 p8
     Legend, op. 12. cf Cantata V, op. 21.
     Requiem for Lajos Kossák, op. 15. cf Cantata V, op. 21.
BALDASSARE
     Sonata, trumpet and organ, no. 1, F major. cf CRD CRD 1008.
BALOGNA, Jacopo da
     Oselleto selvaggio. cf Vanguard VSD 71179.
BANCHIERI, Adriano
     Battaglia, Canzone Italiana dialogo. cf Musical Heritage Society
     MHS 1790.
     Dialogo per organo. cf Columbia M 33514.
     Madrigal. cf Crystal S 201.
BANKS, Don
196 Concerto, horn. MUSGRAVE: Concerto, clarinet. SEARLE: Aubade,
     op. 28. Barry Tuckwell, hn; Gervase de Peyer, clt; NPhO, LSO;
     Norman Del Mar. Argo ZRG 726.
          ++Gr 8-75 p315   ++RR 9-75 p40
          +-HFN 11-75 p149
     Concerto, violin. cf FRICKER: Concerto, violin, no. 11.
BARATI, George
     Concerto, violoncello. cf EGGE: Concerto, piano, no. 2, op. 21.
BARBER, Samuel
197 Adagio, strings. Essay, no. 2, op. 17. Medea's meditation and
     dance of vengeance, op. 23. Overture to the School for Scandal.

NYP; Thomas Schippers. Odyssey Y 33230. (Reissue from Columbia
3211 0006)
    +-Audio 11-75 p97   ++SFC 6-8-75 p23
+MJ 3-75 p24
198 Concerto, piano, op. 38. Concerto, violin, op. 14. John Browning,
pno; Isaac Stern, vln; CO, NYP; Georg Szell, Leonard Bernstein.
CBS 61621. (Reissues from CBS SBRG 72345, Columbia SAX 2575)
    +Gr 3-75 p1639   +-RR 4-75 p21
Concerto, violin, op. 14. cf Concerto, piano, op. 38.
Essay, no. 2, op. 17. cf Adagio, strings.
Hesitation tango. cf Golden Crest CRS 4132.
Medea's meditation and dance of vengeance, op. 23a. cf Adagio,
    strings.
Die Natali, op. 37. cf ADAM: Concerto, violoncello and orchestra.
Overture to the School for Scandal. cf Adagio, strings.
Reincarnations: I got shoes (arr. Parker-Shaw); Sometimes I feel;
    The coolin (arr. Parker-Shaw). cf Wayne State University,
    unnumbered.
199 Sonata, piano. BRITTEN: Diversions on a theme, op. 21. John
Browning, Leon Fleisher, pno; Baltimore Symphony Orchestra;
Sergiu Commissiona. Peerless PRCM 204.
    +-Gr 4-75 p1801
200 Symphony, no. 1, op. 9, in one movement. MAYER: Octagon, piano
and orchestra. William Masselos, pno; Milwaukee Symphony
Orchestra; Kenneth Schermerhorn. Turnabout (Q) QTVS 34564.
    ++MJ 1-75 p49   +SFC 6-22-75 p26
    ++NR 1-75 p5   ++St 5-75 p96
BARBERIIS (16th century)
    Madonna qual certezza. cf L'Oiseau-Lyre 12BB 203/6.
BARBETTA, Giulio
    Moresca detta le Canarie. cf DG Archive 2533 173.
BARBIREAU
    Een vrolic Wesen. cf HMV SLS 5022.
BARBIERI
    Canción de paloma. cf HMV SLS 5012.
    THE BARITONE VOICE, SHERRILL MILNES. cf RCA ARL 1-0851.
BARKIN, Elaine
201 Quartet, strings. BOYKAN: Quartet, strings, no. 1. Contemporary
Quartet, American Quartet. CRI SD 338.
    +NR 9-75 p8
BARLOW, Wayne
    The winter's past, rhapsody for oboe. cf Era 1001.
    BAROQUE BRASS. cf Klavier KS 536.
    BAROQUE FLUTE SONATAS. cf London STS 15198.
    BAROQUE FLUTE SONATAS ON HISTORICAL INSTRUMENTS. cf Pandora PAN 103.
    BAROQUE ORGAN WORKS FROM ALPINE COUNTRIES. cf Philips 6775 006.
    THE BAROQUE SOUND OF THE TRUMPET. cf Argo ZDA 203.
BARRIOS, Agustin
    Medallón antiguo. cf CBS 61654.
BARSANTI, Francesco
    Sonata, recorder, C major. cf Telefunken SMA 25121 T/1-3.
BARTOK, Bela
202 Bagatelles, op. 6 (14). Dance suite. Robert Silverman, pno.
Orion ORS 74152.
    +-HF 7-74 p84   ++RR 7-75 p45
    ++NR 6-74 p11   ++St 8-74 p103

Bagatelle, op. 6, no. 2. cf Piano works (Turnabout THS 65010).
203 Bluebeard's castle (The castle of the Duke Blue Beard), op. 11.
Nina Poliakova, s; Yevgeny Kibkalo, bar; Bolshoi Symphony
Orchestra; Gennady Rozhdestvensky. Westminster-Melodiya WGSO
8219-1.

    +MJ 1973 annual p19   -SFC 5-13-73 p29
    /NR 2-74 p9   -St 9-73 p111
    +-ON 2-22-75 p28

204 Cantata profana (The enchanted stag). Richard Lewis, t; Marko
Rothmuller, bar; New Symphony Orchestra and Chorus; Walter
Susskind. Bartok Records 312.

    +-CJ 9-75 p30

205 Cantata profana. József Réti, t; András Faragó, bar; HRT Orchestra
and Chorus; György Lehel. DG LPM 138873.

    +CJ 9-75 p30

206 Cantata profana. Hungarian folksongs (5). Village scenes.
Choruses (7). Laura Faragó, Anna Adám, s; Julia Hamari, alto;
József Réti, t; András Faragó, bar; Budapest Chorus; Györ
Girls Chorus; Ferenc Liszt Academy Chamber Chorus; Budapest
Symphoy Orchestra, Hungarian State Symphony Orchestra, Budapest
Chamber Ensemble; János Feréncsik, András Kóródy, Antal Doráti.
Hungaroton SLPX 11510.

    +CJ 9-75 p30   +RR 4-74 p73
    ++Gr 7-74 p244

207 Cantata profana. USSR State Philharmonic Orchestra and Chorus;
Rozhoktvensky. Period PRST 2757.

    +-CJ 9-75 p30

208 Cantata profana. Music for strings, percussion and celesta.
Murray Dickie, t; Edmond Hurshell, bar; Vienna Chamber Choir;
VSOO, PH; Heinrich Hollreiser, Thomas Unger. Turnabout
TV 34382S. (Reissue from Vox PL 10480)

    +-CJ 9-75 p30   -HFN 6-73 p1171
    +-Gr 7-73 p187   +-RR 6-73 p86

Choruses (7). cf Cantata profana.
209 Concerto, orchestra. BPhO; Herbert von Karajan. Angel S 37059.
(also HMV ASD 3046)

    +Gr 1-75 p1335   +NYT 11-10-74 pD1
    -HF 2-75 p87   +-RR 1-75 p25
    +NR 12-74 p4   ++St 3-75 p96

210 Concerto, orchestra. BSO; Rafael Kubelik. DG 2530 479.

    ++Gr 6-75 p32   ++NR 5-75 p4
    ++HF 5-75 p69   +SFC 7-27-75 p22
    +HFN 7-75 p73   +-RR 6-75 p30
    +-LJ 12-75 p38   ++St 6-75 p94

211 Concerto, orchestra. The miraculous Mandarin, op. 19: Suite.
Strasbourg Philharmonic Orchestra; Alain Lombard. Erato STU
70835.

    -Gr 12-75 p1022

212 Concerto, violin and orchestra, no. 2, B minor. Itzhak Perlman,
vln; LSO; André Previn. HMV ASD 3014. (also Angel S 37014)

    ++Gr 10-74 p676   +-RR 11-74 p34
    +HF 1-75 p75   ++St 3-75 p97
    +NR 12-74 p6

213 Dance suite. KODALY: Variations on a Hungarian folksong. Hungarian
Radio and Television Orchestra; György Lehel. Hungaroton SHLX
90043.

    +-NR 11-75 p3   -RR 12-75 p36

214 Dance suite. PROKOFIEV: Scythian suite, op. 20. CPhO; Zdeněk
       Košler. Panton 110 374.
           +-HFN 6-75 p73   +-RR 7-75 p22
     Dance suite. cf Bagatelles, op. 6.
215 Divertimento, string orchestra. GINASTERA: Concerto for strings.
       PO; Eugene Ormandy. Columbia M 32874.
           +HF 8-75 p84   +SFC 8-17-75 p22
           ++NR 4-75 p3
216 Duo, 2 violins (44). Lorand Fenyves, Victor Martin, vln. Musical
       Heritage Society MHS 1722.
           +HF 4-75 p80
217 Etudes, op. 18 (3). COPLAND: Fantasy. DALLAPICCOLA: Quaderno
       musicale de Annalibera. Anthony Peebles, pno. Unicorn RHS 323.
           ++Gr 1-75 p1365   +RR 3-75 p45
           ++HF 7-75 p92
218 For children, vols. 1-2. Yida Novik, pno. Toshiba 7032/3.
           +-HF 4-75 p80
219 For children, excerpts. BEETHOVEN: Sonata, violin and piano,
       no. 8, op. 30, no. 3, G major. FRANCK: Sonata, violin and
       piano, A major. Ede Zathureczky, vln; Menahem Pressler, pno.
       Hungaroton SLPX 11641.
           ++NR 8-75 p15
220 Hungarian folksongs (1907-17). Songs, op. 15 (5). Songs, op. 16
       (5). Village scenes (5). Elizabeth Suderburg, s; Bela Siki,
       pno. Turnabout TV 34592.
           ++NR 11-75 p10
     Hungarian folksongs (5). cf Cantata profana.
     Hungarian folksongs. cf Piano works (Turnabout THS 65010).
221 Images, op. 10 (2). Pieces, orchestra, op. 12 (4). Portraits,
       op. 5 (2). Frankfurt Radio Symphony Orchestra; Andreas Röhn,
       vln; Eliahu Inbal. Philips 6500 781.
           +Gr 2-75 p1473   +RR 1-75 p25
     Improvisations on Hungarian peasant songs, op. 20, nos. 1, 2, 6,
       7, 8. cf Piano works (Turnabout THS 65010).
222 Mikrokosmos. Yida Novik, pno. Toshiba TS 7042/3 (2).
           ~ HF 4-75 p80
     Mikrokosmos, excerpts. cf Piano works (Turnabout THS 65010).
223 Mikrokosmos: Pieces (14). Suite, op. 4b. Richard and John
       Contiguglia, pno. Connoisseur CS 2033. Tape (c) E 1012.
           +HF 1-75 p110 tape   ++RR 2-75 p48
224 Mikrokosmos: 4 songs. Slavonic folksongs (5). KODALY: Hungarian
       songs (5). KALLOS: The mountain brigand. Torok, s; Erzsebet
       Tusa, pno. Hungaroton SLPX 11722.
           +SFC 7-20-75 p27
     The miraculous Mandarin, op. 19: Suite. cf Concerto, orchestra.
     Music for strings, percussion and celesta. cf Cantata profana.
     Music for strings, percussion and celesta. cf Toscanini Society
       ATS GC 1201/6.
     Nine little pieces, no. 1: All Ungherese, Preludio. cf Piano
       works (Turnabout THS 65010).
225 Out of doors. BERG: Sonata, piano, op. 1. SCHONBERG: Klavier-
       stück, op. 33a and b. STRAVINSKY: Petrouchka. Adám Fellegri,
       pno. Hungaroton SLPX 11529.
           +-NR 11-75 p12
     Petite suite. cf PIano works (Turnabout THS 65010).
226 Piano works: Nine little pieces, no. 9: All Ungherese, Preludio.
       Bagatelle, op. 6, no. 2. Hungarian folksongs. Improvisations

on Hungarian peasant songs, op. 20, nos. 1, 2, 6, 7, 8.
Mikrokosmos, excerpts.   Petite suite.   Rondo on folk tunes,
no. 1. Bela Bartok, Ditta Pasztory Bartok, pno. Turnabout
THS 65010.
+HF 10-74 p88   +St 3-75 p96
+NR 9-74 p9
Pieces, orchestra, op. 12 (4). cf Images, op. 10.
Portraits, op. 5 (2). cf Images, op. 10.
227 Quartets, strings, nos. 2, 5. Prague Quartet. Supraphon 50645.
(Reissue)
+RR 1-75 p37
228 Rhapsody, violin and piano, no. 1. Rumanian folk dances (2).
Sonata, violin and piano, no. 1. Denés Zsigmondy, vln; Anneliese
Nissen, pno. Klavier KS 535.
+Audio 9-75 p70   +St 10-75 p104
++NR 5-75 p9
Rondo on folk tunes, no. 1. cf Piano works (Turnabout THS 65010).
Rumanian folk dances (2). cf Rhapsody, violin and piano, no. 1.
Rumanian folk dances. cf Connoisseur Society CS 2070.
Slavonic folk songs. cf Mikrokosmos: 4 songs.
229 Sonata, 2 pianos and percussion. Sonata, solo violin. Dezsö Ránki,
Zoltán Kocsis, pno; Ferenc Petz, József Marton, perc, Dénes
Kovács, vln. Hungaroton SLPX 11479.
++Gr 3-75 p1668   ++RR 4-75 p48
230 Sonata, solo violin. Sonata, violin and piano, no. 2. Dénes
Zsigmondy, vln; Anneliese Nissen, pno. Klavier KS 542.
+NR 8-75 p7   +St 10-75 p104
Sonata, solo violin. cf Sonata, 2 pianos and percussion.
231 Sonata, violin and piano, no. 1. PROKOFIEV: Sonata, violin and
piano, no. 1, op. 80, F minor. David Oistrakh, vln; Sviatoslav
Richter, pno. HMV Melodiya ASD 3105.
+Gr 8-75 p341   +RR 8-75 p54
++HFN 8-75 p72   +STL 9-7-75 p37
Sonata, violin and piano, no. . cf Rhapsody, violin and piano,
no. 1.
232 Sonatas, violin and piano, nos. 1 and 2. Gidon Kremer, vln; Yuri
Smirnov, pno. Hungaroton SLPX 11655.
+Gr 1-75 p1361   +-St 7-75 p94
+NR 5-75 p9
Sonata, violin and piano, no. 2. cf Sonata, solo violin.
233 Songs, op. 15 (5). Songs, op. 16 (5). Village scenes. Julia
Hamari, ms; Konrad Richter, pno. DG 2530 405.
+-HF 4-74 p93   +ON 2-22-75 p28
+MJ 7-74 p50   ++SFC 11-3-74 p22
+NR 6-74 p11   ++St 5-74 p96
Songs, op. 15 (5). cf Hungarian folksongs (1907-17).
Songs, op. 16 (5). cf Hungarian folksongs (1907-17).
Songs, op. 16 (5). cf Songs, op. 15 (5).
Suite, op. 4b. cf Mikrokosmos: Pieces.
Village scenes. cf Cantata profana.
Village scenes (5). cf Hungarian folksongs (1907-17).
Village scenes. cf Songs, op. 15.
BASTON, John
Concerto, D major. cf HMV SLS 5022.
BATCHELAR, Daniel
Mounsiers almaine. cf DG Archive 2533 157.

BATTEN, Adrian
    O praise the Lord. cf Argo ZRG 789.
    BATTLE MUSIC FOR ORGAN. cf Musical Heritage Society MHS 1790.
BAX, Arnold
    Fanfare for the wedding of Princess Elizabeth, 1948. cf Decca 419.
234 The garden of Fand. BUTTERWORTH: A Shropshire lad. VAUGHAN
        WILLIAMS: Symphony, no. 8, D minor. Hallé Orchestra; John
        Barbirolli. Pye GSGC 15017. (Reissues from CCT 31000, NCT
        17000, GSGC 14061)
                +Gr 8-75 p335   +RR 7-75 p36
                +-HFN 8-75 p87
    Mediterranean. cf RCA ARM 4-0942/7.
235 Symphony, no. 7, A flat major. LPO; Raymond Leppard. Lyrita
        SRCS 83.
                +Gr 11-75 p801   ++RR 9-75 p28
                +HFN 11-75 p149
BAZELON, Irwin
    Duo, viola and piano. cf CRI SD 342.
    Propulsions. cf Quintet, brass.
236 Quintet, brass. Propulsions. Raymond DesRoches, Richard Fitz,
        Gordon Gottlieb, Herbert Harris, Arnie Lang, Walter Rosenberger,
        Colin Walcott, perc; American Brass Quintet. CRI SD 327.
                +-HFN 11-75 p149   +-RR 7-75 p38
                +NR 1-75 p7
BAZZINI, Antonio
    La ronde des lutins, op. 25. cf Discopaedia MB 1010.
    La ronde des lutins, op. 25 (2). cf RCA ARM 4-0942/7.
BEACH, H. H. A.
    Ballad, op. 6. cf Works, selections (Genesis 1054).
    Hermit thrush at eve, op. 92, nos. 1 and 2. cf Works, selections
        (Genesis 1054).
    Improvisations, op. 148, nos. 1, 2, 4. cf Works, selections
        (Genesis 1054).
    Morceaux caracteristiques, op. 28. cf Works, selections (Genesis
        1054).
    Nocturne, op. 107. cf Works, selections (Genesis 1054).
    Prelude and fugue, op. 81. cf Works, selections (Genesis 1054).
237 Quintet, piano and strings, op. 67, F sharp minor. FOOTE: Quintet,
        piano and strings, op. 38, A minor. Mary Louise Boehm, pno;
        Kees Kooper, Alvin Rodgers, vln; Richard Maximoff, vla; Fred
        Sherry, vlc. Turnabout TVS 34556.
                +Gr 11-75 p845   +NR 10-74 p8
                +-HF 10-74 p88   +RR 12-75 p71
                +HFN 11-75 p147   +St 11-74 p121
    Sketches, op. 15 (4). cf Works, selections (Genesis 1054).
    Valse caprice, op. 4. cf Works, selections (Genesis 1054).
238 Works, selections: Ballad, op. 6. Hermit thrush at eve, op. 92,
        nos. 1 and 2. Improvisations, op. 148, nos. 1, 2, 4. Morceaux
        caracteristiques, op. 28 (3). Nocturne, op. 107. Prelude and
        fugue, op. 81. Sketches, op. 15 (4). Valse caprice, op. 4.
        Virginia Eskin, pno. Genesis GS 1054.
                ++12-75 p85   ++St 12-75 p82
                ++NR 11-75 p12
BEDFORD, David
239 The rime of the ancient mariner. David Bedford, keyboards, record-
        ers, percussion; Mike Oldfield, gtr; Robert Powell, narrator;

Queen's College Girls Choir, London. Virgin V 2038.
  +–HFN 11-75 p150   +RR 12-75 p71
240 Spillihpnerak. BERIO: Sequenza VI. CAGE: Dream. MADERNA: Viola.
     Karen Phillips, vla. Finnadar SR 9007.
       +HHF 6-75 p108   +NYT 4-27-75 pD19
       +MT 7-75 p629
241 Star's end. Mike Oldfield, gtr; Chris Cutler, perc; RPO; Vernon
     Handley. Virgin V 2020. Tape (c) TCV 2020 (ct) TCV 2020.
       +–Gr 3-75 p1639   *RR 1-75 p26
242 The tentacles of the dark nebula. BERKELEY: Ronsard sonnets (4).
     LUTOSLAWSKI: Paroles tissées. Peter Pears, t; London Sinfonietta
     Orchestra; Witold Lutoslawski, Lennox Berkeley, David Bedford.
     Decca Headline HEAD 3. (also London HEAD 3)
       +Gr 5-74 p2063   +HRR 5-74 p20
       +HF 4-75 p87   +SR 2-8-75 p37
       +HNR 10-75 p13
     You asked for it. cf L'Oiseau-Lyre DSLO 3.
BEETHOVEN, Ludwig van
     Adagio, F major. cf Pelca PSRK 41013/6.
     Adagio, op. 150, E flat major. cf HUMMEL: Sonata, mandolin and
     piano, C major.
     Adelaide, op. 46. cf RCA SER 5704/6.
     Andante favori, A major. cf Sonata, piano, no. 21, op. 53, C major.
243 Andante favori, F major. Sonata, piano, no. 21, op. 53, C major.
     Sonata, piano, no. 31, op. 110, A flat major. Alfred Brendel,
     pno. Philips 6500 762.
       +–HFN 10-75 p137
244 Andante favori, F major. Sonata, piano, no. 12, op. 26, A flat
     major. LANNER: Valses viennoises (arr. Landowska). MOZART:
     Sonata, piano, no. 17, K 576, D major. Wanda Landowska, pno.
     Saga 5389. (Taken from piano rolls)
       +–Gr 2-75 p1515   -RR 6-75 p71
       /HFN 6-75 p100
     Andante favori, F major. cf Piano works (Vanguard VSD 71186).
     Bagatelle. cf Piano works (Vanguard VSD 71187).
245 Bagatelle, no. 25, A minor (Für Elise). Rondo a capriccio, op.
     129, G major. Sonata, piano, no. 7, op. 10, no. 3, D major.
     André Watts, pno. Columbia M 33074.
       -NR 2-75 p11
     Bagatelle, no. 25, A minor. cf Piano works (Columbia M 33074).
     Bagatelle, no. 25, A minor. cf Piano works (Vanguard VSD 71187).
     Bagatelle, no. 25, A minor. cf Decca SPA 372.
     Bagatelle, no. 25, A minor. cf DG Heliodor 2548 137.
     Bagatelle, no. 25, A minor. cf Philips 6747 204.
246 Bagatelles, op. 33 (6). Bagatelles, op. 126 (6). Glenn Gould, pno.
     Columbia M 33265. (also CBS 76424).
       +HF 8-75 p80   +–RR 11-75 p73
       +HHFN 11-75 p150   +HSt 8-75 p94
       +HNR 8-75 p12
247 Bagatelles, opp. 33, 119, 126. Stephen Bishop, pno. Philips
     6500 930.
       +HHFN 11-75 p150   +–RR 11-75 p73
248 Bagatelles, opp. 33, 119, 126. Alfred Brendel, pno. Turnabout
     TV 34077. Tape (c) KTVC 34077.
       +–RR 1-75 p68
249 Bagatelles, op. 126 (6). Sonata, piano, no. 31, op. 110, A flat

major. Mihály Bâcher, pno. Hungaroton SLPX 11718.
    +NR 11-75 p14
Bagatelles, op. 126 (6). cf Bagatelles, op. 33.
Bundeslied, op. 122. cf Works, selections (CBS 76404).
250 Concerti, piano, nos. 1-5. Leon Fleisher, pno; CO; Georg Szell.
    CBS 77371 (3). (also Columbia M4X 30052) (Reissues)
    +Gr 1-75 p1336   +-RR 12-74 p25
251 Concerti, piano, nos. 1-5. Wilhelm Kempff, pno; BPhO; Ferdinand
    Leitner. DG 2740 131 (4). (Reissues from SLPM 138774/77)
    +Gr 11-75 p909   +-RR 11-75 p36
252 Concerti, piano, nos. 1-5. Solomon, pno; PhO; Herbert Menges,
    André Cluytens. HMV SLS 5026 (4). (Reissues from ALP 1583,
    1546, 1300, BLP 1024, 1036)
    +Gr 11-75 p802   ++RR 12-75 p37
    +-HFN 10-75 p152
253 Concerti, piano, nos. 1-5. Stephen Bishop, pno; LSO, BBC Symphony
    Orchestra; Colin Davis. Philips 6747 104 (4). (Nos. 1, 3, 5
    reissues from 6500 179, 6500 315, SAL 3787)
    +Gr 10-75 p607   +RR 11-75 p36
    +-HFN 9-75 p96   ++SFC 10-19-75 p33
    +NR 12-75 p6
254 Concerti, piano, nos. 1-5. Jan Panenka, pno; PSO; Václav Smetáček.
    Supraphon 110 1531/34.
    /HFN 7-75 p76   /RR 6-75 p30
255 Concerto, piano, no. 1, op. 15, C major. MOZART: Concerto, piano,
    no. 9, K 271, E flat major. Walter Gieseking, pno; Berlin State
    Opera Orchestra; Hans Rosbaud. Bruno Walter Society RR 411.
    ++NR 9-75 p7
256 Concerto, piano, no. 1, op. 15, C major. Fantasia, op. 80, C
    minor. John Lill, pno; Scottish National Orchestra and Chorus;
    Alexander Gibson. Classics for Pleasure CFP 40232.
    +-Gr 11-75 p802   +RR 11-75 p41
    +HFN 10-75 p136
257 Concerto, piano, no. 1, op. 15, C major. Fantasia, op. 80, C minor.
    Julius Katchen, pno; LSO and Chorus; Pierino Gamba. Decca SDD
    227. Tape (c) KSDC 227.
    +-RR 2-75 p74 tape
258 Concerto, piano, no. 1, op. 15, C major. Sonata, piano, no. 5,
    op. 10, no. 1, C minor. Stephen Bishop, pno; BBC Symphony
    Orchestra; Colin Davis. Philips 6500 179. Tape (c) 7300 116.
    ++ARG 8-72 p600   +NYT 2-6-72 pD32
    -HF 5-72 p85   +RR 2-75 p74 tape
    +MJ 9-72 p60   +SFC 5-14-72 p41
    +NR 4-72 p6   +St 7-72 p68
259 Concerto, piano, no. 2, op. 19, B flat major. Sonata, piano, no.
    21, op. 53, C major. Vladimir Ashkenazy, pno; CSO; Georg
    Solti. Decca SXL 6652.
    +HFN 12-75 p171
260 Concerto, piano, no. 2, op. 19, B flat major. Concerto, piano,
    no. 4, op. 58, G major. Wilhelm Kempff, pno; BPhO; Ferdinand
    Leitner. DG 138 775. Tape (c) 3300 485.
    +HFN 12-75 p173 tape
261 Concerto, piano, no. 3, op. 37, C minor. Sonata, piano, no. 26,
    op. 81a, E flat major. Vladimir Ashkenazy, pno; CSO; Georg
    Solti. Decca SXL 6653. Tape (c) KSXC 6653. (Reissues from
    SXLG 6594/7, 6706)

          +–Gr 11-75 p802   +HFN 12-75 p173 tape
          +HFN 10-75 p152
262 Concerto, piano, no. 3, op. 37, C minor. Sviatoslav Richter, pno;
     VSO; Kurt Sanderling. DG 2535 107. Tape (c) 3335 107.
          +–HFN 11-75 p173   +–RR 9-75 p77 tape
263 Concerto, piano, no. 3, op. 37, C minor. Claudio Arrau, pno; COA;
     Bernard Haitink. Philips 6580 078. (Reissue from SBAL 20).
          +Gr 10-73 p735   +–NR 12-73 p6
          +HLJ 2-75 p36   +RR 10-73 p59
          +MJ 11-73 p8
264 Concerto, piano, no. 4, op. 58, G major. Fantasia, op. 77, G
     minor. Paul Badura-Skoda, pno; Collegium Aureum; Paul Badura-
     Skoda. BASF KHB 21510. (also BASF BAC 3002. Tape (ct) KBACC
     3002)
          +Gr 7-74 p196   -NYT 10-14-73 pD33
          +Gr 12-74 p1237 tape   +RR 6-74 p39
          +–MQ 4-74 p312   +RR 6-75 p90 tape
          +–NR 11-73 p5   /SFC 8-12-73 p32
265 Concerto, piano, no. 4, op. 58, G major. Ivan Moravec, pno; VSO;
     Martin Turnovský. Connoisseur Society CS 163.
          +–SFC 3-2-75 p25
266 Concerto, piano, no. 4, op. 58, G major. Leonore overture, no. 3,
     op. 72b. Vladimir Ashkenazy, pno; CSO; Georg Solti. Decca
     SXL 6654. (Reissues from SXLG 6594/7, SXLP 6684)
          +Gr 5-75 p1962   +HRR 6-75 p30
          +HFN 6-75 p85
267 Concerto, piano, no. 4, op. 58, G major. MOZART: Concerto, violin,
     no. 3, K 216, G major. Emil Gilels, pno; David Oistrakh, vln;
     PhO; Leopold Ludwig, David Oistrakh. HMV Tape (c) TX EXE 156.
          +–HFN 10-75 p155 tape
268 Concerto, piano, no. 4, op. 58, G major: 1st movement. Symphony,
     no. 5, op. 68, C minor. Conrad Hansen, pno; BPhO; Wilhelm
     Furtwängler. Unicorn UNI 106.
          +–NR 9-75 p5
     Concerto, piano, no. 4, op. 58, G major. cf Concerto, piano,
     no. 2, op. 19, B flat major.
     Concerto, piano, no. 4, op. 58, G major. cf MOZART: Concerto,
     violin, no. 3, K 216, G major.
269 Concerto, piano, no. 5, op. 73, E flat major. Alexis Weissenberg,
     pno; BPhO; Herbert von Karajan. Angel S 37062. (also HMV ASD
     3043)
          +HGr 1-75 p1341   +HRR 3-75 p20
          +–HF 2-75 p88   +HSFC 2-9-75 p24
          +HNR 12-74 p6   +HSt 2-75 p106
          +–NYT 11-10-74 pD1
270 Concerto, piano, no. 5, op. 73, E flat major. Malcolm Frager,
     pno; Hamburg Philharmonia Orchestra; Gary Bertini. BASF
     BAC 3093.
          +–Gr 6-75 p35   +–RR 6-75 p31
          -HFN 5-75 p124
271 Concerto, piano, no. 5, op. 73, E flat major. Artur Schnabel, pno;
     CSO; Frederick Stock. Camden CCV 5028.
          +–RR 12-75 p37
272 Concerto, piano, no. 5, op. 73, E flat major. Egmont overture,
     op. 84. Vladimir Ashkenazy, pno; CSO; Georg Solti. Decca SXL
     6655. Tape (c) KSXC 16655. (Reissue from SXLG 6594/7, SXLP 6684)
     (also London 6857)

```
        +Gr 3-75 p1640           -NR 6-75 p7
       ++Gr 6-75 p106 tape       +RR 4-75 p21
      ++HFN 5-75 p124            -RR 7-75 p69 tape
        +HFN 7-75 p90 tape
```

273 Concerto, piano, no. 5, op. 73, E flat major. Christoph Eschenbach,
    pno; BSO; Seiji Ozawa. DG 2530 438. Tape (c) 3300 384.
```
       ++Gr 8-74 p352            +-RR 7-74 p34
        +Gr 9-74 p595 tape       +-RR 8-74 p83 tape
        +-HF 8-74 p85           ++SFC 6-2-74 p22
        +-NR 9-74 p4             +St 1-75 p109
```

274 Concerto, piano, no. 5, op. 73, E flat major. Sonata, piano, no.
    25, op. 79, G major. Andor Foldes, pno; BPhO; Ferdinand Leit-
    ner. DG 2548 206. (Reissue from 138019)
```
        +-Gr 11-75 p840          +-RR 12-75 p37
        +-HFN 10-75 p152
```

275 Concerto, piano, no. 5, op. 73, E flat major. Egmont overture,
    op. 84. Julius Katchen, pno; LSO; Pierino Gamba. London STS
    15210.
```
       ++NR 6-75 p7
```

276 Concerto, piano, no. 5, op. 73, E flat major. Mindru Katz, pno;
    Hallé Orchestra; John Barbirolli. Pye GSGC 15015. (Reissue
    from CSCL 70019)
```
        -Gr 8-75 p335            +-RR 8-75 p28
        +-HFN 8-75 p87
```

277 Concerto, piano, no. 5, op. 73, E flat major. BRAHMS: Concerto,
    piano, no. 2, op. 83, B flat minor. RACHMANINOFF: Concerto,
    piano, no. 3, op. 30, D minor. TCHAIKOVSKY: Concerto, piano,
    no. 1, op. 23, B flat minor. Vladimir Horowitz, pno; NBC Symp-
    hony Orchestra, RCA Symphony Orchestra; Arturo Toscanini, Fritz
    Reiner. RCA CRM 4-0914 (4).
```
        +SFC 5-4-75 p35
```

278 Concerto, piano, no. 5, op. 73, E flat major. Walter Gieseking,
    pno; VPO; Bruno Walter. Turnabout THS 65011. (Reissue from
    Columbia)
```
        +HF 11-74 p99            +St 2-75 p106
```

279 Concerto, piano and orchestra, op. 61, D major (arr. from violin
    concerto). Daniel Barenboim, pno; ECO; Daniel Barenboim. DG
    2530 457. Tape (c) 3300 407.
```
        +-Gr 10-74 p681          +-RR 10-74 p34
        +-HF 10-74 p90           +-RR 1-75 p68 tape
        +HF 4-75 p112 tape      ++St 1-75 p109
```

280 Concerto, violin, op. 61, D major. BUSONI: Concerto, violin, op.
    35a, D major. Adolph Busch, vln; Orchestra; COA; Fritz Busch,
    Bruno Walter. Bruno Walter Society IGI 335.
```
        +-NR 8-75 p6
```

281 Concerto, violin, op. 61, D major. Isaac Stern, vln; NYP; Leonard
    Bernstein. CBS 61598. Tape (c) 40-61598. (Reissue from Fontana
    SCFL 120)
```
       ++Gr 2-75 p1473           +-HFN 12-75 p173 tape
        -Gr 12-75 p1121 tape     +RR 1-75 p26
```

282 Concerto, violin, op. 61, D major. HEUBERGER (Kreisler): Midnight
    bells. KREISLER: Caprice viennois. Fritz Kreisler, vln;
    Michael Raucheisen, Franz Rupp, pno; Berlin State Opera Orches-
    tra; Leo Blech. HMV HLM 7062. (Reissues from Matrix CWR 631/41,
    HMV DB 990/95, BLR 6066, DA 1138, 2RA 1484, DB 3050)
```
        +-Gr 9-75 p451           +RR 9-75 p33
        +-HFN 9-75 p96
```

283   Concerto, violin, op. 61, D major. Henryk Szeryng, vln; COA;
        Bernard Haitink. Philips 6500 531. Tape (c) 7300 275.
                +Gr 10-74 p681   +NR 2-75 p5
                ++Gr 5-75 p2031 tape   +RR 10-74 p36
                +HF 2-75 p88   ++St 8-75 p94
284   Concerto, violin, op. 61, D major. Arthur Grumiaux, vln; COA;
        Colin Davis. Philips 6500 775.
                ++Gr 3-75 p1640   ++RR 3-75 p23
                ++HF 6-75 p83   ++SFC 7-13-75 p21
                ++MJ 10-75 p39   ++St 8-75 p94
                ++NR 5-75 p6
285   Concerto, violin, op. 61, D major. BRAHMS: Hungarian dance, no.
        1, G minor. SIBELIUS: Concerto, violin, op. 47, D minor: 2nd
        movement. Jascha Heifetz, vln; NYP; Orchestra; Artur Rodzinski.
        Rococo 2070.
                +NR 3-75 p12
      Concerto, violin, op. 61, D major. cf HMV SLS 5004.
      Concerto, violin, op. 61, D major. cf RCA CRL 6-0720.
      Concerto, violin, op. 61, D major. cf RCA ARM 4-0942/7.
286   Concerto, violin, violoncello and piano, op. 56, C major. Franz-
        josef Maier, vln; Anner Bylsma, vlc; Paul Badura-Skoda, pno;
        Collegium Aureum. BASF BAC 3097.
                +-Gr 7-75 p174   +MT 10-75 p885
                +-HFN 6-75 p85   +-RR 7-75 p22
287   Concerto, violin, violoncello and piano, op. 56, C major. Josef
        Suk, vln; Josef Chuchro, vlc; Jan Panenka, pno; CPhO; Kurt Masur.
        Supraphon 110 1558.
                -Gr 11-75 p802   +-RR 10-75 p37
                +-HFN 10-75 p136
      Consecration of the house, op. 124: Overture. cf Overtures
        (DG 2548 138).
      Consecration of the house, op. 124: Overture. cf Symphonies, nos.
        1-9 (Philips 6747 135).
      Contradances (12). cf HMV ASD 3017.
      Contradances, WoO 14. cf DG Archive 2533 182.
      Coriolan overture, op. 62. cf Symphonies, nos. 1-9 (DG 2740 115).
      Coriolan overture, op. 62. cf Symphonies, nos. 1-9 (London CPS 9).
      Coriolan overture, op. 62. cf Symphonies, nos. 1-9 (Philips 6747 135
      Coriolan overture, op. 62. cf Symphony, no. 2, op. 36, D major.
      Coriolan overture, op. 62. cf Symphony, no. 3, op. 55, E flat
        major.
      Coriolan overture, op. 62. cf Symphony, no. 6, op. 68, F major.
      Coriolan overture, op. 62, CCV 5023. cf Camden CCV 5000-12, 5014-24.
      Country dances, op. 141 (12). cf Nonesuch H 71141.
288   Duos, violin and violoncello, nos. 1-3. Ruggiero Ricci, vln;
        Mihaly Virizlay, vlc. Orion ORS 7295. (also Ember ECL 9008).
                -Gr 7-75 p200   ++NR 4-73 p8
                +-HF 6-73 p124   +RR 7-75 p45
                -HFN 5-75 p125   +-SFC 9-2-73 p27
289   Egmont overture, op. 84. Leonore overture, no. 3, op. 72.
        BERLIOZ: Les Francs Juges, op. 3. ROSSINI: Il barbiere di
        Siviglia: Overture. WAGNER: Dei Meistersinger von Nürnberg:
        Prelude, Act 1. CSO; Georg Solti. Decca SXLP 6684. Tape (c)
        KSXCP 6684.
                +Gr 10-74 p713   +-RR 10-74 p64
                +Gr 1-75 p1402 tape

290 Egmont overture, op. 84. BRAHMS: Liebeslieder waltzes, op. 52
     (Orchestral version). SCHUMANN: Concerto, piano, op. 54, A
     minor. Shura Cherkassky, pno; Ambrosian Singers; London
     Symphonia; Wyn Morris. Pye TPLS 13063. Tape (c) ZCTPL 13063
     (ct) Y8TPL 13063.
         +-Gr 4-75 p1802   +-HFN 6-75 p100
         +-Gr 5-75 p2037 tape   -RR 4-75 p21
     Egmont overture, op. 84. cf Concerto, piano, no. 5, op. 73, E
     flat major (Decca SLX 6655).
     Egmont overture, op. 84. cf Concerto, piano, no. 5, op. 73,
     E flat major (London 15210).
     Egmont overture, op. 84. cf Leonore overtures, nos. 1-3.
     Egmont overture, op. 84. cf Overtures (DG 2548 138).
     Egmont overture, op. 84. cf Symphonies, nos. 1-9 (DG 2740 115).
     Egmont overture, op. 84. cf Symphonies, nos. 1-9 (London CPS 9).
     Egmont overture, op. 84. cf Symphonies, nos. 1-9 (Philips 6747 135).
     Egmont overture, op. 84. cf Symphony, no. 7, op. 92, A major
     (Decca SPA 327).
     Egmont overture, op. 84. cf Symphony, no. 7, op. 92, A major
     (Decca SXLN 6673).
     Egmont overture, op. 84. cf Symphony, no. 9, op. 25, D minor
     (DG 2700 108).
     Egmont overture, op. 84, excerpts. cf Symphony, no. 9, op. 125,
     D minor (Hungaroton 11736/7).
     Egmont overture, op. 84. cf Philips 6747 204.
     Elegischer Gesang, op. 118. cf Works, selections (CBS 76404).
     Fantasia, op. 77, G minor. cf Concerto, piano, no. 4, op. 58,
     G major.
     Fantasia, op. 77, G minor. cf Sonata, piano, no. 11, op. 22,
     B flat major.
     Fantasia, op. 77, G minor. cf CBS 73396.
     Fantasia, op. 80, C minor. cf Concerto, piano, no. 1, op. 15,
     C major (Classics for Pleasure CFP 40232).
     Fantasia, op. 80, C minor. cf Concerto, piano, no. 1, op. 15,
     C major (Decca 227).
291 Fidelio, op. 72. WAGNER: Götterdämmerung: Brunhilde's immolation.
     Wesendonck Lieder (5). Kirsten Flagstad, Elisabeth Schwarzkopf,
     s; Julius Patzak, Anton Dermota, t; Hans Braun, bar; Paul
     Schöffler, Josef Greindl, bs; Vienna State Opera Chorus; VPO;
     Wilhelm Furtwängler; NYP; Bruno Walter. Bruno Walter Society
     IGI 328.
         ++NR 8-75 p11
292 Fidelio, op. 72. Ingeborg Hallstein, s; Christa Ludwig, ms;
     Gerhard Unger, Jon Vickers, Kurt Wehofschitz, t; Raymond
     Wolansky, bar; Gottlob Frick, Walter Berry, Franz Crass, bs;
     PhO and Chorus; Otto Klemperer. HMV SLS 5006 (3). Tape (c)
     TC SLS 5006. (Reissue from Columbia SAX 2451/3)
         ++Gr 5-75 p2010   ++RR 5-75 p18
         ++HFN 7-75 p76   ++RR 12-75 p99 tape
         ++HFN 12-75 p173 tape
293 Fidelio, op. 72. Birgit Nilsson, Ingeborg Winkler, s; Hans Hopf,
     Gerhard Unger, t; Paul Schöffler, bar; Hans Braun, Gottlob
     Frick, bs; Orchestra and chorus; Erich Kleiber. Rococo 1014 (2).
         ++NR 3-75 p8
     Fidelio, op. 72: ha, Ha, Ha, welch ein Augenblick. cf Decca
     GOSC 666/8.
     Fidelio, op. 72: Leonorenari. cf Discophilia KGP 4.

Fidelio, op. 72: O war ich schön mit der vereint. cf Discophilia
    KGB 2.
294 Fidelio, op. 72: Overtures (4). PhO; Otto Klemperer. Seraphim
    S 60261.
        ++SFC 12-28-75 p30
Fidelio, op. 72: Overture. cf Overtures (DG 2548 138).
Fidelio, op. 72: Overture. cf Symphonies, nos. 1-9 (Philips 6747
    135).
Fidelio, op. 72, Overture. cf Symphony, no. 7, op. 92, A major.
Fidelio, op. 72: Overture. cf Symphony, no. 7, op. 92, A major
    (Philips 7300 024).
German dance, no. 6. cf RCA ARM 4-0942/7.
Die Geschöpfe des Prometheus (The creatures of Prometheus) op. 43:
    Overture. cf Symphonies, nos. 1-9 (Philips 6747 135).
The creatures of Prometheus, op. 43: Overture. cf Symphonies,
    nos. 1-9 (DG 2740 115).
The creatures of Prometheus, op. 43: Overture. cf Symphony, no.
    2, op. 36, D major.
The creatures of Prometheus, op. 43: Overture. cf Symphony, no. 5,
    op. 67, C minor.
The creatures of Prometheus, op. 43: Overture. cf Symphony, no. 4,
    op. 60, B flat major.
The creatures of Prometheus, op. 43: Overture. cf Symphony, no. 6,
    op. 68, F major.
Grand overture (Zur Namensfeier), op. 115. cf Symphony, no. 5,
    op. 67, C major.
Grosse Fuge, op. 133, B flat major. cf Quartets, strings, nos.
    12-16 (Columbia M4 31730).
Grosse Fuge, op. 133, B flat major. cf Quartets, strings, nos.
    12-16 (Hungaroton SLPX 11673/6).
Grosse Fuge, op. 133, B flat major. cf Quartets, strings, nos.
    12-16 (Telefunken SKA 25113 T/1-4).
Grosse Fuge, op. 133, B flat major. cf Symphony, no. 2, op. 36,
    D major.
King Stephen, op. 117. cf Works, selections (CBS 76404).
295 King Stephen, op. 117: Incidental music. Songs: Elegiac song,
    op. 118; Opferlied, op. 121b; Bundeslied, op. 122; Calm sea
    and prosperous voyage, op. 112. Lorna Haywood, s; Ambrosian
    Opera Chorus; LSO; Michael Tilson Thomas. Columbia M 33509.
    (also CBS 76404. Tape (c) 40-76404)
        +Gr 11-75 p871   +-RR 12-75 p87
        +NYT 9-21-75 pD18   +-SFC 10-19-75 p33
296 King Stephen, op. 117: Overture. Wellington's victory, op. 91.
    Viennese dances (11). Paris Philharmonic Orchestra; René
    Leibowitz. Olympic 8113.
        -NR 5-75 p6
King Stephen, op. 117: Overture. cf Leonore overtures, nos. 1-3.
King Stephen, op. 117: Overture. cf Symphonies, nos. 1-9 (Philips
    6747 135).
297 Leonore overtures, nos. 1-3, opp. 72, 138. Egmont overture, op.
    84. King Stephen, op. 117: Overture. CO; Georg Szell. CBS
    Classics 61580. Tape (c) 40-61580. (Reissue from 72689)
        +-Gr 12-74 p1114   +RR 9-74 p43
        +-HFN 12-75 p173 tape
Leonore overtures, nos. 1-3, opp. 72, 138. cf Symphonies, nos.
    1-9 (Philips 6747 135).

Leonore overtures, nos. 2 and 3, op. 72. cf Overtures (DG 2548 138).
Leonore overture, no. 2, op. 72. cf HMV RLS 717.
Leonore overture, no. 3, op. 72. cf Egmont overture, op. 84.
Leonore overture, no. 3, op. 72. cf Symphonies, nos. 1-9 (London
   CPS 9).
Leonore overture, no. 3, op. 72. cf Symphonies, nos. 1, 6, 8, 9.
Leonore overture, no. 3, op. 72. cf Symphony, no. 3, op. 55,
   E flat major.
Leonore overture, no. 3, op. 72. cf Symphony, no. 7, op. 92,
   A major (Decca SPA 327).
Leonore overture, no. 3, op. 72. cf Symphony, no. 7, op. 92,
   A major (London 6668).
Leonore overture, no. 3, op. 72. cf Symphony, no. 8, op. 93, F
   major.
Leonore overture, no. 3, op. 72. cf Symphony, no. 9, op. 125,
   D minor.
Leonore overture, no. 3, op. 72. cf HMV SLA 870.
Lustig-Traurig. cf Piano works (Vanguard VSD 71186).
298 Mass, op. 123, D major. Gundula Janowitz, s; Agnes Baltsa, con;
   Peter Schreier, t; Jose van Dam, bs; Vienna Singverein; BPhO;
   Herbert von Karajan. Angel SB 3821 (2). (also HMV SLS 979)
      +-Gr 7-75 p224   -NYT 9-21-75 pD18
      +HF 8-75 p80   +-RR 7-75 p55
      +HFN 7-75 p75   ++SFC 7-6-75 p16
      +NR 7-75 p9   +-St 12-75 p116
299 Mass, op. 123, D major. Margaret Price, s; Christa Ludwig, ms;
   Wieslaw Ochman, t; Martti Talvela, bs; Vienna State Opera
   Chorus; VPO; Karl Böhm. DG 2707 080 (2).
      ++Gr 6-75 p75   +RR 7-75 p55
      +-HF 11-75 p96   +SFC 10-19-75 p33
      +HFN 7-75 p76   +St 12-75 p116
      +-NYT 9-21-75 pD18
300 Mass, op. 123, D major. Ilona Steingruber, s; Else Scheurhoff,
   con; Erich Majkut, t; Otto Wiener, bs; Vienna Academy Chamber
   Choir; VSO; Otto Klemperer. Turnabout TV 37072. (Reissue from
   Vox PL 6992)
      -Gr 3-75 p1687   +-RR 3-75 p56
Meerestille and Gluckliche Fahrt, op. 112. cf Works, selections
   (CBS 76404).
Minuet, E flat major. cf Piano works (Vanguard VSD 71187).
Minuet, G major. cf Philips 6747 199.
Namensfeier, op. 115: Overture. cf Overtures (DG 2548 138).
Namensfeier, op. 115: Overture. cf Symphonies, nos. 1-9 (Philips
   6747 135).
Mödlinger dances, nos. 1-4, 6, 8. cf Saga 5411.
301 Octet, op. 103, E flat major. Quintet, piano, op. 16, E flat major.
   Rudolf Serkin, pno; Marlboro Festival Winds. Columbia M 33527.
      ++NR 8-75 p7   ++SFC 11-23-75 p26
Opferlied, op. 121b. cf Works, selections (CBS 76404).
302 Overtures: Consecration of the house, op. 124. Egmont, op. 84.
   Fidelio, op. 72. Leonore, nos. 2 and 3, op. 72. Namensfeier,
   op. 115. The ruins of Athens, op. 113. BPhO, Lamoureux
   Orchestra, Bavarian Radio Symphony Orchestra; Eugen Jochum,
   Igor Markevich. DG 2548 138. (Reissues from SLPM 138039,
   138038, 138694, 138032, 136019)
      +-Gr 10-75 p601   /RR 9-75 p28
      +HFN 9-75 p108

Pariser Einzugsmarsche (attrib.). cf DG 2721 077.

303 Piano works: Bagatelle, no. 25, A minor. Rondo a capriccio, op.
     129, G major. Sonata, piano, no. 7, op. 10, no. 3, D major.
     Variations on an original theme, C minor (32). André Watts, pno.
     Columbia M 33074.
  +–HF 4-75 p84   +St 3-75 p108
  -NR 2-75 p11

304 Piano works: Andante favori, F major. Lustig-Traurig. Sonata,
     piano, no. 19, op. 49, no. 1, G minor. Sonata, piano, no. 20,
     op. 49, no. 2, G major. Sonata, piano, no. 21, op. 53, C
     major. Bruce Hungerford, pno. Vanguard VSD 71186.
  +HF 12-74 p98   ++SFC 10-20-74 p28
  ++MJ 2-75 p41   ++St 8-74 p104

305 Piano works: Bagatelle, no. 25, A minor. Minuet, E flat major.
     Rondo, op. 51, no. 1, C major. Sonata, piano, no. 6, op. 10,
     no. 2, F major. Sonata, piano, no. 7, op. 10, no. 3, D major.
     Bruce Hungerford, pno. Vanguard VSD 71187.
  +HF 12-74 p98   ++SFC 10-20-74 p28
  ++MJ 2-75 p41   ++St 8-74 p104

306 Quartet, piano and strings, op. 16, E flat major. Serenade,
     string trio, op. 8, D major. Cantilena Chamber Players.
     Musical Heritage Society MHS 1795.
  +HF 5-75 p72

307 Quartets, strings, nos. 1-6, op. 18. Végh Quartet. Telefunken
     SPA 25128 (3).
  ++NR 12-75 p7

308 Quartet, strings, no. 5, op. 18, no. 5, A major. Quartet, strings,
     no. 6, op. 18, no. 6, B flat major. Quartetto Italiano.
     Philips 6500 647. Tape (c) 7300 348.
  +–Gr 6-75 p56   ++NR 6-75 p9
  +HF 6-75 p84   ++RR 5-75 p47
  ++HFN 6-75 p85   ++RR 7-75 p69 tape

     Quartet, strings, no. 6, op. 18, no. 6, B flat major. cf Quartet,
     strings, no. 5, op. 18, no. 5, A major.

309 Quartets, strings, nos. 7-9, op. 59. Quartetto Italiano. Philips
     6747 139 (2).
  +Gr 12-75 p1062   +RR 11-75 p58
  ++HFN 12-75 p148

310 Quartets, strings, nos. 7-11, opp. 59, 74, 95. Amadeus Quartet.
     DG 2733 005 (3). (Reissues from SLPM 138534/6)
  *Gr 4-75 p1820   +RR 5-75 p47
  +HFN 5-75 p124

311 Quartets, strings, nos. 7-11, opp. 59, 74, 95. Budapest Quartet.
     Odyssey Y3 33316 (3). (Reissues from Columbia SL 173)
  +–HF 12-75 p92

312 Quartet, strings, no. 9, op. 59, no. 3, C major. Quartet, strings,
     no. 11, op. 95, F minor. Drolc Quartet. Classics for Pleasure
     CFP 40089.
  +–Gr 3-75 p1673   +–RR 1-75 p37

313 Quartet, strings, no. 9, op. 59, no. 3, C major. Quartet, strings,
     no. 11, op. 95, F minor. Quartet, strings, no. 15, op. 132,
     A minor. Busch Quartet. World Records SHB 27/1-2. (Reissues
     from HMV DB 2109/12, DB 1799-800, DB 3375-80)
  +Gr 1-75 p1361   ++STL 2-9-75 p37
  +RR 1-75 p37

314 Quartet, strings, no. 11, op. 95, F minor. MOZART: Quartet,

strings, no. 17, K 458, B flat major. SCHUBERT: Quartet,
strings, no. 12, D 703, C minor. Quartetto Italiano. Philips
6599 931. (Reissues from 6500 180, SAL 3633, 3618)
     ++Gr 12-75 p1062   +HFN 11-75 p173
Quartet, strings, no. 11, op. 95, F minor. cf Quartet, strings,
no. 9, op. 59, no. 3, C major.
Quartet, strings, no. 11, op. 95, F minor. cf Quartet, strings,
no. 9, op. 59, no. 3, C major (World SHB 27/1-2).
315 Quartets, strings, nos. 12-16. Grosse Fuge, op. 133, B flat major.
Juilliard Quartet. Columbia M4 31730 (4).
     +HF 9-74 p90   +-St 3-75 p97
     +-NYT 7-14-74 pD23
316 Quartets, strings, nos. 12-16. Grosse Fuge, op. 133, B flat major.
Bartók Quartet. Hungaroton SLPX 11673/6.
     +-HF 5-75 p70   +-RR 5-74 p45
     +-NYT 7-14-74 pD24   ++SFC 8-18-74 p22
317 Quartets, strings, nos. 12-16. Grosse Fuge, op. 133, B flat major.
Végh Quartet. Telefunken SKA 25113 T/1-4 (4).
     +Gr 10-74 p713   ++NR 1-75 p7
     ++HF 8-75 p81   +-RR 10-74 p69
318 Quartet, strings, no. 13, op. 130, B flat major. Busch Quartet.
CBS 61664.
     ++Gr 11-75 p846   +-RR 11-75 p59
     ++HFN 11-75 p151
319 Quartet, strings, no. 15, op. 132, A minor. Collegium Aureum
Quartet. BASF KHB 21509. (also BASF BAC 3071)
     +Gr 1-75 p1362   +-RR 1-75 p38
     +MQ 4-74 p312   +St 12-73 p120
     ++NR 9-73 p7
Quartet, strings, no. 15, op. 132, A minor. cf Quartet, strings,
no. 9, op. 59, no. 3, C major.
Quartet, strings, no. 16, op. 135, F major: Lento assai. cf
Symphony, no. 5, op. 67, C minor.
Quintet, piano, op. 16, E flat major. cf Octet, op. 103, E flat
major.
320 Romance, no. 1, op. 40, G major. Romance, no. 2, op. 50, F major.
MOZART: Adagio, K 261, E major. Rondo, violin, K 373, C major.
SCHUBERT: Rondo, violin and string quartet, D 438, A major.
Josef Suk, vln; AMF; Neville Marriner. Klavier KS 530. (also
HMV SXLP 30179. Reissue from ASD 2725)
     +Gr 4-75 p1979   ++NR 10-74 p13
321 Romances, nos. 1 and 2, opp. 40, 50. SIBELIUS: Concerto, violin,
op. 47, D minor. Pinchas Zukerman, vln; LPO; Daniel Barenboim.
DG 2530 552.
     +-Gr 10-75 p633   +RR 10-75 p55
     +HFN 11-75 p151
322 Romance, no. 1, op. 40, G major. Romance, no. 2, op. 50, F major.
GOLDMARK: Concerto, violin, no. 1, op. 28, A minor.* Nathan
Milstein, vln; PhO; Harry Blech. Seraphim S 60238. (*Reissue
from Capitol SP 8414)
     ++HF 10-75 p71   ++St 9-75 p107
     ++NR 8-75 p15
Romance, no. 1, op. 40, G major. cf RCA ARM 4-0942/7.
Romance, no. 2, op. 50, F major. cf Romance, no. 1, op. 40, G
major (Klavier KS 530).
Romance, no. 2, op. 50, F major. cf Romance, no. 1, op. 40, G
major (Seraphim S 60238).

Romance, no. 2, op. 50, F major. cf BRAHMS: Concerto, violin,
   op. 77, D major.
Romance, no. 2, op. 50, F major. cf RCA ARM 4-0942/7.
323 Rondo, WoO 6, B flat major (edit. Czerny). MOZART: Concerto,
   piano, no. 20, K 466, D minor. Sviatoslav Richter, pno; VSO,
   Warsaw Philharmonic Orchestra; Kurt Sanderling, Stanislaw
   Wislocki. DG 2548 106. (Reissues from 138848, SLPE 133226)
          +Gr 10-75 p607   +RR 12-75 p56
Rondo, op. 51, no. 1, C major. cf Piano works (Vanguard VSD 71187).
Rondo, op. 51, no. 2, G major. cf Sonata, piano, no. 24, op. 78,
   F sharp major.
324 Rondo a capriccio, op. 129, G major. The ruins of Athens, op. 113:
   Turkish march. Sonata, piano, no. 3, op. 2, no. 3, C major.
   Sonata, piano, no. 14, op. 27, no. 2, C sharp minor. Josef
   Hofmann, pno. Saga 5392. (From Duo-Art piano rolls)
          +Gr 7-75 p214   +RR 7-75 p46
Rondo a capriccio, op. 129, G major. cf Bagatelle, no. 25, A
   minor.
Rondo a capriccio, op. 129, G major. cf Piano works (Columbia
   M 33074).
Rondo a capriccio, op. 129, G major. cf DG Heliodor 2548 137.
The ruins of Athens, op. 113: Chorus of dervishes Turkish march.
   cf RCA ARM 4-0942/7.
The ruins of Athens, op. 113: Overture. cf Overtures (DG 2548 138).
Die Ruinen von Athens, op. 113: Overture. cf HMV RLS 717.
The ruins of Athens, op. 113: Overture. cf Symphonies, nos. 1-9
   (Philips 6747 135).
The ruins of Athens, op. 113: Turkish march. cf Rondo a capriccio,
   op. 129, G major.
Die Ruinen von Athens, op. 113: Turkish march. cf Philips
   6580 107.
325 Septet, op. 20, E flat major. Vienna Octet. Decca SDD 200. Tape
   (c) KSDC 200.
          +HFN 7-75 p90 tape   +RR 7-75 p69 tape
Serenade, flute, violin and viola, op. 25, D major: Adagio, Allegro.
   cf Decca SPA 394.
Serenade, string trio, op. 8, D major. cf Quartet, piano and
   strings, op. 16, E flat major.
Serenade, string trio, op. 8, D major. cf Trios, strings, op. 3,
   no. 1, E flat major.
326 Sketchbooks: Symphonies, nos. 1, 2, 7. Denis Matthews, pno and
   speaker. Discourses ABM 4.
          +HFN 9-75 p96   +RR 9-75 p33
327 Sketchbooks: Symphonies, nos. 5, 6. Denis Matthews. Discourse
   ABM 5.
          +HFN 12-75 p149   +RR 12-75 p37
Sonata, C major. cf HUMMEL: Sonata, mandolin and piano, C major.
328 Sonata, horn, op. 17, F major. DANZI: Sonata, horn, op. 28, E
   flat major. SAINT-SAENS: Romance, op. 67. SCHUMANN: Adagio
   and allegro, op. 70, A flat major. Barry Tuckwell, hn; Vladimir
   Ashkenazy, pno. Decca SXL 6717.
          +Gr 8-75 p342   +RR 8-75 p54
          +HFN 8-75 p84
329 Sonatas, piano, complete. Wilhelm Kempff, pno. DG 2740 130 (11).
   (Reissues from SKL 901/11)
          +Gr 11-75 p909   +RR 11-75 p73

Sonata, piano, no. 3, op. 2, no. 3, C major. cf Rondo a capriccio, op. 129, G major.

Sonata, piano, no. 4, op. 7, E flat major. cf BACH: Chaconne (Busoni).

Sonata, piano, no. 5, op. 10, no. 1, C minor. cf Concerto, piano, no. 1, op. 15, C major.

Sonata, piano, no. 6, op. 10, no. 2, F major. cf Piano works (Vanguard VSD 71187).

330 Sonata, piano, no. 7, op. 10, no. 3, D major. Sonata, piano, no. 23, op. 57, F minor. Vladimir Horowitz, pno. RCA VH 012. (also RCA LSC 2366. Tape (c) R8S1040) (Reissue from RB 16230)
    ++Gr 3-75 p1677   +-RR 9-75 p57
    ++HFN 7-75 p76

Sonata, piano, no. 7, op. 10, no. 3, D major. cf Bagatelle, no. 25, A minor.

Sonata, piano, no. 7, op. 10, no. 3, D major. cf Piano works (Columbia M 33074).

Sonata, piano, no. 7, op. 10, no. 3, D major. cf Piano works (Vanguard VSD 71187).

331 Sonatas, piano, nos. 8, 14, 21. Rudolf Firkušny, pno Decca PFS 4341.
      -Gr 12-75 p1075   /-RR 11-75 p73
      -HFN 11-75 p151

332 Sonatas, piano, nos. 8, 14, 17, 21, 23, 26. Wilhelm Kempff, pno. DG 2726 042 (2). (Reissues from SKL 901/11)
    +Gr 8-75 p347   +RR 11-75 p73
    +HFN 9-75 p109

333 Sonatas, piano, nos. 8, 21, 26. Vladimir Ashkenazy, pno. Decca SXL 6706. Tape (c) KSCX 6706.
    +Gr 5-75 p1991   +HFN 7-75 p90 tape
    ++HFN 6-75 p87   +-RR 5-75 p52

334 Sonata, piano, no. 8, op. 13, C minor. Sonata, piano, no. 14, op. 27, no. 2, C sharp minor. Sonata, piano, no. 23, op. 57, F minor. Joseph Cooper, pno. Music for Pleasure MFP 57011. (Reissues)
    +-Gr 4-75 p1835   +-RR 4-75 p48

335 Sonata, piano, no. 10, op. 14, no. 2, G major. Sonata, piano, no. 11, op. 22, B flat major. Sonata, piano, no. 20, op. 49, no. 2, G major. Sviatoslav Richter, pno. Philips 6580 095. (Reissue from SAL 3456/7)
    +-Gr 3-75 p1677   +RR 3-75 p40

336 Sonata, piano, no. 11, op. 22, B flat major. Sonata, piano, no. 24, op. 78, F sharp major. Fantasia, op. 77, G minor. Rudolf Serkin, pno. Columbia M 32294.
    +HF 6-74 p80   ++St 2-75 p106
    +SFC 2-24-74 p25

Sonata, piano, no. 11, op. 22, B flat major. cf Sonata, piano, no. 10, op. 14, no. 2, G major.

Sonata, piano, no. 12, op. 26, A flat major. cf Andante favori, F major.

337 Sonatas, piano, nos. 14, 23, 26. Howard Shelley, pno. Contour 2870 436.
    ++HFN 7-75 p76   +RR 4-75 p48

Sonata, piano, no. 14, op. 27, no. 2, C sharp minor. cf Rondo a capriccio, op. 129, G major.

Sonata, piano, no. 14, op. 27, no. 2, C sharp minor. cf Sonata,
    piano, no. 8, op. 13, C minor.
Sonata, piano, no. 14, op. 27, no. 2, C sharp minor. cf Philips
    6747 199.
Sonata, piano, no. 14, op. 27, no. 2, C sharp minor. cf Turnabout
    TV 37033.
Sonata, piano, no. 14, op. 27, no. 2, C sharp minor: Presto. cf
    Monitor MCS 2143.
Sonata, piano, no. 19, op. 49, no. 1, G minor. cf Piano works
    (Vanguard VSD 71186).
Sonata, piano, no. 19, op. 49, no. 1, G minor. cf Columbia MG
    33202.
Sonata, piano, no. 20, op. 49, no. 2, G major. cf Piano works
    (Vanguard VSD 71186).
Sonata, piano, no. 20, op. 49, no. 2, G major. cf Sonata, piano,
    no. 10, op. 14, no. 2, G major.
Sonata, piano, no. 20, op. 49, no. 2, G major. cf Columbia MG
    33202.
Sonata, piano, no. 20, op. 49, no. 2, G major. cf Connoisseur
    Society (Q) CSQ 2065.
338 Sonata, piano, no. 21, op. 53, C major. Sonata, piano, no. 30,
    op. 109, E major. Antonio Barbosa, pno. Connoisseur Society
    CS 2068 (Q) CSQ 2068.
        /-HF 4-75 p81   +SR 1-11-75 p51
        -NR 2-75 p11   ++St 12-74 p129
        ++SFC 10-20-74 p28
339 Sonata, piano, no. 21, op. 53, C major. Sonata, piano, no. 31,
    op. 110, A flat major. Andante favori, A major. Alfred
    Brendel, pno. Philips 6500 762. Tape (c) 7300 351.
        +Gr 11-75 p859   +-RR 11-75 p74
        ++HFN 7-75 p90 tape
Sonata, piano, no. 21, op. 53, C major. cf Andante favori, F
    major.
Sonata, piano, no. 21, op. 53, C major. cf Concerto, piano, no.
    2, op. 19, B flat major.
Sonata, piano, no. 21, op. 53, C major. cf Piano works (Vanguard
    VSD 71186).
Sonata, piano, no. 21, op. 53, C major. cf International Piano
    Archives IPA 5007/8.
340 Sonatas, piano, nos. 23, 24, 29, 32. Alfred Brendel, pno.
    Philips 6500 138/9 (2).
        +-HF 9-75 p82   ++SFC 10-5-75 p38
341 Sonata, piano, no. 23, op. 57, F minor. Sonata, piano, no. 32,
    op. 111, C minor. Alfred Brendel, pno. Philips 6500 138.
        ++MJ 10-75 p38   +-NR 6-75 p13
Sonata, piano, no. 23, op. 57, F minor. cf Sonata, piano, no. 7,
    op. 10, no. 3, D major.
Sonata, piano, no. 23, op. 57, F minor. cf Sonata, piano, no. 8,
    op. 13, C minor.
342 Sonata, piano, no. 24, op. 78, F sharp major. Sonata, piano, no.
    30, op. 109, E major. Rondo, op. 51, no. 2, G major. Jörg
    Demus, pno. BASF BAC 3063.
        +Gr 1-75 p1365   +-RR 2-75 p49
        ++MT 7-75 p628
343 Sonata, piano, no. 24, op. 78, F sharp major. Sonata, piano, no.
    29, op. 106, B flat major. Alfred Brendel, pno. Philips 6500 139.

```
     ++HFN 2-72 p303            ++SFC 8-10-75 p26
     ++MJ 10-75 p38            +-St 7-75 p94
     +-NR 6-75 p13
```

344 Sonata, piano, no. 24, op. 78, F sharp major. Sonata, piano,
     no. 29, op. 106, B flat major. Claudio Arrau, pno. Philips
     6580 104. (Reissues from SAL 3600, 3494)
```
          +Gr 11-75 p859            +RR 12-75 p78
     ++HFN 10-75 p153
```
     Sonata, piano, no. 24, op. 78, F sharp major. cf Sonata, piano,
     no. 11, op. 22, B flat major.
345 Sonatas, piano, nos. 25-27. Emil Gilels, pno. DG 2530 589.
```
          +Gr 12-75 p1075            ++RR 12-75 p78
     +-HFN 12-75 p149
```
     Sonata, piano, no. 25, op. 79, G major. cf Concerto, piano,
     no. 5, op. 73, E flat major.
     Sonata, piano, no. 26, op. 81a, E flat major. cf Concerto, piano,
     no. 3, op. 37, C minor.
     Sonata, piano, no. 27, op. 90, E minor. cf CBS 73396.
346 Sonatas, piano, nos. 29-32. Wilhelm Kempff, pno. DG 2526 033 (2).
     (Reissue from SKL 901/11)
```
          +Gr 3-75 p1677            +RR 7-75 p45
```
347 Sonatas, piano, nos. 29-32. Claudio Arrau, pno. Philips 6780
     020 (2).
```
          +NR 11-75 p14
```
348 Sonata, piano, no. 29, op. 106, B flat major. Variations on a
     theme by Diabelli, op. 120. Webster Aitkin, pno. Delos DEL
     24101/2.
```
     +-HF 5-75 p72              +-SFC 3-2-75 p24
     ++MJ 3-75 p25             +SR 2-22-75 p47
     -NR 4-75 p12              +-St 7-75 p94
```
349 Sonata, piano, no. 29, op. 106, B flat major. MOZART: Sonata,
     piano, no. 13, K 333, B flat major. Variations on "Ah,
     vous dirai-je Maman", K 265, C major. SCHUMANN: Kinderscenen,
     op. 15. Variations on A-B-E-G-G, op. 1. Christoph Eschenbach,
     pno. DG 2726 044 (2). (Reissues from 2350 080, SLPM 138949,
     139183)
```
          +Gr 9-75 p492            +-RR 12-75 p79
     +HFN 10-75 p153
```
     Sonata, piano, no. 29, op. 106, B flat major. cf Sonata, piano,
     no. 24, op. 78, F sharp major (Philips 6580 104).
     Sonata, piano, no. 29, op. 106, B flat major. cf Sonata, piano,
     no. 24, op. 78, F sharp major (Philips 6500 139).
     Sonata, piano, no. 30, op. 109, E major. cf Sonata, piano, no. 21,
     op. 53, C major (Connoisseur Society CS 2068).
     Sonata, piano, no. 30, op. 109, E major. cf Sonata, piano, no. 24,
     op. 78, F sharp major.
350 Sonata, piano, no. 31, op. 110, A flat major. Sonata, piano, no.
     32, op. 111, C minor. Vladimir Ashkenazy, pno. Decca SXL 6630.
     Tape (c) KSXC 6630. (also London CS 6843)
```
          +Gr 8-74 p375             ++RR 8-74 p46
     ++Gr 10-74 p771 tape      +RR 3-75 p73 tape
     ++HF 3-75 p77             ++St 4-75 p72
```
351 Sonata, piano, no. 31, op. 110, A flat major. Sonata, piano,
     no. 32, op. 111, C minor. Stephen Bishop, pno. Philips 6500 674.
```
          +Gr 6-75 p67              +RR 6-75 p59
     +HFN 8-75 p72             ++SFC 3-2-75 p24
     ++NR 4-75 p12
```

Sonata, piano, no. 31, op. 110, A flat major. cf Andante favori, F major.

Sonata, piano, no. 31, op. 110, A flat major. cf Bagatelles, op. 126.

Sonata, piano, no. 31, op. 110, A flat major. cf Sonata, piano, no. 21, op. 53, C major.

Sonata, piano, no. 32, op. 111, C minor. cf Sonata, piano, no. 23, op. 57, F minor.

Sonata, piano, no. 32, op. 111, C minor. cf Sonata, piano, no. 31, op. 110, A flat major (Decca SXL 6630).

Sonata, piano, no. 32, op. 111, C minor. cf Sonata, piano, no. 31, op. 110, A flat major (Philips 6500 764).

352 Sonatas, violin and piano, nos. 1-3. Arthur Grumiaux, vln; Clara Haskil, pno. Philips 6580 090. (Reissues from ABL 3204, ABL 3199)

    +Gr 4-75 p1819   +RR 3-75 p40

353 Sonatas, violin and piano, nos. 1-10. Josef Suk, vln; Jan Panenka, pno. Supraphon 111 0561/5 (5). (Some reissues from SUAST 50905/50907)

    +-Gr 11-75 p847   +RR 10-75 p69
    +-HFN 10-75 p136

354 Sonata, violin and piano, no. 2, op. 12, A major. Sonata, violin and piano, no. 9, op. 47, A major. Itzhak Perlman, vln; Vladimir Ashkenazy, pno. Decca SXL 6632. (also London CS 6845)

    +Gr 2-75 p1504   +RR 2-75 p49
    +HF 7-75 p71   ++SFC 7-13-75 p21
    ++NR 8-75 p8   +St 8-75 p95

Sonata, violin and piano, no. 3, op. 12, E flat major. cf Discopaedia MB 1002.

Sonata, violin and piano, no. 3, op. 12, E flat major. cf Rococo 2072.

355 Sonata, violin and piano, no. 4, op. 23, A minor. Sonata, violin and piano, no. 6, op. 30, no. 1, A major. Yehudi Menuhin, vln; Wilhelm Kempff, pno. DG 2530 458.

    +NR 1-75 p7

356 Sonata, violin and piano, no. 5, op. 24, F major. BRAHMS: Sonatas, violin and piano, nos. 1-3. Wanda Wilkomirska, vln; Antonio Barbosa, pno. Connoisseur Society CS 2079, 2080 (2).

    +-HF 12-75 p88

357 Sonata, violin and piano, no. 5, op. 24, F major. Sonata, violin and piano, no. 9, op. 47, A major. Wolfgang Schneiderhan, vln; Carl Seemann, pno. DG 2548 135. (Reissue from 138120)

    /-Gr 11-75 p840   -RR 10-75 p70
    -HFN 10-75 p153

Sonata, violin and piano, no. 6, op. 30, no. 1, A major. cf Sonata, violin and piano, no. 4, op. 23, A minor.

358 Sonata, violin and piano, no. 7, op. 30, no. 2, C minor. Sonata, violin and piano, no. 10, op. 96, G major. Yehudi Menuhin, vln; Wilhelm Kempff, pno. DG 2530 346.

    +NR 12-75 p8

359 Sonata, violin and piano, no. 8, op. 30, no. 3, G major. Sonata, violin and piano, no. 9, op. 47, A major. Yehudi Menuhin, vln; Wilhelm Kempff, pno. DG 2530 135. Tape (c) 3300 292. (Reissue from 2720 018)

    ++HFN 2-72 p303   +-RR 7-73 p92 tape
    +HFN 4-73 p792 tape   +SFC 3-10-74 p26
    +-NR 6-74 p6   +St 8-75 p95

360 Sonata, violin and piano, no. 8, op. 30, no. 3, G major. GRIEG:
    Sonata, violin and piano, no. 3, op. 45, C minor. RACHMANINOFF:
    The isle of the dead, op. 29. Vocalise, op. 34, no. 14.
    Symphony, no. 3, op. 44, A minor. SCHUBERT: Sonata, violin and
    piano, op. 162, D 574, A major. Sergei Rachmaninoff, pno;
    Fritz Kreisler, vln; PO; Sergei Rachmaninoff. RCA ARM 3-0295
    (3). (also RCA AVM 3-0295)
            +Gr 3-75 p1682   +NR 2-74 p12
            +HF 2-74 p109   +RR 4-75 p29
            +HFN 6-75 p98   +-St 2-74 p124
    Sonata, violin and piano, no. 8, op. 30, no. 3, G major. cf BARTOK:
    For children, excerpts.
    Sonata, violin and piano, no. 9, op. 47, A major. cf Sonata,
    violin and piano, no. 5, op. 24, F major.
    Sonata, violin and piano, no. 9, op. 47, A major. cf Sonata,
    violin and piano, no. 2, op. 12, A major.
    Sonata, violin and piano, no. 9, op. 47, A major. cf Sonata,
    violin and piano, no. 8, op. 30, no. 3, G major (DG 2530 135).
    Sonata, violin and piano, no. 9, op. 47, A major. cf RCA ARM
    4-0942/7.
    Sonata, violin and piano, no. 10, op. 96, G major. cf Sonata,
    violin and piano, no. 7, op. 30, no. 2, C minor.
    Sonata, violoncello, no. 3, op. 69, A major: 3rd movement. cf
    HMV SEOM 19.
    Sonatina, C major. cf HUMMEL: Sonata, mandolin and piano, C major.
    Sonatina, C minor. cf HUMMEL: Sonata, mandolin and piano, C major.
    Sonatina, G major. cf Connoisseur Society (Q) CSQ 2066.
361 Songs: Ich liebe dich, G 235; Kennst du das Land, op. 75, no. 1;
    Lieder von Gellert, op. 48, no. 1, Bitten; no. 2, Die Liebe
    des Nächsten; no. 3, Vom Tode; no. 4, Die Ehre Gottes aus der
    Natur; no. 5, Gottes Macht und Versehung; no. 6, Busslied;
    Wonne der Wehmut, op. 83, no. 1. SCHUBERT: Songs: An die
    Musik, D 547; Die Allmacht, D 852; Du bist die Ruh, D 776; Der
    Jüngling an der Quelle, D 300; Lachen und Weinen, D 777; Nacht
    und Träume, D 827; Nachtviolen, D 752; Rastlose Liebe, D 138.
    Heather Harper, s; Paul Hamburger, pno. BBC REB 170.
            +Gr 2-75 p1541   +-RR 2-75 p58
    Songs: Elegiac song, op. 118; Opferlied, op. 121b; Bundeslied,
    op. 122; Calm sea and prosperous voyage, op. 112. cf King
    Stephen, op. 117: Incidental music.
    Songs: An die Hoffnung, op. 94; Ich liebe dich, G 235. cf Desto
    DC 7118/9.
    Songs: Mit Mädeln sich vertragen. cf DG Archive 2533 149.
    Songs: Wiedersehen; Though thou so blest. cf RCA ARL 1-5007.
    Song: An die ferne Geliebte, op. 98. cf Telefunken DLP 6-48064.
    Songs from sonatas and symphonies arranged as songs by Friedrich
    Silcher. cf Symphony, no. 8, op. 93, F major.
362 Symphonies, nos. 1-9. Coriolan overture, op. 62. Egmont overture,
    op. 84. The creatures of Prometheus, op. 43: Overture. VPO;
    Carl Böhm. DG 2740 115.
            +HFN 8-75 p87   +-RR 7-75 p22
363 Symphonies, no. 1-9. Soloists; Vienna Singverein; BPhO; Herbert
    von Karajan. DG Tape (c) 3378 001 (6).
            +-RR 7-75 p68 tape
364 Symphonies, nos. 1-9. Coriolan overture, op. 62. Egmont overture,
    op. 84. Leonore overture, no. 3, op. 72. CSO; Georg Solti.

London CPS 9 (9). (also Decca 11BB 188/96 (9). Reissues from
SXL 6655, 6684, 6BB 121/2)
++Gr 9-75 p443   +-RR 9-75 p28
+HFN 9-75 p95   ++SFC 11-23-75 p27

365 Symphonies, nos. 1-9. Overtures: Consecration of the house, op.
124. Coriolan, op. 62. The creatures of Prometheus, op. 43.
Egmont, op. 84. Namensfeier, op. 115. The ruins of Athens,
op. 113. Leipzig Gewandhaus Orchestra; Kurt Masur. Philips
6747 135 (9).
+Gr 9-75 p443   +-RR 9-75 p28
/HFN 9-75 p95

366 Symphonies, nos. 1, 6, 8, 9. Leonore overture, no. 3, op. 72.
Pilar Lorengar, 1; Yvonne Minton, con; Stuart Burrows, t;
Martti Talvela, bs; Vienna State Opera Chorus; CSO; Georg Solti.
Decca Tape (c) K3F 11.
+-RR 11-75 p91 tape

367 Symphony, no. 1, op. 21, C major. Symphony, no. 2, op. 36, D major.
VPO; Hans Schmidt-Isserstedt. Decca SXL 6437. Tape (c) KSXC
6437.
++Gr 5-75 p2031 tape   +-RR 2-75 p74 tape

368 Symphony, no. 1, op. 21, C major. Orchestra; Colin Davis. HMV
Tape (c) TC EXE 138.
+HFN 7-75 p90 tape

369 Symphony, no. 1, op. 21, C major. SCHUBERT: Rosamunde, op. 26,
D 797: Overture; Entr'acte, Act 3; Ballet music, no. 2. COA;
Willem Mengelberg. Pearl HE 301.
+-Gr 8-75 p316   +-RR 6-75 p31

370 Symphony, no. 1, op. 21, C major. Symphony, no. 8, op. 93, F major.
COA; Eugen Jochum. Philips 6500 087. Tape (c) 7300 081.
/HFN 9-75 p110 tape   +-RR 10-75 p96 tape

371 Symphony, no. 1, op. 21, C major. Symphony, no. 2, op. 36, D
major. AMF; Neville Marriner. Philips 6500 113. Tape (c) 7300
087. (Reissue from 6707 013)
+Gr 12-72 p1214 tape   +RR 1-73 p88
-HF 5-75 p72   ++SFC 4-20-75 p23
+HFN 5-72 p913   ++St 5-75 p94
-NR 3-75 p2

372 Symphony, no. 1, op. 21, C major. Symphony, no. 8, op. 93, F
major. LOL; Leslie Jones. Unicorn UNS 200.
+-NR 9-75 p5
Symphony, no. 1, op. 21, C major. cf Toscanini Society ATS GS
1201/6.

373 Symphony, no. 2, op. 36, D major. Coriolan overture, op. 62.
Grosse Fuge, op. 133, B flat major. OSR; Ernest Ansermet.
Decca ECS 739. (Reissues from SXL 2228, SEC 5016, BR 3070)
+Gr 7-75 p173   +RR 6-75 p31
+-HFN 8-75 p87

374 Symphony, no. 2, op. 36, D major. Symphony, no. 4, op. 60, B flat
major. The creatures of Prometheus, op. 43: Overture. VPO;
Karl Böhm. DG 2530 448, 2530 451 (2). (Reissues from 2720 045)
+Gr 10-75 p602   +RR 8-75 p28
+HFN 8-75 p87

375 Symphony, no. 2, op. 36, D major (arr. for piano trio). HAYDN:
Trios, violin, viola and violoncello (Sonata, piano, no. 40,
G major, arr. Haydn). Thomas Brandis, vln; Wolfgang Boettcher,
vlc; Eckart Besch, pno. DG Archive 2533 136.
++NR 5-75 p12   +NYT 8-10-75 pD14

Symphony, no. 2, op. 36, D major. cf Symphony, no. 1, op. 21,
    C major (Decca 6437).
Symphony, no. 2, op. 36, D major. cf Symphony, no. 1, op. 21,
    C major (Philips 6500 113).
376 Symphony, no. 3, op. 55, E flat major. BPhO; Karl Böhm. DG 2535
    101. Tape (c) 3335 101. (Reissue from SLPM 138814)
        +Gr 7-75 p174   +RR 7-75 p22
        +HFN 8-75 p87   +RR 10-75 p96 tape
377 Symphony, no. 3, op. 55, E flat major. BPhO; Ferenc Fricsay.
    DG Heliodor 2548 088. (Reissue from 138 038)
        +Gr 4-75 p1867   +-RR 4-75 p21
378 Symphony, no. 3, op. 55, E flat major. VPO; Pierre Monteux.
    London STS 15190. (Reissue from RCA)
        +St 9-75 p106
379 Symphony, no. 3, op. 55, E flat major. Leonore overture, no. 3,
    op. 72. COA; Pierre Monteux, Eugen Jochum. Philips Tape
    (c) 7317 102.
        +RR 10-75 p96 tape
380 Symphony, no. 3, op. 55, E flat major. San Francisco Symphony
    Orchestra; Seiji Ozawa. Philips 9500 002. Tape (c) 7300 420.
        -SFC 12-28-75 p30
381 Symphony, no. 3, op. 55, E flat major. Coriolan overture, op. 62.
    LSO; Leopold Stokowski. RCA ARL 1-0600. Tape (c) ARK 1-0600
    (ct) ARS 1-0600.
        +-Gr 10-75 p602   +-RR 9-75 p30
        +-HFN 11-75 p150   +-SFC 7-13-75 p21
        +NR 7-75 p4   +St 9-75 p106
382 Symphony, no. 3, op. 55, E flat major. BPhO; Wilhelm Furtwängler.
    Rococo 2050.
        -NR 2-75 p3
Symphony, no. 3, op. 55, E flat major. cf HMV RLS 717.
383 Symphony, no. 4, op. 60, B flat major. Symphony, no. 8, op. 93,
    F major. NYP; Leonard Bernstein. CBS 73427.
        +-Gr 7-75 p174   +-RR 6-75 p32
        -HFN 7-75 p73
384 Symphony, no. 4, op. 60, B flat major. The creatures of Prometheus,
    op. 43: Overture. VPO; Karl Böhm. DG 2530 451. (Reissue from
    2740 115)
        +HFN 8-75 p87   ++NR 12-75 p3
385 Symphony, no. 4, op. 60, B flat major. Symphony, no. 5, op. 67,
    C minor. NBC Symphony Orchestra; Arturo Toscanini. RCA AT 128.
    (Reissue from HMV ALP 1134, ALP 1108)
        +Gr 8-75 p336   +-RR 6-75 p31
386 Symphony, no. 4, op. 60, B flat major. BRAHMS: Academic festival
    overture, op. 80. Bournemouth Symphony Orchestra; Charles
    Groves. Saga 5178. (Reissue from Fidelity STFDY 1902)
        +-Gr 4-75 p1801
Symphony, no. 4, op. 60, B flat major. cf Symphony, no. 2, op. 36,
    D major.
387 Symphony, no. 5, op. 67, C minor. RPO; Antal Dorati. Contour
    2870 482.
        +-Gr 9-75 p444   /RR 9-75 p30
        /HFN 9-75 p96
388 Symphony, no. 5, op. 67, C minor. VPO; Carlos Kleiber. DG 2530
    516. Tape (c) 3300 472.
        ++Gr 6-75 p32   ++RR 6-75 p32

```
            ++HF 11-75 p91              +RR 7-75 p68 tape
            ++HFN 6-75 p85             ++RR 10-75 p96 tape
            ++MT 10-75 p885            +SR 11-1-75 p45
            +NR 12-75 p3              ++St 12-75 p81
            ++NYT 8-10-75 pD14
```

389 Symphony, no. 5, op. 67, C minor. BPhO; Ferenc Fricsay. DG 2548
    028. (Reissue from 138813)
```
            +-Gr 11-75 p840            +-RR 10-75 p37
            -HFN 10-75 p152
```

390 Symphony, no. 5, op. 67, C minor. The creatures of Prometheus,
    op. 43: Overture. LSO; André Previn. HMV ASD 2960. Tape (c)
    TC ASD 2960. (also Angel S 36927)
```
            ++Gr 4-74 p1851            +-RR 4-74 p36
            ++Gr 8-74 p400 tape        -RR 8-74 p83 tape
            -HF 1-75 p76               -SFC 10-20-74 p28
            +-HFN 3-74 p101           +-St 2-75 p107
            +NR 11-74 p4
```

391 Symphony, no. 5, op. 67, C minor. Quartet, strings, no. 16, op.
    135, F major: Lento assai. GLUCK: Alceste: Overture. Salzburg
    Mozarteum Orchestra; Leopold Hager. Pye TPLS 13068.
```
            -HFN 12-75 p148
```

392 Symphony, no. 5, op. 67, C minor. Symphony, no. 8, op. 93, F
    major. BBC Symphony Orchestra; Colin Davis. Philips 6500 462.
    Tape (c) 7300 252.
```
            +-Gr 11-74 p889           +MJ 10-74 p43
            +Gr 9-74 p592 tape        +RR 10-74 p36
            +HF 9-74 p90              +-SFC 9-22-74 p23
            /NR 9-74 p2              ++St 2-75 p106
```

393 Symphony, no. 5, op. 67, C minor. Grand overture (Zur Namens-
    feier), op. 115. COA; Eugen Jochum. Philips Tape (c) 7300 011.
```
            +Gr 5-75 p2032 tape       +-HFN 6-75 p109 tape
```

394 Symphony, no. 5, op. 67, C minor. BRUCKNER: Symphony, no. 7, E
    major. VPO; Georg Szell. Rococo 2081 (2).
```
            +NR 9-75 p2
```
    Symphony, no. 5, op. 67, C minor. cf Concerto, piano, no. 4, op.
    58, G major: 1st movement.
    Symphony, no. 5, op. 67, C minor. cf Symphony, no. 4, op. 60,
    B flat major.
    Symphony, no. 5, op. 67, C minor. cf Camden CCV 5000-12, 5014-24.
    Symphony, no. 5, op. 67, C minor. cf Toscanini Society ATS GC
    1201/6.

395 Symphony, no. 6, op. 68, F major. BPhO; Lorin Maazel. DG 2548 205.
    Tape (c) 3318 045. (Reissue from 138642)
```
            +Gr 11-75 p840            +RR 10-75 p37
            +HFN 10-75 p152           +RR 10-75 p96 tape
```

396 Symphony, no. 6, op. 68, F major. Coriolan overture, op. 62. VPO;
    Karl Böhm. DG Tape (c) 3300 476.
```
            ++RR 10-75 p96 tape
```

397 Symphony, no. 6, op. 68, F major. The creatures of Prometheus,
    op. 43: Overture. LSO; Colin Davis. Philips 6580 050. Tape
    (c) 7317 097.
```
            +Gr 10-72 p752            +-RR 10-72 p59
            +HFN 10-72 p1921          +RR 10-75 p96 tape
```

398 Symphony, no. 6, op. 68, F major. BPhO; Wilhelm Furtwängler.
    Rococo 2077.
```
            +-NR 9-75 p6
```

Symphony, no. 6, op. 68, F major. cf Columbia M2X 33014.

399 Symphony, no. 7, op. 92, A major. Fidelio, op. 72: Overture. CSO;
Fritz Reiner. Camden CCV 5026.
+–RR 10-75 p37

400 Symphony, no. 7, op. 92, A major. Marlboro Festival Orchestra;
Pablo Casals. Columbia M 33788.
+NR 12-75 p3

401 Symphony, no. 7, op. 92, A major. Leonore overture, no. 3, op. 72.
Egmont overture, op. 84. OSR; Ernest Ansermet. Decca SPA 327.
Tape (c) KCSP 327. (Reissues from SXL 2235, SXL 2003)
+–Gr 12-74 p1119   +RR 11-74 p34
+Gr 2-75 p1562 tape

402 Symphony, no. 7, op. 92, A major. Egmont overture, op. 84. LAPO;
Zubin Mehta. Decca SXLN 6673. Tape (c) KSXLN 6673. (also London
CS 6870)
+–Gr 10-74 p681   -NR 9-75 p5
+Gr 1-75 p1403 tape   +RR 10-74 p34
-HF 8-75 p82   +SFC 8-10-75 p26

403 Symphony, no. 7, op. 92, A major. VPO; Karl Böhm. DG 2530 421.
(Reissue from 2720 045)
+–Gr 1-75 p1335   +–RR 11-74 p34
+–NR 1-75 p3

404 Symphony, no. 7, op. 92, A major. BPhO; Ferenc Fricsay. DG 2548
107. (Reissue from SLPM 138757)
+–Gr 10-75 p602   -RR 9-75 p30
+HFN 9-75 p108

405 Symphony, no. 7, op. 92, A major. LSO; André Previn. HMV ASD
3119. (also Angel S 37116)
+–Gr 11-75 p802   -RR 11-75 p41
+HFN 11-75 p150   ++SFC 12-28-75 p30

406 Symphony, no. 7, op. 92, A major. RPO; Colin Davis. HMV SXLP
20038. Tape (c) TC EXE 138. (also Angel S 37027)
+Gr 7-75 p254 tape   +RR 7-75 p69 tape

407 Symphony, no. 7, op. 92, A major. Leonore overture, no. 3, op. 72.
VPO; Hans Schmidt-Isserstedt. London 6668. Tape (c) M 10226
(ct) M 67226.
++LJ 10-75 p47 tape

408 Symphony, no. 7, op. 92, A major. Fidelio, op. 72: Overture.
COA; Eugen Jochum. Philips Tape (c) 7300 024.
+Gr 5-75 p2032 tape   +–HFN 6-75 p109

409 Symphony, no. 7, op. 92, A major. NYP; Arturo Toscanini. RCA AT
153. (Reissue from HMV ALP 1119)
+Gr 10-75 p602   +RR 8-75 p29

410 Symphony, no. 8, op. 93, F major (arr. Liszt). Songs from sonatas
and symphonies arranged as songs by Friedrich Silcher. Hermann
Prey, bar; Leonard Hokanson, pno. DG Archive 2533 121.
+Audio 8-75 p82   +NYT 8-10-75 pD14
+HF 5-75 p73   ++SFC 2-16-74 p24
+NR 5-75 p12

411 Symphony, no. 8, op 93, F major. Leonore overture, no. 3, op. 72.
(also rehearsal of the overture) Stockholm Philharmonic
Orchestra; Wilhelm Furtwängler. Unicorn WFS 5.
+SFC 3-16-75 p25

Symphony, no. 8, op. 93, F major. cf Symphony, no. 1, op. 21,
C major (Unicorn UNS 200).

Symphony, no. 8, op. 93, F major. cf Symphony, no. 1, op. 21, C
major (Philips 6500 087).

Symphony, no. 8, op. 93, F major. cf Symphony, no. 4, op. 60,
   B flat major.

Symphony, no. 8, op. 93, F major. cf Symphony, no. 5, op. 67,
   C minor.

412 Symphony, no. 9, op. 125, D minor (arr. Liszt). LISZT: Festival
   cantata for the unveiling of the Beethoven monument in Bonn,
   G 584. Richard and John Contiguglia, pno. Connoisseur
   Society (Q) CSQ 2052 (2).

   ++Gr 11-75 p847   +-NR 1-74 p13
   /-HF 3-74 p81   ++RR 10-75 p70
   +HFN 11-75 p151   +-SFC 3-10-74 p26
   +MJ 1-75 p48   +St 3-74 p107

413 Symphony, no. 9, op. 125, D minor. Joan Sutherland, s; Norma
   Procter, con; Anton Dermota, t; Arnold van Mill, bs; Du Brassus
   Choir; OSR; Ernest Ansermet. Decca SPA 328. (Reissue from SXL
   2274) (also London STS 15089. Tape (c) A 30689)

   +Gr 3-75 p1640   +RR 3-75 p23

414 Symphony, no. 9, op. 125, D minor. Egmont overture, op. 84.
   Leonore overture, no. 3, op. 72. Irmgard Seefried, s; Maureen
   Forrester, con; Ernst Häfliger, t; Dietrich Fischer-Dieskau,
   bar; St. Hedwig's Cathedral Choir; BPhO; Ferenc Fricsay. DG
   2700 108 (2). (Reissues from SLPM 138002/3)

   +Gr 9-75 p451   +-RR 9-75 p30
   +HFN 9-75 p108

415 Symphony, no. 9, op. 125, D minor. Egmont, op. 84, excerpts. Eva
   Andor, s; Márta Szirmay, ms; György Korondi, t; Sándor Nagy,
   bar; Budapest Chorus; HSO; János Feréncsik. Hungaroton SPLX
   11736/7 (2).

   +-HF 10-75 p70   -NR 9-75 p3

416 Symphony, no. 9, op. 25, D minor. Marita Napier, s; Anna Reynolds,
   con; Helge Brillioth, t; Karl Ridderbusch, bs; Ambrosian
   Singers; NPhO; Seiji Ozawa. Philips 6747 119 (2). Tape (c)
   7505 072.

   +Gr 1-75 p1336   +-RR 7-75 p69 tape
   +HF 10-75 p70   ++SFC 7-6-75 p16
   +-NR 9-75 p3   ++St 10-75 p104
   +RR 1-75 p26

417 Symphony, no. 9, op. 125, D minor. Wilma Lipp, s; Elisabeth
   Höngen, con; Julius Patzak, t; Otto Wiener, bs; Vienna Singverein, Vienna Pro Musica Orchestra; Jascha Horenstein. Turnabout TVS 37074. (Reissue from Vox PL 10000)

   +-Gr 5-75 p1961   +RR 5-75 p27
   +HFN 7-75 p76

Symphony, no. 9, op. 125, D minor, CCV 5021. cf Camden CCV 5000-
   12, 5014-24.

Symphony, no. 9, op. 125, D minor: 3rd movement. cf Philips 6747
   204.

Theme and variations, D major. cf HUMMEL: Sonata, mandolin and
   piano, C major.

Trio, op. 121a, G major. cf Trio, piano, no. 6, op. 97, B flat
   major.

418 Trio, clarinet, op. 11, B flat major. Trio, piano, no. 2, op. 1,
   G major. Wilhelm Kempff, pno; Henryk Szeryng, vln; Karl
   Leister, clt; Pierre Fournier, vlc. DG 2530 408.

   ++Audio 9-75 p70   ++NR 6-75 p9

419 Trios, piano, nos. 1-9. Variations on 'Ich bin der Schneider

Kakadu', op. 121a, G major. Beaux Arts Trio. Philips 6747
142 (4). (Reissue from SAL 3527/30)
  ++Gr 10-75 p640   +RR 10-75 p62
  +HFN 11-75 p173   +SFC 11-23-75 p26
  ++NR 11-75 p8

 Trio, piano, no. 2, op. 1, G major. cf Trio, clarinet, op. 11,
  B flat major.

420 Trio, piano, no. 6, op. 97, B flat major. HAYDN: Trio, piano, no.
  28 E major. Trieste Trio. DG 2538 318. (Reissue from SLPEM
  136220)
   +-Gr 5-75 p1985   +RR 5-75 p48
   ++HFN 6-75 p85

421 Trio, piano, no. 6, op. 97, B flat major*. BRAHMS: Trio, piano,
  no. 1, op. 8, B major**. SCHUBERT: Trio, piano, no. 1, op.
  99, D 898, B flat major. Jascha Heifetz, vln; Emanuel Feuermann,
  vlc; Artur Rubinstein, pno. RCA LRM 2-5093 (2). (Reissues
  from *HMV ALP 1184, **BLP 1056)
   ++Gr 11-75 p846   +RR 9-75 p52
   +HFN 9-75 p97

422 Trio, piano, no. 6, op. 97, B flat major*. Trio, op. 121a, G
  major. Jacques Thibaud, vln; Pablo Casals, vlc; Alfred Cortot,
  pno. Seraphim 60242. (*Reissue from Angel COLH 29)
   +HF 10-75 p71

423 Trio, piano, no. 6, op. 97, B flat major. Suk Trio. Vanguard/
  Supraphon SU 5.
   +HF 5-75 p74   +NR 3-75 p5
   +MJ 10-75 p38   +St 5-75 p102

424 Trio, piano, no. 6, op. 97, B flat major. Variations on 'Ich
  bin der Schneider Kakadu', op. 121a, G major. Jacques Thibaud,
  vln; Pablo Casals, vlc; Alfred Cortot, pno. World Records SH
  230.
   +HFN 12-75 p149   +RR 12-75 p72

425 Trio, piano, clarinet and violoncello, op. 38, E flat major.
  Thomas Brandis, vln; Wolfgang Boettcher, vlc; Eckart Besch, pno.
  DG Archive 2533 118.
   ++NR 6-75 p9   +NYT 8-10-75 pD14

426 Trio, piano, clarinet and violoncello, op. 38, E flat major.
  GOUNOD: Petite symphonie. HAYDN: Trio, piano, flute and violon-
  cello. MOZART: Serenade, no. 11, K 375, E flat major. Rudolf
  Serkin, pno; Michel Dubost, flt; Richard Stoltzman, clt; Peter
  Wiley, Alain Meunier, vlc. Marlboro Recording Society MRS 7/8 (2).
   +HF 10-75 p80   +NYT 5-25-75 pD14

427 Trios, strings, op. 3, no. 1, E flat major; op. 9, nos. 1-3.
  Serenade, string trio, op. 8, D major. Italian String Trio.
  DG 2733 004 (3). (Reissue from 2720 014/1-3)
   *Gr 4-75 p1820   +RR 5-75 p48

428 Trio, strings, op. 9, no. 1, G major. Trio, strings, op. 9, no. 3,
  C minor. Leonid Kogan, vln; Rudolf Barshai, vla; Mstislav
  Rostropovitch, vlc. Saga 5396.
   +RR 11-75 p59

 Trio, strings, op. 9, no. 1, G major. cf Classic Record Library
  SQM 80-5731.

 Trio, strings, op. 9, no. 3, C minor. cf Trio, strings, op. 9,
  no. 1, G major.

429 Trio, violin, violoncello and piano (arr. from Symphony, no. 2,
  op. 36, D major). HAYDN: Trio, strings (arr. from Sonata,

piano, no. 1, op. 53, G major). Thomas Brandis, vln; Siegbert Uberschaef, vla; Wolfgang Boettcher, vlc; Eckart Besch, pno. DG Archive 2533 136.

++Audio 8-75 p82   +-HFN 11-75 p151
+HF 5-75 p73   ++SFC 3-2-75 p24

430 Variations on a theme by Diabelli, op. 120 (33). Hans Petermandl, pno. Supraphon 111 1155.

-Gr 4-75 p1835   +-RR 7-75 p46
+-NR 8-75 p12   ++SFC 3-2-75 p24

431 Variations on a theme by Diabelli, op. 120. Rudolf Buchbinder, pno. Telefunken Tape (c) 4-41891.

+-HFN 9-75 p110 tape

Variations on a theme by Diabelli, op. 120. cf Sonata, piano, no. 29, op. 106, B flat major.

Variations on an original theme, C minor (32). cf Piano works (Columbia M 33074).

432 Variations on Handel's "See the conquering hero comes" (12). Variations on Mozart's "Ein Mädchen oder Weibchen", op. 66 (12). STRAUSS, R.: Sonata, violoncello, op. 6, F major. Mstislav Rostropovich, vlc; Vasso Devetzi, pno. HMV ASD 3066. (also Angel S 37086)

+Gr 4-75 p1820   +RR 4-75 p54
+HF 10-75 p83   +-SR 9-6-75 p40
+NR 8-75 p7

Variations on "Ich bin der Schneider Kakadu", op. 121a, G major. cf Trio, piano, no. 6, op. 97, B flat major.

Variations on "Ich bin der Schneider Kakadu", op. 121a, G major. cf Trios, piano, nos. 1-9.

Variations on Mozart's "Ein Mädchen oder Weibchen, op. 66 (12). cf Variations on Handel's "See the conquering hero comes".

Viennese dances (11). cf King Stephen, op. 117: Overture.

433 Wellington's victory, op. 91. TCHAIKOVSKY: Overture, the year 1812, op. 49. Don Cossack Choir; BPhO; Herbert von Karajan. DG 2538 142.

+NR 5-75 p6   ++SFC 3-9-75 p27

Wellington's victory, op. 91. cf King Stephen, op. 117: Overture.

Wellington's victory, op. 91. cf TCHAIKOVSKY: Overture, the year 1812, op. 49.

434 Works, selections: Bundeslied, op. 122. Elegischer Gesang, op. 118. King Stephen, op. 117. Meerestille und Gluckliche Fahrt, op. 112. Opferlied, op. 121b. Lorna Haywood, s; Ambrosian Singers; LSO; Michael Tilson Thomas. CBS 76404.

+HFN 11-75 p150

York'scher Marsch. cf BASF 292 2116-4 (2).

York'scher Marsch. cf DG 2721 077.

Zapfenstreichen, nos. 1-3. cf Classics for Pleasure CFP 40230.

BEHREND

Eichstatter Hofmühltanz; Riedenburger Tanz. cf DG Archive 2533 184.

Tanz im Aicholdinger Schloss. cf DG Archive 2533 184.

BELLINI, Vicenzo

435 Capuletti ed i Montecchi: O, quante volte, O, quante. La sonnambula: Ah, non credea mirarti. DONIZETTI: Lucia di Lammermoor: Ardon gli incensi. ROSSINI: Il barbiere di Siviglia: Una voce poco fa. La gazza ladra: Di piacer mi balza il cor. VERDI: Rigoletto: Caro nome. Mady Mesplé, s; Paris Opera Orchestra; Gianfranco Masini. Angel S 37095.

+-NR 12-75 p11

I Capuleti ed i Montecchi: Ecco la tomba...O tu bell anima. cf
   Rococo 5368.
436 Norma. Montserrat Caballé, s; Fiorenza Cossotto, Elizabeth Bain-
     bridge, ms; Placido Domingo, Kenneth Collins, t; Ruggero
     Raimondi, bs; LPO; Ambrosian Opera Chorus; Carlo Felice Cillario.
     RCA LSC 6202 (3). (also SER 5658/60)
           +Gr 11-75 p893   -NYT 1-7-73 pD27
           +-HF 4-73 p69   +-ON 3-10-73 p27
           +-HFN 12-75 p149   +RR 11-75 p32
           +-MJ 4-73 p34   +-St 3-73 p115
           ++NR 3-73 p9
     Norma: Arias. cf London S 26277.
     Norma: Deh, con te...Mora o Norma...Si, fino all'ore estreme. cf
        Decca GOSC 666/8.
     Norma: Dormono entrambi. cf Decca SXL 6719.
     Norma: Overture. cf Erato STU 70880.
     Norma: Sgombra è la sacra selva. cf Odyssey Y2 32880.
     Norma: Sgombra è la sacra selva, Deh con te il prendi...Mira O
        Norma...Si fue all'ore. cf Rococo 5368.
437 I puritani. Joan Sutherland, s; Anita Caminada, ms; Luciano Pava-
     rotti, Renato Cazzaniga, t; Piero Cappuccilli, bar; Nicolai
     Ghiaurov, Gian Carlo Luccardi, bs; ROHO Chorus; LSO; Richard
     Bonynge. London OSA 13111 (3). (also Decca SET 587/9)
           +Gr 7-75 p237   ++ON 6-75 p10
           ++HF 6-75 p84   +RR 7-75 p16
           +HFN 7-75 p76   ++SFC 2-23-75 p22
           +NYT 4-6-75 pD18   ++St 5-75 p73
           +ON 4-19-75 p54   +STL 10-5-75 p36
     I puritani: Arias. cf London OS 26373.
     I puritani: A te, o cara. cf Supraphon 012 0789.
     I puritani: Qui la voce. cf ABC ATS 20019.
     I puritani: Qui la voce. cf Decca GOSC 666/8.
     I puritani: Sorgea la notte. cf Rubini RS 300.
     Songs: Bella nice che d'amore; Dolente immagine di fille mia;
        Ma rendi pur contento; Maliconia, ninfa gentile; Vanne o rosa
        fortunata. cf London OS 26391.
438 La sonnambula: Arias. CILEA: Adriana Lecouvreur: Arias. PUCCINI:
     Suor Angelica: Arias. VERDI: Un ballo de maschera: Arias.
     Rigoletto: Arias. Il trovatore: Arias. I vespri siciliani:
     Arias. Montserrat Caballé, s; Barcelona Symphony Orchestra;
     Gianfranco Masini. London OS 26424.
           +-SFC 11-16-75 p32
     La sonnambula: Arias and scenes. cf London OS 26306.
     La sonnambula: Ah, non credea mirarti. cf Capuletti ed i Montecchi:
        O quante volte, O, quante.
     La sonnambula: Ah, se una volta sola...Ah, non credea mirarti...
        Ah, non giunge. cf CILEA: Adriana Lecouvreur: Io sono l'umile
        ancella.
     La sonnambula: Ah, non credea mirarti; Ah, non giunge. cf Pearl
        GEM 121/2.
BELLAMN, Carl
     Song at nightfall. cf Argo ZRG 765.
BEMBERG, Herman
     Ca fait peur aux oiseaux. cf Discophilia KGC 1.
BEN HAIM, Paul
     Sephardic lullaby. cf Da Camera Magna SM 93399.

BENDA, Franz (Frantisek)
439 Symphonies, nos. 1-6, B flat major, D major. Ars Rediviva; Milan
        Munclinger. Supraphon 110 1641/2 (2).
                +-Gr 11-75 p807   +NR 11-75 p3
                +-HFN 10-75 p137   +-RR 9-75 p33
BENDA, Jiři (Georg)
440 Concerto, flute, E minor. GLUCK: Orfeo ed Euridice: Reigen seliger
        Geister. HAYDN: Concerto, flute, D major. Frans Vester, flt;
        Concerto Amsterdam; Jaap Schröder. BASF KMB 20838.
                ++NR 3-74 p3   +SFC 7-13-75 p21
441 Sonata, piano, no. 9, A minor. DUSSEK: Sonata, piano, op. 77,
        F minor. TOMASEK: Eglogue, no. 2, op. 35, F major. VORISEK:
        Impromptu, no. 4, op. 7, A major. Rudolf Firkusny, pno.
        Candide CE 31086.
                +HF 10-75 p88   ++SFC 3-2-75 p24
                +NR 4-75 p11   ++St 12-75 p118
        Sonatina, D major. cf RCA ARL 1-0864.
        Sonatina, D minor. cf RCA ARL 1-0864.
BENDUSI (16th century)
        Cortesa padana e fusta. cf Erato STU 70847.
BENEDICT, Julius
        The gypsy and the moth. cf Decca SPA 394.
BENET, John
        Jacet granum, Sanctus. cf Nonesuch H 71292.
BENNARD
        The old rugged cross. cf RCA ARL 1-0561.
BENNETT, Richard Rodney
442 Calendar. GOEHR: Choruses (2). DAVIES: Leopardi fragments.
        WILLIAMSON: Symphony, voices. Mary Thomas, s; Rosemary Phillips,
        Pauline Stevens, con; Geoffrey Shaw bar; John Alldis Choir;
        Melos Ensemble; John Carewe. Argo ZRG 758. (Reissue from HMV
        ASD 640)
                +-Gr 5-75 p2000   +NR 10-75 p6
                -HFN 5-75 p125   *RR 4-75 p59
                +MT 9-75 p797
        Commedia IV. cf ADDISON: Divertimento, op. 9.
        The house of sleepe. cf HMV EMD 5521.
BENTZON, Niels Viggo
443 Pezzi sinfonici, op. 109. PISTON: Serenata. VAN VACTOR: Fantasia,
        chaconne and allegro. Louisville Orchestra; Jorge Mester.
        Louisville LOU 586.
                *St 6-75 p50
        BERBERIAN, CATHY, SONGS AND OPERATIC ARIAS. cf RCA ARL 1-5007.
BEREZOVSKY, Maximus
        Do not reject me in my old age. cf HMV Melodiya ASD 3102.
BERG, Alban
444 Lulu: Suite. STRAUSS, R.: Salome, op. 54: Final scene. Anja
        Silja, s; VPO; Christoph von Dohnányi. Decca SXL 6657. (also
        London OS 26397)
                +-Audio 10-75 p118   +-Op 11-74 p990
                +-Gr 10-74 p742   -RR 9-74 p38
                +-HF 6-75 p87   +St 9-75 p106
                +NR 6-75 p6
445 Lyric suite, string quartet. Quartet, strings, op. 3. LaSalle
        Quartet. DG 2530 283. (Reissue from 2720 029)
                ++Gr 4-74 p1902   ++NR 4-74 p6

```
      +HFN 3-74 p101              ++RR 4-74 p56
      ++LJ 4-75 p69              +St 6-74 p118
      ++MJ 3-74 p10
```
Lyric suite, string quartet. cf Quartet, strings, op. 3.
446 Lyric suite: Pieces (3). Orchestral pieces, op. 6 (3). SCHONBERG:
      Pelleas und Melisande, op. 5. Verklärte Nacht, op. 4. Varia-
      tions, orchestra, op. 31. WEBERN: Movements, string quartet,
      op. 5 (5). Pieces, op. 6 (6). Passacaglia, op. 1. Symphony,
      op. 21. BPhO; Herbert von Karajan. DG 2711 014 (4). (also
      DG 2530 485/8)
```
      +Gr 3-75 p1640            +RR 3-75 p34
      +-HF 7-75 p61             ++SFC 5-25-75 p17
      ++NR 6-75 p3              +-St 9-75 p120
```
Orchestral pieces, op. 6. cf Lyric suite: Pieces.
Pieces, clarinet and piano, op. 5 (4). cf Crystal S 331.
447 Quartet, strings, op. 3. Lyric suite, string quartet. Alban Berg
      Quartet. Telefunken SAT 22549.
```
      +Gr 12-74 p1158           +RR 1-75 p38
      ++HF 7-75 p71             ++SFC 5-11-75 p23
      +NR 4-75 p7               +St 11-75 p126
```
Quartet, strings, op. 3. cf Lyric suite, string quartet.
448 Sonata, piano, op. 1. SCHONBERG: Chamber symphony, no. 1, op. 9,
      E major. WEBERN: Movements, string quartet, op. 5 (5)*. Idil
      Biret, pno; Chamber Ensemble; Quartetto di Milano; Gunther
      Schuller. Finnadar SR 9008. Tape (ct) TP 9008). (*Reissue
      from Atlantic SD 1420)
```
      +-HF 11-75 p116           +St 9-75 p120
      +NR 8-75 p7
```
Sonata, piano, op. 1. cf BARTOK: Out of doors.
449 Wozzeck. Eileen Farrell, s; David Lloyd, t; Mack Harrell, bar;
      Ralph Herbert, bs; Schola Cantorum; NYP; Dimitri Mitropoulos.
      Odyssey Y2 33126 (2). (Reissue from Columbia SL 118)
```
      +HF 7-75 p72              ++SFC 2-23-75 p22
```
BERGER, Arthur
      Pieces for piano (5). cf Septet.
450 Septet. Pieces for piano (5). SOLLBERGER: Grand quartet, flutes.
      WESTERGAARD: Variations for six players. Robert Miller, pno;
      David Gilbert, Thomas Nyfenger, Harvey Sollberger, Sophie
      Sollberger, flt; Contemporary Chamber Ensemble; Group for Con-
      temporary Music, Columbia University; Arthur Weisberg, Harvey
      Sollberger. DG/Acoustic Research 064 088.
```
      ++HF 2-71 p97             *NYT 1-3-71 pD22
      +MJ 2-75 p29              +SFC 4-18-71 p31
      +-MQ 7-72 p501            +SR 12-26-70 p47
                               +St 5-71 p95
```
BERGER, Jean
      Songs of sadness and gladness; Then I commended mirth; Lets take
      a cat; He that loves a rosy cheek; The seasons; My love is dead;
      Farewell. cf University of Illinois, unnumbered.
BERGER, Ludwig
451 Grande sonate, op. 7. MOSCHELES: Sonate caractèristique, piano,
      op. 27. Frederick Marvin, pno. Genesis GS 1061.
```
      ++NR 10-75 p10
```
BERGSMA, William
      Suite, brass quartet. cf Desto DC 6474/7.
BERIO, Luciano
      Agnus. cf Works, selections (RCA ARL 1-0037).

Air. cf Works, selections (RCA ARL 1-0037).
Chamber music, female voice, clarinet, violoncello and harp. cf
    Works, selections (Philips 6500 631).
Différences, 5 instruments and magnetic tape. cf Works, selections
    (Philips 6500 631).
E vo. cf Works, selections (RCA ARL 1-0037).
El mar la mar. cf Works, selections (RCA ARL 1-0037).
Melodrama. cf Works, selections (RCA ARL 1-0037).
O king. cf Works, selections (RCA ARL 1-0037).
Pieces, violin and piano (2). cf Works, selectons (Philips 6500
    631).
Sequenza. cf Nonesuch HB 73028.
Sequenza III, female voice. cf Works, selections (Philips 6500 631).
Sequenza VI. cf BEDFORD: Spillihpnerak.
Sequenza VII. cf Works, selections (Philips 6500 631).
452 Visage. CAGE: Fontana mix. MIMAROGLU: Agony. Cathy Berberian,
        s. Turnabout Tape (c) KTVC 34046.
            +Gr 2-75 p1562 tape   +RR 9-75 p78 tape
453 Works, selections: Différences, 5 instruments and magnetic tape.
        Chamber music, female voice, clarinet, violoncello and harp.
        Pieces, violin and piano (2). Sequenza III, female voice.
        Sequenza VII. Cathy Berberian, s; Heinz Holliger, ob; Juilliard
        Ensemble; Luciano Berio. Philips 6500 631.
            +HF 8-74 p86   ++NR 7-74 p6
            ++LJ 4-75 p69   ++SFC 11-3-74 p22
            ++MJ 7-74 p46   ++St 9-74 p113
454 Works, selections: Agnus. Air. E vo. El mar la mar. Melodrama.
        O king. Vocalists; London Sinfonietta; Luciano Berio. RCA
        ARL 1-0037.
            +HF 11-75 p98   +-NYT 12-21-75 pD18
            +NR 9-75 p11   +SR 9-6-75 p40
BERKELEY, Lennox
    Concert studies, op. 14, nos. 2-4; op. 48, E flat major. cf
        Piano works (Lyrita RCS 9).
455 Concerto, guitar. RODRIGO: Concierto de Aranjuez. Julian Bream,
        gtr; Monteverdi Orchestra; John Eliot Gardiner. RCA ARL 1-1181.
        Tape (c) ARK 1-1181 (ct) ARS 1-1181.
            ++Gr 11-75 p807   ++RR 12-75 p60
            ++HFN 12-75 p149   +SR 11-29-75 p50
            +NR 12-75 p14
456 Concerto, 2 pianos, op. 30. Symphony, no. 1, op. 16. Garth
        Beckett, Boyd McDonald, pno; LPO; Norman Del Mar. Lyrita
        SRCS 80.
            ++Gr 11-75 p807   ++HFN 11-75 p151
457 Divertimento, op. 18, B flat major. Partita, chamber orchestra,
        op. 66. Serenade, strings, op. 12. Sinfonia concertante, op.
        84: Canzonetta. LPO; Lennox Berkeley. Lyrita SRCS 74.
            +Gr 9-75 p451   +RR 12-75 p44
            ++HFN 10-75 p137
    Impromptu, op. 7, no. 1, G minor. cf Piano works (Lyrita RCS 9).
458 Mont Juic, op. 9 (with Benjamin Britten). BLISS: Mêlée fantasque.
        HOLST: Japanese suite, op. 33. WALTON: Music for children.
        LPO, LSO; Lennox Berkeley, Arthur Bliss, Adrian Boult, William
        Walton. Musical Heritage Society MHS 1919.
            ++HF 2-75 p107
    Nocturne. cf RCA LRL 1-5087.

Partita, chamber orchestra, op. 66. cf Divertimento, op. 18,
     B flat major.
459 Piano works: Concert studies, op. 14, nos. 2-4; op. 48, E flat
     major. Impromptu, op. 7, no. 1, G minor. Preludes, op. 23,
     nos. 1-6. Scherzo, op. 32, no. 2, D major. Sonata, piano, op.
     20, A major. Colin Horsley, pno. Lyrita RCS 9. (Reissue from
     1960)
          +Gr 9-75 p492   +RR 9-75 p57
          +HFN 12-75 p156
Preludes, op. 23, nos. 1-6. cf Piano works (Lyrita RCS 9).
Ronsard sonnets (4). cf BEDFORD: The tentacles of the dark
     nebula.
Scherzo, op. 32, no. 2, D major. cf Piano works (Lyrita RCS 9).
Serenade, strings, op. 12. cf Divertimento, op. 18, B flat major.
Sinfonia concertante, op. 84: Canzonetta. cf Divertimento, op. 18,
     B flat major.
Sonata, piano, op. 20, A major. cf Piano works (Lyrita RCS 9).
460 Songs: Automne; Chinese songs; D'un vanneur de blé; Tant que mes
     yeux. CROSSE: The new world, op. 25. DICKINSON: Extravaganzas.
     Meriel Dickinson, ms; Peter Dickinson, pno. Argo ZRG 788.
          ++RR 9-75 p68   +STL 11-2-75 p38
Symphony, no. 1, op. 16. cf Concerto, 2 pianos, op. 30.
BERLIOZ, Hector
     Absence. cf Desto DC 7118/9.
     Béatrice et Bénédict. cf Overtures (Pye GSGC 15012).
     Béatrice et Bénédict: Vous soupirez, madame. cf Decca DPA 517/8.
     Benvenuto Cellini, op. 23. cf Overtures (Pye GSGC 15012).
     Benvenuto Cellini, op. 23. cf Works, selections (HMV 3080).
     Benvenuto Cellini, op. 23. cf AUBER: Le cheval de bronze: Overture.
461 Le carnaval romain, op. 9*. Les Troyens: Royal hunt and storm*.
     WEBER: Konzertstuck, F major. Nikita Magaloff, pno; LSO, ROHO;
     Colin Davis. Philips 6580 091. (*Reissues from SAL 3753,
     6709 022)
          +-Gr 8-75 p316   +RR 7-75 p37
          ++HFN 7-75 p77
     Le carnaval romain, op. 9. cf Overtures (Odyssey Y 33287).
     Le carnaval romain, op. 9. cf Overtures (Pye GSGC 15012).
     Le carnaval romain, op. 9. cf Works, selections (CBS 77395).
     Le carnaval romain, op. 9. cf Works, selections (HMV 3080).
     Le carnaval romain, op. 9. cf HMV SLS 5019.
     Le carnaval romain, op. 9. cf Philips 6747 199.
     Le Corsaire, op. 21. cf Overtures (Odyssey Y 33287).
     Le Corsaire, op. 21. cf Overtures (Pye GSGC 15012).
     Le Corsaire, op. 21. cf Works, selections (CBS 77395).
462 La damnation de Faust, op. 24. Medith Mathis, Judith Dickison, s;
     Stuart Burrows, t; Donald McIntyre, Thomas Paul, bs; Tangle-
     wood Festival Chorus; Boston Boys Choir; BSO; Seiji Ozawa.
     DG 2709 048 (3). Tape (c) 3371 016.
          +-Gr 12-74 p1208   +-ON 12-21-74 p28
          /Gr 1-75 p1402 tape   +RR 12-74 p68
          +-HF 4-75 p81   +-RR 1-75 p68 tape
          -NR 1-75 p10   +-St 3-75 p97
The damnation of Faust, op. 24, selections. cf RAVEL: Valses
     nobles et sentimentales.
La damnation de Faust, op. 24: D'amour l'ardente flamme. cf RCA
     ARL 1-0844.

The damnation of Faust, op. 24: Dance of the sylphs. cf Decca
  DPA 519/20.
The damnation of Faust, op. 24: Hungarian march. cf DG Heliodor
  2548 148.
La damnation de Faust, op. 24: Hungarian march. cf HMV SLS 5019.
La damnation de Faust, op. 24: Hungarian march. cf Philips 6580 107.
La damnation de Faust, op. 24: Hungarian march, Dance of the sylphs,
  Minuet of the Will-o-the-wisps. cf Works, selections (HMV 3080).
L'Enfance du Christ, op. 25: Shepherds chorus. cf HMV TC MCS
  Tape (c) 14.
Les Francs Juges, op. 3. cf Overtures (Odyssey Y 33287).
Les Francs Juges, op. 3. cf Overtures (Pye GSGC 15012).
Les Francs Juges, op. 3. cf Works, selections (CBS 77395).
Les Francs Juges, op. 3. cf BEETHOVEN: Egmont overture, op. 84.
463 Grande messe des morts, op. 5. Robert Tear, t; Birmingham City
     Symphony Chorus and Orchestra; Louis Frémaux. HMV SLS 982 (2).
     Tape (c) TC SLS 982.
         +Gr 9-75 p499   +-HFN 12-75 p173 tape
         +Gr 12-75 p1121 tape   ++RR 10-75 p81
         +HFN 10-75 p138
464 Harold in Italy, op. 16. Daniel Benyamini, vla; Israel Philharmonic
     Orchestra; Zubin Mehta. Decca SXL 6732.
         +-Gr 8-75 p316   +-RR 8-75 p29
         +HFN 8-75 p73
465 Harold in Italy, op. 16. William Primrose, vla; RPO; Thomas Beecham.
     Odyssey Y 33286.
         +HF 7-75 p89   +St 7-75 p110
     Harold in Italy, op. 16. cf Works, selections (CBS 77395).
King Lear, op. 4. cf Overtures (Odyssey Y 33287).
King Lear, op. 4. cf Works, selections (CBS 77395).
March funebre pour la derniere scene de Hamlet, op. 18, no. 3. cf
  Works, selections (HMV 3080).
466 Les nuits d'été, op. 7. ELGAR: Sea pictures, op. 37. MAHLER:
     Lieder eines fahrenden Gesellen. Kindertotenlieder. Rückert
     Lieder. RAVEL: Schéhérazade. Janet Baker, ms; LSO, NPhO, Hallé
     Orchestra; John Barbirolli. HMV SLS 5013 (3). Reissues from
     ASD 2444, 655, 2518/9, 2338)
         +Gr 9-75 p505
467 Overtures: Le carnaval romain, op. 9. Le Corsaire, op. 21. Les
     Francs Juges, op. 3. King Lear, op. 4. Waverly, op. 2b. RPO;
     Thomas Beecham. Odyssey Y 33287.
         +HF 7-75 p89   +St 7-75 p110
468 Overtures: Béatrice et Bénédict. Benvenuto Cellini, op. 23. Le
     carnaval romain, op. 9. Le Corsaire, op. 2. Les Francs Juges,
     op. 3. Waverly, op. 2. LPO; Adrian Boult. Pye GSGC 15012.
     (Reissues from CCL 30159, GSGC 14084)
         +-Gr 8-75 p335   -RR 8-75 p29
         +-HFN 8-75 p87
469 Roméo et Juliette, op. 17. Christa Ludwig, ms; Michel Sénéchal,
     t; Nicolai Ghiaurov, bs; ORTF Choir; VPO; Lorin Maazel. Decca
     SET 570/2 (2). (also London OSA 12102)
         +Gr 2-74 p1582   +NYT 7-21-74 pD16
         +-HF 9-74 p91   +-ON 12-7-74 p42
         +-HFN 2-74 p334   +-Op 6-74 p518
         +MJ 1-75 p48   +RR 2-74 p16
         ++NR 9-74 p8   ++SFC 6-23-74 p25

Roméo et Juliette, op. 17, Part 1.   cf Columbia M2X 33014.
Songs: Irlande, op. 2: La belle voyageuse, Le coucher du soleil,
   L'Origine de la harpe.   cf BIZET: Songs (Saga 5388).
470 Symphonie fantastique, op. 14.   Toronto Symphony Orchestra; Seiji
     Ozawa.   CBS 61659. (also Odyssey Y 31923)
              -Gr 12-75-p1031   +-RR 11-75 p41
              +HFN 10-75 p152
471 Symphonie fantastique, op. 14.   Lamoureux Orchestra; Igor Markevitch.
     DG 2548 172. (Reissue from 138712)
              /Gr 11-75 p840   +-RR 11-75 p41
              ++HFN 10-75 p152
472 Symphonie fantastique, op. 14.   Strasbourg Philharmonic Orchestra;
     Alain Lombard.   Erato STU 70800.
              -Gr 7-74 p196   +St 7-75 p94
              -RR 12-73 p58
473 Symphonie fantastique, op. 14.   COA; Colin Davis.   Philips 6500 774.
     Tape (c) 7300 313.
              +-Gr 3-75 p1647   +RR 3-75 p23
              +-HF 5-75 p73   +-RR 3-75 p73 tape
              +HF 9-75 p116 tape   +SFC 12-29-74 p20
              ++NR 2-75 p1   +St 7-75 p94
474 Symphonie fantastique, op. 14.   Hallé Orchestra; John Barbirolli.
     Pye GSGC 15010. (Reissue from GSGC 14005)
              /Gr 8-75 p335   +RR 8-75 p29
              /HFN 8-75 p87
     Symphonie fantastique, op. 14.   cf HMV SLS 5003.
     Te Deum, op. 22.   cf Works, selections (CBS 77395).
475 Les Troyens, excerpts.   Berit Lindholm, s; Josephine Veasey, ms;
     Jon Vickers, t; Peter Glossop, bar; ROHO and Chorus; Colin
     Davis.   Philips 6500 16. Tape (c) 7300 050.
              +-RR 2-75 p74 tape   +SFC 2-6-72 p32
476 Les Troyens: Overture, March Troyenne.   MASSENET: La Vièrge: Le
     dernier sommeil de la viérge.   RIMSKY-KORSAKOV: Le coq l'or:
     Wedding march.   SIBELIUS: Karelia suite, op. 11, no. 3: March.
     The tempest, op. 109: Miranda, The Naiads, The storm.   RPO;
     Thomas Beecham.   Odyssey Y 33288. (also CBS 61655. Reissue
     from Philips SBR 6215, Columbia 33CX 1087)
              +Gr 7-75 p199   +-RR 7-75 p27
              +HF 7-75 p89   +St 7-75 p110
              +-HFN 9-75 p108
477 Les Troyens: Royal hunt and storm.   DEBUSSY: Prélude à l'après-
     midi d'un faune.   STRAVINSKY: The firebird: Suite.   LPO; John
     Pritchard.   Pye GSGC 15002. (Reissue from Virtuoso TPLS 13032/3)
              *Gr 8-75 p335   +-RR 8-75 p44
              /HFN 9-75 p109
     Les Troyens: Royal hunt and storm.   cf Le carnaval romain, op. 9.
     Les Troyens: Royal hunt and storm.   cf HMV SLS 5019.
     Les Troyens: Royal hunt and storm, Trojan march.   cf Works,
        selections (HMV 3080).
     Les Troyens: Trojan march.   cf HMV RLS 717.
     Waverly, op. 2b.   cf Overtures (Odyssey Y 33287).
     Waverly, op. 2b.   cf Overtures (Pye GSGC 15012).
     Waverly, op. 2b.   cf Works, selections (CBS 77395).
478 Works, selections: Le carnaval romain, op. 9.   Le Corsaire, op. 21.
     Les Francs Juges, op. 3.   Harold in Italy, op. 16.   King Lear
     op. 4.   Te Deum, op. 22.   Waverly, op. 2b.   William Primrose,

vla; Alexander Young, t; London Philharmonic Chorus; Dulwich
College Boys Choir; Denis Vaughan, org; RPO; Thomas Beecham.
CBS 77395 (3). (Reissues from Philips ABL 3083, Columbia 33CX
1019, Philips ABL 3006)
    +-Gr 10-75 p607   +RR 11-75 p41
    +-HFN 10-75 p152

479 Works, selections: Benvenuto Cellini, op. 23: Overture. Le
carnaval romain, op. 9. La damnation de Faust, op. 24: Hungarian march, Dance of the sylphs, Minuet of the Will-o-the-wisps.
Marche funebre pour la dernier scene de Hamlet, op. 18, no. 3.
Les Troyens: Royal hunt and storm, Trojan march. Birmingham
City Symphony Orchestra and Chorus; Louis Frémaux. HMV ASD
3080.
    +Gr 8-75 p316   +RR 8-75 p29
    +HFN 9-75 p97

BELSTERLING
    March of the steel men. cf Michigan University SM 0002.
BERNEVILLE, Gilebert de
    De moi doleros vos chant. cf Telefunken AW 6-41275.
BERNSTEIN, Charles Harold
480 A Cabris. Amphion suite. Duo, flute and violoncello. Duo, flute
and viola. Elegiac dreams and awakenings. Interlude. The
London flute. CIMAROSA: Quartet, flute and strings, no. 1,
D major. Quartet, flute and strings, no. 4, F major. HAYDN:
Quartet, op. 33, no. 3, C major. Amphion Quartet, Belgium.
Laurel LR 101/2 (2). .
    +-NR 5-75 p10
Amphion suite. cf A Cabris.
Duo, flute and violoncello. cf A Cabris.
Duo, flute and viola. cf A Cabris.
Elegiac dreams and awakenings. cf A Cabris.
Interlude. cf A Cabris.
The London flute. cf A Cabris.
BERNSTEIN, Leonard
481 Dybbuk. David Johnson, bar; John Ostendorf, bs; New York City
Ballet Orchestra; Leonard Bernstein. Columbia M 33082. Tape
(c) MT 33082 (ct) MA 33082 (Q) MQ 33082 Tape (ct) MAQ 33082.
    +HF 11-74 p100   +SFC 10-6-74 p26
    +HF 3-75 p108 Quad tape   +SR 11-30-74 p40
    +NYT 8-4-74 pD20   ++St 11-74 p122
Facsimile. cf Symphony, no. 1.
Jeremiah. cf Symphony, no. 1.
482 Sonata, clarinet and piano. VANHAL: Sonata, clarinet, B flat major.
VAUGHAN WILLIAMS: Studies in English folksong (6). WAGNER:
Adagio. Jerome Bunke, clt; Hidemitsu Hayashi, pno. Musical
Heritage Society MHS 1887.
    ++MJ 3-75 p26   ++St 3-75 p111
483 Symphony, no. 1. Facsimile. Jeremiah. Nan Merriman, ms; St.
Louis Symphony Orchestra, RCA Victor Orchestra; Leonard
Bernstein. RCA SMS 9002.
    +Gr 12-75 p1022   +-RR 9-75 p33
BERNSTEIN AT HARVARD. cf Columbia M2X 33014, 33017, 33020, 33024,
33032.
BERTHIER
    Short pieces. cf BACH: Chorale prelude, O Mensch bewein dein Sünde
grosse, S 622.

BERTOLDO, Sperandio
   Canzon francese. cf Erato STU 70847.
   Petit fleur. cf Erato STU 70847.
BERTONCINI, Mario
   Tune. cf Opus One 22.
BERWALD, Franz
484 Quintets, piano, nos. 1-2. Vienna Philharmonia Quintet; Eduard
        Mrazek. Decca SDD 448.
           ++Gr 6-75 p61   +MT 9-75 p797
           +HFN 6-75 p87   +-RR 6-75 p58
     Symphony capricieuse. cf BLOMDAHL: Sisyphos.
BESARD, Jean
   La bataille de poire. cf Angel S 37015.
BEST, Martin
   Four songs of love's sorrow. cf Argo ZRG 765.
BEVERSDORF, Thomas
485 Sonata, violin and piano. DITTERSDORF: Quartet, E flat major.
        Jacques Israelievitch, vln; Henry Upper, pno; Sinnhoffer Quartet.
        Orion ORS 75170.
           ++NR 2-75 p6
BIANCIARDI, Francesco
   Fantasia terza. cf Pelca PSRK 41013/6.
   Ricercare quinto. cf Pelca PSRK 41013/6.
BIBER, Heinrich
486 Serenade, C major. BOCCHERINI: Quintet, op. 30, no. 6, C major.
        MOZART: Serenade, no. 13, K 525, G major. VIVALDI: Concerto,
        flute and bassoon, op. 10, no. 2, G minor. James Galway, flt;
        Karl Ridderbusch, bs; Lucerne Festival Strings; Rudolf Baum-
        gartner. RCA LRL 1-5085.
           +Gr 5-75 p1979   +RR 4-75 p25
           +-HFN 6-75 p102
     Sonata a 7, C major. cf Nonesuch H 71301.
     Sonata, 2 choirs. cf Nonesuch H 71301.
     Sonata, trumpet, C major. cf Nonesuch H 71301.
     Sonata, violin and harpsichord, no. 10, G minor. cf Strobe SRCM
        123.
BIGAGLIA, Drogenio
   Sonata, recorder, A minor. cf Telefunken SMA 25121 T/1-3.
487 Sonata, recorder, bassoon and harpsichord, A minor. BONONCINI:
        Divertimento, no. 6, C minor. GEMINIANI: Sonata, oboe, bassoon
        and harpsichord, E minor. VERACINI: Sonata prima, F major.
        VIVALDI: Sonata, oboe, bassoon and harpsichord, C minor. Sonata,
        no. 6, G minor. Zurich Ricercare Ensemble. Erato STU 70663.
        (also Musical Heritage Society MHS 1864)
           ++AR 5-75 p51   +Gr 4-72 p1730
BILLINGS, William
   Be glad, then America. cf New England Conservatory NEC 111.
   Chester. cf Michigan University SM 0002.
   Songs: I am come into my garden; I am the rose of Sharon; I charge
        you; An anthem for Thanksgiving: O praise the Lord. cf Nonesuch
        H 71276.
BINCHOIS, Gilles de
   Adieu m'amour. cf Vanguard SRV 316.
   Dueil angoisseus. cf Telefunken ER 6-35257.
   Filles à marier; De plus en plus. cf Telefunken TK 11569/1-2.
   Gloria, laus et honor. cf Telefunken ER 6-35257.

Je ne prise point tels Baisiers, soprano, rebec and 3 viols.
cf Strobe SRCM 123.
BINKERD, Gordon
488 Sonata, violin and piano. COPLAND: Sonata, violin and piano.
IVES: Sonata, violin and piano, no. 4. Jaime Laredo, vln;
Ann Schein, pno. Desto DC 6439.
++HF 11-75 p129
BIRTWISTLE, Harrison
Chronometer. cf The triumph of time.
The fields of sorrow. cf Nenia: The death of Orpheus.
489 Nenia: The death of Orpheus. The fields of sorrow. Verses for
ensembles. Jane Manning, s; The Matrix; London Sinfonietta;
David Atherton. Decca HEAD 7. (also Headline 7).
++Gr 1-75 p1375   +NYT 12-21-75 pD18
++HF 11-75 p98   +RR 11-74 p86
+MT 8-75 p713   ++St 12-75 p117
*NR 11-75 p10   +Te 12-75 p43
490 Tragoedia. CROSSE: Concerto da camera. WOOD: Pieces, piano, op.
6 (3). Manoug Parikian, vln; Susan McGaw, pno; Melos Ensemble;
Edward Downes, Lawrence Foster. Argo ZRG 759. (Reissues)
+-RR 12-75 p72
491 The triumph of time. Chronometer. BBC Symphony Orchestra;
Pierre Boulez. Argo ZRG 790.
+Gr 7-75 p174   +RR 6-75 p32
+-HFN 10-75 p137   +Te 12-75 p43
Verses for ensembles. cf Nenia: The death of Orpheus.
BISHOP, Henry
Grand march, E major. cf Saga 5417.
BIZET, Georges
Agnus Dei. cf Decca ECS 2159.
Agnus Dei. cf RCA ARL 1-0561.
492 L'Arlésienne: Suites, nos. 1 and 2. Carmen: Prelude, Acts 1-4;
Scène des contrebandiers; Habañera; Nocturne; La Garde motante;
Danse bohème. OSR; Ernest Ansermet. Decca ECS 755. (Reissues)
+Gr 3-75 p1711   +-RR 4-75 p22
493 L'Arlésienne: Suites, nos. 1 and 2. Carmen: Suites, nos. 1 and 2.
The Hague Residentie Orchestra; Willem van Otterloo. DG 2548
173. (Reissue from SLPM 138787)
+Gr 11-75 p840
494 L'Arlésienne: Suites, nos. 1 and 2. Carmen: Suite. Detroit
Symphony Orchestra; Paul Paray. Mercury SRI 75060.
++SFC 9-21-75 p34
495 L'Arlésienne: Suites, nos. 1 and 2. Carmen: Suite, no. 1. PhO;
Heinz Wallberg. Music for Pleasure MFP 57015.
+Gr 4-75 p1871   +RR 5-75 p28
-HFN 5-75 p125
496 L'Arlésienne: Suites, nos. 1 and 2. CHABRIER: España. PSA; Václav
Smetáček. Supraphon 110 0694. Tape (c) 04 0694.
+-Gr 6-75 p106   +-HFN 2-72 p304
+Gr 4-75 p1872 tape   +-HFN 5-75 p142 tape
L'Arlésienne: Suites, nos. 1 and 2. cf Camden CCV 5000-12, 5014-24.
L'Arlésienne: Suites, nos. 1 and 2; Intermezzo; Farandole. cf
HMV SLS 5019.
497 Carmen (sung in German). Emmy Destinn, Minnie Nast, Marie Dietrich,
s; Grete Parbs, ms; Karl Jörn, Rudolf Krasa, t; Hermann Bachmann,
Julius Lieban, bar; Felix Dahn, bs; Orchestra and Chorus; Bruno
Seidler-Winkler. Discophilia KS 1/3 (3). (Recorded, 1908)

```
        +-HF 2-75 p89                    +-ON 3-22-75 p33
        +-NR 7-75 p10
```

498 Carmen. Régine Crespin, Jeannette Pilou, Maria Rosa Carminati,
     Nadine Denize, s; Gilbert Py, Rémy Corazza, t; José van Dam,
     Jacques Trigeau, bar; Pierre Thau, Paul Guige, bs; St. Maurice
     Children's Chorus; Rhine Opera Chorus; Strasbourg Philharmonic
     Orchestra; Alain Lombard. Erato STU 70900/2 (3).
```
        +-Gr 8-75 p359                   +-ON 12-13-75 p48
        +-HF 12-75 p85
```
499 Carmen. Risë Stevens, Licia Albanese, s; Jan Peerce, t; Robert
     Merrill, bar; RCA Orchestra; Fritz Reiner. RCA AV 3-0670 (3).
     (Reissue from LM 6012)
```
        +-HF 4-75 p92                    +-SFC 11-24-74 p31
```
500 Carmen: Prelude; Avec la garde montante; L'amour est un oiseau
     rebelle; Près des remparts de Séville; Entre'acte; Les tringles
     des sistres tintaient; Votre toast; Nous avons en tête une
     affaire; La fleur que tu m'avais jetée; Melons, Coupons; Entre'
     act; Finale. Colette Boky, s; Marilyn Horne, Marcia Baldwin,
     ms; James McCracken, Russell Christopher, Andrea Velis, t; Tom
     Krause, bar; Donald Gramm, bs; Manhattan Opera Children's
     Chorus; Metropolitan Opera Orchestra and Chorus; Leonard Bern-
     stein. DG 2530 534. (Reissue from 2740 101)
```
        +-Gr 10-75 p676                  +-RR 8-75 p22
        +-HFN 9-75 p109
```
Carmen: Bolero; Toreador. cf HMV SEOM 20.
Carmen: C'est toi. C'est toi. cf Decca DPA 517/8.
Carmen: La fleur que tu m'avais jetée. cf Decca SDD 390.
Carmen: Le fleur que tu m'avais jetée. cf RCA SER 5704/6.
Carmen: Flower song. cf Decca SXL 6649.
Carmen: Air de fleur. cf Columbia D3M 33448.
Carmen: Habañera. cf HMV SLS 5012.
Carmen: Habañera; Seguidilla; Chanson bohème; Card scene. cf
     Odyssey Y2 32880.
Carmen: Ich sprach dass ich furchtlos. cf Discophilia KGB 2.
Carmen: Ich sprach dass ich furchtlos mich fühle. cf Disophilia
     KGP 4.
Carmen: March of the toreadors, Act 4. cf Decca PFS 4323.
Carmen: Prelude, Acts 1-4; Scène des contrebandiers; Habañera;
     Nocturne; La Garde montante; Danse bohème. cf L'Arlésienne:
     Suites, nos. 1, 2.
Carmen: Smugglers chorus; March and chorus, Act 4. cf Decca ECS
     2159.
Carmen: Suite. cf L'Arlésienne: Suites,nos. 1 and 2.
Carmen: Suite. cf Camden CCV 5000-12, 5014-24.
Carmen: Suite, no. 1. cf L'Arlésienne: Suites, nos. 1 and 2.
Carmen: Suites, nos. 1 and 2. cf L'Arlésienne: Suites, nos. 1 and
     2.

501 Carmen fantasy, op. 25 (arr. Sarasate). SAINT-SAENS: Havanaise,
     op. 83. Introduction and rondo capriccioso, op. 28. SARASATE:
     Zigeunerweisen, op. 20, no. 1. Ruggiero Ricci, vln; LSO;
     Pierino Gamba. Decca SDD 420. (Reissue from SXL 2197)
```
        +-Gr 4-75 p1871                  +-RR 4-75 p22
        +-HFN 6-75 p105
```
502 Carmen for orchestra (arr. Kostelanetz). Orchestra; André Kostel-
     anetz. CBS 61596.
```
        +-Gr 2-75 p1555                  -RR 2-75 p130
```

503 Jeux d'enfants, op. 22. RAVEL: Ma mere l'oye. SAINT-SAENS:
     Le carnaval des animaux. Peter Katin, Philip Fowke, pno;
     Scottish National Orchestra; Alexander Gibson. Classics for
     Pleasure CFP 40086.
          +-Gr 6-75 p33   +-RR 5-75 p38
          ++HFN 7-75 p77
504 Jeux d'enfants, op. 22. DEBUSSY: Nocturnes: Fêtes (arr. Ravel).
     Petite suite. John Ogdon, Brenda Lucas, pno. HMV HQS 1344.
          +-Gr 7-75 p200   +-RR 6-75 p65
          ++HFN 7-75 p77
505 Jeux d'enfants, op. 22 (arr. Odom). CARMICHAEL: Puppet show.
     FAURE: Dolly, op. 56: Berceuse (arr. C. Smith). POLDINI:
     Poupée valsante (arr. Salter). Cyril Smith, Phyllis Sellick,
     pno. Polydor 2460 232.
          ++HFN 6-75 p103
     Jeux d'enfants, op. 22. cf GERSHWIN: Rhapsody in blue.
     La jolie fille de Perth: La, la, la, la...Quand la flamme de
     l'amour. cf Decca SXL 6637.
506 Nocturne, no. 1, F major. Variations chromatiques de concert.
     GRIEG: Sonata, piano, op. 7, E minor. Glenn Gould, pno.
     Columbia 32040. (also CBS 73178)
          +Gr 7-75 p214   +NR 8-73 p15
          +-HF 7-73 p79   /St 7-73 p104
          +HFN 10-75 p139
     Nocturne, no. 1, F major. cf GRIEG: Sonata, piano, op. 7, E minor.
507 Les pêcheurs de perles. Janine Micheau, s; Nicolai Gedda, t;
     Ernest Blanc, bar; Jacques Mars, bs; Paris Opéra Comique Chorus
     and Orchestra; Pierre Dervaux. HMV SLS 877 (1). (Reissue from
     Columbia SAX 2442/3)
          +-Gr 3-75 p1697   +-RR 4-75 p16
          +-MT 11-75 p977
     Les pêcheurs de perles: Au fond du temple saint. cf Discophilia
     KGC 1.
     Les pêcheurs de perles: Au fond du temple saint. cf Ember GVC 41.
     Les pêcheurs de perles: Au fond du temple saint. cf RCA SER 5704/6.
     Les pêcheurs de perles: C'est toi...Au fond du temple saint. cf
     Decca DPA 517/8.
     Les pêcheurs de perles: De mon amie. cf Decca SDD 390.
     Les pêcheurs de perles: De mon amie. cf Discophilia KGC 2.
     The pearl fishers: Je crois entendre. cf Columbua D3M 33448.
     Les pêcheurs de perles: Leila, Leila, Dieu puissant le voilà. cf
     Angèl S 37143.
     Les pêcheurs de perles: Mi par d'udir ancora. cf Bel Canto Club
     no. 3.
     Les pêcheurs de perles: O Dieu Brâhma. cf RCA ARL 1-0844.
     Les pêcheurs de perles: O Nadir, Chanson d'Avril. cf Rococo 5365.
508 Roma, Symphony in C. Symphony, no. 1, C major. Birmingham Symphony
     Orchestra; Louis Frémaux. HMV ASD 3039.
          ++Gr 1-75 p1341   +RR 12-74 p25
509 Songs: Absence; Les adieux de l'hôtesse arabe; Chanson d'Avril;
     Vieille chanson. DEBUSSY: Chansons de Bilitis. FALLA: Spanish
     popular songs(7). NIN: Jesús de Nazareth; Villancico andaluz;
     Villancico asturiano; Villancico castellano. Marilyn Horne,
     ms; Martin Katz, pno. Decca SXL 6577. (also London OS 26301)
          +-Gr 6-74 p96   +-ON 5-75 p36
          +-HF 8-74 p105   +-RR 5-74 p62
          +MJ 12-74 p44   ++SFC 7-21-74 p25
          +NR 8-74 p9   ++St 9-74 p134

510   Songs: Adieux de l'hôtesse arabe; Chanson d'Avril; La chanson de
          la rose; Vous ne priez pas. BERLIOZ: Songs: Irlande, op. 2: La
          belle voyageuese, Le coucher du soleil, L'Origine de la harpe.
          DEBUSSY: Songs: Proses lyriques; Noël des enfants qui n'ont
          plus de maison. Jill Gomez, s; John Constable, pno.   Saga 5388.
              ++Gr 1-75 p1380   +-STL 1-12-75 p36
              +RR 2-75 p58
      Songs: Adieu de l'hôtesse arabe. cf Odyssey Y2 32880.
      Songs: Pastorale. cf Saga 7029.
      Symphony, no. 1, C major. cf Roma, Symphony in C.
      Variations chromatiques de concert. cf Nocturne, no. 1, F major.
      Variations chromatiques de concert. cf GRIEG: Sonata, piano,
          op. 7, E minor.
BLACHER, Boris
      Francesca da Rimini, op. 47. cf Musical Heritage Society MHS 1976.
      Ornaments. cf Sonata, solo violin, op. 40.
511   Sonata, solo violin, op. 40. Ornaments (4). LINKE: Violencia.
          MADERNA: Dedication. Pièce pour Ivry. Christiane Edinger, vln.
          Orion ORS 75171.
              ++HF 9-75 p81   ++NR 10-75 p11
BLAKE (20th century England)
      Variations, piano, cf HMV HQS 1337.
BLANGINI
      Per valli, per boschi. cf Saga 7029.
BLANK, Allan
512   Esther's monologue. Music for solo violin. COPE: Arena. Margins.
          Marlee Sabo, s; Stanley Hoffman, vln; Gerald Stanick, vla;
          Richard Peepo, vlc; Larry Baker, tpt, vlc, perc, pno; David
          Cope, vlc and tape. Orion ORS 75169.
              +NR 3-75 p7
      Music for solo violin. cf Esther's monologue.
513   Rotation. JENNI: Musique printanière. KARLINS: Variations on
          'Obieter dictum'. STOCK: Quintet, clarinet and strings. Allen
          Blustine, clt; Joel Krosnick, vlc; Elizabeth Buccheri, Gilbert
          Kalish, pno; Thomas Siwe, perc; Betty Bang Mather, flt; Contem-
          porary String Quartet; John Simms. CRI SD 329.
              ++NR 1-75 p7   +-RR 7-75 p44
BLANTER, Matvei
      Rostov town. cf HMV ASD 3116.
BLAVET, Michel
      Sonata, flute, no. 2, F major. cf London STS 15198.
      Sonata, flute, no. 2, B minor. cf Pandora PAN 103.
BLISS, Arthur
      Antiphonal fanfare for 3 brass choirs. cf Decca 419.
      Mêlée fantasque. cf BERKELEY: Mont Juic, op. 9.
514   Morning heroes. John Westbrook, orator; Liverpool Philharmonic
          Choir; Royal Liverpool Philharmonic Orchestra; Charles Groves.
          HMV SAN 365.
              +Gr 3-75 p1687   +RR 6-75 p77
              ++HFN 7-75 p77
515   Sonata, piano. LAMBERT: Elegiac blues. Sonata, piano. Rhonda
          Gillespie, pno. Argo ZRG 786.
              +Gr 2-75 p1516   ++RR 2-75 p50
              +MT 10-75 p886
BLITHEMAN, William
      Eterne rerum conditor. cf Philips 6775 006.

BLOCH, Ernest
    Baal shem. cf Da Camera Magna SM 93399.
    Baal Shem suite: Ningun. cf CBS 76420.
516 Concerto, violin and orchestra. Yehudi Menuhin, vln; PhO; Paul
      Kletzki. HMV SXLP 30177. (also Angel S 36192). (Reissue from
      ASD 584)
        ++Gr 3-75 p1648
517 Concerto, violin and orchestra, A minor. Joseph Szigeti, vln;
      OSCCP; Charles Munch. Turnabout THS 65007. (Reissue from EMI,
      1939)
        ++HF 10-74 p90   +St 3-75 p98
    Jewish life: 3 pieces. cf Orion ORS 75181.
    Meditation hébraique. cf Orion ORS 7518.
518 Sinfonia breve. KOUSSEVITZKY: Concerto, double bass. Gary Karr,
      double-bass; Oslo Philharmonic Orchestra, Minneapolis Symphony
      Orchestra; Alfredo Antonini, Antal Dorati. CRI SD 248.
        +-RR 1-75 p28   ++RR 4-75 p28
    Sonata, violin, nos. 1 and 2. cf RCA ARM 4-0942/7.
BLODEK, Vilém
    In the well: The rising of the moon. cf Supraphon 110 1429.
BLOMDAHL, Karl-Birger
519 Sisyphos. BERWALD: Symphony capricieuse. ROSENBERG: Resan till
      America (Voyage to America), excerpts. Stockholm Philharmonic
      Orchestra; Antal Dorati. RCA VICS 1319.
        *St 6-75 p50
520 Symphony, no. 2. PETTERSSON: Symphony, no. 10. Stockholm
      Philharmonic Orchestra, Swedish Radio Symphony Orchestra; Antal
      Dorati. HMV 4E 061 35142.
        +-Gr 12-75 p1061
521 Symphony, no. 3. ROSENBERG: Symphony, no. 6. Stockholm Philharmonic
      Orchestra; Sixten Ehrling, Stig Westerberg. Turnabout TV 34318.
        +Gr 2-72 p1373   +-HFN 2-72 p305
        +HF 11-71 p94   *St 6-75 p50
BLOW, John
    Behold, O God our defender. cf Gemini GM 2022.
522 Coronation and symphony anthems: Blessed is the man. Cry aloud,
      and spare not. God spake sometime in visions. I was glad. O
      sing unto the Lord. Charles Brett, c-t; Philip Langridge, t;
      James Lancelot, org; Kenneth Heath, vlc; Cambridge, King's
      College Chapel Choir; AMF; David Willcocks. Argo ZRG 767.
        +Gr 4-75 p1844   ++RR 4-75 p58
        +NR 8-75 p8   +St 11-75 p126
    Toccata. cf Wealden WS 139.
523 Venus and Adonis. Margaret Ritchie, Margaret Field, Elizabeth
      Cooper, s; Robert Ellis, Michael Cynfelin, t; Gordon Clinton,
      John Frost, bs; L'Oiseau-Lyre Orchestral Ensemble; Anthony Lewis.
      L'Oiseau-Lyre OLS 128. (Reissue from 50004)
        ++ARG 7-72 p558   /NYT 6-4-72 pD26
        +-HF 6-72 p100   ++SFC 1-19-75 p27
        +HFN 1-72 p104
BOCCHERINI, Luigi
524 Concerto, flute, op. 27, D major. MERCADANTE: Concerto, flute, E
      minor. TARTINI: Concerto, flute, a 5, G major. Severino
      Gazzelloni, flt; I Musici. Philips 6500 611. Tape (c) 7300 334.
        +Gr 2-75 p1473   ++RR 7-75 p72 tape
        +HF 4-75 p97   +RR 9-75 p79 tape
        +NR 3-75 p5   +SFC 10-5-75 p38
        ++RR 1-75 p34

525 Concerto, violoncello, B flat major (arr. Grützmacher). HANDEL:
       Sonata, 2 violoncellos and orchestra, G minor (arr. Feuillard
       and P. Tortelier). PAGANINI: Variations on a theme by Rossini,
       D minor (arr. P. Tortelier). VIVALDI: L'Estro armonica: Concerto,
       violoncello, op. 3, no. 9, D major. Paul Tortelier, Maud Martin-
       Tortelier, vlc; ECO; Paul Tortelier, Maud Martin-Tortelier.
       HMV ASD 3015.
               +Gr 10-74 p713   ++SFC 3-2-75 p24
               +-RR 10-74 p38
     Minuet. cf Philips 6580 066.
526 Quintet, guitar, no. 2, C major. Quintet, strings, op. 13, no. 5,
       E major. Alirio Diaz, gtr; Alexander Schneider Quartet. Bach
       (Vanguard) Guild HM 43 SD. (Reissue from Vanguard)
               ++MJ 11-74 p48   ++SFC 8-18-74 p23
               +RR 4-75 p25
     Quintet, strings, op. 13, no. 5, E major. cf Quintet, guitar,
       no. 2, C major.
     Quintet, strings, op. 13, no. 5, E major: Minuet. cf Philips
       6580 098.
     Quintet, op. 30, no. 6, C major. cf BIBER: Serenade, C major.
527 Sonatas, flute and harpsichord, op. 5 (6). Sheridon Stokes, flt;
       Bess Karp, hpd. Orion ORS 75173.
               +NR 5-75 p8
     Sonata, violoncello, no. 6, A major: Allegro. cf Ember GVC 42.
BOCK
     Sunrise, sunset. cf Columbia D3M 33448.
BOELLMAN, Leon
     Suite gothique, op. 25. cf Stentorian SC 1685.
     Suite gothique, op. 25: Toccata. cf Decca SDD 463.
BOHAC, Josef
     March. cf Panton 110 361.
BOHM, Georg
     Chorale prelude, "Wer nur den lieben Gott Lässt walten". cf
       Telefunken DX 6-35265.
     Präludium, Fuge und Postludium, G minor. cf Saga 5402.
     Prelude, D minor. cf Pelca PSRK 41013/6.
     Prelude and fugue, A minor. cf Saga 5374.
     Prelude and fugue, A minor. cf Telefunken DX 6-35265.
     Prelude and fugue, C major. cf Wealden WS 145.
     Suite, no. 1, C minor. cf BACH: Toccata and fugue, S 916, G major.
     Suite, no. 7, F major. cf BACH: Toccata and fugue, S 916, G major.
BOHM, Theobald
528 Introduction and variations on Nel cor piu, op. 4. DEMERSSEMAN:
       Solo de concerto, no. 6, op. 82, F major. DOPPLER: Fantaisie
       pastorale hongroise, op. 26. TULOU: Grand solo, no. 13, op.
       96, A minor. Michel Debost, flt; Christian Ivaldi, pno. Sera-
       phim S 60247.
               ++NR 10-75 p5
     Introduction and variations on Nel cor piu, op. 4. cf Vanguard
       VSD 71207.
BOIELDIEU, François
     Calif of Bagdad: Overture. cf AUBER: Le cheval de bronze: Overture.
529 Concerto, harp, C major. HANDEL: Concerto, harp, B flat major
       (from Concerto, organ, op. 4, no. 6). Susanna Mildonian, hp;
       Luxembourg Radio Orchestra; Louis de Froment. Decca 7234.
               +-HFN 8-75 p73   ++RR 8-75 p30

530 Concerto, harp, C major. RODRIGO: Concierto serenata. Catherine
        Michel, hp; Monte Carlo National Orchestra; Antonio de Almeida.
        Philips 6500 813. Tape (c) 7300 367.
              +-Gr 8-75 p321   ++NR 9-75 p7
              ++HFN 7-75 p77   +RR 7-75 p23
              ++HFN 11-75 p149 tape   +SFC 8-3-75 p30
        La dame blanche: Overture. cf AUBER: Le cheval de bronze: Overture.
BOISDEFFRE
        Au Bord du Ruisseau, op. 52. cf Discopaedia MB 1005.
BOISMORTIER, Joseph Bodin de
        Sonata, bassoon, no. 5, G minor. cf Musical Heritage Society MHS
        1853.
BOITO, Arrigo
        Mefistofele: L'Altra notte. cf Decca GOSC 666/8.
        Mefistofele: Dai campi, dai prati; Se tu mi doni; Lontano, lontano;
            Giunto sul passo estremo. cf Ember GVC 41.
        Mefistofele: Giunto sul passo estremo. cf Supraphon 10924.
        Mefistofele: Lontano, lontano. cf Discophilia KGC 1.
531 Mifestofele: Prologue. VERDI: Pezzi sacri, no. 4: Te Deum. Nicola
        Moscona, bs; Columbus Boys' Choir; Robert Shaw Chorale; NBC
        Symphony Orchestra; Arturo Toscanini. RCA AT 13. (Reissues
        from HMV ALP 1363)
              +Gr 8-75 p336   +RR 6-75 p22
BOLCOM, William
532 Frescoes. Bruce Mather, pno and harmonium; Pierette Le Page, pno
        and hpd. Nonesuch H 71297.
              +HF 10-74 p90   +NR 8-74 p12
              ++HFN 10-75 p138   +St 10-74 p124
              *MT 11-75 p977
        BOLET, JORGE, AT CARNEGIE HALL. cf RCA ARL 2-0512.
BOLLING
533 Suite, flute and jazz piano. Jean-Pierre Rampal, flt; Claude
        Bolling, pno; Max Hédiguer, double bs; Marcel Sabiani, drums.
        Columbia M 33233.
              +NR 10-75 p4
BONONCINI, Giovanni
        Deh più a me non vascondete. cf Decca SXL 6629.
534 Divertimenti da camera, A minor, G minor, E minor, G major, F major,
        D minor, C minor, B flat major. Hans-Martin Linde, rec; Eduard
        Müller, hpd; Konrad Ragossnig, lt; Josef Ulsamer, vla da gamba.
        DG Archive 2533 167.
              +Gr 12-74 p1161   ++RR 12-74 p45
              +HF 7-75 p72   +SFC 7-27-75 p22
              ++NR 5-75 p8
        Divertimento, no. 6, C minor. cf BIGAGLIA: Sonata, recorder,
        bassoon and harpsichord, A minor.
        Per la gloria d'adorarvi. cf London OS 26391.
BORDEN, David
535 Easter. DREWS: Ceres motion. Train. Steve Drews, Linda Fisher,
        synthesizers; David Borden, synthesizers and electric piano.
        Earthquack EQ 0001.
              +St 1-75 p120
BORLET (14th century)
        He, tres doulz roussignol. cf Telefunken TK 11569/1-2.
BORODIN, Alexander
        For the shore of your far-off land. cf London OS 26249.

Prince Igor: Boyars' chorus with Yaroslavna. cf HMV Melodiya SXL
   30190.
536 Prince Igor: Dance of the Polovtsian maidens; Polovtsian dances.
       GLINKA: Russlan and Ludmila: Overture; Waltz fantasy. Jota
       aragonesa. MUSSORGSKY: Khovanschina: Prelude; Dance of the
       Persian slaves. A night on the bare mountain. Lausanne Radio
       Chorus; Choeur de Jeunes de l'Eglise National Vaudoise; OSR;
       Ernest Ansermet. Decca ECS 757. (REissue from SXL 6119)
            +-RR 8-75 p30
537 Prince Igor: Overture. GLINKA: Russlan and Ludmila: Overture.
       MUSSORGSKY (orch. Rimsky-Korsakov): A night on the bare mountain.
       Khovanschina: Prelude and Persian dance. BPhO; Georg Solti.
       Decca SPA 257. (also London CS 6944)
            +Gr 7-73 p193   +RR 7-73 p55
            +-HF 12-75 p111   +-SFC 8-24-75 p28
            +NR 10-75 p2
     Prince Igor: Polovtsian dances, Act 2. cf HMV SLS 5019.
538 Prince Igor: Polovtsian march. PROKOFIEV: Alexander Nevsky, op.
       78. Rosalind Elias, ms; CSO and Chorus; Fritz Reiner. RCA VICS
       1652. Tape (c) MCK 530. (Reissue from SB 6530)
            -HFN 4-72 p705   -RR 1-75 p71 tape
     Prince Igor: Vladimir's recitative and cavatina. cf RCA SER 5704/6.
539 Quartet, strings, no. 2, D major: Nocturne (arr. Sargent). HOLST:
       St. Paul's suite, op. 29, no. 2. TCHAIKOVSKY: Quartet, strings,
       no. 1, op. 11, D major: Andante cantabile (arr. Schmid). WIREN:
       Serenade, strings, op. 11. London Youth String Ensemble;
       Frederick Applewhite. Cameo GOCLP 9004.
            +-Gr 3-75 p1667   +-RR 3-75 p24
     Quartet, strings, no. 2, D major: Nocturne. cf Amon Ra SARB 01.
BORTNIANSKY (Bortnyiansky), Dmitri
     Cherubim hymn, no. 7. cf HMV SEOM 20.
     Cherubim hymn, no. 7. cf HMV Melodiya ASD 3102.
     I will lift up my eyes to the hills. cf HMV Melodiya ASD 3102.
BORTZ, Daniel
     Nightwinds. cf Caprice RIKS LP 46.
BOSSI, Marco
     Etude symphonique. cf Argo ZRG 807.
     Scherzo, G minor. cf Polydor 2460 252.
BOTTEGARI
     Mi stare pone Totesche. cf L'Oiseau-Lyre 12BB 203/6.
BOTTESINI, Giovanni
540 Concerto, double-bass. Duo concertante, double-bass and violin.
       Tarentelle, double-bass. Jean-Marc Rollez, double-bs; Gerard
       Jarry, vln; French National Radio Chamber Orchestra; André
       Girard. Arion ARN 38277.
            +HFN 10-75 p139   +RR 10-75 p37
     Duo concertante, double-bass and violin. cf Concerto, double-bass.
     Tarentelle, double-bass. cf Concerto, double-bass.
BOUIN (17th century France)
     La montauban. cf DG Archive 2533 172.
BOULANGER, Nadia
     Cortege. cf Avant AV 1012.
     Cortege. cf Discopaedia MB 1009.
     Cortege. cf Sicopaedia MB 1010.
     Cortege. cf RCA ARM 4-0942/7.
     Nocturne. cf Discopaedia MB 1010.

Nocturne, F major. cf RCA ARM 4-0942/7.
D'un vieux jardin. cf Avant AV 1012.
BOULEZ, Pierre
541 e. e. cummings (Cummings ist der Dichter). LUTOSLAWSKI: Poèmes
        d'Henri Michaux (3). MESSIAEN: Et exspecto resurrectionem
        mortuorum. STRAVINSKY: Canticum sacrum. Peter Baillie, t;
        Ladislav Illavsky, bar; Vienna ORF Chorus; Vienna ORF Symphony
        Orchestra; Bruno Maderna. Telefunken EK 6-48066 (2).
                +-Gr 7-75 p199   +-RR 7-75 p27
                +-HFN 10-75 p138
BOWEN, York
    Berceuse, op. 83. cf Piano works (Lyrita RCS 17).
    Partita, op. 157, D minor. cf Piano works (Lyrita RCS 17).
542 Piano works: Berceuse, op. 83. Preludes, op. 24, nos. 1, 2, 7, 8,
        10, 15, 16, 19, 20. Partita, op. 157, D minor. Suite Mignonne,
        op. 39: Moto perpetuo. Toccata, op. 155. York Bowen, pno.
        Lyrita RCS 17. (Reissue from 1963)
                +Gr 9-75 p492   /RR 7-75 p46
                +-HFN 12-75 p156
    Preludes, op. 24, nos. 1, 2, 7, 8, 10, 15, 16, 19, 20. cf Piano
        works (Lyrita RCS 17).
    Suite Mignonne, op. 39: Moto perpetuo. cf Piano works (Lyrita RCS
        17).
    Toccata, op. 155. cf Piano works (Lyrita RCS 17).
BOYCE, William
    Voluntary, no. 1, D major. cf Abbey LPB 738.
    Voluntary, no. 1, D major. cf CRD CRD 1008.
BOYKAN, Martin
    Quartet, strings, no. 1. cf BARKIN: Quartet, strings.
BOZAY, Attila
543 Quintet, winds, op. 6. HIDAS: Quintet, winds, no. 2. LANG: Quin-
        tet, winds, no. 2. PETROVICS: Quintet, winds. Hungarian Wind
        Quintet. Hungaroton SLPX 11630.
                +-HFN 6-75 p107   ++RR 5-75 p49
                +NR 5-75 p9
BOZIC
    Kriki, brass quintet. cf Desto DC 6464/7.
BOZZA, Eugene
    Dialogue. cf Armstrong 721-4.
BRADE, William
    Allemande. cf Crystal S 201.
BRAHMS, Johannes
    Academic festival overture, op. 80. cf Concerto, violin and violon-
        cello, op. 102, A minor.
    Academic festival overture, op. 80. cf Symphonies, nos. 1-4.
    Akademische Festouverture, op. 80. cf Symphony, no. 3, op. 90,
        F major.
    Academic festival overture, op. 80. cf Symphony, no. 4, op. 98,
        E minor (Angel 37304).
    Academic festival overture, op. 80. cf Symphony, no. 4, op. 98,
        E minor (Decca 381).
    Academic festival overture, op. 80. cf Symphony, no. 4, op. 98,
        E minor (RCA 1-0719).
    Academic festival overture, op. 80. cf Works, selections (HMV
        SLS 5009).
    Academic festival overture, op. 80. cf BEETHOVEN: Symphony, no. 4,
        op. 60, B flat major.

Alto rhapsody, op. 53. cf Works, selections (HMV SLS 5009).

544 Ballades, op. 10, nos. 1-4. SCHUMANN: Arabeske, op. 18, C major.
Bunte Blätter, op. 99, no. 9. Romances, op. 28, nos. 1-3.
Waldscenen, op. 82, no. 6. Wilhelm Kempff, pno. DG 2530 321.
    +-Gr 11-73 p956   +-NR 7-75 p14
    ++HF 7-75 p83   ++SFC 7-27-75 p22

Ballade, op. 10, no. 3, B minor. cf Concerto, piano, no. 1, op.
15, D minor.

Capriccio, op. 76, no. 1, F major. cf Columbia Special Products
AP 12411.

545 Capriccio, op. 116. Intermezzi, op. 116. SCHUBERT: Fantasia, op.
108, F minor (arr. Kabalevsky). SCHUMANN: Scherzo and presto
passionata. Emil Gilels, pno; RAI Milan Orchestra. Rococo 2087.
    +-NR 10-75 p9

Chorale prelude, no. 8 (arr). cf Kendor KE 22174.

Chorale preludes: Herzlich tut mich verlangen, op. 122, no. 3;
O Welt ich muss dich lassen, op. 122, no. 11. cf Pelca PSRK
41013/6.

Chorale preludes, organ, op. 122, nos. 1, 4, 8, 10. cf Decca
5BBA 1013-5.

546 Concerto, piano, no. 1, op. 15, D minor. Ballade, op. 10, no. 3,
B minor. Julius Katchen, pno; LSO; Pierre Monteux. Decca
SPA 385. (Reissue from SXL 2172, SXL 6160)
    +-Gr 7-75 p175   +-RR 9-75 p34
    +HFN 8-75 p87

547 Concerto, piano, no. 1, op. 15, D minor. Radu Lupu, pno; LPO;
Edo de Waart. Decca SXL 6728. Tape (c) KSXC 6728.
    -Gr 11-75 p808   +HFN 12-75 p173 tape
    +-HFN 11-75 p152   +-RR 10-75 p38

548 Concerto, piano, no. 1, op. 15, D minor. Emil Gilels, pno; BPhO;
Eugen Jochum. DG 2530 258.
    ++NR 12-75 p7

549 Concerto, piano, no. 1, op. 15, D minor. Alexis Weissenberg, pno;
LSO; Carlo Maria Giulini. HMV ASD 2992. (also Angel S 36967)
    +Gr 6-74 p40   +NR 12-74 p6
    -HF 2-75 p88   +RR 5-74 p29

550 Concerto, piano, no. 1, op. 15, D minor.* Concerto, piano, no. 2,
op. 83, B flat major. Bruno-Leonardo Gelber, pno; Munich
Philharmonic Orchestra, RPO; Franz-Paul Decker, Rudolf Kempe.
HMV SXDW 3020 (2). (*Reissue from HQS 1068)
    -Gr 9-75 p452   +RR 8-75 p30
    +-HFN 9-75 p97

551 Concerto, piano, no. 1, op. 15, D minor. Alfred Brendel, pno;
COA; Hans Schmidt-Isserstedt. Philips 6500 623. Tape (c) 7300
281.
    ++Gr 2-75 p1473   ++RR 3-75 p24
    +HF 6-75 p87   ++RR 4-75 p76 tape
    +NR 5-75 p6   ++SFC 5-4-75 p35

552 Concerto, piano, no. 1, op. 15, D minor. Solomon, pno. Rococo 2046.
    ++NR 2-75 p11

553 Concerto, piano, no. 1, op. 15, D minor. Artur Schnabel, pno; LPO;
Georg Szell. World Records SH 223. (also Rococo 2022).
(Reissue from HMV DB 3712/7)
    +Gr 6-75 p35   -NR 12-72 p7

554 Concerto, piano, no. 2, op. 83, B flat major. Bruno-Leonardo
Gelber, pno; RPO; Rudolf Kempe. Connoisseur Society 2088.

(Q) CSQ 2088.

++SFC 11-16-75 p32

555 Concerto, piano, no. 2, op. 83, B flat major. Clifford Curzon,
pno; VPO; Hans Knappertsbusch. Decca ECS 751. (Reissue from
LXT 5434)

    -Gr 3-75 p1648   +-RR 5-75 p28

556 Concerto, piano, no. 2, op. 83, B flat major. Vladimir Ashkenazy,
pno; LAPO; Zubin Mehta. Decca SXL 6309. Tape (c) DSXC 6309.
(also London 6539. Tape (ct) M 10206)

    +Gr 7-75 p254 tape   +-RR 7-75 p70 tape

    ++HFN 7-75 p90 tape

557 Concerto, piano, no. 2, op. 83, B flat major. Emil Gilels, pno;
BPhO; Eugen Jochum. DG 2530 259. Tape (c) 3300 265. (Reissue
from 2707 064)

    +Gr 6-74 p45   ++RR 5-74 p29

    ++NR 7-75 p6   +-RR 7-75 p70 tape

558 Concerto, piano, no. 2, op. 83, B flat major. Alfred Brendel, pno;
COA; Bernard Haitink. Philips 6500 767. Tape (c) 7300 293.

    +-Gr 12-74 p1120   +NR 12-74 p5

    +Gr 5-75 p2031   +RR 11-74 p36

    +-HF 2-75 p89   +RR 4-75 p76 tape

559 Concerto, piano, no. 2, op. 83, B flat major. Edwin Fischer, pno;
BPhO; Wilhelm Furtwängler. Unicorn UNI 102.

    +NR 9-75 p7

Concerto, piano, no. 2, op. 83, B flat major. cf Concerto, piano,
no. 1, op. 15, D minor.

Concerto, piano, no. 2, op. 83, B flat minor. cf BEETHOVEN:
Concerto, piano, no. 5, op. 73, E flat major.

560 Concerto, violin, op. 77, D major. Maurice Hasson, vln; LPO;
James Loughran. Classics for Pleasure CFP 40221.

    /HFN 8-75 p73   /RR 8-75 p31

561 Concerto, violin, op. 77, D major. Christian Ferras, vln; VPO;
Carl Schuricht. Decca ECS 704. (Reissue from LXT 2949)

    ++Gr 2-75 p1474   +-RR 12-74 p26

562 Concerto, violin, op. 77, D major. Nathan Milstein, vln; VPO;
Eugen Jochum. DG 2530 592. Tape (c) 3300 592.

    +-Gr 12-75 p1031   +RR 12-75 p22

563 Concerto, violin, op. 77, D major. David Oistrakh, vln; CO;
Georg Szell. HMV ASD 2525. Tape (c) TC ASD 2525. (Reissue)

    +-Gr 10-75 p721 tape   ++RR 9-73 p69

    +HFN 10-75 p155 tape

564 Concerto, violin, op. 77, D major. Yehudi Menuhin, vln; BPhO;
Rudolf Kempe. HMV SXLP 30186. (Reissue from ASD 264)

    +Gr 6-75 p36   +-RR 7-75 p24

565 Concerto, violin, op. 77, D major. Henryk Szeryng, vln; COA;
Bernard Haitink. Philips 6500 530. Tape (c) 7300 024, 7300 276.

    +-Gr 5-75 p1961   ++NR 12-74 p5

    +-Gr 5-75 p2032 tape   ++RR 4-75 p25

    +HF 12-74 p100   ++RR 6-75 p90 tape

    ++HFN 6-75 p109 tape   /SFC 11-17-74 p31

    ++MJ 2-75 p40   ++St 2-75 p83

566 Concerto, violin, op. 77, D major. Hermann Krebbers, vln; COA;
Bernard Haitink. Philips 6580 087.

    +-RR 3-75 p24

567 Concerto, violin, op. 77, D major. BEETHOVEN: Romance, no. 2, op.
50, F major. Yehudi Menuhin, vln; Lucerne Festival Orchestra;

PhO; Wilhelm Furtwängler. Seraphim 60232. (Reissue from HMV
Victor originals, 1949, 1953)
          +HF 2-75 p80   +St 2-75 p84
Concerto, violin, op. 77, D major. cf HMV SLS 5004.
Concerto, violin, op. 77, D major. cf RCA ARM 4-0942/7.
Concerto, violin, op. 77, D major. cf RCA CRL 6-0720.
568 Concerto, violin and violoncello, op. 102, A minor. Tragic overture, op. 81. David Oistrakh, vln; Pierre Fournier, vlc; PhO;
Alceo Galliera. HMV SXLP 30185. (Reissue from Columbia SAX
2264)
          +-Gr 9-75 p452   -RR 8-75 p30
          +-HFN 9-75 p108
569 Concerto, violin and violoncello, op. 102, A minor. Academic
festival overture, op. 80. Alfredo Campoli, vln; André Navarra,
vlc; Hallé Orchestra; John Barbirolli. Pye GSGC 15016. (Reissue from GSGC 14009)
          -Gr 8-75 p335   -RR 8-75 p30
          +-HFN 8-75 p87
570 Concerto, violin and violoncello, op. 102, A minor. SCHUMANN:
Fantasy, violin and orchestra, op. 131, C major. Ruggiero
Ricci, vln; Giorgio Ricci, vlc; NPhO; Leipzig Gewandhaus
Orchestra; Kurt Masur. Turnabout TVS 34593.
          +-NR 10-75 p3   ++SFC 8-24-75 p28
Concerto, violin, violoncello and orchestra, op. 102, A minor.
cf RCA ARM 4-0942/7.
571 Deutsche Volkslieder (42). Elisabeth Schwarzkopf, s; Dietrich
Fischer-Dieskau, bar; Gerald Moore, pno. Angel B 3675 (2).
          +St 4-75 p68
572 Hungarian dances (16). LSO; Antal Dorati. Mercury SRI 75024.
Tape (r) L 45024. (Reissue from SR 90437)
          *HF 3-74 p98   +SFC 6-29-75 p26 tape
          +SFC 8-11-74 p31   +St 9-74 p113
573 Hungarian dances (21). Michel Beroff, Jean-Philippe Collard, pno.
Connoisseur Society CS 2083.
          ++SFC 10-12-75 p22
574 Hungarian dances, nos. 1-3, 5-6, 10-21. PSO; Dean Dixon. Supraphon 110 1206. Tape (c) 041206.
          +Gr 12-73 p1192   +-RR 12-73 p58
          /HFN 12-73 p2607   /RR 6-75 p90 tape
          /HFN 5-75 p142 tape   +SFC 5-19-74 p29
          -NR 5-74 p4   +St 9-74 p113
575 Hungarian dances, nos. 1, 3, 5, 6. LISZT: Hungarian fantasia,
G 123. Hungarian rhapsodies, nos. 4, 5, G 244. Shura Cherkassky,
pno; BPhO; Herbert von Karajan. DG Tape (c) 3318 005.
          -RR 1-75 p69 tape
576 Hungarian dances, nos. 1, 3, 5-6, 10, 12-13, 19, 21. DVORAK:
Slavonic dances, op. 46, nos. 1, 5-6, 8; op. 72, no. 2, E minor.
LSO; Willi Boskovsky. Decca SXL 6696.
          ++HFN 6-75 p87   /RR 6-75 p33
Hungarian dance, no. 1, G minor. cf BEETHOVEN: Concerto, violin,
op. 61, D major.
Hungarian dance, no. 1, G minor. cf Discopaedia MB 1003.
Hungarian dance, no. 1, G minor. cf Discopaedia MB 1007.
Hungarian dance, no. 1, G minor. cf RCA ARM 4-0942/7.
Hungarian dance, no. 5, G minor. cf Rococo 2035.
577 Hungarian dances, nos. 5 and 6. DVORAK: Slavonic dances, op. 46,

nos. 1-3, op. 72, no. 2, E minor. GRIEG: Norwegian dances,
op. 35, nos. 1-4. SMETANA: The bartered bride: Polka, Furiant.
PhO; Walter Susskind, Charles Mackerras. Classics for Pleasure
CFP 40214.
  +HFN 8-75 p87   +RR 6-75 p41
Hungarian dance, no. 17 (arr. Kreisler). cf Decca SPA 405.
Hungarian dances, nos. 20 and 21. cf Discopaedia MB 1008.
Intermezzi, op. 116. cf Capriccio, op. 116.
Intermezzo, op. 116, no. 4, E major. cf Columbia Special Products
AP 12411.
Intermezzi, op. 117, nos. 1-3. cf DG Heliodor 2548 137.
Intermezzo, op. 117, no. 1, E flat major. cf Sonata, piano, no.3,
op. 5, F minor.
Intermezzo, op. 117, no. 2, B flat minor. cf Decca SPA 372.
Intermezzo, op. 119, no. 3, C major. cf Sonata, piano, no. 3,
op. 5, F minor.
Liebeslieder waltzes, op. 52. cf BEETHOVEN: Egmont overture, op.
84.

Low, how a rose e'er blooming. cf University of Colorado, unnumbered.
578 Motets: Warum ist das Licht gegeben, op. 74, no. 1. BRUCKNER:
Motets: Ave Maria; Christus factus est; Locus iste a Deo factus
est; Os justi; Virga Jesse. VERDI: Motets: Ave Maria; Laudi
alla Vergine Maria; Pater Noster. Saltarello Choir; Richard
Bradshaw. CRD CRD 1009.
  +Gr 4-75 p1859   +RR 3-75 p60
579 Motets: Es ist das Heil uns kommen her, op. 29, no. 1. Schaffe in
mir, Gott, op. 29, no. 2. Warum ist das Licht gegeben dem
Mühseligen, op. 74, no. 1. SCHUBERT: Motets: Christ ist erstan-
den, D 440. Gebet, D 815. Gott im Ungewitter D 985. Psalm,
no. 23: Gott ist mein Hirt, D 706. Cambridge, Kings College
Choir; Philips Ledger. HMV ASD 3091
  +Gr 8-75 p352   +HFN 9-75 p97
580 Quartets, piano and strings, nos. 1-3, opp. 25, 26, 60. Beaux
Arts Trio; Walter Trampler, vla. Philips 6747 068 (3).
  +-Audio 12-75 p105   +NR 5-75 p8
  -HF 9-75 p83   ++SFC 8-24-75 p28
  ++MJ 10-75 p39   ++St 10-75 p106
Quartets, piano and strings, nos. 1-3, opp. 25, 26, 60. cf Works,
selections (DG 2740 117).
581 Quartet, piano and strings, no. 1, op. 25, G minor. Quartet, piano
and strings, no. 2, op. 26, A major. Rudolf Serkin, pno;
Busch Quartet, Members. DaCapo 1C 147 01555/6 (2). (Reissues
from Columbia ML 4296, HMV/Victor 78s)
  +HF 9-75 p83
Quartet, piano and strings, no. 2, op. 26, A major. cf Quartet,
piano and strings, no. 1, op. 25, G minor.
Quartets, strings, nos. 1-3, op. 51, nos. 1 and 2; op. 67. cf
Works, selections (DG 2740 117).
582 Quintet, clarinet, op. 115, B minor. Jack Brymer, clt; Prometheus
Ensemble. Pye GSGC 15004. (Reissue from Virtuoso TPLS 13004)
  +-Gr 8-75 p335   -RR 8-75 p46
  -HFN 8-75 p89
Quintet, clarinet, op. 115, B minor. cf Works, selections (DG 2740
117).
Quintet, piano, op. 34, F minor. cf Works, selections (DG 2740 117).
Quintets, strings, nos. 1 and 2, opp. 88, 111. cf Works, selections
(DG 2740 117).

Rhapsody, op. 79, no. 2, G minor.   cf Sonata, piano, no, 3, op. 5,
    F minor.
Rhapsodie, op. 79, no. 2, G minor.   cf HMV HQS 1353.
Sapphische ode, op. 94, no. 4.   cf Rubini RS 300.
583 Serenade, no. 2, op. 16, A major.   DVORAK: Serenade, op. 44, D
    minor. LSO; István Kertész. Decca SXL 6368. Tape (c)
    KSXC 6368. (also London 6594)
        +Gr 2-75 p1562 tape   ++RR 2-75 p76 tape
    Sextets, strings, nos. 1 and 2, opp. 18 and 36. cf Works, selec-
    tions (DG 2740 117).
    Sonatas, clarinet, nos. 1 and 2, op. 120, nos. 1 and 2.   cf Works,
    selections (DG 2740 117).
584 Sonata, piano, no. 2, op. 2, F sharp minor.   SCHUMANN: Sonata,
    piano, no. 1, op. 11, F sharp minor. István Antal, pno. Hun-
    garoton SLPX 11647.
        -HFN 10-75 p140   -RR 4-75 p54
585 Sonata, piano, no. 3, op. 5, F minor.   Rhapsody, op. 79, no. 2,
    G minor. Bruno-Leonardo Gelber, pno. Connoisseur Society
    2084. (Q) CSQ 2084.
        ++NR 11-75 p14   ++SFC 11-16-75 p32
586 Sonata, piano, no. 3, op. 5, F minor.   Intermezzo, op. 117, no. 1,
    E flat major; op. 119, no. 3, C major. Clifford Curzon, pno.
    London STS 15272. (Reissue from CS 6341)
        ++HF 6-75 p87   ++St 7-75 p95
587 Sonata, viola and piano, no. 1, op. 120, F minor.   Sonata, viola
    and piano, no. 2, op. 120, E flat major. Rainer Moog, vla;
    Werner Genuit, pno. BASF 2521 971/2.
        +Gr 10-75 p640   +RR 11-75 p74
        +HFN 10-75 p139
    Sonatas, viola and piano, nos. 1 and 2, op. 120. cf Sonatas,
    violin, nos. 1-3.
    Sonata, viola and piano, no. 2, op. 120, E flat major. cf Sonata,
    viola and piano, no. 1, op. 120, F minor.
588 Sonatas, violin and piano, nos. 1-3.   Sonatas, viola and piano,
    nos. 1 and 2, op. 120. Pinchas Zukerman, vln and vla; Daniel
    Barenboim, pno. DG 2709 058 (3).
        +-Gr 11-75 p847   +-RR 11-75 p74
        ++HFN 11-75 p152
    Sonatas, violin and piano, nos. 1-3. cf BEETHOVEN: Sonata, violin
    and piano, no. 5, op. 24, F major.
    Sonatas, violin and piano, nos. 1-3, opp. 78, 100, 108. cf Works,
    selections (DG 2740 117).
    Sonata, violin and piano, no. 2, op. 100, A major. cf RCA ARM
    4-0942/7.
    Sonata, violin and piano, no. 3, op. 108, D minor. cf RCA ARM
    4-0942/7.
589 Sonata, violoncello, no. 1, op. 38, E minor.   CHOPIN: Polonaise
    brilliante, op. 3. DVORAK: Rondo, op. 94, G minor. SCHUMANN:
    Fantasiestücke, op. 73. Marius May, vlc; Paul Hamburger, pno.
    Decca SDD 447.
        +/Gr 2-75 p1509   +-RR 2-75 p50
    Sonatas, violoncello, nos. 1 and 2, opp. 38, 99. cf Works,
    selections (DG 2740 117).
590 Songs: Ach, wende diesen Blick, op. 57, no. 4; Am Sonntag Morgen,
    op. 49, no. 1; Es steht ein Lind in jenem Tal; Es traümte mir,
    op. 57, no. 3; Das Mädchen spricht, op. 107, no. 3; Mädchenlied,

op. 107, no. 5; O kühler Wald, op. 72, no. 3; Ständchen, op. 106,
no. 1; Umbewegte, laue Luft, op. 57, no. 8; Vergebliches Ständchen, op. 84, no. 4; Von ewiger Liebe, op. 43, no. 1; Während
des Regens, op. 58, no. 2; Wenn du nur zuweilen lächelst, op. 57,
no. 2. Elly Ameling, s; Norman Shetler, pno. BASF KHG 21021.
(also BASF BAC 3065)
   +–Gr 1-75 p1376   ++RR 1-75 p55
   +–HF 8-73 p78   +SFC 6-17-73 p31
    +NR 6-73 p11   +–St 7-73 p98
   +–ON 4-21-73 p32

591 Songs: Children's folk songs: Beim Ritt auf dem Knie; Dornröschen;
Heidenröslein; Die Henne; Der Jäger im Walde; Das Mädchen und
der Hasel; Der Mann; Marienwürmchen; Die Nachtigall; Sandmännchen;
Das Scharaffenland; Dem Schutzengel; Wiehnachten; Wiegenlied.
German folk songs: Des Abends kann ich; Ach, englische Schäferin;
Ach Gott, wie weh tut Scheiden; Ach könnt ich diesen Abend;
All mein Gedanken; Da unten in Tale; Dort in den Weiden; Du mein
en einzig Licht; Erlaube mir, feins Mädchen; Es ging ein Maidlein
zarte; Es reit ein Herr und auch sein Knecht; Es ritt ein Ritter;
Es steht ein Lind; Es war ein Markgraf über Rhein; Es war eine
schöne Jüdin; Es wohnet ein Fiedler; Feinsliebchen; Gar lieblich
hat sich gesellet; Gunhilde; Guten Abend; Ich stand auf hohem
Berge; Ich wiess mir'n Baidlein; In stiller Nacht; Lungfräulein,
soll ich mit euch gehn; Maria ging aus wandern; Mein Mädel hat
einem Rosenmund; Mir ist ein schöns brauns Madelein; Nur ein
Gesicht auf Erden lebt; Och Moder, ich well; Der Reiter Spreitet
seinen; Sagt mir, o schönste Schäffrin mein; Schöner Augen;
Schönster Schatz, mein Engel; Schwesterlein; So will ich frisch;
So wunsch ich ihr ein gute Nacht; Soll ich der Mond; Die Sonne
scheint nicht mehr; Wach auf mein Herzenschöne; Wach auf, mein
Hort; Wie komm ich; Wo gehst du hin, du Stolze. German folk
songs for chorus: Abschiedlied; Bei nächtlicher Weil; Der Englische
Jäger; In stiller Nacht; Mit Lust tät ich ausreiten; Morgengesang;
Schnitter Tod; Von edler Art; Die Wollust in den Maien. Edith
Mathis, s; Peter Schreier, t; Karl Engel, pno; Leipzig Radio
Chorus; Horst Neumann. DG 2709 057 (3).
   +Gr 10-75 p662   +ON 12-20-75 p38
   +HFN 10-75 p139   +RR 10-75 p81

592 Songs: Der Gang zum Liebchen, op. 31, no. 3; Der Tod ist die kühle
Nacht, op. 96, no. 1; Thérèse, op. 86, no. 1; Zigeunerlieder,
op. 103. SCHUBERT: Songs: Dithyrambe, op. 60, no. 2; Ellens
zweiter Gesang, op. 52; Schwanengesang, no. 11: Die Stadt;
Wiegenlied. WOLF: Songs: Italieniescher Liederbuch: Der Mond
hat eine schwere Klag erhoben; Und willst du deinen Liebsten.
Elena Gerhardt, s; Gerald Moore, pno. Discophilia KGG 4.
   +NR 9-75 p11   +–ON 5-75 p36

593 Songs: Es ist das Heil uns kommen her, op. 29, no. 1; Schaffe in
mir, Gott, op. 29, no. 2; Warum ist das Licht gegeben dem
Mühseligen, op. 74, no. 1. SCHUBERT: Songs: Christ ist erstanden, S 440; Gebet, D 815; Gott im Ungewitter, D 985; Psalm, no.
23, Gott ist mein Hirt, D 706. Cambridge, King's College Choir;
Philip Ledger, pno; Philip Ledger. HMV KASD 3091
   +–RR 9-75 p71

594 Songs: Abenddämmerung, op. 49, no. 5; Abendregen, op. 70, no. 4;
Abschied, op. 69, no. 3; Ach, wende diesen Blick, op. 57, no. 4;
Ade, op. 85, no. 4; Alte Liebe, op. 72, no. 1; Am Sonntag Morgen,
op. 49, no. 1; An den Mond, op. 71, no. 2; An die Nachtigall,

op. 46, no. 4; An die Stolze, op. 107, no. 1; An die Tauben, op. 63, no. 4; An ein Bild, op. 63, no. 3; An ein Veilchen, op. 49, no. 2; An eine Aeolsharfe, op. 19, no. 5; Anklange, op. 7, no. 3; Auf dem Kirchhof, op. 105, no. 4; Auf dem Schiffe, op. 97, no. 2; Auf dem See, op. 59, no. 2; Auf dem See, op. 106, no. 2; Bei dir sind meine Gedanken, op. 95, no. 2; Beim Abschied, op. 95, no. 3; Bitteres zu sagen, op. 32, no. 7; Blinke Kuh, op. 58, no. 1; Botschaft, op. 47, no. 1; Dämmerung senkte sich von oben, op. 59, no. 1; Dein blaues Auge, op. 59, no. 8; Die Dranze, op. 46, no. 1; Du sprichst, dass ich mich täuschte, op. 32, no. 6; Eine gute, gute Nacht, op. 59, no. 6; Entführung, op. 97, no. 3; Erinnerung, op. 63, no. 2; Es hing der Reif, op. 106, no. 3; Es liebt sich so lieblich, op. 71, no. 1; Es schauen die Blumen, op. 96, no. 3; Es träumte mir, op. 57, no. 3. Feldeinsamkeit, op. 86, no. 2; Der Frühling, op. 6, no. 2; Frühlingslied, op. 85, no. 5; Frühlingstrost, op. 63, no. 1; Der Gang zum Liebchen, op. 48, no. 1; Gang zur Liebsten, op. 14, no. 6; Geheimnis, op. 71, no. 3; Heimkehr, op. 7, no. 6; Heimweh, I-III, op. 63, nos. 7-9; Herbstgefühl, op. 48 no. 7; Ich schell mein Horn, op. 43, no. 3; Ich schleich umher betrübt, op. 32, no. 3; Im Garten, op. 70, no. 1; In der Ferne, op. 19, no. 3; In der Fremde, op. 3, no. 5; In der Gasse, op. 58, no. 6; In meiner Nächte Sehnen, op. 57, no. 5; In Waldeseinsamkeit, op. 85, no. 6; Juchhe, op. 6, no. 4; Junge Lieder, I and II, op. 63, nos. 5-6; Kein Haus, keine Heimat, op. 94, no. 5; Klage, op. 105, no. 3; Komm bald, op. 97, no. 5; Die Kränze, op. 46, no. 1; Der Kuss, op. 19, no. 1; Lerchengesang, op. 70, no. 2; Liebe und Frühling, I and II, op. 3, nos. 2-3; Lied, op. 3, nos. 4, 6; Liebesglut, op. 47, no. 2; Mädchenlied, op. 107, no. 5; Magyarisch, op. 46, no. 2; Maienkätzchen, op. 107, no. 4; Die Mainacht, op. 43, no. 2; Meerfahrt, op. 96, no. 4; Mein Herz ist schwer, op. 94, no. 3; Mein wundes Herz verlangt nach Dir, op. 59, no. 7; Meine Lieder, op. 106, no. 4; Minnelied, op. 71, no. 5; Mit vierzig Jahren, op. 94, no. 1; Mondenschein, op. 85, no. 2; Murrays Ermordung, op. 14, no. 3; Nachklang, op. 59, no. 4; Nachtigall, op. 97, no. 1; Nachtigallen schwingen, op. 6, no. 6; Nachtwandler, op. 86, no. 3; Nachwirkung, op. 6, no. 3; Nicht mehr zu dir zu gehen, op. 32, no. 2; O liebliche Wangen, op. 47, no. 4; O komme, holde Sommernacht, op. 58, no. 4; O kühler Wald, op. 72, no. 3; Parole, op. 7, no. 2; Regenlied, op. 59, no. 3; Der Salamander, op. 107, no. 2; Die Schale der Vergangenheit, op. 46, no. 3; Scheiden und Meiden, op. 19, no. 2; Die Schnur, Perl an Perle, op. 57, no. 7; Schön war das ich dir weihte, op. 95, no. 7; Schwermut, op. 58, no. 5; Sehnsucht, op. 14, no. 8; Sehnsucht, op. 49, no. 3; Serenade, op. 58, no. 8; Serenade, op. 70, no. 3; So stehn wir, op. 32, no. 8; Sommerabend, op. 85, no. 1; Sommerfäden, op. 72, no. 2; Ein Sonett, op. 14, no. 4; Sonntag, op. 47, no. 3; Die Spröde, op. 58, no. 3; Ständchen, op. 14, no. 7; Ständchen, op. 106, no. 1; Steig auf geliebter Schatten, op. 94, no. 2; Strahlt zuweilen auch ein mildes Licht, op. 57, no. 6; Der Strom, der neben mir verrauschte, op. 32, no. 4; Tambourliedchen, op. 69, no. 5; Therese, op. 86, no. 1; Der Tod, das ist die kühle Nacht, op. 96, no. 1; Todessehnen, op. 86, no. 6; Trennung, op. 14, no. 5; Trennung, op. 97, no. 6; Treue Liebe, op. 7, no. 1; Trost in Tränen, op. 48, no. 5; Uber die Heide, op. 86, no. 4; Uber die

See, op. 69, no. 7; Unbewegteblaue Luft, op. 57, no. 8; Der
Unberläufer, op. 48, no. 2; Unuberwindlich, op. 72, no. 5;
Vergangen ist mir Glück und Heil, op. 48, no. 6; Verrat, op.
105, no. 5; Versunken, op. 86, no. 5; Verzagen, op. 72, no.
4; Vier ernste Gesänge, op. 121; Volkslied, op. 7, no. 4; Vom
verwundeten Knaben, op. 14, no. 2; Von ewige Liebe, op. 43,
no. 1; Vor dem Fenster, op. 14, no. 1; Vorüber, op. 58, no. 7;
Während des Regens, op. 58, no. 2; Ein Wanderer, op. 106, no.
5; Wehe so willst du mich wieder, op. 32, no. 5; Wenn du nur
zuweilen lächelst, op. 57, no. 2; Wie bist du meine Königin,
op. 32, no. 9; Wie die Wolke nach der Sonne, op. 6, no. 5;
Wie Melodien zieht es, op. 105, no. 1; Wie rafft ich mich auf,
op. 32, no. 1; Wiegenlied, op. 49, no. 4; Willst du dass ich
geh, op. 71, no. 4; Wir wandelten, op. 96, no. 2. Dietrich
Fischer-Dieskau, bar; Gerald Moore, Wolfgang Sawallish, Daniel
Barenboim, pno. HMV SLS 5002 (7). (Some reissues from ASD 630)
        +Gr 2-75 p1524   +RR 3-75 p56
        +-MT 7-75 p628
595 Songs: Die Meere, op. 20, no. 3; Phänomen, op. 61, no. 4; Weg
    der Liebe, op. 20, no. 1. DVORAK: Songs: Der Apfel; Der
    kleine Acker; Möglichkeit; Der Ring; Die Taube auf dem Ahorn.
    MENDELSSOHN: Songs: Abendlied; Herbstlied, op. 63, no. 4;
    Maiglöckchen und Blümelein, op. 63, no. 6. SCHUMANN: Songs:
    Herbstlied, op. 43,no. 2; Schön Blümelein, op. 43, no. 3; So
    wahr die Sonne scheinet, op. 37. Herrad Wehrung, s; Traugott
    Schmohl, bar; Karl-Michael Komma, pno. Musical Heritage
    Society MHS 1896.
        +St 8-75 p109
Songs, alto, piano and viola, op. 91 (2). cf Classic Record
    Library SQM 80-5731.
Songs: We love the place, O God. cf Argo ZRG 785.
Songs: Klänge, op. 66; Klosterfräulein, op. 61, no. 2; Phänomen,
    op. 61, no. 3; Walpurgisnacht, op. 75, no. 4; Weg der Liebe,
    op. 20, nos. 1 and 2. cf Columbia M 33307.
Songs: Botschaft, op. 47, no. 1; Immer leiser wird mein Schlummer,
    op. 105, no. 2; Der Tod das ist die kühle Nacht, op. 96, no.
    1; Von ewiger Liebe, op. 43, no. 1. cf Decca 6BB 197/8.
Songs: Sister dear. cf HMV RLS 714.
Songs: Ständchen, op. 106, no. 1. cf RCA SER 5704/6.
596 Symphonies, nos. 1-4. BPhO; Herbert von Karajan. DG 2721 075 (4).
    Tape (c) 3371 015. (Reissue from SKL 133/9)
        +Gr 7-74 p199   ++RR 2-75 p75 tape
        +RR 7-75 p35
597 Symphonies, nos. 1-4. Academic festival overture, op. 80.
    Tragic overture, op. 81. Janet Baker, ms; John Alldis Choir;
    LPO, LSO; Adrian Boult. HMV SLS 5009 (4). (Reissues from
    ASD 2660, 2746, 2871, 2901)
        +Gr 6-75 p35   ++HFN 5-75 p125
Symphonies, nos. 1-4. cf Works, selections (HMV SLS 5009).
598 Symphony, no. 1, op. 68, C minor. Hallé Orchestra; James Loughran.
    Classics for Pleasure CFP 40096.
        +Gr 3-75 p1648   +RR 3-75 p24
599 Symphony, no. 1, op. 68, C minor. OSR; Ernest Ansermet. Decca
    SPA 378.
        +-HFN 12-75 p171   +RR 12-75 p44
600 Symphony, no. 1, op. 68, C minor. LSO; Leopold Stokowski. Decca
    PFS 4305. Tape (ct) KPFC 4305. (Reissue from OPFS 3/4)

(also London SPC 21131. Tape (c) 521131 (ct) 821131)
   +–Gr 12–74 p1125   +–RR 10–74 p38
   +Gr 12–74 p1237 tape   ++St 6–75 p95
601 Symphony, no. 1, op. 68, C minor. BPhO; Karl Böhm. DG 2535 102.
     Tape (c) 3335 102. (Reissue from SLPM 138113)
   ++Gr 7–75 p175   ++RR 7–75 p23
   +HFN 8–75 p87   +RR 10–75 p97 tape
602 Symphony, no. 1, op. 68, C minor. VPO; István Kertész. London
     CS 6836. (also Decca SXL 6675. Reissue from SXLH 6610/3)
   +–Gr 1–75 p1341   +RR 1–75 p27
   ++HF 3–75 p77   ++SFC 5–4–75 p35
   +NR 6–75 p4   /St 6–75 p95
     Symphony, no. 1, op. 68, C minor. cf Camden CCV 5000–12, 5014–24.
603 Symphony, no. 2, op. 73, D major. Hallé Orchestra; James Loughran.
     Classics for Pleasure CFP 40219.
   +Gr 11–75 p808   +RR 12–75 p45
   +HFN 11–75 p152
604 Symphony, no. 2, op. 73, D major. Tragic overture, op. 81. OSR;
     Ernest Ansermet. Decca SPA 379.
   +HFN 12–75 p171   +RR 12–75 p44
605 Symphony, no. 2, op. 73, D major. VPO; István Kertész. Decca
     SXL 6676. (Reissue from SXL 6172) (also London 6435)
   ++Gr 2–75 p1474   +RR 2–75 p30
606 Symphony, no. 2, op. 73, D major. Variations on a theme by Haydn,
     op. 56a. COA; Bernard Haitink. Philips 6500 375. Tape (c)
     7300 375.
   +Gr 7–75 p175   ++NR 11–75 p2
   +–HF 12–75 p88   +RR 7–75 p23
   +–HFN 7–75 p77
607 Symphony, no. 2, op. 73, D major. NBC Symphony Orchestra; Arturo
     Toscanini. RCA AT 132. (Reissue from HMV ALP 1013)
   +Gr 8–75 p336   ++RR 6–75 p33
     Symphony, no. 2, op. 73, D major. cf HMV RLS 717.
608 Symphony, no. 3, op. 90, F major. Variations on a theme by Haydn,
     op. 56a. OSR; Ernest Ansermet. Decca SPA 380.
   +HFN 12–75 p171   +RR 12–75 p44
609 Symphony, no. 3, op. 90, F major. Tragic overture, op. 81. BPhO;
     Lorin Maazel. DG 2548 192. (Reissue from SLPM 138022)
   +–Gr 11–75 p808   +–RR 9–75 p34
   +–HFN 9–75 p108
610 Symphony, no. 3, op. 90, F major. Akademische Festouverture,
     op. 80. HRT Orchestra; Tamás Pál. Hungaroton SLPX 11734.
   –NR 11–75 p2
611 Symphony, no. 3, op. 90, F major. Variations on a theme by Haydn,
     op. 56a. VPO; István Kertész. London CS 6837. (also Decca
     SXL 6677. Reissue from SXLH 6610/3)
   +–Audio 9–75 p69   +MJ 7–75 p34
   ++Gr 4–75 p1802   +NR 6–75 p4
   ++HF 6–75 p88   ++RR 4–75 p25
612 Symphony, no. 3, op. 90, F major. Tragic overture, op. 81. COA;
     Willem Mengelberg. Rococo 2051.
   –NR 2–75 p3
613 Symphony, no. 4, op. 98, E minor. Academic festival overture,
     op. 80. LPO; Adrian Boult. Angel S 37304.
   +HF 6–75 p88   +–NR 6–75 p4
614 Symphony, no. 4, op. 98, E minor. Munich Philharmonic Orchestra;

Rudolf Kempe. BASF BAC 3064.
+Gr 11-75 p808   +RR 10-75 p38
+-HFN 10-75 p139

615 Symphony, no. 4, op. 98, E minor. BSO; Charles Munch. Camden
CCV 5032.
-RR 10-75 p38

616 Symphony, no. 4, op. 98, E minor. Academic festival overture,
op. 80. OSR; Ernest Ansermet. Decca SPA 381.
+HFN 12-75 p171   +RR 12-75 p44

617 Symphony, no. 4, op. 98, E minor. VPO; István Kertész. London
CS 6838. (also Decca SXL 6678)
+HF 6-75 p88   +-RR 8-75 p31
+HFN 8-75 p87   +St 6-75 p95
+NR 6-75 p4

618 Symphony, no. 4, op. 98, E minor. COA; Bernard Haitink. Philips
6500 389. Tape (c) 7300 238.
+-Gr 10-73 p678   +RR 10-73 p61
+HF 7-73 p78   +-RR 9-75 p79 tape
+NR 6-73 p2   +-SFC 4-22-73 p29

619 Symphony, no. 4, op. 98, E minor. Hallé Orchestra; John Barbirolli.
Pye GSGC 15004. (Reissue from GSGC 14037)
+Gr 8-75 p335   +RR 7-75 p23
+-HFN 8-75 p87

620 Symphony, no. 4, op. 98, E minor. Academic festival overture,
op. 80. NPhO; Leopold Stokowski. RCA ARL 1-0719. Tape (c)
ARK 1-0719 (ct) ARS 1-0719 (Q) ARD 1-0719 Tape (ct) ART 1-0719.
+Gr 12-75 p1031   +NR 11-75 p2
+HF 12-73 p88   +-RR 11-75 p42
+-HFN 11-75 p152

Tragic overture, op. 81. cf Concerto, violin and violoncello,
op. 102, A minor.
Tragic overture, op. 81. cf Symphonies, nos. 1-4.
Tragic overture, op. 81. cf Symphony, no. 2, op. 73, D major.
Tragic overture, op. 81. cf Symphony, no. 3, op. 90, F major.
Tragic overture, op. 81. cf Symphony, no. 3, op. 90, F major
(Rococo 2051).
Tragic overture, op. 81. cf Works, selections (HMV SLS 5009).

621 Trio, clarinet, violoncello and piano, op. 114, A minor. Ralph
McLane, clt; Sterling Hunkins, vlc; Milton Kaye, pno. Grena-
dilla GS 1002. (Reissue)
+NR 12-75 p9

Trio, clarinet, violoncello and piano, op. 114, A minor. cf
Works, selections (DG 2740 117).

622 Trio, horn, violin and piano, op. 40, E flat major. FRANCK: Sonata,
violin and piano, A major. Itzhak Perlman, vln; Barry Tuckwell,
hn; Vladimir Ashkenazy, pno. Decca SXL 6408. Tape (c)
KSXC 6408.
+RR 3-75 p74 tape

623 Trio, horn, violin and piano, op. 40, E flat major. DUVERNOY:
Trio, horn, no. 1. DUKAS: Villanelle. SCHUMANN: Adagio and
allegro, op. 70, A flat major. Adám Friedrich, hn; Miklós
Szenthelyi, vln; Sándor Falvai, pno. Hungaroton SLPX 11672.
+-Gr 5-75 p1991   +-RR 4-75 p39
+-HFN 6-75 p103

Trio, horn, violin and piano, op. 40, E flat major. cf Works,
selections (DG 2740 117).

624 Trios, piano, nos. 1-3. Henryk Szeryng, vln; Pierre Fournier;
    vlc; Artur Rubinstein, pno. RCA LRL 2-7528 (2).
        -Gr 7-75 p205   +-RR 8-75 p51
        ++HFN 8-75 p73
625 Trios, piano, nos. 1-3. SCHUMANN: Trio, piano, no. 1, op. 63, D
    minor. Henryk Szeryng, vln; Pierre Fournier, vlc; Artur
    Rubinstein. RCA ARL 3-0138 (3).
        ++HF 1-75 p76   ++SFC 12-22-74 p24
        ++NR 12-74 p7   ++St 1-75 p109
    Trios, piano, nos. 1-3. cf Works, selections (DG 2740 117).
    Trio, piano, no. 1, op. 8, B major. cf BEETHOVEN: Trio, piano,
    no. 6, op. 97, B flat major.
626 Trio, piano, no. 2, op. 87, C major. SCHUBERT: Fantasia, D 934,
    C major. Rondo brillant, op. 70, D 895, B minor. Sonata,
    violin and piano, op. 162, D 574, A major. Sonatina, violin
    and piano, no. 1, D 384, D major. SCHUMANN: Trio, piano, no.
    1, op. 63, D minor. Joseph Szigeti, vln; Myra Hess, Mieczyslaw
    Horszowski, Andor Foldes, Carlo Bussotti, Joseph Levine, pno;
    Pablo Casals, Rudolf von Tobel, vlc. Bruno Walter Society WAS
    714 (2).
        +-HF 1-75 p76
627 Variations on a theme by Haydn, op. 56a. ELGAR: Enigma variations,
    op. 36. LSO; Eugen Jochum. DG 2530 586. Tape (c) 3300 586.
        +Gr 12-75 p1031   +RR 12-75 p48
        +-HFN 12-75 p151
628 Variations on a theme by Haydn, op. 56a. MOZART: Sonata, 2 pianos,
    K 448 (375a), D major. RAVEL: Ma mere l'oye. Dezső Ranki,
    Zoltán Kocsis, pno. Hungaroton SLPX 11646.
        +-Gr 2-75 p1515   +-RR 1-75 p46
    Variations on a theme by Haydn, op. 56a. cf Symphony, no. 2,
    op. 73, D major.
    Variations on a theme by Haydn, op. 56a. cf Symphony, no. 3, op.
    90, F major (Decca 380).
    Variations on a theme by Haydn, op. 56a. cf Symphony, no. 3, op.
    90, F major (London CS 6837).
    Variations on a theme by Haydn, op. 56a. cf ELGAR: Enigma varia-
    tions, op. 36.
629 Variations on a theme by Paganini, op. 35, A minor. LISZT: Czárdás
    obstiné, G 225. Transcendental etudes, nos. 4, 6, 10-12, G 139.
    Gabriella Torma, pno. Hungaroton SLPX 11572.
        +-NR 5-75 p13
630 Variations on a theme by Paganini, op. 35, A minor. SCHUMANN:
    Carnaval, op. 9. Cécile Oussett, pno. Decca SDDR 477.
        +-Gr 8-75 p347   ++RR 7-75 p53
        +-HFN 12-75 p149
631 Variations on a theme by Paganini, op. 35, A minor. DEBUSSY: Pre-
    ludes, Bk I, no. 10: La cathédrale engloutie; no. 1: Danseuses
    de Delphes; no. 12: Minstrels; no. 4: Les sons et les parfums
    tournent dans l'air du soir. SCHONBERG: Impressions, op. 38.
    SEROCKI: Suita preludiow. Elzbieta Glabowna, pno. Caprice
    RIKS LP 66.
        +RR 10-75 p75
    Variations on a theme by Paganini, op. 35, A minor. cf BACH:
    Chaconne (Busoni).
    Waltz, op. 39, no. 15, A flat major. cf Turnabout TV 37033.

632 Works, selections: Quartets, piano, nos. 1-3, opp. 25, 26, 60.
Quartets, strings, nos. 1-3, op. 51, nos. 1 and 2; op. 67.
Quintet, clarinet, op. 115, B minor. Quintet, piano, op. 34,
F minor. Quintets, strings, nos. 1 and 2, opp. 88, 111. Sex-
tets, strings, nos. 1 and 2, opp. 18 and 36. Sonatas, clarinet,
nos. 1 and 2, opp. 120, nos. 1 and 2. Sonatas, violin, nos. 1-3,
opp. 78, 100, 108. Sonatas, violoncello, nos. 1 and 2, opp.
38, 99. Trio, clarinet, violoncello and piano, op. 114, A minor.
Trio, horn, violin and piano, op. 40, E flat major. Trios,
piano, nos. 1-3, opp. 8, 87, 101. Cecil Aronowitz, Stefano
Passaggio, vla; William Pleeth, Georg Donderer, Pierre Fournier,
vlc; Christoph Eschenbach, Jörg Demus, Pierre Barbizet, Rudolf
Firkušny, pno; Karl Leister, clt; Eduard Drolc, Christian
Ferras, vln; Amadeus Quartet; Trieste Trio. DG 2740 117 (15).
          +-HFN 9-75 p109   +-RR 8-75 p46
633 Works, selections: Academic festival overture, op. 80. Alto
rhapsody, op. 53. Symphonies, nos. 1-4. Tragic overture, op.
81. Janet Baker, ms; John Alldis Choir; LPO, LSO; Adrian Boult.
HMV SLS 5009 (4).
          ++RR 6-75 p33
BRANDL
The old refrain (arr. Kreisler). cf Decca SPA 405.
BRASSART, Jean
O flos fragrans. cf Telefunken ER 6-35257.
BRIAN, Havergal
634 English suite, no. 5. Psalm, no. 23. Symphony, no. 22. Paul
Taylor, t; Brighton Festival Chorus; Leicestershire Schools
Symphony Orchestra; Laszlo Heltay, Eric Pinkett. CBS 61612.
          +Gr 2-75 p1474   +St 6-75 p145
          +MT 9-75 p797   +Te 12-75 p45
          +RR 2-75 p30
Psalm, no. 23. cf English suite, no. 5.
635 Symphonies, nos. 6, 16. LPO; Myer Fredman. Lyrita SRCS 67.
          +Gr 5-75 p1962   ++RR 5-75 p28
          ++HFN 5-75 p125   ++Te 12-75 p45
Symphony, no. 22. cf English suite, no. 5.
BRICCETTI, Thomas
636 The fountain of youth overture. ROREM: Concerto, piano, in six
movements. TUROK: Lyric variations, oboe and strings, op. 32.
Daniel McAninch, ob; Jerome Lowenthal, pno; Louisville Orches-
tra; Jorge Mester. Louisville LS 733.
          +MJ 9-75 p50
BRIDGES
All my hope on God is founded. cf Polydor 2460 250.
BRITISH NATIONAL ANTHEM: GOD SAVE THE QUEEN. cf Decca SXL 6719.
BRITTEN, Benjamin
637 A ceremony of carols, op. 28. VAUGHAN WILLIAMS: Mass, G minor.
Mark Elder, James Finch, soloists; Michael Welles, treble; John
Whitworth, c-t; Gerald English t; Maurice Bevan, bs; Canterbury
Cathedral Choir; Renaissance Chorus; Sidney Campbell. Decca
ECS 748. (Reissue from Argo ZRG 5179)
          +Gr 3-75 p1688   +RR 4-75 p67
A ceremony of carols, op. 28. cf BRUCKNER: Motets (BASF KBB 21232).
638 Death in Venice. Penelope Mackay, Iris Saunders, s; Peter Pears,
Kenneth Bowen, t; James Bowman, c-t; John Shirley-Quirk, bar;
Peter Leeming, bs; Neville Williams, bs-bar; English Opera Group

Chorus; ECO; Steuart Bedford. Decca SET 581/3 (3). (also
London 13109. Tape (c) Q 513109 (r) R 413109)
+Gr 11-74 p955   +ON 12-14-74 p46
++HF 2-75 p79   +RR 11-74 p20
+MJ 2-75 p30   +SR 11-30-74 p40
+NR 2-75 p8   +-St 1-75 p104

Diversions on a theme, op. 21. cf BARBER: Sonata, piano.

Fanfare for St. Edmondsbury. cf Decca 419.

Festival Te Deum. cf Argo ZRG 789.

639 The little sweep, op. 45. David Hemmings, Michael Ingham, Robin
Fairhurst, Lyn Vaughan, treble; Jennifer Vyvyan, April Cantelo,
Marilyn Baker, Gabrielle Soskin, s; Nancy Thomas, con; Peter
Pears, t; Trevor Anthony, bs; Alleyn's School Choir; English
Opera Group Orchestra; Benjamin Britten. Decca ECM 2166.
(Reissue from LXT 5163)
+Gr 7-75 p238   +RR 5-75 p22

Missa brevis, op. 63, D major: Agnus Dei; Gloria in excelsis. cf
Decca ECS 774.

Nocturnal after John Dowland, op. 70. cf L'Oiseau-Lyre DSLO 3.

Peter Grimes: Sea interludes, nos. 1 and 2. cf Decca 396.

Prelude and fugue on a theme by Vittoria. cf Decca 5BBA 1013-5.

640 The rape of Lucretia. Heather Harper, Jenny Hill, s; Elizabeth
Bainbridge, ms; Janet Baker, con; Peter Pears, t; Bryan Drake,
Benjamin Luxon, bar; John Shirley-Quirk, bs; ECO; Benjamin
Britten. London OSA 1288 (2). Tape (r) D 90198. (also Decca
SET 492/3; SET 537)
++Gr 8-72 p396   +NR 2-71 p11
+HF 10-71 p81   +ON 4-8-72 p34
+HFN 6-71 p1128   ++Op 9-71 p808
++LJ 2-75 p38 tape   +RR 8-72 p41
++MJ 2-74 p8   ++St 12-71 p84

641 Saint Nicholas, op. 42. David Hemmings, treble; Peter Pears, t;
Ralph Downes, org; Aldeburgh Festival Orchestra and Chorus;
Benjamin Britten. Decca ECM 765. (Reissue from LXT 5060)
+Gr 6-75 p75   +RR 6-75 p77
++HFN 8-75 p91

Scherzo. cf HMV SLS 5022.

642 Simple symphony, strings, op. 4. Young person's guide to the
orchestra, op. 34. ELGAR: Cockaigne overture, op. 40. Enigma
variations, op. 36. Pomp and circumstances march, op. 39,
no. 1. WALTON: Crown imperial. LSO, COA, I Musici, Eastman
Wind Ensemble; Frederick Fennell, Colin Davis, Antal Dorati,
Bernard Haitink. Philips 6780 753 (2).
+-HFN 12-75 p171

643 Songs: The ash grove; La belle est au jardin d'amour; The bonny
Earl o' Moray; The brisk young widow; Ca' the yowes; Come you
not from Newcastle; The foggy, foggy dew; The Lincolnshire
poacher; Little Sir William; The minstrel boy; O can ye sew
cushions; O waly, waly; Oliver Cromwell; The plough boy; Quand
j'étais chez mon père; Le roi s'en va-t'en chasse; The Sally
gardens; Sweet Polly Oliver; The trees they do grow high.
Robert Tear, t; Philip Ledger, pno. HMV HQS 1341.
+Gr 4-75 p1849   +STL 6-8-75 p36
++RR 5-75 p57

Suite, violoncello, no. 1, op. 72. cf BACH: Suite, solo violoncello, no. 5, S 1011, C minor.

644 War Requiem, op. 66. Jeannine Altmeyer; Douglas Lawrence, bs-bar;
        Michael Sells, t; William Hall Chorale, Columbus Boychoir;
        Vienna Festival Orchestra; William Hall. Klavier KS 544 (2).
               ++Audio 11-75 p96   +NR 8-75 p9
645 Young person's guide to the orchestra (Variations and fugue on a
        theme by Purcell), op. 34. PROKOFIEV: Cinderella, op. 87:
        Introduction; Quarrel; The dancing lesson; Spring fairy; Summer
        fairy; Grasshoppers dance; Winter fairy; The interrupted depart-
        ure; Clock scene; Cinderella's arrival at the ball; Grand valse;
        Cinderella's waltz; Midnight; Apotheosis. LSO; Andrew Davis.
        CBS 76453. Tape (c) 40-76453.
               +Gr 12-75 p1031   /-RR 11-75 p42
               +HFN 12-75 p151
646 Young person's guide to the orchestra, op. 34. PROKOFIEV: Peter
        and the wolf, op. 67. RAVEL: Ma mere l'oye, excerpts. Alec
        McCowan, narrator; COA; Bernard Haitink. Philips Tape (c)
        7317 093.
               +RR 9-75 p77 tape
647 Young person's guide to the orchestra, op. 34. PROKOFIEV: Peter
        and the wolf, op. 67. Will Geer, narrator; ECO; Johannes Somary.
        Vanguard VSD 71189. (Q) VSQ 30033.
               +HF 11-75 p108   +St 8-75 p104
               +NR 4-75 p5
     Young person's guide to the orchestra, op. 34. cf Simple symphony,
        strings, no. 4.
BRIXI, Frantisek
     Prelude and fugue, C major. cf Hungaroton SLPX 11601/2.
BROWN, Earle
648 December 1952. Novara. Octet 1. Times five. Various performers.
        CRI SD 330.
               +HF 6-75 p88   +St 7-75 p95
               +NR 4-75 p7
     Music, violoncello and piano. cf DG 2530 562.
     Novara. cf December 1952.
     Octet 1. cf December 1952.
     Times five. cf December 1952.
BROWNE, John
     Woefully array'd. cf BASF 25 22286-1.
BRUCH, Max
649 Concerto, clarinet, viola and orchestra, op. 88. UHL: Symphonie
        concertante, clarinet and orchestra. Rudolf Irmisch, Ottokar
        Drapal, clt; orchestras and conductors. Grenadilla GS 1005.
               +NR 11-75 p5
650 Concerto, violin, no. 1, op. 26, G minor. KREISLER: Quartet,
        strings, A minor. Scherzo. Fritz Kreisler, vln; Royal Albert
        Hall Orchestra, Kreisler String Quartet; Eugene Goossens.
        Bruno Walter Society IGI 332. (Reissues from HMV 78s, 1924 and
        1935)
               +HF 12-75 p90   ++NR 8-75 p6
651 Concerto, violin, no. 1, op. 26, G minor. MENDELSSOHN: Concerto,
        violin, op. 64, E minor. Ion Voicou, vln; LSO; Rafael Frühbeck
        de Burgos. Decca SDD 443. (Reissue from SXL 6184)
               +-Gr 8-75 p321   +-RR 6-75 p27
               +-HFN 9-75 p108
652 Concerto, violin, no. 1, op. 26, G minor. GLAZUNOV: Concerto,
        violin, op. 82, A minor. Erica Morini, vln; BRSO; Ferenc
        Fricsay. DG 2548 170. (Reissue from 138044)

+Gr 11-75 p840   +HFN 10-75 p152
+HFN 10-75 p152
653 Concerto, violin, no. 1, op. 26, G minor. SIBELIUS: Concerto,
     violin, op. 47, D minor. Zino Francescatti, vln; NYP; Leonard
     Bernstein. Odyssey Y 33522.
          +NR 11-75 p5
654 Concerto, violin, no. 1, op. 26, G minor. Scottish fantasia, op.
     46. Arthur Grumiaux, vln; NPhO; Heinz Wallberg. Philips
     6500 780. Tape (c) 7300 291.
          ++HF 12-75 p90   ++NR 11-75 p5
     Concerto, violin, no. 1, op. 26, G minor. TCHAIKOVSKY: Concerto,
     violin, op. 35, D major.
     Concerto, violin, no. 1, op. 26, G minor. cf Camden CCV 5000-12,
     5014-24.
     Concerto, violin, no. 1, op. 26, G minor. cf HMV SLS 5004.
     Concerto, violin, no. 1, op. 26, G minor. cf RCA ARM 4-0942/7.
     Concerto, violin, no. 1, op. 26, G minor. cf RCA CRL 6-0720.
655 Concerto, violin, no. 2, op. 44, D minor: 2d movement. KORNGOLD:
     Garden scene. MENDELSSOHN: Concerto, violin, op. 64, E minor.
     SAINT-SAENS: Havanaise, op. 83. Jascha Heifetz, vln; Emanuel
     Bay, pno; PhO; Guido Cantelli. Rococo 2071.
          +NR 10-75 p9
     Kol Nidrei, op. 47. cf Discopaedia MB 1005.
     Kol Nidrei, op. 47. cf Orion ORS 75181.
     Scottish fantasia, op. 46. cf Concerto, violin, no. 1, op. 26, G
     minor.
     Scottish fantasia, op. 46. cf RCA ARM 4-0942/7.
BRUCK, Arnold van
     O du armer Judas. cf BASF BAC 3087.
BRUCKNER, Anton
656 Mass, no. 1, D minor. Edith Mathis, s; Marga Schiml, con; Martin
     Ochman, t; Karl Ridderbusch, bs; Bavarian Radio Symphony Orches-
     tra and Chorus; Eugen Jochum. DG 2530 314. (Reissue from
     2720 054)
          +Gr 7-73 p241   ++RR 7-73 p74
          ++NR 6-75 p10   ++SFC 9-14-75 p28
          +NYT 9-21-75 pD18
657 Mass, no. 1, D minor. Pieces, orchestra (4). Barbara Yates, s;
     Sylvia Swan, con; John Steel, t; Colin Wheatley, bs; Alexandra
     Choir; LPO; Hans-Hubert Schönzeler. Unicorn UNS 210.
          +NR 8-75 p8
658 Mass, no. 2, E minor. John Alldis Choir; ECO; Daniel Barenboim.
     HMV ASD 3079.
          +-Gr 10-75 p669   +RR 9-75 p67
          +-HFN 10-75 p140
659 Mass, no. 2, E minor. SCHUBERT: Deutsche Messe, D 872. Berge-
     dorfer Chamber Choir; Hamburg State Philharmonic Orchestra;
     Hellmut Wormsbächer. Telefunken SAT 22545. (also Telefunken
     6-41297)
          ++Gr 12-73 p1232   +MJ 2-75 p33
660 Motets: Ave Maria; Christus factus est; Locus iste; Os justi;
     Virga Jesse floruit. BRITTEN: A ceremony of carols, op. 28.
     Elisabeth Bayer, hp; Vienna Boys' Choir; Hans Gillesberger.
     BASF KBB 21232. (also BASF BAC 3061)
          +Gr 1-75 p1376   +NR 10-73 p8
          /HF 12-74 p95   +-RR 1-75 p55

Motets: Ave Maria; Christus factus est; Locus iste a Deo factus
  est; Os justi; Virga Jesse. cf BRAHMS: Motets (CRD CRD 1009).
Pieces, orchestra (4). cf Mass, no. 1, D minor.
Prelude, D major. cf Pelca PSRK 41013/6.
661 Symphony, no. 2, C minor. VPO; Horst Stein. Decca SXL 6681.
    (also London CS 6879)
        ++Gr 3-75 p1648   +-RR 2-75 p33
         +HF 11-75 p98   +SFC 9-14-75 p28
         +NR 12-75 p4
662 Symphony, no. 2, C minor. VSO; Carlo Maria Giulini. HMV (Q)
    ASD 3146.
        ++Gr 12-75 p1032   -RR 12-75 p45
        +-HFN 12-75 p151
663 Symphony, no. 4, E flat major. BPhO; Eugen Jochum. DG 2535 111.
    Tape (c) 3335 111. (Reissue from SLPM 139134/5)
         +Gr 8-75 p321   +HFN 10-75 p155 tape
         +HFN 8-75 p87   ++RR 6-75 p34
664 Symphony, no. 4, E flat major. Bamberg Symphony Orchestra;
    Heinrich Hollreiser. Turnabout 34107. Tape (c) KTVC 34107.
         +HFN 12-75 p173 tape
665 Symphony, no. 4, E flat major. VSO; Otto Klemperer. Turnabout
    TV 37073. (Reissue from Vox PL 6930)
         -Gr 6-75 p36   +-RR 6-75 p34
        +-HFN 8-75 p87
666 Symphony, no. 5, B flat major. VPO; Lorin Maazel. Decca SXL
    6686/7 (2). (also London CS 2238)
        +-Gr 12-74 p1126   +RR 12-74 p26
         -HF 7-75 p72   +SFC 11-9-75 p22
         +NR 7-75 p4   +St 11-9-75 p22
667 Symphony, no. 6, A major. VPO; Horst Stein. Decca SXL 6682.
    (also London 6880)
        +-Gr 4-75 p1802   +-RR 4-75 p26
668 Symphony, no. 6, A major. BPhO; Joseph Keilberth. Telefunken
    Tape (c) 4-41168.
         -HFN 9-75 p110 tape
669 Symphony, no. 7, E major. WAGNER: Parsifal: Prelude, Good Friday
    music. BPhO, Bavarian Radio Orchestra; Eugen Jochum. DG 2726
    054 (2). (Reissue from SKL 929-39, SLPM 138005)
        +-Gr 8-75 p321   +-RR 6-75 p34
        +-HFN 8-75 p87
    Symphony, no. 7, E major. cf BEETHOVEN: Symphony, no. 5, op. 67,
    C minor.
670 Symphony, no. 8, C minor. LAPO; Zubin Mehta. Decca SXL 6671/2 (2).
    Tape (c) KSXC 27023. (also London CSA 2237)
         -Gr 9-74 p491   +-RR 10-74 p39
        +-Gr 1-75 p1402 tape   -St 4-75 p98
         -HF 2-75 p90
671 Symphony, no. 8, C minor. BPhO; Wilhelm Furtwängler. Rococo 2032.
         +NR 2-75 p3
672 Symphony, no. 8, C minor. VPO; Wilhelm Furtwängler. Unicorn UNI
    109/10 (2).
        ++SFC 9-14-75 p28   ++St 4-75 p98
    Tosterin Musik. cf Wayne State University, unnumbered.
BRUHNS, Nicolaus
    Prelude, E minor. cf Pelca PSRK 41013/6.
673 Prelude and fugue, G minor. Toccatas, organ, nos. 1-3. HANFF:

Chorale preludes: Ach Gott, vom Himmel sich darein; Auf
meinen lieben Gott; Ein fest Burg ist unser Gott; Erbarm dich
mein, O Herre Gott; Helft mir Gott's Güte preisen; Wär Gott
nicht mit uns diese Zeit. Michel Chapuis, org. Telefunken
SAWT 9615.
         ++Gr 12-74 p1173   +RR 1-75 p46
         +MT 7-75 p629
    Toccatas, organ, nos. 1-3. cf Prelude and fugue, G minor.
BRULE, Gace
    Biaus m'est estez. cf Telefunken AW 6-41275.
BRULL, Ignaz
    Scene espagnole. cf Discopaedia MB 1006.
BRUN, Herbert
674 Gestures for eleven. HEISS: Quartet. Four movements for 3 flutes.
    YTTREHUS: Sextet. Ronald Anderson, tpt; Barry Benjamin, hn;
    Thomas Kornacher, vln; Joan Tower, pno; David Walter, bs; Claire
    Heldrich, perc; Efrain Guigui, cond; Boston Musica Viva; Richard
    Pittman, cond; John Heiss, Paul Dunkel, Trix Kout, flt; Univer-
    sity of Illinois Chamber Players. CRI SD 321.
         +-HF 2-75 p108   +-NR 9-74 p6
BRUNNER, Adolf
    Eingangspiele, Weihnacht, Passion, Ostern. cf Pelca PSRK 41013/6.
BRYCE, Frank
675 Promenade. MAURER: Fancies (4) (arr. Gay). ROMAN: Music for the
    royal nuptials. WRIGHT: Concerto, cornet. Gram Gay, cornet;
    Solna Brass; Per Ohlsson. Grosvenor GRS 1031.
         +Gr 4-75 p1871   ++RR 4-75 p26
BULL, John
    Coranto battle. cf Musical Heritage Society MHS 1790.
    English toy. cf BASF BAC 3075.
    Fantasia, D minor. cf BASF BAC 3075.
    The King's hunt. cf BASF BAC 3075.
    The King's hunt. cf BASF BAB 9005.
    The King's hunt. cf Saga 5402.
BULLOCK, Ernest
    Give us the wings of faith. cf Abbey LPB 734.
BULMAN (Bullman) (16th century)
    Pavan. cf DG Archive 2533 157.
BURCKHARDI, J.
    Fantasies, 2 recorders. cf Pelca PSR 40589.
BURGK, Joachim
    The Lord with His disciples. cf BASF BAC 3087.
BURKHART
    Tre advetntssanger. cf BIS LP 2.
BURLEIGH, Cecil
    Characteristic pieces: Indian snake dance. cf Discopaedia MB 1007.
BUSONI, Ferruccio
    Ballet-scène, op. 4, no. 33. cf Elégies, op. 70.
676 Berceuse elégiaque, op. 42. DALLAPICCOLA: Piccola musica notturna.
    Preghiere. Sex carmina alcaei. WOLPE: Piece in 2 parts, 6
    players. Henry Datyner, vln; Colin Bradbury, clt; David Mason,
    tpt; Charles Tunnell, vlc; Michael Jeffries, hp; Katherine Wolpe,
    pno; Heather Harper, s; Barry McDaniel, bar; NPhO, ECO; Frederick
    Prausnitz. Argo ZRG 575.
         +RR 12-75 p45
    Concerto, violin, op. 35a, D major. cf BEETHOVEN: Concerto, violin,
    op. 61, D major.

677 Concerto, violin and orchestra. Nocturne symphonique. Turandot
     suite. Manoug Parikian, vln; RPO; Jascha Horenstein. Rococo
     2036.
          +-NR 7-75 p6
678 Elégies, op. 70. Ballet-scène, no. 4, op. 33. Martin Jones, pno.
     Argo ZRG 741.
               +Gr 12-73 p1222   ++NR 6-74 p9
               +HF 12-74 p102   ++St 2-75 p107
     Nocturne symphonique. cf Concerto, violin and orchestra.
679 Sonata, violin and piano, no. 2, op. 36a, E minor. PADEREWSKI:
     Sonata, violin and piano, op. 13, A minor. Endre Granat, vln;
     Harold Gray, pno. Desmar DSM 1004.
               ++MJ 12-75 p38   ++NR 12-75 p9
     Turandot suite. cf Concerto, violin and orchestra.
BUSSOTTI, Sylvano
     Ultima rara. cf DG 2530 561.
BUTTERLY, Nigel
     The white throated warbler. cf HMV SLS 5022.
BUTTERWORTH, George
680 The banks of green willow. English idylls (2). A Shropshire lad.
     HOWELLS: Elegy, op. 15. Merry eye, op. 20b. Music for a
     prince. Herbert Downes, vla; Desmond Bradley, Gillian Eastwood,
     vln; Albert Cayzer, vla; Norman Jones, vlc; NPhO, LPO; Adrian
     Boult. Lyrita SRCS 69.
               ++HFN 11-75 p153   +RR 12-75 p50
     English idylls. cf The banks of green willow.
     A Shropshire lad. cf The banks of green willow.
     A Shropshire lad. cf BAX: The garden of Fand.
BUXTEHUDE, Dietrich
     Canzonetta, C major. cf Telefunken DX 6-35265.
     Chaconne, E minor. cf Organ works (Cambridge CRS 2515).
     Chorale preludes: Wie schön leuchtet der Morgenstern; Lobt Gott,
        ihr Christen; In dulci jubilo; Puer natus in Bethlehem; Auf
        meinem lieben Gott. cf Organ works (Cambridge CRS 2515).
     Fugue, C major. cf Telefunken DX 6-35265.
681 Organ works: Chaconne, E minor. Chorale preludes: Wie schön leuch-
        tet der Morgenstern; Lobt Gott, ihr Christen; In dulci jubilo;
        Puer natus in Bethlehem; Auf meinem lieben Gott. Prelude, fugue
        and chaconne, C major. Preludes and fugues, G minor, D minor,
        F sharp minor. Lawrence Moe, org. Cambridge CRS 2515.
               +CL 9-75 p16   /HF 3-72 p80
     Partita on "Auf meinen lieben Gott". cf Telefunken DX 6-35265.
     Prelude. cf Kendor KE 22174.
     Prelude and fugue, G minor. cf Vista VPS 1021.
     Preludes and fugues, G minor, D minor, F sharp minor. cf Organ
        works (Cambridge CRS 2515).
     Prelude, fugue and chaconne, C major. cf Organ works (Cambridge
        CRS 2515).
     Toccata and fugue, F major. cf Argo ZRG 783.
     Toccata and fugue, F major. cf Columbia M 33514.
     Toccata and fugue, F major. cf Orion ORS 74161.
BYRD, William
     Alleluia, ascendit Deus. cf Gemini GM 2022.
     Alman. cf Angel SB 3816.
     Ave verum corpus. cf Gemini GM 2022.
     Callina Casturame. cf Angel SB 3816.

Exsurge Domine. cf Argo ZRG 789.
A fancie. cf Angel SB 3816.
French corantos. cf Angel SB 3816.
Haec dies. cf HMV Tape (c) TC MCS 13.
Lavolta, Lady Morley. cf Angel SB 3816.
The leaves be green. cf HMV SLS 5022.
Lullaby. cf Turnabout TV 34569.
682 Mass, 3 parts. Mass, 4 parts. Deller Consort. Vanguard Bach
      Guild HM 6.
            +-HFN 9-75 p109   +RR 10-75 p82
    Mass, 4 parts. cf Mass, 3 parts.
    My Lord of Oxenford's mask. cf Vanguard SRV 316.
    Pavan and galliard, no. 4. cf Angel SB 3816.
    Pavan and galliard of Mr. Peter. cf BASF BAC 3075.
    Rowland or Lord Willoughby's welcome home. cf Angel SB 3816.
    Walsingham variations. cf BASF BAC 3075.
    Wolsey's wilde. cf Saga 5402.
CABANILLES, Juan
    Batalla II. cf Musical Heritage Society MHS 1790.
CABEZON, Antonio de
    Diferencias sobre el canto Llano la alta. cf Argo ZRG 783.
CADMAN, Charles
    At dawning. cf HMV RLS 714.
    Little firefly. cf Discopaedia MB 1005.
CAGE, John
    Amores. cf Opus One 22.
    Dream. cf BEDFORD: Spillihpnerak.
    Fontana mix. cf BERIO: Visage.
    62 mesostics re Merce Cunningham. cf 1750 Arch 1752.
683 Winter music. FLYNN: Wound. George Flynn, pno. Finnadar (Q)
      QD 9006. Tape (ct) QT 0996.
            *Audio 8-75 p84   *NR 3-75 p10
            +-HF 3-75 p77   ++St 10-75 p106
CAIN, David
    Much ado about nothing: Suite. cf BBC REB 191.
CAMBRAI,   de (13th century France)
    Retrowange novelle. cf Telefunken AW 6-41275.
CAMILLERI, Charles
684 Missa mundi. Gillian Weir, org. Argo ZRG 812.
            +Gr 7-75 p214   +MT 12-75 p1070
            +-HFN 8-75 p73   ++RR 8-75 p55
CAMPAGNOLI, Bartolomeo
    Romanza. cf Ember GVC 42.
CAMPANA, Fabio
    M'hai tradito. cf Rococo 5368.
CAMPBELL, Sidney
    Jubilate Deo. cf Argo ZRG 789.
CAMPOS-PARSI
    Plena de concierto. cf International Piano Library IPL 5005/6.
CAMPRA, Andre
685 Cantates françaises de Arion et Didon. LECLAIR: Sonatas, op. 2,
      nos. 5, 8. Ana Marie Miranda, s; Brigitte Haudebourg Trio.
      Arion ARN 38257.
            +RR 10-75 p82
    Rigaudon, A major. cf Columbia M 33514.
CANNABICH, Christian
    Rechenmeister amor. cf Oryx ORYX 1720.

686   Sinfonia concertante, F major.   ROSETTI: Sinfonia, G minor.   VOGLER:
      Comic ballets: 4 movements.   Kuarpfälzische Chamber Orchestra;
      Wolfgang Hofmann.   Oryx ORYX 1722.   (Reissue from Musica Rara
      MUS 28)
           /Gr 11-75 p839   /HFN 10-75 p153

CANOVA
      Fantasia, 2 lutes.   cf Erato STU 70847.
      CANTATA FOR VENICE.   cf Pye GSGC 14153.
      THE CANTELLI LEGACY, vol. 1.   cf Toscanini Society ATS GS 1201/6.
CANTELOUBE, Joseph
687   Chants d'Auvergne: La pastoura als camps; L'antouèno; La past-
      rouletta è lou chibalié; Lo calhé; Lou boussu; Malurous qu'o
      uno fenno; Oï ayaï; Pour l'enfant; Pastorale; Lou coucut; Obal,
      din lo coumbelo; La haut, sur le rocher; Hé, beyla-z-y dáu fè;
      Tè, l'co, tè; Uno jionto postouro.   Victoria de los Angeles, s;
      Lamoureux Orchestra; Jean-Pierre Jacquillat.   HMV (Q) ASD 3134.
      Tape (c) TC ASD 3134.   (also Angel S 36898)
           -Gr 11-75 p875   +NR 12-75 p12
           +HFN 12-75 p151   +-RR 12-75 p93
688   Songs of the Auvergne (31).   Netania Davrath, s; Orchestra; Pierre
      de la Roche.   Vanguard ZCVB 713/4.   Tape (c) M 5713/4 (ct)
      M 8713/4.
           +Gr 6-75 p111 tape
689   Songs of the Auvergne: L'antouèno; Pastourello; L'aïo dè rotso;
      Ballèro; Passo del prat; Malurous qu'o uno fenno; Brezairola.
      RACHMANINOFF: Vocalise, op. 34, no. 14 (arr. Dubensky).   VILLA-
      LOBOS: Bachianas brasileiras, no. 5.   Anna Moffo, s; American
      Symphony Orchestra; Leopold Stokowski.   RCS LSB 4114.
           /Gr 4-75 p1860   +-RR 4-75 p67
           +-HFN 6-75 p97
      Songs of the Auvergne: L'aïo dè rotso; Ballèro; Brezairola; Chut,
      chut; La delaïssádo; La fiolairé; Obal, din lou Limouzi; Ound
      onorén gorda; Passo pel prat.   cf CHAUSSON: Songs: Poeme de
      l'amour et de la mer.
CAPIROLA (16th century Italy)
      Ricercare, nos. 1, 2, 10, 13.   cf DG Archive 2533 173.
CAPUA,   di
      I tel vurria vasa.   cf Musical Heritage Society MHS 1951.
      O sole mio.   cf LEONCAVALLO: I Pagliacci.
      Songs: O sole mio; Sweet dream of love.   cf Ember GVC 41.
      Songs: O sole mio.   cf Polydor 2489 519.
CARDILLO
      Core 'ngrato.   cf Musical Heritage Society MHS 1951.
CARLOS, Walter
      Dialogues, piano and 2 loudspeakers.   cf Columbia CXM 32088.
      Episodes, piano and electronic sound.   cf Columbia CXM 32088.
      Geodesic dance.   cf Columbia CXM 32088.
      Pompous circumstances.   cf Columbia CXM 32088.
CARMICHAEL, John
      Puppet show.   cf BIZET: Jeux d'enfants, op. 22.
CAROSO, Fabrizio
      Laura soave: Gagliarda, Saltarello (Balleto).   cf DG 2530 561.
CAROUBEL (16th century Spain)
      Courante (2).   cf DG Archive 2533 184.
      Volte (2).   cf DG Archive 2533 184.
CARTER
      Lord of the dance.   cf Decca ECS 774.

CARTER, Elliot
690   Concerto, harpsichord, piano and 2 chamber orchestras. Duo,
          violin and piano. Paul Jacobs, hpd; Gilbert Kalish, pno;
          Contemporary Chamber Ensemble; Arthur Weisberg. Nonesuch H 71314.
                +NR 12-75 p7
      Duo, violin and piano. cf Concerto, harpsichord, piano and 2
      chamber orchestras.
      Etudes, woodwind quartet (8). cf Classic Record Library SQM 80-
      5731.
      Heart not so heavy as mine. cf Wayne State University, unnumbered.
691   Quartets, strings, nos. 2 and 3. Juilliard Quartet. Columbia
          M 32738. (Q) MQ 32738 Tape (ct) 32738.
                ++HF 7-74 p73   +NYT 4-41-74 pD26
                ++MQ 1-75 p157   ++SFC 5-12-74 p24
                +NR 7-74 p6   ++St 9-74 p116
CARULLI, Ferdinando
      Serenade, op. 96. cf RCA ARL 1-0456.
CARVALHO, João
      Sonata, organ, D major: Allegro. cf Argo ZRG 783.
CASADESUS, Robert
      Etudes, op. 28. cf Sonata, piano, no. 4, op. 56.
692   Sonata, piano, no. 4, op. 56. Etudes, op. 28 (8). Gaby Casadesus,
          pno. Columbia M 33505.
                ++HF 11-75 p102   ++NR 8-75 p12
CASALS, Pablo
      Preludio de Do mayor. cf International Piano Library IPL 5005/6.
      Les roi mages, 6 celli. cf Sardana, celli.
693   Sardana, celli. Les roi mages, 6 celli. EHRLICH: Short pieces,
          3 celli (6). LINN: Dithyramb, 8 celli. VIVALDI: Concerto,
          op. 3, no. 11, D minor (arr. for cello quartet). I Cellisti;
          Jerome Kessler, vlc. Ember GVC 33.
                +Gr 7-75 p245   +HFN 7-75 p79
      Song of the birds. cf Orion ORS 75181.
CASSADO
      Danse due diable vert. cf Discopaedia MB 1009.
CASSADO, Gaspar
      Requiebros. cf Supraphon 50919.
      Sonata nello stile antico spagnuolo. cf Supraphon 50919.
CASSANOVAS
      Sonata, harp, F major. cf RCA LRL 1-5087.
CASTELNUOVO-TEDESCO, Mario
694   Concertino, harp and chamber orchestra, op. 93. RODRIGO: Fantasia
          para un gentilhombre. VILLA-LOBOS: Concerto, harp. Catherine
          Michel, hp; Monte Carlo Opera Orchestra; Antonio de Almeida.
          Philips 6500 812.
                +Gr 7-75 p175   +RR 6-75 p55
                +HFN 7-75 p79
      Coplas. cf GOLD: Songs of love and parting.
      Etudes d'ondes: Sea murmurs. cf CBS 76420.
      Fandango. cf Pye GSGC 14154.
      La guarda cuydadosa. cf DG 2530 561.
      Sea murmurs (2). cf RCA ARM 4-0942/7.
      Sonatina, flute and guitar. cf Klavier KS 537.
      Tango. cf RCA ARM 4-0942/7.
      Tarantella. cf DG 2530 561.
      Tonadilla on name Andrés Segovia. cf CBS 61654.
      Valse. cf RCA ARM 4-0942/7.

CASTRO, Juan Jose
Tangos. cf Golden Crest CRS 4132.
CATALAN MUSIC FROM MEDIEVAL AND RENAISSANCE SPAIN. cf Candide
CE 31068.
CATALANI, Alfredo
La wally, arias. cf Columbia 33435.
695 La wally: Ebben. Ne andrò lontana. CILEA: Adriana Lecouvreur:
Io son l'umile ancella; Poveri fiori. MASCAGNI: Lodoletta:
Flammen, perdonami. Iris: Un dì (ero piccina) al tempio.
PUCCINI: La bohème: Quando m'en vo' soletta. Gianni Schicchi:
O mio babbino caro. Manon Lescaut: In quelle trine morbide;
Sola, perduta, abbandonata. La rondine: Ch' il bel sogno di
Doretta. Suor Angelica: Senza mamma, o bimbo. Le Villi: Non
ti scordar di me. Renata Scotto, s; LSO; Gianandrea Gavazzeni.
Columbia M 33435. (also CBS 76407) Tape (c) 40-76407)
+Gr 11-75 p904   +-ON 8-75 p28
+HF 9-75 p101   +RR 11-75 p34
+HFN 12-75 p169   ++SFC 8-3-75 p30
+-NR 8-75 p9   +St 9-75 p116
+NYT 7-6-75 pD11
CATCHES AND PARTSONGS. cf BASF BAC 3081.
CATO, Diomedes
Praeludium, Galliardas I, II. cf DG Archive 2533 294.
CAURROY, Eustache de
Fantasia; Prince la France te veut. cf L'Oiseau-Lyre 12BB 203/6.
CAVAZZONI, Marco Antonio
Intabulatura d'organo: Libro secondo. cf Pelca PSRK 41013/6.
Intabulatura cioe recercari canzoni: Hymnus in festo corporis
Christi. cf Pelca PSRK 41013/6.
CAVENDISH, Michael
Sly thief, if so you will believe. cf Harmonia Mundi 593.
Wand'ring in this place. cf L'Oiseau-Lyre 12BB 203/6.
CEELY, Robert
696 Elegia. Mitsyn music. DEL MONACO: Electronic study, no. 2.
Metagrama. RANDALL: Music for the film 'Eakins'. Princeton
University Computer Center; Milan, Studio di Fonologia; Boston
Experimental Electronic-music Projects; Columbia-Princeton
Music Center. CRI SD 328.
+-HFN 10-75 p140   +-RR 7-75 p60
+NR 2-75 p15
Mitsyn music. cf Elegia.
CERTON, Pierre
Psalms and nunc Dimittis (3). cf BIS LP 2.
CESARIS, Johannes
Bonté biauté. cf Telefunken ER 6-35257.
CESTI, Pietro
Orentea: Intorno all'idol mio. cf Pearl SHE 511.
CHABRIER, Emanuel
Air de ballet. cf Piano works (CRD CAL 1828/9).
Bourrée fantasque. cf Piano works (CRD CAL 1828/9).
Capriccio. cf Piano works (CRD CAL 1828/9).
España. cf BIZET: L'Arlésienne: Suites, nos. 1 and 2.
España. cf Decca DPA 519/20.
España: Rapsodie. cf HMV ASD 3008.
Habanera. cf Piano works (CRD CAL 1828/9).
Impromptu. cf Piano works (CRD CAL 1828/9).

Marche des Cipayes. cf Piano works (CRD CAL 1828/9).
Marche joyeuse. cf HMV SLS 5019.
697 Piano works: Air de ballet. Bourrée fantasque. Capriccio. Haba-
        nera. Impromptu. Marche des Cipayes. Pièces pittoresques (10).
        Pièces posthumes (5). Annie d'Arco, pno. CRD CAL 1828/9 (2).
            ++Gr 9-75 p485   +HFN 10-75 p140
    Pièces pittoresques. cf Piano works (CRD CAL 1828/9).
    Pièces pittoresques, no. 10: Scherzo valse. cf Rococo 2035.
    Pièces posthumes. cf Piano works (CRD CAL 1828/9).
690 Le Roi malgré lui: Fête polonaise. LALO: Rapsodie norvégienne.
        MASSENET: Scènes hongroises. Luxembourg Radio Orchestra; Pierre
        Cao. Turnabout (Q)QTV 34570.
            +-NR 3-75 p3
    Scherzo-Valse. cf Columbia M 32070.
699 Songs: Ballad des gros dindons; Les cigales; Chanson pour Jeanne;
        L'Ile heureuse; Lied; Pastorale des petits cochons roses;
        Villanelle des petits canards. LISZT: Songs: Anfangs woll't
        ich fast verzagen, G 311; Ihr Auge, G 310; Kling leise mein
        Lied, G 301; Uber allen Gipfeln ist Ruh, G 306; Die Loreley,
        G 273; Vergiftet sind meine Lieder, G 289; Wieder möcht ich dir
        begegnen, G 322; Die drei Zigeuner, G 320. Paul Sperry, t; Irma
        Vallecillo, pno. Orion ORS 75174.
            +NR 10-75 p7
    Songs: Toutes les fleurs; Les cigales; L'Ile heureuse. cf Rococo
        5365.
    Valses romantiques (3). cf GERSHWIN: Rhapsody in blue.
CHADWICK, George
    Jubilee. cf Century/Advent OU 97 734.
700 Symphonic sketches. PISTON: Incredible flutist. Eastman-Rochester
        Orchestra; Howard Hanson. Mercury SRI 75050. (Reissue)
            ++SFC 8-3-75 p30
CHAGRIN, Francis
    Ballad of County Down, D major. cf HMV SLA 870.
    Concerto, conductor and orchestra: A movement. cf HMV SLA 870.
CHAJES, Julius
    Hechassid, op. 24, no. 1. cf Da Camera Magna SM 93399.
    CHAMBER MUSIC SOCIETY OF LINCOLN CENTER. cf Classic Record Library
        SQM 80-5731.
CHAMBERS
    The boys of the old brigade. cf Michigan University SM 0002.
CHAMBONNIERES, Jacques
    Chaconne, F major. cf Harmonia Mundi HMU 334.
    Rondeau. cf Harmonia Mundi HMU 334.
CHAMINADE, Cecile
    Autrefois, op. 87, A minor. cf L'Oiseau-Lyre DSLO 7.
    Scarf dance. cf Connoisseur Society (Q) CSQ 2065.
    CHANSONS DES TROUVERES. cf Telefunken AW 6-41275.
CHAPIN, Lucius
    Rockbridge. cf Nonesuch H 71276.
CHARPENTIER, Gustave
701 Impressions d'Italie. MASSENET: Scenes alsaciennes: Suite, no. 7.
        OSCCP; Albert Wolff. Decca ECS 773. (Reissues from LXT 5246,
        5100)
            +-Gr 10-75 p608   +RR 8-75 p39
            +HFN 9-75 p108
    Louise: Depuis le jour. cf RCA ARL 1-0844.

CHARPENTIER, Marc-Antoine
    Messe de minuit: Kyrie. cf HMV Tape (c) TC MCS 13.
    Messe de minuit pour Noel: Kyrie. cf University of Colorado,,
        unnumbered.
    Te Deum: Prelude. cf CRD CRD 1008.
CHATTERTON
    Tu sola a me. cf Columbia/Melodiya M 33120.
CHAUN, Frantisek
702 Proces. HAVELKA: Pena. MACHA: Varianty. PAVER: Panychida. CPhO;
        Václav Neumann. Panton 110 264.
        +HFN 6-75 p98
CHAUSSON, Ernest
    Chanson perpetuelle, op. 37. cf Columbia M 33307.
    Concerto, violin, piano and string quartet, op. 21, D major. cf
        RCA ARM 4-0942/7.
703 Poème, op. 25. RAVEL: Tzigane. SAINT-SAENS: Introduction and
        rondo capriccioso, op. 28. Havanaise, op. 83. Itzhak Perlman,
        vln; Orchestre de Paris; Jean Martinon. Angel (Q) S 37118.
        Tape (c) 4XS 37118 (ct) 8XS 37118. (also HMV ASD 3125)
            +HF 12-75 p110   +RR 12-75 p61
            +HFN 12-75 p152
704 Poème de l'amour et de la mer, op. 19. CANTELOUBE (Anon.): Songs
        of the Auvergne: L'Aio de rotso; Ballèro; Brezairola; Chut,
        chut; La delaïssàdo; La fiolairé; Obal, din lou Limouzi;
        Ound onorèn gorda; Passo pel prat. Victoria de los Angeles, s;
        Lamoureux Orchestra; Jean-Pierre Jacquillat. HMV ASD 2826.
        Tape (c) TC ASD 2826. (also Angel S 36897)
            ++Gr 3-73 p1721   +ON 8-73 p30
            ++Gr 12-74 p1237 tape   +RR 3-73 p84
            +-HF 6-73 p74   +-RR 1-75 p72 tape
             +HFN 3-73 p559   +SR 4-73 p72; 3-73 p47
            ++NR 8-73 p11   ++St 6-73 p83
CHEDVILLE, Nicolas
    Musette. cf DG Archive 2533 172.
CHERUBINI, Luigi
705 Requiem, male voices and orchestra, D minor. Ambrosian Singers;
        NPhO; Riccardo Muti. HMV ASD 3073. (Q) SQ 4 ASD 3073. (also
        Angel S 37096)
            +Gr 5-75 p2000   +NR 8-75 p8
            ++HF 12-75 p90   ++NYT 9-21-75 pD18
            +HFN 9-75 p97   +RR 4-75 p58
            +MT 11-75 p977   ++SFC 10-19-75 p33
CHESNOKOV, Pavel
    Spassi, Boje, lioudi tvoi. cf ARKHANGELSKII: Blajen razoumnevai;
        Otche nach.
    Spassi, Bozhe, Lyudi Tvoya (Save, God, our people). cf Monitor
        MFS 757.
CHEVALIER DE SAINT-GEORGES
706 Concerto, violin, no. 2, op. 2, A major. Concerto, violin, no.
        9, op. 8, G major. Jean-Jacques Kantorow, vln; Bernard Thomas
        Chamber Orchestra. Arion ARN 38253.
            +Gr 3-75 p1649
    Concerto, violin, no. 9, op. 8, G major. cf Concerto, violin, no.
        2, op. 2, A major.
CHICHESTER CATHEDRAL, 900 YEARS. cf HMV HQS 1350.

CHIHARA, Paul
707 Ceremony I and III. Grass. Buell Neidlinger, double bs; Peter
     Lloyd, flt; Suenobu Togi, hichi-riki; LSO Members; Neville
     Marriner. Turnabout (Q) QTVS 34572.
          ++HF 9-75 p84   ++NR 4-75 p3
     Grass. cf Ceremony I and III.
CHOPIN, Frederic
708 Andante spianato and grande polonaise, op. 22, E flat major.
     Scherzo, no. 2, op. 31, B flat minor. Sonata, piano, no. 2,
     op. 35, B flat minor. Martha Argerich, pno. DG 2530 530.
          +Gr 6-75 p68   ++NR 7-75 p15
          +HF 7-75 p74   +-RR 6-75 p65
          ++HFN 6-75 p87
     Andante spianato and grande polonaise, op. 22, E flat major. cf
     Piano works (Philips 6500 422).
709 Ballades, nos. 1-4. Milosz Magin, pno. Decca 7167.
          +-HFN 8-75 p74   -RR 8-75 p56
710 Ballade, no. 1, op. 23, G minor. Ballade, no. 2, op. 38, F major.
     Etudes, op. 10. Nocturne, op. 9, no. 3, B major. Dezsö Ránki,
     pno. Hungaroton SLPX 11555.
          ++NR 4-75 p12   ++RR 3-74 p52
711 Ballade, no. 1, op. 23, G minor. Ballade, no. 3, op. 47, A flat
     major. Nocturne, op. 37, no. 2, G major. Polonaise, op. 40,
     no. 1, A major. Mazurka, op. 24, no. 4, B flat minor. Valse,
     op. 34, no. 1, A flat major. DEBUSSY: Images: Reflets dans
     l'eau. MENDELSSOHN: Song without words, op. 67, no. 4, C major
     (Spinning song). SCHUBERT: Hark, hark the lark. Impromptu,
     op. 142, no. 3, D 935, B flat major. Ignace Jan Paderewski,
     pno. Klavier KS 127.
          +-NR 10-75 p8
     Ballade, no. 1, op. 23, G minor. cf Piano works (London SPC 21071).
     Ballade, no. 1, op. 23, G minor. cf PIano works (RCA VH 011).
     Ballade, no. 1, op. 23, G minor. cf Westminster WGM 8309.
     Ballade, no. 2, op. 38, F major. cf Ballade, no. 1, op. 23, G
     minor.
     Ballade, no. 3, op. 47, A flat major. cf Ballade, no. 1, op. 23,
     G minor.
     Ballade, no. 3, op. 47, A flat major. cf Piano works (Decca SXL
     6693).
     Ballade, no. 3, op. 47, A flat major. cf Piano works (RCA VH 018).
     Ballade, no. 3, op. 47, A flat major. cf Piano works (Telefunken
     AW 6-41847).
712 Ballade, no. 4, op. 52, F minor. Impromptu, no. 1, op. 29, A flat
     major. Nocturne, op. 37, no. 1, G minor. Nocturne, op. 62,
     no. 1, B major. Polonaise, op. 71, no. 2, B flat major. Valse,
     op. 18, E flat major. Waltz, op. 64, no. 1, D flat major.
     Ignaz Friedman, pno. Saga 5394.
          +RR 7-75 p47
     Ballade, no. 4, op. 52, F minor. cf Piano works (Decca PFS 4313).
     Ballade, no. 4, op. 52, F minor. cf Piano works (Erato STU 70843).
     Ballade, no. 4, op. 52, F minor. cf International Piano Archives
     IPA 5007/8.
713 Barcarolle, op. 60, F sharp major. Prelude, A flat major. Prelude,
     op. 45, C sharp minor. Preludes, op. 28, nos. 1-24. Garrick
     Ohlsson, pno. Angel S 37087.
          +NR 7-75 p15

714   Barcarolle, op. 60, F sharp major. Impromptu, no. 3, op. 51, G
      flat major. Mazurkas, op. 63, nos. 1-3. Nocturnes, op. 62,
      nos. 1 and 2. Polonaise-Fantaisie, op. 61, A flat major.
      Stephen Bishop, pno. Philips 6500 393. Tape (c) 7300 228.
              +Gr 7-73 p215   +-NYT 4-1-73 pD28
              +HF 6-73 p75   ++RR 6-75 p91 tape
              +MJ 5-73 p37   +SR 4-73 p72
              ++NR 6-73 p12   +-St 5-73 p109
      Barcarolle, op. 60, F sharp major. cf Piano works (Decca SXL 6693).
      Barcarolle, op. 60, F sharp major. cf Piano works (London SPC
      21071).
      Barcarolle, op. 60, F sharp major. cf Preludes, complete.
      Berceuse, op. 57, D flat major. cf Piano works (Classics for
      Pleasure CFP 40061).
      Berceuse, op. 57, D flat major. cf Piano works (Telefunken AW
      6-41847).
      Berceuse, op. 57, D flat major. cf Preludes, op. 28, nos. 1-24.
715   Concerto, piano, no. 1, op. 11, E minor. Krakowiak, op. 14, F
      major. Stefan Askenase, pno; Hague Philharmonic Orchestra;
      Willem van Otterloo. DG 2548 066. (Reissue from SLPM 138085)
              ++Gr 10-75 p608   +RR 9-75 p34
              +HFN 9-75 p108
716   Concerto, piano, no. 1, op. 11, E minor. Stefan Askenase; Hague
      Philharmonic Orchestra; Willem van Otterloo. DG Tape (c) 3318
      013.
              +Gr 1-75 p1402 tape   +RR 12-74 p87 tape
717   Concerto, piano, no. 1, op. 11, E minor. Sándor Falvai, pno;
      Budapest Philharmonic Orchestra; András Koródy. Hungaroton LPS
      11654.
              -SFC 8-10-75 p26
718   Concerto, piano, no. 1, op. 11, E minor. Claudio Arrau, pno;
      LPO; Eliahu Inbal. Philips 6500 255. Tape (c) 7300 109.
              -ARG 6-72 p503   +MJ 6-72 p33
              -Gr 6-72 p45   ++NR 6-72 p5
              -Gr 9-74 p592 tape   ++RR 1-75 p69 tape
              +HF 6-72 p73   -SFC 5-7-72 p46
              +HFN 6-72 p1107
719   Concerto, piano, no. 1, op. 11, E minor. Concerto, piano, no. 2,
      op. 21, F minor. Dinorah Varsi, pno; Monte Carlo Opera Orchestra;
      Jan Krenz. Philips 6580 065.
              +-Gr 5-75 p1807   -RR 5-75 p28
              -HFN 8-75 p74
      Concerto, piano, no. 1, op. 11, E minor. cf Piano works (Connois-
      seur Society CS 2029/30).
720   Concerto, piano, no. 2, op. 21, F minor. Fantasia, op. 49, F minor.
      Witold Malcuzynski, pno; LSO; Walter Susskind. Classics for
      Pleasure CFP 40215.
              +-HFN 8-75 p87   +RR 6-75 p35
721   Concerto, piano, no. 2, op. 21, F minor. Polonaise, op. 40, no. 1,
      A major. Polonaise, op. 35, A flat major. Stefan Askenase,
      pno; BPhO; Leopold Ludwig. DG 2548 124. Tape (c) 3318 014.
      (Reissue from 138791)
              +-Gr 1-75 p1402   +-HFN 10-75 p152
              /Gr 11-75 p840   +-RR 11-75 p38
722   Concerto, piano, no. 2, op. 21, F minor. FALLA: Noches en los
      jardines de España. Alicia de Larrocha, pno; OSR; Sergiu

Comissiona. London CS 6773. Tape (c) M 10252. (also Decca
SXL 16528. Tape (c) KSXC 16528).

++ARG 12-72 p806   ++HFN 7-75 p90 tape
+-Ha 11-72 p77   ++NR 4-72 p7
+-HF 7-72 p77   ++SFC 5-7-72 p46
+HFN 2-72 p306   +-St 6-72 p82

723 Concerto, piano, no. 2, op. 21, F minor. Krakowiak, op. 14, F
     major. Claudio Arrau, pno; LPO; Iliahu Inbal. Philips 6500 309.
     Tape (c) 7300 110.

+Gr 10-72 p689   +NR 7-74 p5
+Gr 7-73 p247 tape   +NYT 8-11-74 pD22
+HF 9-74 p92   +-RR 10-72 p61
+-HFN 10-72 p1905   +-SFC 8-11-74 p34
+-MJ 3-75 p34   +STL 11-19-72 p38

Concerto, piano, no. 2, op. 21, F minor. cf Concerto, piano, no.
1, op. 11, E minor.

Concerto, piano, no. 2, op. 21, F minor. cf Piano works (Decca
SXL 6693).

Concerto, piano, no. 2, op. 21, F minor. cf GINASTERA: Milena,
cantata, no. 3, op. 37, soprano and orchestra.

724 Etudes, opp. 10 and 25 complete. Ilana Vered, pno. Connoisseur
     Society CS 2045. Tape (c) Advent E 1018 (Q) CSQ 2045.

+-HF 10-73 p101   +-MJ 9-73 p44
+HF 7-75 p102 tape   ++NR 7-73 p11
+LJ 2-74 p50

725 Etudes, opp. 10 and 25, complete. Vladimir Ashkenazy, pno. London
     CS 6844. (also Decca SXL 6710)

+Gr 12-75 p1076   +-RR 12-75 p79
+-HF 11-75 p102   ++SFC 8-24-75 p28
+NR 11-75 p12   ++St 12-75 p117

Etudes, op. 10. cf Ballade, no. 1, op. 23, G minor.
Etude, op. 10, no. 3, E major. cf Piano works (Angel S 36088).
Etude, op. 10, no. 3, E major. cf Piano works (London SPX 21071).
Etude, op. 10, no. 3, E major. cf CBS 77513.
Etude, op. 10, no. 3, E major. cf Turnabout TV 37033.
Etudes, op. 10, nos. 3, 4, 12. cf Piano works (Columbia M 32932).
Etudes, op. 10, nos. 5, 12. cf Piano works (Classics for Pleasure
CFP 40061).
Etude, op. 10, no. 10, A flat major. cf Piano works (Connoisseur
Society CS 2029/30).
Fantasia, op. 49, F minor. cf Concerto, piano, no. 2, op. 21, F
minor.
Fantasia, op. 49, F minor. cf Piano works (Telefunken AW 6-41847).
Fantasia, op. 49, F minor. cf Scherzi.
Fantasia, op. 49, F minor. cf Scherzi, nos. 1-4.
Fantasie-Impromptu, op. 66, C sharp minor. cf Piano works (Angel
S 36088.
Fantasie-Impromptu, op. 66, C sharp minor. cf Piano works (Classics
for Pleasure CFP 40061).
Fantasie-Impromptu, op. 66, C sharp minor. cf Piano works (London
SPX 21071).
Fantasie-Impromptu, op. 66, C sharp minor. cf CBS 77513.
Fantasy on Polish airs, op. 13, A major. cf Piano works (Philips
6500 422).
Funeral march (arr. Elgar). cf ELGAR: Works, selections (HMV ASD
3050).

Impromptu, no. 1, op. 29, A flat major. cf Ballade, no. 4, op.
52, F minor.
Impromptu, no. 3, op. 51, G flat major. cf Barcarolle, op. 60,
F sharp major.
Krakowiak, op. 14, F major. cf Concerto, piano, no. 1, op. 11,
E minor.
Krakowiak, op. 14, F major. cf Concerto, piano, no. 2, op. 21,
F minor.
726 Mazurkas, complete. Ronald Smith, pno. HMV SLS 5014 (3).
        +-Gr 5-75 p1992   +-RR 6-75 p60
        ++HFN 5-75 p125
Mazurka, op. 7, no. 1, B flat major. cf Piano works (Decca PFS
4313).
Mazurka, op. 7, no. 3, F minor. cf Piano works (Columbia M 32932).
Mazurka, op. 17, no. 4, A minor. cf Piano works (Decca PFS 4313).
Mazurka, op. 24, no. 4, B flat minor. cf Ballade, no. 1, op.
23, G minor.
Mazurka, op. 24, no. 4, B flat minor. cf Piano works (RCA VH 018).
Mazurka, op. 30, no. 3, D flat major. cf Piano works (Columbia
M 32932).
Mazurka, op. 30, no. 4, C sharp minor. cf Piano works (RCA VH 011).
Mazurka, op. 33, no. 2, D major. cf Piano works (Columbia M 32932).
Mazurka, op. 33, no. 2, D major. cf CBS 77513.
Mazurka, op. 41, no. 2, F minor. cf Piano works (Columbia M 32932).
Mazurka, op. 50, no. 3, C sharp minor. cf Piano works (Columbia
M 32932).
Mazurka, op. 50, no. 3, C sharp minor. cf Piano works (Erato STU
70843).
Mazurka, op. 59, no. 3, F sharp minor. cf Piano works (Columbia
M 32932).
Mazurka, op. 62, no. 2, A minor. cf HMV SLA 870.
Mazurkas, op. 63, nos. 1-3. cf Barcarolle, op. 60, F sharp major.
Mazurka, op. 63, no. 3. cf Piano works (Erato STU 70843).
Mazurkas, op. 67, nos. 2, 4. cf Piano works (Erato STU 70843).
Minuet. cf Piano works (Everest SDBR 3377).
727 Nocturnes (19). Ivan Moravec, pno. Connoisseur CS 1065/1165 (2).
        +RR 2-75 p51
Nocturne, D flat major. cf RCA ARM 4-0942/7.
Nocturne, E flat major. cf RCA ARM 4-0942/7.
Nocturne, E minor. cf RCA ARM 4-0942/7.
Nocturne, op. 9, no. 2, E flat major. cf Piano works (Angel S 36088)
Nocturne, op. 9, no. 2, E flat major. cf Piano works (Classics for
Pleasure CFP 40061).
Nocturne, op. 9, no. 2, E flat major. cf CBS 77513.
Nocturne, op. 9, no. 2, E flat major. cf Discopaedia MB 1006.
Nocturne, op. 9, no. 2, E flat major. cf Discopaedia MB 1003.
Nocturne, op. 9, no. 2, E flat major. cf Discopaedia MB 1009.
Nocturne, op. 9, no. 2, E flat major (arr. Popper). cf Ember
GVC 42.
Nocturne, op. 9, no. 2, E flat major. cf HMV HQS 1353.
Nocturne, op. 9, no. 3, B major. cf Ballade, no. 1, op. 23, G
minor.
Nocturne, op. 9, no. 3, B major. cf Piano works (RCA VH 018).
Nocturne, op. 9, no. 3, B major. cf International Piano Archives
IPA 5007/8.
Nocturne, op. 15, no. 1, F major. cf Piano works (Everest SDBR
3377).

Nocturne, op. 15, no. 1, F major. cf Piano works (RCA VH 018).
Nocturne, op. 15, no. 2, F sharp major. cf Piano works (RCA
    VH 011).
Nocturne, op. 15, no. 2, F sharp major. cf Piano works (Telefunken
    AW 6-41847).
Nocturne, op. 27, no. 1, C sharp minor. cf Piano works (RCA VH 011).
Nocturne, op. 27, no. 2, D flat major. cf Piano works (London
    SPC 21071).
Nocturne, op. 37, no. 1, G minor. cf Ballade, no. 4, op. 52, F
    minor.
Nocturne, op. 37, no. 2, G major. cf Ballade, no. 1, op. 23, G
    minor.
Nocturne, op. 55, no. 1, F minor. cf Piano works (Decca PFS 4313).
Nocturnes, op. 62, nos. 1 and 2. cf Barcarolle, op. 60, F sharp
    major.
Nocturnes, op. 62, nos. 1 and 2. cf Piano works (Erato STU 70843).
Nocturne, op. 62, no. 1, B major. cf Ballade, no. 4, op. 52, F
    minor.
Nocturne, op. 62, no. 1, B major. cf Piano works (Decca SXL 6693).
Nocturne, op. 72, E minor. cf Piano works (RCA VH 018).
728 Piano works: Etude, op. 10, no. 3, E major. Fantasie-Impromptu,
        op. 66, C sharp minor. Nocturne, op. 9, no. 2, E flat major.
        Polonaise, op. 53, A flat major. Preludes, op. 28, nos. 4, 7,
        15, 20. Waltzes, op. 34, no. 2, A minor; op. 64, no. 2, C
        sharp minor; op. 69, no. 1, A flat major; op. 70, nos. 2 and 3.
        Leonard Pennario, pno. Angel S 36088.
            +NR 10-75 p10
729 Piano works: Berceuse, op. 57, D flat major. Etudes, op. 10, nos.
        5, 12. Fantasie-Impromptu, op. 66, C sharp minor. Nocturne,
        op. 9, no. 2, E flat major. Polonaises, op. 26, no. 1, A major;
        op. 40, no. 1, A major. Prelude, op. 28, no. 15, D flat major.
        Scherzo, no. 2, op. 31, B flat minor. Allan Schiller, pno.
        Classics for Pleasure CFP 40061.
            +Gr 2-75 p1516   +-RR 3-75 p45
730 Piano works: Etudes, op. 10, nos. 3, 4, 12. Mazurkas, op. 7, no.
        3; op. 30, no. 3; op. 33, no. 2; op. 41, no. 2; op. 50, no.
        3; op. 59, no. 3. Polonaise, op. 40, no. 1, A major. Prelude,
        op. 28, no. 6, B minor. Waltz, op. 64, no. 2, C sharp minor.
        Vladimir Horowitz, pno. Columbia M 32932. Tape (c) MT 32932
        (ct) MA 32932. (also CBS 76307)
            +Gr 2-75 p1516   ++RR 2-75 p51
            +HF 2-75 p90   +SFC 10-20-74 p26
            +-NR 4-75 p11   +-St 4-75 p98
731 Piano works: Concerto, piano, no. 1, op. 11, E minor. Etude, op.
        10, no. 10, A flat major. Polonaise, op. 44, F sharp minor.
        Scherzo, no. 4, op. 54, E major. Sonata, piano, no. 3, op. 58,
        B minor. Garrick Ohlsson, pno; Warsaw National Philharmonic
        Orchestra; Witold Rowicki. Connoisseur Society CS 2029/30 (3).
            +Gr 3-75 p1677   +RR 2-75 p52
732 Piano works: Ballade, no. 4, op. 52, F minor. Mazurka, op. 17,
        no. 4, A minor. Mazurka, op. 7, no. 1, B flat major. Nocturne,
        op. 55, no. 1, F minor. Polonaise, op. 40, no. 1, A major.
        Sonata, piano, no. 3, op. 58, B minor. Waltz, op. 34, no. 2,
        A minor. Ilana Vered, pno. Decca PFS 4313. Tape (c) KPFC
        4313. (also London SPC 21119. Tape (c) 052119 (ct) 082119) (r)
        E 42119)

         ++Gr 11-74 p922   -RR 11-74 p78
         +Gr 2-75 p1562 tape   ++SFC 11-17-74 p32
         +HF 7-75 p102 tape   +-SR 1-25-75 p50
733 Piano works: Ballade, no. 3, op. 47, A flat major. Barcarolle,
       op. 60, F sharp major. Nocturne, op. 62, no. 1, B major.
       Scherzo, no. 3, op. 39, C sharp minor. Concerto, piano, no. 2,
       op. 21, F minor. Vladimir Ashkenazy, pno; LSO; David Zinman.
       Decca SXL 6693. (Reissues from SXL 6143, 6174, 6215, 6334)
         +Gr 4-75 p1835   ++RR 4-75 p26
         ++HFN 5-75 p125
734 Piano works: Ballade, no. 4, op. 52, F minor. Mazurkas, op. 50,
       no. 3, C sharp minor; op. 63, no. 3; op. 67, nos. 2, 4. Polon-
       aise-Fantasie, op. 61, A flat major. Nocturnes, op. 62 (2).
       Joseph Kalichstein, pno. Erato STU 70843.
         ++RR 2-75 p52
735 Piano works: Minuet. Nocturne, op. 15, no. 1, F major. Polish
       song, op. 74, no. 1: Maiden's wish. Scherzo, no. 2, op. 31,
       B flat minor. Waltz, op. 18, E flat major. RACHMANINOFF:
       Barcarolle, op. 10, no. 3, G minor. Elegie, op. 3, no. 1.
       Melodie, op. 3, no. 3, E major. Polichinelle, op. 3, no. 4,
       F sharp minor. Polka de W. R. Prelude, op. 3, no. 2, C sharp
       minor. Prelude, op. 23, no. 5, G minor. Sergei Rachmaninoff,
       pno. Everest SDBR 3377.
         +NR 11-75 p15
736 Piano works: Ballade, no. 1, op. 23, G minor. Barcarolle, op. 60,
       F sharp major. Etude, op. 10, no. 3, E major. Fantasie-
       Impromptu, op. 66, C sharp minor. Nocturne, op. 27, no. 2,
       D sharp major. Scherzo, no. 3, op. 39, C sharp minor.
       Valses, op. 64, nos. 1 and 2. Ivan Davis, pno. London SPC
       21071.
         ++MJ 3-75 p26
737 Piano works: Andante spianato and grande polonaise, op. 22, E flat
       major. Fantasy on Polish airs, op. 13, A major. Variations on
       "Là ci darem la mano", op. 2. Claudio Arrau, pno; LPO; Eliahu
       Inbal. Philips 6500 422. Tape (c) 7300 198.
         ++Audio 8-75 p80   +MJ 7-74 p48
         +-Gr 7-73 p194   ++NR 6-74 p10
         +-HF 9-74 p92   +RR 7-73 p46
         ++HFN 7-75 p90 tape   ++SFC 3-17-74 p25
738 Piano works: Ballade, no. 1, op. 23, G minor. Mazurka, op. 30,
       no. 4, C sharp minor. Nocturne, op. 27, no. 1, C sharp minor.
       Nocturne, op. 15, no. 2, F sharp major. Polonaise, op. 53,
       A flat major. Scherzo, no. 1, op. 20, B minor. Waltz, op. 34,
       no. 2, A minor. Waltz, op. 64, no. 2, C sharp minor. Vladimir
       Horowitz, pno. RCA VH 011. (Reissues from HMV DB 6688, DB
       10131, APL 1111, 1430, RCA RB 16064)
         +-Gr 3-75 p1678   +-RR 4-75 p49
739 Piano works: Ballade, no. 3, op. 47, A flat major. Mazurka, op.
       24, no. 4, B flat minor. Nocturnes, op. 9, no. 3, B major;
       op. 72, E minor; op. 15, no. 1, F major. Scherzo, no. 2, op.
       31, B flat minor. Scherzo, no. 3, op. 39, C sharp minor.
       Vladimir Horowitz, pno. RCA VH 018.
         +-RR 12-75 p80
740 Piano works: Ballade, no. 3, op. 47, A flat major. Berceuse, op.
       57, D flat major. Fantasia, op. 49, F minor. Nocturne, op.
       15, no. 2, F sharp major. Polonaise, op. 53, A flat major.

Scherzo, no. 1, op. 20, B minor. Nelson Freire, pno. Telefunken AW 6-41847. Tape (c) CX 4-41847.
    +Gr 6-75 p68   ++HFN 9-75 p110 tape
    +Gr 8-75 p376 tape   +NR 7-75 p14
    -HF 10-75 p72   +RR 5-75 p52
    /HFN 5-75 p126

741 Polish songs (Chants polonaise), op. 74 (17). Songs, posth. (2).
Annette Celine, s; Felicja Blumenthal, pno. Everest 3370.
    +NR 4-75 p8   +-St 7-75 p95
    +SFC 8-31-75 p20

742 Chants polonaise, op. 74, no. 1: A maiden's wish (arr. Liszt).
LISZT: Hungarian rhapsodies, nos. 2, 10, G 244. PADEREWSKI:
Caprice, op. 14, no. 3, G major. Mélodie, op. 8, no. 2.
Légende, op. 16, no. 1. Minuet, op. 14, no. 1, G major.
Nocturne, op. 16, no. 4, B flat major. Ignace Paderewski, pno.
Ember GVC 43.
    +HFN 12-75 p160   +-RR 9-75 p61
Polish songs, op. 74, no. 1: Maiden's wish. cf Piano works
(Everest SDBR 3377).
Chants polonaise, op. 74, no. 1: The maiden's wish; no. 12, My
joy. cf Scherzi, nos. 1-4.
Polonaise, op. 3. cf BRAHMS: Sonata, violoncello, no. 1, op. 38,
E minor.
Polonaise, op. 26, no. 1, A major. cf Piano works (Classics for
Pleasure CFP 40061).
Polonaise, op. 26, no. 2, E flat minor. International Piano
Archives IPA 5007/8.
Polonaise, op. 35, A flat major. cf Concerto, piano, no. 2, op.
21, F minor.
Polonaise, op. 40, no. 1, A major. cf Ballade, no. 1, op. 23, G
minor.
Polonaise, op. 40, no. 1, A major. cf Concerto, piano, no. 2,
op. 21, F minor.
Polonaise, op. 40, no. 1, A major. cf Piano works (Classics for
Pleasure CFP 40061).
Polonaise, op. 40, no. 1, A major. cf Piano works (Columbia
32932).
Polonaise, op. 40, no. 1, A major. cf Piano works (Decca PFS 4313).
Polonaise, op. 40, no. 1, A major. cf CBS 77513.
Polonaise, op. 40, no. 1, A major. cf HMV HQS 1353.
Polonaise, op. 44, F sharp minor. cf Piano works (Connoisseur
Society CS 2029/30).
Polonaise, op. 53, A flat major. cf Piano works (Angel S 36088).
Polonaise, op. 53, A flat major. cf Piano works (RCA VH 011).
Polonaise, op. 53, A flat major. cf Piano works (Telefunken AW
6-41847).
Polonaise, op. 53, A flat major. cf CBS 77513.
Polonaise, op. 53, A flat major. cf International Piano Library
IPL 5005/6.
Polonaise, op. 71, no. 2, B flat major. cf Ballade, no. 4, op.
52, F minor.
Polonaise-Fantaisie, op. 61, A flat major. cf Barcarolle, op. 60,
F sharp major.
Polonaise-Fantaisie, op. 61, A flat major. cf Piano works (Erato
STU 70843).
Prelude, A flat major. cf Barcarolle, op. 60, F sharp major.

743 Preludes, complete. Barcarolle, op. 60, F sharp major. Garrick
      Ohlsson, pno. HMV HQS 1338. (also Angel S 37087)
              +—Gr 1-75 p1366   ++SFC 6-29-75 p26
              -HF 10-75 p72   +—St 11-75 p126
              +RR 1-75 p47
      Preludes, complete. cf Camden CCV 5000-12, 5014-24.
744 Preludes, op. 28, nos. 1-24. Berceuse, op. 57, D flat major.
      Alicia de Larrocha, pno. Decca SXL 6733.
              +Gr 11-75 p860   +—RR 11-75 p76
              +—HFN 11-75 p152
745 Preludes, op. 28, nos. 1-24. Carol Rosenberger, pno. Delos DEL
      15311.
              -HF 1-75 p80   -NR 6-74 p10
              ++MJ 2-74 p13   +—NYT 8-11-74 pD22
746 Preludes, op. 28, nos. 1-24. Maurizio Pollini, pno. DG 2530 550.
              +Gr 12-75 p1076   ++RR 12-75 p80
747 Preludes, op. 28, nos. 1-24. Prelude, op. 45, C sharp minor.
      Prelude, op. posth., A flat major. Ruth Slenczynska, pno.
      Musical Heritage Society MHS 1841.
              +HF 1-75 p80
      Preludes, op. 28, nos. 1-24. cf Barcarolle, op. 60, F sharp major.
748 Preludes, op. 28, nos. 1-24; op. 45, C sharp minor; op. posth.,
      no. 26. Caludio Arrau, pno. Philips 6500 622. Tape (c)
      7300 335.
              +Gr 6-75 p68   ++NR 9-75 p12
              +—Gr 10-75 p721 tape   ++RR 6-75 p65
              ++HF 10-75 p72   ++SFC 8-10-75 p26
              ++HFN 6-75 p87   ++St 11-75 p126
              +HFN 7-75 p90
      Preludes, op. 28. cf RCA ARL 2-0512.
      Preludes, op. 28, nos. 4, 7, 15, 20. cf Piano works (Angel S 36088).
      Prelude, op. 28, no. 6, B minor. cf Piano works (Columbia M 32932).
      Prelude, op. 28, no. 7, A major. cf CBS 77513.
      Prelude, op. 28, no. 15, D flat major. cf Piano works (Classics
          for Pleasure CFP 40061).
      Prelude, op. 28, no. 15, D flat major. cf Decca SPA 372.
      Prelude, op. 45, C sharp minor. cf Barcarolle, op. 60, F sharp
          major.
      Prelude, op. 45, C sharp minor. cf Preludes, op. 28, nos. 1-24
          (Musical Heritage Society MHS 1841).
      Prelude, op. 45, C sharp minor. cf Preludes, op. 28, nos. 1-24
          (Philips 6500 622).
      Prelude, op. posth., A flat major. cf Preludes, op. 28, nos. 1-24
          (Musical Heritage Society MHS 1841).
      Prelude, op. posth., no. 26. cf Preludes, op. 28, nos. 1-24 (Phil-
          ips 6500 622).
749 Scherzi, nos. 1-4. Fantasia, op. 49, F minor. François Duchable,
      pno. Connoisseur Society CSQ 2086.
              -SFC 11-16-75 p32
750 Scherzi, nos. 1-4. Chants polonaise, op. 74, no. 1: The maiden's
      wish; no. 12, My joy. Antonio Barbosa, pno. Connoisseur
      Society CS 2071 (Q) CSQ 2071.
              +HF 3-75 p77   ++SFC 7-27-75 p22
              +—NR 10-75 p10   +St 3-75 p98 Quad
751 Scherzi, nos. 1-4. Felicja Blumenthal, pno. Everest SDBR 3384.
              -NR 12-75 p13
752 Scherzi, nos. 1-4. Fantasia, op. 49, F minor. Garrich Ohlsson,

pno. HMV HQS 1328. (also Angel S 37017)
```
        +-Gr 7-74 p233                    +-RR 7-74 p61
        -HF 8-75 p82                      +-SR 3-22-75 p36
        ++NR 4-75 p11                     +St 6-74 p95
```
Scherzo, no. 1, op. 20, B minor. cf Piano works (RCA VH 011).

Scherzo, no. 1, op. 20, B minor. cf Piano works (Telefunken AW
6-41847).

Scherzo, no. 2, op. 31, B flat minor. cf Andante spianato and
grande polonaise, op. 22, E flat major.

Scherzo, no. 2, op. 31, B flat minor. cf Piano works (Classics
for Pleasure CFP 40061).

Scherzo, no. 2, op. 31, B flat minor. cf Piano works (Everest
SDBR 3377).

Scherzo, no. 2, op. 31, B flat minor. cf Piano works (RCA VH 018).

Scherzo, no. 3, op. 39, C sharp minor. cf Piano works (Decca
SXL 6693).

Scherzo, no. 3, op. 39, C sharp minor. cf Piano works (London
SPC 21071).

Scherzo, no. 3, op. 39, C sharp minor. cf Piano works (RCA VH 018).

753 Scherzo, no. 4, op. 54, E major. PROKOFIEV: Cinderella, op. 87:
Gavotte. Visions fugitives, op. 22 (5). RACHMANINOFF: Preludes
(3). RAVEL: Jeu d'eau. Miroirs: La vallee des cloches.
Sviatoslav Richter, pno. RCA AGL 1-1279. (Reissue)
+SFC 11-16-74 p32

Scherzo, no. 4, op. 54, E major. cf Piano works (Connoisseur
Society CS 2029/30).

754 Sonata, piano, no. 2, op. 35, B flat minor. Sonata, piano, no.
3, op. 58, B minor. Daniel Barenboim, pno. HMV ASD 3064.
```
        +Gr 8-75 p347                     ++RR 8-75 p56
        ++HFN 8-75 p74
```
755 Sonata, piano, no. 2, op. 35, B flat minor. LISZT: Sonata, piano,
G 178, B minor. Tedd Joselson, pno. RCA ARL 1-1010. Tape
(ct) ARS 1-1010.
```
        +-HF 9-75 p84                     +-SR 9-6-75 p40
        +NR 7-75 p14                      +St 10-75 p110
        -SFC 7-27-75 p22
```
756 Sonata, piano, no. 2, op. 35, B flat minor. CLEMENTI: Sonata,
piano, B flat major. Arturo Benedetti Michelangeli, pno.
Rococo 2088.
+-NR 10-75 p9

Sonata, piano, no. 2, op. 35, B flat minor. cf Andante spianato
and grande polonaise, op. 22, E flat major.

757 Sonata, piano, no. 3, op. 58, B minor. LISZT: Gnomenreigen, G 145.
Etude d'exécution transcendente d'après Paganini, no. 6, A
minor, G 140. Paraphrases, SCHUBERT: Horch. Horch. Die Lerch,
G 588/9; Der Muller und der Bach; Liebesbotschaft; Das Wandern.
Emanuel Ax, pno. RCA ARL 1-1030.
```
        +Gr 12-75 p1085                   ++RR 12-75 p79
        +HF 11-75 p128                    +SR 9-6-75 p40
        +NR 9-75 p11                      +St 10-75 p72
```
758 Sonata, piano, no. 3, op. 58, B minor. LISZT: Sonata, piano,
G 178, B minor. Alfred Cortot, pno. Seraphim M 60241. (Re-
issue).
+NR 9-75 p12

Sonata, piano, no. 3, op. 58, B minor. cf Piano works (Connoisseur
Society CS 2029/30).

Sonata, piano, no. 3, op. 58, B minor. cf Piano works (Decca
    PFS 4313).
Sonata, piano, no. 3, op. 58, B minor. cf Sonata, piano, no. 2,
    op. 35, B flat minor.
Sonata, violoncello, op. 65, G minor: Largo. cf Orion ORS 75181.
Sonata, violoncello, op. 65, G minor: 3rd movement. cf HMV SEOM 19.
Songs, posth. (2). cf Polish songs, op. 74.
Les sylphides: Prélude; Valse; Mazurka; Valse; Grand valse
    brillante. cf Decca SDDJ 393/95.
Les sylphides: Prélude; Valse; Mazurka; Valse; Grand valse bril-
    lante (arr. Douglas). cf Decca DPA 515/6.
Variations on "Là ci darem la maon", op. 2. cf Piano works
    (Philips 6500 422).
759 Waltzes, complete. Felicja Blumenthal, pno. Everest SDBR 3378.
        +Gr 12-75 p1076   +NR 11-75 p7
760 Waltzes (18). Aldo Ciccolini, pno. Seraphim S 60252.
        /NR 12-75 p13
761 Waltzes (19). Abbey Simon, pno. Turnabout TVS 34580.
        +HF 11-75 p102   +NR 9-75 p12
762 Waltzes, nos. 1-4. Stefan Askenase, pno. DG Heliodor 2548 146.
    (Reissue from DG 136396)
        +Gr 4-75 p1867   +-RR 4-75 p50
        +HFN 10-75 p152
763 Waltzes, nos. 1-4. Dinu Lipatti, pno. HMV HLM 7075. (Reissue
    from Columbia LX 1341/6)
        +-Gr 12-75 p1076   +RR 12-75 p80
        +HFN 12-75 p152
Valse, op. 18, E flat major. cf Ballade, no. 4, op. 52, F minor.
Waltz, op. 18, E flat major. cf Piano works (Everest SDBR 3377).
Waltz, op. 18, E flat major. cf CBS 77513.
Waltz, op. 18, E flat major. cf International Piano Archives
    IPA 5007/8.
Waltz, op. 18, E flat major. cf Philips 6747 204.
Valse, op. 34, no. 1, A flat major. cf Ballade, no. 1, op. 23,
    G minor.
Valse, op. 34, no. 1, A flat major. cf Rococo 2049.
Waltz, op. 34, no. 2, A minor. cf Piano works (Angel S 36088).
Waltz, op. 34, no. 2, A minor. cf Piano works (Decca PFS 4313).
Waltz, op. 34, no. 2, A minor. cf Piano works (RCA VH 011).
Waltz, op. 34, no. 2, A minor. cf Connoisseur Society (Q) CSQ 2066.
Valses, op. 64, nos. 1 and 2. cf Piano works (London SPC 21071).
Waltz, op. 64, no. 1, D flat major. cf Ballade, no. 4, op. 52,
    F minor.
Waltz, op. 64, no. 1, D flat major. cf CBS 77513.
Waltz, op. 64, no. 1, D flat major. cf Discopaedia MB 1008.
Waltz, op. 64, no. 1, D flat major. cf International Piano Archives
    IPA 5007/8.
Waltz, op. 64, no. 1, D flat major (arr. and orch. Gerhardt). cf
    RCA LRL 1-5094.
Waltz, op. 64, no. 2, C sharp minor. cf Piano works (Angel S 36088).
Waltz, op. 64, no. 2, C sharp minor. cf Piano works (Columbia M
    32932).
Waltz, op. 64, no. 2, C sharp minor. cf Piano works (RCA VH 011).
Waltz, op. 64, no. 2, C sharp minor. cf CBS 77513.
Waltz, op. 64, no. 2, C sharp minor. cf HMV HQS 1353.
Waltz, op. 69, no. 1, A flat major. cf Piano works (Angel S 36088).

Waltz, op. 69, no. 1, A flat major. cf Connoisseur Society
    (Q) CSQ 2065.
Waltz, op. 69, no. 2, B minor. cf Discopaedia MB 1009.
Waltz, op. 70, no. 1, G flat major. cf CBS 77513.
Waltz, op. 70, no. 1, G flat major. cf Discopaedia MB 1008.
Waltzes, op. 70, nos. 2 and 3. cf Piano works (Angel S 36088).
Waltz, op. 70, no. 3, D flat major. cf Turnabout TV 37033.
Zyczenie. cf Discophilia KGS 3.
CHRISTMAS AT COLORADO STATE UNIVERSITY. cf University of Colorado,
    unnumbered.
CHRISTOV, Dobri
    Velikaia ekteniia; Vo tsarstvii tvoem; Herouvimska; Tebe poem;
        Dostoine est. cf ARKHANGELSKII: Blajen razoumnevai; Otche nach.
    Velika ekteniya (Great litany); Vo Tsarstve tvoyem (In Thy king-
        dom); Kheruvimska, no. 2 (Hymn of the cherubim, no. 2); Tebe
        poyem (To Thee we sing); Dostoyno yest (It is fitting). cf
        Monitor MFS 757.
CILEA, Francesco
764 Adriana Lecouvreur: Io sono l'umile ancella. BELLINI: La sonnam-
        bula: Ah, se una volta sola...Ah, non credea mirarti...Ah,
        non giunge. PUCCINI: Suor Angelica: Senza mamma. VERDI: Un
        ballo in maschera: Morrò, ma prima in grazia. Rigoletto: Caro
        nome. Il trovatore: D'amor sull'ali rosee. I vespri Siciliani:
        Arrigo...Ah parli a un core; Mercè, dilette amiche. Montserrat
        Caballé, s; Barcelona Symphony Orchestra; Gianfranco Masini.
        Decca SXL 6690.
                +Gr 5-75 p2019   +RR 4-75 p19
                +HFN 6-75 p103
    Adriana Lecouvreur: Arias. cf BELLINI: La sonnambula: Arias.
    Adriana Lecouvreur: Io son l'umile ancella; Poveri fiori. cf
        CATALANI: La wally: Ebben, Ne andrò lontana.
    Adriana Lecouvreur: Ma dunque e vero. cf London OS 26315.
CIMA, Giovanni
    Sonatas, recorder, D minor, G minor. cf Telefunken SAWT 9589.
    Sonatas, recorder, D major, G major. cf Telefunken SMA 25121 T/1-3.
CIMAROSA, Domenico
    Concertante, 2 flutes and orchestra, G major. cf RCA CRL 2-7003.
    Concerto, 2 flutes and orchestra, G major: Allegro. cf Decca SPA
        394.
    Concerto, oboe, C minor. cf ALBINONI: Concerto, oboe, op. 7, no.
        3, B flat major.
    Quartet, flute and strings, no. 1, D major. cf BERNSTEIN: A
        Cabris.
    Quartet, flute and strings, no. 4, F major. cf BERNSTEIN: A
        Cabris.
CINQUE
    Trobadorica. cf LEONCAVALLO: I Pagliacci.
CIOFFI
    Na sera 'e maggio. cf LEONCAVALLO: I Pagliacci.
    Na sera 'e maggio. cf Musical Heritage Society MHS 1951.
CLARKE, Herbert
    The bride of the waves. cf Nonesuch H 71298.
    The débutante. cf Nonesuch H 71298.
    From the shores of the mighty Pacific. cf Nonesuch H 71298.
    Sounds from the Hudson. cf Nonesuch H 71298.

CLARKE, Jeremiah
    Prince of Denmark's march. cf Saga 5417.
    Suite, D major. cf Argo ZDA 203.
    Trumpet voluntary. cf CRD CRD 1008.
    Trumpet voluntary. cf Decca SDD 463.
    Trumpet voluntary. cf Philips 6580 066.
    Trumpet voluntary. cf Philips 6747 204.
CLAYTON
    Come, come ye saints. cf CBS 61599.
CLEMENS NON PAPA
    Vox in Roma. cf Turnabout TV 34569.
CLEMENTI, Muzio
    Sonata, piano, B flat major. cf CHOPIN: Sonata, piano, no. 2,
        op. 35, B flat minor.
    Sonata, piano, op. 14, no. 3, F minor. cf Sonata, piano, op. 34,
        no. 2, G minor.
    Sonata, piano, op. 26, no. 2, F sharp minor. cf Sonata, piano,
        op. 34, no. 2, G minor.
765 Sonata, piano, op. 34, no. 2, G minor. Sonata, piano, op. 14,
      no. 3, F minor. Sonata, piano, op. 26, no. 2, F sharp minor.
      Vladimir Horowitz, pno. RCA VH 007. (Reissue from HMV ALP
      1340)
          +Gr 9-75 p544   ++RR 9-74 p73
          +MT 12-75 p1070
766 Sonatinas, op. 36, nos. 1-6. GRIEG: Concerto, piano, op. 16, A
      minor. Marie-Aimée Varro, pno; Moravian Philharmonic Orchestra;
      Ludovic Rajter. Orion ORS 75196.
        -HF 12-75 p94   +NR 9-75 p7
    Sonatinas, op. 36, nos. 1-6. cf Columbia MG 33202.
CLERAMBAULT, Louis
    Basse et dessus de trompette. cf Vista VPS 1021.
    Largo on G string. cf RCA ARM 4-0942/7.
    Suite du deuxième ton: Plain jeu; Duo, Trio, Basse de cromarne,
        Caprice sur les grands jeux. cf University of Iowa Press,
        unnumbered.
CLERISSE
    Caravane. cf Canon CNN 4983.
    Introduction et scherzo. cf Canon CNN 4983.
COATES, Eric
    The dambusters march. cf Decca 419.
COLEMAN
    Pieces, sackbuts and cornets (3). cf Crystal S 202.
COLERIDGE-TAYLOR, Samuel
    Danse negre. cf STILL: Afro-American symphony.
    Hiawatha's wedding feast: Awake, beloved. cf STILL: Afro-American
        symphony.
    Onaway. cf STILL: Afro-American symphony.
COMPERE, Louis
    Virgo celesti. cf L'Oiseau-Lyre 12BB 203/6.
CONRADI, Johann
    Suite, lute, C major. cf BACH: Suite, lute, S 995, G minor.
COOKE, Arnold
    Jabez and the devil. cf Symphony, no. 3, D major.
767 Symphony, no. 3, D major. Jabez and the devil. LPO; Nicholas
      Braithwaite. Lyrita SRCS 78.
          ++Gr 11-75 p817   ++RR 12-75 p46
          ++HFN 10-75 p141

COOLIDGE, Clark
    Preface. cf 1750 Arch 1752.
COOPER, Robert
    I have been a foster. cf L'Oiseau-Lyre SOL 329.
COPE, David
    Arena. cf BLANK: Esther's monologue.
    Margins. cf BLANK: Esther's monologue.
COPLAND, Aaron
    Appalachian spring. cf Works, selections (Columbia D3M 33720).
    Billy the kid. cf Works, selections (Columbia D3M 33720).
768 Dance panels. Danzón cubano. Latin American sketches (3). El
      salón Mexico. LSO, NPhO; Aaron Copland. Columbia M 33269.
      (also CBS 73451)
            +Gr 9-75 p452   +RR 9-75 p34
            +HF 8-75 p82   ++SFC 6-15-75 p24
            +HFN 9-75 p97   +St 7-75 p95
            +NR 6-75 p2   +Te 12-75 p40
    Dance panels. cf Works, selections (Columbia D3M 33720).
769 Dance symphony. STEVENS: Symphony, no. 1. Japan Philharmonic
      Symphony Orchestra; Akeo Watanabe. CRI SD 129.
            +-RR 12-75 p46
    Danzón cubano. cf Dance panels.
770 Down a country lane. John Henry. Letter from home. Music for
      movies. The red pony. NPhO; Aaron Copland. Columbia M 33586.
            ++SFC 12-21-75 p39
    Fanfare for the common man. cf Works, selections (Columbia D3M
      33720).
    Fantasy. cf BARTOK: Etudes, op. 18.
    John Henry. cf Down a country lane.
    Latin American sketches (3). cf Dance panels.
    Letter from home. cf Down a country lane.
    Lincoln portrait. cf Works, selections (Columbia D3M 33720).
    Music for movies. cf Down a country lane.
    Our town. cf Works, selections (Columbia D3M 33720).
    Passacaglia. cf Scherzo humoristique: The cat and the mouse.
    Piano blues (4). cf Scherzo humoristique: The cat and the mouse.
771 Quartet, piano. IVES: Trio, piano. Cardiff Festival Ensemble.
      Argo ZRG 794.
            +Gr 8-75 p341   +-RR 8-75 p52
            +HFN 8-75 p75   ++Te 12-75 p40
    Quiet city. cf Era 1001.
    The red pony. cf Down a country lane.
    Rodeo. cf Works, selections (Columbia D3M 33720).
    El salón Mexico. cf Dance panels.
    El salón Mexico. cf Works, selections (Columbia D3M 33720).
772 Scherzo humoristique: The cat and the mouse. Passacaglia. Piano
      blues (4). Sonata, piano. Robert Silverman, pno. Orion ORS
      7280.
            +HF 9-72 p74   ++RR 9-75 p57
            +MJ 4-73 p34   ++SFC 3-4-73 p33
            ++NR 8-72 p13   ++St 10-72 p123
    Serenade, ukulele. cf Orion ORS 74160.
    Sonata, piano. cf Scherzo humoristique: The cat and the mouse.
    Sonata, violin and piano. cf BINKERD: Sonata, violin and piano.
773 Works, selections: Appalachian spring. Billy the kid. Dance panels.
      Fanfare for the common man. Lincoln portrait. Our town.

Rodeo. El salón Mexico. LSO, NPhO; Henry Fonda, narrator;
Aaron Copland. Columbia D3M 33720 (3).
+SFC 12-21-75 p39
CORDERO, Roque
774 Concerto, violin and orchestra. Miniatures, small orchestra (8).
Sanford Allen, vln; Detroit Symphony Orchestra; Paul Freeman.
Columbia M 32784.
+HF 6-74 p71   +-NYT 5-12-74 pD26
++MQ 10-75 p645   *St 7-74 p105
+NR 7-74 p14
Miniatures, small orchestra (8). cf Concerto, violin and orchestra.
CORDIGIANI
La via del pastore. cf Rococo 5368.
CORELLI, Arcangelo
775 Concerti grossi, op. 6. AMF; Neville Marriner. Argo ZRG 773/5 (3).
+Gr 9-74 p492   ++NR 2-75 p2
++HF 1-75 p82   ++RR 9-74 p44
776 Concerti grossi, op. 6, nos. 6-8. AMF; Neville Marriner. Argo
ZRG 828.
+HFN 12-75 p171   ++RR 12-74 p46
777 Concerto grosso, op. 6, no. 8, C minor. GLUCK: Chaconne. PACHEL-
BEL: Canon, D minor. RICCIOTTI: Concerto, no. 2, G major
(attrib.). Stuttgart Chamber Orchestra; Karl Münchinger.
London CS 6206.
++SFC 12-28-75 p30
Concerto grosso, op. 6, no. 8, C minor. cf Decca SDD 411.
Concerto grosso, op. 6, no. 8, C minor. cf Grange SGR 1124.
Sonata, recorder, F major. cf Telefunken SMA 25121 T/1-3.
Sonata, recorder and continuo, op. 5, no. 4, F major. cf Tele-
funken SAWT 9589.
778 Sonatas, violin and continuo, op. 5 (12). Eduard Melkus, vln;
Huguette Dreyfus, hpd, org; Karl Scheit, lt; Garo Altmacayan,
vlc; Vienna Capella Academica. DG 2533 132/3 (2).
+Gr 10-75 p640   ++RR 9-75 p58
+-HF 6-73 p74   ++St 6-73 p118
++HFN 9-75 p98   +STL 10-5-75 p36
779 Sonatas, violin and continuo, op. 5 (12). Sonya Monosoff, vln;
James Weaver, hpd and org; Judith Davidoff, vlc and vla da
gamba. Musical Heritage Society MHS 1690/3 (3).
+HF 5-75 p75
780 Suite, strings (arr. Pinelli). HANDEL: The royal fireworks music:
Suite (arr. Harty). Water music: Suite (arr. Ormandy). PO;
Eugene Ormandy. CBS 61639. (Reissues from ABE 10057, SABE
2020)
-Gr 7-75 p175   +-RR 6-75 p41
+HFN 8-75 p87
Variations on 'La follia', recorder. cf Telefunken SMA 25121 T/1-3.
CORNAGO, Juan
Songs: Señora, qual soy venido; Gentil dama; Pues que Dios. cf
Telefunken SAWT 9620/1.
CORNELIUS, Peter
781 Der Barbier von Bagdad. Sylvia Geszty, s; Trudeliese Schmidt, ms;
Adalbert Kraus, Gerhard Unger, t; Bernd Weikl, bar; Karl Ridder-
busch, bs; Bavarian Broadcasting Orchestra and Chorus; Heinrich
Hollreiser. Eurodisc 86 830 (2).
+HF 10-74 p92   +St 3-74 p114
+ON 2-1-75 p29

Songs: Die Könige; Simeon. cf Telefunken AJ 6-41867.
CORNYSHE, William
    Adieu mes amours. cf L'Oiseau-Lyre SOL 329.
    Ah, Robin. cf BASF 25 22286-1.
    Ah, Robin; Hoyda jolly Rutterkin. cf Harmonia Mundi 204.
    Blow thy horn, hunter. cf BASF 25 22286-1.
    Blow thy horn, hunter. cf L'Oiseau-Lyre SOL 329.
    Fa la sol. cf L'Oiseau-Lyre SOL 329.
    Hoyda, jolly Rutterkin. cf BASF 25 22286-1.
    While life or breath. cf L'Oiseau-Lyre SOL 329.
CORRETTE, Michel
    Menuets, nos. 1, 2. cf DG Archive 2533 172.
COSTE, Napoleon
    Study, no. 22, op. 38, A major. cf Amberlee ACL 501X.
COSTELEY, Guillaume
    Hélas, hélas, que de mal. cf L'Oiseau-Lyre 12BB 203/6.
COUPERIN, François
    Air tendre. cf Klavier KS 536.
    L'apothéose de Corelli. cf Works, selections (Philips 6747 174).
    L'apothéose de Lully. cf Works, selections (Philips 6747 174).
    L'art de toucher le clavecin. cf Livres de clavecin, Bk II, Ordre
        nos. 6, 7, 8.
782 L'art de toucher le clavecin: Preludes (8). Livres de clavecin,
        Bk I, Ordre nos. 3, 4; Bk II, Ordre nos. 6, 11; Bk III, Ordre
        nos. 13, 16, 18; Bk IV, Ordre nos. 24, 26. Huguette Dreyfus,
        hpd. Telefunken EK 6-35276 (4).
            +-RR 12-75 p80
    Les barricades mysterieuses. cf Angel S 37015.
783 Concerts royaux, nos. 1-4. Joachim Starke, flt; Rainer Kussmaul,
        vln; Helmut Lissok, vlc; Klaus Preis, hpd. Oryx ORYX 1719.
        (Reissue from Musica Rara MUS 3)
            +-Gr 8-75 p341   -HFN 11-75 p153
    Concerts royaux, nos. 1-4. cf Works, selections (Philips 6747 174).
784 Concerts royaux, nos. 2 and 3. Livres de clavecin, Bk III, Ordres
        nos. 13-19. Kenneth Gilbert, Martha Brickman, hpd; Gian Lyman-
        Silbiger, vla da gamba. RCA SER 5720/3 (4).
            +-Gr 4-75 p1836   +-RR 4-75 p50
    Couplet on Domine Deus, Rex Coelestis. cf Wealden WS 160.
    Les goûts réunis: Concerto, no. 9, E major. cf BACH: Sonata,
        flute (oboe) and harpsichord, S 1030, G minor.
    Livres de clavecin, Bk I, Ordre nos. 3, 4; Bk II, Ordre nos 6, 11;
        Bk III, Ordre nos. 13, 16, 18; Bk IV, Ordre nos. 24, 26.
        cf L'art de toucher le clavecin: Preludes.
785 Livres de clavecin, Bk II, Ordre nos. 6, 7, 8. L'art de toucher
        le clavecin. Kenneth Gilbert, hpd. RCA LHL 1-5048/9.
            +-Gr 1-75 p1366   +-STL 1-12-75 p36
            +-RR 12-74 p56
786 Livres de clavecin, Bk II, Ordres nos. 9, 10, 11, 12. Kenneth
        Gilbert, hpd. RCA LHL 1-5050/1 (2).
            +-Gr 2-75 p1523   +-RR 12-74 p56
            +-RR 1-75 p47
    Livres de clavecin, Bk III, Ordres, nos. 13-19. cf Concerts
        royaux, nos. 2 and 3.
    Livres de clavecin, Bk III, Ordre no. 17: Les petits moulins à
        vent. cf Discopaedia MB 1010.
    Livres de clavecin, Bk IV, Ordre no. 17: Les petits moulins à
        vent. cf RCA ARM 4-0942/7.

787 Livres de clavecin, Bk IV, Ordres 20-27. Kenneth Gilbert, hpd;
       Gian Lyman-Silbiger, vla da gamba. RCA LHL 4-5906 (4).
         +HFN 12-75 p152   ++RR 10-75 p22
     Livres de clavecin, Bk IV: Ordre no. 26. cf Draco DR 1333.
788 Messe pour les couvents. Messe pour les paroisses. André Isoir,
       org. CRD CAL 1907/8 (2)
         ++Gr 9-75 p486
     Messe pour le couvents: Et in terra pax. cf Hungaroton SLPX 11548.
     Messe pour les paroisses. cf Messe pour les couvents.
     Muséte de choisi. cf HMV SLS 5022.
     Muséte de taverni. cf HMV SLS 5022.
789 Les Nations. Quadro Amsterdam. Telefunken TK 11550/1-2 (2).
       (Reissue from SAWT 9476, 9546)
         ++HF 4-75 p81   ++RR 8-74 p44
         ++NR 10-74 p8   ++St 2-75 p107
     Nouveaux concerts, nos. 5-7, 9-12, 14. cf Works, selections
       (Philips 6747 174).
     Offertoire sur les grands jeux. cf BIS LP 7.
     La steinquerque. cf Works, selections (Philips 6747 174).
     La sultane, D minor. cf Works, selections (Philips 6747 174).
     La superbe. cf Works, selections (Philips 6747 174).
     La tromba. cf Klavier KS 536.
790 Works, selections: L'apothéose de Corelli. L'apothéose de Lully.
       Concerts royaux, nos. 1-4. Nouveaux ceoncerts, nos. 5-7, 9-12,
       14. La sultane, D minor. La superbe. La steinquerque.
       Sigiswald Kuijken, baroque violin, treble viol, bass viola da
       gamba; Lucy van Dael, Janine Rubinlicht, baroque violins;
       Wieland Kuijken, baroque violoncello, bass viola da gamba;
       Adelheid Glatt, bass viola da gamba; Barthold Kuijken, German
       flute, recorder; Oswald van Olmen, Frans Brüggen, German flt;
       Bruce Haynes, Jürg Schaeftlein, Paul Dombrecht, baroque oboes;
       Hans Jürg Lange, Milan Turkovíc, baroque bassoons, Robert
       Kohnen, hpd, speaker. Philips 6747 174 (6).
         +-Gr 10-74 p645   +RR 9-75 p58
         +HFN 10-75 p141
COUPERIN, Louis
     Passacaille, C major. cf Harmonia Mundi HMU 334.
     Pavane. cf Suite, harpsichord, D minor.
     Pavane, D minor. cf Angel S 37015.
     La Piémontaise, A minor. cf Suite, harpsichord, D minor.
     La Piémontaise, A minor. cf Harmonia Mundi HMU 334.
     Prelude, D minor. cf BASF BAB 9005.
791 Suite, harpsichord, D minor. Suite, harpsichord, C major. Suite,
       harpsichord, F major. La Piémontaise, A minor. Pavane. Blan-
       dine Verlet, hpd. Telefunken SAWT 9605.
         +Gr 6-74 p79   +NR 7-74 p12
         +HF 7-74 p88   ++RR 4-74 p64
         +MJ 3-75 p34
     Suite, harpsichord, C major. cf Suite, harpsichord, D minor.
     Suite, harpsichord, F major. cf Suite, harpsichord, D minor.
     Le tombeau de M. Blancrocher. cf Angel S 37015.
     Le tombeau de M. Blancrocher. cf Saga 5402.
     COURTLY PASTIMES OF 16TH CENTURY ENGLAND. cf L'Oiseau-Lyre SOL 329.
COWELL, Henry
     Symphony, no. 16. cf ISOLFFSON: Passacaglia.
CRECQUILLON, Thomas
     Le corps absent. cf Hungaroton SLPX 11549.

Un gay bergier. cf Hungaroton SLPX 11549.
CRESTON, Paul
    Suite, op. 18. cf Orion ORS 74160.
CROFT
    O God, our help in ages past. cf RCA ARL 1-0562.
CROFT, William
    Ground. cf RCA LRL L-5087.
    Sarabande. cf RCA LRL 1-5087.
CROSS, Lowell
    Etudes, magnetic tape. cf CRI SD 342.
CROSSE, Gordon
    Concerto da camera. cf BIRTWISTLE: Trajoedia.
    The new world, op. 25. cf BERKELEY: Songs (Argo ZRG 788).
792 Purgatory. Glenville Hargreaves, bar; Peter Bodenham, t; Royal
        Northern School of Music Orchestra and Chorus; Michael Lankester.
        Argo ZRG 810.
                +Gr 12-75 p1093   ++RR 11-75 p32
                +HFN 12-75 p152
CRUMB, George
793 Black angels, electric string quartet. RAXACH: Quartet, strings,
        no. 2. LEEUW: Quartet, strings, no. 2. Gaudeamus Quartet.
        Philips 6500 881.
                +-HF 6-75 p106   +-SR 6-14-75 p46
                ++NR 6-75 p8   ++St 11-75 p127
794 Eleven echoes of autumn, 1965. WOLPE: Trio, flute, violoncello
        and piano. Group for Contemporary Music, Columbia University;
        Aeolian Chamber Players. CRI SD 233.
                +Gr 7-75 p200   *RR 1-75 p45
                +-HFN 5-75 p126
795 Madrigals. Elizabeth Suderburg, s; David Shrader, perc; Felix
        Skowronek, flt; Pamela Vokolek, hp; W. Ring Warner, double-bs.
        Turnabout TVS 34523.
                +HR 11-74 p100   +St 1-75 p106
                +-NR 2-75 p10
    Madrigals, Bks. I-IV. cf LONDON: Portraits of three ladies
        (American).
796 Makrokosmos, vol. 1. David Burge, pno. Nonesuch H 71293.
                +Gr 5-75 p1995   +NR 5-74 p13
                ++HF 6-74 p80   +RR 3-75 p46
                ++LJ 5-75 p43   +St 8-74 p108
797 Music for a summer evening (Makrokosmos III). Gilbert Kalish,
        James Freeman, pno; Raymond DesRoches, Richard Fitz, perc.
        Nonesuch H 71311.
                ++HF 10-75 p73   +-RR 10-75 p75
                ++HFN 12-75 p152   ++St 12-75 p117
                +NR 10-75 p14
798 Sonata, solo violoncello. HINDEMITH: Sonata, solo violoncello,
        no. 3, op. 25. WELLESZ: Sonata, solo violoncello, op. 30.
        YSAYE: Sonata, solo violoncello, op. 28. Robert Sylvester, vlc.
        Desto DC 7169.
                ++HF 6-75 p108   ++St 10-75 p120
CRUSELL, Bernard
799 Quartet, clarinet, no. 2, op. 4, C minor. HUMMEL: Quartet, clari-
        net, E flat major. The Music Party. L'Oiseau-Lyre Florilegium
        DSLO 501.
                +Gr 11-74 p915   +RR 11-74 p17
                +-HF 11-75 p104   +SFC 12-28-75 p30
                ++NR 5-75 p8

CUI, Cesar
    Kaleidoscope, op. 50, no. 9. cf Discopaedia MB 1007.
    Kaleidoscope, op. 50, no. 9. cf Discopaedia MB 1008.
    Statue in Tsarskoye Selo. cf RCA ARL 1-5007.
CUMMING, Richard
    Songs: Go lovely rose; The little black boy; Memory, hither come.
        cf Duke University Press DWR 7306.
CURTIS, E. de
    Carme. cf Ember GVC 46.
    Torna a Surriento; Voce e notte. cf Musical Heritage Society
        MHS 1951.
    Tu c'nun chiagne. cf LEONCAVALLO: I Pagliacci.
    Vergiss mein nicht. cf Polydor 2489 519.
CUTTING, Francis
    Almain. cf DG Archive 2533 157.
    Galliard. cf Vanguard SRV 316.
    Greensleeves. cf DG Archive 2533 157.
    Greensleeves. cf L"Oiseau-Lyre SOL 336.
    The squirrel's toy. cf DG Archive 2533 157.
    Walsingham. cf DG Archive 2533 157.
CZERNY, Carl
    Duo concertante, flute and piano, op. 129. cf BACH: Partita,
        flute and harpsichord, S 987, C minor.
    Fantasia on Scottish airs, op. 471. cf International Piano
        Library IPL 5005/6.
    Fantasy and variations on Persiani's "Inez de Castro", op. 377.
        cf Sonata, piano, no. 1, op. 7, A flat major.
800 Sonata, piano, no. 1, op. 7, A flat major. Fantasy and variations
        on Persiani's "Inez de Castro", op. 377. Hilde Somer, pno.
        Genesis GS 1057.
            +Audio 9-75 p70   +-SFC 3-2-75 p25
            +-HF 6-75 p88   +-St 12-75 p118
            ++NR 9-75 p12
DAHL, Ingolf
    Music, brass instruments. cf Crystal S 202.
    Music, brass instruments. cf Desto DC 6476/7.
DALL'ABACO, Evaristo
801 Concerto da chiesa, a 4, op. 2. MARCELLO: Concerto, oboe and
        strings, D minor. VIVALDI: Concerto, violoncello, D major.
        Concerto, flute and strings, op. 10, no. 3, G major. Kurt
        Hausmann, ob; Jörg Metzger, vlc; Kurt Redel, flt; Würzburg
        Camerata Accademica, Pro Arte Orchestra; Hans Reinartz, Kurt
        Redel. Pye GSGC 14142.
            +Gr 2-75 p1504   +-RR 2-75 p33
DALLAPICCOLA, Luigi
    Piccola musica notturna. cf BUSONI: Berceuse elégiaque, op. 42.
    Preghiere. cf BUSONI: Berceuse elégiaque, op. 42.
802 Il Prigionero. Giulia Barrera, con; Maurizio Mazzieri, bar; Romano
        Emile, Gabor Carelli, Ray Harrell, t; University of Maryland
        Chorus; Washington National Symphony Orchestra; Antal Dorati.
        Decca HEAD 10. (also London OSA 1166)
            +Gr 5-75 p2010   +RR 5-75 p23
            +HFN 7-75 p79   ++SFC 11-9-75 p22
            ++MT 10-75 p885   +Te 6-75 p48
            ++NYT 12-21-75 pD18
    Quaderno musicale de Annalibera. cf BARTOK: Etudes, op. 18.

Sex carmina alcaei. cf BUSONI: Berceuse elégiaque, op. 42.
Sicut umbra. cf Tempus destruendi, Tempus aedificandi.
803 Studi, violin and piano (2). HINDEMITH: Sonata, violin, op. 31,
     no. 2. SCHONBERG: Fantasia, violin and piano, op. 47. SWIFT:
     Sonata, solo violin, op. 15. WEBERN: Pieces, op. 7 (4). Robert
     Gross, vln; Richard Grayson, pno. Orion ORS 74147.
          +HF 9-74 p108   ++St 12-74 p151
          +NR 3-75 p12
804 Tempus destruendi, Tempus aedificandi. Sicut umbra. SHAW: A
     lesson from Ecclesiastes. Music when soft voices die. Peter
     and the lame man. To the Bandusian spring. Sybil Michelow, ms;
     BBC Singers; London Sinfonietta Orchestra; Gary Bertini. Argo
     ZRG 791.
          +Gr 12-75 p1086   +-RR 11-75 p84
          +NYT 12-21-75 pD18
DALZA, Joan Ambrosio
     Calata ala Spagnola. cf DG Archive 2533 111.
     Calata ala Spagnola. cf DG Archive 2533 184.
     Enfermo estaba Antioco. cf HMV SLS 5012.
     Recercar; Suite ferrarese; Tastar de corde. cf L'Oiseau-Lyre
     12BB 203/6.
DAMASE, J. M.
     Sonata, G major. cf Musical Heritage Society MHS 1345.
DANCE MUSIC OF THE HIGH BAROQUE. cf DG Archive 2533 172.
DANCE MUSIC OF THE RENAISSANCE. cf DG Archive 2533 111.
DANDRIEU, Jean
     Armes, amours. cf Telefunken ER 6-35257.
     Noëls: Le Roy des cieux vient de naître; Adam fut un pauvre homme;
     A minuit fut fait un reveil; Chrétien que suivez l'eglise;
     Joseph est bien Marié. cf Argo ZRG 783.
DANKWORTH, John
805 Tom Sawyer's Saturday. HANDEL: Concerto, organ, no. 4, op. 4, F
     major: 3rd movement. Solomon: Entry of the Queen of Sheba.
     ROSSINI: La cenerentola: Overture. TRAD.: Ding dong. The first
     Nowell. God rest you merry gentlemen. Merrily on high. The
     national anthem. O come all ye faithful. It came upon the mid-
     night clear. BBC Singers; BBC Academy; John Birch, org;
     Edward Heath. HMV CSD 3763.
          +Gr 1-75 p1379   +-RR 2-75 p38
DANYEL, John
     A fancy. cf L'Oiseau-Lyre SOL 336.
     Passymeasures galliard. cf L'Oiseau-Lyre SOL 336.
     Rosamunde pavan. cf L'Oiseau-Lyre SOL 336.
DANZI, Franz
     Quartet, flute, op. 56, no. 2, D minor. cf Oryx ORYX 1720.
     Sonata, horn, op. 28, E flat major. cf BEETHOVEN: Sonata, horn,
     op. 17, F major.
     Songs: Ach was ist die Liebe; Ich denke dein; Dir ruh o die ich
     liebe; Ich hab ein Mädchen; Ich liebe dich; In des Lebens
     Maien; Oft am Rande stiller Flöten; Wann erwacht der Knabe
     Wieder. cf Oryx ORYX 1720.
DAQUIN, Louis
     The Noël, G major. cf University of Colorado, unnumbered.
     Noël X. cf Hungaroton SLPX 11548.
DARE, Elkanah Kelsey
     Babylonian captivity. cf Nonesuch H 71276.

DARGOMIJSKY, Alexander (also Dargomizhsky, Alexander)
    Romance. cf Desto DC 7118/9.
    Songs: Nocturnal breeze; The old corporal; The worm. cf London
        OS 26249.
    DARWIN, SONG FOR A CITY. cf Decca SXL 6719.
DAVICO, Vicenzo
    Songs: O lune che fa'lume; Japanese songs. cf Pearl SHE 511.
DAVID, Félicien
    La perle de Breseil: Charmant oiseau. cf Rubini RS 300.
DAVID, Gyula
806 Concerto, horn. Festive overture. RANKI: Cantus Urbis. Margit
    László, s; Márta Szirmay, con; György Korondi, t; Endre Utö,
    bs; Gábor Lehotka, org; János Sebestyén, hpd; Ferenc Tarjáni,
    hn; Debrecen Kodály Chorus; HSO, HRT Orchestra; András Kórody,
    János Feréncsik, György Lehel. Hungaroton SLPX 11699.
        +NR 7-75 p8   +RR 9-75 p71
    Festive overture. cf Concerto. horn.
DAVIDOVSKY, Mario
807 Chacona (1972), violin, violoncello and piano. Inflexions, 14
    players. STREET: Quartet, strings. TRYTHALL: Coincidences,
    piano. Jeanne Benjamin, vln; Joel Krosnick, vlc; Robert Miller,
    Richard Trythall, pno; Concord Quartet; Orchestra; David Gilbert.
    CRI SD 305.
        +-HF 8-74 p111   +-NR 3-74 p6
        +LJ 4-75 p70   ++St 11-74 p125
    Inflexions, 14 players. cf Chacona (1972), violin, violoncello
        and piano.
    Junctures. cf Nonesuch HB 73028.
808 Synchronisms, no. 1. KORTE: Remembrances. KUPFERMAN: Superflute.
    Samuel Baron, flt; with prerecorded tapes. Nonesuch H 71289.
        ++HF 6-74 p106   +NR 4-74 p14
        +LJ 2-75 p36   ++St 6-74 p128
DAVIES
    Psalm 121. cf Pye GH 589.
DAVIES, Peter Maxwell
    Ave Maria. cf HMV HQS 1350.
    Fantasia on John Taverner's 'In Nomine', no. 2. cf Taverner:
        Points and dances.
    Leopardi fragments. cf BENNETT: Calendar.
    Lullaby for Ilian Rainbow. cf L'Oiseau-Lyre DSLO 3.
    O magnum mysterium. cf Decca 5BBA 1013-5.
809 Taverner: Points and dances. Fantasia on John Taverner's 'In
    Nomine', no. 2. Fires of London; Peter Maxwell Davies; NPhO;
    Charles Groves. Argo ZRG 712.
        +-Gr 5-73 p2048   +HFN 5-73 p983
        +HF 6-75 p89   +RR 5-73 p25
DAVIES, Wolford
    Solemn melody. cf Decca SDD 463.
DAVIS
    The West's awake. cf Philips 6599 227.
DAY, E.
    Psalm 84: O how amiable. cf Polydor 2460 250.
DEBUSSY, Claude
    Arabesques. cf Saga 5356.
    Berceuse héroïque. cf Piano works (Telefunken SMA 25110 T/1-3).
    Berceuse héroïque. cf Works, selections (HMV SLS 893).

Berceuse héroïque. cf Works, selections (Angel S 37064).
810 La boîte à joujoux. Printemps. French National Radio Orchestra;
     Jean Martinon. Angel S 37124.
          ++NR 12-75 p4   ++SFC 9-21-75 p34
     La boîte à joujoux. cf Piano works (Telefunken SMA 25110 T/1-3).
     La boîte à joujoux. cf Works, selections (HMV SLS 893).
     Chansons de Bilitis. cf BIZET: Songs (Decca SXL 6577).
     Chansons de Bilitis (3). cf Desto DC 7118/9.
     Chansons de Bilitis. cf Odyssey Y2 32880.
     Chanson de Bilitis: La chevelure. cf RCA ARM 4-0942/7.
     Children's corner suite. cf Piano works (Telefunken SMA 25110 T/1-3).
     Children's corner suite. cf Works, selections (Angel S 37064).
     Children's corner suite. cf Works, selections (HMV SLS 893).
     Children's corner suite, no. 6: Golliwog's cakewalk. cf Canon
          CNN 4983.
     Children's corner suite, no. 6: Golliwog's cakewalk. cf Connoisseur
          Society (Q) CSQ 2065.
     Children's corner suite: Golliwog's cakewalk; Snowflakes are danc-
          ing. cf Works, selections (RCA ARL 1-0488).
     Children's corner suite: Serenade for the doll. cf RCA VH 020.
     Danse. cf Works, selections (Angel S 37064).
811 Danse sacrée et profane. MEHUL: The two blind men of Toledo: Over-
          ture. RAVEL: Pavane pour une infante défunte. Tzigane. SAINT-
          SAENS: Havanaise, op. 83. Introduction and rondo capriccioso,
          op. 28. Yan Pascal Tortelier, vln; John Manson, hp; London
          Symphonia; Wyn Morris. Pye TYPLS 13062. Tape (c) ZCTPL 13062
          (ct) Y8TPL 13062.
          +-Gr 5-75 p1980   +-Gr 5-75 p2032 tape
     Danses sacrée et profane. cf Works, selections (Angel S 37065).
     Danses sacrée et profane. cf Works, selections (Columbia D3M 32988).
     Danses sacrée et profane. cf Works, selections (HMV SLS 893).
     L'Enfant prodigue: Prelude. cf RCA ARM 4-0942/7.
812 Estampes, no. 3: Jardins sous la pluie. Pour le piano. Préludes,
          Bk 2: La puerta del vino; Ondine; Feuilles mortes. RAVEL:
          Sonatine. Ivan Moravec, pno. Connoisseur Society CS 2010.
          +RR 2-75 p55
     Estampes, no. 3: Gardens in the rain. cf Works, selections (RCA
          ARL 1-0488).
813 Etudes, complete. Anthony di Bonaventura, pno. Connoisseur Society
          (Q) CSQ 2074.
          ++HF 11-75 p102   ++SFC 4-20-75 p23
          +NR 11-75 p13   +St 9-75 p107
     Etudes, Bks 1-2. cf Piano works (Telefunken SMA 25110 T/1-3).
     Fantaisie, piano. cf Works, selections (Angel S 37065).
     Fantaisie, piano. cf Works, selections (HMV SLS 893).
814 Fantaisie, piano and orchestra. Rhapsody, clarinet and orchestra.
          Rhapsody, saxophone and orchestra. Marylène Dosse, pno; Guy
          Dangain, clt; Jean-Marie Londeix, sax; Luxembourg Radio Orches-
          tra; Louis de Froment. Candide CE 31069. (also Vox STGBY 679)
          +Gr 4-75 p1962   +NR 10-73 p13
          +-HF 3-74 p84   +RR 5-75 p29
          /HFN 7-75 p79
     Fetes galantes, Set I: En sourdine; Fantoches; Clair de lune;
          Mandoline. cf Songs (Cambridge CRS 2774).
     Fetes galantes, Set I: Fantoches. cf Columbia M 32231.
     Hommage à Haydn. cf Piano works (Telefunken SMA 25110 T/1-3).
     Images (1894, unpublished). cf Piano works (Telefunken SMA 25110 T/1-3).

Images. cf Works, selections (HMV SLS 893).

815 Images, Bks. 1 and 2. RAVEL: Gaspard de la nuit. Arturo Benedetti
    Michelangeli, pno. Rococo 2073.
    ++NR 2-75 p11

816 Images, Bk 1. TAVEL: Gaspard de la nuit. REZAC: Sisyfova nedele
    (The Sunday of Sisyfos). Boris Krajný, pno. Supraphon 110 227.
    +-RR 6-74 p72

Images: Ibéria, gigues, rondes de printemps. cf Works, selections
    (Columbia D3M 32988).

Images: Reflets dans l'eau. cf CHOPIN: Ballade, no. 1, op. 23,
    G minor.

Images: Reflets dans l'eau. cf Columbia Special Products AP 12411.

817 Images for orchestra. Jeux. ORTF: Jean Martinon. Angel S 37066.
    +-HF 6-75 p89   +St 8-75 p95
    ++NR 6-75 p4

818 Images pour orchestra: Ibéria. Nocturnes, nos. 1-3. Prélude à
    l'après midi d'un faune. Netherlands Radio Chorus; Netherlands
    Philharmonic Orchestra; Jean Fournet. Decca PFS 4317. Tape
    (c) KPFC 4317.
      -Gr 10-75 p608   -RR 9-75 p35
    +HFN 12-75 p153

819 Images pour orchestra; Ibéria, Rondes de printemps. FALLA: El
    sombrero de tres picos: Three dances. CPhO; Jean Fournet.
    Supraphon SUAST 50614. Tape (c) 04 50614.
    +Gr 6-75 p106 tape   +-RR 7-75 p70 tape
    +HFN 5-75 p142 tape

Jeux. cf Images for orchestra.

Jeux. cf Works, selections (Columbia D3M 32988).

Jeux. cf Works, selections (HMV SLS 893).

Jeux: Poeme danse. cf SABATA: Juventus.

Khamma. cf Works, selections (Angel 37067/8).

Khamma. cf Works, selections (HMV SLS 893).

Marche écossaise. cf Works, selections (Angel 37067/8).

Marche écossaise. cf Works, selections (HMV SLS 893).

820 La mer. RAVEL: Daphnis and Chloe: Suite, no. 2. Pavane pour une
    infante défunte. CO; Georg Szell. CBS 61075. Tape (c) 40-
    61075.
    +-HFN 12-75 p173 tape

821 La mer. RAVEL: Daphnis et Chloe: Suite, no. 2. La valse. Hallé
    Orchestra; John Barbirolli. Pye GSGC 15013. (Reissue from
    GSGC 14010)
      +Gr 8-75 p335   -RR 8-75 p31
    +-HFN 8-75 p87

822 La mer. Prélude à l'près midi d'un faune. RAVEL: Daphnis and
    Chloe: Suite, no. 2. PO; Mendelssohn Club Chorus, Philadelphia;
    Eugene Ormandy. RCA ARL 1-0029. Tape (c) ARS 1-0029 (ct) ART
    1-0029 (r) ARK 1-0029 (Q) ARD 1-0029.
      ++Gr 6-73 p44   /MJ 1973 annual p16
      ++Gr 6-75 p36   -NR 9-73 p3
      +-HF 9-73 p101   +-RR 7-73 p56
      +-HFN 6-73 p1175   ++RR 6-75 p35
      +HFN 8-75 p87   +-St 9-73 p116
      -LJ 5-74 p58

La mer. cf Preludes, Bk 1, no. 10: La cathédrale engloutie.

La mer. cf Works, selections (Angel 37067/8).

La mer. cf Works, selections (Columbia D3M 32988).

La mer. cf Works, selections (HMV SLS 893).
La mer. cf RAVEL: Rapsodie espagnole.
La mer: Dialogue between the wind and the sea. cf Decca 396.
Musiques pour le Roi Lear. cf Works, selections (HMV SLS 893).
823 Nocturnes (arr. Ravel). RACHMANINOFF: Symphonic dances, op. 45.
    Anne Shasby, Richard McMahon, pno. Argo ZRG 808.
        ++Gr 6-75 p61   ++RR 6-75 p72
        ++HFN 6-75 p91
Nocturnes. cf Works, selections (Angel 37067/8).
Nocturnes. cf Works, selections (HMV SLS 893).
Nocturnes. cf RAVEL: Daphnis et Chloe: Suite, no. 2.
Nocturnes, nos. 1-3. cf Images pour orchestra: Ibéria.
Nocturnes: Fêtes. cf BIZET: Jeux d'enfants, op. 22.
Nocturnes: Nuages, fêtes, sirènes. cf Works, selections (Columbia
    D3M 32988).
824 Pelléas et Mélisande. Irène Joachim, Leila Ben-Sedira, s; Germaine
    Cernay, ms; Jacques Jansen, t; H. Etcheverry, bar; A. Narçon,
    Paul Cabanel, bs; Yvonne Gouverne Choeurs; Roger Désmormière.
    EMI Odeon 2C 153 12513/5 (3).
        +HF 4-75 p76
Petite suite. cf Works, selections (Angel S 37064).
Petite suite. cf Works, selections (HMV SLS 893).
Petite suite. cf BIZET: Jeux d'enfants, op. 22.
Petite suite: En bateau. cf Connoisseur Society CS 2070.
825 Piano works: Berceuse héroïque. La boîte à joujoux. Children's
    corner suite. Etudes, Bks 1-2. Hommage à Haydn. Images (1894,
    unpublished). La plus que lente. Préludes, Bk 2. Noël Lee,
    pno. Telefunken SMA 25110 T/103 (3). (Reissue from Oryx 6XLC 4)
        -Gr 1-75 p1366   +-RR 12-74 p57
La plus que lente. cf Piano works (Telefunken SMA 25110 T/1-3).
La plus que lente. cf Works, selections (Angel S 37064).
La plus que lente. cf Works, selections (HMV SLS 893).
La plus que lente. cf CBS 76420.
La plus que lente. cf Connoisseur Society CS 2070.
La plus que lente. cf RCA ARM 4-0942/7.
Pour le piano. cf Estampes, no. 3: Jardins sous la pluie.
Preludes. cf Draco DR 1333.
826 Preludes, Bk 1. Livia Rev, pno. Saga 5391.
        +Gr 3-75 p1678   +-RR 4-75 p50
827 Preludes, Bks 1 and 2. Monique Haas, pno. DG 2726 038.
        +HFN 9-75 p109   /RR 8-75 p56
Preludes, Bk 1, no. 2: Voiles. Preludes, Bk 2, no. 24: Feux
    d'artifice. cf Columbia Special Products AP 12411.
Preludes, Bk 1, no. 8: La fille aux cheveux de lin. cf Orion ORS
    75181.
Prelude, Bk 1, no. 8: La fille aux cheveux de lin (2). cf RCA
    ARM 4-0942/7.
Preludes, Bk 1, no. 8: The girl with the flaxen hair; no. 10: The
    engulfed cathedral; no. 6: Footprints in the snow. cf Works,
    selections (RCA ARL L-0488).
Preludes, Bk 1, no. 8: The girl with the flaxen hair. cf Columbia
    M 32070.
828 Preludes, Bk 1, no. 10: La cathédrale engloutie. La mer. Prélude
    à l'après midi d'un faune. LSO, NPhO; Leopold Stokowski. London
    21109. Tape (c) M5 21109 (ct) M8 21109 (r) L 475109. (Reissues
    from SPC 21059, 21006, 21090/1) (also Decca SDD 455. Reissues

from PFS 4220, 4095, OPFS 3/4)
+–Gr 6-75 p90   +–RR 6-75 p35
+–HF 11-74 p91   +SFC 4-28-74 p29
–HFN 8-75 p87

Préludes, Bk 1, op. 10: La cathédrale engloutie; no. 1: Danseuses
de Delphes; no. 12: Minstrels; no. 4: Les sons et les parfums
tounent dans l'air du soir. cf BRAHMS: Variations on a theme
by Paganini, op. 35, A minor.
Préludes, Bk 2. cf Piano works (Telefunken SMA 25110 T/1-3).
Préludes, Bk 2: La puerta del vino; Ondine; Feuilles mortes. cf
Estampes, no. 3: Jardins sous la pluie.
Prélude à l'apres-midi d'un faune. cf Prelude, Bk 1, no. 10: La
cathédrale engloutie.
Prélude à l'après-midi d'un faune. cf Images pour orchestra: Ibéria.
Prélude à l'après-midi d'un faune. cf La mer.
Prélude à l'après-midi d'un faune. cf Works, selections (Angel
37067/8).
Prélude à l'après-midi d'un faune. cf Works, selections (Columbia
D3M 32988).
Prélude à l'après-midi d'un faune. cf Works, selections (HMV SLS
893).
Prélude à l'après-midi d'un faune. cf BERLIOZ: Les Troyens: Royal
hunt and storm.
Prélude à l'après-midi d'un faune. cf Columbia M2X 33014.
Prélude à l'après-midi d'un faune. cf Decca DPA 519/20.
Prélude à l'après-midi d'un faune. cf HMV ASD 3008.
Printemps. cf La boîte à joujoux.
Printemps. cf Works, selections (Columbia D3M 32988).
Printemps. cf Works, selections (HMV SLS 893).
829 Le promenoir des deux amants. RESPIGHI: Il tramonto. SANTOLIQUIDO:
Poesi persiane (3). WOLF-FERRARI: Quattro rispetti, op. 11,
no. 4. Howard Thain, t; Epsilon Quartet; Hubert Dawkes, Mag-
daleine Panzera-Baillot, pno; Yannis Daris. Pearl SHE 521.
–RR 12-75 p91
830 Quartet, strings, op. 10, G minor. RAVEL: Quartet, strings, F
major. Guarneri Quartet. RCA ARL 1-0187.
+–Gr 3-75 p1673   +RR 3-75 p37
+–HF 5-75 p75   +–SFC7-13-75 p21
+NR 2-75 p6   +–St 6-75 p147
831 Quartet, strings, op. 10, G minor. RAVEL: Quartet, strings, F
major. Danish Quartet. Telefunken SAT 22541. (Reissue from
Valois)
+–HF 11-74 p101   /–RR 10-73 p88
+HFN 11-73 p2314   +St 1-75 p110
+NR 9-74 p6
Quartet, strings, op. 10, G minor. cf Works, selections (DG 2533
007).
Rêverie. cf Works, selections (RCA ARL 1-0488).
Rêverie. cf Columbia M 32070.
Rhapsody, clarinet and orchestra. cf Fantaisie, piano and orches-
tra.
Rhapsody, clarinet and orchestra. cf Works, selections (Angel
S 37065).
Rhapsody, clarinet and orchestra. cf Works, selections (HMV SLS
893).
Rhapsody, clarinet and orchestra, no. 1. cf Works, selections
(Columbia D3M 32988).

Rhapsody, saxophone and orchestra. cf Fantaisie, piano and
    orchestra.
Rhapsody, saxophone and orchestra. cf Works, selections (Angel
    S 37065).
Rhapsody, saxophone and orchestra. cf Works, selections (HMV SLS
    893).
Le Roi Lear: Fanfare, Le sommeil de Lear. cf Works. selections
    (Angel 37067/8).
Sonata, flute, viola and harp. cf Works, selections (DG 2533 007).
Sonata, flute, viola and harp. cf Decca SPA 394.
Sonata, violin. cf Works, selections (DG 2533 007).
832 Sonata, violin and piano. FAURE: Sonata, violin and piano, no. 1,
    op. 13, A major. Maurice Hasson, vln; Michael Isador, pno.
    Classics for Pleasure CFP 40210.
        ++Gr 7-75 p200   +RR 5-75 p52
        +HFN 9-75 p98
833 Sonata, violin and piano, no. 3, G minor. FRANCK: Sonata, violin
    and piano, A major. RAVEL: Tzigane. Denés Zsigmondy, vln;
    Anneliese Nissen, pno. Klavier KS 534.
        +-NR 5-75 p9
834 Sonata, violin and piano, no. 3, G minor. MILHAUD: Sonata, violin
    and piano, no. 2. PROKOFIEV: Sonata, violin and piano, no. 2,
    op. 94 bis, D major. Ion Voicou, vln; Monique Haas, pno.
    London STS 15175.
        +HF 9-75 p104
Sonata, violin and piano, no. 3, G minor. cf BACH: Sonata and
    partita, solo violin, S 1003, A minor.
Sonata, violoncello, D minor. cf Works, selections (DG 2533 007).
835 Sonata, violoncello and piano, no. 1, D minor. PROKOFIEV: Sonata,
    violoncello and piano, op. 119. WEBERN: Kleine Stücke, violon-
    cello, op. 11 (3). Lynn Harrell, vlc; James Levine, pno. RCA
    ARL 1-1262.
        ++SR 11-1-75 p46
836 Songs: Ariettes oubliées: C'est l'extase; Chevaux de Bois; Green;
    L'ombres des arbres; Il pleure dans mon coeur; Spleen. Fêtes
    galantes, Set 1: En sourdine; Fantoches; Clair de lune; Mando-
    line. DUPONT: Mandoline. FAURE: Songs: Clair de lune, op. 46,
    no. 2; Melodies, op. 58; Spleen, op. 51, no. 3. SZULC: Clair
    de lune. Carole Bogard, s; John Moriarty, pno. Cambridge CRS
    2774.
        +HF 6-75 p106   +ON 5-75 p36
        -NR 3-75 p9   ++St 6-75 p108
837 Songs: Chansons de Bilitis: La flûte de Pan; Le chevelure; Le
    tombeau des naiades. Fêtes galantes: En sourdine; Fantoches;
    Clair de lune. Poèmes de Baudelaire: Le balcon; Harmonie du
    soir; Le jet d'eau; Recueillement; La mort des amants. Mando-
    line. Romance. Voici que le printemps. Anna Moffo, s; Jean
    Casadesus, pno. RCS SB 6890.
        +-Gr 11-75 p875   -RR 9-75 p68
        +-HFN 10-75 p141
838 Songs: Ballades de François Villon (3); Chansons de France (3);
    Fêtes galantes, 2nd series; Noël des enfants qui n'ont pas de
    maison; Poèmes de Stéphane Mallarmé (3); Le promenoir de deux
    amants. Bernard Kruysen, bar; Noël Lee, pno. Telefunken SAT
    22540.
        ++HF 5-74 p83   ++NR 3-74 p10
        +-MQ 1-75 p171   ++St 5-74 p100

Song: Ariettes oubliées, no. 2: Il pleure dans mon coeur. cf
     Command COMS 9006.
Song: Ariettes oubliées, no. 2: Il pleure dans mon coeur. cf
     RCA ARM 4-0942/7.
Songs: Ariettes oubliées; Beau soir. cf Seraphim S 60251.
Songs: Proses lyriques: Noël des enfants qui n'ont plus de maison.
     cf BIZET: Songs (Saga 5388).
Suite bergamasque: Clair de lune. cf Connoisseur Society (Q) CSQ
     2066.
Suite bergamasque: Clair de lune. cf Decca SPA 372.
Suite bergamasque: Clair de lune. cf London STS 15160.
Suite bergamasque: Claire de lune; Passepied. cf Works, selections
     (RCA ARL 1-0488).
Suite bergamasque: Clair de lune. cf Turnabout TV 37033.
Syrinx, flute. cf Works, selections (DG 2533 007).
Tarantelle styrienne. cf Works, selections (HMV SLS 893).
839 Works, selections: Berceuse héroïque. Children's corner suite.
     Petite suite. Danse. La plus que lente. French National Radio
     Orchestra; Jean Martinon. Angel S 37064.
          +HF 4-75 p82   ++St 4-75 p99
          ++SFC 12-8-74 p36
840 Works, selections: Danses sacrée et profane. Fantaisie, piano.
     Rhapsody, clarinet and orchestra. Rhapsody, saxophone and
     orchestra. Marie-Claire Jamet, hp; Aldo Ciccolini, pno; Guy
     Dangain, clt; Jean-Marie Londeix, sax; ORTF; Jean Martinon.
     Angel S 37065.
          +HF 4-75 p82   +SFC 4-4-75 p22
          +NR 4-75 p2   +St 8-75 p95
841 Works, selections: Khamma. Marche écossaise. La mer. Nocturnes.
     Prélude à l'après-midi d'un faune. Le Roi Lear: Fanfare; Le
     Sommeil de Lear. ORTF: ORTF Choir; Jean Martinon. Angel S
     37067/8 (2).
          +HF 10-75 p74   ++SFC 8-3-75 p30
          ++NR 8-75 p5
842 Works, selections: Prélude à l'après-midi d'un faune. Danses
     sacrée et profane. Images: Ibéria, gigues, rondes de printemps.
     Jeux. La mer. Nocturnes: Nuages, fêtes, sirènes. Printemps.
     Rhapsody, clarinet and orchestra, no. 1. NPhO; Pierre Boulez.
     Columbia D3M 32988 (3).
          ++SFC 4-20-75 p23   ++St 4-75 p100
843 Works, selections: Quartet, strings, op. 10, G minor. Sonata,
     flute, viola and harp. Sonata, violin. Sonata, violoncello,
     D minor. Syrinx, flute. RAVEL: Introduction and allegro.
     Quartet, strings, F major. Sonata, violin and piano. Trio,
     piano, A minor. Joseph Silverstein, Max Rostal, Monique Frasca-
     Colombier, Marguerite Vidal, vln; Michael Tilson Thomas, Monique
     Haas, pno; Jule Eskin, Hamisa Dor, vlc; Doriot Dwyer, Christian
     Lardé, flt; Burton Fine, Anka Moraver, vla; Ann Hobson, Nicanor
     Zabaleta, hp; Guy Deplus, clt; Drolc Quartet; Triest Trio. DG
     2733 007. (Reissues from 2530 049, SLPM 139369, 138016, 139394,
     138054)
          +-Gr 9-75 p480   -RR 7-75 p47
          +-HFN 8-75 p89
844 Works, selections: Berceuse héroïque. La boîte à joujoux (orch.
     Caplet). Children's corner suite (orch. Caplet). Tarantelle
     styrienne (orch. Ravel). Danses sacrée et profane. Fantaisie,

piano. Images. Jeux. Khamma. Marche écossaise. La mer.
Musiques pour le Roi Lear. Nocturnes. La plus que lente.
Petite suite (orch. Busser). Prélude à l'après-midi d'un faune.
Printemps. Rhapsody, clarinet and orchestra. Rhapsody, saxo-
phone and orchestra. Aldo Ciccolini, Fabienne Boury, pno; John
Leach, cimbalom; Guy Dangain, clt; Jean-Marie Londeix, sax;
Marie-Claire Jamet, hp; ORTF: Jean Martinon. HMV SLS 893 (5).

    +Gr 2-75 p1479   +-MT 7-75 p629
    ++HFN 5-75 p126   +RR 3-75 p25

845 Works, selections: Children's corner suite: Golliwog's cakewalk;
    Snowflakes are dancing. Estampes, no. 3: Gardens in the rain.
    Préludes, Bk 1, no. 8: The girl with the flaxen hair; no. 10,
    The engulfed cathedral; no. 6; Footprints in the snow. Rêverie.
    Suite bergamasque: Clair de lune; Passepied. (arr. by Tomita).
    Isao Tomita, Moog synthesizer. RCA ARL 1-0488. (Q) ARD 1-0488.

    -Audio 8-75 p83   +-NR 7-74 p14
    -Gr 4-75 p1867   -St 8-74 p109

DEGTIAREV
    Preslavnia dnes. cf Harmonia Mundi HMU 133.

DELIBES, Leo
846 Coppelia. Minneapolis Symphony Orchestra; Antal Dorati. Philips
    6780 253 (2). (Reissue from Mercury AMS 16018/9)
        +-Gr 11-75 p817   +RR 12-75 p46

847 Coppelia: Ballet music. Sylvia: Ballet music. GOUNOD: Faust:
    Ballet music. ROSSINI: William Tell: Ballet music. NPhO;
    Charles Mackerras. Classics for Pleasure CFP 40229.
        +HFN 10-75 p152   +RR 10-75 p49

848 Coppelia: Prelude and mazurka; Scène et valse de Swanhilde; Czardas;
    Scène et valse de la poupée; Ballade; Thème slav varie. Sylvia:
    Prélude; Les chassesses; Intermezzo et valse lente; Pas des
    Ethiopiens; Chant bacchique; Pizzicati polka; Cortège de Bacchus.
    BSO Members; Pierre Monteux. Camden CCV 5030.
        +-RR 11-75 p42
    Coppelia: Prélude and mazurka; Valse lente; Thème slav varié;
    Czardas. cf Decca SDDJ 393/95.
    Coppelia: Prelude and mazurka; Valse lente, Act 1; Thème slav varié;
    Czardas, Act 1. cf Decca DPA 515/6.
    Les filles de Cadiz. cf RCA ARL 1-5007.
    Le Roi s'amuse: Passepied. cf Discopaedia MB 1006.
849 Sylvia, ballet. Desmond Bradley, vln; NPhO; Richard Bonynge.
    Decca SXL 6635/6 (2). (also London CSA 2236)
        +Gr 6-74 p45   +NYT 3-9-75 pD23
        ++HF 1-75 p82   ++RR 5-74 p32
        +NR 2-75 p3   ++SFC 10-20-74 p26
    Sylvia: Ballet music. cf Coppelia: Ballet music.
    Sylvia: Prelude; Les chassesses; Intermezzo et valse lente; Pas
    des Ethiopiens; Chant bacchique; Pizzicati polka; Cortège de
    Bacchus. cf Coppelia: Prelude and mazurka; Scène et valse
    de Swanhilde; Czardas; Scène et valse de la poupée; Ballade;
    Thème slav varié.

DELIUS, Frederick
850 Appalachia. North country sketches. RPO: Thomas Beecham. CBS
    61354. (Reissue from HMV Columbia 33 CX 1112, Columbia LX
    1399/01) (also Odyssey Y 33283)
        ++Gr 5-73 p2097   ++RR 5-73 p53
        +HF 7-75 p89   +St 7-75 p110
        ++HFN 5-73 p981

851 Aquarelles (2). (orch. Fenby). Fennimore and Gerda: Intermezzo.
       On hearing the first cuckoo in spring. Summer night on the river.
       VAUGHAN WILLIAMS: Fantasia on "Greensleeves". The lark ascend-
       ing. WALTON: Death of Falstaff: Passacaglia. Touch her soft
       lips and part. Pinchas Zukerman, vln; ECO; Daniel Barenboim.
       DG 2530 505.
                    +Audio 11-75 p95   ++RR 4-75 p37
                    +Gr 4-75 p1807   ++SFC 10-5-75 p38
                    +HF 8-75 p102   +STL 6-8-76 p36
                    ++NR 8-75 p2
852 Brigg Fair. Hassan: Intermezzo and serenade. In a summer garden.
       Irmelin: Prelude. On hearing the first cuckoo in spring. PO,
       RPO, CO; Eugene Ormandy, Thomas Beecham, Georg Szell. CBS
       Harmony 30056. (Reissues from CBS SBDG 72086, Columbia 33C 1017,
       Fontana CF 1020)
                    +Gr 5-75 p1962   -HFN 7-75 p79
853 Brigg Fair. Dance rhapsody, no. 2. In a summer garden. On hear-
       ing the first cuckoo in spring. PO; Eugene Ormandy. CBS 61426.
       Tape (c) 40-61426. (Reissue from SBRG 72086)
                    +Gr 6-74 p46   -RR 5-74 p32
                    +-HFN 12-75 p173 tape
854 Caprice and elegy. Concerto, piano. Concerto, violin. Albert
       Sammons, vln; Benno Moiseiwitsch, pno; Beatrice Harrison, vlc;
       Liverpool Philharmonic Orchestra, PhO, CO; Malcolm Sargent,
       Constant Lambert. World Records SH 224. (Reissues from Colum-
       bia DX 1160/2, HMV 3533/5, B 3721)
                    +Gr 8-75 p321   +RR 7-75 p24
                    +HFN 7-75 p79
     Concerto, piano. cf Caprice and elegy.
     Concerto, violin. cf Caprice and elegy.
     Dance rhapsody, no. 2. cf Brigg Fair.
     Eventyr. cf Koanga: Closing scene.
     Fennimore and Gerda: Intermezzo. cf Aquarelles.
     Hassan: Intermezzo and serenade. cf Brigg Fair.
     In a summer garden. cf Brigg Fair (CBS 30056).
     In a summer garden. cf Brigg Fair (CBS 61426).
     Irmelin: Prelude. cf Brigg Fair.
855 Koanga: Closing scene. Eventyr. Paris, The song of a great city.
       RPO and Chorus; Thomas Beecham. Odyssey Y 33284. (Reissues).
       (also CBS 61271. Reissues from Philips ABL 3088; Fontana CFL
       1042, CFE 15022, CFL 1033)
                    +HF 7-75 p89   +St 7-75 p110
                    /HFN 2-72 p306
856 Life's dance. North country sketches. A song of summer. RPO;
       Charles Groves. HMV ASD 3139. (also Angel S 37140)
                    +HFN 11-75 p154   ++RR 12-75 p47
                    ++NR 12-75 p3
     North country sketches. cf Appalachia.
     North country sketches. cf Life's dance.
     On hearing the first cuckoo in spring. cf Aquarelles.
     On hearing the first cuckoo in spring. cf Brigg Fair (CBS 30056).
     On hearing the first cuckoo in spring. cf Brigg Fair (CBS 61426).
     Paris, The song of a great city. cf Koanga: Closing scene.
857 Sonatas, violin and piano, nos. 1-3. Wanda Wilkomirska, vln; David
       Garvey, pno. Connoisseur Society CS 2069. (Q) CSQ 2069.
                    +-Gr 3-75 p1673   ++SFC 10-27-74 p10
                    -HF 2-75 p90   ++St 1-75 p110
                    ++NR 3-75 p12

858   Sonatas, violin and piano, nos. 1-3.   Ralph Holmes, vln; Eric
         Fenby, pno.   Unicorn RHS 310.
                  +Gr 5-73 p2063   +RR 5-73 p84
                  +HF 2-75 p90   +St 12-75 p118
                  +-HFN 5-73 p982
859   Sonata, violin and piano, no. 1.   Sonata, violin and piano, B
         minor.   David Stone, vln; Robert Threlfall, pno.   Pearl SHE 522.
                  +-Gr 10-75 p645   +-RR 12-75 p81
                  +-HFN 12-75 p153
      Sonata, violin and piano, B major.   cf Sonata, violin and piano,
         no. 1.
      A song of summer.   cf Life's dance.
      Summer night on the river.   cf Aquarelles.
DELLO JOIO, Norman
860   Homage to Haydn.   WELCHER: Concerto, flute and orchestra.   Francis
         Fuge, flt; Louisville Orchestra; Leonard Slatkin, Jorge Mester.
         Louisville LS 742.
                  +HF 8-75 p82
DEL MONACO, Alfredo
      Electronic study, no. 2.   cf CEELY: Elegia.
      Metagrama.   cf CEELY: Elegia.
DEMANTIUS, Johann
      St. John Passion.   cf BASF BAC 3087.
DEMERSSEMAN
      Solo de concerto, no. 6, op. 82, F major.   cf BOHM: Introduction
         and variations on Nel cor piu, op. 4.
DENZA, Luigi
      Funiculi funicula.   cf Polydor 2489 519.
DERING, Richard
      Country cries.   cf Turnabout TV 37079.
DESMARETS, Henri
      Menuet.   cf DG Archive 2533 172.
      Passepied.   cf DG Archive 2533 172.
DIABELLI, Anton
861   Variations über einen Walzer von Diabelli.   Jörg Demus, hammer-
         flugel.   DG Archive 2708 025 (2).
                  ++Audio 8-75 p85   +NR 5-75 p13
      DIABELLI VARIATIONS.   cf DG Archive 2708 025.
DIBDIN, Charles
      The Warwickshire lad.   cf Argo ZRG 765.
DICKINSON, Peter
      Extravaganzas.   cf BERKELEY: Songs (Argo ZRG 788).
      Recorder music.   cf HMV SLS 5022.
      Winter afternoons.   cf HMV EMD 5521.
DIEGO DE CONCEICAO, Fray
      Batalha de 5.   cf Musical Heritage Society MHS 1790.
DIEMER, Louis
      Valse de concerto.   cf Rococo 2049.
DIEUPART, Charles
      Sarabande, Gavotte, Menuet en Rondeau.   cf Argo ZRG 746.
      Suite, recorder and harpsichord, G major.   cf Telefunken Tape (c)
         CX 4-41203.
DIJON, Guiot
      Chanterai por mon coraige.   cf Telefunken AW 6-41275.
DILETZKY, Nikolai
      Glorify the name of the Lord.   cf HMV Melodiya ASD 3102.

DINICU, Dimitri
> Hora staccato (arr. and orch. Gerhardt).   cf RCA LRL 1-5094.
> Hora staccato.   cf RCA ARM 4-0942/7.

DIRUTA
> Toccata in Ionian mode.   cf Hungaroton SLPX 11601/2.

DISTLER, Hugo
> Weinachts, Geschichte: Den Dei hirten lobten sehre; Frohlich soll
>   mein Herze springen; Zu Bethlehem geboren.   cf University of
>   Colorado, unnumbered.

DITTERSDORFF, Karl Ditters von
> Quartet, E flat major.   cf BEVERSDORF: Sonata, violin and piano.
> DIVISIONS ON A GROUND: An introduction to the recorder and its
>   music.   cf Transatlantic TRA 292.

DLUGORAJ, Adalbert
> Carola Polonesa.   cf DG Archive 2533 294.
> Fantasia.   cf DG Archive 2533 294.
> Finale (2).   cf DG Archive 2533 294.
> Kowaly.   cf DG Archive 2533 294.
> Vilanella (2).   cf DG Archive 2533 294.

DODGE, Charles
> Speech songs.   cf 1750 Arch 1752.

DODGSON, Stephen
> Sonata, brass.   cf ADDISON: Divertimento, op. 9.

DOHNANYI, Ernst von
> Ruralia Hungarica: Gypsy andante.   cf RCA ARM 4-0942/7.
> Treasure waltz.   cf Musical Heritage Society MHS 1959.

DONIZETTI, Gaetano
> Anna Bolena: Debole io fu.   cf ABC ATS 20018.
> Anna Bolena: Pianget voi.   cf ABC ATS 20019.
> Belisario: A si tremendo annunzio.   cf Bel Canto Club no. 3.

862   Don Pasquale. Anna Maccianti, s; Ugo Benelli, t; Mario Basiola,
> bar; Alfredo Mariotti, Augusto Frato, bs; Maggio Musicale Chorus
> and Orchestra; Ettore Gracis. DG 2705 039 (2). (Reissue from
> SLPM 138971/2)
>> +Gr 2-75 p1543   +RR 2-75 p28

> Don Sebastiano: Deserto in terra.   cf Bel Canto Club no. 3.
> Il Duca d'Alba: Angelo casto e bel.   cf Bel Canto Club no. 3.
> Il Duca d'Alba: Angelo casto e bel.   cf Discophilia KGC 2.
> Il Duca d'Alba: Angelo casto e bel.   cf Saga 7206.

863   L'Elisir d'amore: Quanto è bella; Della crudele Isotta; Chiedi
> all aura lusinghiera; Udite, udite, o rustici...Voglio dire;
> La nina gondoliera; Venti scudi; Quanto amore...Una furtiva
> lagrima; Prendi, per me sei libero; Ei corregge ogni difetto.
> Joan Sutherland, s; Maria Casula, ms; Luciano Pavarotti, t;
> Domonic Cossa, bar; Spiro Malas, bs; Ambrosian Opera Chorus;
> ECO; Richard Bonynge. Decca SET 564. (Reissue from SET 503/5)
> (also London OS 26343. Reissue from London 13101)
>> +-Gr 12-73 p1252   +-Op 4-74 p421
>> +HFN 12-73 p1252   +RR 12-73 p44
>> +OC 12-75 p49

> L'Elisir d'amore: Una furtiva lagrima.   cf RCA SER 5704/6.
> L'Elisir d'amore: Una furtiva lagrima.   cf Supraphon 10924.
> L'Elisir d'amore: Una furtiva lagrima.   cf Supraphon 012 0789.
> L'Elisir d'amore: una furtiva lagrima.   cf Telefunken AJ 6-41867.
> L'Elisir d'amore: Quanto è bella, quanto e cara.   cf Decca GOSC
>   666/8.

Emilia di Liverpool: Cavatina and rondo finale. cf Lucia di
    Lammermoor: Sulla tomba to end of Act 1.
La favorita: Arias. cf London OS 26373.
La favorita: Una vergine. cf BASF BC 21723.
La favorita: Una vergine, un angel di Dio; Spirto gentil. cf
    Supraphon 012 0789.
La fille du régiment: Arias. cf London OS 26373.
La fille du régiment: Ah, mes amis...Que dire, que faire...Pour
    mon ame. cf Decca GOSC 666/8.
La fille du régiment: Au bruit. cf Rubini RS 300.
La fille du régiment: Chacun le sait, chacun le dit. cf RCA ARL
    1-0844.
864 Linda di Chamounix: Ah, tardai troppo...O luce di quest'anima.
        Lucia di Lammermoor: Il dolce suono mi corpi sua voce...Ardon
        gl'incensi; Ancor non giunse...Regnava nel silenzio. VERDI:
        Ernani: Surta è la notte...Ernani, Ernani, involami. I vespri
        Siciliani: Mercé, diletti amiche. Joan Sutherland, s; Nadine
        Sautereau, s; Paris Opera Chorus; OSCCP; Nello Santi. Decca
        SDD 146. Tape (c) KSDC 146. (Reissues)
            +-RR 3-75 p74 tape
     Linda di Chamounix: Arias and scenes. cf London OS 26306.
     Linda di Chamounix: Ah, tardai troppo, O luce di quest'anima. cf
        RCA ARL L-0702.
865 Lucia di Lammermoor: Sulla tomba to end of Act 1; Chi mi frena to
        end of Act 2; Tombe degli avi miei to end of opera. Emilia
        di Liverpool: Cavatina and rondo finale. ROSSINI: La fioraria
        fiorentina. SPOHR: Zemira et Azor: Rose softly blooming.
        Renata Scotto, Joan Sutherland, s; Stefania Malagù, ms; Giuseppe
        di Stefano, Franco Ricciardi, t; Ettore Bastianini, bar; Ivo
        Vinco, bs; Richard Bonynge, pno; La Scala Orchestra and Chorus;
        Nino Sanzogno. Ember GVC 45. (Reissues from DG SLPM 138704/5;
        Belcantodisc LR 1)
            +-Gr 2-75 p1542   +RR 12-74 p22
     Lucia di Lammermoor: Alfin son tua. cf Saga 7029.
     Lucia di Lammermoor: Ardon gli incensi. cf BELLINI: Capuletti ed
        i Montecchi: O, quante volte, O, quante.
     Lucia di Lammermoor: Il dolce suono. cf ABC ATS 20019.
     Lucia di Lammermoor: Il dolce suono mi corpi sua voce...Ardon
        gl'incensi; Ancor non giunse...Regnava nel silenzio. cf
        Linda di Chamounix: Ah, tardai troppo...O luce di quest'anima.
     Lucia di Lammermoor: Regnava nel silenzio, Quando rapito in
        estasi. cf RCA ARL L-0702.
     Lucia di Lammermoor: Sulla tomba. cf ABC ATS 20018.
     Lucia di Lammermoor: Sulla tomba. cf Supraphon 012 0789.
     Lucrezia Borgia: Arias. cf London OS 26277.
DOPPLER, Franz
     Fantaisie pastorale hongroise, op. 26. cf BOHM: Introduction and
        variations on Nel cor piu, op. 4.
     Fantaisie pastorale hongroise, op. 26. cf RCA LRL 1-5094.
DORAN
     Andante, flute and guitar. cf Klavier KS 537.
     Suite, flute and guitar: Finale. cf Klavier KS 537.
     DOUCE DAME/MUSIC OF COURTLY LOVE FROM MEDIEVAL FRANCE AND ITALY.
        cf Vanguard VSD 71179.
DOWLAND, John
     Captain Digorie piper's galliard. cf CRD CRD 1019.

Dances (4). cf Desto DC 6474/7.
The Earl of Essex galliard. cf DG Archive 2533 157.
Fantasia. cf DG Archive 2533 157.
Forlorne hope fancy. cf DG Archive 2533 157.
The King of Denmark's galliard. cf CRD CRD 1019.
The King of Denmark's galliard. cf DG Archive 2533 157.
Lachrimae antiquae pavan. cf DG Archive 2533 157.
Master Piper's pavan. cf L'Oiseau-Lyre SOL 336.
Melancholy galliard. cf DG Archive 2533 157.
Melancholy galliard. cf Strobe SRCM 123.
Mrs. Winter's leap. cf DG Archive 2533 157.
My Lady Hunsdon's puffe. cf DG Archive 2533 157.
My Lady Hunsdon's puffe, lute solo. cf Strobe SRCM 123.
My Lord Willoughbie's welcome home. cf L'Oiseau-Lyre SOL 336.
Semper Dowland semper dolens. cf DG Archve 2533 157.
Semper Dowland semper dolens. cf L'Oiseau-Lyre SOL 336.
What if I never speede. cf Advent 5012.

DOWNES, Bob
866 Episodes at 4 am. Bob Downes, Wendy Benka, flt, electronics,
        perc. Openian BDOM 002.
            +-RR 1-75 p40

DRAPER
All creatures of our God and King. cf CBS 61599.

DRDLA, Franz
Souvenir. cf Discopaedia MB 1004.

DRESDEN, Sem
867 Concerto, oboe. PIJPER: Concerto, piano. VAN HEMEL: Concerto,
        violin, no. 2. Herman Krebbers, vln; Theo Bruins, pno; Koen
        van Slogteren, ob; COA, RPO; Bernard Haitink, Willem van Otter-
        loo, Francis Travis. Donemus 7374/2.
            +-RR 2-75 p40

DREW
Bless the Lord, O my soul. cf RCA ARL 1-0561.

DREWS, Steve
Ceres motion. cf BORDEN: Easter.
Train. cf BORDEN: Easter.

DRIGO, Riccardo
Les millions d'Arlequin: Serenade. cf Discopaedia MB 1004.
Les millions d'Arlequin: Serenade. cf Discopaedia MB 1008.
Les millions d'Arlequin: Serenade (arr. and orch. Gamley). cf
    RCA LRL 1-5094.
Valse bluette (2). cf RCA ARM 4-0942/7.

DRISCHNER, Max
Chorale prelude on 'O heiliger Geist, O heiliger Gott'. cf BACH:
    Chorale prelude, O Mensch bewein dein Sünde grosse, S 622.
Choralfantasia 'Lobe den Herren'. cf BACH: Chorale prelude, O
    Mensch bewein dein Sünde grosse, S 622.
Nordische Fantasie. cf BACH: Chorale prelude, O Mensch bewein
    dein Sünde grosse, S 622.
Partita on 'Wilhemnus von Naussauen'. cf BACH: Chorale prelude,
    O Mensch bewein dein Sünde grosse, S 622.

DUBEN
Partiten uber "Erstanden ist der heil'ge Christ". cf Pelca PSRK
    41013/6.

DUBOIS
Preludes faciles (9). cf Armstrong 721-4.

DUETS WITH THE SPANISH GUITAR. cf Angel S 36076.
DUFAY, Guillaume
    L'alta bellezza. cf Telefunken ER 6-35257.
    Ave virgo. cf Telefunken ER 6-35257.
    Bien veignes vous. cf Telefunken ER 6-35257.
    Bon jour, bon mois. cf Telefunken ER 6-35257.
    C'est bien raison. cf Telefunken ER 6-35257.
    Credo. cf Telefunken ER 6-35257.
    La dolce vista. cf Telefunken ER 6-35257.
    Dona i ardente ray. cf Telefunken ER 6-35257.
868 Ecclesiae militantis. Missa Sancti Jacobi. Rite Majorem.
      Capella Cordina; Alejandro Planchart. Lyrichord LLST 7275.
        +NR7-75 p8
    Ecclesie militantis. cf Telefunken ER 6-35257.
    Gloria. cf Telefunken ER 6-35257.
    Helas mon dueil. cf Telefunken ER 6-35257.
    J'ay mis mon cuer. cf Telefunken ER 6-35257.
    Je vous pri. cf Telefunken ER 6-35257.
    Kyrie. cf Telefunken ER 6-35257.
    Lamentatio Sanctae matris ecclesiae Constantinopolitanae. cf
      Telefunken ER 6-35257.
869 Las, que Feray. Ma bella dame souveraine. Missa se la face ay
      pale. Vergine bella. Capella Cordina; Alejandro Planchart.
      Lyrichord LLST 7274.
        +NR 7-75 p8
    Ma bella dame souveraine. cf Las, que Feray.
    Missa Sancti Jacobi. cf Ecclesiae militantis.
    Missa se la face ay pale. cf Las, que Feray.
    Mon chier amy. cf Telefunken ER 6-35257.
    Moribus et genere Christo. cf Telefunken ER 6-35257.
870 Motets: Supremum est mortalibus. Flos forum. Ave virgo quae de
      caelis. Vasilissa, ergo gaude. Alma redemptoris mater (II).
      DUNSTABLE: Motets: Beata mater. Preco proheminencie. Salve
      regina misericordie. Veni sancti spiritus. Pro Cantione
      Antiqua; Hamburg Wind Ensemble, Members; Bruno Turner. DG
      Archive 2533 291.
        ++Gr 8-75 p352   +-HFN 8-75 p75
    Qui latuit. cf Telefunken ER 6-35257.
    Rite Majorem. cf Ecclesiae militantis.
    Sanctus. cf Telefunken ER 6-35257.
    Songs: C'est bien raison de devoir essaucier; Je me complains
      piteusement; Invidia nimica; Malheureux cuer que veux to faire;
      Par droit je puis bien complaindre. cf 1750 Arch S 1753.
    Veni creator spiritus. cf Telefunken ER 6-35257.
    Vergine bella. cf Las, que Feray.
    Vergine bella. cf BIS LP 2
DUGGER, Edwin
871 Music for synthesizer and six instruments. ERICKSON: Ricercar a
      5 for trombones. HOFFMANN: Orchestra piece, 1961. Stuart
      Dempster, trom; Instrumental Ensemble; Oberlin College Conserv-
      tory Orchestra; David Epstein, Robert Baustain. DG/Acoustic
      Research 0654 084.
        +HF 2-71 p97   *NYT 1-3-71 pD22
        +MJ 2-75 p29   +SR 12-26-70 p47
        +MQ 7-72 p501   +St 5-71 p95

DUKAS, Paul
872 The sorcerer's apprentice. RESPIGHI: La boutique fantasque. Israel
      Philharmonic Orchestra; Georg Solti. London STS 15005. Tape
      (c) A 30605. (also Decca SPA 376. Tape (c) KCSP 376)
              +Gr 6-75 p90   ++HFN 7-75 p91 tape
             +HFN 8-75 p89   +RR 6-75 p46
      The sorcerer's apprentice. cf Decca DPA 519/20.
      L'Apprenti sorcier. cf HMV ASD 3008.
      L'Apprenti sorcier. cf HMV (Q) ASD 3131.
      La Plainte, au loin, du faune. cf Variations on a theme by
      Rameau.
873 Variations on a theme by Rameau. La Plainte, au loin, du faune.
      d'INDY: Sonata, piano, E minor. Vladimir Pleshakov, pno.
      Orion ORS 7266.
              +CL 12-75 p5   +SFC 5-7-72 p46
             ++NR 5-72 p12
      Villanelle. cf BRAHMS: Trio, horn, violin and piano, op. 40,
      E flat major.
DUKE, John
      Songs: I carry your heart; In just spring; The mountains are
      dancing. cf Duke University Press DWR 7306.
DU MAGE, Pierre
      Livre d'orgue: Recit and basse de trompette. cf Wealden WS 139.
DUMONT, Henry
      Pavane, D minor. cf Harmonia Mundi HMU 334.
DUNSTABLE, John
      Beata mater. cf Telefunken ER 6-35257.
      Motets: Beata mater. Preco proheminencie. Salve regina miseri-
      cordie. Veni sancti spiritus. cf DUFAY: Motets (DG 2533 291).
      O rosa bella. cf BIS LP 3.
      O rosa bella; Hastu mir. cf L'Oiseau-Lyre 12BB 203/6.
DUPARC, Henri
      La vie antérieure. cf Columbia M 32231.
      Songs: L'Invitation au voyage; Phidylé. cf Seraphim S 60251.
DUPHLY, Jacques
      Chaconne. cf La Félix.
874 La Félix. La Forqueray. Chaconne. FORQUERAY: La Laborde. La
      Bellemont. La Couperin. MARCHAND: Suite. Kenneth Gilbert,
      hpd. Harmonia Mundi HMU 940.
             ++RR 11-75 p76
      La Forqueray. cf La Félix.
      Pièces de clavecin (8). cf BACH: Sonata, harpsichord, no. 1,
      A minor.
875 Pièces de clavecin: Allemande, courante; Chaconne; La damanzy (2);
      La Félix; Les graces; La Forqueray; Menuets; La de belombre;
      La pothoütin. Gustav Leonhardt, hpd. Philips Seon 6575 017.
              +Gr 1-75 p1371   +RR 1-75 p47
DUPONT, Gabriel
      Mandoline. cf DEBUSSY: Songs (Cambridge CRS 2774).
DUPRE, Marcel
      Annonciation: Méditations, op. 56, no. 1, E minor; no. 2, G major.
      cf Organ works (RCA Victrola LVL 1-5018).
876 Carillon. Cortege et Litainie, op. 19. Fileuse. In dulci jubilo.
      Preludes and fugues, op. 7 (3). Robert Noehren, org. Delos
      DELS 24201.
             ++MJ 2-75 p39   ++SFC 2-23-75 p23
             ++NR 2-75 p13

Cortege et Litainie, op. 19. cf Carillon.
Esquisses, op. 41, no. 1, E minor; no. 2, B flat minor. cf Organ
    works (RCA LVL 1-5018).
Fileuse. cf Carillon.
In dulci jubilo. cf Carillon.
877 Organ works: Annonciation: Méditations, op. 56, no. 1, E minor;
    no. 2, G major. Esquisses, op. 41, no. 1, E minor; no. 2,
    B flat minor. Preludes and fugue, op. 7, no. 1, B major; no.
    2, F minor; no. 3, G minor. Variations sur un Noël, op. 20.
    Graham Steed, org. RCA Victrola LVL 1-5018.
        +Gr 7-74 p234   +RR 7-74 p67
        +MT 9-75 p797
    Preludes and fugues, op. 7. cf Carillon.
    Preludes and fugues, op. 7, no. 1, B major; no. 2, F minor; no. 3,
        G minor. cf Organ works (RCA LVL 1-5018).
    Prelude and fugue, op. 7, no. 1, B major. cf Wealden WS 145.
    Prelude and fugue, op. 7, no. 3, G minor. cf Pye TPLS 13066.
    Variations sur un Noël, op. 20. cf Organ works (RCA LVL 1-5018).
DURANTE, Francesco
878 Concertos, strings and continuo (4). Collegium Aureum; Rolf
    Reinhardt. BASF KHB 21681. (also BASF BAC 3060)
        +Gr 12-74 p1133   +-RR 1-75 p27
        ++NR 1-74 p4   +St 2-74 p115
DURKO, Zsolt
879 Altamira, chamber choir and orchestra. Fire music. Iconography,
    no. 1, 2 bass viols (cellos) and harpsichord. Iconography,
    no. 2, horn and chamber ensemble. Attila Lajos, flt; Béla
    Kovács, clt; Ferenc Tarjáni, hn; László Almásy, pno; Károly
    Duska, vln; Gábor Fias, vla; Janos Devich, Ede Banda, László
    Mëzo, vlc; János Sebestyén, hpd; Budapest Chamber Ensemble;
    HRT Chamber Orchestra and Chorus; András Mihály, György Lehel.
    Hungaroton LSPX 11607.
        ++RR 1-75 p40
    Fire music. cf Altamira, chamber choir and orchestra.
    Iconography, no. 1, 2 bass viols (cellos) and harpsichord. cf
        Altamira, chamber choir and orchestra.
    Iconography, no. 2, horn and chamber ensemble. cf Altamira,
        chamber choir and orchestra.
DURUFLE, Maurice
880 Prelude et fugue sur le nom d'Alain, op. 7. Requiem, op. 9.
    Robert King, treble; Christopher Keyte, bar; Stephen Cleobury,
    org; St. John's College Chapel Choir, Cambridge; George Guest.
    Argo ZRG 787.
        +-Gr 5-75 p2000   +NR 10-75 p6
        ++HFN 6-75 p88   +RR 5-75 p58
        +MT 12-75 p1070
    Prelude et fugue sur le nom d'Alain, op. 7. cf BIS LP 7.
    Requiem, op. 9. cf Prelude et fugue sur le nom d'Alain, op. 7.
    Suite, organ, op. 5: Toccata. cf Pye GSGC 14153.
DUSSEK (Dusik or Dessek), Johann (Jan) Ladislaus
    La chasse. cf Piano works (Musical Heritage Society MHS 1966).
    La consolation. cf Piano works (Musical Heritage Society MHS 1966).
    Minuet. cf Discopaedia MB 1003.
    Partante pour la Syrie. cf Piano works (Musical Heritage Society
        MHS 1966).
881 Piano works: La chasse. La consolation. Partante pour la Syrie.
    Sonata, op. 77, F minor. Edward Gold, pno. Musical Heritage

Society MHS 1966.
        +—St 12-75 p118
Sonata, piano, op. 77, F minor. cf BENDA: Sonata, piano, no. 9,
    A minor.
Sonata, piano, op. 77, F minor. cf Piano works (Musical Heritage
    Society MHS 1966).
Sonatina, op. 20, no. 1. cf Columbia MG 33202.
Within a mile of Edinburgh. cf Angel SB 3816.
DUTILLEUX, Henri
882 Concerto, violoncello. LUTOSLAWSKI: Concerto, violoncello.
        Mstislav Rostropovitch, vlc; Orchestre de Paris; Serge Baudo,
        Witold Lutoslawski. HMV ASD 3145.
            ++HFN 12-75 p153   ++RR 12-75 p47
883 Symphony, no. 2. ROUSSEL: Suite, op. 33, F major. Lamoureux
        Orchestra; Charles Munch. Musical Heritage Society MHS 3022.
        (Reissue from Westminster WST 17119)
            +HF 5-75 p70   ++St 12-75 p122
DUVERNOY, Frederic
Trio, horn, no. 1. cf BRAHMS: Trio, horn, violin and piano, op.
    40, E flat major.
DVORAK, Antonin
Berceuse, op. posth., G major.   cf Piano works (Supraphon 111
    1179).
Capriccio, op. posth., G minor. cf Piano works (Supraphon 111
    1179).
884 Carnival overture, op. 92. Slavonic dances, opp. 46, 72. CO;
        Georg Szell. Odyssey Y2 33524 (2). (Reissue from Epic)
            +—NR 10-75 p10   +SFC 8-31-75 p20
Carnival overture, op. 92. cf Symphony, no. 5, op. 76, F major.
Carnival overture, op. 92. cf Camden CCV 5000-12, 5014-24.
885 Concerto, violin and piano, op. 53, A minor. SIBELIUS: Concerto,
        violin, op. 47, D minor. Ruggiero Ricci, vln; LSO; Malcolm
        Sargent, Ølvin Fjeldstad. Decca SPA 398. (Reissues from SXL
        2279, 2077)
            +—Gr 7-75 p176   +RR 7-75 p35
            +—HFN 9-75 p108
886 Concerto, violin and piano, op. 53, A minor. Romance, violin and
        orchestra, op. 11, F minor. Itzhak Perlman, vln; LPO; Daniel
        Barenboim. HMV ASD 3120. (also Angel S 37069)
            +Gr 10-75 p611   +RR 10-75 p38
            +HFN 10-75 p141   ++SFC 11-16-75 p32
            ++NR 12-75 p7
887 Concerto, violin and piano, op. 53, A minor. Romance, violin and
        orchestra, op. 11, F minor. Josef Suk, vln; CPhO; Karel Ančerl.
        Supraphon 50181. Tape (c) 045181D. (also Vanguard Supraphon
        SU 3. Reissue from Artia ALPS 193)
            +Audio 8-75 p83   -RR 3-74 p72 tape
            ++HF 5-75 p74   ++SFC 2-2-75 p26
            ++NR 3-75 p4   +St 5-75 p102
            +RR 3-74 p35
Concerto, violin and piano, op. 53, A minor. cf RAVEL: Tzigane.
888 Concerto, violoncello, op. 104, B minor. FAURE: Elégie, op. 24,
        C minor. Janos Starker, vlc; PhO; Walter Susskind. Classics
        for Pleasure CFP 40070. (Reissue from Columbia SAX 2263)
            -Gr 1-75 p1342   +St 6-75 p147
            +—RR 1-75 p27

889 Concerto, violoncello, op. 104, B minor. Pierre Fournier, vlc;
     BPhO; Georg Szell. DG 2535 106. Tape (c) 3335 106. (Reissue
     from SLPM 138755)
              /Gr 7-75 p176   +HFN 10-75 p155 tape
              +HFN 9-75 p108   +RR 6-75 p35
890 Concerto, violoncello, op. 104, B minor. Anja Thauer, vlc; CPhO;
     Zdeněk Mácal. DG Heliodor 2548 134. (Reissue from 139392)
              +-Gr 4-75 p1867   +RR 4-75 p27
891 Concerto, violoncello, op. 104, B minor. Paul Tortelier, vlc; PhO;
     Malcolm Sargent. HMV SXLP 30018. Tape (c) TC EXE 158.
              +-HFN 12-75 p173 tape
892 Concerto, violoncello, op. 104, B minor. Mstislav Rostropovich,
     vlc; RPO; Adrian Boult. HMV SXLP 30176. (Reissue from ALP
     1545) (also Seraphim S 60136)
              ++Gr 1-75 p1341
893 Concerto, violoncello, op. 104, B minor. TCHAIKOVSKY: Variations
     on a rococo theme, op. 33. Christine Walevska, vlc; LPO;
     Alexander Gibson. Philips 6500 224.
              +Gr 2-75 p1479   +RR 2-75 p33
              /HF 5-72 p92   ++SFC 1-7-73 p23
              +NR 4-72 p5   +St 6-72 p82
894 Concerto, violoncello, op. 104, B minor. Lynn Harrell, vlc; LSO;
     James Levine. RCA ARL 1-1155. Tape (c) ARK 1-1155 (ct) ARS
     1-1155.
              ++NR 11-75 p7   +SR 11-1-75 p46
     Dumka and furiant, op. 12, nos. 1-2, C minor. cf Piano works
     (Supraphon 111 1179).
     The golden spinning wheel, op. 109. cf Symphonic variations,
     op. 78.
     Humoresque. cf Rococo 2035.
     Humoresque. cf Supraphon 110 1429.
     Humoresque, F sharp major. cf Piano works (Supraphon 111 1179).
     Humoresque, op. 101. cf Connoisseur Society (Q) CSQ 2066.
     Humoresque, op. 101, no. 7, G flat major. cf Decca SPA 372.
     Humoresque, op. 101, no. 7, G flat major. cf Discopaedia MB 1006.
     Humoresque, op. 101, no. 7, G flat major. cf Philips 6747 199.
     Impromptu, D minor. cf Piano works (Supraphon 111 1179).
895 In nature's realm overture, op. 91. My home overture, op. 62.
     SMETANA: Ma Vlast: Vltava; From Bohemia's woods and fields.
     Supraphon 110 1589.
              +-HFN 10-75 p141   +RR 9-75 p47
              +NR 9-75 p4
     Indian lament. cf Ember GVC 46.
896 Legends, op. 59. Brno State Philharmonic Orchestra; Jiří Pinkas.
     Supraphon 110 1395.
              ++Gr 4-75 p1807   +RR 3-75 p25
              +NR 8-75 p2   +SFC 2-23-75 p23
897 Legends, op. 59, nos. 1-4. Slavonic dances, op. 46, nos. 1-8; op.
     72, nos. 9-16. Vlastimil Lejsek, Věra Lejsková, pno. Supra-
     phon 111 1301/2 (2).
              -Gr 7-75 p205   +NR 10-75 p10
              -HFN 7-75 p80   +RR 7-75 p24
     March. cf Panton 110 361.
898 Mass, op. 86, D major. Christ Church Cathedral Choir; Simon
     Preston. Argo ZRG 781.
              +Gr 12-74 p1200   +-RR 12-74 p71
              +HF 7-75 p74   +-SFC 7-6-75 p16
              +-NR 6-75 p10   ++St 7-75 p96

      +HF 7-75 p74   +-SFC 7-6-75 p16
      +-NR 6-75 p10   ++St 7-75 p96

899 Mass, op. 86, D major. Mavis Beattie, s; Christina Stephenson, ms;
    John Dudley, t; Mark Rowlinson, bar; Ralph Downes, org; London
    Oratory Choir; John Hoban. Discourses All About Music AMB 18.
        +Gr 2-75 p1523   -RR 2-75 p59

Mazurkas, op. 56. cf Piano works (Supraphon 111 1179).

900 My home overture, op. 61. Slavonic dances, op. 72. Bavarian
    Radio Symphony Orchestra; Rafael Kubelik. DG 2530 593. Tape
    (c) 3300 593.
        ++Gr 12-75 p1032   ++RR 12-75 p47

My home overture, op. 62. cf In nature's realm overture, op. 91.

My home overture, op. 62. cf Symphony, no. 5, op. 76, F major.

901 Piano works: Berceuse, op. posth., G major. Capriccio, op. posth.,
    G minor. Dumka and furiant, op. 12, nos. 1-2. Humoresque,
    F sharp major. Impromptu, D minor. Mazurkas, op. 56. Radoslav
    Kvapil, pno. Supraphon 111 1179.
        +Gr 5-75 p1995   +RR 3-75 p46
        ++NR 4-75 p12

902 Quartet, strings, no. 2, op. 34, D minor. MARTINU: Quartet, strings,
    no. 4. Smetana Quartet. Supraphon 50529. (Reissue)
        ++LJ 10-75 p47   ++RR 5-74 p46

903 Quartet, strings, op. 51, E flat major. Quartet, strings, no. 7,
    op. 105, A flat major. Gabrieli Quartet. Decca SDD 479.
        +RR 12-75 p72

904 Quartet, strings, no. 4, op. 61, C major. Terzetto, 2 violins and
    viola, op. 74, C major. Guarneri Quartet. RCA ARL 1-0082.
    Tape (r) ERP 1-0082.
        +HF 8-73 p75   ++NR 9-73 p9
        +HF 4-75 p112 tape   ++St 9-73 p116
        +MJ 10-73 p19

905 Quartet, strings, no. 5, op. 87, E flat major. Artur Rubinstein,
    pno; Guarneri Quartet, Members. RCA LSC 3340. Tape (r) ERPA
    3340. (also RCA SB 6884)
        +Gr 3-74 p1711   ++MJ 9-73 p44
        +HF 8-73 p75   +NR 5-73 p7
        +HF 4-75 p112 tape   +RR 3-74 p48
        +LJ 10-74 p52   +St 9-73 p117

906 Quartet, strings, no. 5, op. 87, E flat major. Dvořák Quartet.
    Supraphon 50528.
        ++LJ 10-75 p47   +RR 5-74 p46

907 Quartet, strings, no. 6, op. 96, F major. Quartet, strings, no. 7,
    op. 105, A flat major. Prague Quartet. Supraphon 50816.
        ++LJ 10-75 p47   ++RR 3-74 p48

Quartet, strings, no. 7, op. 105, A flat major. cf Quartet, strings,
    no. 6, op. 96, F major.

Quartet, strings, no. 7, op. 105, A flat major. cf Quartet,
    strings, op. 51, E flat major.

908 Quartet, strings, no. 13, op. 106, G major. Prague Quartet. DG
    2530 480.
        +Gr 4-75 p1820   ++RR 4-75 p39
        -HF 6-75 p93   +STL 5-4-75 p37
        ++NR 6-75 p9

909 Quartet, strings, no. 13, op. 106, G major. Vlach Quartet. Supra-
    phon SUAST 50172.
        +LJ 10-75 p47

910 Quintet, strings, no. 1, op. 1, A minor. Quintet, strings, no. 3,

op. 97, E flat major. Sextet, strings, op. 48, A major. Trio,
piano, op. 26, G minor. Trio, piano, op. 65, F minor. Waltz,
op. 54, no. 4. Irmgard Schuster, vla; Dankwahrt Gahl, vlc;
Dumka Trio; Austrian Quartet. Vox SVBX 588 (3).
   +-HF 9-75 p84   +NR 9-75 p9
Quintet, strings, no. 3, op. 97, E flat major. cf Quintet, strings,
no. 1, op. 1, A minor.
Romance, violin and orchestra, op. 11, F minor. cf Concerto, violin
and piano, op. 53, A minor (HMV 3120).
Romance, violin and orchestra, op. 11, F minor. cf Concerto,
violin and piano, op. 53, A minor (Supraphon 50181).
Rondo, op. 94, G minor. cf BRAHMS: Sonata, violoncello, no. 1,
op. 38, E minor.
Rusalka, op. 114: Arias. cf London OS 26381.
911 Scherzo capriccioso, op. 66, D flat major. Slavonic dances, op.
46. Bavarian Radio Symphony Orchestra; Rafael Kubelik. DG
2530 466.
   +-Gr 11-75 p817   +RR 11-75 p42
   +HFN 11-75 p153   ++SFC 11-16-75 p32
Serenade, op. 44, D minor. cf BRAHMS: Serenade, no. 2, op. 16,
A major.
912 Serenade, strings, op. 22, E major. TCHAIKOVSKY: Serenade, strings,
op. 48, C major. Israel Philharmonic Orchestra; Georg Solti,
Rafael Kubelik. Decca SPA 375. Tape (c) KCSP 375.
   +-Gr 4-75 p1871   +RR 5-75 p45
   +-HFN 6-75 p88   +-RR 6-75 p93 tape
913 Serenade, strings, op. 22, E major. TCHAIKOVSKY: Serenade, strings,
op. 48, C major. Hamburg Radio Orchestra; Dresden Staatskapelle
Orchestra; Hans Schmidt-Isserstedt, Otmar Suitner. DG Heliodor
2548 121. (Reissue from DG 136481, 135109)
   +Gr 4-75 p1867   +RR 4-75 p36
914 Serenade, strings, op. 22, E major. TCHAIKOVSKY: Serenade, strings,
op. 48, C major. ECO; Daniel Barenboim. HMV ASD 3036. (also
Angel S 37045)
   +Gr 11-74 p890   -NR 8-75 p2
   +HF 11-75 p102   +-RR 11-74 p36
915 Serenade, strings, no. 22, E major. JANACEK: Idyll, string orches-
tra. South West German Chamber Orchestra; Paul Angerer. Turna-
bout TV 34532.
   +-Gr 7-75 p176   +NR 9-74 p4
   +HF 10-74 p95   +RR 5-75 p33
   +HFN 5-75 p126
Sextet, strings, op. 48, A major. cf Quintet, strings, no. 1, op.
1, A minor.
Slavonic dance, no. 2. cf RCA ARM 4-0942/7.
Slavonic dances, op. 46. cf Scherzo capriccioso, op. 66, D flat
major.
916 Slavonic dances, op. 46 (8). Bamberg Symphony Orchestra; Antal
Dorati. Turnabout (Q) QTV 34582.
   +NR 11-75 p1
Slavonic dances, opp. 46, 72. cf Carnival overture, op. 92.
Slavonic dances, op. 46, nos. 1-3. cf BRAHMS: Hungarian dances,
nos. 5, 6.
Slavonic dances, op. 46, nos. 1, 5-6, 8. cf BRAHMS: Hungarian
dances, nos. 1, 3, 5-6, 10, 12-13, 19, 21.
Slavonic dances, op. 46, nos. 1-8; op. 72, nos. 9-16. cf Legends,
op. 59, nos. 1-4.

Slavonic dance, op. 46, no. 1, C major (arr. Kreisler). cf
Decca SPA 405.
Slavonic dance, op. 46, no. 1, C major. cf Decca DPA 519/20.
Slavonic dances, op. 72. cf My home overture, op. 62.
Slavonic dance, op. 72, no. 1, B major. cf HMV (Q) ASD 3131.
917 Slavonic dance, op. 72, no. 2, E minor. RIMSKY-KORSAKOV: Capriccio
espagnol, op. 34. SCRIABIN: Poème de l'extase, op. 54. CPhO,
NPhO; Leopold Stokowski. Decca PFS 4333. (also London 21117)
    +Gr 7-75 p176   +NR 11-75 p2
    +-HF 11-75 p124   ++RR 6-75 p48
    ++HFN 6-75 p102   +SFC 9-21-75 p34
Slavonic dance, op. 72, no. 2, E minor. cf RCA ARM 4-0942/7.
Slavonic dance, op. 72, no. 2, E minor. cf BRAHMS: Hungarian
dances, nos. 5, 6.
Slavonic dance, op. 72, no. 2, E minor. cf BRAHMS: Hungarian
dances, nos. 1, 3, 5-6, 10, 12-13, 19, 21.
Slavonic dance, op. 72, no. 8. cf RCA ARM 4-0942/7.
918 Salvonic dances, op. 72, nos. 9, 10, 15. Symphony, no. 9, op. 95,
E minor: Largo. SMETANA: The bartered bride: Overture, Polka,
Furiant. Ma Vlast: Vltava. Various orchestras and conductors.
Supraphon ST 50547.
    +-LJ 9-75 p37   +-NR 10-74 p2
919 Songs: Folk songs, op. 73: Good night; The mower; Maiden's lament;
Loved and lost. Gypsy songs, op. 55: My song of love; Hey,
Ring out; All round about; Songs my mother taught me; Come and
join the dancing; Wide the sleeves and trousers; Give a hawk
a fine cage. Love songs, op. 83: Never will love lead us;
Death reigns in many a human breast; I wander oft; I know that
on my love to thee; Nature lies peaceful; In deepest forest
glade; When thy sweet glances; Thou only dear one. Modern
Greek songs, op. 50: Kolias; The Nereids; Parga's lament. Jindřich Jindrák, bar; Alfred Holeček, pno. Supraphon 112 1349.
    +-Gr 7-75 p227   +NR 11-75 p11
    +-HFN 7-75 p81   +RR 7-75 p57
Songs; Der Apfel; Der kleine Acker; Möglichkeit; Der Ring; Die
Taube auf dem Ahorn. cf BRAHMS: Songs (Musical Heritage
Society MHS 1896).
Songs: Die Bescheidene, op. 32, no. 8; Die Gefangene, op. 32, no.
11; Scheiden ohne leiden, op. 32, no. 4; Die verlassene, op.
32, no. 6; Die Zuversicht, op. 32, no. 10. cf BIS LP 17.
920 Symphonic variations, op. 78. The golden spinning wheel, op. 109.
LSO; István Kertész. London CS 6721. (also Decca SXL 6510.
Tape (c) KSXC 6510)
    +Gr 12-74 p1237 tape   ++RR 1-75 p69 tape
    ++HF 8-72 p70   ++SFC 12-31-72 p22
    ++NR 7-72 p3   ++St 7-72 p78
921 Symphonies, nos. 1-9. CPhO; Václav Neumann. Supraphon 110 1621/8 (8
    +-Gr 10-75 p608   +NR 9-75 p4
    +HFN 9-75 p98   +-RR 9-75 p35
922 Symphony, no. 5, op. 76, F major. Carnival overture, op. 92. My
home overture, op. 62. LSO; István Kertész. Decca 6273. Tape
(c) KSXC 16273.
    -HFN 7-75 p90 tape   /RR 7-75 p70
923 Symphony, no. 6, op. 60, D major. BPhO; Rafael Kubelik. DG
2530 425. (Reissue from 2720 066)
    +Gr 12-74 p1133   ++NR 1-75 p3

924 Symphony, no. 7, op. 70, D minor. LSO; István Kertész. Decca
        Tape (c) KSXC 6155.
            -Gr 1-75 p1403 tape   +-RR 2-75 p75 tape
925 Symphony, no. 7, op. 70, D minor. CPhO; Václav Neumann. Vanguard
        SU 7.
            +NR 11-75 p1
926 Symphony, no. 8, op. 88, G major. VPO: Herbert von Karajan.
        Decca SDD 440. (Reissue from SXL 6169) (also London 6443)
            -Gr 2-75 p1479   +RR 2-75 p33
927 Symphony, no. 8, op. 88, G major. Columbia Symphony Orchestra;
        Bruno Walter. Odyssey Y 33231. (Reissue from Columbia MS 6361)
            ++HF 6-75 p93
928 Symphony, no. 8, op. 88, G major. CPhO; Václav Neumann. Vanguard/
        Supraphon SU 2.
            +HF 5-75 p74   ++St 5-75 p102
            +NR 3-75 p3
929 Symphony, no. 9, op. 95, E minor. SMETANA: Ma Vlast: Vltava.
        VPO; Rafael Kubelik. Decca ECS 771. (Reissues from SXL 2005,
        2064/5)
            \ -Gr 11-75 p817   -RR 8-75 p31
            -HFN 9-75 p108
930 Symphony, no. 9, op. 95, E minor. NBC Symphony Orchestra; Arturo
        Toscanini. RCA AT 114. Tape (c) MCK 580. (Reissue from RCA
        VICS 1187)
            +Gr 4-73 p1926   +-RR 4-73 p52
            +Gr 6-75 p111 tape   -RR 6-75 p91 tape
            -HFN 4-73 p780
931 Symphony, no. 9, op. 95, E minor. CPhO; Václav Neumann. Supra-
        phon 110 1334.
            +NR 9-75 p4   +RR 9-75 p36
932 Symphony, no. 9, op. 95, E minor. CPhO; Václav Neumann. Vanguard
        SU 8.
            ++NR 11-75 p1
     Symphony, no. 9, op. 95, E minor. cf Camden CCV 5000-12, 5014-24.
     Symphony, no. 9, op. 95, E minor. cf HMV SLS 5003.
     Symphony, no. 9, op. 95, E minor: Largo. cf Slavonic dances, op.
        72, nos. 9, 10, 15.
     Terzetto, 2 violins and viola, op. 74, C major. cf Quartet,
        strings, no. 4, op. 61, C major.
     Trios, piano, op. 26, G minor. cf Quintet, strings, no. 1, op. 1,
        A minor.
933 Trio, piano, op. 65, F minor. Yuval Trio. DG 2530 371
            ++Gr 4-75 p1820   +-RR 3-75 p38
            +HF 6-74 p82   ++SFC 5-19-74 p29
            ++NR 8-74 p6   ++St 5-74 p101
     Trio, piano, op. 65, F minor. cf Quintet, strings, no. 1, op. 1,
        A minor.
934 Trio, piano, op. 90, E minor. SUK: Elegie, op. 23. Smetana Trio.
        Panton 110 234.
            +-Gr 7-75 p200   +RR 6-75 p58
            -HFN 7-75 p80
     Waltzes, op. 54 (2). cf Quartet, strings, no. 5, op. 87, E flat
        major.
     Waltz, op. 54, no. 1, A major. cf Supraphon 110 1429.
     Waltz, op. 54, no. 4. cf Quintet, strings, no. 1, op. 1, A minor.
DYKES
     Holy, holy, holy, Lord God almighty. cf RCA ARL 1-0562.

Lead kindly light. cf CBS 61599.
EARLS, Paul
    Songs: Arise my love; Entreat me not to leave you. cf Duke
      University Press DWR 7306.
    EARLY AMERICAN VOCAL MUSIC: New England anthems and southern folk
      hymns. cf Nonesuch H 71276.
EAST, Michael
    Desperavi. cf Crystal S 202.
    Desperavi. cf Desto DC 6474/7.
    Triumphavi. cf Desto DC 6474/7.
EBERLIN, Johann
    Toccata and fugue, A minor. cf Hungaroton SLPX 11601/2.
    Toccata and fugue tertia. cf Philips 6775 006.
    Toccata e Fughe per l'Organo: Fugue, G minor. cf Pelca PSRK
      41013/6.
    Toccata sexta. cf Philips 6775 006.
ECCARD, Johann
    Nun komm der Heiden Heiland. cf Turnabout TV 34569.
EGGE, Klaus
935 Concerto, piano, no. 2, op. 21. BARATI: Concerto, violoncello.
    Robert Riefling, pno; Oslo Philharmonic Orchestra; Ølvin
    Fjeldstad. CRI 184.
        +St 6-75 p50
EHRLICH, Jesse
    Short pieces, 3 cellos. cf CASALS: Sardana, celli.
EISMA, Will
936 Le gibet Little lane. RAXACH: Imaginary landscape.   Paraphrase.
    Ileana Melita, con; Rien de Reede, flt; Willy Goudswaard,
    perc; Cor Coppens, ob; Radio Chamber Orchestra; Pro-Hontra
    Ensemble; Paul Hupperts. Donemus Audio-Visual DAVS 7475/2.
      +RR 10-75 p67
    Little lane. cf Le gibet.
EKLUND, Hans
    Pezzi per organi (3). cf BIS LP 7.
ELGAR, Edward
    Allegretto on a theme of five notes. cf Works, selections (Pearl
      523).
937 The apostles, op. 49. Sheila Armstrong, s; Helen Watts, con;
    Robert Tear, t; Benjamin Luxon, bar; Clifford Grant, John
    Carol Case, bs; Downe House School Choir; LPO and Choir; Adrian
    Boult. HMV SLS 976 (3). Tape (c) TC SLS 976.
        +Gr 11-74 p935   +RR 11-74 p18
        +Gr 3-75 p1712 tape   +RR 6-75 p91 tape
    Ave verum, op. 2, no. 1. cf ARgo ZRG 785.
    The banner of St. George, op. 33: Epilogue. cf Works, selections
      (HMV RLS 713).
    Bavarian dances, op. 27, nos. 1-3. cf Works, selections (HMV RLS
      713).
    Beau Brummel: Minuets (2). cf Works, selections (HMV RLS 713).
    Bizarrerie, op. 13, no. 2. cf Works, selections (Pearl 523).
    Caprice, op. 51, no. 2, A minor. cf Discopaedia MB 1005.
    La capricieuse, op. 17. cf Works, selections (Pearl 523).
    La capricieuse, op. 17 (2). cf RCA ARM 4-0942/7.
    Caractacus, op. 35. cf Works, selections (HMV ASD 3050).
    Caractacus, op. 35: Woodland interlude; Triumphal march. cf Works,
      selections (HMV RLS 713).

Carillon, op. 75. cf Works, selections (HMV ASD 3050).
Carissima. cf Works, selections (CBS 76423).
Carissima. cf Works, selections (HMV RLS 713).
Chanson de matin, op. 15, no. 2. cf Works, selections (CBS 76423).
Chanson de matin, op. 15, no. 2. cf Works, selections (HMV RLS 713).
Chanson de matin, op. 15, no. 2. cf Works, selections (Pearl 523).
Chanson de nuit, op. 15, no. 1. cf Works, selections (CBS 76423).
Chanson de nuit, op. 15, no. 1. cf Works, selections (HMV RLS 713).
Chanson de nuit, op. 15, no. 1. cf Works, selections (Pearl 523).
938   Choral works: Ecce sacerdos magnus (edit. Tozer). From the Bavar-
       ian highlands, op. 27, no. 1, The dance; No. 2, False love; No.
       3, Lullaby; No. 4, Aspiration; No. 5, On the Alm; No. 6, The
       marksman. The light of life, op. 29: Doubt not thy Father's
       care (arr. H. A. Chambers); Light of the world. O salutaris
       hostia; 3 settings. Tantum ergo. Worcester Cathedral Choir;
       Frank Wibaut, pno; Harry Bramma, org; Christopher Robinson.
       Polydor 2460 239.
            +-Gr 2-75 p1529
939   Cockaigne overture, op. 40. Enigma variations, op. 36. Pomp and
       circumstance marches, op. 39, nos. 1, 4. Imperial march, op.
       32. NYP, LPO, RPO, PO; Leonard Bernstein, Daniel Barenboim,
       Thomas Beecham, Eugene Ormandy. CBS 30055. (Reissues from
       61111, 76248, 72982, Philips ABL 3053).
            -Gr 6-75 p39   +-HFN 7-75 p81
940   Cockaigne overture, op. 40. Enigma variations, op. 36. Serenade,
       strings, op. 20, E minor. RPO; Thomas Beecham. CBS 61660.
       (Reissue from Philips ABL 3053)
            +Gr 9-75 p459   +RR 10-75 p43
            -HFN 11-75 p153
Cockaigne overture, op. 40. cf Works, selections (HMV RLS 713).
Cockaigne overture, op. 40. cf BRITTEN: Simple symphony, strings,
       op. 4.
Concerto, violin, op. 61, B minor. cf RCA ARM 4-0942/7.
941   Concerto, violoncello, op. 85, E minor. Introduction and allegro,
       op. 47. Serenade, strings, op. 20, E minor. Paul Tortelier,
       vlc; Rodney Friend, John Willison, vln; John Chambers, vla;
       Alexander Cameron, vlc; LPO; Adrian Boult. HMV ASD 2906.
       Tape (c) TC ASD 2906. (also Angel 37029)
            +Gr 8-73 p330   ++RR 8-73 p41
            ++Gr 6-75 p106 tape   ++RR 6-75 p91 tape
            +-HF 8-75 p84   ++SFC 10-5-75 p38
            +HFN 5-75 p142 tape   ++St 6-75 p95
            ++NR 4-75 p5
942   Concerto, violoncello, op. 85, E minor. Enigma variations, op.
       36. André Navarra, vlc; Hallé Orchestra; John Barbirolli. Pye
       GSGC 15005. (Reissue from CCL 30101, 30103, GSGC 14057)
            +Gr 8-75 p335   +-RR 8-75 p32
            +-HFN 8-75 p87
Concerto, violoncello, op. 85, E minor: 3rd movement. cf HMV SEOM
       19.
Contrasts, op. 10, no. 3. cf Works, selections (HMV RLS 713).
Crown of India suite, op. 66. cf Imperial march, op. 32.
Crown of India suite, op. 66. cf Works, selections (HMV RLS 713).
Dream children, op. 43. cf Works, selections (HMV ASD 3050).
Dream children, op. 43, nos. 1 and 2. cf London STS 15160.
943   The dream of Gerontius, op. 38. Gladys Ripley, con; Heddle Nash, t;

Dennis Noble, bar; Norman Walker, bs; Huddersfield Choral
Society; Royal Liverpool Philharmonic Orchestra; Malcolm Sargent.
HMV RLS 709 (2).   (Reissue from C 3435/6)
+Gr 4-75 p1844   +RR 4-75 p59

The dream of Gerontius, op. 38: Kyrie eleison...All ye saints;
Rescue him, O Lord...O Lord, into they hands; Go in the name...
Through the same; Praise to the Holies...Praise to the Holiest;
And now the threshold...Most sure in all His ways; Jesu, by
that shuddering dread...To that glorious home; Take me away...
Praise to the Holiest. cf Works selections (HMV RLS 713).

The dream of Gerontius, op. 38: Praise to the holiest. cf Decca
ECS 2159.

The dream of Gerontius, op. 38: Praise to the holiest. cf HMV TC
MCS Tape (c) 14.

944 Elegy, strings, op. 58. Introduction and allegro, op. 47. Sere-
nade, strings, op. 20, E minor. Sospiri, op. 70. The Spanish
lady: Suite. AMF; Neville Marriner. Argo ZRG 573. Tape (c)
KZRC 573.
+HFN 2-74 p347 tape   +RR 1-74 p77 tape
+HFN 5-75 p142 tape

Elegy, strings, op. 58. cf Enigma variations, op. 36.
Elegy, strings, op. 58. cf Works, selections (CBS 76423).
Elegy, strings, op. 58. cf Works, selections (HMV RLS 713).
Elegy, strings, op. 58. cf Works, selections (HMV ASD 3050).

945 Enigma variations, op. 36. Pomp and circumstance marches, op. 39,
nos. 1-5. RPO; Phillip Moore, org; Norman del Mar. Contour
2870 440.
+Gr 6-75 p39   +RR 7-75 p25
+HFN 6-75 p88

946 Enigma variations, op. 36. Elegy, strings, op. 58. Serenade,
strings, op. 20, E minor. CPhO, LSO; Leopold Stokowski, Ainslee
Cox. Decca PFS 4338.
+Gr 8-75 p322   +RR 8-75 p32
+HFN 8-75 p75

947 Enigma variations, op. 36. BRAHMS: Variations on a theme by
Haydn, op. 56a. LSO; Pierre Monteux. London STS 15188.
++NR 11-74 p2   ++St 1-75 p110

948 Enigma variations, op. 36. Falstaff, op. 68. NPhO; Andrew Davis.
Lyrita SRCS 77.
+Gr 9-75 p457   +RR 9-75 p36
+HFN 9-75 p98

949 Enigma variations, op. 36. STRAUSS, R.: Don Juan, op. 20. COA;
Bernard Haitink. Philips 6500 481. Tape (c) 7300 344.
+Gr 3-75 p1649   ++NR 2-75 p3
+Gr 6-75 p111 tape   ++RR 4-75 p27
++HF 4-75 p84   +RR 6-75 p91 tape
++HFN 7-75 p90 tape   -SFC 4-4-75 p22
+MJ 2-75 p39   ++St 5-75 p94

Enigma variations, op. 36. cf Cockaigne overture, op. 40 (CBS
30055).

Enigma variations, op. 36. cf Cockaigne overture, op. 40 (CBS
61660).

Enigma variations, op. 36. cf Concerto, violoncello, op. 85, E
minor.

Enigma variations, op. 36. cf BRITTEN: Simple symphony, strings,
op. 4.

Enigma variations, op. 36. cf IVES: Symphony, no. 1, D minor.
Enigma variations, op. 36: Nimrod. cf Vista VPS 1021.
950 Falstaff, op. 68. The sanguine fan, op. 8. Fantasia and fugue,
    op. 86, C minor (Bach trans. Elgar). LPO; Adrian Boult. HMV
    ASD 2970. Tape (c) TC ASD 2970.
         +Gr 3-74 p1694   +-HFN 9-75 p110 tape
         +Gr 8-75 p376 tape   ++RR 5-74 p33
    Falstaff, op. 68. cf Enigma variations, op. 36.
    Falstaff, op. 68: Interludes. cf Works, selections (HMV RLS 713).
    Fantasia and fugue, op. 86, C minor. cf Falstaff, op. 68.
    Froissart overture, op. 19. cf Works, selections (HMV RLS 713).
    Gavotte. cf Works, selections (Pearl 523).
    Grania and Diarmid, op. 42: Incidental music, Funeral march. cf
       Works, selections (HMV ASD 3050).
    Une idylle, op. 4, no. 1. cf Works, selections (Pearl 523).
951 Imperial march, op. 32. Pomp and circumstances marches, op. 39,
    nos. 1-5. Crown of India suite, op. 66. LPO; Daniel Barenboim.
    Columbia M 32936. Tape (c) MT 32926 (ct) MA 32926 (Q) MQ 32936
    Tape (ct) MAQ 32936. (also CBS 76248. Tape (c) 40-76248)
         -Gr 9-74 p495   +-HFN 12-75 p173 tape
         +-Gr 12-75 p1121 tape   +-RR 10-74 p45
         -HF 4-75 p112 tape   ++SFC 9-22-74 p22
         +-HF 4-75 p112 Quad
    Imperial march, op. 32. cf Cockaigne overture, op. 40.
    Imperial march, op. 32. cf Decca 419.
    In the south overture, op. 50. cf Works, selections (HMV RLS 713).
952 Introduction and allegro, op. 47. Serenade, strings, op. 20, E
    minor. TIPPETT: Fantasia concertante on a theme by Corelli.
    Little music for strings. St. John's Symphony Orchestra; John
    Lubbock. Pye (Q) TPLS 13069.
         -Gr 12-75 p1032
    Introduction and allegro, op. 47. cf Concerto, violoncello, op.
       85, E minor.
    Introduction and allegro, op. 47. cf Elegy, strings, op. 58.
    The land of hope and glory. cf Works, selections (HMV RLS 713).
    The light of life, op. 29: Meditation. cf Works, selections (HMV
       RLS 713).
    The light of life, op. 29: Meditation. cf Works, selections (HMV
       ASD 3050).
    May song. cf Works, selections (HMV RLS 713).
    Mazurka, op. 10, no. 1. cf Works, selections (HMV RLS 713).
    Minuet, op. 21. cf Works, selections (HMV RLS 713).
    Mot d'amour, op. 13, no. 1. cf Works, selections (Pearl 523).
    The national anthem. cf Works, selections (HMV RLS 713).
    Nursery suite. cf Works, selections (HMV RLS 713).
    O God, our help in ages past. cf Works, selections (HMV RLS 713).
    Offertoire. cf Works, selections (Pearl 523).
    Pastourelle, op. 4, no. 2. cf Works, selections (Pearl 523).
    Piano improvisations (5). cf Works, selections (HMV RLS 713).
    Polonia, op. 76. cf Works, selections (HMV ASD 3050).
    Pomp and circumstance marches, op. 39. cf Decca 419.
    Pomp and circumstance marches, op. 39, nos. 1-5. cf Enigma varia-
       tions, op. 36.
    Pomp and circumstance marches, op. 39, nos. 1-5. cf Imperial march,
       op. 32.
    Pomp and circumstance marches, op. 39, nos. 1-4. cf Works, selec-
       tions (HMV RLS 713).

Pomp and circumstance marches, op. 39, nos. 1, 4. cf Cockaigne
overture, op. 40.
Pomp and circumstance marches, op. 39, nos. 1, 2, 4. cf Works,
selections (HMV RLS 713).
Pomp and circumstance march, op. 39, no. 1. cf BRITTEN: Simple
symphony, strings, op. 4.
Reminiscences, op. 1. cf Works, selections (Pearl 523).
Romance, op. 1. cf Works, selections (Pearl 523).
Romance, bassoon, op. 62. cf Works, selections (CBS 76423).
Rosemary. cf Works, selections (CBS 76423).
Rosemary. cf Works, selections (HMV RLS 713).
Salut d'amour, op. 12. cf Works, selections (CBS 76423).
Salut d'amour, op. 12. cf Works, selections (HMV RLS 713).
Salut d'amour, op. 12. cf Works, selections (Pearl 523).
The sanguine fan, op. 8. cf Falstaff, op. 68.
953 Sea pictures, op. 37. MAHLER: Rückert songs (5). Janet Baker, ms;
LSO, NPhO; John Barbirolli. Angel S 36796.
+ARG 5-72 p410   +-ON 3-4-72 p35
+Audio 12-75 p104   ++SFC 1-23-72 p30
+NR 1-72 p12   ++St 3-72 p84
Sea pictures, op. 37. cf BERLIOZ: Les nuits d'été, op. 7.
Serenade, strings, op. 20, E minor. cf Cockaigne overture, op. 40.
Serenade, strings, op. 20, E minor. cf Concerto, violoncello,
op. 85, E minor.
Serenade, strings, op. 20, E minor. cf Elegy, strings, op. 58.
Serenade, strings, op. 20, E minor. cf Enigma variations, op. 36.
Serenade, strings, op. 20, E minor. cf Introduction and allegro,
op. 47.
Serenade, strings, op. 20, E minor. cf Works, selections (CBS
76423).
Serenade, strings, op. 20, E minor. cf Works, selections (HMV RLS
713).
Serenade lyrique. cf Works, selections (HMV RLS 713).
Severn suite, op. 87. cf Works, selections (HMV RLS 713).
Sonata, organ, op. 28, G major. cf Decca 5BBA 1013-5.
Sonata, organ, op. 28, G major: 1st movement. cf HMV HLM 7065.
Sospiri, op. 70. cf Elegy, strings, op. 58.
Sospiri, op. 70. cf Works, selections (CBS 76423).
Sospiri, op. 70. cf Works, selections (Pearl 523).
The Spanish lady: Suite. cf Elegy, strings, op. 58.
954 Symphony, no. 1, op. 55, A flat major. Symphony, no. 2, op. 63,
E flat major. LPO; Daniel Barenboim. CBS 78289 (2).
-RR 12-75 p48
955 Symphony, no. 2, op. 63, E flat major. LPO; Georg Solti. Decca
SXL 6723. Tape (c) KSXC 16723.
+Gr 6-75 p36   ++RR 6-75 p35
++HFN 6-75 p88   +RR 12-75 p173 tape
+-MT 9-75 p798   +STL 9-7-75 p37
956 Symphony, no. 2, op. 63, E flat major. LPO; Adrian Boult. Pye
GSGC 15008. (Reissue from GSGC 14002)
+-Gr 8-75 p335   /RR 8-75 p32
+HFN 8-75 p87
Symphony, no. 2, op. 63, E flat major. cf Symphony, no. 1, op.
55, A flat major.
Virelai, op. 4, no. 3. cf Works, selections (Pearl 523).
Wand of youth suites, nos. 1-2. cf Works, selections (HMV RLS 713).

957 Works, selections: Carissima. Chanson de matin, op. 15, no. 2.
       Chanson de nuit, op. 15, no. 1. Elegy, strings, op. 58. Romance
       bassoon, op. 62. Rosemary. Salut d'amour, op. 12. Serenade,
       strings, op. 20, E minor. Sospiri, op. 70. Martin Gatt, bsn;
       ECO; Daniel Barenboim. CBS 76423. Tape (c) 40-76423.
          +—Gr 11-75 p818   +RR 11-75 p43
          +—HFN 11-75 p153
958 Works, selections: The banner of St. George, op. 33: Epilogue.
       Bavarian dances, op. 27, nos. 1-3. Beau Brummel: Minuets (2).
       Caractacus, op. 35: Woodland interlude; Triumphal march. Car-
       issima. Chansons, op. 15, nos. 1-2. Cockaigne overture, op.
       40 (2). Contrasts, op. 10, no. 3. Crown of India suite, op.
       66. The dream of Gerontius, op. 38: Kyrie eleison...All ye
       saints; Rescue him, O Lord...O Lord, into thy hands; Go in the
       name...Through the same; Praise to the Holiest...Praise to the
       Holiest; And now the threshold...Most sure in all His ways;
       Jesu, by that shuddering dread...To that glorious home; Take
       me away...Praise to the Holiest. Elegy, strings, op. 58.
       Falstaff, op. 68: Interludes. Froissart overture, op. 19. In
       the south overture, op. 50. The land of hope and glory. The
       light of life, op. 29: Meditation. May song. Mazurka, op. 10,
       no. 1. Minuet, op. 21. The national anthem. Nursery suite.
       O God, our help in ages past. Rosemary. Piano improvisations
       (5). Pomp and circumstance marches, op. 39, nos. 1, 2, 4.
       Pomp and circumstance marches, op. 39, nos. 1-4. Rosemary.
       Salut d'amour, op. 12. Serenade, strings, op. 20, E minor.
       Severn suite, op. 87. Wand of youth suites, nos. 1-2. Film
       "Land of hope and glory", Elgar speaking and conducting. Var-
       ious artists and orchestras; Edward Elgar. HMV RLS 713 (6).
          ++Gr 2-75 p1480   +RR 2-75 p34
959 Works, selections: Caractacus, op. 35. Carillon, op. 75. Dream
       children, op. 43. Elegy, strings, op. 58. Grania and Diarmid,
       op. 42: Incidental music, Funeral march. The light of life,
       op. 29: Meditation. Polonia, op. 76. CHOPIN: Funeral march
       (arr. Elgar). LPO; Adrian Boult. HMV ASD 3050.
          ++Gr 2-75 p1491   ++RR 3-75 p26
960 Works, selections: Allegretto on a theme of five notes. Bizarrerie,
       op. 13, no. 2. La capricieuse, op. 17. Chanson de matin, op.
       15, no. 2. Chanson de nuit, op. 15, no. 1. Gavotte. Une idylle,
       op. 4, no. 1. Mot d'amour, op. 13, no. 1. Offertoire. Past-
       ourelle, op. 4, no. 2. Reminiscences. Romance, op. 1. Salut
       d'amour, op. 12. Sospiri, op. 70. Virelai, op. 4, no. 3.
       John Georgiadis, vln; John Parry, pno. Pearl SHE 523.
          ++Gr 11-75 p848   +RR 12-75 p81
ELLINGTON, Duke
     Daybreak express. cf New England Conservatory NEC 111.
     Koko. cf New England Conservatory NEC 111.
ELWYN-EDWARDS, Dilys
     Caneuom y tri aderun. cf Argo ZRG 769.
ENCINA, Juan del
     Fata la parte. cf Telefunken TK 11569/1-2.
     Romerico. cf DG 2530 504.
     Songs: Ay triste que vengo; Si abrá en este baldrés; Qu'es de ti;
       Levanta Pascual. cf Telefunken SAWT 9620/1.
     Todos los bienes. cf BIS LP 3.

ENESCO, Georges
Rumanian rhapsody, op. 11, no. 1, A major. cf HMV SEOM 14.
961 Songs, op. 15 (7). ROUSSEL: Songs: Adieu; A flower given to my
daughter; Jazz dans la nuit; Light; Mélodies, op. 20 (2). Odes
anacréontiques, nos. 1, 5; Odelette; Poèmes chinois, op. 12 (2);
Poemès chinois, op. 35 (2). Yolanda Marcoulescou, s; Katja
Phillabaum, pno. Orion ORS 75184.
++HF 10-75 p78   +NYT 6-8-75 pD19
+NR 7-75 p12
ENGEL
Chabad melody and Freilachs, op. 20, nos. 1 and 2. cf Da Camera
Magna SM 93399.
Sea shell. cf Orion ORS 74160.
ENGELMANN
Paduana and galliarda. cf Crystal S 202.
THE ENGLISH HARPSICHORD. cf Angel SB 3816.
ENGLISH MADRIGALS AND FOLKSONGS. cf Harmonia Mundi 593.
ENGLISH VIRGINALISTS. cf BASF BAC 3075.
EPSTEIN, David
962 Quartet, strings. The seasons. Trio, strings. Vent-ures.
Jan DeGaetani, ms; Robert Freeman, pno; Pacific Trio, Philha-
delphia Quartet, Eastman Wind Ensemble; Donald Hunsberger.
Dest DC 7148.
+/HF 5-75 p75
The seasons. cf Quartet, strings.
Trio, strings. cf Quartet, strings.
Vent-ures. cf Quartet, strings.
ERBACH, Christian
Canzon in the Phrygian mode. cf University of Iowa Press, un-
numbered.
Introitus sexti toni. cf Pelca PSRK 41013/6.
ERICKSON, Robert
963 End of the mime. FERRITTO: Oggi, op. 9. IVEY: Hera, hung from
the sky. RANDALL: Improvisation. Bethany Beardslee, s; Neva
Pilgrim, Elaine Bonazzi, ms; Allen Blustine, clt; Ursula Oppens,
pno; New Music Choral Ensemble; Instrumental Ensemble; Kenneth
Gaburo, David Gilbert, Andrew Thomas. CRI SD 325.
+MQ 10-75 p636   +NR 2-75 p10
Ricercar a 5 for trombones. cf DUGGER: Music for synthesizer and
six instruments.
ERTL, D.
Hoch-und Deutschmeister Marsch. cf DG 2721 077.
ESPEJO
Airs tziganes, op. 11. cf Discopaedia MB 1006.
ESPLA, Oscar
Prego. cf HMV SLS 5012.
ESTEVE Y GRIMAU, Pablo
Songs: Alam, sintamos; Ojos, llorar. cf DG 2530 598.
EUSTACE
Les grenadiers de la Vieille Garde à Waterloo. cf Classics for
Pleasure 40230.
EVANS, Merle
Symphonia. cf Columbia 33513.
EVETT, Robert
964 Quintet, piano and strings. PARRIS: The book of imaginary beings.
Robert Parris, pno; University of Maryland Trio, University of
Maryland Quartet; Dorothy Skidmore, flt; Ronald Barnett,

Thomas Jones, perc.   Turnabout TVS 34568.
   +-HF 9-75 p88   ++St 7-75 p102
   +NR 4-75 p7

EWALD, Victor
   Quintet, op. 5, B flat minor.   cf Desto DC 6474/7.

EYBLER, Joseph
   Polonaise.   cf DG Archive 2533 182.

EYCK, Jacob van
   Pavane lachrymae.   cf Telefunken Tape (c) CX 4-41203.
   Variations on 'Amarilli mia bella'.   cf Transatlantic TRA 292.

FALLA, Manuel de
965   El amor brujo.   WAGNER: Tristan und Isolde: Love music, Acts 2
     and 3 (arr. Stokowski.   Shirley Verrett, ms; LPO; Leopold
     Stokowski).   Odyssey Y 32368.   (Reissue from Columbia MS 6147).
     (also CBS 61288)
       +Gr 3-75 p1649   +SFC 4-28-74 p29
       ++HF 11-74 p91   +St 7-74 p103
       +-RR 9-74 p44

    El amor brujo: Danza del terror.   cf Works, selections (Musical
     Heritage Society MHS 1929).
    El amor brujo: Pantomime.   cf RCA ARM 4-0942/7.
    El amor brujo: Ritual fire dance.   cf Connoisseur Society (Q)
     CSQ 2066.
    El amor brujo: Ritual fire dance (arr. Piatigorsky); Pantomime
     (arr. Sadlo).   cf Supraphon 50919.
966   El amor brujo: Suite.   Fantasia baetica.   Piezas españolas: Aragon-
     ese, Cubana, Montanesca, Andaluza.   El sombrero de tres picos:
     Three dances.   Alicia de Larrocha, pno.   Decca SXL 6683.   (also
     London CS 6881)
       ++Gr 4-75 p1836   ++SFC 6-29-75 p26
       ++HF 12-75 p92   ++St 9-75 p112
       ++RR 4-75 p50

    Canción.   cf Angel S 36076.
967   Concerto, harpsichord and chamber orchestra.   Nights in the gardens
     of Spain.   GRANADOS: Goyescas: Intermezzo; The maiden and the
     nightingale.   Spanish National Orchestra; Ataulfo Argenta.
     Decca SDD 446.   (Reissues)
       +HFN 8-75 p87   +-RR 7-75 p25

    Fantasia baetica.   cf El amor brujo: Suite.
    Fantasia baetica.   cf Works, selections (Musical Heritage Society
     MHS 1929).
    Homenaje.   cf CBS 72950.
    Homenaje a Debussy.   cf ALBENIZ: Suite española, no. 1, Granada;
     no. 3, Sevillanas.
    Homenaje a Debussy.   cf Nonesuch H 71233.
968   Mélodies (3).   Spanish popular songs (7).   GRANADOS: Tonadilla: La
     maja dolorosa.   Tonadillas al estilo antiguo: Amor y odio; El
     majo discreto; El majo timido; El mirar de la Maja; El tra la
     la y el punteado.   TURINA: Poema en forma de canciones, op. 19.
     Jill Gomez, s; John Constable, pno.   Sage 5409.
       ++RR 12-75 p88

    Night in the gardens of Spain.   cf Concerto, harpsichord and
     chamber orchestra.
    Noches en los jardines de España.   cf CHOPIN: Concerto, piano, no.
     2, op. 21, F minor.
    Piezas españolas (4).   cf El amor brujo: Suite.

Piezas españolas; Aragonesa, Cubana, Montañesa, Andaluza. cf
Works, selections (Musical Heritage Society MHS 1929).
El sombrero de tres picos: Danza del molinero. cf ALBENIZ: Suite
española, no. 1, Granada; no. 3, Sevillanas.
El sombrero de tres picos: Danza de los vecinos; Danza de la moli-
nera. cf Works, selections (Musical Heritage Society MHS 1929).
The three-cornered hat: Farruca. cf Angel S 36076.
El sombrero de tres picos: The miller's dance. cf London STS
15224.
El sombrero de tres picos: The neighbours; The miller's dance;
Final dance. cf Decca SDDJ 393/95.
El sombrero de tres picos: Suite, no. 2. cf Decca SDDJ 393/95.
El sombrero de tres picos: Three dances. cf El amor brujo: Suite.
El sombrero de tres picos: Three dances. cf DEBUSSY: Images pour
orchestra: Ibéria, Rondes de printemps.
Spanish popular songs (7). cf Mélodies.
Spanish popular songs (7). cf BIZET: Songs (Decca SXL 6577).
Spanish popular songs (7). cf HMV SLS 5012.
Canciones populares españolas: Jota. cf Discopaedia MB 1010.
Spanish popular songs: Jota. cf RCA ARM 4-0942/7.
Spanish popular songs: Nana. cf CBS Classics 61579.
Spanish popular songs: Nana. cf CBS 76420.
La vida breve: Danse espagnole. cf Argo ZRG 805.
La vida breve: Danza, no. 1. cf RCA ARM 4-0942/7.
La vida breve: Spanish dance, no. 1. cf Saga 5356.
La vida breve: Danza, no. 2. cf Works, selections (Musical Heri-
tage Society MHS 1929).
La vida breve: Vivan los que rien. cf HMV SLS 5012.
969 Works, selections: El amor brujo: Danza del terror. Fantasia
baetica. Piezas españolas: Aragonesa, Cubana, Montañesa, Anda-
luza. El sombrero de tres picos: Danza de los vecinos (Seguid-
illas); Danza de la molinera. La vida breve: Danza, no. 2.
Alicia de Larrocha, pno. Musical Heritage Society MHS 1929.
(Reissue from Hispavox)
++St 9-75 p112

FALVO
Dicitencello vuie. cf Columbia D3M 33448.
Dicitencello vuie. cf Musical Heritage Society MHS 1951.
FARINA, Carlo
Pavane. cf Grange SGR 1124.
FARKAS
Pieces, guitar (6). cf Hungaroton SLPX 11629.
FARMER, John
A little pretty bonny lass. cf Harmonia Mundi 593.
A little pretty bonnie lass. cf University of Illinois, un-
numbered.
FARNABY, Giles
Farnaby's conceit. cf Angel SB 3816.
Giles Farnaby's dream. cf Angel SB 3816.
His humour. cf Angel SB 3816.
His rest. cf Angel SB 3816.
Loath to depart. cf Saga 5402.
Maske, G minor. cf BASF BAC 3075.
The new Sa-hoo. cf Angel SB 3816.
The old spagnoletta. cf Angel SB 3816.
Tell me, Daphne. cf Angel SB 3816.

A toye. cf Angel SB 3816.
Up tails all. cf Angel SB 3816.
FARRANT, Richard
    Call to remembrance. cf Argo ZRG 789.
FARRAR
    Bombasto. cf Michigan University SM 0002.
FASCH, Johann Friedrich
    Sonata, bassoon, C major. cf Musical Heritage Society MHS 1853.
FASOLO (17th century Italy)
        Cangia, cangia tue voglie. cf Pearl SHE 511.
FAURE, Gabriel
    Après un rêve, op. 7, no. 1. cf HMV SEOM 19.
    Après un rêve, op. 7, no. 1. cf Orion ORS 75181.
970 Barcarolles, nos. 1-3. Jean-Philippe Collard, pno. French EMI
        C 065 12048. (also Connoisseur Society CS 2078)
            ++NR 11-75 p13   +SR 11-29-75 p50
            ++SFC 1-7-73 p23   ++St 12-75 p125
            ++SFC 8-10-75 p26
    Berceuse, op. 16. cf Rococo 2035.
971 Chant funéraire, op. 117. KOECHLIN: Quelques chorals pour des
        fêtes populaires. SCHMITT: Dionysiaques, op. 62, no. 1.
        Musique de Gardiens de la Paix; Désiré Dondeyne. CRD CAL 1839.
            ++Gr 10-75 p611   ++RR 11-75 p24
    La charité. cf Rubini RS 300.
    Dolly, op. 56. cf Classic Record Library SQM 80-5731.
    Dolly, op. 56: Berceuse. cf BIZET: Jeux d'enfants, op. 22.
972 Elégie, op. 24, C minor. Sonata, violoncello, no. 1, op. 109, D
        minor. Sonata, violoncello, no. 2, op. 117, G minor. Sicilienne,
        op. 78. Thomas Igloi, vlc; Clifford Benson, pno. CRD 1016.
            +-Gr 12-75 p1062   +-HFN 9-75 p98
973 Elégie, op. 24, C minor. Sonatas, violoncello and piano, nos. 1,
        2. Miklós Perényi, vlc; Loránd Szücs, pno. Hungaroton SLPX
        11658.
            +-Gr 7-75 p205   ++RR 1-75 p48
            +-NR 5-75 p9
    Elégie, op. 24, C minor. cf DVORAK: Concerto, violoncello, op.
        104, B minor.
    Elégie, op. 24, C minor. cf Command COMS 9006.
    Elégie, op. 24, C minor. cf Orion ORS 75181.
974 Fantaisie, piano and orchestra, op. 111. RAVEL: Concerto, piano,
        for the left hand, D major. Concerto, piano, G major. Alicia
        de Larrocha, pno; LPO; Rafael Frühbeck de Burgos, Lawrence
        Foster. Decca SXL 6680. (also London CS 6787)
            ++Gr 11-74 p890   ++RR 10-74 p53
            +HF 6-75 p102   ++SFC 4-27-75 p23
            +NR 7-75 p6   ++St 7-75 p106
    Fantasy, op. 79. cf Classic Record Library SQM 80-5731.
    Impromptu, no. 2, op. 31, F minor. cf Decca SPA 372.
    Impromptu, no. 3, op. 34, A flat major. cf Columbia M 32070.
    Nocturne. cf Decca SPA 405.
975 Nocturnes, op. 33, nos. 1-3; op. 36; op. 37; op. 63; op. 74; op.
        84, no. 8; op. 97; op. 99; op. 104; op. 107; op. 119. Theme
        and variations, op. 73, C minor. Jean-Philippe Collard, pno.
        Connoisseur Society/ Pathé Marconi CS 2072 (2).
            +HF 6-75 p94   ++St 7-75 p69
            ++SFC 4-6-75 p22

Nocturne, op. 63, D flat major. cf Columbia M 32070.
Papillon, op. 77. cf Command COMS 9006.
Pavane. cf London STS 15160.
Pavane, op. 50. cf Requiem, op. 48.
Pavane, op. 50. cf Decca DPA 519/20.
Pelléas et Mélisande, op. 80. cf RAVEL: Valses nobles et senti-
 mentales.
976 Quartet, piano, op. 15, C minor. FRANCK: Quintet, piano and strings,
 F minor. Jesús María Sanroma, Clifford Curzon, pno; Budapest
 Quartet. Odyssey Y 33315.
   +—HF 12-75 p92   ++NR 8-75 p7
977 Quartet, piano, op. 15, C minor. Quartet, strings, op. 121, E
 minor. Artur Rubinstein, pno; Guarneri Quartet. RCA ARL
 1-0761.
   +Gr 5-75 p1985   +RR 3-75 p38
   +—HF 3-75 p78   ++SFC 11-24-74 p32
   +MT 9-75 p798   +St 4-75 p100
   ++NR 1-75 p8
Quartet, strings, op. 121, E minor. cf Quartet, piano, op. 15,
 C minor.
978 Requiem, op. 48. Pavane, op. 50. Sheila Armstrong, s; Dietrich
 Fischer-Dieskau, bar; Henriette Puig-Roget, org; Edinburgh
 Festival Chorus; Orchestre de Paris; Daniel Barenboim. Angel
 S 37077. (also HMV (Q) SQ Q4ASD 3065)
   +Gr 6-75 p76 Quad   +—NYT 9-21-75 pD18
   +—HF 5-75 p76   +RR 6-75 p79 Quad
   +HFN 6-75 p88 Quad +—St 7-75 p96
   +—NR 4-75 p6
979 Requiem, op. 48. Kyoko Ito, s; Norio Ohga, bar; Takashi Sakai,
 org; Tokyo Metropolitan Choir and Symphony Orchestra; Kazuo
 Yamada. Columbia MQ 32883. Tape MAQ (Q) 32883.
   +HF 10-74 100   +St 11-74 p125
   +HF 1-75 p110 Quad tape
980 Requiem, op. 48. Orchestra; Ernest Ansermet. Decca SDD 154.
 Tape (c) KSDC 154.
   +—HFN 7-75 p90 tape
Requiem, op. 48: Agnus Dei. cf HMV Tape (c) TC MCS 13.
Requiem, op. 48: Sanctus. cf HMV Tape (c) TC MCS 14.
Sicilienne, op. 78. cf Elégie, op. 24, C minor.
Sicilienne, op. 78. cf Classic Record Library SQM 80-5731.
Sicilienne, op. 78. cf Command COMS 9006.
981 Sonata, violin and piano, no. 1, op. 13, A major. Sonata, violin,
 no. 2, op. 108, E minr. Clara Bonaldi, vln; Sylvaine Billier,
 pno. Arion ARN 38267.
   ++Gr 11-75 p848   +RR 10-75 p76
Sonata, violin and piano, no. 1, op. 13, A major. cf RCA ARM
 4-0942/7.
Sonata, violin and piano, no. 1, op. 13, A major. cf DEBUSSY:
 Sonata, violin and piano.
Sonata, violin and piano, no. 2, op. 108, E minor. cf Sonata,
 violin and piano, no. 1, op. 13, A major.
Sonatas, violoncello and piano, nos. 1, 2. cf Elégie, op. 24,
 C minor.
Sonata, violoncello, no. 1, op. 109, D minor. cf Elégie, op. 24,
 C minor.
Sonata, violoncello, no. 2, op. 117, G minor. cf Elégie, op. 24,
 C minor.

982 Songs: L'Absent, op. 5, no. 3; Après un rêve, op. 7, no. 1; Au
      bord de l'eau, op. 8, no. 1; Aubade, op. 6, no. 1; Aurore, op.
      39, no. 4; Barcarolle, op. 7, no. 3; Chanson du pêcheur, op. 4,
      no. 1; Chant d'automne, op. 5, no. 1; Dans les ruins d'un abbaye;
      Hymne, op. 7, no. 2; Ici-bas, op. 8, no. 3; Lydia, op. 4, no.
      2; Mai, op. 1, no. 2; Les matelots, op. 2, no. 2; Le papillon
      et la fleur, op. 2, no. 1; Le rançon, op. 8, no. 2; Rêve d'amour,
      op. 5, no. 2; Sylvia, op. 6, no. 3; Tristesse, op. 6, no. 2.
      Jacques Herbillon, bar; Theodore Paraskivesco, pno. CRD CAL
      1841.
            +RR 10-75 p24
983 Songs: Accompagnement, op. 85, no. 3; Arpège, op. 76, no. 2; La
      bonne chanson, op. 61; Dans la forêt de septembre, op. 85, no.
      1; La fleur qui va sur l'eau, op. 85, no. 2; Le perfum im-
      périssable, op. 76, no. 1; Le plus doux chemin, op. 87, no.
      1; Prison, op. 83, no. 1; Le ramier, op. 87, no. 2; Sérénade
      du bourgeois gentilhommbre; Soir, op. 83, no. 2. Jacques Herbillon,
      bar; Theodore Paraskivesco, pno. CRD CAL 1844.
            +RR 10-75 p24
984 Songs: L'Horizon chimérique, op. 118; Nocturne, op. 43, no. 2;
      Poème d'un jour, op. 21. LULLY: Alceste: Air de Caron. POULENC:
      L'Anguille; La belle jeunesse; Priez pour paix; Serenade.
      TIERSOT: Chants de la vieille France (4). Martian Singher, bar;
      Alden Gilchrist, hpd or pno. 1750 Arch S 1754.
            +NR 12-75 p12
985 Songs: La bonne chanson, op. 71; Two songs, op. 76; Three songs,
      op. 85; Mirages, op. 113. Bernard Kruysen, bar; Noël Lee, pno.
      Telefunken SAT 22546.
            +-Gr 7-74 p250   +-MQ 1-75 p171
            -HF 8-74 p93
     Songs: Claire de lune, op. 46, no. 2; Mélodies, op. 58; Spleen,
      op. 51, no. 3. cf DEBUSSY: Songs (Cambridge CRS 2774).
     Songs: Chanson Lorraine; Les rameaux. cf Discophilia KGC 1.
     Songs: Les bercaux, op. 23, no. 1; La chanson du pêcheur, op. 4,
      no. 1; Mai, op. 1, no. 2. cf Seraphim S 60251.
     Theme and variations, op. 73, C minor. cf Nocturnes, op. 33,
      nos. 1-3 (Connoisseur Society CS 2072).
     FAVORITE DUETS WITH TENORS. cf ABC ATS 20018.
     FAVORITE SPANISH ENCORES. cf RCA ARL 1-0485.
     FAVORITE TENOR ARIAS. cf London OS 26384.
FAYRFAX, Robert
     I love, loved; Thatt was my woo. cf L'Oiseau-Lyre 12BB 203/6.
FELD, Jindrich
986 Quartet, strings, no. 4. MATEJ: Quartet, strings, no. 2. Smetana
      Quartet; Vlach Quartet. Supraphon 111 0970.
            +Gr 5-75 p1985   ++RR 4-75 p40
            +NR 8-73 p6
987 Sonata, piano. LISZT: Sonata, piano, G 178, B minor. Božena
      Steinerová, pno. Panton 110 379.
            +-RR 9-75 p60
FENNELLY, Brian
988 Evanescences. HIBBARD: Quartet, strings. Da Capo Chamber Players,
      members; Stradivari Quartet. CRI SD 322.
            +HF 1-75 p83   +-RR 10-75 p62
            ++NR 10-74 p9

FERGUSON
    The lark in the clear air. cf Philips 6599 227.
FERGUSON, Howard
    Discovery. cf Decca 6BB 197/8.
    O Jesus, I have promised. cf Abbey LPB 734.
FERNANDEZ, Oscar Lorenzo
    Brasileira, no. 2: Ponteio, Moda, Cataretè. cf Angel S 37110.
FERNSTROM, John
989 Concertino, flute, women's chorus and chamber orchestra, op. 52.
        LARSSON: Concerto, violin, op. 42. KOCH: Oxberg variations.
        Erik Holmstedt, flt; André Gertler, vln; Stockholm Radio Orches-
        tra, Stockholm Symphony Orchestra; Sten Frykberg, Stig Wester-
        berg. Turnabout TVS 34498.
           +-NR 1-75 p4   ++St 3-75 p105
          +St 6-75 p51
FERRABOSCO, Alfonso II
    Four note pavan. cf Turnabout TV 37071.
    Pavane, no. 4. cf Grange SGR 1124.
FERRITTO, John
    Oggi, op. 9. cf ERICKSON: End of the mime.
    FESTIVAL OF FLUTE CONCERTOS. cf RCA CRL 2-7003.
    FESTIVE MARCHES. cf Panton 110 361.
FEVIN, Antoine de
    Faulte d'argent. cf Telefunken TK 11569/1-2.
FIALA, Josef
    Concerto, cor anglais, E flat major. cf ALBINONI: Concerto, oboe,
        op. 9, no. 2, D minor.
FIBICH, Zdenĕk
    At twilight: Poem. cf Supraphon 110 1429.
FIELD, John
    Nocturne, fortepiano, A major. cf Strobe SRCM 123.
    FIFE AND DRUM TUNE. cf Old North Bridge Records ONB 1775.
FILLMORE, Henry
    Americans we. cf Michigan University SM 0002.
    The footlifter. cf Columbia 33513.
FILTZ, Anton
    Sonata, 2 flutes, op. 2, no. 4, D major: Allegro and minuetto. cf
        Oryx ORYX 1720.
FINCK
    Greiner, zanner. cf Crystal S 202.
FINGER, Gottfried
    Divisions on a ground. cf Transatlantic TRA 292.
    Sonata, flute, D minor. cf Orion ORS 75199.
FINZI, Gerald
    Dies natalis, op. 8. cf HOLST: Choral fantasia, op. 51.
990 Intimations of immortality, tenor, chorus and orchestra. Ian
        Partridge, t; Guildford Philharmonic Orchestra and Choir;
        Vernon Handley. Lyrita SRCS 75.
           +Gr 5-75 p2000   /RR 5-75 p59
           +HFN 5-75 p127   +Te 12-75 p43
FIOCCO, Joseph-Hector
    Andante. cf CRD CRD 1008.
FIRENZE, Lorenzo di
    Opposte messe. cf Vanguard VSD 71179.
FISCHER, Johann
    Bouree. cf DG Archive 2533 172.

Gigue. cf DG Archive 2533 172.
Praeludium and fugue, C major. cf Telefunken TK 11567/1-2.
Prelude and fugue, D minor. cf Telefunken TK 11567/1-2.
Suite, no. 9, D minor: Passacaglia. cf BACH: Toccata and fugue,
    S 916, G major.
FISCHER, Johann Kaspar
    Preludes and fugues, B minor, D major, E flat major, C minor.
        cf Philips 6775 006.
FISER, Luboš
991 Requiem. JIRASEK: Stabat mater. Musica Nova Bohemica. Supraphon
        112 1537.
            +-HFN 8-75 p75   +RR 7-75 p58
            +NR 10-75 p6
992 Sonata, piano, no. 3. MUSSORGSKY: Pictures at an exhibition.
        Richard Kratzmann, pno. Panton 110 283.
            +-RR 12-75 p84
FISHER
    I heard a cry. cf Discophilia KGS 3.
    FIVE CENTURIES AT ST. GEORGE'S. cf Argo ZRG 789.
FLORENTIA,   de
    Come da lupo. cf Telefunken TK 11569/1-2.
FLOTOW, Friedrich
    Martha: M'appari. cf Decca SXL 6649.
    Martha: M'appari tutt'amor. cf RCA SER 5704/6.
    Martha: M'appari. cf Supraphon 10924.
    Martha: M'appari. cf Supraphon 012 0789.
    Martha: Ja was nun. cf Discophilia KGM 2.
    Martha: Letzte Rose. cf Discophilia KGP 4.
    Martha: Nancy, Julia verweille. cf Discophilia KGB 2.
    FLUTE MUSIC OF THE ROMANTIC ERA. cf Vanguard VSD 71207.
FLYNN, George
    Wound. cf CAGE: Winter music.
FOGLIANO, Lodovico
    L'amor donna. cf Telefunken TK 11569/1-2.
FONTANA, Giovanni Battista
993 Sonatas, trumpet, nos. 1-6. FRESCOBALDI: Canzoni, trumpet, nos.
        1-5. Gerard Schwarz, tpt; Julie Feves, bsn; Helen Katz, hpd.
        Peerless PRCM 203.
            +Gr 4-75 p1820   +HFN 6-75 p97
    Sonata, violin. cf L'Oiseau-Lyre 12BB 203/6.
FONTENAILLES
    Obstination. cf Pearl GEM 121/2.
FOOTE, Arthur
    Quintet, piano and strings, op. 38, A minor. cf BEACH: Quintet,
        piano and strings, op. 67, F sharp minor.
    FOOTLIFTER: A century of American marches in authentic versions.
        cf Columbia 33513.
FORBES, Sebastian
    Gracious spirit, Holy Ghost. cf Gemini GM 2022.
FORQUERAY, Antoine
    La Bellemont. cf DUPHLY: La Félix.
    La Couperin. cf DUPHLY: La Félix.
    La Laborde. cf DUPHLY: La Félix.
FORSTER, Josef Bohuslav
    Deborah: Polka. cf Supraphon 110 1429.
994 Symphony, no. 4, op. 54, C minor. PSO; Václav Smetáček. Nonesuch

H 71267. Tape (r) NST 71267. (also Supraphon 110 0617)
+-Gr 4-72 p1704   +NR 8-72 p5
+HF 9-72 p78   +SFC 6-11-72 p34
+HF 6-75 p122 tape   +St 12-72 p124
+-HFN 4-72 p709

Vitrum nostrum glorosum. cf L'Oiseau-Lyre 12BB 203/6.

FOSTER, Donald
At the round earth's imagined corner. cf Wayne State University,
unnumbered.

FOSTER, Stephen
Songs: We are coming Father Abraham, 300,000 more; Willie has gone
to war; Jenny June; Wilt thou be true; Katy Bell. cf Vox SVBX
5304.

FRACKENPOHL
Pop suite. cf Kendor KE 22174.

FRANCAIX, Jean
995 Concerto, piano, D major. HAHN: Concerto, piano, no. 1, E minor.
LAMBERT: The Rio Grande, piano, chorus and orchestra. MILHAUD:
Scaramouche. cf Jean Françaix, Magda Tagliafero, Hamilton
Harty, Marcelle Meyer, Darius Milhaud, pno; Paris Philharmonic
Orchestra, Hallé Orchestra; Orchestra; St. Michael's Singers;
Nadia Boulanger, Reynaldo Hahn, Constant Lambert. World Records
SH 227.
+HFN 10-75 p140   +RR 10-75 p85
996 Concerto, piano and orchestra. Rhapsody, viola and chamber orches-
tra. Suite, violin and orchestra. Claude Paillard-Fançaix;
pno; Susanne Lautenbacher, vln; Ulrich Koch, vla; Luxembourg
Radio Orchestra; Jean Françaix. Turnabout TV 34552.
++NR 1-75 p4   ++SFC 12-8-74 p36
Petit quatuor. cf Canon CNN 4983.
Quartet, flute, oboe, clarinet and bassoon. cf ARNOLD: Diverti-
mento, flute, oboe and clarinet, op. 37.
Rhapsody, viola and chamber orchestra. cf Concerto, piano and
orchestra.
Suite, violin and orchestra. cf Concerto, piano and orchestra.

FRANCESCHINI, Petronio
Sonata, 2 trumpets and strings, D major. cf HMV ASD 2938.

FRANCHETTI, Alberto
Germania: Ah, vieni qui...No, non chiuder gli occhi; Studenti
udite. cf Saga 7206.

FRANCK, Cesar
Cantabile, B major. cf Organ works (Saga 5390).
997 Le chasseur maudit. Rédemption. LALO: Rapsodie norvegienne.
Scherzo, orchestra. MASSENET: Scènes pittoresques, no. 4*.
OSCCP; Albert Wolff. Decca ECS 772. (*Reissue from LXT 5100)
+-Gr 10-75 p611   +RR 8-75 p39
+HFN 9-75 p108
998 Chorales, nos. 1-3. Thomas Murray, org. Nonesuch H 71310.
++NR 10-75 p11
999 Chorales, nos. 1-3. Pièces pour grand orgue: Cantabile. Marie-
Claire Alain, org. Supraphon 50823.
+-Gr 4-75 p1836   +-RR 3-75 p47
Chorale, no. 1. cf Argo ZRG 807.
Chorale, no. 3, A minor. cf Pye TPLS 13066.
Fantasie, A major. cf Organ works (Saga 5390).
1000 Organ works: Cantabile, B major. Fantasie, A major. L'Organiste:

Poco allegretto; Tres lent; Priere (quasi lento); Andantino;
Poco allegretto; Poco lento; Quasi allegro; Non troppo lento.
Pièce héroïque. Pierre Cochereau, org. Saga 5390.
    -Gr 5-75 p1995   +-RR 6-75 p66
    /HFN 5-75 p127
L'Organiste: Poco allegretto; Tres lent; Priere (quasi lento);
  Andantino; Poco allegretto; Poco lento; Quasi allegro; Non
  troppo lento. cf Organ works (Saga 5390).
Panis Angelicus. cf RCA ARL 1-0562.
Pastorale, op. 19. cf Hungaroton SLPX 11548.
Pastorale, op. 19. cf Polydor 2460 252.
Pièce héroïque. cf Organ works (Saga 5390).
Pièce héroïque. cf Decca 5BBA 1013-5.
Pièce héroïque. cf Stentorian SC 1685.
Pièces pour grand orgue: Cantabile. cf Chorales, nos. 1-3.
Prelude, fugue and variations, op. 18. cf Decca 5BBA 1013-5.
1001 Quintet, piano and strings, F minor. Samson François, pno; Bernède
  Quartet. Connoisseur Society CS 2077.
    ++SFC 8-31-75 p20   ++St 10-75 p107
Quintet, piano and strings, F minor. cf FAURE: Quartet, piano,
  op. 15, C minor.
Rédemption. cf Le chasseur maudit.
1002 Sonata, flute, A major. PROKOFIEV: Sonata, flute and piano, op.
  94, D major. James Galway, flt; Martha Argerich, pno. RCA LRL
  1-5095.
    ++Gr 11-75 p848   ++RR 10-75 p78
    ++HFN 12-75 p153
1003 Sonata, violin and piano, A major. BACH: Partita, violin, no. 2,
  S 1004, D minor. Arthur Rubinstein, pno; Jascha Heifetz, vln.
  Seraphim M 60230. (Reissue from RCA LCT 1120)
    +-HF 1-75 p83   +-St 1-75 p110
    ++NR 11-74 p11
Sonata, violin and piano, A major. cf BRAHMS: Trio, horn, violin
  and piano, op. 40, E flat major.
Sonata, violin and piano, A major. cf BARTOK: For children, excerpts.
Sonata, violin and piano, A major. cf DEBUSSY: Sonata, violin and
  piano, no. 3, G minor.
Sonata, violin and piano, A major. cf SYZMANOWSKI: Mythes, op. 30.
1004 Symphonic variations, piano and orchestra. LISZT: Totentanz,
  G 126. Andre Watts, pno; LSO; Erich Leinsdorf. Columbia M
  33072. Tape (c) MT 33072 (ct) MA 33072 (Q) MQ 33072 Tape (ct)
  MAQ 33072.
    +HF 4-75 p84   +SFC 11-17-74 p31
    +HF 7-75 p102 tape and Quad ++St 3-75 p108
    +NR 5-75 p13
1005 Symphonic variations, piano and orchestra. KHACHATURIAN: Concerto,
  piano, D flat major. Alicia de Larrocha, pno; LPO; Rafael
  Frühbeck de Burgos. Decca SXL 6599. Tape (c) KSXC 6599. (also
  London CS 6818).
    +-Gr 6-73 p49   ++NR 2-74 p4
    +Gr 7-75 p254 tape   ++RR 6-73 p52
    ++HF 4-74 p98   +RR 7-75 p71 tape
    +-HFN 6-73 p1177   +-St 6-74 p121
    +HFN 7-75 p90 tape
Symphonic variations, piano and orchestra. cf HMV SLS 5033.
1006 Symphony, D minor. PhO; Constantin Silvestri. Classics for

Pleasure CFP 40090. (Reissue from HMV ASD 408)
+Gr 1-75 p1342   ++RR 1-75 p27
1007 Symphony, D minor. Dresden Staatskapelle Orchestra; Kurt Sander-
     ling. DG 2548 132. (Reissue from 135036)
             /Gr 10-75 p612   +-RR 10-75 p43
             +HFN 9-75 p108
1008 Symphony, D minor. Rotterdam Philharmonic Orchestra; Charles
     Munch. Olympic 8133.
             +-NR 4-75 p4
FRANZL, Ferdinand
     Andenken an Elisen. cf Oryx ORYX 1720.
     FRENCH DUETS. cf Angel S 37143.
     FRENCH ORCHESTRAL WORKS. cf HMV ASD 3008.
FRESCOBALDI, Girolamo
     Bergamasca. cf Telefunken AW 6-41890.
     La bernadina. cf Telefunken SAWT 9589.
     La bernadina. cf Telefunken SMA 25121 T/1-3.
     Canzona VI. cf Works, selections (BASF BAC 3077).
     Canzoni, trumpet, nos. 1-5. cf FONTANA: Sonatas, trumpet, nos.
         1-6.
     Canzona, viola da gamba and cembalo. cf Pelca PSR 40589.
     Capriccio seconda sopra la, sol, fa, mi, re, ut. cf Works, selec-
         tions (BASF BAC 3077).
     Capriccio sopra la battaglia. cf Musical Heritage Society MHS
         1790.
     Courante (2). cf Angel S 37015.
     Fantasia sesta sopra doi soggetti. cf Works, selections (BASF
         BAC 3077).
     Fiori musicali, op. 12: Canzon dopo l'Epistola. cf Pelca PSRK
         41013/6.
     Gagliarda. cf Angel S 37015.
     Partite sopra passacagli. cf Works, selections (BASF BAC 3077).
     Recercar III. cf Works, selections (BASF BAC 3077).
     Il secondo libro di toccata: Toccata sesta. cf Pelca PSRK 41013/6.
     Toccata. cf L'Oiseau-Lyre 12BB 203/6.
     Toccata, Toccata VIII, Toccata IX. cf Works, selections (BASF BAC
         3077).
     Toccata cromatica per l'elevazione. cf Telefunken AW 6-41890.
     Toccata per l'elevazione. cf Works, selections (BASF BAC 3077).
     Toccata per l'elevazione. cf Pelca PSRK 41013/6.
1009 Works, selections: Canzona VI. Capriccio seconda sopra la, sol,
     fa, mi, re, ut. Partite sopra passacagli. Fantasia sesta doi
     soggetti. Recercar III. Toccata, Toccata IX, Toccata VIII.
     Toccata per l'elevazione. Gustav Leonhardt, hpd and org.
     BASF BAC 3077.
             +Gr 3-75 p1678   +RR 4-75 p51
FRICKER, Peter Racine
1010 Concerto, violin, op. 11. BANKS: Concerto, violin. Yfrah Neaman,
     vln; RPO; Norman del Mar. Argo ZRG 517.
             +Gr 1-75 p1335   +Te 3-75 p44
             +-RR 1-75 p28
     Pastorale. cf Abbey LPB 738.
     Pastorale. cf Vista VPS 1021.
     Waltz, restricted orchestra. cf HMV SLA 870.
FROBERGER, Johann
     Canzona, F major. cf Telefunken TK 11567/1-2.

Capriccio, C major, G major. cf Telefunken TK 11567/1-2.
Capriccio, no. 8. cf Philips 6775 006.
Fantasy, no. 2, E minor. cf Pandora PAN 101.
Libro secondo: Toccata da sonarsi alla levatione. cf Pelca PSRK
    41013/6.
Ricercar, G minor. cf Telefunken TK 11567/1-2.
Ricercar, no. 1. cf Philips 6775 006.
Suite, C major: Allemande. cf Pandora PAN 101.
Toccata, A minor. cf BASF BAB 9005.
FUCIK, Julius
    Der alte Brummbär. cf Telefunken DX 6-35262.
    Einzug der Gladiatoren. cf BASF 292 2116-4.
    Florentiner Marsch. cf DG 2721 077.
    Marches: Entry of the gladiators; Florentine; Hertzegovatz;
        Sempre avanti; Il soldato. cf Marinarello overture.
    Marches: Herzegovatz; Entry of the gladiators; Florentine. cf
        Telefunken DX 6-35262.
1011 Marinarella overture. Marches: Entry of the gladiators; Florentine;
        Hertzegovatz; Sempre avanti; Il soldato. Winter storms waltz.
        Czechoslovak Army Band; Rudolf Urbanec. Suraphon 54839.
            ++Gr 2-75 p1710   +RR 3-75 p26
    Marinarella overture. cf Telefunken DX 6-35262.
    Regimentskinder. cf DG 2721 077.
    Vom Donauufer. cf Rediffusion 15-16.
    Vom Donauufer. cf Supraphon 114 1458.
    Waltzes: Winter storms; Tales of the Danube. cf Telefunken DX
        6-35262.
    Winter storms waltz. cf Marinarella overture.
FUENLLANA, Miguel de
    De los álamos vengo. cf HMV SLS 5012.
    Perdida de Antequera. cf DG 2530 504.
FUKUSHIMA, Kazuo
    Pieces from Chu-u. cf Nonesuch HB 73028.
FURGEOT
    La favorite, Marche des pupilles de la garde. cf Classics for
        Pleasure 40230.
    March des bonnets à poils. cf Classics for Pleasure CFP 40230.
FUSSELL, Robert
1012 Processionals, orchestra (3). SCHUBEL: Fracture. STRANDBERG:
        Sea of tranquility. Springfield Symphony Orchestra; Robert
        Gutter. Opus One 21.
            +St 7-75 p96
FUSTE-VILA
    Háblamo de amores. cf HMV SLS 5012.
FUX, Johann
    Sonata, organ, no. 5. cf Philips 6775 006.
GABRIELI, Andrea
    Canzona francese. cf L'Oiseau-Lyre 12BB 203/6.
GABRIELI, Domenico
    Sonata, trumpet. cf Nonesuch H 71301.
GABRIELI, Giovanni
    Canzona a 5, a 6, a 7. cf Erato STU 70847.
    Canzona per sonare, nos. 1-4. cf Crystal S 201.
    Canzona per sonare, no. 2. cf Klavier KS 536.
    Canzoni septimi toni. cf University of Colorado, unnumbered.
    Sanctus Dominus Deus. cf L'Oiseau-Lyre 12BB 203/6.
    La spiritata. cf Wealden WS 139.

1013 Symphonie sacrae, nos. 1-8. New York Brass Ensemble; Samuel Baron.
     Orion ORS 7270. (also Ember ECL 9039)
              -ARG 7-72 p562   /NR 4-72 p9
           +-Gr 11-75 p818   +-RR 8-75 p52
              -HFN 12-75 p153   ++SFC 7-9-72 p33
1014 Symphoniae sacrae: Quis est iste; Sonata piano e forte; Maria
     Virgo; Sancta Maria, succurre miseris. LASSUS: Praeter rerum
     seriam: Magnificat. Alma redemptoris mater. Ave Maria. Pro
     Cantione Antiqua; Tölzer Boys' Choir, Members; Hamburg Old
     Music Wind Ensemble, Collegium Aureum, Members; Bruno Turner.
     BASF BAC 3080.
              +Gr 5-75 p2005   +-RR 5-75 p59
              +MT 11-75 p977
GADE, Niels
1015 Echoes of Ossian. NIELSEN: Helios overture, op. 17. Saga-dream,
     op. 39. RIISAGER: Qartsiluni. Etude. Royal Danish Orchestra;
     Jerzy Semkov, Igor Markevitch, Johan Hye-Knudsen. Turnabout
     TVS 34085.
              *St 6-75 p52
     Tre tonestukker, op. 22. cf Abbey LPB 738.
GAGLIANO, Marco da
1016 La Dafne. Mary Rawcliffe, Maurita Thornburgh; Robert White, Dale
     Terbeek, Hayden Blanchard, Jonathan Mack, t; Chamber Chorus
     and Instrumental Ensemble; Paul Vorwerk. ABC COMS 9004 (2).
              ++ON 12-13-75 p48
1017 La Dafne. Elizabeth Humes, Christine Whittlesey, s; Daniel
     Collins, c-t; Ray DeVoll, t; New York Pro Musica Antiqua; George
     Houle. Musical Heritage Society MHS 1953/4 (2).
              +-St 7-75 p96
GAGNON, Roland
     Sillsiana. cf International Piano Library IPL 5005/6.
GALILEI, Vicenzo
     Capriccio a due voci. cf Erato STU 70847.
     Contrapunto, 2 lutes. cf Erato STU 70847.
GALLIARD, Ernst
     Sonata, bassoon, no. 3, F major. cf Musical Heritage Society MHS
     1853.
GALLIARD, Johann E.
     Suite, bassoon and continuo, no. 2. cf Strobe SRCM 123.
GALUPPI, Baldessare
     Sonata, organ. cf Telefunken AW 6-41890.
GALYNIN, Herman
1018 Concerto, piano. KHACHATURIAN: Rhapsody concerto, violoncello and
     orchestra. Dmitri Bashkirov, pno; Natalia Shakhovskaya, vlc;
     MRSO: Yevgeny Svetlanov, Aram Khachaturian. Ember ECL 9009.
              +-Gr 3-75 p1650   +-RR 12-74 p30
GANASSI
     Ricercar. cf Telefunken TK 11569/1-2.
GANGI, Mario
     Ut fabulae ferunt 1970. cf Pye GSGC 14154.
GARDINER, Henry Balfour
     Evening hymn. cf Polydor 2460 250.
GARDNER, John
     Theme and variations, op. 7. cf ADDISON: Divertimento, op. 9.
GARLAND, Peter
     Apple blossom. cf Opus One 22.

GARRETA, Juli
    Sonata, piano, C minor. cf BACK: Sonata alla ricercare.
GASTOLDI, Giovanni
    Capriccio a due voci (2). cf Erato STU 70847.
    Intradas a 5 (2). cf Erato STU 70847.
GAUBERT, Philippe
    Nocturne and allegro scherzando. cf Vanguard VSD 71207.
GAULTIER, Denys
    Suite, G minor. cf London STS 15198.
GAYNOR/BROWNIEL
    The slumber song; Four leaf clover. cf Rubini RS 300.
GEIJER
    Dansen. cf BIS LP 17.
GEMINIANI, Francesco
    Sonata, flute, D major. cf Orion ORS 75199.
    Sonata, oboe, bassoon and harpsichord, E minor. cf BIGAGLIA:
        Sonata, recorder, bassoon and harpsichord, A minor.
GENIN
    Carnival of Venice variations. cf Vanguard VSD 71207.
GENTIAN (15th century France)
    Je suis Robert. cf Harmonia Mundi 204.
GERDALIA
    Little evening; When beauty is young. cf Rubini RS 300.
GERGELY, Ferenc
    Improvisation on Hungarian Christmas tunes. cf Hungaroton SLPX
        11548.
GERHARD, Roberto
    Don Quixote: Dances. cf Symphony, no. 1.
1019 The plague. Alec McCowen, narrator; Washington National Symphony
        Orchestra and Chorus; Antal Dorati. Decca HEAD 6. (also Head-
        line HEAD 6)
                +Gr 12-74 p1200   +NYT 12-21-75 pD18
                ++HF 10-75 p63   +RR 11-74 p86
                +NR 10-75 p6   ++SFC 8-17-75 p22
1020 Symphony, no. 1. Don Quixote: Dances. BBC Symphony Orchestra;
        Antal Dorati. Argo ZRG 752. (Reissue from HMV ASD 613)
                ++Gr 10-74 p687   ++RR 11-74 p37
                +NR 1-75 p4   ++SFC 8-17-75 p22
GERMANI, Fabio
    Cantata for Venice. cf Pye GSGC 14153.
GERSHWIN, George
1021 An American in Paris. Cuban overture. Rhapsody in blue. Ivan
        Davis, pno; Daniel Majeski, vln; CO; Lorin Maazel. Decca SXL
        6727. (also London CS 6946. Tape (c) 56946 (ct) 86946)
                ++Gr 9-75 p458   ++RR 8-75 p32
                +HFN 8-75 p75   ++SFC 9-21-75 p34
                ++NR 12-75 p13   ++St 12-75 p125
1022 An American in Paris. Cuban overture. Porgy and Bess. Monte
        Carlo National Orchestra; Edo de Waart. Philips 6500 290.
        Tape (c) 7300 189.
                *Gr 2-73 p1502   *NR 11-72 p4
                -HF 1-73 p114   +RR 2-73 p58
                +HFN 2-73 p337   ++SFC 11-12-72 p38
                ++HFN 5-75 p142 tape
1023 An American in Paris. GROFE: Grand Canyon suite. NBC Symphony
        Orchestra; Arturo Toscanini. RCA TA 129. (Reissues from HMV

        ALP 1107, ALP 1232)
            +Gr 8-75 p336   +-RR 6-75 p41
    An American in Paris. cf Works, selections (Vox (Q) QSVBX 5132).
    An American in Paris. cf HMV SEOM 14.
    An American in Paris, rehearsal excerpt. cf HMV SEOM 14.
    Clap yo' hands. cf Works, selections (Monmouth/Evergreen MES 7071).
1024 Concerto, piano, F major. Rhapsody in blue. Variations on "I
    got rhythm". Werner Haas, pno; Monte Carlo Opera Orchestra;
    Edo de Waart. Philips 6500 118. Tape (c) 7300 096.
            -ARG 1-72 p210   +-MJ 4-72 p70
            +Gr 11-71 p827   /NR 12-71 p8
            +Gr 9-74 p592 tape   +-RR 1-75 p69 tape
            /HF 3-72 p92   ++St 2-72 p82
1025 Concerto, piano, F major. Preludes, piano (3). Rhapsody in blue
    (original version for piano and band). Eugene List, pno; BeSO;
    Samuel Adler. Turnabout TV 34457. Tape (c) KTVC 34457.
            +Gr 12-72 p1197   +RR 1-74 p40
            +Gr 1-75 p1402 tape   +RR 4-75 p77
            +HFN 12-73 p2610   ++St 11-72 p112
            ++NR 4-73 p5

    Concerto, piano, F major. cf Works, selections (Vox (Q) QSVBX
    5132).
1026 Cuban overture. Do it again. Variations on "I got rhythm". The
    man I love. Preludes (3). Rhapsody, no. 2. Somebody loves
    me. Francis Veri, Michael Jamanis, pno. Connoisseur Society
    (Q) CSQ 2067.
            +Gr 11-75 p848   ++SFC 10-27-74 p3
            +HFN 10-75 p142   +-SR 2-22-75 p47
            +NR 3-75 p10   +St 12-74 p130
            +RR 10-75 p76
    Cuban overture. cf An American in Paris (Decca 6727).
    Cuban overture. cf An American in Paris (Philips 6500 290).
    Cuban overture. cf Works, selections (Vox (Q) QSVBX 5132).
    Do do do. cf Works, selections (Monmouth/Evergreen MES 7071).
    Do it again. cf Cuban overture.
1027 George Gershwin's songbooks (Improvisations on 18 songs). Impromp-
    tu, in two keys. Merry Andrew. Piano playin' Jazzbo Brown.
    Preludes (3). Promenade. Rialto ripples. Three-quarter blues.
    William Bolcom, pno. Nonesuch H 71284. Tape (c) ZCH 71284
    (Q) HQ 1284 Tape (r) NSTQ 1284.
            +Gr 12-73 p1197   +RR 10-73 p98
            +HF 11-73 p106   ++RR 10-74 p97 tape
            /HF 1-74 p84 Quad   ++SFC 10-21-73 p28
            +HF 7-75 p102 Quad tape   ++St 9-73 p79
            +NR 9-73 p11
1028 George Gershwin's songbooks (Improvisations on 18 songs). MAYERL:
    Pieces (8). Richard Rodney Bennett, piano. Polydor 2460 245.
            +HFN 10-75 p142
    Impromptu, in two keys. cf George Gershwin's songbooks (Improvi-
    sations on 18 songs).
    Lullaby. cf Works, selections (Vox (Q) QSVBX 5132).
    Lullaby. cf CBS 76267.
    The man I love. cf Cuban overture.
    Maybe. cf Works, selections (Monmouth/Evergreen MES 7071).
    Merry Andrew. cf George Gershwin's songbooks (Improvisations on
    18 songs).

Piano playin' Jazzbo Brown. cf George Gershwin's songbooks (Improvisations on 18 songs).

Porgy and Bess. cf An American in Paris.

1029 Porgy and Bess: Act 1, Introduction; Summertime; A woman is a sometime thing; Where is Brudder Robbins; My man's gone now; Oh, the train is at the station; Act 2, It takes a long pull to get there; I got plenty o' nuttin'; Buzzard song; Bess, you is my woman now; It ain't necessarily so; Oh, what you want wid Bess; O Doctor Jesus...Street cry; Strawberry woman; Street cry; Crab man; I love you, Porgy; O de Lawd shake de heavens; A red-headed woman; Act 3, Clara, Clara, don't you be downhearted; The fight and murder of Crown by Porgy; There's a boat dat's leavin' soon for New York; Good mornin'...Sistuh; Oh, Bess, Oh where's my Bess; Ain't you say Bess gone to New York...Oh Lawd, I'm on my way. Camilla Williams, June McMechen, Inez Matthews, s; Helen Dowdy, ms; Harrison Cattenhead, Avon Long, Ray Yeats, t; Edward Matthews, Lawrence Winters, Warren Colman, bar; Orchestra and chorus; Lehman Engel. CBS 61622. (Reissue from Philips NBL 5016/8)

    +-Gr 3-75 p1698   +-RR 4-75 p16

Porgy and Bess: Suite. cf Works, selections (Vox (Q) QSVBX 5132).

Preludes, piano (3). cf Concerto, piano, F major.

Preludes, piano (3). cf Cuban overture.

Preludes, piano (3). cf George Gershwin's songbooks (Improvisations on 18 songs).

Preludes, piano (3). cf Works, selections (Monmouth/Evergreen MES 7071).

Primrose: Songs (12). cf Works, selecdtions (Monmouth/Evergreen MES 7071).

Promenade. cf George Gershwin's songbooks (Improvisations on 18 songs).

Promenade. cf Works, selections (Vox (Q) QSVBX 5132).

Rhapsody, no. 2. cf Cuban overture.

Rhapsody, no. 2. cf Works, selections (Vox (Q) QSVBX 5132).

1030 Rhapsody in blue. MILHAUD: Scaramouche. CHABRIER: Valses romantiques (3). BIZET: Jeux d'enfants, op. 22. Frances Veri, Michael Jamanis, pno. Connoisseur Society CS 2054. (Q) CSQ 2054.

       +Gr 2-75 p1510   +NR 2-74 p13
       +MJ 1-74 p40   +St 3-74 p111
       +-MJ 7-74 p50

1031 Rhapsody in blue. PROKOFIEV: Concerto, piano, no. 3, op. 26, C major. RAVEL: Concerto, piano, for the left hand, D major. Julius Katchen, pno. Decca SXL 6411. Tape (c) KSXC 6411. (also London 6633)

       +-Gr 6-75 p111 tape   ++HFN 7-75 p90 tape

Rhapsody in blue. cf An American in Paris.

Rhapsody in blue. cf Concerto, piano, F major.

Rhapsody in blue. cf Works, selections (Vox (Q) QSVBX 5132).

Rhapsody in blue. cf RAVEL: Concerto, piano, for the left hand, D major.

Rhapsody in blue (original version for piano and band). cf Concerto, piano, F major.

Rhapsody in blue. cf Century/Advent OU 97 734.

Rhapsody in blue: Andante. cf Works, selections (Monmouth/Evergreen MES 7071).

Rialto ripples. cf George Gershwin's songbooks (Improvisations
   on 18 songs).
Somebody loves me. cf Cuban overture.
Someone to watch over me. cf Works, selections (Monmouth/Evergreen
   MES 7071).
Three-quarter blues. cf George Gershwin's songbooks (Improvisations
   on 18 songs).
Variations on "I got rhythm". cf Concerto, piano, F major.
Variations on "I got rhythm". cf Cuban overture.
Variations on "I got rhythm". cf Works, selections (Vox (Q) QSVBX
   5132).

1032 Works, selections: Do do do. Clap yo' hands. Primrose: Songs (12).
   Preludes, piano (3). Rhapsody in blue: Andante. Maybe. Someone to watch over me. Soloists; Winter Gardens Theatre Orchestra; J. Ansell; George Gershwin, pno. Monmouth/Evergreen MES
   7071.
           ++NR 10-75 p8
1033 Works, selections: An American in Paris. Concerto, piano, F major.
   Cuban overture. Lullaby. Porgy and Bess: Suite. Promenade.
   Rhapsody, no. 2. Rhapsody in blue. Variations on "I got
   rhtyhm". Jeffrey Siegel, pno; St. Louis Symphony Orchestra;
   Leonard Slatkin. Vox (Q) QSVBX 5132 (3). (also Turnabout TV
   37080/2)
           +-Gr 7-75 p176   +RR 6-75 p36
           /HF 4-75 p85   +SFC 10-27-74 p3
           +-HFN 6-75 p88   ++St 2-75 p110
           +-NR 12-74 p2

GERVAISE, Claude
   Allemande. cf L'Oiseau-Lyre SOL 329.
   Basse danse. cf L'Oiseau-Lyre SOL 329.
   Bransles (2). cf RCA CRL 2-7001.
   Branle de Bourgogne. cf DG Archive 2533 111.
   Branle de Bourgogne. cf DG Archive 2533 184.
   Branle de Champaigne. cf DG Archive 2533 111.
   Branle de Champaigne. cf DG ARchive 2533 184.
   Dances (4). cf RCA CRL 2-7001.
   La volunte. cf L'Oiseau-Lyre SOL 329.

GESUALDO, Carlo
   Canzona francese; Mille volte il dir moro. cf L'Oiseau-Lyre
      12BB 203/6.
   Io tacero. cf Wayne State University, unnumbered.
1034 Madrigals: Volgi, mia luce; O dolorosa gioia; Non t'umo, o voce
   ingrata; Che fai meco; Questa crudele; Dolcissima mia vita;
   T'amo mia vita; O che in gioia; O sempre crudo amore; Deh
   coprite il bel seno; Cor mio deh non piangete; Dunque non
   m'offendete. Sacred music: Psalmi delle Compliete (In te
   Domine speravi); Responsoria, no. 8 (Aestimatus sum). Gesualdo
   Madrigal Singers; Robert Craft. Odyssey Y 32886.
           +St 1-75 p110
1035 Responsoria et alia ad officium Sabbati Sancti. Prague Madrigal
   Singers; Miroslav Venhoda. Telefunken SAWT 9613.
           +-Gr 12-74 p1200   +-RR 10-74 p84
           +-HF 7-75 p74   +St 1-75 p110
           ++NR 12-74 p9

GIBBONS, Orlando
   Do not repine, fair sun. cf Turnabout TV 37079.

The cries of London. cf Harmonia Mundi 204.
Fancy, D minor. cf BASF BAC 3075.
Fantasia, D minor. cf BASF BAC 3075.
In nomine. cf Turnabout TV 37071.
Now each flowery bank. cf L'Oiseau-Lyre 12BB 203/6.
O clap your hands. cf Argo ZRG 789.
Pavane, G minor. cf BASF BAC 3075.
1036 Songs: Ah, dear heart; The silver swan; What is our life. TOMKINS:
      Music divine; See, see the shepherds queen; When I observe;
      Adieu, ye city-prisoning towers; When David heard; Fusca, in
      thy starry eyes; Weep no more, thou sorry boy. WILBYE: Flora
      gave me fairest flowers; Adieu sweet Amaryllis; Away, thou shalt
      not love me; When shall my wretched life; Lady, when I behold;
      Unkind, O stay thy flying; Lady, your words do spite me; Thus
      saith my Cloris bright. Ursula Connors, Elaine Barry, s; Marg-
      aret Cable, con; Nigel Rogers, Ian Partridge, t; Geoffrey Shaw,
      bs; Wilbye Consort; Peter Pears. Decca SXL 6639.
            +-Gr 9-74 p572   +RR 9-74 p83
            ++MT 8-75 p716
GIDEON, Miriam
1037 The condemned playground. Questions on nature. WEISGALL: End of
      summer. Jan DeGaetani, ms; Phyllis Bryn-Julson, s; Charles
      Bressler, Constantine Cassolas, t; New York Chamber soloists;
      Instrumentalists. CRI SD 343.
            ++NYT 12-21-75 pD18
      Questions on nature. cf The condemned playground.
GIGOUT, Eugene
      Scherzo. cf Argo ZRG 807.
GINASTERA, Alberto
      Concerto for strings. cf BARTOK: Divertimento, string orchestra.
1038 Milena, cantata no. 3, op. 37, soprano and orchestra. CHOPIN:
      Concerto, piano, no. 2, op. 21, F minor. Phyllis Curtin, s;
      Nerine Barrett, pno; Denver Symphony Orchestra; Brian Priestman.
      Desto DC 7171.
            ++SFC 11-4-73 p27   +-Te 3-75 p47
            +St 4-74 p111
      Triste. cf Odyssey Y2 32880.
GIORDANI, Giuseppe
      Caro mio ben. cf Columbia D3M 33448.
      Caro mio ben. cf Ember GVC 41.
GIORDANO, Umberto
      Andrea Chénier: Colpito que m'avete...Un dì, all'azzurro spacio;
      Come un bel dì di maggio. cf Decca SDD 390.
      Andrea Chénier: Come un bel dì di maggio. cf RCA SER 5704/6.
      Andrea Chénier: Come un bel dì di maggio. cf Supraphon 10924.
      Andrea Chénier: Come un bel dì di maggio; Vicino a te. cf Supra-
      phon 012 0789.
      Andrea Chénier: Nemico della patria. cf Ember GVC 37.
      Andrea Chénier: Un dì all'azzurro spazio. cf Columbia D3M 33448.
GIORNO, John
      Give it to me, baby. cf 1750 Arch 1752.
GIULIANI, Mauro
1039 Concerto, guitar, op. 30, A major. RODRIGO: Concierto madrigal,
      2 guitars and orchestra. Pepe and Angel Romero, gtr; AMF;
      Neville Marriner. Philips 6500 918. Tape (c) 7300 369.
            +-RR 12-75 p48

Grande ouverture, op. 61. cf DG 2530 561.

1040 Le Rossiniane, op. 121 and op. 119. SOR: Sonata, guitar, op. 25,
     C major. Julian Bream, gtr. RCA ARL L-0711. Tape (c) 1-0711
     (ct) ARS 1-0711.

    ++Gr 3-75 p1678   ++RR 3-75 p47
    ++MJ 3-75 p25   ++SFC 3-2-75 p24
    +NR 3-75 p14   ++St 4-75 p106

1041 Sonata, flute and guitar, op. 85. LOEILLET: Sonata, flute, op. 1,
     A minor. VISEE: Suite, guitar. Jean-Pierre Rampal, flt;
     René Bartoli, gtr. Harmonia Mundi HMU 711.
    -RR 12-75 p83

Sonata, guitar, op. 15: 1st movement. cf CBS 72526.

1042 Sonata eroica, op. 150, A major. PONCE (Weiss): Suite, A minor
     (Suite antica). SOR: Fantasia elegiaca, op. 59: Marcha fúnebra.
     Sonata, op. 22, C major. Eric Hill, gtr. Saga 5406.
    +Gr 12-75 p1081   ++RR 11-75 p78
    +HFN 12-75 p169

Variazioni concertante, op. 130. cf RCA ARL 1-0456.

GIUSTINI DI PISTOIA, Lodovico

1043 Sonata, piano, no. 1, G minor. Sonata, piano, no. 7, G major.
     HOVHANESS: Prayer of Saint Gregory. Symphony, no. 6. Mieczy-
     slaw Horszowski, pno; Polyphonia Orchestra; Alan Hovhaness.
     Poseidon 1017.
    +NR 6-75 p2

Sonata, piano, no. 7, G major. cf Sonata, piano, no. 1, G minor.

GLAZUNOV, Alexander

Concerto, violin, op. 82, A minor. cf BRUCH: Concerto, violin,
    no. 1, op. 26, G minor.
Concerto, violin, op. 82, A minor. cf RCA ARM 4-0942/7.
Concerto, violin, op. 82, A minor. cf RCA CRL 6-0720.
In modo religioso, op. 38. cf Desto DC 6474/7.
Meditation, op. 32. cf HMV SXLP 30193.
Meditation, op. 32 (2). cf RCA ARM 4-0942/7.
Morceaux, op. 49, no. 3: Gavotte. cf Ember GVC 40.
Raymonda, op. 57: Valse. cf Discopaedia MB 1010.
Raymonda, op. 57: Valse fantastique. cf HMV SEOM 20.
Raymonda, op. 57: Valse grande adagio. cf RCA ARM 4-0492/7.
Waltz, op. 42, no. 3, D major. cf L'Oiseau-Lyre DSLO 7.

GLIERE, Reinhold

The red poppy, op. 70: Russian sailors' dance. cf HMV SEOM 20.

GLINKA, Mikhail

Barcarolle, G major. cf Piano works (Musical Heritage Society
    MHS 1973).
Jota aragonesa. cf BORODIN: Prince Igor: Dance of the Polovtsian
    maidens; Polovtsian dances.
Mazurkas, C minor, A minor. cf Piano works (Musical Heritage
    Society MHS 1973).

1044 Nocturne. Variations on a theme by Mozart. LISZT: Etude de con-
     cert, no. 3, G 144, D flat major: Un sospiro (arr. Renié). The
     nightingale, G 250. RENIE: Contemplation. Legende. Pièce
     symphonique. SPOHR: Fantasie, op. 35, A flat major. Susanna
     McDonald, hp. Klavier KS 543.
    ++NR 10-75 p4

Nocturne, F minor. cf Piano works (Musical Heritage Society MHS
    1973).

1045 Piano works: Barcarolle, G major. Mazurkas, C minor, A minor.

Nocturne, F minor. Waltz, G major. Trio pathétique, D minor.
Variations on Alabiev's song "The nightingale". Thomas Hrynkiv,
pno; Esther Lamneck, clt; Michael McCraw, bsn; New American
Trio. Musical Heritage Society MHS 1973.
    ++HF 9-75 p85
Russlan and Ludmila: March of the sorcerer. cf Philips 6580 107.
Russlan and Ludmila: Mysterious Lel. cf HMV Melodiya SXL 30190.
Russlan and Ludmila: Overture. cf BORODIN: Prince Igor: Overture.
Russlan and Ludmila: Overture; Waltz fantasy. cf BORODIN: Prince
    Igor: Dance of the Polovtsian maidens; Polovtsian dances.
Russlan and Ludmila: Persian song. cf Discopaedia MB 1008.
Songs: Doubt; Vain temptation. cf Desto DC 7118/9.
Songs: The midnight review. cf London OS 26249.
Trio pathétique, D minor. cf Piano works (Musical Heritage
    Society MHS 1973).
Variatons on a theme by Mozart. cf Nocturne.
Variations on Alabiev's song "The nightingale". cf Piano works
    (Musical Heritage Society MHS 1973).
Waltz, G major. cf Piano works (Musical Heritage Society MHS
    1973).
GLOGAUER LIEDERBUCH
    Zwe Lieder. cf BIS LP 3.
    THE GLORIOUS VOICE OF FRITZ WUNDERLICH. cf Polydor 2489 519.
GLUCK, Christoph
    Alceste: Divinités du Styx. cf Decca SXL 6629.
    Alceste: Overture. cf BEETHOVEN: Symphony, no. 5, op. 67, C minor.
    Chaconne. cf CORELLI: Concerto grosso, op. 6, no. 8, C minor.
    Chaconne. cf Decca SDD 411.
    Don Juan: Allegretto. cf DG Archive 2533 182.
1046 Orfeo ed Euridice. Gundula Janowitz, Edda Moser, s; Dietrich
    Fischer-Dieskau, bar; Munich Bach Orchestra and Choir; Karl
    Richter. DG 2726 043 (2). (Reissue from SLPM 139268/9)
         +-Gr 10-75 p679   +RR 7-75 p22
         +HFN 9-75 p109
1047 Orfeo ed Euridice, Act 2. Nan Merriman, ms; Barbara Gibson, s;
    Robert Shaw Chorale; NBC Symphony Orchestra; Arturo Toscanini.
    RCA AT 127. (Reissue from HMV ALP 1357)
         +Gr 8-75 p336   +RR 6-75 p22
    Orfeo ed Euridice: Arias. cf London OS 26277.
    Orfeo ed Euridice: Ballet. cf DG Archive 2533 182.
    Orpheus and Eurydice: Dance of the blessed spirits. cf London
        SPC 21100.
    Orfeo ed Euridice: Dance of the blessed spirits. cf Decca SPA 394.
    Orfeo ed Euridice: Dance of the blessed spirits. cf London STS
        15160.
    Orfeo ed Euridice: Dance of the blessed spirits. cf RCA LRL 1-5094.
    Orfeo ed Euridice: Reigen seliger Geister. cf BENDA: Concerto,
        flute, E minor.
    Orphée et Eurydice: Viens, viens, Eurydice, suis-moi. cf Angel
        S 37143.
    Paride ed Elena: O del mio dolce ardor. cf Decca GOSC 666/8.
    Paride ed Elena: O del mio dolce ardor. cf Decca SXL 6629.
    Paride ed Elena: O del mio dolce ardor. cf Columbia D3M 33448.
GNAZZO
    The population explosion. cf 1750 Arch 1752.

GODARD, Benjamin
    Jocelyn: Berceuse. cf Discophilia KGC 1.
    Jocelyn: Berceuse. cf Ember GVC 46.
    Pieces, op. 116: Waltz. cf RCA LRL 1-5094.
    Suite de trois morceaux, op. 116. cf Vanguard VSD 71207.
GODOWSKY, Leopold
    Alt Wien. cf L'Oiseau-Lyre DSLO 7.
    Alt Wien. cf RCA ARM 4-0942/7.
    Waltz, D major. cf RCA ARM 4-0942/7.
    Waltz-poem, no. 4, for the left hand. cf L'Oiseau-Lyre DSLO 7.
GOEHR, Alexander
    Choruses (2). cf BENNETT: Calendar.
    Pieces, op. 18 (3). cf HMV HSQ 1337.
    GOETHE LIEDER. cf DG Archive 2533 149.
GOLD
    The Exodus song. cf Columbia D3M 33448.
GOLD, Ernest
1048 Songs of love and parting. CASTELNUOVO-TEDESCO: Coplas. Marni
    Nixon, s; Vienna Volksoper Orchestra; Ernest Gold. Crystal
    S 501.
      +-NR 4-75 p9   ++St 6-75 p98
    THE GOLDEN AGE OF VIENNESE MUSIC. cf Olympic 8136.
    THE GOLDEN AGE OF VIENNESE WALTZES. cf Rediffusion 15-16.
    GOLDEN DANCE HITS OF 1600. cf DG Archive 2533 184.
    THE GOLDEN PARADE OF VIENNESE WALTZES. cf Supraphon 114 1458.
GOLDFADEN
    Rozshinkes mit Mandlin. cf Columbia D3M 33448.
GOLDMARK, Carl
    Concerto, violin, A minor: Andante. cf RCA ARM 4-0942/7.
    Concerto, violin, no. 1, op. 28, A minor. cf HMV SXLP 30193.
    Concerto, violin, no. 1, op. 28, A minor. cf BEETHOVEN: Romance,
      no. 1, op. 40, G major.
    Die Königin von Saba: Lockruf. cf Pearl GEM 121/2.
    Regina di Saba: Magiche Note. cf Discophilia KGC 2.
    Die Königin von Saba: Magische Töne. cf Saga 7206.
GOLTERMANN, Georg
    Concerto, violoncello, op. 14, A minor: Cantilena. cf Ember GVC 42.
GOMBERT, Nicolas
    Caeciliam cantate. cf L'Oiseau-Lyre 12BB 203/6.
GOMES, Antonio
    Guarany: Sento una forza. cf Discophilia KGC 2.
GORDIGIANI
    Santa Lucia. cf Polydor 2489 519.
GOSS, John
    Praise, my soul, the King of Heaven. cf Decca ECS 774.
GOTTSCHALK, Louis
    Bamboula, op. 2. cf Works, selections (Angel S 36077).
    Bamboula, op. 2. cf Works, selections (Vanguard VSD 723/4).
    Le bananier, op. 5. cf Works, selections (Angel S 36077).
    Le bananier, op. 5. cf Works, selections (Vanguard VSD 723/4).
    The banjo, op. 15. cf Works, selections (Angel S 36077).
    Le banjo, op. 15. cf Works, selections (Vanguard VSD 723/4).
    Battle cry of freedom, op. 55. cf Piano works (Angel S 36090).
    Berceuse, op. 47. cf Piano works (Angel S 36090).
    Columbia, op. 34. cf Piano works (Angel S 36090).
    Creole eyes (Ojos criollos), op. 37. cf Works, selections (Angel
      S 36077).

Ojos criollos, op. 37. cf Works, selections (Vanguard VSD 723/4).
Ojos criollos. cf Columbia 31726.
Danza, op. 33. cf Works, selections (Angel S 36077).
Dying poet. cf Works, selections (Vanguard VSD 723/4).
La gallina, op. 53. cf Piano works (Angel S 36090).
La gallina, op. 53. cf Columbia 31726.
Grand scherzo, op. 57. cf Piano works (Angel S 36090).
Grand tarantelle, piano and orchestra, op. 67. cf Works, selections
    (Vanguard VSD 723/4).
The last hope, op. 16. cf Works, selections (Angel S 36077).
The last hope, op. 16. cf Works, selections (Vanguard VSD 723/4).
Maiden's blush. cf Works, selections (Vanguard VSD 723/4).
Marguerite. cf Piano works (Angel S 36090).
Mazurka, F sharp minor. cf Piano works (Angel S 36090).
Midnight in Seville, op. 30. cf Works, selections (Angel S 36077).
O, ma charmante, epargnez-moi, op. 44. cf Piano works (Angel
    S 36090).
Pasquinade, op. 59. cf Works, selections (Angel S 36077).
Pasquinade, op. 59. cf Works, selections (Vanguard VSD 723/4).
1049 Piano works: Battle cry of freedom, op. 55. Berceuse, op. 47.
        Columbia, op. 34. La gallina, op. 53. Grand scherzo, op. 57.
        Marguerite. Mazurka, F sharp minor. O, ma charmante, eparg-
        nez-moi, op. 44. Polka, B flat major. Suis-moi, op. 45.
        Tournament galop. Leonard Pennario, pno. Angel S 36090. Tape
        (c) 4XS 36090 (ct) 8XS 36090.
            +St 12-75 p120
Polka, B flat major. cf Piano works (Angel S 36090).
La Savane. cf Works, selections (Vanguard VSD 723/4).
Souvenir of Puerto Rico. cf Works, selections (Angel S 36077).
Souvenir de Porto Rico. cf Works, selections (Vanguard VSD 723/4).
Suis-moi, op. 45. cf Piano works (Angel S 36090).
Suis-moi, op. 45. cf Works, selections (Vanguard VSD 723/4).
Symphony, no. 1 (Night in the tropics). cf Works, selections
    (Vanguard VSD 723/4).
Tournament galop. cf Piano works (Angel S 36090).
Tournament galop. cf Works, selections (Vanguard VSD 723/4).
The Union, op. 48. cf Works, selections (Angel S 36077).
Variations on the Brazilian national anthem. cf International
    Piano Library IPL 5005/6.
1050 Works, selections: Bamboula, op. 2. Le bananier, op. 5. The banjo,
        op. 15. Creole eyes, op. 37. Danza, op. 33. The last hope,
        op. 16. Midnight in Seville, op. 30. Pasquinade, op. 59.
        Souvenir of Puerto Rico. The Union, op. 48. Leonard Pannario,
        pno. Angel S 36077. Tape (c) 4XS 36077 (ct) 8XS 36077.
            +HF 3-75 p78   +St 11-74 p130
            ++NR 5-75 p13
1051 Works, selections: Bamboula, op. 2. Le bananier, op. 5. Le banjo,
        op. 15. Dying poet. Grand tarantelle, piano and orchestra,
        op. 67. The last hope, op. 16. Maiden's blush. Ojos criollos,
        op. 37. Pasquinade, op. 59. La Savane. Souvenir de Porto
        Rico. Suis-moi. Symphony, no. 1 (Night in the tropics).
        Tournament galop. Eugene List, Ried Nibley, pno; Orchestra;
        Utah Symphony Orchestra; Maurice Abravanel. Vanguard VSD 723/4.
        (Reissues)
            +LJ 3-75 p33   +SFC 6-30-74 p39
            +MJ 7-74 p30

GOULD, Morton
    Symphonietta, no. 2: Pavane.   cf Pye GSGC 14154.
GOUNOD, Charles
    Ave Maria.   cf Pye GH 589.
1052 Fantasy on the Russian national hymn. MASSENET: Concerto, piano.
    SAINT-SAENS: Africa, op. 89. Marylène Dosse, pno; Westphalian
    Symphony Orchestra; Siegfried Landau. Candide (Q) QCE 31088.
        +NR 3-75 p3
1053 Faust, excerpts. Sylvia Sass, Magda Kalmár, s; György Korondi,
    Lajos Miller, bar; Kolos Kováts, bs; HRT Orchestra and Choir;
    Ervin Lukács. Hungaroton SLPX 11712.
        /NR 11-75 p9
    Faust: Ah, je rie. cf Saga 7029.
    Faust: Ballet music. cf DELIBES: Coppelia: Ballet music.
    Faust: Fantasie sur la valse. cf Rococo 2072.
    Faust: Love duet. cf Ember GVC 41.
    Faust: O Dieu, que de bijoux. cf HMV SLS 5012.
    Faust: Quel trouble...Salut, demeure. cf Decca SDD 390.
    Faust: Salut, demeure chase et pure. cf BASF BC 21723.
    Faust: Salut, demeure. cf Decca SXL 6649.
    Faust: Salut, demeure. cf RCA SER 5704/6.
    Faust: Salve, dimora casta e pura. cf Supraphon 10924.
    Faust: Soldiers chorus. cf Decca PFS 4323.
    Faust: Soldiers chorus. cf Philips 6747 204.
    Faust: Vous que faîtes l'endormie. cf Decca SXL 6637.
    Funeral march of a marionette. cf Philips 6580 107.
    Messe solennelle a St. Cecîle: Domine salvum. cf HMV Tape TC MCS
        (c) 14.
    Mireille: Vincenette à votre âge. cf Angel S 37143.
    Petite symphonie. cf BEETHOVEN: Trio, piano, clarinet and violon-
        cello, op. 38, E flat major.
    Roméo et Juliette: Ange adorable. cf Discophilia KGC 1.
    Roméo et Juliette: Je veux vivre. cf RCA ARL L-0844.
    Roméo et Juliette: Je veux vivre dans ce rêve. cf Saga 7029.
    Roméo et Juliette: Madrigal of Juliet and Roméo. cf Angel S 37143.
    Serenade. cf Pearl GEM 121/2.
1054 Songs: Absence; L'absent; Aimons-nous; Ce que je suis sans toi;
    Chanson de printemps; Les deux pigeons; Envoi de fleurs; Ma
    belle amie est morte; O ma belle rebelle; Où voulez-vous aller;
    Prière; Quanti mai; Sérénade; Le soir; Venise; Viens les gazons
    sont verts. Gérard Souzay, bar; Dalton Baldwin, pno. HMV ASD
    3083.
        ++Gr 9-75 p500   +RR 9-75 p69
        +HFN 9-75 p99   +STL 10-5-75 p36
    Song: Biondina bella. cf Rococo 5365.
    Songs: Aimons-nous; Òu voulez-vous aller. cf Seraphim S 60251.
GOW, David
    Ave maris stella. cf Gemini GM 2022.
GRAFULLA
    Washington grays. cf Michigan University SM 0002.
GRAINGER, Percy
    Hill song, no. 2. cf Philips 6747 177.
1055 Paraphrase on TCHAIKOVSKY: Nutcracker: Waltz of the flowers.
        LISZT: Paraphrase on TCHAIKOVSKY: Eugen Onegin: Polonaise, G
        429. TCHAIKOVSKY: Concerto, piano, no. 2, op. 44, G major.
    Michael Ponti, pno. Turnabout TV 34560.
        +NR 1-75 p5

Scotch strathspey and reel. cf Decca 396.

GRANADOS, Enrique

Allegro de concierto, C major. cf Works, selections (Musical Heritage Society MHS 1870).

Danse espagnole. cf Argo ZRG 805.

Danzas españolas: Danza lenta. cf Works, selections (Musical Hertage Society MHS 1870).

Spanish dance, op. 5, no. 2 (arr. Piatigorsky); op. 5, no. 5 (arr. Saleski). cf Supraphon 50919.

Spanish dance, op. 37, no. 5: Andaluza. cf CBS 72950.

Danza española, op. 37, no. 5: Andaluza. cf Nonesuch H 71233.

Danza española, op. 37, no. 5: Andaluza. cf RCA ARM 4-0942/7.

Danza española, op. 37, no. 5: Andaluza. cf ALBENIZ: Suite española, no. 1, Granada, no. 3, Sevillanas.

Danza española, op. 37, nos. 6 and 11. cf RCA ARL 1-0456.

Exquise: Valse Tzigane. cf International Piano Library IPL 5005/6.

Goyescas: Intermezzo. cf HMV SLS 5019.

Goyescas: Intermezzo. cf Supraphon 50919.

Goyescas: Intermezzo; The maiden and the nightingale. cf FALLA: Concerto, harpsichord and chamber orchestra.

Goyescas: The maiden and the nightingale. cf Decca SPA 372.

Goyescos: La maja y el ruiseñor. cf HMV SLS 5012.

Piezas sobre cantos populares españoles (6). cf Works, selections (Musical Heritage Society MHS 1870).

Tonadilla. cf Supraphon 50919.

Tonadilla: La maja dolorosa. cf FALLA: Mélodies.

Tonadilla: La maja dolorosa. cf HMV SLS 5012.

Tonadillas: La maja dolorosa, nos. 1-3; El majo discreto; El tra-la-la y el punteado; El majo tímido. cf DG 2530 598.

Tonadillas al estilo antiguo: La maja de Goya. cf ALBENIZ: Suite española, no. 1, Granada; no. 3, Sevillanas.

Tonadillas al estilo antiguo: Amor y odio; El majo discreto; El majo tímido; El mirar de la Maja; El tra-la-la y el punteado. cf FALLA: Mélodies.

Tonadillas al estilo antiguo: La maja de Goya. cf Nonesuch H 71233.

Valses poéticas. cf Works, selections (Musical Heritage Society MHS 1870).

1056 Works, selections: Allegro de concierto, C major. Danzas españolas: Danza lenta. Piezas sobre cantos populares españoles (6). Valses poéticas. Alicia de Larrocha, pno. Musical Heritage Society MHS 1870.
++St 4-75 p100

GRAND OPERA GALA. cf Decca GOSC 666/8.

GREAT HITS YOU PLAYED WHEN YOU WERE YOUNG, volume 3. cf Connoisseur (Q) CSQ 2065.

GREAT HITS YOU PLAYED WHEN YOU WERE YOUNG, volume 4. cf Connoisseur Society (Q) CSQ 2066.

GREAT TRUMPET CONCERTOS. cf RCA CRL 2-7002.

GREENE, Maurice

Introduction and trumpet tune. cf CRD CRD 1008.

Lord let us know mine end. cf Argo ZRG 789.

GREGSON, Edward

Prelude and capriccio. cf RCA LRL 1-5072.

GRENON, Nicholas

La plus jolie. cf Telefunken ER 6-35257.

GREP
  Paduana.   cf Crystal S 201.
GRETCHANINOV, Alexander
  Verouiou; Slava...Edinorodnii.   cf ARKANGELSKII: Blajen Razoumnevai;
    Otche nach.
  Veruyu, op. 29, no. 8 (Credo); Slava I Edinorodni (Glory and only
    begotten).   cf Monitor MFS 757.
GRETRY, Andre
  Richard, Coeur de Lion: O Richard, O mon Roi.   cf RCA ARL 1-0851.
GRIEG, Edvard
  Album leaf, op. 47, no. 2.   cf Lyric pieces (DG 2530 476).
  Arietta, op. 12, no. 1.   cf Lyric pieces (DG 2530 476).
  At the cradle, op. 68, no. 5.   cf Lyric pieces (DG 2530 476).
  At your feet, op. 68, no. 3.   cf Lyric pieces (DG 2530 476).
  Ballade, op. 65, no. 5.   cf Lyric pieces (DG 2530 476).
  Berceuse, op. 38, no. 1.   cf Lyric pieces (DG 2530 476).
  Brooklet, op. 62, no. 4.   cf Lyric pieces (DG 2530 476).
  Butterfly, op. 43, no. 1.   cf Lyric pieces (DG 2530 476).
1057 Concerto, piano, op. 16, A minor.   SCHUMANN: Concerto, piano, op.
    54, A minor.   Walter Gieseking, pno; Orchestra; NPhO; Wilhelm
    Furtwängler.   Bruno Walter Society IGI 348.
        +HF 12-75 p94   +NR 9-75 p7
1058 Concerto, piano, op. 16, A minor.   RACHMANINOFF: Concerto, piano,
    no. 2, op. 18, C minor.   Clifford Curzon, pno; LPO, LSO; Anatole
    Fistoulari, Adrian Boult.   Decca ECS 753.   (Reissues from LXT
    5165, 5178)
        /-Gr 6-75 p39
1059 Concerto, piano, op. 16, A minor.   SCHUMANN: Concerto, piano, op.
    54, A minor.   Radu Lupu, pno; LSO; André Previn.   Decca SXL
    6624.   Tape (c) KSXC 6624.   (also London CS 6840)
        +-Gr 2-74 p1552   +MJ 1-75 p48
        +HF 11-74 p122   ++NR 11-74 p5
        +HFN 3-74 p105   +RR 2-74 p33
        +HFN 9-75 p110   +-SFC 8-11-74 p34
1060 Concerto, piano, op. 16, A minor.   SCHUMANN: Concerto, piano, op.
    54, A minor.   Sviatoslav Richter, pno; Monte Carlo National
    Orchestra; Lovro von Matačič.   HMV (Q) ASD 3133.   (also Angel
    (Q) S 36899.   Tape (c) 4XS 36889 (ct) 8XS 36899)
        +-Gr 11-75 p818   ++NR 11-75 p6
        +-HF 12-75 p94   -RR 12-75 p49
        +-HFN 12-75 p153   ++St 12-75 p126
1061 Concerto, piano, op. 16, A minor.   RAVEL: Concerto, piano, G major.
    Jenö Jandó, pno; HRT Orchestra; Antal Jancsovics.   Hungaroton
    SLPX 11710.
        +-NR 11-75 p6   +-RR 12-75 p58
1062 Concerto, piano, op. 16, A minor.   SCHUMANN: Concerto, piano, op.
    54, A minor.   Claudio Arrau, pno; COA; Christoph von Dohnanyi.
    Philips 5680 108.   (Reissue from SAL 3452)
        +Gr 8-75 p322   ++RR 9-75 p38
1063 Concerto, piano, op. 16, A minor.   SCHUMANN: Concerto, piano, no.
    54, A minor.   Claudio Arrau, pno; COA; Christoph von Dohnanyi.
    Philips 6833 020.
        +HFN 8-75 p87
  Concerto, piano, op. 16, A minor.   cf CLEMENTI: Sonatinas, op. 36,
    nos. 1-6.
  Concerto, piano, op. 16, A minor.   cf Works, selections (Classics
    for Pleasure 40225).

Concerto, piano, op. 16, A minor. cf Camden CCV 5000-12, 5014-24.
Concerto, piano, op. 16, A minor. cf CBS 77513.
Concerto, piano, op. 16, A minor. cf HMV SLS 5033.
1064 Elegiac melodies, op. 34. Peer Gynt: Suite, no. 1, op. 46. Sym-
   phonic dances, op. 64. Hallé Orchestra; John Barbirolli. Pye
   GSGC 15018. (Reissues from CCL 30126, GSGC 14077).
      +-Gr 8-75 p335   +RR 8-75 p33
   Elegaic melodies, op. 34. cf Works, selections (Classics for
   Pleasure 40225).
   Elegaic melodies, op. 34, no. 2: The last spring. cf WIREN:
   Serenade, strings, op. 11.
   Gone, op. 71, no. 6. cf Lyric pieces (DG 2530 476).
   Grandmother's minuet, op. 68, no. 2. cf Lyric pieces (DG 2530 476).
1065 Af Haugtussa, op. 67. RANGSTROM: Songs: Flickan under Nymanen;
   Pan; Semele; Sköldmön; Villemo. SIBELIUS: Songs: Blackroses,
   op. 36, no. 1; The kiss, op. 37, no. 1; The tryst, op. 37, no.
   5; Varen flyktar hastigt, op. 13, no. 4. Siv Wennberg, s;
   Geoffrey Parsons, pno. HMV HQS 1345.
      +-Gr 10-75 p676   +-RR 9-75 p69
      +HFN 9-75 p104
1066 Holberg suite, op. 40. TCHAIKOVSKY: Serenade, strings, op. 48,
   C major. Netherlands Chamber Orchestra; David Zinman. Philips
   6580 102.
      +Gr 4-75 p1807   ++RR 4-75 p36
      +NR 12-75 p2
   Holberg suite, op. 40. cf Works, selections (Classics for Pleasure
   40225).
   Holberg suite, op. 40. cf WIREN: Serenade, strings, op. 11.
   Holberg suite, op. 40: Prelude. cf HMV SEOM 19.
   Home-sickness, op. 57, no. 6. cf Lyric pieces (DG 2530 486).
   Homeward, op. 62, no. 6. cf Lyric pieces (DG 2530 476).
   Humoresques, op. 6, nos. 1-4. cf Piano works (Arion 38268).
   Lonely wanderer, op. 43, no. 2. cf Lyric pieces (DG 2530 476).
1067 Lyric pieces: Album leaf, op. 47, no. 2. Arietta, op. 12, no. 1.
   At the cradle, op. 68, no. 5. At your feet, op. 68, no. 3.
   Ballade, op. 65, no. 5. Berceuse, op. 38, no. 1. Brooklet,
   op. 62, no. 4. Butterfly, op. 43, no. 1. Gone, op. 71, no.
   6. Grandmother's minuet, op. 68, no. 2. Home-sickness, op.
   57, no. 6. Homeward, op. 62, no. 6. Lonely wanderer, op. 43,
   no. 2. Melody, op. 47, no. 3. Nocturne, op. 54, no. 4. Norwegian dance, op. 47, no. 4. Once upon a time, op. 71, no. 1.
   Puck, op. 71, no. 3. Remembrances, op. 71, no. 7. Scherzo,
   op. 54, no. 5. Emil Gilels, pno. DG 2530 476.
      +Gr 3-75 p1681   +RR 3-75 p47
      ++HF 9-75 p85   ++SFC 7-27-75 p22
      ++NR 9-75 p12   ++St 12-75 p126
   Lyric pieces, op. 12, no. 5; op. 38, no. 4; op. 43, no. 1; op. 47,
   nos. 3, 6; op. 71, no. 3. cf Piano works (Arion 38268).
   Lyric pieces, op. 43, no. 6: To the spring. cf Connoisseur Society
   (Q) CSQ 2065.
   Lyric pieces, op. 54: March of the dwarfs. cf CBS 77513.
   Lyric pieces, op. 54, no. 6: Scherzo. cf RCA ARM 4-0942/7.
   Lyric pieces, op. 65, no. 6: Wedding day at Troldhaugen. cf Decca
   SPA 372.
   Lyric pieces, op. 71, no. 3: Puck. cf Discopaedia MB 1010.
   Lyric pieces, op. 71, no. 3: Puck. cf RCA ARM 4-0942/7.

Melody, op. 47, no. 3. cf Lyric pieces (DG 2530 476).
Nocturne, op. 54, no. 4. cf Lyric pieces (DG 2530 476).
Norwegian dances, op. 35. cf Works, selections (Classics for
    Pleasure 40225).
Norwegian dances, op. 35, nos. 1-4. cf BRAHMS: Hungarian dances,
    nos. 5, 6.
Norwegian dance, op. 35, no. 2. cf CBS 77513.
Norwegian dance, op. 47, no. 4. cf Lyric pieces (DG 2530 476).
Norwegian dances and songs, op. 17, nos. 1, 4-6, 12, 15. cf
    Piano works (Arion ARN 38268).
Norwegian folk melodies, op. 66, nos. 1, 5, 8, 10, 13, 16, 18. cf
    Piano works (Orion 38268).
Once upon a time, op. 71, no. 1. cf Lyric pieces (DG 2530 476).
Peer Gynt, op. 46, no. 1: In the hall of the mountain king. cf
    DG Heliodor 2548 148.
Peer Gynt, op. 46: Suites. cf Camden CCV 5000-12, 5014-24.
1068 Peer Gynt, op. 46: Suite, no. 1; op. 55: Suite, no. 2. Sigurd
    Jorsalfar suite, op. 56. BPhO; Herbert von Karajan. DG 2530
    243. Tape (ct) 89 466 (c) 3300 314.
          +-Gr 11-73 p920   +RR 11-73 p36
          +HF 11-73 p106   +RR 1-75 p69 tape
          ++HFN 11-73 p2315   ++SFC 5-19-74 p29
          ++NR 3-74 p2   +St 11-73 p121
Peer Gynt, op. 46: Suite, no. 1. cf Elegiac melodies, op. 34.
Peer Gynt, op. 46: Suite, no. 1. cf ALFVEN: Midsommarvaka, no. 1,
    op. 19.
Peer Gynt, op. 46: Suite, no. 1. cf CBS 77513.
1069 Piano works: Humoresques, op. 6, nos. 1-4. Lyric pieces, op. 12,
    no. 5; op. 38, no. 4; op. 43, no. 1; op. 47, nos. 3, 6; op. 71,
    no. 3. Norwegian dances and songs, op. 17, nos. 1, 4-6, 12,
    15. Norwegian folk melodies, op. 66, nos. 1, 5, 8, 10, 13,
    16, 18. Françoise Thinat, pno. Arion ARN 38268.
          +-Gr 1-75 p657   +-RR 10-75 p76
Puck, op. 71, no. 3. cf Lyric pieces (DG 2530 476).
Remembrances, op. 71, no. 7. cf Lyric pieces (DG 2530 476).
Scherzo, op. 54, no. 5. cf Lyric pieces (DG 2530 476).
Sigurd Jorsalfar, op. 56: cf Peer Gynt, op. 46: Suite, no. 1;
    op. 55: Suite, no. 2.
Sigurd Jorsalfar, op. 56, excerpts. cf Works, selections (Classics
    for Pleasure CFP 40225)
Sigurd Jorsalfar, op. 56: Homage march. cf CBS 77513.
Sigurd Jorsalfar, op. 56: Homage march. cf DG Heliodor 2548 148.
Sigurd Jorsalfar, op. 56: Homage march. cf Philips 6580 107.
1070 Sonata, piano, op. 7, E minor. BIZET: Nocturne, no. 1, F major.
    Variations chromatiques de concert. Glenn Gould, pno. Colum-
    bia M 32040. (also CBS 73178)
          +-HF 7-73 p79   +-RR 6-75 p66
          +NR 8-73 p15   /St 7-73 p104
Sonata, piano, op. 7, E minor. cf BIZET: Nocturne, no. 1, F major.
Sonata, violin and piano, no. 2, op. 13, G minor. cf RCA ARM
    4-0942/7.
Sonata, violin and piano, no. 3, op. 45, C minor. cf BEETHOVEN:
    Sonata, violin and piano, no. 8, op. 30, no. 3, G major.
1071 Sonata, violoncello, op. 36, A minor. RHEINBERGER: Sonata, violon-
    cello, op. 92, C major. Ludwig Hoelscher, vlc; Kurt Rapf, pno.
    BASF 212 2397-3.
          +-HFN 11-75 p154   +-RR 11-75 p78

Songs: Ich liebe dich, op. 5, no. 3. cf CBS 77513.
Songs: To a waterlily; A dream. cf HMV RLS 714.
Songs: Last-spring. cf Rubini RS 300.
Symphonic dances, op. 64. cf Elegiac melodies, op. 34.
To the spring. cf Discopedia MB 1005.
1072 Works, selections: Concerto, piano, op. 16, A minor. Holberg
      suite, op. 40. Elegaic melodies, op. 34. Norwegian dances,
      op. 35 (2). Sigurd Jorsalfar, op. 56, excerpts. PhO; George
      Weldon. Classics for Pleasure CFP 40225. (Reissue from SCX
      3416)
            +HFN 9-75 p108   +RR 9-75 p38
GRIGNY, Nicolas de
1073 La messe. LEBEGUE: Elévation, G major. Symphonie sur le bémol
      fa. Suite du deuxième ton. André Isoir, org. CRD CAL 1910/11
      (2).
            +Gr 10-75 p657
GRIMACE, Magister
      A l'arme a l'arme. cf 1750 Arch S 1753.
GROFE, Ferdé
1074 Death Valley suite. Grand Canyon suite. Capitol Symphony Orches-
      tra; Ferde Grofé. Angel S 36089. (Reissue)
            -NR 10-75 p2
      Grand Canyon suite. cf Death Valley suite.
      Grand Canyon suite. cf GERSHWIN: An American in Paris.
GROVLEZ, Gabriel
      Les petites litanies de Jesus. cf Canon CNN 4983.
GRUNFELD
      Emperor waltz. cf Musical Heritage Society MHS 1959.
      Paraphrase on themes from Aschenbrödel. cf Musical Heritage
      Society MHS 1959.
      Voices of spring. cf Musical Heritage Society MHS 1959.
GUAMI, Gioseffo
      La brillantina. cf L'Oiseau-Lyre 12BB 203/6.
GUARNIERI, Camargo
      Dansa brasileira. cf Angel S 37110.
      Dansa negra. cf Angel S 37110.
GUBAJDULINA, Sofia
      Concordanza, chamber ensemble. cf ARISTAKESIAN: Sinfonietta,
      piano, xylophone and strings.
GUERRERO, Francisco
      Virgen santa. cf Turnabout TV 34569.
GULIELMUS, M.
      Bassa danza à 2. cf DG Archive 2533 111.
GURIDI, Jesus
      Jota. cf HMV SLS 5012.
      Songs: Llámle con el pañuelo; No quiero tus avellanas; Cómo
      quieres que advine. cf DG 2530 598.
GURLITT
      Toy symphony, op. 169, C major. cf Angel S 36080.
GUYARD
      M'y levay par ung matin. cf Telefunken TK 11569/1-2.
GYSIN, Bryon
      Come to free the words. cf 1750 Arch 1752.
HABA, Alois
1075 The mother, op. 35: Scene 1, excerpt. Nonet, no. 2, op. 41. The
      path of life, op. 46, excerpt. Quartet, strings, no. 11, op. 87.

Suite, solo bass clarinet, op. 96.   Josef Horák, bs clt; Novák
Quartet, CPhO, Prague National Theatre Orchestra; Václav Smetá-
cek, Jiří Jirous.   Supraphon 111 1418.
            +NR 9-75 p4
Nonet, no. 2, op. 41.   cf The mother, op. 35: Scene 1, excerpt.
The path of life, op. 46, excerpt.   cf The mother, op. 35: Scene 1,
excerpt.
Quartet, strings, no. 11, op. 87.   cf The mother, op. 35: Scene 1,
excerpt.
Suite, solo bass clarinet, op. 96.   cf The mother, op. 35: Scene 1,
excerpt.
HADZIDAKIS, Manos
1076 For a little white seashell, op. 1.   PONIRIDY: Rythemes Grecs.
SKALKOTTAS: Suite, piano, no. 3.   Nicolas Constantinidis, pno.
Musical Heritage Society MHS 3055.
            +St 8-75 p95
HAHN, Reynaldo
1077 Concerto, piano, E major.   Concerto provençal, flute, clarinet
basson, horn and strings.   PIERNE: Giration.   Magda Tagliafero,
pno; Various orchestras and conductors.   Rococo 2053.
            +NR 3-75 p3
Concerto, piano, no. 1, E minor.   cf FRANCAIX: Concerto, piano,
D major.
Concerto provençal, flute, clarinet, bassoon, horn and strings.
cf Concerto, piano, E major.
Etudes latines: Phyllis, L'Enamourée, L'Heure exquise; Le plus
beau present; Paysage triste; Le temps d'Aimer: Le chien fidèle,
Lettre d'amour; Je me mets en votre mercy; Offrande; Venezia-
Ghè-pecà.   cf Rococo 5365.
L'Offrande.   cf RCA ARL L-5007.
Si mes vers avaient des ailes.   cf Desto DC 7118/9.
HAKIM, Talib Rasul
Visions of Ishwara.   cf ANDERSON: Squares.
HALEVY, Jacques
La Juive: Rachel, quand du Seigneur.   cf BASF BC 21723.
La Juive: Rachel, quand du Seigneur.   cf Discophilia KGC 2.
La Juive: Rachel, quand du seigneur.   cf Columbia D3M 33448.
HALFFTER, Ernesto
1078 Concerto, guitar.   BACARISSE: Concertino, guitar and orchestra,
op. 72, A minor.   Narciso Yepes, gtr; Spanish Radio and Tele-
vision Orchestra; Odón Alonso.   DG 2530 326.
            +-Gr 6-73 p187   +-RR 7-73 p42
            -NR 6-75 p8   +SR 6-14-75 p46
Danza de la gitana (Escriche).   cf RCA ARM 4-0942/7.
HALLAM, Percy
Prelude, F major.   cf Wealden WS 110.
HAMILTON, Iain
1079 Epitaph for this world and time.   Voyage, horn and chamber orches-
tra.   Larry King, David Agler, Jack Jones, org; Cathedral of
St. John the Divine Choir, New York City, Trinity Church Choir,
Princeton, N. J., Trinity Church Choir, New York City; Alec
Wyton, Barry Tuckwell, hn; London Sinfonietta; David Atherton.
CRI SD 280.
            +ARG 6-72 p413   *NR 5-72 p8
            +Gr 2-75 p1530   +RR 2-73 p81
            +HF 11-72 p82   ++RR 2-75 p59
            +HFN 3-73 p562   ++St 12-72 p125

1080 Sonata, piano, op. 13. TIPPETT: Sonata, piano, no. 1. Margaret
        Kitchin, pno. Lyrita RCS 5. (Reissue from 1960)
            ++Gr 9-75 p492   +RR 7-75 p54
            +HFN 12-75 p156
     Voyage, horn and chamber orchestra. cf Epitaph for this world and
        time.
HANBY
     Nelly Gray. cf Vox SVBX 5304.
HANDEL, Georg Friedrich
     Acis and Galatea: O ruddier than the cherry. cf HMV SLS 5022.
     Agrippina. cf Overtures (DG 2530 342).
1081 Agrippina condotta a morire, cantata no. 14. Armida abbandonata,
        cantata no. 13. Pensieri notturni di filli, cantata no. 17.
        Agnes Giebel, s; Helma Elsner, hpd; Alfred Mann, rec; Helmut
        Reimann, vlc; Tonstudio Orchestra, Stuttgart; Rudolf Lamy.
        Olympic 8116.
            -NR 4-75 p9
1082 Ah, che troppo inegali. Joseph: Overture. Look down, harmonious
        Saint. Nel dolce dell'oblio. Silete venti. Elly Ameling, s;
        Theo Altmeyer, t; Collegium Aureum of Original Instruments.
        BASF BAC 3058/59 (2).
            +-Gr 1-75 p1376
     Alcina: Dream music. cf HMV RLS 717.
     Alcina: Overture. cf Overtures (DG 2530 342).
     Alcina: Verdi prati. cf Decca SXL 6629.
     Allegro, oboe, F major. cf Works, selections (Philips 6747 096).
     Alessandro: Lusinghe piu care. cf Discophilia KGS 3.
     Andante, flute, B minor. cf Works, selections (Philips 6747 096).
1083 Arias: Ariodante: Dopo notte. Atalanta: Care selve. Hercules:
        Where shall I fly. Joshua: O had I Jubal's lyre. Lucrezia,
        complete. Rodelinda: Pompe vane di morte...Dove sei, amato
        bene. Serse: Ombra mai fù. Janet Baker, ms; ECO; Raymond Lep-
        pard. Philips 6500 523. Tape (c) 7300 345.
            +Audio 12-75 p104   +NR 12-74 p13
            +-Gr 10-74 p737   +Op 12-74 p1087
            ++Gr 6-75 p111 tape   ++RR 11-74 p87
            +HF 3-75 p80   ++St 3-75 p101
            ++MJ 2-75 p39
     Ariodante: Dopo notte. cf Arias (Philips 6500 523).
     Armida abbandonata, cantata no. 13. cf Agrippina condotta a
        morire, cantata no. 14.
     Atalanta: Care selve. cf Arias (Philips 6500 523).
     Atalanta: Care selve. cf London OS 26391.
     Aylesford pieces: Fugue, G major; Saraband; Impertinence. cf
        Columbia M 33514.
     Belshazzar. cf Overtures (DG 2530 342).
     Berenice: Minuet. cf HMV ASD 3017.
     Chaconne, G major. cf Harpsichord works (Erato STU 70906).
1084 Chandos anthems, nos. 2, 5. Caroline Friend, s; Philip Langridge,
        t; Cambridge, King's College Chapel Choir; AMF; David Willcocks.
        Argo ZRG 766.
            +Gr 3-75 p1688   +RR 3-75 p61
            ++NR 8-75 p9   /SFC 10-26-75 p24
     Concerto, harp, B flat major (From concerto, organ, op. 4, no. 6).
        cf BOIELDIEU: Concerto, harp, C major.
     Concerto, harp, op. 4, no. 5, F major. cf RCA CRL 2-7003.

Concerto, harp, op. 4, no. 6, B flat major. cf BACH: Concerto,
    violin and oboe, S 1060, D minor.
Concerto, harp, op. 4, no. 6, B flat major. cf Saga 5356.
1085 Concerto, 2 lutes, strings and recorders, op. 4, no. 6, B flat
    major. KOHAUT: Concerto, guitar (lute), F major. VIVALDI:
    Concerto, lute, D major. Julian Bream, 1t; Robert Spencer,
    chitarrone; Nicholas Kraмer, hpd; Marilyn Sansom, vlc; John
    Gray, violone; Monteverdi Orchestra; John Eliot Gardiner. RCA
    ARL 1-1180.
            ++Gr 11-75 p839   ++NR 12-75 p7
            ++HFN 12-75 p155   +RR 12-75 p70
1086 Concerti, oboe, nos. 1-3. HAYDN, J.: Concerto, trumpet, E flat
    major. HAYDN, M.: Concerto, trumpet, D major. Maurice André,
    tpt; Hilde Noé, hpd; Munich Bach Orchestra, Munich Chamber Orch-
    estra; Karl Richter, Hans Stadlmair. DG 2538 319. (Reissues
    from SLPEM 136517, SAPM 198341).
            ++Gr 7-75 p199   +RR 5-75 p29
            ++HFN 6-75 p107
1087 Concerto, oboe, no. 3, G minor. MARAIS: Alcyone: Suite. VIVALDI:
    Concerto, flute and strings, op. 10, no. 2, G minor. VCM;
    Nikolaus Harnoncourt. Telefunken SAW 9626.
            +-Gr 4-75 p1818   +RR 3-75 p36
1088 Concerti, oboe, nos. 3, 4. Concerti, organ, nos. 13, 14. Gustav
    Leonhardt, org; Hans Kamésch, ob; Vienna State Opera Chamber
    Orchestra; Ernest Kuyler. Olympic 8119.
            +-NR 4-75 p5
Concerti, oboe, nos. 8, 9. cf ALBINONI: Concerto, oboe, op. 7,
    no. 6, D major.
1089 Concerti, organ, nos. 1-6, op. 4; nos. 13-18, op. 7. Karl Richter,
    org; Chamber Orchestra; Karl Richter. Decca SDD 470/2 (3).
    (Reissues from SXL 2115, 2187, 2201)
            +Gr 11-75 p818   +HFN 10-75 p153
1090 Concerti, organ, nos. 1-6, op. 4; nos. 7-8; nos. 13-16, 18, op.
    7; nos. 19-20. E. Power Biggs, org; LPO; Adrian Boult. Colum-
    bia D3M 33716. (also CBS 77358. Reissues from Philips SABL
    148-8, ABL 3326/7)
            +Gr 3-74 p1765   ++RR 11-73 p36
            +HFN 11-73 p235   +SFC 11-30-75 p34
Concerto, organ, no. 4, op. 4, F major: Allegro. cf Stentorian
    SC 1685.
Concerto, organ, no. 4, op. 4, F major: 3rd movement. cf DANKWORTH:
    Tom Sawyer's Saturday.
Concerti, organ, nos. 13-14. cf Concerti, oboe, nos. 3, 4.
1091 Concerto, recorder and strings, B flat major (arr. from Concerto,
    organ, op. 4, no. 6). SAMMARTINI: Concerto, recorder and
    strings, F major. TELEMANN: Suite, recorder and strings, A
    minor. David Munrow, rec; Christopher Hogwood, hpd; AMF;
    Neville Marriner. Angel S 37019. (also HMV ASD 3028)
            ++Ar 5-75 p52   ++NR 10-74 p6
            +Gr 11-84 p890   ++RR 11-74 p68
            +HF 3-75 p89
Concerto, trumpet, G minor. cf Erato STU 70871
Converto, violin, no. 3, G minor: Arioso, Sarabande. cf Discopaedia
    MB 1003.
1092 Concerto, 2 wind choirs and strings, B flat major. Royal fireworks
    music. Water music. La Grande Ecurie & La Chambre du Roy;

Jean-Claude Malgoire. Columbia MG 32813 (2).
+AR 5-75 p52

1093 Concerto à due cori (double concerto), F major. Royal fireworks
music (original version). Pro Arte Orchestra; Charles Mackerras.
Pye GSGC 15009. (Reissues from CML 33005, GSGC 14003)
+Gr 8-75 p335   ++RR 7-75 p25
+-HFN 8-75 p87

1094 Concerti grossi, op. 6 (12). South West German Chamber Orchestra;
Paul Angerer. Vox QSVBX 558 (3).
+NR 10-75 p1

Coronation anthem: Zadok the Priest. cf Works, selections (Decca
PFS 4295).

Deidamia: Overture. cf Overtures (DG 2530 342).

1095 Delirio amoroso. Nel dolce del'oblio (Pensiere notturni di Filli).
Magda Kalmár, s; Ference Liszt Academy Chamber Orchestra; Frig-
yes Sándor. Hungaroton LSPX 11653.
++HFN 7-75 p81   ++SFC 6-1-75 p21
++NR 6-75 p12   ++St 8-75 p96
+RR 4-75 p59

Dixit Dominus: Judicabit in nationibus. cf HMV Tape (c) TC MCS 13.

1096 The faithful shepherd suite (arr. Beecham). HAYDN: Symphony, no.
93, D major. RPO; Thomas Beecham. Odyssey Y 33285.
+HF 7-75 p89   +St 7-75 p110

Fantasia, C major. cf Harpsichord works (Erato STU 70906).

Forest music (arr. Kipnis). cf Angel SB 3816.

Fugue, no. 4, B minor. cf Saga 5374.

1097 Giulio Cesare: Hast du mich ganz berauscht, Es blaut die Nacht,
Breite aus die gänd'gen Hande, Weine nur klage nur, Heil und
sicher kam mein Nachen. MOZART: Cosí fan tutte, K 588: Per
pieta ben mio. Don Giovanni, K 527: In quali eccessi...Mi tradi,
Ah, fuggi, il traditor, Crudele...Non mir dir. Le nozze di
Figaro, K 492: Dove sono. Lisa della Casa, s; VPO; Richmond
SR 33200.
+NR 12-75 p10

Giulio Cesare: E pur cosí...Piangerò la sorte mia. cf Philips
6833 105.

1098 Harpsichord works: Chaconne, G major. Fantasia, C major. Partita,
A major. Prelude and lesson, A minor. Suite, harpsichord, D
minor. Luciano Sgrizzi, hpd. Erato STU 70906.
+-Gr 12-75 p1081

Hercules: Where shall I fly. cf Arias (Philips 6500 523).

Israel in Egypt: He spake the word; He gave them hailstones. cf
Works, selections (Decca PFS 4295).

Jephtha: Overture. cf Overtures (DG 2530 342).

Jephtha: When His loud voice. cf Works, selections (Decca PFS
4295).

Joseph: Overture. cf Ah, che troppo inegali.

Joshua: O had I Jubal's lyre. cf Arias (Philips 6500 523).

Joshua: O had I Jubal's lyre. cf HMV RLS 714.

1099 Judas Maccabaeus. Martina Arroya, s; Mary Davenport, ms; Jan
Peerce, Lawrence Avery, t; David Smith, bar; Vienna Academy
Chorus; VSOO; Thomas Scherman. Desto DST 6452/4 (3).
-Gr 6-75 p76   +HFN 5-75 p127

1100 Judas Maccabaeus, highlights. Heather Harper, s; Helen Watts, con;
Alexander Young, t; John Shirley-Quirk, bar; Amor Artis Chorale,
Wandsworth School Boys' Choir: ECO; Johannes Somary. Vanguard

(Q) VSQ 30029.
  +-Audio 8-75 p80
Judas Maccabaeus: See the conquering hero comes. cf Works,
  selections (Decca PFS 4295).
Larghetto. cf Rococo 2072.
Look down, harmonious Saint. cf Ah, che troppo inegali.
Lucrezia, complete. cf Arias (Philips 6500 523).
1101 Messiah (arr. Mozart). Edith Mathis, s; Birgit Finnila, con;
  Peter Schreir, t; Theo Adam, bs; Austrian Radio Choir; Symphony
  Orchestra; Charles Mackerras. DG Archive 2723 019 (3). (also
  DG Archive 2710 016)
              ++Gr 11-74 p941   ++RR 11-74 p88
              +-HF 3-75 p78   ++St 3-75 p101
               +NR 1-75 p8
1102 Messiah. Margaret Price, s; Yvonne Minton, con; Alexander Young,
  t; Justino Diaz, bs; Amor Artis Chorale; ECO; Johannes Somary.
  Vanguard C 10090/2. (Tape (c) ZCVX 10092.
               +RR 3-75 p74 tape
1103 Messiah, excerpts. Heather Harper, s; Helen Watts, con; John
  Wakefield, t; John Shirley-Quirk, bs; Ralph Downs, org; Leslie
  Pearson, hpd; LSO and Chorus; Colin Davis. Philips 6833 144.
               ++MJ 1-75 p49
1104 Messiah: Glory to God in the highest; And the glory of the Lord;
  Lift up your heads; Behold the Lamb of God; His yoke is easy;
  Worthy is the lamb; For unto us a child is born; But thanks be
  to God; Surely He hath borne our griefs; And with his stripes
  we are healed; And we like sheep have gone astray; Hallelujah.
  Mormon Tabernacle Choir; RPO; Richard Condie. Columbia M 32935.
  Tape (c) MT 32935 (ct) MA 32935 (Q) Tape (ct) MAQ 32935. (also
  CBS 61582. Tape (c) 40-61582)
              -HF 7-75 p102 tape   +RR 11-74 p88
              +-NR 1-75 p11   +-RR 12-75 p173 tape
Messiah: For unto us a child is born. cf HMV Tape (c) TC MCS 13.
Messiah: For unto us a child is born. cf HMV Tape (c) TC MCS 14.
Messiah: Hallelujah; For unto us a child is born; Worthy is the
  Lamb...Amen. cf Works, selections (Decca PFS 4295).
Messiah: I know that my Redeemer liveth. cf Pye GH 589.
Messiah: Pastoral symphony. cf HMV ASD 3017.
Messiah: Rejoice greatly, O daughter of Zion; Come unto Him; How
  beautiful are the feet; I know that my Redeemer liveth; If God
  be for us. cf HMV RLS 714.
Minuet, flute, E minor. cf Works, selections (Philips 6747 096).
Movement, recorder, D minor (2). cf Works, selections (Philips
  6747 096).
Nel dolce dell'oblio. cf Ah, che troppo inegali.
Nel dolce dell'oblio (Pensieri notturno di Filli). cf Delirio
  amoroso.
1105 Overture, D major. Royal fireworks music. Water music, excerpts.
  Schola Cantorum Basiliensis; August Wenzinger. DG 2538 100.
  Tape (c) 3318 011.
               +-RR 1-75 p69 tape
1106 Overtures: Alcina. Agrippina. Belshazzar. Deidamia. Jephtha.
  Radamisto. Rinaldo. Rodelinda. Susanna. Hedwig Bilgram,
  hpd; LPO; Karl Richter. DG 2530 342. Tape (r) L 43342.
              +-Gr 4-74 p1851   +NR 3-74 p4
              ++HF 3-74 p86   +-RR 3-74 p38
              +HF 6-74 p124 tape   +SFC 1-27-74 p31
              +HFN 3-74 p105   +SFC 6-29-75 p26 tape

Partita, A major. cf Harpsichord works (Erato STU 70906).
Pensieri notturni di filli, cantata no. 17. cf Agrippina condotta
    a morire, cantata no. 14.
Il pensieroso: Sweet bird. cf Pearl GEM 121/2.
Prelude and lesson, A minor. cf Harpsichord works (Erato STU 70906).
Radamisto: Overture. cf Overtures (DG 2530 342).
Rinaldo: Overture. cf Overtures (DG 2530 342).
Rodelinda: Art thou troubles. cf HMV RLS 714.
Rodelinda: Overture. cf Overtures (DG 2530 342).
Rodelinda: Pompe vane di morte...Dove sei, amato bene. cf Arias
    (Philips 6500 523).
1107 Royal fireworks music. Water music: Suite. Schola Cantorum;
    Archive Wind Ensemble; August Wenzinger. DG 2548 169. (Reissue
    from Archive 198365, 198146)
        ++Gr 11-75 p840   +-RR 10-75 p43
        +HFN 10-75 p153
1108 Royal fireworks music. Water music. Ars Rediviva Orchestra; Milan
    Munclinger. Supraphon 110 1556.
        +-Gr 10-75 p612   +NR 9-75 p6
        /HFN 7-75 p81   +-RR 7-75 p26
        +-HFN 10-75 p143
1109 Royal fireworks music. Water Music: Overture, Adagio e staccato;
    Allegro in 3/4; Air; Minuet; Bourrée; Hornpipe; Andante, D minor;
    Allegro; Bourrée; Alla hornpipe. ECO Wind Ensemble (augmented);
    Johannes Somary. Vanguard VSD 71176. Tape (c) ZCVSM 71176 (Q)
    VSQ 30020.
        +HF 9-73 p106   +RR 11-73 p37
        -RR 3-75 p73 tape   ++St 7-73 p104
Royal fireworks music. cf Concerto à due cori, F major.
Royal fireworks music. cf Concerto, 2 wind choirs and strings,
    B flat major.
Royal fireworks music. cf Overture, D major.
Royal fireworks music. cf Camden CCV 2000-12, 5014-24.
Royal fireworks music: Minuet and allegro. cf Vista VPS 1021.
1110 Royal fireworks music: Suites. Water music: Suites. RPO; George
    Weldon. HMV SXLP 20033. Tape (c) TC EXE 137 (ct) 8X EXE 137.
        -Gr 8-75 p376 tape   +-RR 7-75 p70 tape
        -HFN 7-75 p90 tape
Royal fireworks music: Suite. cf CORELLI: Suite, strings.
Samson: Arias and scenes. cf London OS 26306.
Samson: Awake the trumpets's lofty sound. cf Columbia M 33514.
Samson: Let the bright seraphim. cf HMV RLS 714.
Sarabande, D major. cf Philips 6833 159.
Saul: Gird on thy sword. cf Works, selections (Decca PFS 4295).
1111 Semele. Sheila Armstrong, Felicity Palmer, s; Helen Watts, con;
    Mark Deller, c-t; Robert Tear, t; Justino Diaz, bs; Amor Artis
    Chorale; Harold Lester, hpd and org; ECO; Johannes Somary.
    Vanguard VSD 71180/2 (3). (Q) VSQ 30013/5 (3).
        ++AR 5-74 p56   +ON 3-30-74 p28
        +Gr 7-75 p238   +SFC 2-24-74 p25
        ++HF 1-74 p60   ++St 1-74 p109
        -HF 1-74 p84 Quad
Semele: Arias. cf London OS 26277.
Serse: Largo. cf Decca SDD 463.
Serse: Ombra mai fù. cf Arias (Philips 6500 523).
Silete venti. cf Ah, che troppo inegali.

Sing unto God, anthem.   cf BACH: Cantata, no. 131, Aus de Tiefe
    rufe ich, Herr.

1112 Solomon.   Sheila Armstrong, Felicity Palmer, s; Robert Tear, t;
    Justino Diaz, Michael Rippon, s; Amor Artis Chorale; ECO;
    Johannes Somary.   Vanguard VSD 71204/6 (3).   (Q) VSQ 30041/3.
          +-HF 10-75 p67   ++SFC 10-26-75 p24
          +NR 8-75 p10   ++St 12-75 p126
          ++NYT 9-21-75 pD18

Solomon: Arrival of the Queen of Sheba.   cf London SPC 21100.
Solomon: Arrival of the Queen of Sheba.   cf Philips 6580 066.
Solomon: Entry of the Queen of Sheba.   cf DANKWORTH: Tom Sawyer's
    Saturday.
Solomon: May no rash intruder.   cf Works, selections (Decca PFS
    4295).
Sonata, flute, A minor.   cf Works, selections (Philips 6747 096).
Sonata, flute, B minor.   cf Works, selections (Philips 6747 096).
Sonata, flute, op. 1, no. 1b, E minor.   cf Works, selections
    (Philips 6747 096).
Sonata, flute, op. 1, no. 5, G major.   cf Works, selections
    (Philips 6747 096).
Sonata, flute, op. 1, no. 5, G major.   cf Decca SPA 394.
Sonata, flute, op. 1, no. 5, G major.   cf London STS 15198.
Sonata, flute, op. 1, no. 11, F major.   cf Argo ZRG 746.
Sonata, oboe, B flat major.   cf Works, selections (Philips 6747
    096).
Sonata, oboe, op. 1, no. 8, C minor.   cf Works, selections
    (Philips 6747 096).
Sonata, oboe, op. 1, no. 8, C minor.   cf ALBINONI: Concerto, oboe,
    op. 9, no. 2, D minor.
Sonata, recorder, B flat major.   cf Works, selections (Philips
    6747 096).
Sonata, recorder, D minor.   cf Works, selections (Philips 6747 096).
Sonatas, recorder, op. 1, nos. 2, 4, 7, 11.   cf Works, selections
    (Philips 6747 096).
Sonata, recorder, op. 1, no. 11, F major.   cf Transatlantic TRA 292.
Sonata, trumpet, A major.   cf ALBINONI: Concertos, trumpet, F
    major, D minor.

1113 Sonatas, violin, op. 1, nos. 10, 13, 14, 15.   Josef Suk, vln;
    Zuzana Ruzičkova, hpd.   Supraphon 111 0864.
          ++SFC 10-26-75 p24

Sonata, violin, op. 1, no. 13, D major.   cf RCA ARM 4-0942/7.
Sonata, violin, no. 5, G major: Minuet.   cf Discopaedia MB 1003.
Sonata, violin, no. 9, B minor.   cf Discopaedia MB 1002.
Sonata, 2 violoncellos and orchestra, G minor.   cf BOCCHERINI:
    Concerto, violoncello, B flat major.
Suite, D major.   cf Argo ZDA 203.
Suite, harpsichord, D minor.   cf Harpsichord works (Erato STU
    70906).

1114 Suites, harpsichord, nos. 1-8.   Colin Tilney, hpd.   DG Archive
    2533 168/9 (2).
          ++Gr 2-75 p1516   +NR 1-75 p15
          +HF 4-75 p86   +-RR 2-75 p56

1115 Suites, harpsichord, nos. 1-8.   Blandine Verlet, hpd.   Telefunken
    SAWT 9623/4 (2).
          +Gr 3-75 p1681   +RR 3-75 p47

Suite, harpsichord, no. 7: Passacaglia.   cf Saga 5356.

Suite, harpsichord, no. 8, F minor. cf Angel SB 3816.
Susanna: Overture. cf Overtures (DG 2530 342).
1116 Trio sonatas, op. 2, nos. 1, 2, 4-5, 7-9; op. 5, nos. 1, 2, 4-6.
    Ars Rediviva Orchestra. Supraphon 111 1251/3 (3).
        +Gr 12-74 p1161   ++RR 12-74 p45
        ++HF 6-75 p94   ++SFC 1-19-75 p27
        ++NR 2-75 p7
1117 Water music. Collegium Aureum. BASF BHM 20341. Tape (c) KBACC
    3021.
        +-Gr 5-75 p2032 tape   +NR 7-73 p3
        +-HF 8-73 p82   +RR 7-75 p70 tape
        +-HFN 2-73 p337   ++SFC 5-27-73 p27
        +HFN 5-75 p142 tape
1118 Water music. Virtuosi of England; Arthur Davison. Classics for
    Pleasure CFP 40092.
        /Gr 2-75 p1492   +-RR 1-75 p28
1119 Water music. NYP; Pierre Boulez. Columbia M 33436. (Q) MQ 33436.
    Tape (ct) MAQ 33436. (also CBS 76440)
        +-Gr 9-75 p457   +NR 8-75 p4
        +HF 10-75 p74B   ++RR 9-75 p38
        +-HFN 9-75 p99
1120 Water music. COA; Eduard van Beinam. Philips Tape (c) 7317 012.
        +RR 9-75 p77 tape
    Water music. cf Concerto, 2 wind choirs and strings, B flat major.
    Water music. cf Royal fireworks music.
    Water music. cf Camden CCV 5000-12, 5014-24.
    Water music, excerpts. cf Overture, D major.
    Water music: Air. cf Decca SDD 463.
    Water music: Overture; Adagio e staccato; Allegro in 3/4; Air;
        Minuet; Bourrée; Hornpipe; Andante, D minor; Allegro; Bourrée;
        Alla hornpipe. cf Royal fireworks music.
    Water music: Pomposo. cf Columbia M 33514.
    Water music: Sarabande, Rigaudon. cf Philips 6580 098.
    Water music: Suites. cf Royal fireworks music: Suites.
    Water music: Suite. cf Royal fireworks music.
    Water music: Suite. cf CORELLI: Suite, strings.
    Water music: Suite, no. 3. cf Philips 6747 204.
1121 Works, selections: Coronation anthem: Zadok the Priest. Israel
    in Egypt: He spake the word; He gave them hailstones. Jephtha:
    When His loud voice. Judas Maccabaeus: See the conquering hero
    comes. Messiah: Hallelujah; For unto us a child is born; Worthy
    is the lamb...Amen. Saul: Gird on thy sword. Solomon: May no
    rash intruder. Handel Society Orchestra and Chorus; Charles
    Farncombe. Decca PFS 4295. (also London Tape (c) 0 521106 (ct)
    0 821106 (r) E 421106)
        +-Gr 12-74 p1203   +-RR 11-74 p88
        -HF 7-75 p102 tape   +SFC 10-26-75 p24
1122 Works, selections: Allegro, oboe, F major. Andante, flute, B minor.
    Minuet, flute, E minor. Movement, recorder, D minor (2). Son-
    ata, flute, op. 1, no. 1b, E minor. Sonata, flute, op. 1, no.
    5, G major. Sonata, flute, B minor. Sonata, flute, A minor.
    Sonata, oboe, op. 1, no. 8, C minor. Sonata, oboe, B flat major.
    Sonatas, recorder, op. 1, nos. 2, 4, 7-11. Sonata, recorder,
    B flat major. Sonata, recorder, D minor. Frans Brüggen, rec
    and flt; Bruce Haynes, ob; Hans Jürg Lange, bsn; Anner Bylsma,
    vlc; Bob van Asperen, hpd and org. Philips 6747 096 (3).

        ++Gr 10-74 p714   ++RR 10-74 p18
        +-MT 8-75 p713
    Xerxes: Ombra mai fù. cf Decca SXL 6629.
    Xerxes: Ombra mai fù. cf Supraphon 10924.

HANFF, Johann
    Ach, Gott, vom Himmel sich darein. cf Pelca PSRK 41013/6.
    Chorale preludes: Ach, Gott vom Himmel sich darein; Auf meinen
        lieben Gott; Ein fest Burg ist unser Gott; Erbarm dich mein, O
        Herre Gott; Helft mir Gott's Güte preisen; Wär Gott nicht mit
        uns diese Zeit. cf BRUHNS: Prelude and fugue, G minor.

HANSON
    Pastorale, oboe, strings and harp. cf Era 1001.
    Serenade, flute, strings and harp, op. 35. cf Era 1001.

d'HARDELOT
    Say yes, Mignon. cf Rubini RS 300.

HARRIS, Roy
1123 Symphony, no. 4. Utah Chorale; Utah Symphony Orchestra; Maurice
        Abravanel. Angel (Q) S 36091
             +HF 12-75 p94   +SR 9-6-75 p40
             ++NR 11-75 p3

HARRIS, William
    Behold now praise the Lord. cf Argo ZRG 789.

HARRISON, John
1124 Grand defilé, organ and percussion. MERRICK: Celtic suite. STAN-
        FORD: Irish dances. John Harrison, org; Orchestra. Rare
        Recorded Editions SRRE 147.
           /HFN 7-75 p81

HARRISON, Lou
    Fugue. cf Opus One 22.

HARTLEY
    Midnight sun. cf Canon CNN 4983.

HARTY, Hamilton
    Songs: The stranger's grave; Grace for light. cf HMV RLS 714.
    HASLEMERE FESTIVAL GOLDEN JUBILEE, 1925-1974. cf Strobe SRCM 123.

HASPROIS, Jehan
    Puisque je voy. cf Telefunken ER 6-35257.

HASSLER
    Agnus Dei. cf Wayne State University, unnumbered.

HAUBIEL, Charles
1125 Gothic variations. Nuances. Shadows. Sonata, violin and piano.
        1865 A. D. Endre Granat, vln; Carol Roberts, pno; Westphalian
        Symphony Orchestra; Paul Freeman. Orion ORS 74158.
           +NR 3-75 p12   +St 5-75 p94
    1865 A. D. cf Gothic variations.
1126 Metamorphoses. LEGINSKA: Victorian portraits. Jeanane Dowis, pno.
        Orion ORS 75188.
           +NR 10-75 p10
1127 Miniatures. Pioneers. Suite passacaille: Minuet. Graunke Symph-
        ony Orchestra, Hamburg Philharmonia Orchestra; James Swift,
        Hans Jürgen Walther. Orion ORS 75197.
           +NR 9-75 p4
    Nuances. cf Gothic variations.
    Pioneers. cf Miniatures.
    Shadows. cf Gothic variations.
    Sonata, violin and piano. cf Gothic variations.
    Suite passacaille: Minuet. cf Miniatures.

HAUSSMANN, Valentin
    Catkanei. cf DG Archive 2533 184.
    Galliard. cf DG Archive 2533 184.
    Tantz. cf DG Archive 2533 184.
HAVELKA, Svatopluk
    Pena. cf CHAUN: Proces.
HAWTHORNE
    Listen to the mocking bird. cf Vox SVBX 5304.
HAYDN, Josef
    Andante and minuet, op. 104. cf CBS 76267.
1128 Arianna a Naxos. Berenice che fai. MOZART: La clemenza di Tito,
    K 621: Parto, parto, ma tu ben mio; Deh, per questo instante
    solo. Songs: Abendempfindung, K 523; Das Veilchen, K 476.
    Janet Baker, ms; Raymond Leppard, fortepiano; ECO; Raymond
    Leppard. Philips 6500 660. Tape (c) 7300 350.
        ++Audio 12-75 p104   ++NR 6-75 p11
        ++Gr 3-75 p1688   ++ON 8-75 p28
        ++Gr 6-75 p111 tape   ++RR 3-75 p61
        ++HF 6-75 p106   ++St 6-75 p70
        +HFN 7-75 p90 tape
    Concerto, flute, D major. cf BENDA: Concerto, flute, E minor.
1129 Concerto, 2 flutes and orchestra, F major. Concerto, violin,
    harpsichord and strings. Jeanette Dwyer, Claude Legrand, flt;
    Jacques-Francis Manzone, vln; Françoise Petit, hpd; Mozart
    Society Orchestra; Orchestra; Henri-Claude Fantapie, Guido
    Bozzi. Orion ORS 75198.
        ++NR 10-75 p4
1130 Concerto, harpsichord, op. 21, D major. MOZART: Concerto, piano,
    no. 21, K 467, C major. Emil Gilels, pno; Moscow Chamber
    Orchestra; Rudolf Barshai. Columbia M 33098.
        +NR 2-75 p5
1131 Concerto, harpsichord, op. 21, D major. MOZART: Concerto, piano,
    no. 12, K 414, A major. Vasso Devetzi, pno; Moscow Chamber
    Orchestra; Rudolf Barshai. HMV SXLP 30184. (Reissue from CSD
    3516)
        /Gr 7-75 p181   +-RR 7-75 p26
        +-HFN 7-75 p81
    Concerto, harpsichord, op. 21, D major. cf BACH, J. C.: Concerto,
    harpsichord, A major.
    Concerto, harpsichord, op. 21, D major. cf MOZART: Concerto,
    piano, no. 21, K 467, C major.
1132 Concerto, horn, no. 1, D major. MOZART: Concerto, horn, no. 4,
    K 495, E flat major. STRAUSS, R.: Concerto, horn, no. 2, E
    flat major. Barry Tuckwell, hn; AMF, LSO; Neville Marriner,
    Peter Maag, István Kertesz. Decca SPA 393. (Reissue from Argo
    ZRG 5498, Decca SWL 8011, SXL 6285)
        +Gr 7-75 p199   +RR 7-75 p28
        +HFN 8-75 p89
    Concerti, organ and strings, nos. 1-3, C major. cf MOZART: Son-
    atas, organ and orchestra.
    Concerto, piano, no. 4, G major. cf Concerto, piano, no. 11, D
    major.
1133 Concerto, piano (clavier), no. 11, D major. Concerto, piano, no.
    4, G major. Arturo Benedetti Michelangeli, pno; Zurich Chamber
    Orchestra; Edmond de Stoutz. HMV ASD 3128. Tape (c) TC ASD
    3128. (also Angel S 37136)

```
            -Gr 9-75 p458                  +NR 11-75 p5
          +-HFN 9-75 p99                   +RR 9-75 p38
```
Concerto, trumpet, E flat major. cf HANDEL: Concerti, oboe, nos.
    1-3.
Concerto, trumpet, E flat major. cf Works, selections (Vanguard
    703/4).
Concerto, trumpet, E flat major. cf HMV ASD 2938.
Concerto, trumpet, E flat major. cf RCA CRL 2-7002.
Concerto, trumpet, E flat major: Allegro. cf Philips 6580 066.
Concerto, trumpet, E flat major: 3rd movement. cf Sound Superb
    SPR 90049.
1134 Concerto, violin, no. 1, C major. Sinfonia concertante, op. 84,
    B flat major. Franzjosef Maier, vln; Collegium Aureum. BASF
    KHC 21799.
```
            ++HF 9-75 p86                   -SFC 7-27-75 p22
            +NR 2-75 p5
```
Concerto, violin, harpsichord and strings. cf Concerto, 2 flutes
    and orchestra, F major.
1135 Concerto, violoncello, op. 101, D major. Concerto, violoncello,
    C major. Frédéric Lodéon, vlc; Bournemouth Sinfonietta Orches-
    tra; Theodor Guschlbauer. Erato STU 70869.
```
            +Gr 3-75 p1650
```
The creation (Die Schöpfung): On mighty pens. cf HMV RLS 714.
1136 Divertimenti, 2 oboes, 2 horns and 2 bassoons (7). London Wind
    Soloists. London STS 15078. (also Decca SDD 450. Reissue
    from SXL 6338)
```
            +Gr 10-75 p646                  +HFN 8-75 p89
            +HF 4-72 p104                   +RR 8-75 p33
```
1137 Divertimenti, 2 oboes, 2 horns and 2 bassoons, nos. 1-5. Péter
    Pongrácz, Bertalan Hock, ob; András Medveczky, Dezsö Mesterházy,
    hn; Tibor Fülemile, András Nagy, bsn. Hungaroton SLPX 11719.
```
            ++NR 11-75 p8                   +RR 12-75 p73
```
Fantasia, C major. cf Sonatas, piano, nos. 6, 10, 18, 33, 38-39,
    47, 50, 52, 60.
Feldpartita, B flat major. cf Saga 5417.
German dances, nos. 4, 10-11. cf Saga 5411.
Marches for the Derbyshire Cavalry Regiment. cf Classics for
    Pleasure 40230.
March for the Prince of Wales. cf Classics for Plesure 40230.
Mass, no. 7, C major: Credo, excerpts. cf HMV Tape (c) TC MCS 13.
1138 Mass, no. 9, D minor (Missa solemnis, Nelson Mass). Benita Valente,
    s; Ingeborg Russ, con; Karl Markus, t; Michael Schopper, bs;
    Stuttgart Kantatenchor; Werner Keltsch Instrumental Ensemble;
    August Langenbeck. BASF KHB 20351. (also BAG 3076)
```
            /Gr 2-75 p1530                  +ON 3-2-74 p28
          +-HFN 5-75 p128                   -RR 6-75 p79
            +NR 5-73 p8
```
1139 Mass, no. 12, B flat major. Judith Blegen, s; Frederica von Stade,
    con; Kenneth Riegel, t; Simon Estes, bs; Westminster Symphonic
    Choir; NYP; Leonard Bernstein. Columbia M 33267. (Q) MQ 33267.
    (also CBS 76410)
```
            +-Gr 12-75 p1089               ++NYT 9-21-75 pD18
            +-HF 9-75 p85                  ++RR 10-75 p84
            ++HFN 10-75 p142              ++SFC 7-6-75 p16
            ++NR 8-75 p9                   ++St 9-75 p84
```
Minuets (2). cf DG Archive 2533 182.

1140 Piano works: Sonata, no. 31, A flat major. Sonata, piano, no. 58,
        C major. MOZART: Adagio, K 540, B minor. Andante, mechancical
        organ, K 616, F major. Eine Kleine Gigue, K 574, G major.
        Minuet, K 355, D major. Rondo, K 511, A minor. Renée Sándor,
        pno. Hungaroton SLPX 11638.
            +—NR 9-75 p11   +—RR 8-75 p57
     Pieces for mechanical clock, nos. 4, 11, 12, 23. cf Vista VPS 1021.
     Quartet, strings, D major: Vivace. cf RCA ARM 4-0942/7.
     Quartet, strings, op. 3, no. 5, F major: Serenade. cf BACH, J. C.:
        Symphony, no. 3, op. 3, E flat major.
     Quartet, strings, op. 3, no. 5, F major: Serenade. cf Amon Ra
        SARB 01.
     Quartet, strings, op. 3, no. 5, F major: Serenade. cf Philips
        6580 066.
1141 Quartets, strings, op. 9, nos.1-6; op. 17, nos. 1-6. Aeolian
        Quartet. Argo HDNQ 61/6 (6).
            +—Gr 5-75 p1985   +—MT 9-75 p798
            +HFN 5-75 p127   +RR 5-75 p48
     Quartets, strings, op. 17, nos. 1-6. cf Quartets, strings, op. 9,
        nos. 1-6.
1142 Quartets, strings, op. 20, nos. 1-6. Lenox String Quartet. Desto
        DC 7152/4 (3).
            +Gr 3-75 p1674   ++MJ 11-73 p9
            +—HF 1-74 p73   ++SFC 6-30-74 p36
            -HFN 5-75 p127   ++St 10-73 p140
            ++LJ 3-75 p33
     Quartet, op. 33, no. 3, C major. cf BERNSTEIN: A Cabris.
1143 Quartets, strings, op. 54, nos. 1-3. Juilliard Quartet. CBS
        61549.
            +Gr 4-75 p1825   -RR 5-75 p49
            +—HFN 5-75 p128
1144 Quartets, strings, op. 54, nos. 1-3; op. 55, nos. 1-3. Aeolian
        Quartet. Argo HDNS 67/69 (3).
            +Gr 11-75 p853   +RR 11-75 p59
            +HFN 12-75 p155
     Quartet, strings, op. 54, no. 1, G major: Allegretto. cf Amon Ra
        SARB 01.
1145 Quartet, strings, op. 74, no. 3, G minor. Quartet, strings, op.
        76, no. 3, C major. Berg Quartet. Telefunken SAT 22550.
            +Gr 1-75 p1362   ++RR 12-74 p46
            +—HF 8-75 p86   ++SFC 8-10-75 p26
            -NR 6-75 p8   ++St 8-75 p72
1146 Quartet, strings, op. 74, no. 3, G minor. Quartet, strings, op.
        76, no. 3, C major. Alban Berg Quartet. Telefunken 6-41302.
        Tape (c) 4-41302.
            +HFN 9-75 p110 tape
1147 Quartet, strings, op. 74, no. 3, G minor. MARTINU: Quartet,
        strings, no. 2. MICA: Quartet, C major: Rondo. STRAVINSKY:
        Concertino. Talich Quartet. Panton 110 362.
            +HFN 7-75 p83   +RR 7-75 p38
     Quartet, strings, op. 74, no. 3, G minor. cf Works, selections
        (Vanguard 703/4).
1148 Quartets, strings, op. 76, nos. 1-6; op. 77, nos. 1-2. Amadeus
        Quartet. DG 2734 001 (4).
            +—Gr 2-75 p1509   +—RR 11-74 p71
1149 Quartets, strings, op. 76, nos. 1-6. Budapest Quartet. Odyssey

Y3 33324 (3). (Reissue from Columbia SL 203)
    +-HF 12-75 p92
1150 Quartet, strings, op. 76, no. 1, G major. Quartet, strings, op.
    76, no. 4, B flat major. Amadeus Quartet. DG 2530 089.
        ++HF 8-75 p86   +St 8-75 p97
        ++NR 6-75 p9
    Quartet, strings, op. 76, no. 3, C major. cf Quartet, strings,
        op. 74, no. 3, G minor (Telefunken SAT 22550).
    Quartet, strings, op. 76, no. 3, C major. cf Quartet, strings,
        op. 74, no. 3, G minor (Telefunken 6-41302).
    Quartet, strings, op. 76, no. 3, C major: Menuetto. cf Canon CNN
        4983.
    Quartet, strings, op. 76, no. 4, B flat major. cf Quartet, strings,
        op. 76, no. 1, G major.
    Raccolta de menuetti ballabili, nos. 1, 14. cf Nonesuch H 71141.
1151 Il ritorno di Tobia. Klára Takács, alto, Zsolt Bende, bar; Buda-
    pest Madrigal Choir; HSO; Ferenc Szekeres. Hungaroton SLPX
    11660/3 (4).
        /Gr 4-75 p1849   ++NYT 8-25-74 pD20
        +HF 6-75 p81   +RR 3-75 p62
1152 Sinfonia concertante, op. 84, B flat major. MOZART: Sinfonia
    concertante, K 364, E flat major. István Engl, ob; László
    Baranyai, bsn; Igor Ozim, Igor Oistrakh, vln; David Oistrakh,
    vla; Zoltán Rácz, vlc; Ph, MPO; Antal Dorati, Kiril Kondrashin.
    Decca SDD 445. (Reissues from HDNH 3540, SXL 6088)
        +Gr 4-75 p1808
    Sinfonia concertante, op. 84, B flat major. cf Concerto, violin,
        no. 1, C major.
1153 Sonatas, piano, nos. 1-26, 29-34. Rudolf Buchbinder, pno. Tele-
    funken FK 6-35088 (6).
        +-Gr 5-75 p1995   +-RR 3-75 p48
1154 Sonatas, piano, nos. 6, 10, 18, 33, 38-39, 47, 50, 52, 60. Fan-
    tasia, C major. Variations, F minor. John McCabe, pno. Decca
    HSN 100/2 (3).
        +-Gr 10-75 p657   +-RR 10-75 p76
        +HFN 10-75 p142
    Sonata, piano, no. 9, D major: Adagio. cf CBS Classics 61579.
    Sonata, piano, no. 31, A flat major. cf Piano works (Hungaroton
        11638).
1155 Sonatas, piano, nos. 35-53. Rudolf Buchbinder, pno. Telefunken
    FK 6-35249 (6).
        +-Gr 9-75 p486   +-RR 6-75 p67
        ++HFN 7-75 p83
    Sonata, piano, no. 35, C major. cf Columbia MG 33202.
    Sonata, piano, no. 58, C major. cf Piano works (Hungaroton 11638).
    Symphony, no. A, B flat major. cf Symphonies, nos. 93-104.
    Symphony, no. B, B flat major. cf Symphonies, nos. 93-104.
    Symphony, no. 22, E flat major (2d version). cf Symphonies, nos.
        93-104.
1156 Symphonies, nos. 25-28. Ph; Antal Dorati. Decca SDD 457.
        +HFN 11-75 p173   +-RR 11-75 p43
1157 Symphony, no. 45, F sharp minor. Symphony, no. 55, E flat major.
    The Hague Residentie Orchestra; Willem van Otterloo. DG
    2548 105. (Reissue from SLPM 138825)
        ++Gr 11-75 p840
    Symphony, no. 45, F sharp minor. cf Works, selections (Vanguard
        703/4).

1158 Symphony, no. 48, C major. Symphony, no. 92, G major. PCO;
     Dean Dixon. Supraphon 110 1202.
          +HFN 11-75 p154   +–RR 9-75 p38
     Symphony, no. 48, C major: Menuetto. cf Philips 6580 098.
1159 Symphony, no. 51, B flat major. Symphony, no. 55, E flat major.
     PH; Antal Dorati. Decca SDD 415. (Reissue from HDND 19/22)
          +Gr 9-75 p458   +RR 8-75 p53
          +HFN 8-75 p89
     Symphony, no. 53, D major (3 alternate finales). cf Symphonies,
     nos. 93-104.
     Symphony, no. 53, D major: Menuetto. cf Philips 6580 098.
     Symphony, no. 55, E flat major. cf Symphony, no. 45, F sharp minor.
     Symphony, no. 55, E flat major. cf Symphony, no. 51, B flat major.
     Symphony, no. 63, C major (1st version). cf Symphonies, nos.
     93-104.
1160 Symphony, no. 82, C major. Symphony, no. 84, E flat major. Menu-
     hin Festival Orchestra; Yehudi Menuhin. HMV ASD 3136.
          +HFN 12-75 p155   /RR 11-75 p43
1161 Symphony, no. 83, G minor. Symphony, no. 101, D major. LPO;
     Gaetano Delogu. Classics for Pleasure CFP 40222.
          +HFN 8-75 p76   +RR 8-75 p33
1162 Symphony, no. 83, G minor. Symphony, no. 85, B flat minor. South
     German Philharmonic Orchestra; Alexander von Pitamic. Pye
     GSGC 15006.
          -Gr 8-75 p322   -RR 8-75 p33
          -HFN 8-75 p75
     Symphony, no. 84, E flat major. cf Symphony, no. 82, C major.
     Symphony, no. 85, B flat major. cf Symphony, no. 83, G minor.
     Symphony, no. 88, G major. cf Toscanini Society ATC GC 1201/6.
1163 Symphony, no. 91, E flat major. Symphony, no. 92, G major. VPO;
     Karl Böhm. DG 2530 524. Tape (c) 3300 470.
          +–Gr 7-75 p181   +–NR 12-75 p5
          +HF 12-75 p97   ++RR 7-75 p26
          ++HFN 7-75 p81   +RR 10-75 p97 tape
1164 Symphony, no. 91, E flat major. Symphony, no. 103, E flat major.
     Bavarian Radio Symphony Orchestra; Eugen Jochum. DG Heliodor
     2548 147. (Reissue from 138007)
          +Gr 4-75 p1867   ++RR 4-75 p27
     Symphony, no. 92, G major. cf Symphony, no. 48, C major.
     Symphony, no. 92, G major. cf Symphony, no. 91, E flat major.
1165 Symphonies, nos. 93-104. PH; Antal Dorati. Decca HDNJ 41-6 (6).
     Tape (c) K4D 8. (also London STS 15319/24)
          +Gr 9-74 p496   +RR 11-75 p92 tape
          +RR 9-74 p45
1166 Symphonies, nos. 93-104. LOL; Leslie Jones. Nonesuch/Advent
     Tape (c) E 1001/3 (3).
          +HF 3-75 p108 tape
1167 Symphonies, nos. 93-104. Symphony, no. A, B flat major. Symphony,
     no. B, B flat major. Symphony, no. 22, E flat major (2d version).
     Symphony, no. 63, C major (1st version). Symphony, no. 53,
     D major (3 alternate finales). Symphony, no. 103, E flat major
     (alternate finale). PH; Antal Dorati. London STS 15319/24 (6),
     STS 15316/7 (2).
          ++HF 5-75 p76   ++SFC 2-9-75 p24
          ++NR 4-75 p4   ++St 6-75 p96
     Symphony, no. 93, D major. cf HANDEL: The faithful shepherd suite.

Symphony, no. 94, G major: Andante.   cf HMV SLA 870.
1168 Symphony, no. 95, C minor.   Symphony, no. 96, D major.   NYP; Leonard
    Bernstein.   Columbia M 32598.   (Q) MQ 32598. Tape (ct) MAQ 32598.
              +JF 2-75 p92   +SFC 9-15-74 p28
              ++NR 11-74 p4   +St 6-75 p96
              +NYT 8-4-74 pD20
1169 Symphony, no. 95, C minor.   Symphony, no. 96, D major.   LPO; Eugen
    Jochum.   DG 2530 420.   (Reissue from DG 2720 064)
              +Gr 12-74 p1134   +-RR 11-74 p38
              ++NR 1-75 p3   +St 6-75 p97
    Symphony, no. 96, D major.   cf Symphony, no. 95, C minor (Columbia
        M 32598).
    Symphony, no. 96, D major.   cf Symphony, no. 95, C minor (DG 2530
        420).
1170 Symphony, no. 99, E flat major.   Symphony, no. 100, G major.   LPO;
    Eugen Jochum.   DG 2530 459.   Tape (c) 3300 402.   (Reissue from
        2720 064)
              +Gr 8-74 p358   +-RR 8-74 p35
              +HF 3-75 p108 tape   ++RR 11-74 p99 tape
              +NR 12-74 p3   +St 6-75 p96
    Symphony, no. 100, G major.   cf Symphony, no. 99, E flat major.
    Symphony, no. 100, G major.   cf Works, selections (Vanguard 703/4).
1171 Symphony, no. 101, D major.   Symphony, no. 103, E flat major.   NYP;
    Leonard Bernstein.   Columbia M 33531.
              +-NR 12-75 p5
1172 Symphony, no. 101, D major.   Symphony, no. 104, D major.   LOL;
    Leslie Jones.   Oryx Tape (c) BRL 18.
              +-RR 2-75 p75 tape
    Symphony, no. 101, D major.   cf Symphony, no. 83, G minor.
1173 Symphony, no. 103, E flat major.   Symphony, no. 104, D major.   LPO;
    Eugen Jochum.   DG 2530 525.   Tape (c) 3300 471.   (Reissue from
        DG 2720 064)
              ++Gr 7-75 p181   +NR 12-75 p5
              /HFN 8-75 p89   +-RR 6-75 p42
              +HFN 10-75 p155 tape   +-RR 7-75 p70
    Symphony, no. 103, E flat major.   cf Symphony, no. 91, E flat major.
    Symphony, no. 103, E flat major (alternate finale).   cf Symphonies,
        nos. 93-104.
    Symphony, no. 103, E flat major.   cf Symphony, no. 101, D major.
    Symphony, no. 104, D major.   cf Symphony, no. 101, D major.
    Symphony, no. 104, D major.   cf Symphony, no. 103, E flat major.
    Trio, piano, no. 28, E major.   cf BEETHOVEN: Trio, piano, no. 6,
        op. 97, B flat major.
    Trio, piano, flute and violoncello.   cf BEETHOVEN: Trio, piano,
        clarinet, violoncello, op. 38, E flat major.
    Trio, strings (arr. from Sonata, piano, no. 1, op. 53, G major).
        cf BEETHOVEN: Trio, violin, violoncello and piano.
    Trio, strings, op. 53, no. 1, G major.   cf Classic Record Library
        SQM 80-5731.
    Trios, violin, viola and violoncello.   cf BEETHOVEN: Symphony, no.
        2, op. 36, D major.
    Variations, F minor.   cf Sonatas, piano, nos. 6, 10, 18, 33, 38-39,
        50, 52, 60.
1174 Works, selections: Concerto, trumpet, E flat major.   Quartet,
    strings, op. 74, no. 3, G minor.   Symphony, no. 45, F sharp
    minor.   Symphony, no. 100, G major.   Zingarese, nos. 1, 6, 8.

VSOO; Zagreb Radio Symphony Orchestra; Helmut Wobisch, tpt;
Griller Quartet, Willi Boskovsky Ensemble; Mogens Wöldlike,
Antonio Janigro. Vanguard 703/4. Tape (c) ZCVB 703/4.
            -RR 5-75 p77 tape
Zingarese, nos. 1, 6, 8.   cf Works, selections (Vanguard 703/4).
HAYDN, Michael
1175 Choral works: Missa Sancti Aloysii. Offertorium pro festo cuius-
    cunque St. Virginis et Martyris (Diffusa est gratia). Offer-
    torium pro festo SS Innocentium (Anima nostra). Sequentia ad
    festum S. P. Augustini (De profundis tenebrarum). Eva Marton,
    Katalin Szökeflavy-Nagy, s; Zsuzsa Németh, con; Imre Sulyok, hpd;
    János Sebestyén, org; Györ Girls' Choir; Györ Philharmonic
    Orchestra; Miklós Szabó. Hungaroton SLPX 11678.
            -Gr 5-75 p2005   ++RR 5-75 p60
        +-HFN 5-75 p128
Coburger-Marsch. cf DG 2721 077.
Concerto, trumpet, D major. cf HANDEL: Concerti, oboe, nos. 1-3.
Pappenheimer-Marsch. cf DG 2721 077.
HAYNE VAN GHIZEGHEM
    Gentil gallans. cf Vanguard SRV 316.
HEADINGTON (20th century England)
    Toccata. cf HMV HQS 1337.
HEATH, Fenno
    Beat, beat, drums; Thy word is a lantern. cf Wayne State University
    unnumbered.
HECKEL, Wolf (Wolfgang)
    Mille regretz; Nach willen dein. cf L'Oiseau-Lyre 12BB 203/6.
    THE HEIFETZ COLLECTION, 1917-55. cf RCA ARM 4-0942/7.
HEINICHEN, Johann
    Concerto, 4 alto recorders, strings and continuo: Pastorell. cf
    Strobe SRCM 123.
HEINTZ
    Da truncken sie. cf BIS LP 3.
HEISS, John
    Four movements for 3 flutes. cf BRUN: Gestures for eleven.
    Quartet. cf BRUN: Gestures for eleven.
HELLER, Stephen
    L'Avalanche. cf Connoisseur Society (Q) CSQ 2066.
HEMY
    Faith of our fathers. cf RCA ARL 1-0561.
    Living still. cf RCA ARL 1-0561.
HENRION, R.
    Dreutzritter Fanfare. cf BASF 292 2116-4.
    Fehrbelliner Reitermarsch. cf BASF 292 2116-4.
    Fehrbelliner Reitermarsch. cf DG 2721 077.
    Kreuzritter Fanfare. cf DG 2721 077.
HENRY VIII, King
    Green groweth holly. cf Oryx BRL 1.
    If love now reigned. cf L'Oiseau-Lyre SOL 329.
    King Harry's pavane. cf Oryx BRL 1.
    Pastime with good company. cf BASF 25 22286-1.
    Pastime with good company. cf Oryx BRL 1.
    Tho' some saith. cf L'Oiseau-Lyre SOL 329.
    The time of youth. cf L'Oiseau-Lyre SOL 329.
HENZE, Hans Werner
1176 Compases para preguntas ensemismadas, viola and 22 players.

Concerto, violin, tape, voices and 13 instrumentalists, no. 2.
Hirofumi Fukai, vla; Brenton Langbein, vln; London Sinfonietta
Orchestra; Hans Werner Henze. Decca Headline HEAD 5.
                +Gr 11-74 p895   +RR 11-74 p43
                +MT 8-75 p713
    Concerto, violin, tape, voices and 13 instrumentalists, no. 2. cf
        Compases para preguntas ensemismadas, viola and 22 players.
1177 In memoriam: Die Weisse Rose. Kammermusik. Philip Langridge, t;
        Timothy Walker, tr; London Sinfonietta Orchestra; Hans Werner
        Henze. L'Oiseau-Lyre DSLO 5.
                +Gr 11-75 p875   +−NYT 12-21-75 pD18
                +HFN 12-75 p155   +RR 9-75 p69
    Kammermusik. cf In Memoriam: Die Weisse Rose.
    HEROINES FROM GREAT FRENCH OPERAS. cf RCA ARL 1-0844.
HEROLD, Louis Joseph Ferdinand
    La fille mal gardée: Simone; Clog dance; Maypole dance; Storm and
        finale; Spinning; Tambourine dance; Harvesters. cf Decca SDDJ
        393/95.
    La fille mal gardée: Simone; Clog dance; Maypole dance; Storm and
        finale, Act 1; Spinning; Tambourine dance; Harvesters, Act 2.
        (arr. Lanchbery). cf Decca DPA 515/6.
1178 Zampa: Overture. OFFENBACH: Orphée aux enfers: Overture. ROSSINI:
        Guglielmo Tell: Overture. SUPPE: Light cavalry: Overture. Poet
        and peasant: Overture. PO; Eugene Ormandy. RCA ARL 1-0453.
        (Q) ARD 1-0453.
                +NR 11-74 p3   +St 7-75 p108
                -SFC 10-20-74 p26
    Zampa: Overture. cf AUBER: Le cheval de bronze: Overture.
HERON, Henry
    Cornet voluntary. cf Argo ZRG 783.
HERRMANN, Bernard
1179 The devil and Daniel Webster. Welles raises Kane. LPO; Bernard
        Herrmann. Unicorn UNS 237. (Reissue from Pye TLPS 13010)
                +Gr 7-75 p181   +HFN 8-75 p89
    Welles raises Kane. cf The devil and Daniel Webster.
1180 Wuthering Heights. Pamela Bowden, Morag Beaton, s; Elizabeth
        Bainbridge, ms; Joseph Ward, t; John Kitchiner, David Kelly,
        bar; Michael Rippon, bs; Pro Arte Orchestra; Elizabethan Sing-
        ers; Bernard Herrmann. Unicorn UNB 400 (4).
                +Op 10-72 p918   +−St 5-75 p94
                +SFC 2-23-75 p23
HERTEL, Johan Wilhelm
    Concerto, trumpet, 2 oboes and 2 bassoons. cf ALBINONI: Concerto,
        trumpet, C major.
HERZER
    Hoch Heidecksburg. cf BASF 292 2116-4.
HEUBERGER, Richard
    In our secluded rendezvous. cf Columbia D3M 33448.
    Midnight bells. cf BEETHOVEN: Concerto, violin, op. 61, D major.
    Der Opernball: In chambre separée. cf Angel S 35696.
    The opera ball: Im chambre separée. cf Pye NSPH 6.
    Der Opernball: Overture. cf Telefunken DX 6-35262.
HEURTEUR
    Troys jeunes bourgeoises. cf HMV SLS 5022.
HEWITT-JONES
    O clap your hands together. cf Gemini GM 2022.

HIBBARD, William
    Quartet, strings. cf FENNELLY: Evanescences.
    Trio. cf CRI SD 324.
HIDAS, Frigyes
    Quintet, winds, no. 2. cf BOZAY: Quintet, winds, op. 6.
HILLER, Ferdinand
1181 Konzertstück, piano, op. 113. MOSONYI: Concerto, piano. Jerome
       Rose, pno; Luxembourg Radio Orchestra; Pierre Cao. Candide (Q)
       QCE 31090.
       +NR 7-75 p6
HILLER, Lejaren
1182 Quartet, strings, no. 6. JOLAS: Quatuor III. Concord String
       Quartet. CRI SD 332.
       ++NR 4-75 p7   ++St 11-75 p127
1183 Sonatas, piano, nos. 4 and 5. Frina Boldt, Kenwyn Boldt, pno.
       Orion ORS 75176.
       ++NR 10-75 p10
HINDEMITH, Paul
1184 Mathis der Maler symphony. STRAUSS, R.: Death and transfiguration,
       op. 24. LSO; Jascha Horenstein. Nonesuch H 71307.
       ++NR 6-75 p6   ++St 10-75 p107
       ++SFC 10-5-75 p38
    Morgenmusik. cf Desto DC 6474/7.
1185 Pieces, string orchestra, op. 44, no. 4 (5). SCHONBERG: Verklärte
       Nacht, op. 4. WEBERN: Movements, string quartet, op. 5 (5).
       AMF; Neville Marriner. Argo ZRG 763.
       ++Gr 1-75 p1342   +RR 1-75 p33
       +-HF 9-75 p90   +St 9-75 p120
       ++NR 6-75 p2
    Rondo, 3 guitars. cf Trio for soprano and 2 alto recorders.
    Septet, wind instruments. cf ARNOLD: Divertimento, flute, oboe
       and clarinet, op. 37.
1186 Sonata, English horn and piano. Sonata, solo viola, op. 25, no.
       1. William Kosinski, English hn; Sven Reher, vla; Lincoln
       Mayorga, pno. GSC 4.
       +NR 8-75 p7
    Sonata, harp. cf Trio for soprano and 2 alto recorders.
    Sonata, organ, no. 3. cf Decca 5BBA 1013-5.
    Sonata, solo viola, op. 25, no. 1. cf Sonata, English horn and
       piano.
    Sonata, violin, op. 31, no. 2. cf DALLAPICCOLA: Studi, violin and
       piano.
    Sonata, solo violoncello, no. 3, op. 25. cf CRUMB: Sonata, solo
       violoncello.
1187 Sonata, violoncello and piano, op. 11, A minor. KODALY: Sonata,
       violoncello and piano, op. 4. George Isaac, vlc; Martin Jones,
       pno. Argo ZRG 762.
       +Gr 4-75 p1825   ++RR 3-75 p49
1188 Symphonic metamorphoses on a theme by Carl Maria von Weber.
       JANACEK: Sinfonietta. LSO; Claudio Abbado. Decca Tape (c)
       KSXC 6398.
       +Gr 2-75 p1562 tape
1189 Trio for soprano and 2 alto recorders (Plöner Musiktag). Rondo,
       3 guitars. Sonata, harp. Trio, viola, heckelphone and piano,
       op. 47. Myra Kestenbaum, vla; John Ellis, heckelphone; Delores
       Stevens, pno; Laurindo Almeida, gtr; Robin Howell, rec; Gail
       Laughton, hp. GSC 5.
       ++NR 6-75 p10

Trio for soprano and 2 alto recorders. cf HMV SLS 5022.
Trio, viola, heckelphone and piano, op. 47. cf Trio for soprano
    and 2 alto recorders.
HISTORIC ORGANS OF AUSTRIA. cf Telefunken TK 11567/1-2.
HISTORY OF SCOTTISH MUSIC, 1250-1625. cf Scottish Records SRSS 1-2.
HODDINOTT, Alun
1190 Divertimento, op. 32. Septet, op. 10. Sonata, violin, no. 1, op.
    63. Clarence Myerscough, vln; Martin Jones, pno; Nash Ensemble.
    Argo ZRG 770.
            +Gr 8-75 p341   +RR 6-75 p70
            +HFN 8-75 p76
Septet, op. 10. cf Divertimento, op. 32.
Sonata, violin and piano, no. 1, op. 63. cf Divertimento, op. 32.
HOFFMAN
    Miniatures, solo trumpet (4). cf BACH: Brandenburg concerto, no.
    2, S 1047, F major.
HOFFMANN, Johann (also called Giovanni Hoffmann)
1191 Concerto, mandoline and orchestra, D major. HUMMEL: Concerto,
    mandoline and orchestra, G major. Edith Bauer-Slais, Elfriede
    Kunschak, mand; Vienna Pro Musica Orchestra; Hladky. Turna-
    bout Tape (c) KTVC 34003.
            +RR 11-75 p92 tape
HOFFMANN, Richard
    Orchestra piece 1961. cf DUGGER: Music for synthesizer and six
    instruments.
    HOFFNUNG'S MUSIC FESTIVALS, November 13, 1956; Interplanetary music
    festival, November 21-22, 1958; Astronautical music festival,
    November 28, 1961. cf HMV SLA 870.
HOFHEIMER, Paul
    Nach willen dein. cf L'Oiseau-Lyre 12BB 203/6.
    Recordare. cf Hungaroton SLPX 11601/2.
HOFMANN, Josef
    Kaleideskop, op. 40. cf International Piano Archives IPA 5007/8.
    Kaleideskop, op. 40. cf L'Oiseau-Lyre DSLO 7.
    Penguine. cf International Piano Archives IPA 5007/8.
HOHNE, Carl
    Slavische fantasie. cf Nonesuch H 71298.
HOLBORNE, Anthony
    Honey suckle. cf Crystal S 201.
    Night watch. cf Crystal S 201.
    Pieces (3). cf Crystal S 202.
    The choice. cf HMV SLS 5022.
    Galliard. cf DG Archive 2533 157.
    Galliard. cf Turnabout TV 37071.
    Muylinda. cf HMV SLS 5022.
    Pavan and galliard. cf HMV SLS 5022.
    Romeo and Juliet: Heartsease. cf BBC REB 191.
    Sic semper soleo. cf HMV SLS 5022.
HOLCOMBE, Henry
    Airs for German flutes. cf Argo ZRG 765.
HOLDEN
    All hail the power of Jesus' name. cf RCA ARL 1-0562.
HOLLOWAY
    Wood-up quick-step. cf Michigan University SM 0002.
HOLMBOE, Vagn
1192 Symphony, no. 8, op. 56. NØRGARD: Constellations, 12 strings, op.

op. 22. Royal Danish Orchestra; Jerzy Semkov. Turnabout TVS
34168.
    *St 6-75 p50
HOLST, Gustav
    Beni Mora suite, op. 29, no. 1, E minor. cf Works, selections
    (Pearl GEM 126).
1193 Choral fantasia, op. 51. Psalm LXXXVI., FINZI: Dies natalis, op.
        8. Janet Baker, s; Ian Partridge, Wilfred Brown, t; The Purcell
        Singers; Ralph Downes, org; ECO; Imogen Holst, Christopher Finzi.
        Everest SDBR 3365. (Reissues)
            +Audio 12-75 p104   +NR 10-74 p13
1194 Choral symphony, op. 41. Felicity Palmer, s; LPO and Chorus;
        Adrian Boult. HMV SAN 354. (also Angel S 37039)
            ++Gr 10-74 p738   ++SFC 11-24-74 p31
            +HF 3-75 p80   +SR 1-11-75 p51
            +NR 1-75 p10   ++St 3-75 p104
            ++RR 10-74 p46   ++Te 3-75 p44
    Egdon Heath, op. 47. cf The tale of the wandering scholar, op. 50.
    Festival chorus, op. 36, no. 2: Turn back, O man. cf HMV HQS 1350.
    Hammersmith, op. 52: Prelude and scherzo. cf Philips 6747 177.
    The heart worships. cf RCA ARL 1-0562.
    Japanese suite, op. 33. cf BERKELEY: Mont Juic, op. 9.
    A Moorside suite. cf RCA LSL 1-5072.
    The perfect fool, op. 39. cf The tale of the wandering scholar,
        op. 50.
1195 The planets, op. 32. Ambrosian Singers; LSO; André Previn. Angel
        S 36991. (also HMV ASD 3002. (Q) Q4 ASD 3002)
            +Gr 9-74 p206   +RR 7-74 p43
            +-HF 2-75 p92   -SFC 9-29-74 p28
            +NR 11-74 p2   ++St 3-75 p104
1196 The planets, op. 32. Bournemouth Symphony Orchestra; Bournemouth
        Municipal Choir; George Hurst. Contour 2870 367. Tape (c)
        3470 367.
            +Gr 4-74 p1857   -RR 7-75 p70 tape
            +-RR 5-74 p34
1197 The planets, op. 32. St. Louis Symphony Orchestra; Walter Susskind.
        Turnabout (Q) QTVS 34598.
            ++NR 12-75 p3   ++SFC 11-16-75 p32
    Psalm LXXXVI. cf Choral fantasia, op. 51.
    St. Paul's suite, op. 29, no. 2. cf Works, selections (Pearl GEM
        126).
    St. Paul's suite, op. 29, no. 2. cf BORODIN: Quartet, strings,
        no. 2, D major: Nocturne.
1198 Songs: English folk songs: Matthew, Mark, Luke and John; I sowed
        the seeds of love; I love my love; Jig. Welsh folk songs: The
        lively pair; Lisa Lan; My sweetheart's like Venus; The mother-
        in-law; 2 folksong fragments; Oh, I hae seen the roses blaw;
        The shoemaker; Adar mân ymynydd (The nightingale and the linnet);
        Lliw gwyn rhosyn yr haf (White summer rose); The first love;
        Green grass; AWake, awake; The dove; O twas on a Monday morning;
        The lover's complaint. Two-part canons: If twere the time of
        lilies; Evening on the Moselle. Nocturne. Toccata. Newburn
        lads. This have I done for my true love. Keith Swallow, pno;
        BBC Northern Singers; Stephen Wilkinson. Abbey LPB 726.
            +-HFN 9-75 p99   +RR 4-75 p60

1199 Songs: Before sleep; David's lament for Jonathan; A dirge for two
     veterans; Drinking song; The fields of sorrow; Good Friday;
     The homecoming; Hymn to Manas, op. 26 (Group 4, no. 3); I
     love my love, op. 36, no. 5; I sowed the seeds of love, op. 36,
     no. 1; Intercession; A love song; Matthew, Mark, Luke and John,
     op. 36, no. 2; The song of the blacksmith, op. 36, no. 3; Swan-
     sea town, op. 36, no. 4; Truth of all truth. Baccholian Singers,
     London; Philip Jones Brass Ensemble, ECO; HMV CSD 3764.
          ++HFN 8-75 p76   ++RR 8-75 p62
          +MT 12-75 p1070
     Songs, voice and violin, op. 35 (4). cf Works, selections (Pearl
          GEM 126).
     Songs, voice and violin, op. 35 (4). cf Musical Heritage Society
          MHS 1976.
     Songs without words, op. 22, nos. 1, 2. cf Works, selections
          (Pearl GEM 126).
     Suite, no. 1, op. 28, E flat major. cf Philips 6747 177.
     Suite, no. 2, op. 28, F major. cf Philips 6747, 177.
1200 The tale of the wandering scholar, op. 50. The perfect fool, op.
     39. Egdon Heath, op. 47. Norma Burrowes, s; Robert Tear, t;
     Michael Rippon, Michael Langdon, bs; ECO, LSO; Steuart Bedford,
     André Previn. HMV ASD 3097.
          +-Gr 9-75 p506   +MT 12-75 p1070
          +HFN 9-75 p99   +RR 9-75 p18
1201 Works, selections: Beni Mora suite, op. 29, no. 1, E minor. St.
     Paul's suite, op. 29, no. 2. Songs, voice and violin, op. 35
     (4). Songs without words, op. 22, nos. 1, 2. Dora Labette, s;
     W. H. Reed, vln; LSO; String Orchestra; Gustav Holst. Pearl
     GEM 126 (Reissues)
          +Gr 3-75 p1650   -RR 3-75 p27
HOLYOKE, Samuel
     Processional march. cf New England Conservatory NEC 111.
HOLZMANN
     Feuert los. cf BASF 292 2116-4.
HONEGGER, Arthur
1020 Pacific 231. ROUSSEL: Le festin de l'araignée, op. 17. Petite
     suite, op. 39. OSR; Ernest Ansermet. Decca ECS 756. (Reissues
     from SXL 6065, LXT 5035)
          +Gr 7-75 p182   +RR 6-75 p46
          +HFN 8-75 p89
1203 Pacific 231. IBERT: Divertissement. POULENC: Les biches, Ballet
     suite. SATIE (orch. Debussy): Gymnopédies: no. 1, Lent et
     grave; no. 3, Lent et douloureux. Birmingham (City) Symphony
     Orchestra; Louis Frémaux. HMV ASD 2989. (Q) Q4 ASD 2989. Tape
     (c) TC ASD 2989.
          ++Gr 5-74 p2032   +-HFN 7-75 p90 tape
          +-Gr 4-75 p1819 Quad   +RR 5-74 p35
          +Gr 8-75 p376 tape   +RR 7-75 p71 tape
     Sonata, violin and piano. cf Sonata, violin and piano, no. 2.
1204 Sonata, violin and piano, no. 2. Sonata, violin and piano. Sona-
     tina, violoncello and piano. Primož Novšak, vln; Susanne Basler,
     vlc; Annette Weisbrod, pno. Musical Heritage Society MHS 1869.
          +HF 3-75 p81
     Sonatina, violoncello and piano. cf Sonata, violin and piano.
1205 Symphony, no. 1. Symphony, no. 4. CPhO; Serge Baudo. Supraphon
     110 1536.
          ++NR 9-75 p4   +RR 5-75 p29

Symphony, no. 4. cf Symphony, no. 1.
HOPKINSON, Francis
    Beneath a weeping willow's shade; My generous heart disdains. cf
        New England Conservatory NEC 111.
HOROVITZ, Joseph
    Horrortorio. cf HMV SLA 870.
    Metamorphosis on a bedtime theme. cf HMV SLA 870.
HOROWITZ, Vladimir
    Variations on themes from Bizet's Carmen. cf RCA VH 020.
HORSLEY
    There is a green hill far away. cf Decca ECS 774.
HOTTETERRE, Jean
    Bourrée. cf DG Archive 2533 172.
    Suite, D major. cf Pandora PAN 103.
HOVHANESS, Alan
1206 Fra Angelico, op. 220. Requiem and resurrection. RPO; North
        Jersey Wind Symphony Orchestra; Alan Hovhaness. Poseidon 1002.
            +–HF 9-72 p98   +SFC 8-17-75 p22
            +NR 2-71 p2   +St 3-73 p116
1207 Fra Angelico, op. 220. Symphony, no. 11, op. 186. RPO; Alan
        Hovhaness. Unicorn UNS 240. (also UNS 243)
            +Gr 4-72 p1709   ++NR 9-75 p3
            ++HFN 5-72 p923
    Hercules, op. 56, no. 4. cf Musical Heritage Society MHS 1976.
    Prayer to Saint Gregory. cf GIUSTINI DI PISTOIA: Sonata, piano,
        no. 1, G minor.
    Requiem and resurrection. cf Fra Angelico, op. 220.
    Symphony, no. 6. cf GIUSTINI DI PISTOIA: Sonata, piano, no. 1,
        G minor.
    Symphony, no. 11, op. 186. cf Fra Angelico, op. 220.
1208 Symphony, no. 24, op. 273. Martyn Hill, t; John Wilbraham, tpt;
        Sidney Sax, vln; John Alldis Choir; London, National Philhar-
        monic Orchestra; Alan Hovhaness. Poseidon 1016.
            +NR 5-75 p4
1209 Triptych. HUSA: Mosaiques. STRAIGHT: Development. Benita Valente,
        s; Bamberg Symphony Orchestra, Members, Singers; Stockholm
        Radio Orchestra, LPO; Alfredo Antonini, Karel Husa, Russell
        Stanger. CRI SD 221.
            ++RR 6-75 p79
1210 Tumburu, op. 264, no. 1. Varuna, op. 264, no. 2. WEIGL: Nature
        moods. New England suite. George Shirley, t; Stanley Drucker,
        clt; Kenneth Gordon, vln; Ilse Sass, pno; Kermit Moore, vlc;
        Macalester Trio. CRI SD 326.
            ++NR 12-74 p8   ++St 5-75 p95
    Varuna, op. 264, no. 2. cf Tumburu, op. 264, no. 1
HOWELLS, Herbert
    Elegy, op. 15. cf BUTTERWORTH: The banks of green willow.
    Magnificat. cf HMV HQS 1350.
    Magnificat and nunc dimittis, G major. cf Polydor 2460 250.
    Merry eye, op. 20b. cf BUTTERWORTH: The banks of green willow.
    Music for a prince. cf BUTTERWORTH: The banks of green willow.
    Psalm prelude, Set 2. cf ALCOCK: Introduction and passacaglia.
1211 Quartet, piano, op. 21, A minor. Quartet, strings, op. 25. Quin-
        tet, clarinet and strings, op. 31. Thea King, clt; Richards
        Piano Quartet, Richards Ensemble. Lyrita SRCS 68.
            ++Gr 9-75 p480   +RR 9-75 p52
            +HFN 9-75 p100

Quartet, strings, op. 25. cf Quartet, piano, op. 21, A minor.
Quintet, clarinet and strings, op. 31. cf Quartet, piano, op. 21,
    A minor.
Rhapsody, no. 1, G flat major. cf Vista VPS 1021.
HOY, Bonnee
1212 Songs: The Freeman celebration songs and dances; The Verlaine
        songs; The winter cycle. Kathryn Bouleyn, s; Bailus Webb, bar;
        David Barg, flt; Peter Wiley, vlc; Bonnee Hoy, pno. Encore
        EN 2002.
            +NR 2-75 p10
HUBO, Jeno
    Blumenleben, op. 30: Der Zephir. cf Discopaedia MB 1008.
    Hejre Kati, violin, op. 32. cf Discopaedia MB 1005.
    Hejre Kati, violin, op. 32. cf Discopaedia MB 1007.
HUBER
    In te Domine speravi. cf Pelca PSRK 41013/6.
    Invention über den Choral "In dich hab ich gehoffet, Herr". cf
        Pelca PSRK 41013/6.
HUFFER
    Black Jack. cf Michigan University SM 0002.
HUFFINE
    Them basses. cf Columbia 33513.
HUGHES, Herbert
    The stuttering lover. cf Philips 6599 227.
HUME, Tobias
1213 Captain Hume's galliard. MARAIS: Les folies d'Espagne: Couplets.
        Suite, G minor. Peggie Sampson, vla da gamba; Susanne Shapiro,
        hpd. Orion ORS 74162.
            +NR 3-75 p6
    Musick and mirth. cf L'Oiseau-Lyre 12BB 203/6.
HUMFREY, Pelham
    A hymne to God the Father. cf Pye GH 589.
HUMMEL, Johann
    Andante, A flat major. cf Pelca PSR 40598.
    Concerto, mandoline and orchestra, G major. cf HOFFMANN: Concerto,
        mandoline and orchestra, D major.
1214 Concerto, piano, op. 85, A minor. Rondo brillant, op. 56, A major.
        Ivan Palovic, Rudolf Macudzinski, pno; Slovak Philharmonic Orch-
        estra; Ladislav Slovak. Supraphon 110 1008.
            /HFN 2-72 p308   +SFC 1-5-75 p23
1215 Concerto, trumpet, E flat major. JOLIVET: Concerto, trumpet,
        strings and piano. TELEMANN: Concerto, trumpet and strings,
        D major. TOMASI: Concerto, trumpet. Pierre Thibaud, tpt; ECO;
        Marius Constant. DG 2530 289.
            +-Gr 12-73 p1217   ++RR 11-73 p38
            ++NR 6-75 p6
1216 Concerto, trumpet, E flat major (ed. Oubradous). MOZART, L.:
        Concerto, trumpet, D major (ed. Seiffert). TELEMANN: Concerto,
        trumpet, D major (ed. Grebe). VIVALDI: Concerto, trumpet, A
        flat major (ed. Thilde). Maurice André, tpt; BPhO; Herbert von
        Karajan. HMV ASD 3044. (also Angel S 37063)
            +Gr 6-75 p55   ++RR 5-75 p30
            +HFN 6-75 p107   ++STL 6-8-75 p36
            +NR 2-75 p4
    Concerto, trumpet, E flat major. cf RCA CRL 2-7002.
1217 Etudes, op. 125 (24). Mary Louise Boehm, pno. Turnabout TVS

34562.
        ++HF 11-75 p104   ++SFC 1-5-75 p23
        ++NR 4-75 p11
    Prelude and fugue, C minor. cf Hungaroton SLPX 11601/2.
    Quartet, clarinet, E flat major. cf CRUSELL: Quartet, clarinet,
        no. 2, op. 4, C minor.
    Rondo, op. 11, E flat major. cf RCA ARM 4-0942/7.
    Rondo brillant, op. 56, A major. cf Concerto, piano, op. 85, A
        minor.
    Sonata, flute, D major. cf Vanguard VSD 71207.
1218 Sonata, mandolin and piano, C major. BEETHOVEN: Adagio, op. 150,
        E flat major. Sonata, C major. Sonatina, C major. Sonatina,
        C minor. Theme and variations, D major. Hugo d'Alton, mand;
        John Beckett, fortepiano. Saga 5350. Tape (c) CA 5350.
        +RR 4-74 p67   ++ 9-75 p77
    Zur Logenfeier. cf DG Archive 2533 149.
HUMPERDINCK, Engelbert
1219 Hansel and Gretel, excerpts. Anna Moffo, Helen Donath, Arleen
        Auger, Lucia Popp, s; Charlotte Berthold, Christa Ludwig, ms;
        Dietrich Fischer-Dieskau, bar; Tölzer Boys' Choir; Bavarian
        Radio Orchestra; Kurt Eichorn. RCA Tape (c) ARK 1-0792 (ct)
        ARS 1-0792.
        +HF 8-75 p116 tape
1220 Hänsel und Gretel, excerpts. MOZART, L.: Toy symphony. OCHS:
        's kommt ein Vogerl geflogen. Berlin State Opera Orchestra,
        Berlin Chamber Orchestra; Hans van Benda, Richard Muller-
        Lampertz, Wolfgang Martin. Telefunken AG 6-41334.
        +RR 11-75 p46
    Hänsel und Gretel: Overture. cf HMV (Q) ASD 3131.
    Konigskinder: O du liebheilige Einfalt; Wohen bist du gegangen.
        cf Rubini RS 300.
HURFORD, Peter
    Laudate Dominum. cf Argo ZRG 807.
HUSA, Karel
1221 Apotheosis of the earth. Music for Prague, 1968. Michigan Univ-
        ersity Symphony Band; Karel Husa. Golden Crest CRS 4134.
        +MJ 10-75 p54
    Mosaiques. cf HOVHANESS: Triptych.
    Music for Prague, 1968. cf Apotheosis of the earth.
    Nocturne. cf Symphony, no. 1.
    Serenade, woodwind quintet with strings, harp and xylophone. cf
        Symphony, no. 1.
1222 Symphony, no. 1. Serenade, woodwind quintet with strings, harp
        and xylophone. Nocturne. PSO; Soloistes de Paris; Karel Husa.
        CRI SD 261.
        ++ARG 5-71 p562   -SFC 3-26-72 p43
        +NR 5-71 p2   +St 10-71 p91
        ++RR 2-75 p35
HUZELLA, Elek
    Dances (3). cf Hungaroton SLPX 11629.
IBERT, Jacques
    Divertissement. cf HONEGGER: Pacific 231.
    Entr'acte. cf Musical Heritage Society MHS 1345.
    Little white donkey. cf Monitor MCS 2143.
1223 Paris 32. MILHAUD: Symphonies, small orchestra, nos. 1-4. POULENC:
        Marches et un intermède (2). SATIE: Petites pièces montées (3).

Leningrad Chamber Orchestra; Gennady Rozhdestvensky. Westminster
WGS 8310.
+HF 9-75 p98

IMBRIE, Andrew
1224 Symphony, no. 3. SCHUMAN: Credendum (Article of faith)*. LSO, PO;
Harold Farberman, Eugene Ormandy. CRI SD 308. (*Reissue from
Columbia ML 5185)
+HF 11-73 p109   ++RR 4-75 p27
+HFN 5-75 p128   ++St 8-74 p114
+NR 11-73 p2   ++Te 3-75 p48

THE INCOMPARABLE VICTORIA DE LOS ANGELES, songs and operatic
arias. cf HMV SLS 5012.

d'INDY, Vincent
Sonata, piano, E minor. cf DUKAS: Variations on a theme by
Rameau.
1225 Symphony on a French mountain air, op. 25. POULENC: Aubade, piano
and 18 instruments. Joela Jones, pno; Westphalian Symphony
Orchestra, LSO; Paul Freeman. Ember ECL 9036. (also Orion
74139)
+-Gr 9-75 p452   +-RR 8-75 p41
+-HFN 7-75 p79

INGALLS, Jeremiah
Northfield. cf Nonesuch H 71276.

IRELAND, John
Decorations. cf Piano works (Lyrita SRCS 87).
Ex ore innocentium. cf Argo ZRG 785.
Greater love hath no man. cf HMV HQS 1350.
London pieces. cf Piano works (Lyrita SRCS 87).
Merry Andrew. cf Piano works (Lyrita SRCS 87).
1226 Piano works: Decorations. London pieces. Merry Andrew. Preludes
(4). Prelude, E flat major. Rhapsody. The towing path. Eric
Parkin, pno. Lyrita SRCS 87.
+Gr 12-75 p1082   ++RR 11-75 p78
+HFN 11-75 p154

Preludes (4). cf Piano works (Lyrita SRCS 87).
Prelude, E flat major. cf Piano works (Lyrita SRCS 87).
Rhapsody. cf Piano works (Lyrita SRCS 87).
Sea fever. cf Decca 396.
1227 Songs: Houseman settings: We'll to the woods no more; In boyhood;
Spring will not wait. Marigold: Youth's spring-tribute; Penum-
bra; Spleen. Poems to Thomas Hardy: Beckon to me to come; In
my sage moments; It was what you bore with you, woman; The
tragedy of the moment; Dear, think not that they will forget
you. Songs of a wayfarer: Memory; When daffodils begin to
peer; English May; I was not sorrowful; When lights go rolling
round the sky. Hope the hornblower. Sea fever. Songs: Love
and friendship; Friendship in misfortune; The one hope. Benjamin Luxon, bar; Alan Rowlands, pno. Lyrita SRCS 65.
+Gr 5-75 p2005   +RR 5-75 p61
++HFN 5-75 p128   +-STL 6-8-76 p36

1228 Songs: The advent; All in a garden green; An aside; The bells of
San Marie; Blow out, you bugles; During music; Great things;
Hymn for a child; I have twelve oxen; If there were dreams to
sell; If we must part; The journey; The merry month of May; My
fair; A report song; The Salley gardens; Santa Chiara; A scape-
goat; The soldier's return; Spring sorrow; The sweet season;

A thanksgiving; Tryst; Tutto è sciolot; Vagaband; When I am
dead, my dearest. Benjamin Luxon, bar; Alan Rowlands, pno.
Lyrita SRCS 66.
    +-Gr 7-75 p227   ++RR 7-75 p57
    ++HFN 7-75 p83
The towing path. cf Piano works (Lyrita SRCS 87).
IRISH SONGS. cf Philips 6599 227.
ISAAC, A.
    Ne più bella di queste; Palle, palle; Quis dabit pacem. cf
    L'Oiseau-Lyre 12BB 203/6.
ISAAC, Heinrich
    La la hö hö. cf L'Oiseau-Lyre 12BB 12BB 203/6.
ISOLFFSON, Pall
1229 Passacaglia. LEIFS: Icelandic overture, op. 9. COWELL: Symphony,
    no. 16. Iceland Symphony Orchestra; William Strickland. CRI
    179.
        *St 6-75 p50
ISTVAN Miloslav
1230 Isle of toys. KUCERA: The kenetic ballet: The labyrinth; The
    spiral. VOSTRAK: Scales of light. Electronic tape. Supraphon
    111 1423.
        /HFN 11-75 p155   +-RR 7-75 p60
Refrains fro string trio. cf Panton 110 253.
ITALIAN RECORDER SONATAS. cf Telefunken SAWT 9589.
ITALIAN RENAISSANCE LUTE MUSIC. cf DG Archive 2533 173.
ITALIAN SONG RECITAL. cf Peal SHE 511.
IVES, Charles
    Allegretto (Inventions). cf Piano works (Desto DST 6458/61).
    Bad resolutions and good. cf Piano works (Desto DST 6458/61).
    Baseball take-off. cf Piano works (Desto DST 6458/61).
1231 The celestial country. Hazel Holt, s; Alfreda Hodgson, alto; John
    Elwes, t; John Noble, bar; Schütz Choir; LSO; Harold Farberman.
    CRI SD 314.
        +-Gr 11-74 p942   -NR 3-74 p7
        +-HF 2-74 p92   *NYT 10-20-74 pD26
        +-LJ 1-75 p51   +RR 3-75 p62
        +MJ 1-74 p10   +SR 5-18-74 p6
        ++MJ 3-74 p47   +-St 3-74 p112
        +MQ 7-74 p500
Celestial railroad. cf Piano works (Desto DST 6458/61).
1232 Charle Rutlage. Chromatimelodtune. Country band march. Evening.
    Fugue in 4 keys on "The shining shore". Gyp the blood or
    Hearst. Holiday quickstep. March II. March III. Mists. An
    old song deranged. Overture and march: 1776. Remembrance.
    The swimmers. Which is worst. Yale Theater Orchestra; James
    Sinclair. Columbia M 32969. (Q) Tape (c) MAQ 32969.
        +HF 4-75 p86   +NR 12-74 p4
        +HF 9-75 p116 Quad tape   +St 5-75 p96
Chromatimelodtune. cf Charlie Rutlage. cf Columbia M 32969.
Country band march. cf Charlie Rutlage.
Evening. cf Charlie Rutlage.
Fugue in 4 keys on "The shining shore". cf Charlie Rutlage.
Gyp the blood or Hearst. cf Charlie Rutlage.
Holiday quickstep. cf Charlie Rutlage.
1233 Largo. KHACHATURIAN: Trio, clarinet, violin and piano. KRENEK:
    Trio, clarinet, violin and piano. STRAVINSKY: L'histoire du

soldat. Roy d'Antonio, clt; Myron Sandler, vln; Delores Stevens,
 pno. Laurel LR 103.
  ++NR 6-75 p8
March II. cf Charlie Rutlage.
March II. cf New England Conservatory NEC 111.
March III. cf Charlie Rutlage.
March in G and D, "Here's to good old Yale". cf Piano works
 (Desto DST 6458/61).
March intercollegiate. cf Columbia 33513.
March intercollegiate. cf New England Conservatory NEC 111.
Mists. cf Charlie Rutlage.
An old song deranged. cf Charlie Rutlage.
Omega Lambda Chi. cf Columbia 33513.
Overture and march: 1776. cf Charlie Rutlage.
1234 Piano works: Allegretto (Invention). Bad resolutions and good.
 Baseball take-off. Celestial railroad. March in G and D,
 "Here's to good old Yale". Processional, anthem. Rough and
 ready. Scene episode. Seen and unseen: Processional. Sonatas,
 piano, nos. 1, 2. Song with (good) words. Storm and distress.
 Studies, nos. 2, 5-8, 9, 15, 18, 20-22. Three page sonata.
 Varied air and variations (six protests). Waltz-rondo. Alan
 Mandel, piano. Desto DST 6458/61 (4).
  +-Gr 4-75 p1836   -HFN 5-75 p129
Processional, anthem. cf Piano works (Desto DST 6458/61).
1235 Quartets, strings, nos. 1 and 2. Concord Quartet. Nonesuch H
 71306.
  ++HF 8-75 p88   ++NR 6-75 p10
  ++HFN 7-75 p83   ++SFC 7-20-75 p27
  +MT 11-75 p977   ++St 10-75 p107
Remembrance. cf Charlie Rutlage.
Rough and ready. cf Piano works (Desto DST 6458/61).
Scene episode. cf Piano works (Desto DST 6458/61).
Seen and unseen: Processional. cf Piano works (Desto DST 6458/61).
Serenity. cf New England Conservatory NEC 111.
Sonatas, piano, nos. 1, 2. cf Piano works (Desto DST 6458/61).
Sonata, violin and piano, no. 4. cf BINKERD: Sonata, violin and
 piano.
Song without (good) words. cf Piano works (Desto DST 6458/61).
1236 Songs: The greatest man; At the river; The circus band. SCHONBERG:
 Pierrot Lunaire, op. 21. Anthony Hymas, pno; Cleo Laine, voca-
 list; Nash Ensemble; Elgar Howarth. RCA LRL 1-5058.
  +-Gr 9-74 p535   -RR 9-74 p35
  -HF 11-75 p116   ++SFC 8-3-75 p30
  +NR 9-75 p11   -SR 9-6-75 p40
  +-ON 9-75 p60
Songs: An old flame; Circus band; A Civil War memory; In the
 alley; Karen; Romanzo di Central Park; A son of a gambolier.
 cf Vox SVBX 5304.
Studies, nos. 2, 5-8, 9, 15, 18, 20-22. cf Piano works (Desto
 DST 6548/61).
The swimmers. cf Charlie Rutlage.
1237 Symphony, no. 1, D minor. ELGAR: Enigma variations, op. 36. LAPO:
 Zubin Mehta. London CS 6816. (also Decca SXL 6592. Tape (c)
 KSXC 6592)
  +Gr 9-73 p483   ++RR 9-73 p70
  ++Gr 10-74 p766 tape +RR 9-75 p71 tape

```
         +-HF 2-74 p92                +SFC 12-2-73 p32
         +NR 2-74 p3                 ++St 4-74 p112
```
1238 Symphony, no. 2. PO; Eugene Ormandy. RCA ARL 1-0663. (Q)
     ARD 1-0663.
```
         +-HF 10-74 p81               +NR 11-74 p4
         -HF 2-75 p110 Quad          ++SFC 10-6-74 p26
         -MJ 2-75 p41                +-St 11-74 p134
```
1239 Symphony, no. 4. LPO; José Serebrier. RCA ARL 1-0589. (Q)
     ARD 1-0589.
```
         +Gr 10-74 p688              ++NYT 10-20-74 pD26
         +HF 10-74 p81               +-RR 10-74 p47
         +HF 2-75 p110 Quad          ++SFC 10-6-74 p26
         +MJ 12-74 p45               +St 11-74 p134
         +NR 11-74 p2
```

Three page sonata. cf Piano works (Desto DST 6458/61).

Trio, piano. cf COPLAND: Quartet, piano.

1240 Trio, violin, violoncello and piano. KORNGOLD: Trio, piano, op.
     1. Pacific Art Trio. Delos DEL 25402.
```
         +NR 9-75 p8                 +-SFC 7-20-75 p27
```

The unanswered question. cf Columbia M2X 33014.

Variations on America. cf Century/Advent OU 97 734.

Varied air and variations (six protests). cf Piano works (Desto
     DST 6458/61).

Waltz-rondo. cf Piano works (Desto DST 6458/61).

Which is worst. cf Charlie Rutlage.

IVEY, Jean Eichelberger

1241 Aldebaran, viola and tape. Cortege for Charles Kent. Songs of
     night, soprano, 5 instruments and tape: The astronomer; I
     dreamed of Sappho; Heraclitus. Terminus, mezzo and tape.
     Catherine Rowe, s; Elaine Bonazzi, ms; Jacob Glick, vla;
     Peabody Conservatory Contemporary Music Ensemble; Electronic
     piece. Ethnic Folkways Library FTS 33439.
```
         ++MJ 10-74 p43              ++NR 6-74 p13
         +MQ 10-75 p638
```

Cortege for Charles Kent. cf Aldebaran, viola and tape.

Hera, hung from the sky. cf ERICKSON: End of the mime.

Songs of night, soprano, 5 instruments and tape: The astronomer;
     I dreamed of Sappho; Heraclitus. cf Aldebaran, viola and tape.

Terminus, mezzo and tape. cf Aldebaran, viola and tape.

JACOB, Gordon

1242 Divertimento, harmonica and string quartet. MOODY: Quintet, harm-
     onica. Tommy Reilly, harmonica; Hindar Quartet. Argo ZDA 206.
```
         +Gr 6-75 p61               +RR 8-75 p52
```

Variations on "Annie Laurie". cf HMV SLA 870.

William Byrd suite. cf Philips 6747 177.

JACOTIN, Jacques

Voyant souffrir. cf HMV SLS 5022.

JACQUET DE LA GUERRE, Elizabeth

Suite, D minor. cf Avant AV 1012.

JANACEK, Leos

1243 By overgrown tracks, Bks I and II. Radoslav Kvapil, pno. Panton
     110 212.
```
         /Gr 7-75 p219              +RR 6-75 p70
         /HFN 7-75 p84
```

By overgrown tracks, Bk I. cf Works, selections (Supraphon 111
     1481/2).

Capriccio, piano, left hand and wind ensemble. cf Works, selections
   (Supraphon 111 1481/2).
Concertino, piano and chamber ensemble. cf Works, selections
   (Supraphon 111 1481/2).
1244 Fairy tale, violoncello and piano. Sonata, violin. Youth, wind
   sextet. Petr Messiereur, vln; Jarmila Kozderková, Radoslav
   Kvapil, pno; Stanislav Apolin, vlc; Josef Horák, bass clt;
   Förster Wind Quintet. Panton 110 214.
         +-Gr 7-75 p206   ++RR 7-75 p47
         +HFN 7-75 p83
1245 Glagolitic mass. Teresa Kubiak, s; Anne Collins, ms; Robert Tear,
   t; Wolfgang Schöne, bs; John Birch, org; Brighton Festival
   Chours; RPO; Rudolf Kempe. Decca SXL 6600. (also London OS
   26338)
         +-Gr 2-74 p1589   ++NYT 8-18-74 pD20
         +-HF 11-74 p106   +ON 12-21-74 p28
         +HFN 2-74 p339   +RR 2-74 p18
         ++MJ 1-75 p48   ++St 1-75 p112
         +NR 11-74 p7
1246 Idyll, string orchestra. MARTINU: Concerto, 2 orchestras, piano
   and timpani. PCO: Hans-Hubert Schönzeler. RCA LHL 1-5086.
         +-Gr 3-75 p1653   +RR 3-75 p27
Idyll, string orchestra. cf DVORAK: Serenade, op. 22, E major.
In the mists. cf Works, selections (Supraphon 111 1481/2).
1247 Katya Kabanova. Drahomirá Tikalova, Eva Hlobilová, s; Ivana
   Mixová, Marcela Leomoriová, ms; Ludmila Komancová, con; Beno
   Blachut, Viktor Kočí, Bohumir Vích, t; Zdenek Kroupa, Rudolf
   Jedlicka, bs; Prague National Theatre Orchestra and Chorus;
   Jaroslav Krombholc. Supraphon 50781/82 (2). (Reissue from SUAST
   50005/6)
         ++Gr 4-74 p1909   ++NR 4-75 p8
         +-HFN 2-74 p337   +RR 3-74 p28
Lachian dances: The saws. cf Supraphon 110 1429.
1248 Sbory z Mladi: Ploughing; True love; I wonder at my lover; The
   drowned wreath; No escape from fate; The wild duck; Autumn
   song; Our song; In the pine tree; On the ferry; Love's uncer-
   tainty; Alone without comfort; Rest in peace; Choral elegy.
   Prague Philharmonic Orchestra; Josef Veselka. Panton 110 400.
         +-HFN 7-75 p83   +-RR 7-75 p58
Sinfonietta. cf HINDEMITH: Symphonic metamorphoses on a theme
   by Carl Maria von Weber.
Sonata, piano. cf Works, selections (Supraphon 111 1481/2).
Sonata, violin. cf Fairy tale, violoncello and piano.
Sonata, violin and piano, op. 21. cf PROKOFIEV: Sonata, violin
   and piano, no. 1, op. 80, F minor.
1249 Suite, strings. STRAUSS, R.: Capriccio, op. 85: Sextet (arr. for
   string orchestra). SUK: Serenade, strings, op. 6, E flat major.
   Los Angeles Chamber Orchestra; Neville Marriner. Argo ZRG 792.
         +Gr 3-75 p1653   +-RR 3-75 p36
         +HF 9-75 p92   ++St 11-75 p139
         +NR 7-75 p5
1250 Works, selections: By overgrown tracks, Bk I. Capriccio, piano,
   left hand and wind ensemble. Concertino, piano and chamber
   ensemble. In the mists. Sonata, piano (October 1, 1905).
   Josef Pálenícek, pno; Czech Philharmonic Wind Ensemble. Supraphon 111 1481/2 (2).

```
    +-Gr 11-75 p860              ++NR 11-75 p11
    +-HFN 10-75 p142            +RR 9-75 p58
```
    Youth, wind sextet. cf Fairy tale, violoncello and piano.
JANNEQUIN, Clement
1251 Chansons: Je ne congnois femme en ceste contrée; Ce petit dieu qui
        vole; Dur acier et diamant; Le chant de l'alouette; Or veit
        mon cueur en grand tristesse; Ma peine n'est pas grande; Sus
        approchez ces lebvres vermeillettes; Il ferait bon planter le
        may; Je liz au cueur de ma mye; Si le coqu; Ce may nous dit la
        verdure; Aussi tost que je voy ma mye; Le Guerre (La bataille
        de Marignan); Quand contremont; O crualté logée en grand
        beauté; Est-il possible o ma maistresse; Le chant des oyseaulx;
        A ce joly moys de may; Cent baysers; Je veulx que ma mye soit
        telle. ORTF Polyphonic Ensemble; Charles Ravier. Musical Heri-
        tage Society MHS 1872.
            +St 7-75 p101
1252 Songs: A ce joly moys de may; Aussi que je voy ma mye; Ce may nous
        dit la verdure; Ce petit dieu que vole; Cent baysers; Le chant
        des oyseaulx; Le chant de l'alouette; Dur acier et diamant; Est-
        il possible o ma maitresse; Il feroit con planter le may; Le
        guerre, la bataille de Marignan; Je liz au cueur de ma mye; Je
        ne congnois femme en ceste contrée; Je veulx que ma mye soit
        telle; Ma peine n'est pas grande; O crualté logée en grand
        beauté; O veir mon cueur; Quand contremont; Si le coqu; Sus
        approchez ces lebvres. ORTF Polyphonic Ensemble; Charles
        Ravier. Telefunken AW 6-41877. (Reissue from Valois MB 928)
            ++Gr 7-75 p228   +NR 9-75 p10
            ++HF 11-75 p105   /RR 6-75 p80
            ++HFN 7-75 p84
    Le chant des oyseaux. cf Caprice RIKS LP 46.
    Les cris de Paris. cf L'Oiseau-Lyre 12BB 203/6.
    Il était une fillette. cf Vanguard SRV 316.
    Ma peine n'est pas grande. cf Vanguard SRV 316.
    Or vien ca vien. cf Hungaroton SLPX 11549.
JANOWSKI
    Avinu Malkeynu. cf RCA ARL 1-0561.
JEANJEAN, Paul
    Carnival of Venice. cf Crystal S 331.
    Theme and variations. cf Crystal S 331.
JENEY, Zoltán
1253 Alef: Homage à Schonberg. Round. Soliloquim, flute, no. 1.
        SARY: Catocoustics, 2 pianos. Immaginario, no. 1. Incanto.
        István Matuz, flt; Margit Bognár, hp; Zsuzsa Pertis, hpd; Nora
        Schmidt, István Nagy, Zoltan Benkö, pno; Gesualdo Vocal Quintet,
        Hungarian Radio and Television Orchestra, Györ Philharmonic
        Orchestra; Peter Eötvös, János Sándor. Hungaroton SLPX 11589.
            ++NR 10-75 p13   ++RR 9-75 p59
    Round. cf Alef: Homage à Schonberg.
    Soliloquium, flute, no. 1. cf Alef: Homage à Schonberg.
JENNI, Donald
    Cucumber music. cf CRI SD 324.
    Musique printanière. cf BLANK: Rotation.
    JENNIE TOUREL AT ALICE TULLY HALL. cf Desto DC 7118/9.
JEREMIAS, Otakar
    March. cf Panton 110 361.

JEUNE, Claude le
      Fière cruelle. cf L'Oiseau-Lyre 12BB 203/6.
      Revecy venir du printemps. cf Crystal S 202.
JIMENEZ, José
      Batalla de sexto tono. cf Musical Heritage Society MHS 1790.
JIRASEK, Ivo
      Serenade, flute, bass clarinet and tape. cf Panton 110 253.
      Stabat mater. cf FISER: Requiem.
      JOAN SUTHERLAND, coloratura spectacular. cf London OS 26306.
JOHNSON, John
      The delight pavan and galliard. cf L'Oiseau-Lyre SOL 336.
      Dump, no. 3. cf L'Oiseau-Lyre SOL 336.
      Laveche's galliard. cf Argo ZRG 765.
      Rogero. cf L'Oiseau-Lyre SOL 336.
JOHNSON, Robert
      Alman. DG Archive 2533 157.
      Care-charming sleep. cf Harmonia Mundi 204.
JOLAS, BEtsy
      Quatuor III. cf HILLER: Quartet, strings, no. 6.
JOLIVET, André
      Concerto, trumpet, strings and piano. cf HUMMEL: Concerto, trumpet,
            E flat major.
JONES, Jeff
1254 Ambiance. ROCHBERG: Blake songs. WOLPE: Quartet, trumpet, tenor
            saxophone, percussion and piano. Phyllis Bryn-Julson, s; Jan
            DeGaetani, ms; Contemporary Chamber Ensemble; Arthur Weisberg.
            Nonesuch H 71302.
                  ++HF 4-75 p98   +RR 12-75 p74
                  +NR 1-75 p6   +St 6-75 p110
JONGEN, Joseph
      Choral. cf Stentorian SC 1685.
JOPLIN, Scott
      Combination march (arr. Schuller). cf Columbia 33513.
      Maple leaf rag. cf Columbia 31726.
1255 The prodigal son, ballet (arr. Hossack). Michael Bassett, pno;
            London Festival Ballet Orchestra; Grant Hossack. CBS 73363.
                  +Gr 10-74 p688   /RR 10-74 p48
                  +Gr 5-75 p2037 tape
JOSQUIN DES PRES (also Des Pres, Depres)
1256 Dominus regnavit, psalm, no. 92. In exitu Israel. Missa Gaudeamus.
            Nymphes des Bois. Capella Cordina; Alejandro Planchart. Lyri-
            chord LLST 7265.
                  +NR 2-75 p8
      In exitu Israel. cf Dominus regnavit, psalm, no. 92.
      Mille regretz. cf L'Oiseau-Lyre 12BB 203/6.
      Missa Gaudeamus. cf Dominus regnavit, psalm, no. 92.
      Nymphes des Bois. cf Dominus regnavit, psalm, no. 92.
JOUBERT, John
1257 Dance suite, op. 21. Sonata, piano, no. 1, op. 24. Sonata, piano,
            no. 2, op. 71. John McCabe, pno. Pearl SHE 520.
                  +Gr 12-75 p1082   +RR 12-75 p82
      Sonata, piano, no. 1, op. 24. cf Dance suite, op. 21.
      Sonata, piano, no. 2, op. 71. cf Dance suite, op. 21.
      JUDITH BLEGEN AND FREDERICA VON STADE, recital. cf Columbia 33307.
JUON, Paul
      Berceuse. cf RCA ARM 4-0942/7.

Pieces, violin and piano, op. 28, no. 3. cf Discopaedia MB 1010.
KABALEVSKY, Dmitri
1258 The comedians, op. 26. LINDE: Concerto, violin, op. 18. Karl-Ove
Mannberg, vln; Gävleborg Symphony Orchestra; Rainer Miedel. HMV
4E 055 34649.
+Gr 12-75 p1061
1259 Concerto, piano, no. 3, op. 50, D major. Concerto, violin, op. 48,
C major. Overture pathétique, op. 64, B minor. Spring, op. 65.
Vladimir Fetsman, pno; Victor Pikaizen, vln; MPO; Dimitri Kaba-
levsky, Fuat Mansurov. HMV Melodiya ASD 3078.
+-Gr 8-75 p322   +RR 8-75 p34
+-HFN 8-75 p76
1260 Concerto, piano, no. 3, op. 50, D major. RUBINSTEIN: Concerto,
piano, no. 3, op. 45, G major. Robert Preston, pno; Westphalian
Symphony Orchestra; Paul Freeman. Orion ORS 74149. (also Ember
ECL 9037)
+-Gr 7-75 p182   +NR 9-74 p5
+-HF 11-74 p118   +RR 6-75 p46
+-HFN 8-75 p76   +St 10-74 p136
    +MJ 12-74 p46
Concerto, violin, op. 48, C major. cf Concerto, piano, no. 3,
op. 50, D major.
Good night. cf Works, selections (Westminster 83038).
Overture pathétique, op. 64, B minor. cf Concerto, piano, no. 3,
op. 50, D major.
Overture pathétique, op. 64, B minor. cf Works, selections (West-
minster 83038).
School years. cf Works, selections (Westminster 83038).
1261 Sonata, piano, no. 3, op. 46. MUCZYNSKI: Suite, op. 13. PROKOFIEV:
Pieces, op. 96 (3). RACHMANINOFF: Variations on a theme by
Chopin, op. 22. Paulina Drake, pno. Orion ORS 75168.
+HF 11-75 p128   +-NR 6-75 p13
Songs of morning, spring and peace, op. 57. cf Works, selections
(Westminster 83038).
Spring, op. 65. cf Concerto, piano, no. 3, op. 50, D major.
Spring, op. 65. cf Works, selections (Westminster 83038).
The unit of young pioneers. cf Works, selections (Westminster
83038).
1262 Works, selections: Good night. Overture pathétique, op. 64, B
minor. School years. Songs of morning, spring and peace, op.
57. Spring, op. 65. The unit of young pioneers. Central House
for Railwaymen's Children Chorus; Moscow State Orchestra;
Dmitri Kabalevsky. Westminster WGC 83038.
+St 10-75 p109
KAGEL, Mauricio
Unguis incarnatus est. cf DG 2530 562.
KALASHNIKOV, Nikolai
Concerto, 12 voice choir: Cherubic hymn. cf HMV Melodiya ASD 3102.
KALLIWODA, Johann Wenzel (also Kalivoda, Jan Václav)
1263 Symphony, no. 1, op. 7, F minor. TOMASEK: Concerto, piano, op. 18,
C major. Petr Toperczer, pno; PSO; Jindřich Rohán. Candide
CE 31073. (also Vox STGBY 677)
+-Gr 6-75 p39   +-NR 8-74 p5
+HF 11-74 p108   +-RR 5-75 p33
+-HFN 5-75 p129   ++SFC 9-22-74 p22
+-MT 11-75 p978   ++St 9-74 p124

KALLOS
    The mountain brigand. cf BARTOK: Mikrokosmos: 4 songs.
KALMAN, Emmerich
1264 Gräfin Mariza, abridged. Rita Zorn, Vera Westhoff, Ursula Richter,
        s; Martin Ritzman, Gerd Pallesche, t; Richard Westemayer, bar;
        Leipzig Radio Symphony Orchestra and Chorus; Herbert Kegel.
        Saga 5397. (Reissue from Urania)
            +Gr 5-75 p2010   +—HFN 8-75 p77
    Grandioso. cf Olympic 8136.
KALMAR, László
    Pieces, piano (3). cf Hungaroton SLPX 11692.
KAMINSKI
    Recitative and dance. cf Da Camera Magna SM 93399.
KANITZ, Ernest
1265 Sinfonietta da camera. Sonata, violin and piano, no. 2. Visions
        at twilight. Israel Baker, vln; Harvey Pittel, sax; Delores
        Stevens, pno; Ensemble; Jan Popper. Orion ORS 75190.
            ++NR 9-75 p8
    Sonata, violin and piano, no. 2. cf Sinfonietta da camera.
    Visions at twilight. cf Sinfonietta da camera.
KAPLAN, Nathan Ivan
1266 Concert etudes (18). Lawrence Sobol, clt. Grenadilla S 1001.
            +NR 12-75 p15
KARG-ELERT, Sigfried
    Kaleidoscope, op. 144. cf Organ works (Polydor 2460 231).
    Nun danket alle Gott, op. 65/69. cf Decca SDD 463.
1267 Organ works: Kaleidoscope, op. 144. Pastels from Lake Constance,
        op. 96, no. 4: The reed-grown waters. Sonatina, no. 1, op. 74,
        A minor. Triptych, organ, op. 141, no. 1: Legend. Michael
        Austin, org. Polydor 2460 231.
            +Gr 1-75 p1371   ++HFN 5-75 p132
    Pastels from Lake Constance, op. 96, no. 4: The reed-grown waters.
        cf Organ works (Polydor 2460 231).
    Sonatina, no. 1, op. 74, A minor. cf Organ works (Polydor 2460
        231).
    Triptych, organ, op. 141, no. 1: Legend. cf Organ works (Polydor
        2460 231).
    Vom Himmel hoch, chorale prelude. cf Wealden WS 145.
KARKOFF, Maurice
1268 Symphony, no. 4, op. 69. LARSSON: Orchestral variations, op. 51.
        Swedish Radio Symphony Orchestra; Sixten Ehrling, Stig Wester-
        berg. Swedish Discofil SLT 33164.
            +St 6-75 p51
KARLINS, M. William
    Variations on 'Obiter dictum'. cf BLANK: Rotation.
KATZ
    The Briansk forest. cf HMV ASD 3116.
KAY, Ulysses
1269 Dances for string orchestra (6). STILL: From the black belt.
        Darker America. Music for Westchester Symphony Orchestra,
        Westphalian Symphony Orchestra; Siegfried Landau, Paul Freeman.
        Turnabout TVS 34546.
            +St 10-75 p118
1270 Markings. WALKER: Concerto, trombone and orchestra. Denis Wick,
        trom; LSO; Paul Freeman. Columbia M 32783.
            +HF 6-74 p71   +—NR 7-74 p14
            ++MQ 10-75 p645   *St 7-74 p105

KEDROV, N., Sr.
Otche Nash (Our Father).   cf Monitor MFS 757.
KEETMAN, Gunild
1271 The Christmas story (Weihnachtgeschichte).   Tobi Reiser Shepherd
     Boys Choir, Tölzer Boys' Choir; Instrumental Ensemble; Carl
     Orff.   BASF BAC 3062.
          ++RR 1-75 p57
KELLER, Homer
Serenade, clarinet and strings.   cf Era 1001.
KENNAN, Kent
Night soliloquy, flute and orchestra.   cf Era 1001.
KERLL, Johann
Canzona, G minor.   cf Philips 6775 006.
Canzona, G minor, D minor.   cf Telefunken TK 11567/1-2.
Capriccio, cembalo.   cf Pelca PSR 40589.
Feldschlacht.   cf Musical Heritage Society MHS 1790.
Passacaglia, D minor.   cf Pelca PSRK 41013/6.
Toccata con durezza e ligature.   cf Philips 6775 006.
KEURIS, Tristan
1272 Concerto, saxophone.   LOEVENDIE: Scaramuccia.   PORCELIJN: 10-5-6-5
     (a).   Continuations.   SCHAT: Thema.   Hans de Vries, ob; Nether-
     lands Wind Ensemble, Radio Philharmonic Orchestra, Radio Chamber
     Orchestra; Ed Bogaard, sax; Piet Honigh, clt; Peter Schat,
     Diego Masson, Roelof Krol, David Atherton.   Donemus 7374/4.
          +RR 4-75 p30
KHACHATURIAN, Aram
Concerto, piano, D flat major.   cf FRANCK: Symphonic variations,
     piano and orchestra.
1273 Concerto, violin.   PROKOFIEV: Concerto, violin, no. 1, op. 19, D
     major.   David Oistrakh, vln; National Philharmonic Orchestra;
     Aram Khachaturian, Sergei Prokofiev.   Everest 3367.
          +LJ 12-75 p38   +NR 4-75 p5
Gayaneh: Sabre dance.   cf Monitor MCS 2143.
1274 Gayaneh suite.   Masquerade suite.   Brno State Philharmonic Orches-
     tra; Jiři Belohlávek.   Supraphon 110 1226.
          -NR 3-75 p4
Masquerade suite.   cf Gayaneh suite.
1275 Rhapsody concerto, violoncello and orchestra.   Symphony, no. 3,
     C major.   Mstislav Rostropovich, vlc; Harry Grodberg, org;
     Bolshoi Theatre Trumpeters' Ensemble; USSR, MPO; Kiril Kondrash-
     in, Yevgeny Svetlanov.   HMV Melodiya ASD 3108.
          +-Gr 9-75 p458   +-MT 12-75 p1071
          +-HFN 8-75 p77   +RR 8-75 p34
Rhapsody concerto, violoncello and orchestra.   cf GALYNIN: Concerto,
     piano.
1276 Spartacus.   Bolshoi Theatre Orchestra; Algis Zhuratis.   Columbia/
     Melodiya D4M 33493 (4).
          +-HF 11-75 p105   +St 9-75 p107
          -NR 8-75 p4
Symphony, no. 3, C major.   cf Rhapsody concerto, violoncello and
     orchestra.
1277 Trio, clarinet, violin and piano.   MOZART: Quintet, clarinet, K 581,
     A major.   Simeon Bellison, clt; Bela Urban, vln; Virginia Urban,
     pno; Roth Quartet.   Grenadilla GS 1003.
          ++NR 11-75 p8

Trio, clarinet, violin and piano. cf IVES: Largo.
KICKHAM
     She lived beside the Anner. cf Philips 6599 227.
KIENZL, Wilhelm
     Kahn Szene, neuer Walzer. cf Rococo 2049.
KILPINEN, Yrjö
1278 Lieder und Tunturilauluja: Kirkkorabbassa, op. 54, no. 2; Kesäyö,
     op. 23, no. 3; Laululle, op. 52, no. 3; Tunturille, op. 52,
     no. 4; Vanhakirkko, op. 54, no. 1; Bannalta, I, op. 23, no. 1.
     SCHUMANN: Songs: Zwolf Gedichte von Justinus Kerner, op. 35:
     No. 1, Lust der Sturmnacht; No. 2, Stirb, Lieb und Freud; No.
     3, Wanderlied; No. 4, Erstes Grün; No. 5, Sehnsucht nach der
     Waldegegend; No. 6, Auf das Trinkglas; No. 7, Wanderung; No. 8,
     Stille Liebe; No. 9, Frage; No. 10, Stille Tränen; No. 11, Wer
     machte dich so krank; No. 12, Alte Laute. Martti Talvela, bs;
     Irwin Gage, pno. Decca SXL 6522. (also London OS 26240)
          +Gr 3-72 p1567   +Op 1-25-75 p32
          +-HF 3-73 p106   +SR 1-73 p51
          +-HFN 3-72 p507   ++St 4-73 p122
          +NR 1-73 p7   +STL 5-7-72 p37
KING
     Hosts of freedom. cf Michigan University SM 0002.
KITTRIDGE, Walter
     Tenting on the old campground. cf Vox SVBX 5304.
KJERULF, Halfdan
1279 Albumblatt, op. 24, no. 1. Allegro, op. 24, no. 2. Caprice, op.
     12, no. 4. Impromptu, op. 12, no. 9. Menuett, op. 12, no. 2.
     Pieces, op. 28 (6). Scherzo, op. 29. Springtanz, op. 27, no.
     2. Wiegenlied, op. 4, no. 3. SJOGREN: Erotikon, op. 10.
     Gerald Robbins, pno. Genesis GS 1017.
          +NR 8-75 p12
     Allegro, op. 24, no. 2. cf Albumblatt, op. 24, no. 1.
     Caprice, op. 12, no. 4. cf Albumblatt, op. 24, no. 1.
     Impromptu, op. 12, no. 9. cf Albumblatt, op. 24, no. 1.
     Menuett, op. 12, no. 2. cf Albumblatt, op. 24, no. 1.
     Pieces, op. 28 (6). cf Albumblatt, op. 24, no. 1.
     Scherzo, op. 29. cf Albumblatt, op. 24, no. 1.
     Springtanz, op. 27, no. 2. cf Albumblatt, op. 24, no. 1.
     Wiegenlied, op. 4, no. 3. cf Albumblatt, op. 24, no. 1.
KJORLING
     Evening mood. cf RCA SER 5719.
KLING
     Kitchen symphony, op. 445. cf Angel S 36080.
KLOHR
     The billboard. cf Michigan University SM 0002.
KLUSAK, Jan
     Invention, no. 5. cf Panton 110 253.
KMOCH, Frantisek
1280 Marches: Andulka; Below the mill; Black horses; Jara Mláďá; Jara-
     báček; Kolin; The moon is shining; Hoj Marenko; Green groves;
     Music music; My little horse; Czech music. Czechoslovak Army
     Band; Rudolf Urbanec. Supraphon 54993.
          +RR 3-75 p26
1281 March songs: Andulka Safárova; Ceska Musika; Hoj Marenko; Koline
     Koline; Jarabáček; Jara Mláďá; Měsícek Svítí; Muj Koníček;
     Muziky Muziky; Pode Mlejnem; Vraný Kone; Zelený Hajové. FOK

Men's Choir; Czech Army Band; Jindřich Brejšek, Eduard Kudelásek.
Panton 110 330.
+RR 9-75 p70
KOCH, Erland von
Nordic capriccio, op. 26. cf NYSTROEM: Sinfonia concertante,
violoncello and orchestra.
Oxberg variations. cf FERNSTROM: Concertino, flute, women's chorus
and chamber orchestra, op. 52.
KOCSAR, Miklós
Improvvisazioni. cf Hungaroton SLPX 11692.
1282 Lonely song of poems by Attila József. Repliche, flute and cimb-
alom. Variations, wind quintet. PAPP: Dialogue, piano and
orchestra. Meditations in memory of Milán Füst. István Matuz,
flt; Márta Fábián, cimb; Loránd Szücs, pno; Erika Sziklay, s;
Hungarian Wind Quintet, Budapest Chamber Ensemble, Hungarian
Radio and Television Orchestra; András Mihály, Miklós Erdélyi.
Hungaroton SLPX 11635.
+NR 4-75 p14   +RR 4-75 p51
Repliche, flute and cimbalom. cf Lonely song of poems by Attila
József.
Variations, wind quintet. cf Lonely song on poems by Attila József.
KODALY, Zoltan
Adagio. cf Orion ORS 74160.
Ballet music. cf Works, selections (Decca SXLM 6665-7).
1283 Concerto, orchestra. Dances of Galánta. Dances of Marosszék.
Theatre overture. PH; Antal Dorati. Decca SXL 6712. (Re-
issue from SXLM 6665/7)
+-Gr 12-75 p1032   +-RR 12-75 p50
+HFN 12-75 p171
Concerto, orchestra. cf Works, selections (Decca SXLM 6665-7).
Dances of Galánta. cf Concerto, orchestra.
Dances of Galánta. cf Works, selections (Decca SXLM 6665-7).
Dances of Marosszék. cf Concerto, orchestra.
Dances of Marosszék. cf Piano works (Candide CE 31077).
Dances of Marosszék. cf Works, selections (Decca SXLM 6665-7).
1284 Háry János: Suite. PROKOFIEV: Lietutenant Kijé suite, op. 60.
CO; Georg Szell. CBS 61193. Tape (c) 40-61193.
+-HFN 12-75 p173   +-RR 12-75 p99 tape
Háry János: Suite. cf Works, selections (Decca SXLM 6665-7).
Hungarian songs. cf BARTOK: Mikrokosmos: 4 songs.
Hungarian tunes. cf Works, selections (Decca SXLM 6665-7).
Meditation sur un motif de Claude Debussy. cf Piano works (Candide
CE 31077).
Minuetto serio. cf Works, selections (Decca SXLM 6665-7).
1285 Piano works: Dances of Marosszék. Meditation sur un motif de
Claude Debussy. Pieces, op. 3 (9). Pieces, op. 11 (7).
Valsette. György Sándor, pno. Candide CE 31077. (also Vox
STGBY 680)
+-Gr 4-75 p1839   +NR 7-74 p11
+HF 9-74 p97   +RR 5-75 p55
+-HFN 5-75 p132   +St 10-74 p131
Pieces, op. 3 (9). cf Piano works (Candide CE 31077).
Pieces, op. 11 (7). cf Piano works (Candide CE 31077).
Sonata, violoncello and piano, op. 4. cf HINDEMITH: Sonata, violon-
cello and piano, op. 11, A minor.
Songs: Csillagoknak teremtöje; Kiolvaso. cf BIS LP 17.

Summer evening. cf Works, selections (Decca SXLM 6665-7).
Symphony, C major. cf Works, selections (Decca SXLM 6665-7).
Theatre overture. cf Concerto, orchestra.
Theatre overture. cf Works, selections (Decca SXLM 6665-7).
Valsette. cf Piano works (Candide CE 31077).
Variations on a Hungarian folk song (Peacock). cf BARTOK: Dance
    suite.
Variations on a Hungarian folk song. cf Works, selections (Decca
    SXLM 6665-7).
1286 Works, selections: Ballet music. Concerto, orchestra. Dances of
    Galánta. Dances of Marosszék. Háry János: Suite. Hungarian
    tunes. Minuetto serio. Summer evening. Symphony, C major.
    Theatre overture. Variations on a Hungarian folk song (Peacock).
    PH; Antal Dorati. Decca SXLM 6665-7 (3)/ (also London CSA
    2313)
        +Gr 9-74 p499   ++RR 9-74 p48
        +HF 8-75 p88   ++St 8-75 p97
        +NR 8-75 p3
KOECHLIN, Charles
1287 Etudes, alto saxophone and piano, op. 188. Paul Brodie, sax;
    Antonín Kubalek, pno. Classic Editions 16.
        ++St 1-75 p112
Miniatures (4). cf Klavier KS 537.
Quelques chorals pour des fêtes populaires. cf FAURE: Chant
    funéraire, op. 117.
KOHAUT, Carl
Concerto, guitar, F major. cf HANDEL: Concerto, 2 lutes, strings
    and recorders, op. 4, no. 6, B flat major.
KOLOSS
Partita. cf Hungaroton SLPX 11548.
KOMZAK, Karel
An der schönen grünen Narenta, op. 227. cf HMV SLS 5017.
Bäd'ner Mäd'ln, op. 252. cf Olympic 8136.
Bäd'ner Mäd'ln, op. 252. cf Rediffusion 15-16.
Bäd'ner Mäd'ln, op. 252. cf Supraphon 114 1458.
Echtes Wiener Blut, op. 189. cf HMV SLS 5017.
Erzherzog-Albrecht-Marsch, op. 136. cf DG 2721 077.
Erzherzog-Albrecht-Marsch, op. 136. cf HMV SLS 5017.
Fidels Wien, op. 190. cf HMV SLS 5017.
Vindobona-Marsch. cf DG 2721 077.
KORNGOLD, Erich
Fair pictures (Märchenbilder), op. 3. cf Sonata, piano, no. 2,
    op. 2, E major.
Garden scene. cf BRUCH: Concerto, violin, no. 2, op. 44, D
    minor: 2d movement.
1288 Much ado about nothing, op. 11 (4). Sonata, violin and piano,
    op. 4. Endre Granat, vln; Harold Gray, pno. Orion ORS 74166.
        /HF 6-75 p97   ++NR 11-74 p6
Much ado about nothing, op. 11: Garden scene. cf Decca SPA 405.
Much ado about nothing, op. 11: Holzapfel und Schlehwein, Garden
    scene. cf RCA ARM 4-0942/7.
1289 Quintet, piano and strings, op. 15, E major. Sonata, piano, no. 3,
    op. 25, C major. Endre Granat, Sheldon Sanov, vln; Milton
    Thomas, vla; Douglas Davis, vlc; Harold Gray, pno. Genesis GS
    1063.
        ++NR 10-75 p4

1290 Sonata, piano, no. 2, op. 2, E major. Fair pictures (Märchenbilder),
     op. 3. Antonín Kubalek, pno. Genesis GS 1055.
           +HF 6-75 p97   ++St 10-75 p110
           ++NR 5-75 p13
     Sonata, piano, no. 3, op. 25, C major. cf Quintet, piano and
     strings, op. 15, E major.
     Sonata, violin and piano, op. 4. cf Much ado about nothing, op. 11.
1291 Symphony, op. 40, F sharp major. Munich Philharmonic Orchestra;
     Rudolf Kempe. RCA ARL 1-0443. Tape (c) ARK 1-0443
           +Gr 6-75 p111 tape
1292 De tote Stadt, op. 12. Carol Neblett, s; Rose Wagemann, ms; René
     Kollo, t; Hermann Prey, bar; Benjamin Luxon, bs; Bavarian Radio
     Chorus; Munich Radio Orchestra; Erich Leinsdorf. RCA ARL
     3-1199 (3).
           +-NYT 11-16-75 pD1
     Die tote Stadt, op. 12: Arias. cf London OS 26381.
     Trio, piano, op. 1. cf IVES: Trio, violin, violoncello and piano.
KORTE, Karl
     Remembrances. cf DAVIDOVSKY: Synchronisms, no. 1.
KOSA, György
1293 Miniatures, harp trio (12). Orpheus, Eurydike, Hermes. Sonata,
     violoncello and piano. Katalin Székelfalvi-Nagy, s; Klára
     Takács, ms; Gabriella Zsigmond, con; Atilla Fülöp, t; Gábor
     Németh, bar; Peter Kovács, bs; Péter Lukacs, vla; Arpád Szász,
     Ede Banda, vlc; László Som, double bs; Henrik Pröhle, alto flt;
     Hédy Lubik, hp; András Gartner, timpani; György Kósa, pno;
     Hungarian Harp Trio; Miklós Erdélyi. Hungaroton SLPX 11628.
           +Gr 7-75 p182   +RR 4-75 p60
           +NR 5-75 p9
     Orpheus, Eurydike, Hermes. cf Miniatures, harp trio.
     Sonata, violoncello and piano. cf Miniatures, harp trio.
KOUGUELL
     Berceuse. cf Da Camera Magna SM 93399.
KOUSSEVITZKY, Serge
     Concerto, double bass. cf BLOCH: Sinfonia breve.
KOVAROVIC, Karel
     Mr. Broucek's excursion to the exhibition: Miners' polka. cf
     Supraphon 110 1429.
KOVATS, Barna
     Three movements for guitar. cf Hungaroton SLPX 11629.
KRAFFT, Frans Jozef
1294 Missa di requiem. Kortrijks Gemengd Koor; Herman Roelstraete.
     Arion ARN 38246.
           -RR 10-75 p84
KRAMAR
     Quartet, flute and strings, op. 75, D major. cf REICHA: Quartet,
     flute and strings, op. 98, no. 1, G minor.
KRAMAR-KROMMER, Frantisek
1295 Harmonie, op. 57, F major. MYSLIVECEK: Octets, no. 1, E flat major;
     no. 2, E flat major. Prague Chamber Orchestra, Members.
     Supraphon 59763.
           +Gr 12-75 p1065   +RR 12-75 p73
KRAPF, Gerhard
     Fantasia on a theme by Frescobaldi. cf University of Iowa Press,
     unnumbered.

KREBS, Johann
  Chorales (6). cf RCA CRL 2-7001.
  Fugue on B-A-C-H. cf Columbia M 33514.
  Klavierübung: Praeambulum sopra "Jesu meine Freude". cf Philips
    6775 006.
  Suite, G minor. cf Orion ORS 73114.
  Suite, flute and harpsichord, G major. cf Ember ECL 9040.
  Trio, C minor. cf BACH: Chorale preludes, S 669-671.
KREIN, Michael
  Valse caprice. cf Canon CNN 4983.
KREISLER, Fritz
  Allegretto (Boccherini). cf Works, selections (Philips 6833 164).
  Caprice viennois, op. 2. cf BEETHOVEN: Concerto, violin, op. 61,
    D major.
  Caprice viennois, op. 2. cf Works, selections (Philips 6833 164).
  Caprice viennois, op. 2. cf Decca SPA 405.
  Caprice viennois, op. 2. cf Discopaedia MB 1007.
  Caprice viennois, op. 2. cf Ember GVC 46.
  Caprice viennois, op. 2. cf Pearl GEM 121/2.
  Caprice viennois, op. 2. cf Rococo 2035.
  Chanson Louis XIII and pavane (Couperin). cf Works, selections
    (Philips 6833 164).
  Chanson Louis XIII and pavane. cf Discopaedia MB 1006.
  La chasse (in the style of Cartier). cf CBS 76420.
  Dances: Liebesfreud, Liebeslied, Schön Rosmarin. cf Argo ZRG 805.
  Liebesfreud. cf Works, selections (Philips 6833 164).
  Liebesfreud. cf Discopaedia MB 1004.
  Liebesfreud. cf Ember GVC 46.
  Liebeslied. cf Works, selections (Philips 6833 164).
  Liebeslied. cf Discopaedia MB 1004.
  Londonderry air (arr.). cf Connoisseur Society CS 2070.
  Minuet. cf RCA ARM 4-0942/7.
  Minuet (Porpora). cf Works, selections (Philips 6833 164).
  The old refrain. cf Works, selections (Philips 6833 164).
  Polichinelle sérénade. cf Argo ZRG 805.
  Praeludium and allegro (Pugnani). cf Works, selections (Philips
    6833 164).
  Quartet, strings, A minor. cf BRUCH: Concerto, violin, no. 1, op.
    26, G minor.
  Recitative and scherzo-caprice, solo violin. cf Works, selections
    (Philips 6833 164).
  Rondino on a theme by Beethoven. cf Works, selections (Philips
    6833 164).
  Rondino on a theme by Beethoven. cf Argo ZRG 805.
  Rondino on a theme by Beethoven. cf Discopaedia MB 1006.
  Scherzo. cf BRUCH: Concerto, violin, no. 1, op. 26, G minor.
  Schön Rosmarin. cf Works, selections (Philips 6833 164).
  Schön Rosmarin. cf Decca SPA 405.
  Sicilienne et Rigaudon. cf Discopaedia MB 1006.
  Sicilienne et Rigaudon. cf RCA ARM 4-0942/7.
  Syncopation. cf Argo ZRG 805.
  Tambourin chinois. cf Works, selections (Philips 6833 164).
  Tambourin chinois. cf Argo ZRG 805.
  Tambourin chinois. cf Ember GVC 46.
  Tempo di minuetto (Pugnani). cf Works, selections (Philips 6833
    164).

1296 Works, selections: Allegretto (Boccherini). Caprice viennois, op.
     2. Chanson Louis XIII and pavane (Couperin). Liebesfreud.
     Liebeslied. Minuet (Porpora). The old refrain. Praeludium
     and allegro (Pugnani). Recitative and scherzo-caprice, solo
     violin. Rondino on a theme by Beethoven. Schön Rosmarin.
     Tambourin chinois. Tempo di minuetto (Pugnani). Henryk
     Szeryng, vln; Charles Reiner, pno. Philips 6833 164.
            +HFN 11-75 p173   +RR 11-75 p79
KREMSER
    We gather together. cf RCA ARL 1-0562.
KRENEK, Ernst
    Trio, clarinet, violin and piano. cf IVES: Largo.
KRESTYANIN, Feodor
    Befittingly. cf HMV Melodiya ASD 3102.
KRETTNER
    Tölzer Schütenmarsch. cf BASF 292 2116-4.
KREUTZER, Konradin
    Ein Bettler vor dem Tor. cf DG Archive 2533 149.
KRIEGER
    Schlacht. cf Musical Heritage Society MHS 1790.
KRIEGER, Johann
    Fantasia, D minor. cf Pelca PSR 40598.
    Fantasia, D minor. cf Telefunken TK 11567/1-2.
    Praeludium and ricercar, A minor. cf Telefunken TK 11567/1-2.
    Toccata, D major. cf Telefunken TK 11567/1-2.
KRIEGER, Johann Philippe
    Toccata and fugue, A minor. cf Telefunken TK 11567/1-2.
KRUMPHOLZ, Johann B.
    Sonata, flute and harp, F major. cf Musical Heritage Society
        MHS 1345.
KRUYF, Ton de
    Quatre pas de deux, flute and orchestra, op. 30. cf KUNST:
        Trajectoire, 16 singers and 11 instrumentalists.
KUCERA, Vaclav
    The kenetic ballet: The labyrinth; The spiral. cf ISTVAN: Isle
        of toys.
    Scenario, flute and string trio. cf Panton 110 253.
KUHLAU, Friedrich
1297 Grand solo, piano and flute, op. 57, no. 3. REINECKE: Sonata,
     flute and piano, op. 167, E minor. RIES: Sonata, flute and
     piano, op. 169, E flat major. Louise di Tullio, flt; Virginia
     di Tullio, pno. Genesis GS 1048.
            +-Gr 6-75 p61   +NR 11-74 p7
            +MJ 12-74 p45   +-SFC 9-22-74 p22
     Sonatinas, op. 20, nos. 1-3. cf Columbia MG 33202.
     Sonatinas, op. 55, nos. 1-3. cf Columbai MG 33202.
1298 Biblical sonatas (6). Christopher Bowers-Broadbent, org. Pearl
     SHE 518/9 (2).
            -Gr 3-75 p1681   ++RR 3-75 p49
KUKUCK
    Die Brücke. cf BIS LP 2.
KUNST, Jos
1299 Trajectoire, 16 singers and 11 instrumentalists. KRUYF: Quatre
     pas de deux, flute and orchestra, op. 30. LEEUW: Quartet,
     strings. VLIJMEN: Omaggio a Gesualdo, violin and 6 instrumen-
     tal groups. Theo Olof, vln; Rien de Reede, flt; Amsterdam

Quartet, Radio Wind Ensemble, Radio Chamber Orchestra and Choir;
Ernest Bour, Hans Vonk, Paul Hupperts. Donemus 7374/3.
          ++RR 2-75 p59
KUPFERMAN, Meyer
    Superflute. cf DAVIDOVSKY: Synchronisms, no. 1.
KURKA, Robert
1300 The good soldier Schweik: Suite. WEILL: The threepenny opera:
    Suite. Music for Westchester Symphony Orchestra; Siegfried
    Landau. Candide CE 31089.
          +NR 6-75 p5   ++St 8-75 p107
LA FORGE
    To a messenger. cf Discophilia KGS 3.
LAJTHA, László
1301 Symphony, no. 4, op. 52. Symphony, no. 9, op. 67. HSO; János
    Ferencsik. Hungaroton SLPX 11564.
          +Gr 12-74 p1134   ++RR 1-75 p29
    Symphony, no. 9, op. 67. cf Symphony, no. 4, op. 42.
LALANDE, Michel
1302 Caprice, no. 2, G minor. Concert de trompettes pour les festes
    sur le canal de Versailles. Symphonies pour les soupers du
    roi, no. 2. L'Oiseau-Lyre Orchestral Ensemble; Louis de Froment.
    L'Oiseau-Lyre OLS 170. (Reissue from OL 50152)
          +Gr 6-74 p51   +RR 5-74 p30
          ++NR 6-75 p6
    Concert de trompettes pour les festes sur le canal de Versailles.
    cf Caprice, no. 2, G minor.
    Symphonies pour les soupers du roi, no. 2. cf Caprice, no. 2, G
    minor.
1303 Symphonies pour les soupers du roi, no. 2. MOURET: Suites of
    symphonies. Adolf Scherbaum, tpt; Chamber Orchestra; Paul
    Kuentz. DG Archive 198333.
          -Audio 9-75 p70   ++NR 6-75 p6
LALO, Edouard
    Rapsodie norvégienne. cf CHABRIER: Le Roi malgré lui: Fête polo-
    naise.
    Rapsodie norvégienne. cf FRANCK: Le chasseur maudit.
    Le Roy d'Ys: Cher mylio. cf Angel S 37143.
    Le Roy d'Ys: Vainement ma bien aimee. cf Discophilia KGC 1.
    Scherzo, orchestra. cf FRANCK: Le chasseur maudit.
1304 Symphonie espagnole, op. 21. TCHAIKOVSKY: Sérénade mélancolique,
    op. 26. Souvenir d'un lieu cher, op. 42: Meditation. Leonid
    Kogan, vln; PhO, OSCCP; Kyril Kondrashin, Constantin Silvestri.
    Classics for Pleasure CFP 40040. (Reissue from Columbia SAX
    2323, 2329)
          +-Gr 4-75 p1808   ++RR 2-75 p35
1305 Symphonie espagnole, op. 21. PAGANINI: Fantaisie on 'Dal tuo
    stellato sogio". Variations on 'Nel cor più non mi sento'.
    Salvatore Accardo, vln; André Collard, pno; Paris, Association
    des Concerts Pasdeloup Orchestra; Herbert Albert. Saga 5398.
          +Gr 7-75 p187   ++RR 6-75 p42
          +-HFN 8-75 p89
    Symphonie espagnole, op. 21, D minor: Andante. cf RCA ARM 4-0942/7.
    Symphonie espagnole, op. 21, D minor: 4th movement. cf Discopaedia
    MB 1010.
1306 Symphony, G minor. LISZT: A Faust symphony, G 108. Alexander
    Young, t; Beecham Choral Society; RPO, French National Radio
    Orchestra; Thomas Beecham. HMV SXLP 3022 (2). Reissues from

ASD 388, 317/8)
    +Gr 12-75 p1032   +-RR 12-75 p55
    +-HFN 12-75 p171
LAMBERT, Constant
    Elegiac blues. cf BLISS: Sonata, piano.
    The Rio Grande, piano, chorus and orchestra. ' cf FRANCAIX: Concerto,
        piano, D major.
    Sonata, piano. cf BLISS: Sonata, piano.
LAMPUGNANI, Giovanni
    Meraspe: Arias. cf London OS 26277.
LANDINI, Francesco
    Ecco la primavera. cf Telefunken TK 11569/1-2.
    Ecco la primavera. cf Vanguard VSD 71179.
    Gran piant. cf Telefunken TK 11569/1-2.
    Songs: Se la nimica mie; Adiu adiu. cf 1750 Arch S 1753.
LANG, C. S.
    Prelude on Leoni. cf Wealden WS 139.
LANG, István
    Intermezzi. cf Hungaroton SLPX 11692.
    Quintet, winds, no. 2. BOZAY: Quintet, winds, op. 6.
LANGFORD, Gordon
    A London scherzo. cf RCA LRL 1-5072.
    A west country fantasy. cf RCA LRL 1-5072.
LANGLAIS, Jean
    Hymne d'actions de graces: Te deum. cf Wealden WS 110.
    Te Deum. cf Argo ZRG 807.
LANIERE, Nicholas
    Though I am young. cf BASF BAC 3081.
LANNER, Josef
    Abendsterne, op. 180. cf HMV SLS 5017.
    Jagd, op. 82. cf HMV SLS 5017.
    Mitternachtswalzer, op. 8. cf Nonesuch H 71141.
    The parting of the ways. cf Saga 5411.
    Regata-Galopp, op. 134. cf Nonesuch H 71141.
    Die Schönbrunner, op. 200. cf Olympic 8136.
    Die Schönbrunner, op. 200. cf Pye NSPH 6.
    Die Schönbrunner, op. 200(arr. Urbanec). cf Rediffusion 15-16.
    Die Schönbrunner, op. 200. cf Supraphon 114 1458.
    Summer night's dream. cf Saga 5411.
    Valses viennoises. cf BEETHOVEN: Andante favori, F major.
1307 Waltzes: Hofballtänze; Pesther Walzer; Die Schönbrunner, op. 200.
        Steirische Tänze. ZIEHRER: Fächer-Polonaise; Hereinspaziert;
        Samt und Seide; Singen, Lachen, Tanzen; Der Zauber der Montur.
        Vienna Volksoper Orchestra; Josef Drexler. Rediffusion 15-28.
            +Gr 8-75 p363
LANSKY, Paul
    Modal fantasy. cf CRI SD 342.
LANTINS, Arnold de
    Puisque je voy. cf Telefunken ER 6-35257.
LANTINS, Hugo de
    Gloria. cf Telefunken ER 6-35257.
LARA, Agustin
    Granada. cf LEONCAVALLO: I Pagliacci.
LARSSON, Lars-Erik
    Concerto, violin, op. 42. cf FERNSTROM: Concertino, flute, women's
        chorus and chamber orchestra, op. 52.

1308 Förklädd Gud (A God in disguise), op. 24. RANGSTROM: Songs:
     Vingar i natten (Wings in the night); Vinden och trädet (The
     wind and the tree); Sköldmön (Amazon). Catarina Ligendza, s;
     Ingvar Wixell, bar; Max von Sydow, narrator; Stockholm Radio
     Orchestra and chorus; Stig Westerberg. HMV 4D 071 35149.
          +Gr 12-75 p1061
     Orchestral variations, op. 51. cf KARKOFF: Symphony, no. 4, op.
     69.
LASERNA, Blas
     Jilguerillo con pico de oro; Las majas de Paris. cf HMV SLS 5012.
LASSUS, Roland de
     Alma redemptoris mater. cf GABRIELI, G.: Symphonie sacrae: Quis
          est iste; Sonata pian e forte; Maria Virgo; Sancta Maria,
          succurre miseris.
     Ave Maria. cf GABRIELI, G.: Symphoniae sacrae: Quis est iste;
          Sonata, pian e forte; Maria Virgo; Sancta Maria, succurre
          miseris.
1309 Bell Amfitrit altera mass. Psalmus Poenitentalis VII. Christ
     Church Cathedral Choir; Simon Preston. Argo ZRG 735.
          +Gr 8-74 p382   +RR 7-74 p75
          *NR 2-75 p7
     Cathalina, apra finestra; Matona mia cara. cf L'Oiseau-Lyre
          12BB 203/6.
1310 Lagrime di San Pietro. Raphaël Passaquet Vocal Ensemble; Raphaël
     Passaquet. Harmonia Mundi HMU 961.
          +-Gr 11-75 p875
1311 Motets: Ave Regina caelorum; Salve Regina; O mors, quam amara est.
     Penitential psalms: Miserere mei, Deus; Domine, ne in furore
     tuo. Roderick Skeaping, Trevor Jones, viol; Pro Cantione Antiqua,
     Early Music Wind Ensemble; Bruno Turner. DG Archive 2533 290.
          +Gr 10-75 p669   +RR 10-75 p85
          +HFN 11-75 p155
     Penitential psalms: Miserere mei, Deus; Domine, ne in furore tuo.
          cf Motets (DG Archive 2533 290).
     Praeter rerum seriem: Magnificat. cf GABRIELI, G.: Symphonie
          sacrae: Quis est iste; Sonata pian e forte; Maria Virgo; Sancta
          Maria, succurre miseris.
     Psalmus Poenitentalis VII. cf Bell Amfitrit altera mass.
1312 Sacrae lectiones ex propheta Job. Prague Madrigal Singers (Prager
     Madrigalisten); Miroslav Venhoda. Telefunken 6-41274.
          +NR 10-75 p6
     Tibi Laus. cf Wayne State University, unnumbered.
     LATE 16TH CENTURY MUSIC, Part 2: The Harvard University Press
          historial anthology of music. cf Pleiades P 255.
LAURENTIUS, A.
     Mij heeft een piperken. cf Telefunken TK 11569/1-2.
LAURO, Antonio
1313 Danza negra. Suite venezolana: Valse. Valse, no. 3. PONCE:
     Campo. SAINZ DE LA MAZA: Homenaje à la guitarra. VILLA-LOBOS:
     Chôro typico. Preludes, nos. 1-5. Julian Byzantine, gtr.
     Classics for Pleasure CFP 40209.
          +Gr 10-75 p658   +RR 9-75 p62
          +HFN 11-75 p155
     Suite venezolana: Valse. cf Danza negra.
     Valse, no. 3. cf Danza negra.
LAVRY, Marc
     Jewish dances (3). cf Da Camera Magna SM 93399.

LAW, Andrew
    Bunker Hill. cf Nonesuch H 71276.
LAWES, Henry
    Songs: Gather your rosebuds; See how in gathering. cf BASF BAC
       3081.
LAYOLLE, Francesco de
1314 Missa Ces fascheux Sotz. Motets, madrigals and songs. ANON.: Ces
       fascheux Sotz. Capella Cordina; Alejandro Planchart. Lyrichord
       LLST 7266.
          +NR 2-75 p7
    Motets, madrigals and songs. cf Missa Ces fascheux Sotz.
LEATHERLAND, Thomas
    Pavan, 6 viols, G minor. cf Strobe SRCM 123.
LEBEGUE, Niclas
    Elévation, G major. cf GRIGNY: La messe.
    Symphonie sur le bemól fa. cf GRIGNY: La messe.
    Suite du deuxième ton. cf GRIGNY: La messe.
LECHNER, Leonard
    Allein zu dir, Herr Jesu Christ. cf BASF BAC 3087.
LECLAIR, Jean Marie
    Gigue. cf Klavier KS 537.
    Musette. cf Klavier KS 537.
    Sonatas, op. 2, nos. 5, 8. cf CAMPRA: Cantates françaises de
       Arion et Didon.
1315 Sonata, 2 violins, op. 3. Claire Bernard, Annie Jodry, vln.
       Arion ARN 38269.
          +-RR 10-75 p78
LEDENEV, Roman
    Pieces, harp and string quartet, op. 16 (6). cf ARISTAKESIAN:
       Sinfonietta, piano, xylophone and strings.
LEEUW, Reinbert de
    Quartet, strings. cf KUNST: Trajectoire, 16 singers and 11 instru-
       mentalists.
    Quartet, strings, no. 2. cf CRUMB: Black angels, electric string
       quartet.
LEGINSKA, Ethel
    Victorian portraits. cf HAUBIEL: Metamorphoses.
LEHAR, Franz
1316 Eva: Wär es auch nichts als ein Traum vom Gluck. Die lustige Witwe:
       Viljalied. MOZART: Le nozze di Fagaro: Overture; Deh vieni non
       tardar; Venite, inginocchiateve. Il Re Pastore: L'amerò sarò
       costante. STRAUSS, J. II: Die Fledermaus: Overture; Mein Herr
       Marquis. Rita Streich, s; London Symphonia; Wyn Morris. Pye
       TPLS 13064. Tape (c) ZCTPL 13064 (ct) Y8 TPL 13064.
          +Gr 5-75 p2019   +RR 6-75 p24
    Friederike: Warum hast du mich wachgeküsst. cf Works, selections
       (London OSA 26220).
    Giuditta: Meine Lippen, sie küssen so heiss. cf Angel S 35696.
1317 Gold and silver waltz, op. 79. STRAUSS, J. II: Die Fledermaus,
       op. 363: Overture. Leichtes Blut, op. 319. Tales from the
       Vienna woods, op. 325. STRAUSS, Joseph: Sphärenklange, op. 235.
       SUPPE: Morning, noon and night in Vienna. Dresden Staatskapelle
       Orchestra; Rudolf Kempe. RCA LRL 1-5044.
          ++Gr 3-75 p1710   +RR 4-75 p38
    Gold and silver waltz, op. 79. cf Decca SXL 6572.
    Gold und Silber, op. 79. cf Olympic 8136.

Der Graf von Luxemburg: Hoch, Evoë, Angèle Didier; Heut noch werd
ich Ehefrau. cf Angel S 35696.
The Count of Luxemburg: Fragrance of May; Polka dance; Love and
age. cf Pye NSPH 6.
The Count of Luxembourg: Lieber Freund...Bist du's lachendes Glück.
cf Works, selections (London OSA 26220).

1318 Das Land des Lächelns: Overture; Immer nur Lächeln; Bei einem Tee
à Deux; Von Apfelbluten einen Kranz; Act 1, finale; Im Salon
zur blau'n Pagode; Wer hat die Liebe uns ins Herz gesenkt; Meine
Liebe, deine Liebe; Dein ist mien Ganzes Herz; Ich möcht wieder
Einmal die Heimat sehn; Zig, zig, zig; Wie rasch verwelkte doch;
Act 3, finale. Soloists; Vienna Volksoper Orchestra; Heinz
Lambrecht. Decca SDD 459.
        +RR 9-75 p18
The land of smiles: Bei einem Tee en deux; Dein ist mien ganzes
Herz; Wer hat die Liebe uns ins Herz gesenkt. cf Works, selec-
tions (London OSA 26220).

1319 Die lustige Witwe (The merry widow): Verehrteste Damen und Herren;
Ich bin eine anstand'ge Frau; Bitte meine Herr'n; Da geh ich
zu Maxim; O kommet doch, o dommet, ihr Ballsirenen; Vilja;
Lied vom dummer Reiter; Wie die Weiber; Wie eine Rosenknospe--
sieh dort den kleinen Pavillon; Act 2, finale; Grisetten Lied;
Lippen Schweigen; Ja, das Studium der Weiber is schwer. Soloists; VSOO and Chorus; Robert Stolz. Decca SDD 460.
        +-RR 9-75 p18
1320 Die lustige Witwe: Act 1, Prelude; Ich bin eine anständ'ge Frau;
Hab'in Paris mich noch...Gar oft hab' ich's gehört; O Vaterland
du machst bei Tag...Da geh' ich zu Maxim; Act 2, Introduction
and dance; Vilja, o Vilja; Heia, Mädel aufgeschaut...Dummer,
Dummer, dummer, Reitersmann; Interlude; Wie die Weiber man
behandelt...Ja, das Studium der Weiber; Wie eine Rosenknospe...
Sieh dort den kleinen Pavillon; Act 3, Cake walk; Grisettenlied;
Lippen schweigen...Finale. Mimi Coertse, Friedl Loor, s; Karl
Terkal, t; Vienna Volksoper Orchestra and Chorus; Hans Hagen.
Saga 5365. (Reissue from Vox VX 1310)
        /Gr 1-75 p1383   -RR 1-75 p24
The merry widow: Lippen Schweigen. cf Works, selections (London
OSA 26220).
Die lustige Witwe: Viljalied. cf Eva: Wär es auch nichts als ein
Traum vom Gluck.
Paganini: Gern hab' ich die Frau'n geküsst; Niemand liebt Dich;
Liebe, du Himmel auf Erden. cf Works, selections (London OSA
26220).
Paganini: Love live for ever. cf Pye NSPH 6.
Schön ist die Welt: Frei und jung dabei; Schön ist die Welt; Ich
bin verliebt. cf Works, selections (London OSA 26220).
Wiener Frauen: Overture. cf Telefunken DX 6-35262.

1321 Works, selections: The best of Franz Lehar: The Count of Luxembourg:
Lieber Freund...Bist du's lachendes Glück. Friederike: Warum
hast du mich wachgeküsst. The land of smiles: Bei einem Tee
en deux; Dein ist mein ganzes Herz; Wer hat die Liebe uns ins
Herz gesenkt. The merry widow: Lippen Schweigen. Paganini:
Gern hab' ich die Frau'n geküsst; Niemand liebt Dich; Liebe du
Himmel auf Erden. Schön ist die Welt: Frei und jung dabei;
Schön ist die Welt; Ich bin verliebt. Der Zarewitsch: Wolgalied;
Kosende Wellen. Werner Krenn, t; Renate Holm, s; Vienna Volks-

oper Orchestra; Anton Paulik.   London OSA 26220.   (also Decca
SXL 6711)
     +Gr 12-75 p1098   ++RR 12-75 p28
     +NR 5-73 p12   +-SR 3-73 p48
    +-ON 12-73 p44   +St 5-73 p117
  Yours is my heart alone.   cf Columbia D3M 33448.
1322 Der Zarewitsch: Es steht ein Soldat; Heil erklingt ein liebliches
    frohes Heimatlied; Einer wird kommen; Allein...Es steht ein
    Soldat am Wolgastrand; Champagner ist ein Feuerwein; Bleib bei
    mir...Hab nur dich allein; Kosende wellen; Finale. Der Graf
    von Luxemburg: Karneval; Mein Ahnherr war der Luxemburg; Ein
    Scheck auf die Englische Bank; Heut noch werd ich Ehefrau; Frau
    Grafin--Sie geht links, er geht rechts; Bist du's lachendes
    Glück; Trêfle incarnat; Sind sie von sinnen; Lieber Freund.
    Soloists; Vienna Volksoper Chorus and Orchestra; Max Schonherr.
    Decca SDD 461.
      +-RR 9-75 p23
  Der Zarewitsch: Einer wird kommen.   cf Angel S 35696.
  Der Zarewitsch: Wolgalied; Kosende wellen.   cf Works, selections
    (London OSA 26220).
LEHMANN, L.
  There are fairies at the bottom of our garden.   cf RCA ARL 1-5007.
LEIFS, Jon
  Icelandic overture, op. 9.   cf ISOLFFSON: Passacaglia.
LEIGHTON, Kenneth
  Et resurrexit, op. 49.   cf Wealden WS 145.
  Fanfare.   cf Wealden WS 139.
  Give me the wings of faith.   cf HMV HQS 1350.
LEKEU, Guillaume
1323 Sonata, violin and piano, G major.   VIEUXTEMPS: Ballade et polonaise,
    op. 38.   YSAYE: Reve d'enfant, op. 14.   Arthur Grumiaux, vln;
    Dinorah Varsi, pno. Philips 6500 841.
      +HF 7-75 p76   +SR 3-22-75 p36
     +-NR 5-75 p9   ++St 7-75 p70
LEMMENS, Nicolas
  Fanfare.   cf Vista VPS 1021.
LENNON, John
  Eleanor Rigby.   cf Columbia CXM 32088.
  Yesterday.   cf L'Oiseau-Lyre DSLO 3.
LEONARD, Lawrence
  Mobile, 7 orchestras.   cf HMV SLA 870.
LEONCAVALLO, Ruggiero
  La bohème: Io non ho che una povera stanzetta.   cf Bel Canto Club
    no. 3.
  La bohème: Non qui...Testa adorata.   cf Saga 7206.
  La bohème: Testa adorata.   cf Discophilia KGC 2.
  Manon Lescaut: Tu, tu amore.   cf London OS 26315.
  Mattinata.   cf Polydor 2489 519.
1324 I Pagliacci.   Lucine Amara, s; Richard Tucker, Thomas Hayward, t;
    Giuseppe Valdengo, Clifford Harvuot, bar; Metropolitan Opera
    Orchestra and Chorus; Fausto Cleva. CBS 61658. (Reissue from
    Philips ABL 3041/2).
      +-Gr 10-75 p679   -RR 10-75 p28
     +-HFN 10-75 p153
1325 I Pagliacci.   Italian songs: CINQUE: Trobadorica. CIOFFI: Na sera
    'e maggio. CURTIS: Tu ca'nun chiagne. CAPUA: O sole mio. LARA:

Granada. MAINARDI: Varca d' 'o primo ammore. Gabriella Tucci,
s; Mario del Monaco, Piero de Palma, t; Cornell MacNeil, Renato
Capecchi, bar; Rome, Santa Cecilia Orchestra and Chorus; Fran-
cesco Molinari-Pradelli; Orchestra; Ernesto Nicelli. Decca GOS
658/9 (2). (Reissue from SXL 2185/6)
        +—Gr 4-75 p1863
1326 I Pagliacci. MASCAGNI: Cavalleria rusticana. Lucine Amara,
Thelma Votipka, Margaret Hershaw, s; Mildred Miller, ms; Richard
Tucker, Thomas Hayward, t; Frank Guarrera, Giuseppe Valdengo,
Cifford Harvuot, bar; Metropolitan Opera Orchestra and Chorus;
Fausto Cleva. Odyssey Y3 33122 (3). (Reissue from Columbia
SL 123, 113)
        +—HF 4-75 p92
I Pagliacci: Bell chorus, Act 1. cf Decca PFS 4323.
I Pagliacci: Prologue. cf Columbia/Melodiya M 33120.
I Pagliacci: Qual fiamma avea nel guardo. cf RCA ARL 1-0702.
I Pagliacci: Vesti la giubba. cf Columbia D3M 33448.
I Pagliacci: Vesti la giubba. cf Decca SXL 6649.
I Pagliacci: Vesti la giubba. cf RCA SER 5704/6.
LERDAHL, Fred
    Wake. cf BABBITT: Philomel, soprano, recorded soprano, synthesized
        sound.
Le ROY, A.
    Branle de Bourgogne. cf DG Archive 2533 111.
LESCHITZKY, Theodor
    Gavotte. cf Rococo 2049.
LEVY, Burt
    Orbs with flute. cf Nonesuch HB 73028.
LEWER
    Fidelia. cf Nonesuch H 71276.
LEWIS, Peter
    Gestes. cf CRI SD 324.
LEYBACH
    Nocturne, no. 5, op. 52, A flat major. cf Discopaedia MB 1005.
LIBAEK, Sven Erik
    Musical pictures, guitar, nos. 2 and 3. cf CBS 61654.
LIDDLE
    How lovely are thy dwellings. cf RCA ARL 1-0561.
LIEB, R.
    Feature suite. cf Kendor KE 22174.
LIGETI, Gyorgy
    Musica ricercata. cf BIS LP 18.
LINDE, Bo
    Concerto, violin, op. 18. cf KABALEVSKY: The comedians, op. 26.
LINDEMANN
    Unter dem Grillenbanner. cf DG 2721 077.
LINKE, Norbert
    Violencia. cf BLACHER: Sonata, solo violin, op. 40.
LINN, Robert
    Dithyramb, 8 celli. cf CASALS: Sardana, celli.
LISZT, Franz
    (G refers to Grove's number, 5th edition)
1327 Ad nos, ad salutarem undam, Fantasia and fugue, G 259. Introduction,
        fugue and magnificat (arr. Gottschlat). Prelude and fugue on the
        name B-A-C-H, G 260. Peter Le Huray, org. Saga 5401.
            +Gr 8-75 p348   +RR 9-75 p59
            +HFN 8-75 p77

Ad nos ad salutarem undam, fantasia and fugue, G 259. cf Organ
    works (Vox SVBX 5328/9).
Am grabe Richard Wagners, G 267. cf Organ works (Vox SVBX 5328/9).
Andante maestoso, G 668. cf Organ works (Vox SVBX 5328/9).
Andante religioso, G 261a. cf Organ works (Vox SVBX 5328/9).
1328 Années de pelerinage, 2nd year, G 161: No. 1, Sposalizio; No. 2,
    I penseroso; No. 3, Canzonetta del Salvator Rosa; No. 4, Sonet÷
    to del Petrarca, no. 47; No. 5, Sonetto del Petrarca, no. 123;
    No. 8, Gondoliera. Legends, G 175, No. 1, St. Francis of Assisi;
    No. 2, St. Francis of Paola. Wilhelm Kempff, pno. DG 2530 560.
        ++Gr 11-75 p860   ++RR 11-75 p79
        ++HFN 12-75 p156
    Années de pelerinage, 3rd year, G 163: Angelus. cf Organ works
        (Vox SVBX 5328/9).
Ave Maria, G 659. cf Organ works (Vox SVBX 5328/9).
Ave maris stella, G 669, no. 2. cf Organ works (Vox SVBX 5328/9).
1329 Ballade, no. 2, G 171, B minor. Paraphrase, DONIZETTI: Dom
    Sebastien, Funeral march, G 402. THALBERG: Fantasy on Rossini's
    "Moise", op. 33. Fantasy on Rossini's "Barber of Seville", op.
    63. Raymond Lewenthal, pno. Angel S 36079.
        +NR 11-75 p12
1330 Concerto, piano, no. 1, G 124, E flat major. SCHUMANN: Kreisler-
    iana, op. 16. Walter Gieseking, pno; Maastricht Municipal Orch-
    estra; Henri Heimans. Bruno Walter Society IGI 341.
        +-NR 9-75 p8
1331 Concerto, piano, no. 1, G 124, E flat major. Concerto, piano, no.
    2, G 125, A major. Philippe Entremont, pno; Zurich Radio
    Symphony Orchestra; Walter Goehr. Olympic 8135.
        +NR 4-75 p6
1332 Concerto, piano, no. 1, G 124, E flat major. Concerto, piano, no.
    2, G 125, A major. Sviatoslav Richter, pno; LSO; Kyril Kondrash-
    in. Philips 835 474. Tape (c) PCR 490 0000, 18008 CAA (4),
    PC 890 0000, Ampex L 5474. (also 6580 071. Reissue from SABL
    207)
        ++Gr 2-75 p1492   ++RR 2-75 p35
        +HF 4-71 p66   ++SFC 5-7-72 p46
1333 Concerto, piano, no. 2, G 125, A major. Etudes d'execution trans-
    cendente, no. 7, G 139. Sonata, piano, G 178, B minor. Gyula
    Kiss, pno; HSO; Tamás Pál. Hungaroton SHLX 90005.
        -RR 1-75 p29
1334 Concerto, piano, no. 2, G 125, A major. Sonata, piano, G 178,
    B minor. Gyula Kiss, pno; HSO; Tamás Pál. Qualiton LPX 11368.
        +LJ 3-75 p33
    Concerto, piano, no. 2, G 125, A major. cf Concerto, piano, no. 1,
        G 124, E flat major (Olympic 8135).
    Concerto, piano, no. 2, G 125, A major. cf Concerto, piano, no. 1,
        G 124, E flat major (Philips 835474).
1335 Consolations, G 172. Harmonies poètiques et réligieuses, G 173.
    Legendes, G 175. Weihnachtsbaum, G 186. Jerome Rose, pno.
    Vox SVBX 5475 (3).
        ++NR 2-75 p12
    Consolations, G 172 (6). cf Etudes d'execution transcedente, G 139.
    Consolation, no. 3, G 172, D flat major. cf Piano works (Saga
        5405).
    Czárdás obstiné, G 225. cf BRAHMS: Variations on a theme by Paga-
        nini, op. 35, A minor.

Dante. cf Organ works (Vox SVBX 5328/9).
1336 Etudes de concerto, G 144 (3). Etudes d'exécution transcendente,
     G 139. Etudes d'exécution transcendente d'après Paganini, G 140.
     Rhapsodie espagnole, G 254. Rumanian rhapsodie, G 242. Louis
     Kentner, pno. Vox SVBX 5353 (3).
     Etude de concert, no. 3, G 144, D flat major: Un sospiro (arr.
     Renié). cf GLINKA: Nocturne.
     Etude de concerto, no. 3, G 144, D flat major: Un sospiro. cf
     HMV HQS 1353.
1337 Etudes d'exécution transcendente, G 139. Russell Sherman, pno.
     Advent Tape (c) D 1010.
          +—HF 6-75 p122 tape   +NYT 3-30-75 pD22
1338 Etudes d'exécution transcendente, G 139. Joseph Banowetz, pno.
     Educo 3084/5 (2).
          +NYT 3-30075 pD22
1339 Etudes d'exécution transcendente, G 139. Consolations, G 172 (6).
     Liebestraum, no. 3, G 541, A flat major. Jorge Bolet, pno.
     RCA CRL 2-0446 (2). (also Pye EnSayo NELD 701)
          +—Gr 7-74 p234   + NR 7-74 p12
          -HF 9-74 p98   -NYT 3-30-74 pD22
          +MJ 9-74 p65   +—St 8-74 p115
1340 Etudes d'exécution transcendente, nos. 4, 5, 8, 12, G 139. Etudes
     d'exécution transcendente d'après Paganini, nos. 1-6, G 140.
     France Clidat, pno. French Decca 7205.
          +Gr 9-75 p491   +RR 8-75 p57
          +HFN 8-75 p77
     Transcendental etudes, nos. 4, 6, 10-12, G 139. cf BRAHMS:
     Variations on a theme by Paganini, op. 35.
     Etudes d'exécution transcendente, no. 7, G 139. cf Concerto,
     piano, no. 2, G 125, A major.
     Etudes d'exécution transcendente, G 139. cf Etudes de concert,
     G 144.
     Etudes d'exécution transcendente d'après Paganini, G 140. cf
     Etudes de concert, G 144.
     Etudes d'exécution transcendente d'après Paganini, no. 1-6, G 140.
     cf Etudes d'exécution transcendente, nos. 4, 5, 8, 12, G 139.
     Etudes d'exécution transcendente d'après Paganini, no. 3, G 140,
     A flat minor. cf CBS 73396.
     Etudes d'exécution transcendente d'après Paganini, no. 3, G 140,
     A flat minor. cf Monitor MCS 2143.
     Etudes d'exécution transcendente d'après Paganini, no. 5, G 140,
     E major. cf Westminster WGM 8309.
     Etudes d'exécution transcendente d'après Paganini, no. 6, G 140,
     A minor. cf CHOPIN: Sonata, piano, no. 3, op. 58, B minor.
     Evocation à la Chapelle Sixtine, G 658. cf Organ works (Vox
     SVBX 5328/9).
     Fantasia on Wagner's Sancta Spirita cavaliere, G 439. cf Piano
     works (Saga 5405).
     A Faust symphony, G 108. cf LALO: Symphony, G minor.
     Festival cantata for the unveiling of the Beethoven monument in
     Bonn, G 584. cf BEETHOVEN: Symphony, no. 9, op. 125, D minor
     (arr. Liszt).
     Festklänge, G 101/2. cf Symphonic poems (Philips 6500 191).
1341 Funerailles, G 173. Gnomenreigen, G 145. Liebestraum, G 541.
     Sonata, piano, G 178, B minor. Paraphrase on GOUNOD: Faust:
     Waltz, G 407. Simon Barere, pno. Turnabout THS 65001. (Re-
     issues)

```
    +HF 11-74 p110              +St 6-75 p98
    +NR 9-74 p9
```
Gebet, G 265. cf Organ works (Vox SVBX 5328/9).
Gnomenreigen, G 145. cf Funerailles, G 173.
Gnomenreigen, G 145. cf CHOPIN: Sonata, piano, no. 3, op. 58,
    B minor.
Harmonies poètiques et réligieuses, G 173. cf Consolations, G 172.
Harmonies poètiques et réligieuses, G 173: Funerailles. cf
    Piano works (Saga 5405).
Harmonies poètiques et réligieuses, G 173: Funerailles. cf
    Columbia Special Products AP 12411.
Heroïde funèbre, G 102. cf Mephisto waltz, no. 1, G 514.
Hosannah, G 677. cf Organ works (Vox SVBX 5328/9).
1342 Hungaria, G 103. Symphonic poem: Tasso, lamento e trionfo, G 96.
    HSO; János Ferencsik. Hungaroton SLPX 11683.
```
    +-Gr 9-75 p465              -NR 7-75 p2
    +HFN 10-75 p143             -RR 8-75 p39
```
Hungarian fantasia, G 123. cf BRAHMS: Hungarian dances, nos. 1,
    3, 5, 6.
Hungarian fantasia, G 123. cf HMV SLS 5033.
1343 Hungarian rhapsodies, nos. 2 and 4, G 244. Mazeppa, G 138. Les
    preludes, G 97. BPhO; Herbert von Karajan. DG Tape (c) 3335
    110.
```
    +HFN 10-75 p155 tape
```
1344 Hungarian rhapsodies, nos. 2, 5, 9, 14, 15, 19. Roberto Szidon,
    pno. DG 2530 441. Tape (c) 3300 386. (Reissue from 2720 072)
```
    +Gr 7-74 p234              +RR 7-74 p69
    ++NR 8-75 p13             +RR 9-74 p96 tape
```
Hungarian rhapsody, no. 2, G 244, C sharp minor. cf Piano works
    (Saga 5405).
Hungarian rhapsody, no. 2, G 244, C sharp minor. cf CHOPIN: Polish
    songs, op. 74: Maiden's wish.
Hungarian rhapsody, no. 2, G 244, C sharp minor. cf HMV SLS 5019.
Hungarian rhapsody, no. 2, G 244, C sharp minor. cf Monitor MCS
    2143.
Hungarian rhapsodies, nos. 4, 5, G 244. cf BRAHMS: Hungarian
    dances, nos. 1, 3, 5, 6.
Hungarian rhapsody, no. 6, G 244, D flat major. cf Sonata, piano,
    G 178, B minor.
1345 Hungarian rhapsodies, nos. 8-11, 13, G 244. Claudio Arrau, pno.
    Desmar DSM 1003.
```
    ++MJ 12-75 p38              +NR 12-75 p13
```
Hungarian rhapsody, no. 9, G 244, E flat major. cf Westminster
    WGM 8309.
Hungarian rhapsody, no. 10, G 244, E minor. cf CHOPIN: Polish
    songs, op. 74: Maiden's wish.
Hungarian rhapsody, no. 11, G 244, A minor. cf HMV HQS 1353.
Hungarian rhapsody, no. 13, G 244, A minor. cf Rococo 2049.
Hungarian rhapsody, no. 14, G 244, F minor. cf Piano works (Saga
    5405).
Die Ideale, G 106. cf Symphonic poems (Philips 6500 191).
Introduction, fugue and magnificat. cf Ad nos, ad salutarem undam,
    fantasia and fugue, G 259.
Introduction, fugue and magnificat. cf Hungaroton SLPX 11601/2.
Introitus, G 268, no. 1. cf Organ works (Vox SVBX 5328/9).
1346 Legend of Saint Elizabeth, op. 2. Eva Andor, s; Erzsébet Komlóssy,

ms; Sándor Nagy, Lajos Miller, György Bordás, bar; Kolos Kováts, József Gregor, bs; Czech Radio Children's Choir; Slovak Philharmonic Orchestra and Chorus; János Feréncsik. Hungaroton SLPX 11650/52.

      +Gr 12-75 p1089   +RR 10-75 p84
      +NR 11-75 p10   ++SFC 10-19-75 p33

Legendes, G 175. cf Consolations, G 172.

Legend, G 175: St. Francis of Assisi. cf Draco DR 1333.

Legendes, G 175, no. 1, St. Francis of Assisi; No. 2, St. Francis of Paola. cf Années de pelerinage, 2nd year, G 161.

Liebestraum, G 541. cf Funerailles. G 173.

Liebestraum, G 541. cf Connoisseur Society (Q) CSQ 2065.

Liebestraum, no. 3, G 541, A flat major. cf Etudes d'exécution transcendente, G 139.

Liebestraum, no. 3, G 541, A flat major. cf Piano works (Saga 5405).

Liebestraum, no. 3, G 541, A flat major. cf Rococo 2049.

Liebestraum, no. 3, G 541, A flat major. cf Turnabout TV 37033.

Mazeppa, G 138. cf Hungarian rhapsodies, nos. 2 and 4, G 244.

Mephisto waltz. cf HMV RLS 717.

1347 Mephisto waltz, no. 1 (Der Tanz in der Dorfschenke), G 514. Symphonic poems: Tasso, lamento e trionfo, G 96; Von der Wiege bis zum Grabe, G 107. Orchestre de Paris; Georg Solti. London CS 6925. (also Decca SXL 6709)

     ++Gr 5-75 p1967   +RR 6-75 p42
     ++HFN 5-75 p132   ++SFC 6-29-75 p26
      +NR 8-75 p5   ++St 11-75 p128

1348 Mephisto waltz, no. 1, G 514. Symphonic poems: Heroïde funèbre, G 102. Prometheus, G 99. LPO; Bernard Haitink. Philips 6500 190. (Reissue from 6709 005)

     ++Gr 8-74 p363   +RR 7-74 p44
      +MJ 3-75 p33   ++St 9-74 p125

Missa pro organo, G 264. cf Organ works (Vox SVBX 5328/9).

The nightingale, G 250. cf GLINKA: Nocturne.

Nun danket alle Gott, G 61. cf Organ works (Vox SVBX 5328/9).

Offertorium, G 667. cf Organ works (Vox SVBX 5328/9).

Oh, Quand je dors, G 282. cf Columbia M 32231.

Operatic transcriptions. cf Paraphrases.

Ora pro nobis, G 262. cf Organ works (Vox SVBX 5328/9).

1349 Organ works: Ad nos ad salutarem undam, fantasia and fugue, G 259. Am grabe Richard Wagners, G 267. Andante maestoso, G 668. Andante religioso, G 261a. Années de pelerinage, 3rd year, G 163: Angelus. Ave Maria, G 659. Ave maris stella, G 669, no. 2. Dante. Evocation à la Chapelle Sixtine, G 658. Gebet, G 265. Hasannah, G 677. Introitus, G 268, no. 1. Missa pro organo, G 264. Nun danket alle Gott, G 61. Offertorium, G 667. Ora pro nobis, G 262. Prelude and fugue on the name B-A-C-H, G 260. Preludium. Prière aux anges gardiens, G 163. Requiem, organ, G 266. Resignazione, G 263 (Adagio, D flat major). Rosario, G 670. Salve Regina, no. 1, G 669. San Francesco, G 665. Trauerode, G 268. Tu es Petrus, G 664. Ungarns Gott, G 674. Weimars Volkslied, G 672. Variations on Bach's "Weinen, Klagen, Sorgen, Zagen", G 673. Zur Trauung, G 60. János Sebestyén, org. Vox SVBX 5328/9.

      +-HF 5-75 p78   +NR 9-74 p11

1350 Paraphrases: Bellini: Norma, G 655. VERDI: Rigoletto, G 434. Il trovatore: Miserere, G 433. WAGNER: Der Mesitersinger von

Nürnberg: Am stillen Herd, G 448. Tannhäuser: Overture, G 442.
Tristan und Isolde; Liebestod, G 447. Craig Sheppard, pno.
Classics for Pleasure CFP 40206.
+-HFN 9-75 p100   +RR 9-75 p60
1352 Paraphrases: BELLINI: Sonnambula: Fantaisie, G 393. Norma, G 394.
GLINKA: Russlan and Ludmila: Tscherkessenmarsch, G 406. MOZART:
Don Juan, G 418. Richard and John Contiguglia, pno. Connois-
seur Society CS 2039. Tape (c) E 1027.
++Gr 12-75 p1065   +NYT 10-28-73 pD21
+HF 7-72 p84   ++RR 10-75 p78
+HF 1-75 p110 tape   +SR 5-20-72 p50
++MJ 2-75 p39   ++St 10-72 p1065
+NR 2-73 p11
1352 Paraphrases: BEETHOVEN: Ruins of Athens, G 649. BELLINI: Norma,
G 655. MOZART: Don Juan, G 656. ROSSINI: Soirées musicales,
no. 9: La danza, G 424. Bracha Eden, Alexander Tamir, pno.
Decca SXL 6708.
/Gr 9-75 p48   +-RR 8-75 p57
/HFN 8-75 p77
1353 Paraphrases: BELLINI: I puritani: Hexameron, G 392. MEYERBEER:
Robert le Diable: Valse infernale, G 413. TCHAIKOVSKY: Eugene
Onegin: Polonaise, G 429. WAGNER: Der fliegende Holländer:
Spinnenlied, G 440. Sylvia Kersenbaum, pno. HMV HQS 1342.
+-Gr 4-75 p1839   +-RR 4-75 p51
+-HFN 5-75 p132   +SR 1-25-75 p50
1354 Paraphrases: BELLINI: Norma, G 394. GOUNOD: Faust: Waltz, G 407.
MEYERBEER: Le prophète, no. 2: Les patineurs, scherzo, G 414.
TCHAIKOVSKY: Eugene Onegin: Polonaise, G 429. Michele Campa-
nella, pno. Philips 6500 310.
+-HF 9-75 p86   +SFC 6-29-75 p26
++NR 8-75 p13   -St 11-75 p127
Paraphrase: DONIZETTI: Dom Sebastien: Funeral march, G 402. cf
Ballade, no. 2, G 171, B minor.
Paraphrases: DONIZETTI: Lucia di Lammermoor, G 397. VERDI: Rigo-
letto, G 434. cf International Piano Library IPL 5005/6.
Paraphrase: GOUNOD: Faust: Waltz, G 407. cf Funerailles, G 173.
Paraphrases: SCHUBERT: Horch, Horch, Die Lerch, G 558/9; Der
Muller und der Bach; Liebesbotschaft; Das Wandern. cf CHOPIN:
Sonata, piano, no. 3, op. 58, B minor.
Paraphrase: TCHAIKOVSKY: Eugene Onegin: Polonaise, G 429. cf
GRAINGER: Paraphrase on TCHAIKOVSKY: Nutcracker: Waltz of the
flowers.
Paraphrase: VERDI: Rigoletto, G 434. cf Rococo 2049.
Paraphrase: VERDI: Rigoletto, G 514. cf Philips 6747 199.
Paraphrase: WAGNER: Tannhäuser: Overture, G 422. cf Classics for
Pleasure CFP 40205.
1355 Piano works: Consolation, no. 3, G 172, D flat major. Fantasia on
Wagner's Rienzi: Sancta Spirito cavaliere, G 439. Harmonies
poètiques et réligieuses, G 173: Funerailles. Hungarian rhap-
sody, no. 2, G 244, C sharp minor. Hungarian rhapsody, no. 14,
G 244, F minor. Liebestraum, no. 3, G 541, A flat major. David
Wilde, pno. Saga 5405.
+-RR 12-75 p83
Les préludes, G 97. cf Hungarian rhapsodies, nos. 2 and 4, G 244.
Les préludes, G 97. cf HMV RLS 717.
Les préludes, G 97. cf HMV SLS 5019.

Prelude and fugue on the name B-A-C-H, G 260. cf Ad nos, ad salut-
   arem undam, fantasia and fugue, G 259.
Prelude and fugue on the name B-A-C-H, G 260. cf Organ works
   (Vox SVBX 5328/9).
Prelude and fugue on the name B-A-C-H, G 260. cf Abbey LPB 738.
Prelude and fugue on the name B-A-C-H, G 260. cf Decca 5BBA 1013-5.
Prelude and fugue on the name B-A-C-H, G 260. cf Pelca PSRK 41013/6.
Preludium. cf Organ works (Vox SVBX 5328/9).
Prière aux anges gardiens, G 163. cf Organ works (Vox SVBX 5328/9).
1356 Prometheus, G 99. Eva Andor, s; Erzsébet Komlóssy, con; József
   Réti, Béla Turpinsky, t; Lajos Miller, bar; József Gregor, bs;
   Adolf Peter Hoffmann, narrator; Budapest Choir; HSO; Miklós
   Forrai. Hungaroton SLPX 11604.
       +-Gr 2-75 p1535   +NYT 8-25-74 pD20
       +NR 4-75 p6   +/RR 1-75 p56
Prometheus, G 99. cf Mephisto waltz, no. 1, G 514.
1357 Psalms, 13, 18, 23, 125, 129, G 13-16. Jószef Réti, t; László
   Jambór, bar; Anna Lelkes, hp; Sándor Margittay, org; Budapest
   Chorus; Hungarian People's Army Male Chorus; Hungarian State
   Orchestra; Miklós Forrai. Qualiton LPX 1261.
       +-LJ 10-75 p48
Reminiscences. cf Paraphrases.
Requiem, organ, G 266. cf Organ works (Vox SVBX 5328/9).
Resignazione, G 263 (Adagio, D flat major). cf Organ works (Vox
   SVBX 5328/9).
Rhapsodie espagnole, G 254. cf Etudes de concert, G 144.
Rosario, G 670. cf Organ works (Vox SVBX 5328/9).
Rumanian rhapsodie, G 242. cf Etudes de concert, G 144.
Salve Regina, no. 1, G 669. cf Organ works (Vox SVBX 5328/9).
San Francesco, G 665. cf Organ works (Vox SVBX 5328/9).
1358 Sonata, piano, G 178, B minor. Hungarian rhapsody, no. 6, G 244,
   D flat major. Alexander Slobodyanik, pno. Columbia/Melodiya
   M 33119.
       +-HF 7-75 p79   +-St 6-75 p98
       +-NR 2-75 p12
Sonata, piano, G 178, B minor. cf Concerto, piano, no. 2, G 125,
   A major (Hungaroton SHLX 90005).
Sonata, piano, G 178, B minor. cf Concerto, piano, no. 2, G 125,
   A major (Qualiton 11368).
Sonata, piano, G 178, B minor. cf Funerailles, G 173.
Sonata, piano, G 178, B minor. cf CHOPIN: Sonata, piano, no. 2,
   op. 35, B flat minor.
Sonata, piano, G 178, B minor. cf CHOPIN: Sonata, piano, no. 3,
   op. 58, B minor.
Sonata, piano, G 178, B minor. cf FELD: Sonata, piano.
1359 Songs: Liebestraum (cycle): Du bist wie eine Blume (Heine); Einst
   (Bodenstedt); Es war ein König in Thule (Goethe); Gestroben war
   ich (Uhland); Hohe Liebe (Uhland); Jeanne d'Arc au bücher (Dumas);
   O Lieb...(Freiligrath); S'il est un charmant garzon (Hugo); La
   tome et la rose (Hugo); Verlassen (Mitchell); Wer nie sein Brot
   mit Tränen ass (Goethe); Wilhelm Tell, Der Fischerknabe; Der
   Hirt; Der Alpenjäger (Schiller). Margit László, Erika Sziklay,
   s; Marta Szirmay, con; György Melis, bar; Tibor Webner; György
   Miklós, pno. Qualiton LPX 1272.
       +-ARG 3-72 p316   +SFC 3-31-74 p29
       +LJ 5-75 p44

Songs: Anfangs woll't ich fast verzagen, G 311; Ihr Auge, G 310;
    Kling leise mein Lied, G 301; Über allen Gipfeln ist Ruh, G 306;
    Die Loreley, G 273; Vergiftet sind meine Lieder, G 289; Wieder
    möcht ich dir begegnen, G 322; Die drei Zigeuner, G 320. cf
    CHABRIER: Songs (Orion ORS 75174).
Songs: Benedetto sia'l giorno; I'vidi in terra angelici costumi;
    Pace non trovo. cf DG 2530 332.
Songs: Comment disaient-ils, G 276; O, quand je dors, S 282; Über
    allen Gipfeln ist Ruh, G 306; Vergiftet sind meine Lieder, G
    289; Mignon's Lied, G 275. cf Desto DC 7118/9.
1360 Symphonic poems: Festklänge, G 101/2. Die Ideale, G 106. LPO;
    Bernard Haitink. Philips 6500 191. (Reissue from 6709 005)
        ++Gr 8-74 p363   ++RR 7-74 p43
        +MJ 3-75 p33   ++St 9-74 p125
    Symphonic poems: Tasso, lamento e trionfo, G 96; Von der Wiege bis
    zum Grabe, G 107. cf Mephisto waltz, no. 1 (Der Tanz in der
    Dorfschenke), G 514.
    Symphonic poem: Tasso, lamento e trionfo, G 96. cf Hungaria, G
    103.
1361 A symphony on Dante's "Divina Commedia", G 109. Luxembourg Radio
    Orchestra; La Psallette Vocal Ensemble; Pierre Cao. Candide
    (Q) QCE 31082.
        +HF 7-75 p96 Quad   -SFC 3-9-75 p26 Quad
        /NR 6-75 p4   /St 8-75 p100 Quad
1362 Totentanz, G 126. SCHUBERT: Fantasia, op. 15, D 760, C major
    (arr. Liszt). Alfred Brendel, pno. Turnabout TV 34265. Tape
    (c) KTVC 34265.
        +Gr 5-75 p2031 tape   +HFN 5-75 p142 tape
    Totentanz, G 126. cf FRANCK: Symphonic variations, piano and
    orchestra.
    Trauerode, G 268. cf Organ works (Vox SVBX 5328/9).
    Tu es Petrus, G 664. cf Organ works (Vox SVBX 5328/9).
    Ungarns Gott, G 674. cf Organ works (Vox SVBX 5328/9).
    Valse oubliée, no. 1, G 215. cf Decca SPA 372.
    Variations on Bach's "Weinen, Klagen, Sorgen, Zagen", G 673. cf
    Organ works (Vox SVBX 5328/9).
    Weihnachtsbaum, G 186. cf Consolations, G 172.
    Weimars Volkslied, G 672. cf Organ works (Vox SVBX 5328/9).
LISZYNYAI-SZABO
    Két magyar pasztorál. cf Hungaroton SLPX 11548.
LITOLFF, Henri
    Concerto symphonique, no. 4, op. 102, D minor: Scherzo. cf HMV
    SLS 5033.
    Concerto symphonique, no. 4, op. 102, D minor: Scherzo. cf Sound
    Superb SPR 90049.
LLOYD, Richard
    View me, Lord, a work of thine. cf Abbey LPB 734.
LOCATELLI, Pietro
    Sonata, flute and harpsichord. cf Klavier KS 537.
LOEILLET, Jean-Baptiste
    Corente. cf DG Archive 2533 172.
    Gigue. cf DG Archive 2533 172.
    Sarabande. cf DG Archive 2533 172.
    Sonata, flute, F major. cf London STS 15198.
    Sonata, flute, op. 1, A minor. cf GIULIANI: Sonata, flute and
    guitar, op. 85.

Sonata, flute, recorder and harpsichord, op. 1, G minor. cf Orion
    ORS 75199.
Sonata, recorder and harpsichord, C minor. cf Telefunken Tape (c)
    CX 4-41203.
Sonata, recorder, op. 3, no. 3, G minor. cf Transatlantic TRA 292.
Sonata, trumpet and organ, C major. cf RCA CRL 2-7001.
LOEVENDIE, Theo
    Scaramuccia. cf KEURIS: Concerto, saxophone.
LOEWE, Karl
    Tom der Reimer, op. 135. cf RCA ARL 1-5007.
LOFFELHOLTZ, Christoph
    Die kleine Schlacht. cf Musical Heritage Society MHS 1790.
LOGY, Johann
    Aria. cf Pelca PSR 40589.
    Sarabande and gigue. cf Pelca PSR 40589.
LONDON, Edwin
1363 Portraits of three ladies (American). CRUMB: Madrigals, Bks I-IV.
    Marilyn Coles, s; Royal MacDonald, narrator; University of
    Illinois Contemporary Chamber Ensemble; Edwin London; Jan De-
    Gaetani, ms; University of Pennsylvania Contemporary Chamber
    Players; Richard Wernick. DG/Acoustic Research 0654 085.
            +HF 2-71 p97   *NYT 1-3-71 pD22
            +MJ 2-75 p29   *SR 12-26-70 p48
            +-MQ 7-72 p501   +-St 5-71 p95
LOQUEVILLE, Richard de
    Sanctus. cf Telefunken ER 6-35257.
LORTZING, Gustav
    Der Waffenschmied: Man wird ja einmal nur geboren; Ach, er fuhlt
    nicht wie ich...er ist so gut; War einst ein junger Springin-
    feld; Wir armen, armen Mädchen; Auch ich war ein Jüngling. cf
    Zar und Zimmermann: O sancta justitia; Zimmermannslied; Die
    Eifersucht ist eine Plage; Lebe wohl, mein Flanrisch Mädchen;
    Den hohen Herrscher wurdig zu empfangen; Holzschutanz; Sonst
    spielt ich mit Zepter.
1364 Zar und Zimmermann: O sancta justitia; Zimmermannslied; Die Eifer-
    sucht ist eine Plage; Lebe wohl, mein Flandrisch Mädchen; Den
    hohen Herrscher wurdig zu emfangen; Holzschutanz; Sonst spielt
    ich mit Zepter. Der Waffenschmied: Man wird ja einmal nur ge-
    boren; Ach, er fuhlt nicht wie ich...er ist so gut; War einst
    ein junger Springinsfeld; Wir armen, armen Mädchen; Auch ich
    war ein Jüngling. Soloists; Vienna State Opera Chorus; Vienna
    Volksoper Orchestra; Peter Ronnefeld. Decca SDD 462.
            +RR 9-75 p23
    Zar und Zimmermann: Van Bett's aria. cf Rubini RS 300.
LOTTI, Antonio
    Pur dicesti, o bocca bella. cf Pearl SHE 511.
    LOVE, LUST, PIETY AND POLITICS. cf BASF 25 22286-1.
LOVER, Samuel
    The low-back'd car. cf Philips 6599 227.
LUBBERT
    Helenen Marsch. cf BASF 292 2116-4.
LUBECK, Vincenz
    Prelude and fugue, E major. cf Pelca PSR 40598.
    Prelude and fugue, E major. cf Telefunken DX 6-35265.
LUBLINA, Jn Z (Lublin, Jan van)
    Dances (4). cf Saga 5374.

LUCIANO PAVAROTTI, King of the high C's. cf London OS 26373.
LUENING, Otto
1365 Fugue and chorale fantasy with electronic doubles for organ and
       tape. Sonata, piano. Ursula Oppens, pno; Alec Wyton, org;
       tape part realized at the Columbia Princeton Electronic Music
       Studio. CRI SD 335.
            ++NR 9-75 p14
     Sonata, piano. cf Fugue and chorale fantasy with electronic doubles
       for organ and tape.
LULLY, Jean
1366 Alceste. Felicity Palmer, Anne-Marie Rodde, Sonia Nigoghossian, s;
       Bruce Brewer, John Elwes, t; Max von Egmond, Pierre-Yves Le
       Maiget, bs; Raphael Passaquet Vocal Ensemble; Le Grand Ecurie
       et La Chambre du Roy; Jean-Claude Malgoire. CBS 79301 (3).
            +HFN 12-75 p156   ++RR 11-75 p32
     Alceste: Air de Caron. cf FAURE: Songs (1750 Arch S 1754).
     Amadis: Bois épais. cf Rococo 5365.
     Au clair du la lune. cf Discophilia KGC 1.
1367 Le bourgeois gentilhomme. Rachel Yakar, Dorothea Jungmann, s;
       René Jacobs, c-t; Norbert Lohmann, Klaus Heider, Michel Lecocq,
       t; Siegmund Nimsgern, Dirk Schortemeier, Franz Müller-Heuser,
       bar; Tölzer Boys' Choir; La Petite Bande; Gustav Leonhardt.
       BASF BAC 3078/9 (2).
            +Gr 5-75 p2010   +RR 4-75 p16
            +MT 9-75 p714
     Gigas. cf Westminster WGM 8309.
1368 Music from Moliere's plays: Le bourgeois gentilhomme: Turkish scene;
       Menuet; Ballet des nations; Chaconne. Les amants magnifiques:
       Menuets. George Dandin: Overture. Pastorale comique: Airs
       des Egyptiens et des Egyptiennes; Peasants fighting and recon-
       ciled. Psyché: Air de trompette; Prélude; Rondeau pour les
       Enseignes; Air pour les Enseignes, no. 2; Symphonie; Lamenti;
       Dishevelled women; Lamento. Suzanne Simonka, s; John Elwes, t;
       Luis Masson, bar; La Grande Ecurie et la Chambre du Roy, Male
       Chorus; Jean-Claude Malgoire. CBS 76184.
            +-Gr 5-75 p2006   +-RR 6-75 p23
     Unce noce de village: Dernière entrée. cf DG Archive 2533 172.
LUNDEN, Lennart
     Lilltåa och 9 till. cf BIS LP 2.
LURANO, Filipp de
     Se me grato. cf Telefunken TK 11569/1-2.
     LUTE MUSIC OF THE RENAISSANCE, ENGLAND. cf DG Archive 2533 157.
LUTOSLAWSKI, Witold
     Concerto, violoncello. cf DUTILLEUX: Concerto, violoncello.
     Paroles tissées. cf BEDFORD: The tentacles of the dark nebula.
     Poèmes d'Henri Michaux. cf BOULEZ: e. e. cummings.
LUZZASCHI
     O dolcezze. cf Telefunken TK 11569/1-2.
McBAIN
     Brother James' air (arr. Jacob). cf Decca ECS 774.
McCABE, John
1369 The Chagall windows. Variations on a theme by Karl Amadeus Hart-
       mann. Hallé Orchestra; James Loughran. HMV (Q) ASD 3096.
            +Gr 10-75 p621   ++RR 9-75 p39
            ++HFN 10-75 p144

McCABE (cont.)   240

Variations on a theme by Karl Amadeus Hartmann. cf The Chagall
    windows.
McCALL, J. P.
    Songs: Kelly, the boy from Killane; Boolavogue. cf Philips 6599
        227.
MacDERMID, James
    He that dwelleth in the secret place. cf RCA ARL 1-0561.
    In my Father's house are many mansions. cf RCA ARL 1-0562.
MacDOWELL, Edward
    Concerto, piano and orchestra, no. 2, op. 23, D minor. cf Suite,
        orchestra, no. 2, op. 48.
1370 Etudes, op. 39, Bk 2, no. 2: Shadow dance. Fantastic pieces, op.
        17, no. 2: Witches' dance. Modern suite, op. 14, no. 2. Sea
        pieces, op. 55. Andrea Anderson Swem, pno. Orion ORS 75175.
            +NR 9-75 p12   +St 12-75 p128
    Fantastic pieces, op. 17, no. 2: Witches' dance. cf Etudes, op.
        39, Bk 2, no. 2: Shadow dance.
    Modern suite, op. 14, no. 2. cf Etudes, op. 39, Bk 2, no. 2:
        Shadow dance.
    Sea pieces, op. 55. cf Etudes, op. 39, Bk 2, no. 2: Shadow dance.
    Sea pieces, op. 55, no. 1. cf HMV HLM 7065.
1371 Sonata, piano, no. 2, op. 50, G minor. Sonata, piano, no. 3, op.
        57, D minor. Yoriko Takahashi, pno. Orion ORS 75183.
            ++HF 12-75 p98   ++NR 10-75 p10
    Sonata, piano, no. 3, op. 57, D minor. cf Sonata, piano, no. 2,
        op. 50, G minor.
1372 Suite, orchestra, no. 2, op. 48. Concerto, piano and orchestra,
        no. 2, op. 23, D minor. Eugene List, pno; Westphalian Symphony
        Orchestra; Siegfried Landau. Turnabout TVS 34535.
            +Gr 3-75 p1653   +RR 2-75 p36
            +HF 4-74 p98   +-St 5-74 p102
            ++NR 3-74 p5
    To a wild rose. cf Connoisseur Society (Q) CSQ 2066.
McLEAN, Barton
1373 Dimensions II. Genesis. The sorcerer revisited. Robert Hamilton,
        pno. Orion ORS 75192.
            -NR 10-75 p11
    Genesis. cf Dimensions II.
    The sorcerer revisited. cf Dimensions II.
    Spirals. cf McLEAN, P.: Dance of dawn.
McLEAN, Priscilla
1374 Dance at dawn. McLEAN, B.: Spirals. Tape realized at the
        Indiana University Electronic Music Studio. CRI SD 335.
            *Audio 11-75 p89   +NR 9-75 p14
MACHA, Otmar
    March. cf Panton 110 361.
    Varianty. cf CHAUN: Proces.
MACHAUT, Guillaume
    Ballad and plus dure. cf BIS LP 2.
    Comment qu'à moy. cf Vanguard VSD 71179.
    Douce dame jolie. cf Vanguard VSD 71179.
    Je sui aussi. cf Vanguard VSD 71179.
    Rose, liz, printemps. cf Vanguard VSD 71179.
1375 Songs: Ce qui soutient; Chanson roiale; Dame a vous; Dame de qui
        toute; Douce dame; Douce dame jolie; Hoquet David; Ma fin est
        mon commencement; Mes esperis de combat; Plourez dame; Plus

dure; Rose liz; Sans coeur dolens; Tels rit au main; Tuit me
penser. Paris Ars Antiqua; Michel Sanvoisin.   Arion ARN 38252.
   -RR 8-75 p63
Voys porter. cf Vanguard VSD 71179.
MACHE, Françoise-Bernard
1376 Canzona, brass quintet, no. 2. Kemit, darbuka or zarb. Korward,
    harpsichord and magnetic tape. Temes Nevinbür, 2 pianos, per-
    cussion and magnetic tape. Elisabeth Chojnacka, hpd; Jean-
    Pierre Drouet, zarb, percussion; Katia and Marielle Labeque,
    pno; Ars Nova Brass Quintet. Erato STU 70860.
      +Gr 4-75 p1825   +RR 4-75 p40
    Kemit, darbuka or zarb. cf Canzona, brass quintet, no. 2.
    Korward, harpsichord and magnetic tape. cf Canzona, brass quintet,
      no. 2.
    Temes Nevinbür, 2 pianos, percussion and magnetic tape. cf Canzona,
      brass quintet, no. 2.
    THE MAD SCENES. cf ABC ATS 200019.
MADERNA, Bruno
    Dedication. cf BLACHER: Sonata, solo violin, op. 40.
    Pièce pour Ivry. cf BLACHER: Sonata, solo violin, op. 40.
    Viola. cf BEDFORD: Spillihpnerak.
    THE MAGIC OF VIENNA. cf Pye NSPH 6.
MAHLER, Gustav
1377 Kindertotenlieder. Symphony, no. 10, F sharp major: Adagio. Janet
    Baker, ms; Israel Philharmonic Orchestra, NYP; Leonard Bernstein.
    Columbia M 33532. (Q) MQ 33532.
      +HF 10-75 p76   ++SFC 9-21-75 p34
      +NR 11-75 p11   ++St 11-75 p127
    Kindertotenlieder. cf Symphony, no. 5, C minor.
    Kindertotenlieder. cf BERLIOZ: Les nuits d'été, op. 7.
1378 Das klagende Lied. Marta Boháčová, s; Věra Soukupová, con; Ivo
    Zidek, t; Czech Philharmonic Chorus; PSO; Herbert Ahlendorf.
    Supraphon 112 1329.
      /Gr 10-74 p738   +NR 11-74 p10
      +-HF 8-75 p89   /RR 9-74 p32
1379 Des Knaben Wunderhorn (Youth's magic horn): Das irdische Leben;
    Wo die schönen Trompeten; Urlicht. Rückert Lieder: Liebst du
    um Schönheit; Ich bin der Welt abhanden gekommen. SCHUBERT:
    Songs: Ave Maria, D 839; Jäger, ruhe von der Jagd, D 838; Raste,
    Krieger, D 837; Schwestergruss, D 762; Der Zwerg, D 771. Jessye
    Norman, s; Irwin Gage, pno. Philips 6500 412.
      ++Gr 10-73 p713   /NYT 1-21-73 pD26
      -HF 4-73 p100   +ON 5-75 p36
      +MJ 11-73 p8   +RR 10-73 p115
      +-NR 5-73 p12   +-St 3-73 p124
1380 Des Knaben Wunderhorn: Revelge; Der Tambourg'sell. Lieder und
    Gesange aus der Jugendzeit: Frühlingsmorgen; Erinnering; Hans
    und Grethe; Don Juan Serenade; Phantasie aus Don Juan; Um
    schlimme Kinder artig zu machen; Ich ging mit Lust durch einen
    grünen Wald; Aus, Aus; Starke Einbildungskraft; Zu Strassburg
    auf der Schanz; Ablösung im Sommer; Scheiden und Meiden; Nicht
    Wiedersehen; Selbstgefühl. Roland Hermann, bar; Geoffrey Par-
    sons, pno. HMV HQS 1346.
      +-Gr 10-75 p669   +-RR 11-75 p86
      +-HFN 10-75 p143
1381 Das Knaben Wunderhorn: Un schlimme Kinder artig; Ich ging mit Lust;

Des Antonius von Padua; Rheinlegendchen. Lieder und Gesange aus
der Jugendzeit: Frühlingsmorgen; Hans und Grethe. Rückert
Lieder: Ich atmet' einen Linden Duft; Ich bin der Welt abhanden.
SCHUBERT: An di Musik, D 547; Auf dem Wasser zu singen, D 774;
Ellens Gesang III, Ave Maria, D 839; Die Forelle, D 550; Der
Hirt auf dem Felsen, D 965; Der Musensohn, D 764. Christa Lud-
wig, ms; Gerald Moore, Geoffrey Parsons, pno. HMV SXLP 30182.
(Reissues from Columbia SAX 2358, 5272, 5274)
    +-Gr 4-74 p1850   +RR 3-75 p65

1382 Das Lied von der Erde. Christa Ludwig, ms; René Kollo, t; Israel
Philharmonic Orchestra; Leonard Bernstein. Columbia KM 31919.
Tape (c) MT 31919 (ct) MA 31919 (Q) KMQ 31919. (also CBS 76105)
    +Gr 2-75 p1535   +-RR 12-74 p72
    +HF 5-75 p78   +-SR 6-14-75 p46
    +NR 4-75 p9   +-St 7-75 p101

1383 Das Lied von der Erde. Rückert Lieder (5). Christa Ludwig, ms;
René Kollo, t; BPhO; Herbert von Karajan. DG 2707 082 (2).
    ++Gr 12-75 p1089   +RR 12-75 p88

1384 Das Lied von der Erde. Lieder eines fahrenden Gesellen. Nan
Merriman, con; Ernst Häfliger, t; COA; Eduard van Beinum. Philips
6780 013 (2). (Reissue from AOO 410)
    +Gr 4-75 p1250   +RR 4-75 p60
    +-HFN 5-75 p134

Lieder eines fahrenden Gesellen. cf Das Lied von der Erde.
Lieder eines fahrenden Gesellen. cf Symphony, no. 5, C minor.
Lieder eines fahrenden Gesellen. cf BERLIOZ: Les nuits d'été,
op. 7.
Lieder und Gesange aus der Jugendzeit: Frühlingsmorgen; Hans
und Grethe. cf Das Knaben Wunderhorn (HMV SXLP 30182).
Lieder und Gesange aus der Jugendzeit: Frühlingsmorgen; Erinnering;
Hans und Grethe; Don Juan Serenade; Phantasie aus Don Juan; Um
schlimme Kinder artig zu machen; Ich ging mit Lust durch einen
grünen Wald; Aus, Aus; Starke Einbildungskraft; Zu Strassburg
auf der Schanz; Ablösung im Sommer; Scheiden und Meiden; Nicht
Wiedersehen; Selbstgefühl. cf Des Knaben Wunderhorn: Revelge;
Der Tambourg'sell.
Rückert Lieder (5). cf Das Lied von der Erde.
Rückert Lieder. cf BERLIOZ: Les nuits d'été, op. 7.
Rückert songs. cf ELGAR: Sea pictures, op. 37.
Rückert Lieder: Ich atmet' einen Linden Duft; Ich bin der Welt
abhanden. cf Das Knaben Wunderhorn (HMV SXLP 30182).
Rückert Lieder: Liebst du um Schönheit; Ich bin der Welt abhanden
gekommen. cf Des Knaben Wunderhorn: Das irdische Leben; Wo
die schönen Trompeten; Urlicht.

1385 Symphonies, nos. 1-3. Heather Harper, s; Helen Watts, con; Ambrosian Opera Chorus, Wandsworth School Boys' Chorus; LSO and Chorus;
Georg Solti. Decca 7BB 173/7 (5). (Reissues from SXL 6113, SET
325/6, 385/6)
    +Gr 5-75 p1967   +-RR 5-75 p34
    ++HFN 5-75 p132

1386 Symphony, no. 1, D major. Columbia Symphony Orchestra; Bruno
Walter. CBS 61116. Tape (c) 40-61116.
    +-HFN 12-75 p173 tape   +-RR 12-75 p99 tape

1387 Symphony, no. 1, D major. LSO; Jascha Horenstein. Nonesuch 71240.
Tape (c) D 1019. (also Unicorn RHS 301)
    +HF 12-74 p146 tape   ++NR 8-75 p5

1388 Symphony, no. 1, D major. LSO; James Levine. RCA ARL 1-0894.
     Tape (c) ARK 1-0894 (ct) ARS 1-0894.
                +Gr 5-75 p1967   ++RR 5-75 p33
                -HF 6-75 p98   ++SFC 5-11-75 p23
                ++HFN 5-75 p134   +-SR 5-3-75 p33
                -NR 5-75 p5   +St 5-75 p96
1389 Symphony, no. 1, D major. Symphony, no. 4, G major. Judith Blegen,
     s; LSO, CSO; James Levine. RCA (Q) CRD 3-1040 (3).
                +Audio 11-75 p89 Quad   +-NR 11-75 p4 Quad
1390 Symphony, no. 1, D major. Symphony, no. 5, C sharp minor. Symph-
     ony, no. 6, A minor. Symphony, no. 10, F sharp major: Adagio.
     Utah Symphony Orchestra; Maurice Abravanel. Vanguard SRV 320/4
     (5). (Q) VSQ 30044.
                +HF 6-75 p98   +St 10-75 p105
                +NR 7-75 p3
1391 Symphony, no. 2, C minor. Janet Baker, Sheila Armstrong, s; Edin-
     burgh Festival Chorus; LSO; Leonard Bernstein. Columbia M2
     32681 (2). Tape (c) M2T 32681 (ct) M2A 32681 (Q) M2Q 32681
     Tape (ct) QMA 32681. (also CBS 78249)
                +-Gr 11-74 p895   +RR 12-74 p31
                +HF 1-75 p71   -SFC 9-29-74 p26
                +-NR 12-74 p3   +-St 1-75 p113
                +NYT 8-4-74 pD20
1392 Symphony, no. 2, C minor. Ileana Cotrubas, s; Christa Ludwig, con;
     Vienna State Opera Chorus; VPO; Zubin Mehta. Decca SXL 6744/5.
     Tape (c) KSXC 2-7037.
                ++Gr 12-74 p1037   +RR 12-75 p55
1393 Symphony, no. 2, C minor. Mimi Coertse, s; Lucretia West, ms;
     Vienna Academy Choir; VSOO; Hermann Scherchen. Westminster WGS
     8262-2 (2).
                +-HF 1-75 p71
1394 Symphony, no. 4, G major. Symphony, no. 8, E flat major. Symphony,
     no. 9, D major. Sylvia Stahlman, Lucia Popp, Heather Harper,
     Arleen Auger, s; Yvonne Minton, Helen Watts, con; René Kollo, t;
     John Shirley-Quirk, bar; Martti Talvela, bs; COA, CSO, LSO;
     Vienna State Opera Chours; Vienna Singverein, Vienna Boys'
     Choir; Georg Solti. Decca 7BB 183/7 (5). (Reissues from SXL
     2276, SET 534/5, 360/1)
                +Gr 11-75 p818   +-RR 9-75 p39
                +-HFN 9-75 p108
1395 Symphony, no. 4, G major. Elsie Morison, s; Rudolf Koeckert, vln;
     Bavarian Radio Symphony Orchestra; Rafael Kubelik. DG 2535 119.
     (Reissue from 139339) (also DG 139339. Tape (c) 923-082)
                +-Gr 8-75 p325   +RR 7-75 p26
                +HFN 8-75 p89
1396 Symphony, no. 4, G major. Judith Blegen, s; CSO; James Levine.
     RCA ARL 1-0895. Tape (c) ARK 1-0895 (ct) ARS 1-0895.
                ++Gr 10-75 p612   +NR 5-75 p5
                ++HF 6-75 p98   ++RR 11-75 p43
                -HFN 11-75 p155   ++St 8-75 p100
1397 Symphony, no. 4, G major. CPhO; Hans Swarowsky. Supraphon 110
     1346.
                +HFN 10-75 p143   +-RR 10-75 p44
                +NR 11-75 p4
     Symphony, no. 4, G major. cf Symphony, no. 1, D major.
1398 Symphony, no. 5, C sharp minor. Symphony, no. 6, A minor. Symphony,

no. 7, D minor. CSO; Georg Solti. Decca 7BB 178/182 (5).
(Reissues from SET 471/2, 469/70; 518/9)
    +-Gr 8-75 p325   +-RR 7-75 p26
    ++HFN 8-75 p89

1399 Symphony, no. 5, C minor. Kindertotenlieder. Christa Ludwig, ms;
    BPhO; Herbert von Karajan. DG 2707 081 (2). Tape (c) 3370 006.
        +Gr 6-75 p40   ++NR 12-75 p5
        +HFN 9-75 p100   ++RR 6-75 p43
        +-MT 11-75 p978   +-RR 10-75 p97 tape

1400 Symphony, no. 5, C minor. Lieder eines fahrenden Gesellen.
    Hermann Prey, bar; COA; Bernard Haitink. Philips Tape (c) 7505
    069 (2).
        +HF 8-75 p116 tape   +RR 4-75 p77 tape

1401 Symphony, no. 5, C sharp minor. Symphony, no. 10, F sharp major:
    Adagio. Utah Symphony Orchestra; Maurice Abravanel. Vanguard
    SRV 321/2 (2).
        /Audio 11-75 p89
  Symphony, no. 5, C sharp minor. cf Symphony, no. 1, D major.

1402 Symphony, no. 6, A minor. (Includes Horenstein interview with
    Alan Blyth) Stockholm Philharmonic Orchestra; Jascha Horenstein.
    Nonesuch HB 73029 (2).
        +-HF 10-75 p76   +NR 9-75 p2
        +HFN 9-75 p100   +St 12-75 p129

1403 Symphony, no. 6, A minor. Stockholm Philharmonic Orchestra; Jascha
    Horenstein. Unicorn RHS 320/1 (2).
        ++Gr 9-75 p465   +RR 9-75 p40
        +HFN 9-75 p100

1404 Symphony, no. 6, A minor. Utah Symphony Orchesta; Maurice Abravanel.
    Vanguard SRV 323/4 (2).
        +Audio 11-75 p88
  Symphony, no. 6, A minor. cf Symphony, no. 1, D major.
  Symphony, no. 6, A minor. cf Symphony, no. 5, C sharp minor.
  Symphony, no. 7, B minor. cf Symphony, no. 5, C sharp minor.

1405 Symphony, no. 8, E flat major. Symphony, no. 10, F sharp major:
    Adagio. Martina Arroyo, Erna Spoorenberg, s; Julia Hamari,
    Norma Procter, con; Donald Grobe, t; Dietrich Fischer-Dieskau,
    bar; Franz Crass, bs; Bavarian Radio, NDR Radio and WDR Radio
    Choruses, Regensburg Cathedral Boys' Choir, Munich Motet Choir
    Women's Chorus; Eberhard Kraus, org; Bavarian Radio Symphony
    Orchestra; Rafael Kubelik. DG 2726 053 (2). (Reissue from
    2707 062, 2707 037)
        +Gr 9-75 p465   +-RR 8-75 p39
        +HFN 8-75 p89

1406 Symphony, no. 8, E flat major. Joyce Barker, Elisabeth Simon,
    Norma Burrowes, s; Joyce Blackham, Alfreda Hodgson, ms; John
    Mitchinson, t; Raymond Myers, bar; Gwynne Howell, bs; Orpington
    Junior Singers, Highgate School Choir, Finchley Children's
    Music Group, Ambrosian Singers, New Philharmonia Chorus, Bruchner-
    Mahler Choir; London Symphonia; Wyn Morris. RCA CRL 2-0359 (2).
    Tape (c) CRK 2-0359 (ct) CRS 2-0359.
        +-HF 5-74 p73   +-ON 1-26-74 p36
        +LJ 4-75 p69   -SFC 5-19-74 p29
        ++NR 2-74 p3   +St 8-74 p117
        +-NYT 2-10-74 pD26
  Symphony, no. 8, E flat major. cf Symphony, no. 4, G major.

1407 Symphony, no. 9, D major. NPhO; Otto Klemperer. HMV SXLP 3021/2.

(Reissue). (also Angel S 3708)
        ++HFN 12-75 p171   +RR 12-75 p56
Symphony, no. 9, D major. cf Symphony, no. 4, G major.
Symphony, no. 9, D major: Adagio. cf Columbia M2X 33014.
1408 Symphony, no. 10, F sharp major (performing edition by Deryck
        Cooke). PO; Eugene Ormandy. CBS 61447. Tape (c) 40-61447.
        +-HFN 12-75 p173   +RR 12-75 p99 tape
Symphony, no. 10, F sharp major: Adagio. cf Kindertotenlieder.
Symphony, no. 10, F sharp major: Adagio. cf Symphony, no. 1,
        D major.
Symphony, no. 10, F sharp major: Adagio. cf Symphony, no. 5,
        C sharp minor.
Symphony, no. 10, F sharp major: Adagio. cf Symphony, no. 8,
        E flat major.
MAINARDI
        Varca d' 'o promo ammore. cf LEONCAVALLO: I Pagliacci.
MAINERIO (17th century Italy)
        Primo libro de balli: Suite. cf Erato STU 70847.
        Schiarazula marazula. cf DG Archive 2533 184.
        Ungarescha-Saltarello. cf DG Archive 2533 184.
MALOTTE
        The Lord's prayer. cf RCA ARL 1-0562.
MANCINELLI, Luigi
        Cleopatra: Overture. cf Erato STU 70880.
MANN-REIZENSTEIN-WETHERELL
        Let's fake an opera or "The tales of Hoffnung". cf HMV SLA 870.
MANNEY
        Consecration. cf Discophilia KGS 3.
MANOLOV, Hristo
        Verouiou. cf Harmonia Mundi HMU 133.
MANZONI
        Quadruplum. cf Desto DC 6474/7.
MARAIS, Marin
        Alcyone: Suite. cf HANDEL: Concerto, oboe, no. 3, G minor.
        Allemande l'asmatique. cf Works, selections (Delos DEL 25403).
        Les folies d'Espagne: Couplets. cf Works, selections (Delos DEL
        25403).
        Les folies d'Espagne: Couplets. cf BACH: Sonata, flute (oboe) and
        harpsichord, S 1030, G minor.
        Les folies d'Espagne: Couplets. cf HUME: Captain Hume's galliard.
1409 Pièces de viole, Bk I, 1686: Suite, no. 1, D minor, excerpts; Suite
        no. 2: Chaconne, D major; Suite, no. 3, G minor (minus a prelude);
        Suite, no. 4: Rondeau, A major. John Hsu, vla da gamba; Judith
        Davidoff, flt; Louis Bagger, hpd. Musical Heritage Society
        MHS 1809.
        +-St 1-75 p113
        Prelude, D minor. cf Works, selections (Delos DEL 25403).
        La sincope. cf Works, selections (Delos DEL 25403).
        Suite, G minor. cf HUME: Captain Hume's galliard.
        Suites, no. 4, A minor; no. 5, A major. cf Works, selections
        (Delos DEL 25403)
1410 Suite, viola da gamba and harpsichord, E minor. SAINTE-COLOMBE:
        Concerts à duex voiles esgales (3). Catharina Meints, James
        Caldwell, vla da gamba; James Weaver, hpd. Cambridge CRS 2201.
        ++NR 3-75 p6
        Tombeau de M. de Ste. Colombe. cf Works, selections (Delos DEL 25403).

Tombeau pour M. de Lully. cf Works, selections (Delos DEL 25403).
1411 Works, selections: Allemande l'asmatique. Les folies d'Espagne:
     Couplets. Prelude, D minor. La sincope. Suites, no. 4, A
     minor; no. 5, A major. Tombeau de M. de Ste. Colombe. Tombeau
     pour M. de Lully. Eva Heinitz, vla da gamba; Malcolm Hamilton,
     hpd. Delos DEL 25403.
          +--Audio 12-75 p104   +NR 10-75 p13
MARBECK, John
     Credo. cf Argo ZRG 789.
MARCELLO, Alessandro
     Concerto, 2 guitars. cf ALBINONI: Adagio, organ and strings, G
     minor.
     Concerto, oboe, C minor. cf ALBINONI: Concerto, oboe, op. 7, no.
     3, B flat major.
1412 Concerto, oboe and strins. SAMMARTINI: Concerto, soprano recorder
     and strings. SCARLATTI, A.: Concerto, alto recorder and 2
     violins. STRADELLA: Sonata, trumpet and strings. Hermann
     Sauter, tpt; Gunther Holler, rec; Helmut Hucke, ob; South West
     German Chamber Orchestra; Paul Angerer. Turnabout (Q) QTV 34573.
          +-+NR 4-75 p5
     Concerto, oboe and strings, D minor. cf DALL'ABACO: Concerto da
     chiesa, a 4, op. 2.
     Sonata, flute, op. 2, B minor. cf Orion ORS 73114.
     Sonata, flute and harpsichord, G major. cf Armstrong 721-4.
MARCELLO, Benedetto
1413 Concerti a cinque, op. 1 (12). I Solisti di Milano; Angelo Ephrik-
     ian. Telefunken SAWT 9601/2 (2).
          +-+AR 2-75 p22   +HF 10-74 p107
          +Gr 12-73 p1198   +HFN 12-73 p2612
     Sonata, flute and harpsichord, B major. cf Ember ECL 9040.
     Sonata, recorder, D minor. cf Telefunken SMA 25121 T/1-3.
     Sonata, recorder and continuo, op. 2, no. 11, D minor. cf Tele-
     funken SAWT 9589.
MARCHAND
     Suite. cf DUPHLY: La Félix.
MARCHETTI, Filippo
     Ruy Blas: O dolce volutta. cf Rubini RS 300.
MARELLA, Giovanni Batista
     Suite, no. 1, A major. cf Pye GSGC 14154.
MARENZIO, Luca
     O voi che sospirate; Occhi lucenti. cf L'Oiseau-Lyre 12BB 203/6.
     Tirsir morir volea. cf Erato STU 70847.
MAREV, Krustiu
     Otche nach. cf Harmonia Mundi HMU 133.
MARGIS
     Valse bleue. cf Discopaedia MB 1007.
MARSHALL
     Suite catalonia: Foc-Follets. cf International Piano Library IPL
     5005/6.
MARTINI, Giovanni
     Sonata al postcommunio. cf ALBINONI: Concertos, trumpet, F major,
     D minor.
MARTINI IL TEDESCO, Johann
     Plaisir d'amour. cf Decca SXL 6629.
MARTINO, Donald
     Notturno. cf WUORINEN: Speculum speculi.

1414 Paradiso choruses. PINKHAM: For evening draws on. Liturgies.
     Toccatas for the vault of heaven. Soloists; New England Conserv-
     atory Symphony Orchestra and Chorus; Lorna Cooke de Varon.
     Golden Crest NEC 114.
          +NYT 12-21-75 pD18
MARTINU, Bohuslav
1415 Concerto, piano, no. 4. Sinfonietta giocosa. Stanislav Knor,
     Josef Pálenícek, pno; Prague Symphony Orchestra, Brno State
     Philharmonic Orchestra; Václav Smetácek, Jirí Pinkas. Supra-
     phon 58591.
          +Gr 1-75 p1342   -RR 1-75 p29
     Concerto, 2 orchestras, piano and timpani. cf JANACEK: Idyll,
     string orchestra.
1416 Concerti, violin, nos. 1, 2. Josef Suk, vln; CPhO; Václav Neumann.
     Supraphon 110 1535.
          ++Gr 7-75 p187   +RR 5-75 p34
          +NR 11-75 p7
1417 Concerto, violin, piano and orchestra. Concerto, violoncello, no.
     1. Josef Chuchro, vlc; Nora Grumliková, vla; Jaroslav Kolar,
     pno; CPhO; Zdenek Kosler. Supraphon 110 1348.
          +Gr 3-75 p1654   +RR 2-75 p36
          +-NR 1-75 p4   ++SFC 6-8-75 p23
     Concerto, violoncello, no. 1. cf Concerto, violin, piano and
     orchestra.
1418 Esquisses (3). Jazz suite. Le jazz. Le revue de cuisine. Sextet,
     wind instruments and piano. Who is the most powerful in the
     world. Shimmy foxtrot. Zdenek Jílek, Frantisek Rauch, Jan
     Panenka, pno; Karel Dlouhy, clt; Jiri Formácek, bsn; Václav
     Junek, tpt; Bruno Belcik, vln; Milos Sádlo, vlc; Prague Symphony
     Orchestra, Prague Wind Quintet; Lubomir Panek Singers; Zbynek
     Vostrák. Supraphon 110 1014.
          +Gr 7-73 p200   +LJ 3-75 p34
          +HF 9-73 p110   +RR 7-73 p50
     Etudes and polkas. cf BACH: Prelude and fugue, B flat minor.
     Le jazz. cf Esquisses.
     Jazz suite. cf Esquisses.
     Quartet, strings, no 2. cf HAYDN: Quartet, strings, op. 74, no. 3,
     G minor.
     Quartet, strings, no. 4. cf DVORAK: Quartet, strings, no. 2, op.
     34, D minor.
     Le revue de cuisine. cf Esquisses.
     Sextet, wind instruments and piano. cf Esquisses.
     Shimmy foxtrot. cf Esquisses.
     Sinfonietta giocosa. cf Concerto, piano, no. 4.
     Who is the most powerful in the world. cf Esquisses.
MARTIRANO, Salvatore
     Chansons innocentes. cf CRI SD 324.
MASCAGNI, Pietro
     L'Amico Fritz: Intermezzo, Act 3. cf HMV SLS 5019.
1419 Cavalleria rusticana. Margaret Harshaw, s; Thelma Votipka, Mildred
     Miller, con; Richard Tucker, t; Frank Guarrera, bar; Metropoli-
     tan Opera Orchestra and Chorus; Fausto Cleva. CBS 61640. Tape
     (c) 40-61640. (Reissue from Philips ABR 4000/1)
          -Gr 8-75 p359   +-HFN 12-75 p173 tape
          /HF 8-75 p77   +-RR 7-75 p22
     Cavalleria rusticana. cf LEONCAVALLO: I Pagliacci.

1420 Cavalleria rusticana: Duets and solos. VERDI: Il trovatore: Duets
        and solos. Irina Arkhipova, ms; Vladislav Piavko, t; Bolshoi
        Theatre Orchestra; Mark Ermler. Columiba/Melodiya M 33099.
            +-NR 1-75 p12   +-ON 7-75 p30
        Cavalleria rusticana: Als euer Sohn. cf Discophilia KGP 4.
        Cavalleria rusticana: O Lola; Addio alla madre. cf RCA SER 5704/6.
        Cavalleria rusticana: Trinklied. cf Bel Canto Club No. 3.
        Cavalleria rusticana: Tu qui, Santuzza. cf Supraphon 10924.
        Cavalleria rusticana: Voi lo sapete. cf HMV SLS 5012.
        Iris: Apri la tua finestra. cf Rubini RS 300.
        Iris: Un dì (ero piccina) al tempio. cf CATALANI: La wally: Ebben.
            Ne andrò lontana.
        Lodoletta: Flammen, perdonami. cf CATALANI: La wally: Eben. Ne
            andrò lontana.
MASCHERA, Florentio
        Canzona quarta. cf L'Oiseau-Lyre SOL 329.
        Canzona seconda. cf L'Oiseau-Lyre SOL 329.
MASSENET, Jules
1421 Le Cid: Ballet music. MEYERBEER: Les patineurs: Ballet suite
        (arr. Lambert). Netherlands Radio Philharmonic Orchestra;
        Stanley Black.   Decca PFS 4322.
            +-Gr 5-75 p2023   +RR 5-75 p34
            +-HFN 8-75 p77
        Le Cid: O Souverain, O juge, O père. cf Columbia D3M 33448.
        Le Cid: O Souverain. cf Discophilia KGC 2.
        Le Cid: Pleurez, mes yeux. cf Saga 7029.
        Concerto, piano. cf GOUNOD: Fantasy on the Russian national hymn.
        Don César de Bazan: Sevillana. cf Saga 7029.
        Elégie. cf Desto DC 7118/9.
        Hérodiade: Dors, ô cité perverse...Astres étincellants. cf Decca
            SXL 6637.
        Hérodiade: Il est doux, il est bon. cf RCA ARL 1-0844.
        Hérodiade: Vision fuggitiva. cf Ember GVC 37.
        Le jongleur de Notre Dame: La Vierge entend fort bien. cf Decca
            SXL 6637.
1422 Manon. Janine Micheau, Libero de Luca, t; Roger Bourdin, bar;
        Julien Giovanetti, bs; Paris, Opéra-Comique Orchestra and Chorus;
        Albert Wolff. Richmond RS 63023 (3). (Reissue from London
        LLA 7, A 4305)
            +-HF 7-75 p79
        Manon: Ay fuyez, douce image. cf BASF BC 21723.
        Manon: En fermant les yeux. cf Decca SDD 390.
        Manon: En fermant les yeux. cf Saga 7206.
        Manon: J'ai marqué l'heure du départ. cf Angel S 37143.
        Manon: Je ne suis que faiblesse...Adieu, notre petite table. cf
            HMV SLS 5012.
        Manon: On l'appelle Manon. cf Discophilia KGC 2.
        Manon: Oui, Je fus cruelle et coupable. cf ABC ATS 20018.
        Manon: Le rêve. cf Discophilia KGC 1.
1423 La navarraise. Lucia Popp, s; Alain Vanzo, Michel Sénéchal, t;
        Gérard Souzay, Vicenzo Sardinero, Claude Meloni, bar; Ambrosian
        Opera Chorus; LSO; Antonio de Almeida. CBS 76403. (also Colum-
        umbia M 33506. (Q) MQ 33506)
            +-Gr 6-75 p83   +NYT 8-10-75 pD14
            +HF 8-75 p74   +OC 12-75 p49
            +HFN 6-75 p89   +ON 9-75 p60

+MJ 11-75 p21   +RR 6-75 p23
+MT 12-75 p1071   +St 8-75 p71
+NR 8-75 p11

1424 La navarraise. Marilyn Horne, ms; Placido Domingo, Ryland Davies,
    Leslie Fyson, t; Sherrill Milnes, Gabriel Bacquier, bar; Nicolas
    Zaccaria, bs; Ambrosian Opera Chorus; LSO; Henry Lewis. RCA ARL
    1-1114. Tape (c) ARK 1-1114 (ct) ARS 1-1114 (Q) ARD 1-1114 Tape
    (c) ART 1-1114.
            +Gr 11-75 p1894   +-RR 11-75 p33
            +HF 12-75 p99   ++SFC 10-12-75 p22
            +MJ 11-75 p21

Le Roi de Lahore: Aux troupes du Sultan...Promesse de mon avenir.
    cf RCA ARL 1-0851.
Romeo et Juliette: Ach gehe auf. cf Bel Canto Club No. 3.
Sapho: Ce que j'appelle beau. cf Rubini RS 300.
Scènes alsaciennes: Suite, no. 7. cf CHARPENTIER: Impressions
    d'Italie.
Scènes hongroises. cf CHABRIER: Le Roi malgré lui: Fête polonaise.
Scènes pittoresques, no. 4. cf FRANCK: Le chasseur maudit.
Songs: Ouvre tes yeux bleus (2). cf Discophilia KGS 3.
1425 Thaïs. Anna Moffo, s; José Carreras, t; Gabriel Bacquier, bs-bar;
    Justino Díaz, bs; Ambrosian Opera Chorus; NPhO; Julius Rudel.
    RCA ARD 3-0842 (3). (Q) ARD 3-0842.
            +-Gr 5-75 p2013   -NYT 2-16-75 pD19
            -HF 5-75 p65   +-OC 9-75 p48
            +HFN 5-75 p134   +-ON 3-8-75 p34
            -MT 12-75 p1071   +SR 3-22-75 p36
            +-NR 3-75 p7   -St 6-75 p98
Thaïs: Méditation. cf Connoisseur Society (Q) CSQ 2065.
Thaïs: Méditation. cf Ember GVC 46.
Thaïs: Méditation. cf London STS 15160.
Thaïs: Voila donc la terrible cité. cf Rubini RS 300.
1426 Thérèse. Huguette Tourangeau, ms; Ryland Davies, Ian Calley, t;
    Louis Quilico, Neilson Taylor, Alan Opie, bar; Linden Singers;
    NPhO; Richard Bonynge. London OSA 1165. Tape (r) L 41165.
    (also Decca SET 572)
            ++Gr 8-74 p394   +Op 9-74 p800
            /HF 10-74 p107   ++SFC 8-11-74 p31
            +-MJ 1-75 p48   ++SFC 12-29-74 tape
            +NR 11-74 p8   +SR 11-30-74 p41
            +-NYT 7-28-74 pD22   +St 9-74 p126
            +ON 9-74 p64

La vièrge: Le dernier sommeil de la vièrge (The last sleep of the
    virgin). cf BERLIOZ: Les Troyens: Overture, March Troyenne.
La vièrge: Dernier sommeil de la vièrge. cf London STS 15160.
Werther: Letter scene. cf RCA ARL 1-0844.
Werther: O wie süss hier zu weilen. cf Bel Canto Club No. 3.
Werther: Pourquoi me réveiller. cf Decca SDD 390.
Werther: Pourquoi me réveiller. cf Discophilia KGC 1.
MASTERS OF THE BOW, Albert Spalding. cf Discopaedia MB 1009.
MASTERS OF THE BOW, Efrem Zimbalist. cf Discopaedia MB 1008.
MASTERS OF THE BOW, Ferenc von Vecsey. cf Discopaedia MB 1002.
MASTERS OF THE BOW, Jascha Heifetz. cf Discopaedia MB 1010.
MASTERS OF THE BOW, Leopold Auer, Willy Burmester and Pablo de
    Sarasate.
MASTERS OF THE BOW, Maud Powell. cf Discopaedia MB 1005.

MASTERS OF THE BOW, Mischa Elman.   cf Discopaedia MB 1006.
MASTERS OF THE BOW, Toscha Seidel.   cf Discopaedia MB 1007.
MASTERS OF THE BOW, Vasa Prihoda.   cf Discopaedia MB 1004.
MATEJ, Josef
    Quartet, strings, no. 2.   cf FELD: Quartet, strings, no. 4.
MATHIAS, William
1427 Invocations, op. 35.   Jubilate, op. 67, no. 2.   Partita, op. 19:
        Chorale, Postlude, Processional.   Toccata giocosa, op. 36, no.
        2.   Variations on a hymn tune, op. 20.   Christopher Herrick,
        org.   L'Oiseau-Lyre SOL 342.
            +Gr 7-75 p219   +RR 7-75 p48
            +HFN 8-75 p78
    Jubilate, op. 67, no. 2.   cf Invocations, op. 35.
    O sing unto the Lord a new song.   cf Abbey LPB 734.
    Partita, op. 19: Chorale, Postlude, Processional.   cf Invocations,
        op. 35.
    Processional.   cf Abbey LPB 738.
    Processional.   cf Argo ZRG 807.
    Toccata giocosa, op. 36, no. 2.   cf Invocations, op. 35.
    Variations on a hymn tune, op. 20.   cf Invocations, op. 35.
MATTEIS, Nicola
    Ground after the Scotch humour.   cf Argo ZRG 746.
MATTHESON
    Sonata, flute, A minor.   cf Orion ORS 75199.
MATTHYSZ
    Variations from "Der Gooden Fluyt Hemel".   cf Transatlantic TRA 292.
MATTULATH
    Calm as the night.   cf Ember GVC 46.
MAUERSBERGER, Rudolf
    Dresden requiem: De profundis.   cf Telefunken AJ 6-41867.
MAURER, Ludwig
    Fancies (4).   cf BRYCE: Promenade.
    Lied.   cf Crystal S 201.
    Scherzo.   cf Crystal S 201.
MAURO-COTTONE, Melchiarre
    Ninna Nanna.   cf Pye GSGC 14153.
MAY
    Ein Lied geht um die Welt.   cf Polydor 2489 519.
MAY, Frederick
1428 Quartet, strings, C minor.   Aeolian Quartet.   Claddagh CSM 2.
            +Gr 4-75 p1825
MAYER, William
    Octagon, piano and orchestra.   cf BARBER: Symphony, no. 1, op. 9,
        in one movement.
MAYERL, Billy
    Pieces.   cf GERSHWIN: George Gershwin's songbooks.
MEAUX,   de (13th century France)
    Trop est mes maris jalos.   cf Telefunken AW 6-41275.
    MEDIEVAL AND CONTEMPORARY LITURGICAL MUSIC: Plainsong mass for
        the Epiphany; Eucharistic liturgy in English.   cf Pleiades P 150.
MEDTNER, Nicolai
1429 Concerto, piano, no. 3, op. 50.   Sonata, piano, op. 22, G minor.
        Sonata tragica, op. 39, C minor.   Michael Ponti, pno; Luxembourg
        Radio Orchestra; Pierre Cao.   Candide CE 31092.
            +NR 10-75 p3
    Fairy tale, B flat minor.   cf RCA ARM 4-0942/7.

Sonata, piano, op. 22, G minor. cf Concerto, piano, no. 3, op. 50.
Sonata tragica, op. 39, C minor. cf Concerto, piano, no. 3, op. 50.
MEHUL, Etienne
    Le chant du départ. cf Classics for Pleasure 40230.
    Joseph: Champs paternels. cf Columbia D3M 33448.
    Overture burlesque. cf Angel S 36080.
    The two blind men of Toledo: Overture. cf DEBUSSY: Danse sacrée
      et profane.
MENDELSSOHN, Felix
    Albumblatt, op. 117, E minor. cf Gondellied, op. 102, no. 7.
    Andante and scherzo. cf CBS 76267.
    Andante with variations, D major. cf Pelca PSR 40598.
    Athalie, op. 74: War march of the priests. cf Philips 6580 107.
    Auf Flugeln des Gesanges, op. 34, no. 2. cf Discopaedia MB 1009.
1430 Concerto, piano, no. 1, op. 25, G minor. Concerto, piano, no. 2,
      op. 40, D minor. Murray Perahia, pno; AMF; Neville Marriner.
      Columbia M 33207. (Q) MQ 33207 Tape (ct) MAQ 33207. (also
      CBS 76376. Tape (c) 40-76376)
          ++Gr 7-75 p187   +-HFN 12-75 p173 tape
          ++Gr 12-75 p1121 tape   ++NR 7-75 p6
          +HF 7-75 p79   +St 9-75 p108
          +HFN 7-75 p84
    Concerto, piano, no. 1, op. 25, G minor. cf HMV SLS 5033.
    Concerto, piano, no. 2, op. 40, D minor. cf Concerto, piano, no.
      1, op. 25, G minor.
1431 Concerto, violin, op. 64, E minor. TCHAIKOVSKY: Concerto, violin,
      op. 35, D major. Ruggiero Ricci, vln; Netherlands Radio Orches-
      tra; Jean Fournet. Decca PFS 4345.
          +-RR 12-75 p68
    Concerto, violin, op. 64, E minor. cf BRUCH: Concerto, violin,
      no. 1, op. 26, G minor.
    Concerto, violin, op. 64, E minor. cf BRUCH: Concerto, violin,
      no. 2, op. 44, D minor: 2d movement.
    Concerto, violin, op. 64, E minor. cf Camden CCV 5000-12, 5014-24.
    Concerto, violin, op. 64, E minor. cf RCA CRL 6-0720.
    Concerto, violin, op. 64, E minor: Finale. cf RCA ARM 4-0942/7.
    Concerto, violin, op. 64, E minor: Finale. cf Rococo 2035.
    Concerto, violin, op. 64, E minor: 2d movement. cf Works, selec-
      tions (Decca SPA 433).
1432 Denn er hat seinen Engeln befohlen (Psalm 91). Mein Herz erhebet
      Gott, den Herrn, op. 69, no. 3. Prelude and fugue, op. 37, no.
      2, G major. Richte mich, Gott (Psalm 43), op. 78. Sonata,
      organ, op. 65, no. 3, A major. Basel Knabenkantorei; Markus
      Ulbrich, Hans Peter Aeschlimann, org. Pelca PSR 40585.
          +NR 11-75 p15
    Elijah, op. 70: And then shall your light break forth. cf HMV
      Tape (c) TC MCS 14.
    Elijah, op. 70: Höre, Israel, höre. cf Philips 6833 105.
    Elijah, op. 70: O rest in the Lord. cf Works, selections (Decca
      SPA 433).
    Elijah, op. 70: Thanks be to God; He that shall endure to the end;
      And then shall your light break forth. cf Decca ECS 2159.
    Elijah, op. 70: What have I to do with thee; Hear ye Israel. cf
      HMV RLS 714.
1433 Die erste Walpurgisnacht (The first Walpurgis night), op. 60.
      Infelice, op. 94. Edda Moser, Annelies Burmeister, con; Eber-
      hard Büchner, t; Siegfried Lorenz, bar; Siegfried Vogel, bs;

Leipzig Radio Chorus; Leipzig Gewandhaus Orchestra; Kurt Masur.
HMV ASD 3009. (also Angel S 37016)

+Gr 9-74 p571   +ON 1-11-75 p35
+HF 11-74 p114   +RR 8-74 p58
/NR 12-74 p9   ++St 12-74 p136
+NYT 9-8-74 pD36

1434 Gondellied, op. 102, no. 7. Albumblatt, op. 117, E minor. Kinderstücke, op. 72, nos. 1-6. Klavierstücke, op. posth., nos. 1, 2.
Songs without words, op. 19, nos. 1-6; po. 30, nos. 1-6; op. 38,
nos. 1-6; op. 53, nos. 1-6; op. 62, nos. 1-6; op. 67, nos. 1-6;
op. 85, nos. 1-6; op. 102, nos. 1-6. Daniel Barenboim, pno.
DG 2740 104 (3). (also DG 2709 052)

+-Gr 12-74 p1181   +NYT 1-26-75 pD26
+-HF 3-75 p81   +-RR 4-75 p52
++NR 2-75 p12   ++SFC 1-12-75 p26

Hebrides overture, op. 26. cf Symphony, no. 3, op. 56, A minor.
Hebrides overture, op. 26. cf Works, selections (Decca SPA 433).
Hebrides overture, op. 26. cf Philips 6747 199.
Kinderstücke, op. 72, nos. 1-6. cf Gondellied, op. 102, no. 7.
Klavierstücke, op. posth., nos. 1, 2. cf Gondellied, op. 102,
no. 7.
Mein Herz erhebet Gott, den Herrn, op. 69, no. 3. cf Denn er hat
seinen Engeln befohlen (Psalm 91).

1435 A midsummer night's dream, op. 61/21: Overture, nocturne and
scherzo. TCHAIKOVSKY: The nutcracker, op. 71a: Suite. PhO;
Igor Markevitch, Heinz Wallberg. Music for Pleasure MFP 57019.

+RR 4-75 p35

A midsummer night's dream, op. 21: Overture. cf Works, selections
(Decca SPA 433).
A midsummer night's dream, op. 21: Overture. cf SCHUMANN: Symphony,
no. 3, op. 97, E flat major.
A midsummer night's dream, op. 61: Scherzo. cf Philips 6580 066.
A midsummer night's dream, op. 61: Scherzo. cf Philips 6747 199.
A midsummer night's dream, op. 61: Wedding march. cf Sound Superb
SPR 90049.
A midsummer night's dream, op. 61: Wedding march and variations.
cf RCA VH 020.
A midsummer night's dream, op. 61: Wedding march; Fairies march.
cf DG Heliodor 2548 148.

1436 Octet, strings, op. 20, E flat major. RIMSKY-KORSAKOV: Quintet,
piano and winds, op. posth., B flat major. Vienna Octet.
Decca SDD 389. Tape (c) KSDC 378.

+Gr 12-73 p1218   +RR 9-75 p78 tape
+HFN 7-75 p90 tape   ++STL 1-6-74 p29
+-RR 1-74 p52

1437 Octet, strings, op. 20, E flat major. Janáček Quartet, Smetana
Quartet. Vanguard/Supraphon SU 4. (Reissue from Westminster,
1959)

+-HF 5-75 p74   ++SFC 8-3-75 p30
++NR 3-75 p5   +St 5-75 p102

Octet, op. 20, E flat major. cf Symphony, strings, no. 10, B minor.
Octet, op. 20, E flat major: Scherzo. cf Works, selections (Decca
SPA 433).
Octet, op. 20, E flat major: Scherzo. cf HMV ASD 3017.
On wings of song, op. 34, no. 2 (2). cf RCA ARM 4-0942/7.
Pieces, string quartet, op. 81 (4). cf Quartet, strings, no. 4,
op. 44, no. 2, E minor.

Prelude and fugue, C minor. cf Wealden WS 139.
Prelude and fugue, op. 37, no. 2, G major. cf Denn er hat seinen
    Engeln fohlen.
Psalm no. 55: O for the wings of a dove. cf Works, selections
    (Decca SPA 433).
Quartet, strings, no. 1, op. 12, E flat major: Canzonetta. cf Amon
    Ra SARB 01.
Quartet, strings, no. 1, op. 12, E flat major: Canzonetta. cf
    Swedish Society SLT 33189.
1438 Quartet, strings, no. 2, op. 13, A minor. SCHUMANN: Quartet,
    strings, no. 3, op. 41, A major. Alberni Quartet. CRD CRD
    1017.
          +Gr 11-75 p853   ++RR 11-75 p65
          +HFN 12-75 p157
Quartet, strings, no. 2, op. 13, A minor: Intermezzo. cf Amon
    Ra SARB 01.
1439 Quartet, strings, no. 4, op. 44, no. 2, E minor. Pieces, string
    quartet, op. 81 (4). Gabrieli Quartet. Decca SDD 469.
          +Gr 4-75 p1826   +RR 5-75 p49
1440 Quintet, strings, no. 2, op. 87, B flat major. VERDI: Quartet,
    strings, E minor. New Vienna String Quartet, Vienna Philharmonia
    Quintet. Musical Heritage Society MHS 1865.
          ++St 9-75 p115
Richte mich, Gott (Psalm 43), op. 78. cf Denn er hat seinen Engeln
    Befohlen.
Rondo brillant, op. 29, E flat major. cf HMV SLS 5033.
Rondo capriccioso, op. 14, E major. cf Rococo 2049.
Scherzo a capriccio. cf Canon CNN 4983.
Sonata, organ, no. 2. cf Orion ORS 74161.
Sonata, organ, op. 65, no. 3, A major. cf Denn er hat seinen
    Engeln befohlen.
Sonata, organ, op. 65, no. 4, B flat major. cf Polydor 2460 252.
Song: Auf Flügeln des Gesanges, op. 34, no. 2. cf Works, selections
    (Decca SPA 433).
Songs: Abendlied; Herbstlied, op. 63, no. 4; Maiglöckchen und
    Blümelein, op. 63, no. 6. cf BRAHMS: Songs (Musical Heritage
    Society 1896).
Song: Auf Flügeln des Gesange, op. 34, no. 2. cf Discopaedia MG
    1010.
Song: Hear my prayer. cf Decca ECS 774.
Songs: Hear my prayer, op. 39, no. 1; Oh, for the wings of a dove.
    cf Pye GH 589.
Song without words (arr. Kreisler). cf Decca SPA 405.
Songs without words, op. 19, nos. 1-6; op. 30, nos. 1-6; op. 38,
    nos. 1-6; op. 53, nos. 1-6; op. 62, nos. 1-6; op. 67, nos. 1-6;
    op. 85, nos. 1-6; op. 102, nos. 1-6. cf Gondellied, op. 102,
    no. 7.
Songs without words, op. 38, no. 6. cf Westminster WGM 8309.
Songs without words, op. 62, no. 3: Spring song. cf Turnabout
    TV 37033.
Songs without words: Spring song, op. 62, no. 6; The bees wedding,
    op. 67, no. 4. cf Works, selections (Decca SPA 433).
Songs without words, op. 62, no. 6; op. 67, no. 5; op. 85, no. 4.
    cf RCA VH 020.
Song without words, op. 67, no. 4, C major. cf Decca SP 372.
Song without words, op. 67, no. 4, C major (Spinning song). cf
    CHOPIN: Ballade, no. 1, op. 23, G minor.

Song without words, op. 67, no. 6, E major. cf Discopaedia MB
    1006.

Sweet remembrance. cf RCA ARM 4-0942/7.

1441 Symphonies, nos. 1-5. BPhO; Herbert von Karajan. DG 2740 128 (4).
    (Reissues from 2720 068, 2530 126)
            +Gr 12-75 p1037   ++RR 10-75 p44
            +HFN 10-75 p152

1442 Symphony, no. 1, op. 11, C minor. Symphony, no. 4, op. 90, A
    major. Hallé Orchestra; John Barbirolli. Olympic 8134.
            +-NR 4-75 p4

1443 Symphony, no. 3, op. 56, A minor. WAGNER: Siegfried Idyll. RPO,
    LPO; Felix Weingartner. Bruno Walter Society IGI 336.
            +NR 8-75 p5

1444 Symphony, no. 3, op. 56, A minor. Hebrides overture, op. 26. LSO;
    Peter Maag. Decca SDD 145. Tape (c) KSDC 145.
            +RR 9-75 p78 tape

1445 Symphony, no. 3, op. 56, A minor. Symphony, no. 4, op. 90, A
    major. LSO; Claudio Abbado. Decca SXL 6363. Tape (c) KSXC
    6363. (also London 6587)
            +HFN 5-75 p142 tape

1446 Symphony, no. 3, op. 56, A minor. Prague Symphony Orchestra; Dean
    Dixon. Supraphon 110 1124. Tape (c) 041124.
            /HFN 5-75 p142 tape   -RR 8-73 p47
            +NR 11-73 p3   /RR 6-75 p92 tape

1447 Symphony, no. 4, op. 90, A minor. Symphony, no. 5, op. 107, D
    minor. BSO; Charles Munch. Camden CCV 5035.
            +RR 10-75 p45

1448 Symphony, no. 4, op. 90, A major. Symphony, no. 5, op. 107, D
    minor. BPhO; Lorin Maazel. DG 138 684. Tape (c) 923 013 (ct)
    88684. (also DG 2538 329. Reissue from SLPM 138684)
            +Gr 1-75 p1342   +-RR 12-74 p34

1449 Symphony, no. 4, op. 90, A major. Symphony, no. 5, op. 107, D
    minor. BPhO; Herbert von Karajan. DG 2530 416. Tape (c) 3300
    418. (Reissue from 2720 068)
            ++Gr 9-74 p500   +NR 4-74 p1
            +HF 9-75 p116 tape   +-RR 9-74 p50

1450 Symphony, no. 4, op. 90, A major. SCHUMANN: Symphony, no. 4, op.
    120, D minor. PhO; Otto Klemperer. HMV SXLP 30178. (Reissue
    from Columbia SAX 2398) (also Angel 35629)
            ++Gr 1-75 p1492   ++RR 1-75 p29

1451 Symphony, no. 4, op. 90, A major. SCHUBERT: Symphony, no. 8,
    D 759, B minor. RPO, PhO; Paul Kletzki, Heinz Wallberg.
    Music for Pleasure MFP 57020.
            +-RR 3-75 p28

Symphony, no. 4, op. 90, A major. cf Symphony, no. 1, op. 11,
    C minor.

Symphony, no. 4, op. 90, A major. cf Symphony, no. 3, op. 56,
    A minor.

Symphony, no. 4, op. 90, A major: 1st movement. cf Works, selec-
    tions (Decca SPA 433).

Symphony, no. 5, op. 107, D minor. cf Symphony, no. 4, op. 90,
    A major (Camden 5035).

Symphony, no. 5, op. 107, D minor. cf Symphony no. 4, op. 90,
    A major (DG 138 684).

Symphony, no. 5, op. 107, D minor. cf Symphony, no. 4, op. 90,
    A major (DG 2530 416).

1452 Symphony, strings, no. 10, B minor. Symphony, strings, no. 12,
        G minor. Octet, op. 20, E flat major.* I Musici. Philips
        6580 103. (*Reissue from SAL 3640)
            +Gr 6-75 p40   +RR 5-75 p49
            +-HFN 6-75 p89
    Symphony, strings, no. 12, G minor. cf Symphony, strings, no. 10,
        B minor.
    Trio, piano, no. 1, op. 49, D minor: Scherzo. cf RCA ARM 4-0942/7.
    Venetian boat song, op. 19, no. 6. cf Connoisseur Society (Q)
        CSQ 2066.
1453 Works, selections: Song: Auf Flügeln des Gesanges, op. 34, no. 2.
        Concerto, violin, op. 64, E minor: 2d movement. Elijah, op.
        70: O rest in the Lord. Hebrides overture, op. 26. A mid-
        summer night's dream, op. 21: Overture. Octet, op. 20, E flat
        major: Scherzo. Psalm no. 55: O for the wings of a dove. Songs
        without words: Spring song, op. 62, no. 6; The bees wedding,
        op. 67, no. 4. Symphony, no. 4, op. 90, A major: 1st movement.
        Joan Sutherland, s Alastair Roberts, treble; Kathleen Ferrier,
        con; Ruggiero Ricci, vln; Wilhelm Backhaus, pno; St. John's
        College Chapel Choir; OSR, NPhO, LSO, Vienna Octet; Ernest
        Ansermet, Richard Bonynge, Peter Maag, Pierino Gamba, George
        Guest, Boyd Neel. Decca SPA 433.
            +RR 10-75 p44
MENNIN, Peter
1454 Symphony, no. 3. SESSIONS: Symphony, no. 2. NYP; Dmitri Mitro-
        poulos. CRI SD 278. (Reissues)
            +HFN 8-75 p78   +SFC 3-26-74 p43
            +-LJ 7-72 p2369   -St 6-72 p90
            -RR 6-75 p53
MENOTTI, Gian Carlo
1455 Sebastian. LSO; José Serebrier. Peerless PRCM 202. (also Desto
        6432)
            +Gr 4-75 p1808
MERCADANTE, Guiseppe
    Concerto, flute, E minor. cf BOCCHERINI: Concerto, flute, op. 27,
        D major.
MERIKANTO, Aare
1456 Juha. Raili Kostia, s; Maiju Kuusoja, con; Hendrik Krumm, t;
        Matti Lehtinen, bar; Finnish National Opera Orchestra and
        Chorus; Ulf Soderblom. Musical Heritage Society MHS 3079/31 (3).
            ++St 9-75 p108
    Preludio. cf BIS LP 18.
MERILAINEN
    Opusculum. cf BIS LP 18.
MERRICK
    Celtic suite. cf HARRISON: Grand defilé, organ and percussion.
MERULA, Tarquinio
    Un cromatico ovvero capriccio. cf Philips 6775 006.
MERULO, Claudio
    Canzon à 4. cf Telefunken AW 6-41890.
    Canzona françese. cf L'Oiseau-Lyre 12BB 203/6.
MESSIAEN, Oliver
    Apparition de l'eglise eternelle. cf Organ works (CRD CAL 1925/30).
    L'ascension. cf Organ works (CRD CAL 1925/30).
    L'ascension. cf Decca 5BBA 1013-5.
    Le banquet céleste. cf Organ works (CRD CAL 1925/30).

1457 Catalogue d'oiseaux: La bouscarle. Préludes: Les sons impalpables
     du rêve; Cloches d'angoisse et larmes d'adieu. Vingt regards
     sur l'enfant-Jésus: Première communion de la Vierge; Noël; Re-
     gard du silence; Regard e l'esprit de joie. Paul Crossley, pno.
     L'Oiseau-Lyre DSLO 6.
          ++Gr 4-75 p1839   +RR 3-75 p50
1458 Catalog d'oiseaux: La bouscarle. Oiseaux exotiques. Reveil des
     oiseaux. Yvonne Loriod, pno; CPhO; Václav Neumann. Supraphon
     SUAST 50749. Tape (c) 04 50749.
          ++HFN 5-75 p142 tape   +RR 9-75 p78 tape
1459 Catalogue d'oiseau: La rousserolle effarvatte. Vingt regards sur
     l'enfant Jésus. Peter Serkin, pno. RCA CRL 3-0759.
          ++HF 10-75 p80   ++SFC 8-17-75 p22
          ++NR 9-75 p12
     Chants de terre et de ciel. cf Poèmes pour mi.
     Les corps glorieux, nos. 1-7. cf Organ works (CRD CAL 1925/30).
     Les corps glorieux: Joie et clarté des corps glorieux. cf Orion
     ORS 74161.
     Les corps glorieux: Joie et clarté des corps glorieux. cf Pye
     TPLS 13066.
     Diptyque. cf Organ works (CRD CAL 1925/30).
     Et exspecto resurrectionem mortuorum. cf BOULEZ: e. e. cummings.
     Livre d'orgue: Chants d'oiseau; Les mains de l'abîme; Pièce de en
     trio; Reprises par interversions; Soixantequatre durées; Les
     yeux dans les roues. cf Organ works (CRD CAL 1925/30).
1460 Meditations sur le mystère de la Sainte Trinité. Olivier Messiaen,
     org. Erato STU 70750/1 (2). (also Musical Heritage Society
     MHS 1797/8)
          ++Gr 3-73 p1717   +RR 2-73 p77
          +-HF 3-75 p82   +St 3-75 p105
          +HFN 3-73 p565
     Messe de la Pentecoste. cf Organ works (CRD CAL 1925/30).
     La nativité du Seigneur, nos. 1-9. cf Organ works (CRD CAL 1925/30).
     La nativité du Seigneur: Les anges. cf Monitor MCS 2143.
     Oiseaux exotiques. cf Catalog d'oiseaux: La bouscarle.
1461 Organ works: Apparition de l'eglise eternelle. L'ascension. Les
     corps glorieux, nos. 1-7. Le banquet céleste. Diptyque. Messe
     de la Pentecoste. La nativité du Seigneur, nos. 1-9. Livre
     d'orgue: Chants d'oiseaux; Les mains de l'abîme; Pièce de en
     trio (2); Reprises par interversions; Soixantequatre durées;
     Les yeux dans les roues. Verset pour la fête de la dédicace.
     Louis Thiry, org. CRD CAL 1925/30 (6).
          ++Gr 11-75 p863   ++STL 10-5-75 p36
1462 Poèmes pour mi. Chants de terre et de ciel. Noëlle Barker, s;
     Robert Sherlaw Johnson, pno. Argo ZRG 699.
          +-Gr 11-72 p940   *NR 6-73 p12
          +HF 10-73 p104   -SFC 4-7-74 p25
          +HFN 8-72 p1465   +St 10-73 p141
          /LJ 12-75 p38   +STL 6-11-72 p38
     Préludes: Les sons impalpables du rêve; Cloches d'angoisse et
     larmes d'adieu. cf Catalogue d'oiseaux: La bouscarle.
     Reveil des oiseaux. cr Catalog d'oiseaux: La bouscarle.
     Rondeau. cf Catalogue d'oiseaux: La bouscarle.
1463 La transfiguration de notre Seigneur Jésus-Christ. Michael Syl-
     vester, t; Paul Aquino, bar; Yvonne Loriod, pno; Janos Starker,
     vlc; Wallace Mann, flt; Loren Kitt, clt; Frank A. Ames, marimba;
     John A. C. Kane, xylorimba; Ronald Barnett, vibraphone; West-

mister Symphonic Choir; Washington National Symphony Orchestra;
Antal Dorati. Decca Headline HEAD 1-2 (2). (also London HEAD
1-2)
        ++Gr 5-74 p2050   ++NR 1-75 p10
        +-HF 4-75 p88   +-RR 5-74 p19
        ++MJ 10-75 p45   ++SFC 11-3-74 p22
Verset pour la fête de la dédicace. cf Organ works (CRD CAL 1925/30).
Vingt regards sur l'enfant Jésus. cf Catalogue d'oiseau: La rous-
    serolle effarvatte.
Ving regards sur l'enfant Jésus: Première communion de la Vierge;
    Noël; Regard du silence; Regard e l'esprit de joie. cf Catalogue
    d'oiseaux: La bouscarle.
1464 Visions de l'amen. Katia and Marielle Labèque, pno. Musical Heri-
    tage Society MHS 1762.
        /HF 3-75 p82
    METROPOLITAN OPERA MADRIGAL SINGERS: Simple gifts. cf Advent 5012.
MEYERBEER, Giacomo
    L'Africana: Ballata; Adamastor, Re dell'acque profonde. cf Ember
        GVC 37.
    L'Africana: Deh ch'io ritorno. cf Discophilia KGC 2.
    L'Africana: Mi batte il core...O paradiso. cf Saga 7206.
    L'Africaine: O paradiso. cf Columbia D3M 33448.
    L'Africaine: O paradiso. cf RCA SER 5704/6.
    L'Africana: O paradiso. cf Supraphon 10924.
    L'Africana: Schlummerarie. cf Discophilia KGM 2.
1465 Les Huguenots. Joan Sutherland, Martina Arroya, s; Anastasios
    Vrenios, t; Gabriel Bacquier, bar; Nicola Ghiuselev, bs; Ambros-
    ian Opera Chorus; NPhO; Richard Bonynge. London OSA 1437 (4).
    Tape (r) 1-90175.
        /Ha 1-71 p97   +ON 1-2-71 p35
        -HF 1-71 p73   /SR 11-28-70 p75
        +LJ 3-75 p35 tape   /St 1-71 p74
        +MJ 1-71 p68
    Les Huguenots: Duet of Marguerite and Raoul. cf Angel S 37143.
    Les Huguenots: Un vieil air Huguenot...Piff, paff. cf Decca SXL
        6637.
    Les patineurs: Ballet suite. cf MASSENET: Le Cid: Ballet music.
    Le prophète: Ach mein Sohn, Gebt O Gebt. cf Discophilia KGM 2.
    Le prophète: Coronation march, Act 4. cf Philips 6580 107.
    Le prophète: Coronation march, Act 4. cf Philips 6747 204.
    Robert de teufel: Du rendezvous; Le bonheur. cf Discophilia KGC 1.
    Robert le diable: Robert, toi que j'aime. cf RCA ARL 1-0844.
1466 Songs: Cantique du Trappiste; Le chant du dimanche; Der Garten des
    Herzens; Hör ich das Liedchen klingen; Komm; Menschenfeindlich;
    Mina; Le poète mourant; Die Rose, die Lilie, die Taube; Die
    Rosenblätter; Scirocco; Sicilienne; Sie und ich; Ständchen.
    Dietrich Fischer-Dieskau, bar; Karl Engel, pno. DG Archive
    2533 295.
        +Gr 9-75 p500   +RR 9-75 p70
        +HFN 9-75 p101   +SR 11-29-75 p50
        +NR 12-75 p12
MIASKOVSKY, Nikolai
    Concerto, violoncello, op. 66, C minor. cf PROKOFIEV: Concerto,
        violin, no. 2, op. 63, G minor.
    Grillen, op. 25, nos. 1, 6. cf Ember GVC 40.
1467 Symphony, no. 22, op. 54, B minor. SVETLANOV: Festive poem, op.
    9. USSR Symphony Orchestra; Yevgeny Svetlanov. HMV Melodiya

ASD 3062.
+-Gr 4-75 p1808   +-RR 4-75 p28
+MT 9-75 p799

MICA, Frantisek
Quartet, C major: Rondo. cf HAYDN: Quartet, strings, op. 74, no. 3, G minor.

MICHEELSEN
Introduction and chorale on "Erhalt uns, Herr dei deinem Wort". cf BACH: Chorale prelude, O Mensch bewein dein Sünde grosse, S 622.

Toccata on "Lobe den Herren". cf BACH: Chorale prelude, O Mensch bewein dein Sünde grosse, S 622.

MIGNONE, Francisco
Passarinho está cantando. cf Angel S 36076.

MIGUEZ, Leopoldo Amérigo
Nocturne. cf Angel S 37110.

MILAN, Luis
Fantasias, nos. 10-12, 16. cf Pavanas, nos. 1-6.
Pavan. cf DG Archive 2533 184.

1468 Pavanas, nos. 1-6. Fantasias, nos. 10-12, 16. MUDARRA: Diferencias sobre El Conde claros. Fantasia que contrahaza la harpa en la manera de Ludovico. Gallarda. O guárdame las vacas. Pavana de Alexandre. NARVAEZ: Baxa de contrapunto. Diferencias sobre guárdame las vacas. Fantasia. Mille regres. Konrad Ragossnig, lt. DG Archive 2533 183.
++Gr 7-75 p223   ++RR 7-75 p48
++HFN 7-75 p87

Pavana I and II. cf DG Archive 2533 111.
Songs: Aquel caballero, madre; Toda mi vida hos amé. cf DG 2530 504.
Sospirastes baldovinos. cf Telefunken SAWT 9620/1.

MILANO, Francesco Canova de
La canzon delli Uccelli. cf Angel S 37015.
Fantasia. cf DG Archive 2533 173.
Ricercare. cf Angel S 37015.

MILES
Anchor's away. cf BASF 292 2116-4.
In the garden. cf RCA ARL 1-0562.

MILHAUD, Darius
Caramel-Mou. cf Golden Crest CRS 4132.
Le carnaval d'Aix: Le capitaine Cartuccia. cf DG Heliodor 2548 148.

1469 Chansons de Ronsard, op. 223 (4). Symphony, no. 6, op. 343. Paula Seibel, s; Louisville Orchestra; Jorge Mester. Louisville LS 744.
++HF 8-75 p90   ++SFC 5-25-75 p17

1470 Concerto, piano and orchestra, no. 2, op. 225. La muse ménagère, op. 245. Suite cisalpine sur des airs populaires piémontais, op. 332. Grant Johannesen, pno; Thomas Blees, vlc; Luxembourg Radio Orchestra; Bernard Kontarsky. Turnabout TVS 34496.
+Gr 7-75 p187   +NR 8-74 p4
+HF 7-74 p96   ++SFC 6-9-74 p28

1471 La création du monde. WEILL: Kleine Dreigroschenmusik. Contemporary Chamber Ensemble; Arthur Weisberg. Nonesuch H 71281. Tape (c) ZCH 71281 (Q) HQ 1281.
+-Gr 3-74 p1711   -RR 9-74 p96 tape
+-HF 11-73 p128   ++SFC 10-14-73 p32
-HF 1-74 p84 Quad   +-St 10-73 p157
-NR 11-73 p2   +Te 6-75 p51

La muse ménagère, op. 245. cf Concerto, piano and orchestra, no.
2, op. 225.
1472 Le printemps, Bks 1 and 2. Rag-caprices (3). Saudades do Brazil.
William Bolcom, pno. Nonesuch H 71316.
+NR 12-75 p13
Rag-caprices (3). cf Le printemps, Bks 1 and 2.
Rag-caprices (3). cf Golden Crest CRS 4132.
1473 Sabbath morning service. Heinz Rehfuss, bar; Paris Opera Chorus
and Orchestra; Darius Milhaud. Westminster Gold WGS 8281.
(Reissue from WST 17052)
+St 6-75 p99
1474 Saudades do Brazil. SATIE: Les aventures de Mercure. La belle
excentrique. Jack in the box (orch. Milhaud). London Festival
Players; Bernard Herrmann. Decca PFS 4286. (also London SPC
21094)
-Gr 5-74 p2031   -St 6-75 p99
++RR 5-74 p39
Saudades do Brazil. cf Le printemps, Bks 1 and 2.
Saudades do Brasil, no. 7: Corcovado. cf RCA ARM 4-0942/7.
Saudades do Brasil, no. 10: Sumaré. cf RCA ARM 4-0942/7.
Scaramouche. cf FRANCAIX: Concerto, piano, D major.
Scaramouche. cf GERSHWIN: Rhapsody in blue.
Sonata, violin and piano, no. 2. cf DEBUSSY: Sonata, violin and
piano, no. 3, G minor.
Sonnets composes au secret par Jean Cassou (6). cf Caprice RIKS
LP 46.
Suite cisalpine sur des airs populaires piémontais, op. 332. cf
Concerto, piano and orchestra, no. 2, op. 225.
Symphonies, small orchestra, nos. 1-4. cf IBERT: Paris 32.
Symphony, no. 6, op. 343. cf Chansons de Ronsard, op. 223.
Tango des Fratellini. cf Golden Crest CRS 4132.
MILLER, Edward
Quartet, variations. cf Opus One 22.
MILLOCKER, Karl
Der Bettelstudent: A slap in the face; The alligator and the
Brahmin's daughter. cf Pye NSPH 6.
Carlotta. cf HMV SLS 5017.
The Dubarry: I gave my heart. cf Pye NSPH 6.
Die Dubarry: Ich schenk mein Herz; Was ich im Leben beginne. cf
Angel S 35696.
Gasparone: Tarantella. cf Rubini RS 300.
Traumwälzer. cf HMV SLS 5017.
MILNER, Anthony
1475 Roman spring. Salutatio Angelica. Felicity Palmer, s; Alfreda
Hodgson, con; Robert Tear, t; London Sinfonietta Chorus and
Orchestra; David Atherton. Decca SXL 6699.
+-Gr 3-75 p1691   ++STL 5-4-75 p37
+RR 3-75 p65   +Te 6-75 p52
MIMAROGLU, Ilhan
Agony. cf BERIO: Visage.
1476 Tract. Tuly Sand, speaking and singing with electronic music.
Folkways FTS 33441.
-NR 10-75 p11
MIYAGI
Haru no umi (arr. and orch. Gerhardt). cf RCA LRL 1-5094.

MODENA
    Ricercare a 4.   cf L'Oiseau-Lyre 12BB 203/6.
MOERAN, Ernest
1477 Symphony, G minor.   NPhO; Adrian Boult.   Lyrita SRCS 70.
        ++Gr 7-75 p187   +RR 7-75 p28
        ++HFN 7-75 p84
MOLINARO, Simone
    Ballo detto il Conte Orlando.   cf DG Archive 2533 173.
    Fantasias, nos. 1, 9, 10.   cf DG Archive 2533 173.
    Saltarello (2).   cf DG Archive 2533 173.
MOLTER, Johann
    Concerto, flute and strings, G major.   cf RCA CRL 2-7003.
    Concertos, trumpet, nos. 1 and 2.   cf ARUTANIAN: Concerto, trumpet.
    Symphony, 4 trumpets, C major.   cf Nonesuch H 71301.
MOLTKE
    Des grossen Kurfürsten Reitermarsch.   cf BASF 292 2116-4.
    Des grossen Kurfürsten Reitermarsch.   cf DG 2721 077.
MOMPOU, Frederico
    El combat del Somni: Damunt de tu nomes.   cf HMV SLS 5012.
MONIOT DE PARIS
    Vadurie.   cf Vanguard SRV 316.
MONIUSZKO, Stanislaw
    Halka: Aria, Act 1: Jontek's dream.   cf Rubini RS 300.
MONN, Georg
    Fugue, F major.   cf Hungaroton SLPX 11601/2.
MONNIKENDAM, Marius
    Toccata.   cf BACH: Chorale preludes, S 669-671.
MONSIGNY, Pierre
    Roe et Colas: La Sagesse est un trésor.   cf Desto DC 7118/9.
    MONSTER CONCERT: 10 pianos, 16 pianists.   cf Columbia 31726.
MONTE, Philipp de
    O suavitas; Stella del nostro mar a l'apparir del sol.   cf Panton
        010335.
MONTEVERDI, Claudio
    Adoramus te, Cantate Domino; Zefiro torno.   cf Wayne State Univ-
        ersity, unnumbered.
    Baci soavi e cari; Lamento dell Ninfa.   cf Harmonia Mundi 204.
    Beatus vir.   cf Abbey LPB 734.
1478 L'Incoronazione di Poppea.   Cathy Berberian, Maria Minetto, Mar-
        garet Baker, ms; Carlo Gaifa, Kurt Equiluz, t; Enrico Fissore,
        bar; Giancarlo Luccardi, bs; VCM; Nikolaus Harnoncourt. Tele-
        funken HS 6-35247 (5).
        +-Gr 3-75 p1697   ++NR 2-75 p10
        +HF 2-75 p84   +ON 8-75 p28
        +-MT 8-75 p715   +St 6-75 p99
1479 L'Incoronazione di Poppea: Concert suite.   L'Orfeo: Concert suite.
        Sue Harmon, s; Michael Sells, t; Robert Rodriguez, hpd; Emanuel
        Gruber, vlc; Orion Chamber Orchestra and Singers.   Ember ECL
        9038.
        -HFN 12-75 p157
    Lamento d'Olimpia.   cf L'Oiseau-Lyre 12BB 203/6.
1480 Madrigals: Altri canti d'amore; Hor ch'el ciel e la terra; Gira
        il nemico; Se vittorie si belle; Armato il cor; Ogni amante è
        guerrier; Ardo avvampo; Il ballo per l'Imperatore Ferdinando.
        Luigi Alva, Ryland Davies, Robert Tear, Alexander Oliver, t;

Clifford Grant, Stafford Dean, bs; Glyndebourne Festival Chorus;
Raymond Leppard, Leslie Pearson, Henry Ward, hpd; Joy Hall, vlc;
Robert Spencer, lt; ECO; Raymond Leppard. Philips 6500 663.
(Reissue from 6799 006)
             +-Gr 1-75 p1379   +-RR 1-75 p56
1481 Madrigals: Madrigali guerrieri, Sinfonia; Altri canti d'amor; Hor
     che' el ciel e la terra; Così sol d'una chiara fonte; Gira il
     nemico insidioso; Ardo avvampo. Madrigali amorosi, Altri canti
     di Marte; Due belli occhi; Lamento della Ninfa; Non havea Febo
     ancora...Amor...Non partir ritrosetta; Dolcissimo uscignolo;
     Vago augelletto. Prague Madrigal Singers; Miroslav Venhoda.
     Supraphon 112 1306.
          -Gr 6-75 p76   +NR 10-75 p6
          /HFN 5-75 p134   -RR 4-75 p63
1482 Madrigali amorosi: Altri canti di Marte; Vago augelletto; Ardo e
     scoprir; O sia tranquillo il mare; Ninfa che scalza il piede;
     Dolcissimo uscignnolo; Chi vol haver felice; Non havea Febo
     ancora; Lamento dell Ninfa...Si tra sdegnosi pianti; Perchè t'en
     fuggi, O Fillide; Non partir, ritrosetta; Su, su pastorelli
     vezzosi. Yvonne Fuller, Angela Bostok, Lillian Watson, Sheila
     Armstrong, s; Alfreda Hodgson, ms; Anne Collins, con; Luigi
     Alva, Ryland Davies, Robert Tear, Alexander Oliver, t; Stafford
     Dean, Clifford Grant, bs; Joy Hall, vlc; Robert Spencer, lt;
     Osian Ellis, hp; Raymond Leppard, Henry Ward, hpd. Glyndebourne
     Festival Chorus, Members; ECO; Raymond Leppard. Philips 6500
     846. (Reissue from 6799 006)
          +Gr 7-75 p228   +-RR 6-75 p80
          +HFN 8-75 p91
     Madrigals: Qui rise Tirsi: O Rosetta. cf Klavier KS 536.
     Madrigali a cinque voci, Libro V: Ecco Silvio...Ma se con la pieta...
          Dorinad, ah dirò...Ecco piegando...Ferir quel petto. cf Caprice
          RIKS LP 46.
     Magnificat. cf Vespro della beata vergine. cf Archive 2723 043.
1483 Missa In illo tempore. SCHUTZ: Deutsches Magnificat, S 949.
     Eleanor Smith, Joanne Roberts, s; Myra Brown, con; David Rhodes,
     Neil Mackie, t; Charles Stewart, Angus MacIntyre, trom; Colin
     Tipple, chamber org; Scottish Chamber Choir; George McPhee.
     Decca ECS 764.
          -Gr 7-75 p228   +RR 6-75 p80
          -HFN 8-75 p78
     Missa, In ill tempore. cf Vespro della beata vergine.
1484 L'Orfeo. Emilia Petrescu, s; Anna Reynolds, ms; Nigel Rogers,
     Ian Partridge, John Elwes, t; James Bowman, c-t; Stafford Dean,
     Alexander Malta, bs; Hamburg Monteverdi Choir; Hamburg Camerata
     Accademica; Jürgen Jürgens. DG Archive 2723 018 (3). (also
     DG Archive 2710 015)
          +Gr 11-74 p956   +ON 8-75 p28
          +-HF 2-75 p82   ++RR 11-74 p27
          ++NR 1-75 p14   -St 6-75 p101
          +OC 4-75 p49
     L'Orfeo: Concert suite. cf L'Incoronazione di Poppea: Concert suite.
1485 Vespro della beata Vergine. Felicity Palmer, Jill Gomez, s; James
     Bowman, c-t; Robert Tear, Philip Langridge, t; John Shirley-
     Quirk, bar; Michael Rippon, bs; Salisbury Cathedral Boys' Choir;
     Philip Jones Brass Ensemble; Monteverdi Orchestra and Choir;
     John Eliot Gardiner. Decca SET 593/4 (2).

                    +Gr 4-75 p1850   +RR 4-75 p63
                    +HFN 5-75 p133   ++STL 5-4-75 p37
                    +MT 11-75 p978
1486 Vespro della beata vefgine. Missa, In illo tempore. Magnificat.
     Paul Esswood, Kevin Smith, c-t; Ian Partridge, John Elwes, t;
     David Thomas, Christopher Keyte, bs; Edward H. Tarr, Ralph Bryant,
     Richard Cook, cor; Fritz Brodersen, Harald Strutz, Walfried
     Kohlert, trom; Sebastian Kelber, Klaus Holsten, flt and Renais-
     sance rec; Eduard Melkus, Spiros Rantos, Thomas Weaver, vln;
     Lilo Gabriel, David Becker, vla; Klaus Storck, Eugene Eicher,
     vlc; Laurenzius Strehl, vla da gamba and violone; Dieter Kirsch,
     lt; Hubert Gumz, Gerd Kaufmann, org; Regensburg Domspatzen;
     Hanns-Martin Schneidt. DG Archive 2723 043 (3).
                    +Gr 11-75 p876   +RR 11-75 p86
                    ++HFN 11-75 p155   +STL 11-2-75 p38
MONTSALVATGE, Xaver
     Canciones negras. cf DG 2530 598.
     Canciones negras: Canción de cuna para dormir. cf HMV SLS 5012.
MOODY, James
     Quintet, harmonica. cf JACOB: Divertimento, harmonica and string
     quartet.
MOORE
     Songs: Believe me, if all those endearing young charms; The minstrel
     boy; The young May moon. cf Philips 6599 227.
MORALES, Cristobal de
1487 Missa l'homme armé. Capella Cordina; Alejandro Planchart. Lyri-
     chord LLST 7267.
                    +NR 4-75 p6
MORGAN, Justin
     Amanda. cf Nonesuch H 71276.
     Judgment anthem. cf Nonesuch H 71276.
MORLEY, Thomas
     As you like it: It was a lover and his lass. cf BBC REB 191.
     My bonny lass she smileth. cf Harmonia Mundi 593.
1488 Dances for broken consort. SUSATO: Danserye: Dances (12). London,
     Early Music Consort; David Munrow. HMV Tape (c) TC EXE 104.
                    +RR 1-75 p72 tape
     Now is the month of maying. cf Advent 5012.
     Pavan. cf DG Archive 2533 157.
     The turtle dove. cf Argo ZRG 765.
     Twelfth night: O mistress mine. cf BBC REB 191.
MORTON, Robert
     La perontina. cf Telefunken ER 6-35257.
MOSCHELES, Ignaz
1489 Grande sonate symphonique, op. 112. PIXIS: Concerto, violin,
     piano and strings. Mary Louis Boehm, Pauline Boehm, pno;
     Kees Kooper, vln; Westphalian Symphony Orchestra; Siegfried
     Landau. Turnabout TV 34590.
                    +NR 12-75 p7
     Sonata caractèristique, piano, op. 27. cf BERGER: Grande sonate,
     op. 7.
MOSONYI, Mihaly
     Concerto, piano. cf HILLER: Konzertstück, piano, op. 113.
MOSS, Lawrence
     Omaggio. cf Desto DC 7131.

MOSSOLOV, Alexander
  Music of the machines. cf SABATA: Juventus.
MOSZKOWSKI, Mortiz
  Caprice espagnol, op. 37. cf L'Oiseau-Lyre DSLO 7.
  En automne, op. 36, no. 4. cf Virtuoso etudes, op. 72.
  Etincelles. cf RCA VH 020.
  Etincelles, op. 36, no. 6. cf Virtuoso etudes, op. 72.
  Etude, A flat major. cf RCA VH 020.
  La jongleuse, op. 52, no. 4. cf RCA ARL 2-0512.
  Scherzino. cf Canon CNN 4983.
  Serenata. cf Discopaedia MB 1005.
  Siciliano, op. 42, no. 2. cf Virtuoso etudes, op. 72.
  Stücke, op. 45, no. 2: Guitarre. cf Virtuoso etudes, op. 72.
  Stücke, op. 45, no. 2: Guitarre. cf Discopaedia MB 1002.
  Stücke, op. 45, no. 2: Guitarre. cf Discopaedia MB 1009.
  Stücke, op. 45, no. 2: Guitarre. cf RCA ARM 4-0942/7.
  Stücke, op. 45, no. 2: Guitarre. cf Rococo 2072.
  Suite, 2 violins and piano, op. 71, G minor. cf Classic Record
    Library SQM 80-5731.
  Valse brillante, A flat major. cf Virtuoso etudes, op. 72.
1490 Virtuoso etudes, op. 72. Siciliano, op. 42, no. 2. Valse brillante,
    A flat major. Etincelles, op. 36, no. 6. En automne, op. 36,
    no. 4. Stücke, op. 45, no. 2: Guitarre. Ilana Vered, pno.
    Connoisseur Society CS 2023.
        +-ARG 5-71 p586   ++RR 3-75 p50
        +Ha 11-71 p145   ++SFC 2-13-72 p34
        +-MJ 2-72 p14   +SR 3-27-71 p71
        ++NR 6-71 p11   ++St 6-71 p92
MOURET, Jean Joseph
  Rondeau. cf Klavier KS 536.
  Suites of symphonies. cf LALANDE: Symphonies pour les soupers du
    roi, no. 2.
MOUTON, Charles
  La, la, la l'oysillon du bois. cf L'Oiseau-Lyre 12BB 203/6.
  Noel. cf Turnabout TV 34569.
MOZART, Leopold
  Concerto, hose-pipe and strings: 3rd movement. cf HMV SLA 870.
  Concerto, trumpet, D major. cf HUMMEL: Concerto, trumpet, E flat
    major.
1491 Divertimenti, nos. 2 and 3, B flat major and D major. WRANITZKY:
    Symphony, C major. HCO: Vilmos Tátrai. Hungaroton SLPX 11656.
        /Gr 10-75 p611   +NR 8-75 p4
        +HFN 8-75 p78   +RR 8-75 p46
  Divertimento, no. 3, D major. cf Divertimento, no. 2, B flat
    major.
  Toy symphony. cf HUMPERDINCK: Hänsel und Gretel, excerpts.
MOZART, Wolfgang Amadeus
  Adagio. cf Columbia M 33514.
1492 Adagio, K 261, E major. Concerto, violin, no. 2, K 211, D major.
    Rondo, violin, K 373, C major. Rondo, violin, K 269, B flat
    major. Pinchas Zukerman, vln; ECO; Daniel Barenboim. Columbia
    M 33206. (Q) MQ 33206.
        -Audio 11-75 p95   +-St 6-75 p102
        -NR 5-75 p7
1493 Adagio, K 261, E major. Concerto, violin, no. 3, K 216, G major.
    Singonia concertante, K 364, E flat major. Wolfgang Schneiderhan,

Thomas Brandis, vln; Giusto Cappone, vla; BPhO; Karl-Böhm. DG
Tape (c) 3318 046.
　　　+RR 9-75 p78 tape
　Adagio, K 261, E major. cf Concerti, violin, nos. 1-5 (CBS 77381).
　Adagio, K 261, E major. cf Concerti, violin, nos. 1-5 (DG 2740 116).
　Adagio, K 261, E major. cf Concerto, violin, no. 2, K 211, D major.
　Adagio, K 261, E major. cf BEETHOVEN: Romance, no. 1, op. 40, G
　　　major.
1494 Adagio, K 540, B minor. Sonata, piano, no. 9, K 311, D major.
　　　Sonata, piano, no. 13, K 333, B flat major. Variations on
　　　"Salve tu Domine", K 398, F major. Michael Cave, pno. Orion
　　　ORS 75185.
　　　　+NR 9-75 p11
　Adagio, K 540, B minor. cf HAYDN: Piano works (Hungaroton 11638).
1495 Adagio and allegro, K 594, F minor. Sonata, 2 pianos, K 19d, C
　　　major. Sonata, 2 pianos, K 381, D major. Sonata, 2 pianos,
　　　K 358, B flat major. Christoph Eschenbach, Justus Frantz, pno.
　　　DG 2530 529.
　　　　+-Gr 5-75 p1986　　　　　　　　+-RR 5-75 p56
　　　　+-HFN 5-75 p135
　Adagio and allegro, K 594, F minor. cf Organ works (Philips 6500
　　　598).
1496 Adagio and fugue, K 546, C minor. Divertimenti, nos. 1-3, K 136-
　　　138. Quartets, strings, nos. 1-13. Quartetto Italiano.
　　　Philips 6500 644/5 (2). (Reissue from 6749 097)
　　　　+Gr 10-74 p717　　　　　　　　++RR 10-74 p71
　　　　++HF 1-75 p84　　　　　　　　++SFC 4-13-75 p23
　　　　+NR 12-74 p8　　　　　　　　++St 2-75 p110
　Adagio and fugue, K 546, C minor. cf Serenade, no. 13, K 525, D
　　　major.
　Adagio and fugue, K 546, C minor. cf SCHUBERT: Quartet, strings,
　　　no. 14, K 810, D minor.
　Ah, se in ciel, benigne stelle, K 538. cf Arias (Crystal S 902).
　Allegro, K 3. cf Connoisseur Society (Q) CSQ 2065.
　Andante, flute, K 315, C major. cf Concerto, flute, no. 1, K 313,
　　　G major.
　Andante, mechanical organ, K 616, F major. cf Organ works (Philips
　　　6500 598).
　Andante, mechanical organ, K 616, F major. cf HAYDN: Piano works
　　　(Hungaroton 11638).
1497 Andante and allegretto, K 404, C major. Sonatas, violin and piano,
　　　nos. 23, 26, 27, K 306, 378, 379. Sonata, violin and piano,
　　　K 372, B flat major. Oleg Kagaan, vln; Sviatoslav Richter, pno.
　　　HMV SLS 5020 (2).
　　　　+Gr 9-75 p485　　　　　　　　+RR 8-75 p58
　　　　+HFN 8-75 p79
1498 Arias: Ah, se in ciel, benigne stelle, K 538. Vesperae de dominica,
　　　K 321. Delcina Stevenson, s; Bonnie Hurwood, ms; Keith Wyatt,
　　　t; Orrin Nehls, bs; Los Angeles Camerata Chorus and Orchestra;
　　　H. Vincent Mitzelfelt. Crystal S 902.
　　　　+-HF 5-75 p84　　　　　　　　-RR 1-75 p56
　　　　++NR 12-74 p13
1499 Arias: Don Giovanni: Non mi dir, bell'idol mio. Die Entführung aus
　　　dem Serail, K 384: Martern aller Arten. Popoli di Tessaglia...
　　　Io non chiedo, eterni dei, K 316. Ma che vi fece, o stelle...
　　　Sperai vicino il lido, K 368. Die Zaüberflote, K 680: O zittre
　　　nicht, mien lieber Sohn; Der Hölle Rache. Edda Moser, s;

Bavarian State Opera Orchestra; Wolfgang Sawallisch, Leopold
Hager. EMI Electrola C 063-29082.
    +OC 12-75 p49   ++St 12-74 p97

1500 Arias: Così fan tutte, K 588 (2). Don Giovanni, K 527 (2). Die
Entführung aus dem Serail, K 384 (4). The magic flute, K 680
(2). Le nozze di Figaro, K 492 (1). József Réti, t; Budapest
Philharmonic Orchestra; Adám Medveczky. Hungaroton SLPX 11679.
    ++SFC 1-26-74 p26

1501 Arias: Ch'io mi scordi di te, K 505. Chi sà, chi sà qual sia, K 582.
Così fan tutte, K 588: Temerari...Come scoglio. Don Giovanni,
K 527: Batti, batti; Vedrai, carino. Misera dove son, K 369.
Le nozze di Figaro, K 492: Non sò più cosa son; Giunse al fin...
Deh vieni, non tardar; Voi che sapete. Vado, ma dove, K 583.
Elly Ameling, s; Dalton Baldwin, pno; ECO; Edo de Waart. Phil-
ips 6500 544.
    +Gr 10-74 p745   +ON 2-1-75 p29
    +-HF 11-74 p114   +Op 12-74 p1086
    +MJ 10-74 p43   +RR 10-74 p24
    +NR 10-74 p13   +-St 12-74 p138

1502 Arias (concert): Al desio di chi t'adora, K 577. Bella mia fiamma...
Resta, oh cara, K 528. Ch'io mi scordi di te...Non temer, amato
bene, K 505. Nehmt meinen Dank, K 383. Non più, tutto ascolta
...Non temer, amato bene, K 490. Vado, ma dove, K 583. Vorrei
spiegarvi, oh Dio, K 418. Margaret Price, s; LPO; James Lock-
hart. RCA LRL 1-5077.
    +Gr 10-75 p670   +RR 11-75 p87
    +HFN 11-75 p159

Ave verum corpus, K 618. cf Mass, no. 16, K 317, C major.
Ave verum corpus, K 618. cf Pye GH 589.

1503 Cassation, no. 1, K 63, G major. March for a cassation, K 62, D
major. Serenade, K 62a, D major. Dresden Philharmonic Orchestra; Günther Herbig. Philips 6500 704.
    +Gr 6-75 p45   ++MT 9-75 p799
    ++HFN 5-75 p135   +-RR 5-75 p34

Cassation, no. 1, K 63, G major: Adagio. cf Works, selections
(Philips 6775 012).

1504 Cassation, no. 2, K 99, B flat major. Divertimento, K 63, G major.
Vienna Mozart Ensemble; Willi Boskovsky. London STS 15302.
    +NR 7-75 p2

La clemenza di Tito, K 621: Parto, parto, ma tu ben mio; Deh per
questo instante solo. cf HAYDN: Arianna a Naxos

1505 Concerto, bassoon, K 191, B flat major. Concerto, clarinet, K 622,
A major. Concerto, oboe, K 314, C major. John de Lancie, ob;
Anthony Gigliotti, clt; Bernard Garfield, bsn; PO; Eugene Ormandy.
CBS 61657. (Reissue from 72128, 72127)
    +Gr 9-75 p465   +RR 8-75 p39
    +HFN 9-75 p108

Concerto, bassoon, K 191, B flat major. cf BACH, J. C.: Symphony,
no. 3, op. 3, E flat major.

1506 Concerto, clarinet, K 622, A major. Concerto, oboe, K 314, C
major. Derek Wickens, ob; Thea King, clt; London, Little Orchestra; Leslie Jones. Oryx Tape (c) BRL 21.
    +RR 1-75 p70 tape

1507 Concerto, clarinet, K 622, A major. NIELSEN: Concerto, clarinet.
John McCaw, clt; NPhO; Raymond Leppard. Unicorn Tape (c) ZCUN
239.
    +RR 3-75 p74 tape   ++St 10-75 p111

Concerto, clarinet, K 622, A major. cf Concerto, bassoon, K 191,
    B flat major.
Concerto, clarinet, K 622, A major. cf Works, selections (Columbia
    D3M 33261).
Concerto, clarinet, K 622, A major. cf Camden CCV 5000-12, 5014-24.
1508 Concerto, flute, no. 1, K 313, G major. Concerto, flute, no. 2,
    K 314, D major. Andante, flute, K 315, C major. Karl-Heinz
    Zöller, flt; ECO; Bernhard Klee. DG 2530 344.
         ++Gr 4-75 p1811   +RR 3-75 p28
         ++HF 6-75 p99   ++SFC 7-6-75 p16
         +NR 5-75 p7
1509 Concerto, flute, no. 1, K 313, G major. Concerto, oboe, K 314,
    C major. Werner Tripp, flt; Gerhard Turetschek, ob; VPO; Karl
    Böhm. DG 2530 527.
         +Gr 7-75 p191   +NR 12-75 p6
         ++HFN 7-75 p85   ++RR 7-75 p28
    Concerto, flute, no. 2, K 314, D major. cf Concerto, flute, no. 1,
    K 313, G major.
    Concerto, flute, no. 2, K 314, D major: 3rd movement. cf Decca
    SPA 394.
1510 Concerto, flute and harp, K 299, C major. Sinfonia concertante,
    oboe, clarinet, horn, bassoon and orchestra, K 297b, E flat
    major. Claude Monteux, flt; Osian Ellis, hp; Neil Black, ob;
    Jack Brymer, clt; Alan Civil, hn; Michael Chapman, bsn; AMF;
    Neville Marriner. Philips 6500 380. Tape (c) 7300 301. (Re-
    issue from 6707 020)
         ++AR 2-74 p24   ++NR 1-74 p6
         +Gr 12-74 p1137   +NYT 1-27-74 pD24
         +Gr 9-74 p592 tape   ++RR 10-74 p50
         +HF 4-74 p100   +-RR 1-75 p70
         +HF 1-75 p110 tape   +St 3-74 p115
         ++MJ 2-74 p42
    Concerto, flute and harp, K 299, C major. cf RCA CRL 2-7003.
    Concerto, harp, K 545, C major. cf Saga 5356.
1511 Concerti, horn and strings, nos. 1-4, K 412, K 417, K 447, K 495.
    Concerto, horn and strings, no. 5, K 494a, E major (fragment).
    Barry Tuckwell, hn; LSO; Peter Maag. Decca SDD 364. Tape (c)
    KCDC 364. (Reissue from SXL 6108, 2238, SWL 8011)
         +Gr 4-73 p1918   ++HFN 4-73 p783
         +Gr 1-75 p1402 tape   ++RR 4-73 p60
1512 Concerti, horn and strings, nos. 1-4, K 412, K 417, K 447, K 495.
    Hermann Baumann, hn; VCM; Nikolaus Harnoncourt. Telefunken
    SAWT 9627.
         ++Audio 9-75 p77   ++NR 5-75 p7
         /Gr 12-74 p1137   +RR 12-74 p34
         +-HF 7-75 p80   ++St 7-75 p102
1513 Concerto, horn and strings, no. 1, K 412, D major. Concerto,
    horn and strings, no. 2, K 417, E flat major. Concerto, horn
    and strings, no. 3, K 447, E flat major. Concerto, horn and
    strings, no. 4, K 495, E flat major. Rondo, horn, K 371, E flat
    major. Alan Civil, hn; AMF; Neville Marriner. Philips 6500 325.
    Tape (c) 7300 199. (REeissue from 6706 020)
         ++AR 2-74 p24   ++RR 4-74 p47
         ++Gr 5-74 p2073   +RR 1-75 p70 tape
         ++Gr 1-74 p1430 tape   ++SFC 1-20-74 p26
         ++HF 6-73 p86   ++St 6-73 p120
         ++NR 6-73 p5

1514 Concerti, horn and strings, nos. 1-2, 4, K 412, K 417, K 495.
       Rondo, horn, K 371, E flat major. Ferenc Tarjani, hn; Györ
       Philharmonic Orchestra; Janos Sándor. Hungaroton SLPX 11707.
              -HFN 5-75 p135   +RR 3-75 p28
       Concerto, horn and strings, no. 2, K 417, E flat major. cf Concerto,
       horn and strings, no. 1, K 412, D major.
       Concerto, horn and strings, no. 3, K 447, E flat major. cf Con-
       certo, horn and strings, no. 1, K 412, D major.
1515 Concerto, horn and strings, no. 4, K 495, E flat major. The
       marriage of Figaro, K 492: Overture. Symphony, no. 40, K 550,
       G minor. Alan Civil, hn; RPO; Lawrence Foster. London SPC
       21093. (also Decca PFS 4314)
              +-Gr 7-75 p188   +MJ 3-75 p26
              +-HFN 9-75 p101   +-RR 7-75 p28
       Concerto, horn and strings, no. 4, K 495, E flat major. cf Concerto,
       horn and strings, no. 1, K 412, D major.
       Concerto, horn and strings, no. 4, K 495, E flat major. cf HAYDN:
       Concerto, horn, no. 1, D major.
       Concerto, horn and strings, no. 4, K 495, E flat major: Rondo. cf
       Works, selections (Philips 6833 163).
       Concerto, horn and strings, no. 5 K 494a, E major, fragment. cf
       Concertos, horn, nos. 1-4, K 412, K 417, K 447, K 495.
       Concerto, oboe, K 314, C major. cf Concerto, bassoon, K 191, B
       flat major.
       Concerto, oboe, K 314, C major. cf Concerto, clarinet, K 622, A
       major.
       Concerto, oboe, K 314, C major. cf Concerto, flute, no. 1, K 313,
       G major.
       Concerto, oboe, K 314, C major. cf ALBINONI: Concerto, oboe, op.
       7, no. 6, D major.
1516 Concerti, piano, nos. 1-4. Géza Anda, pno; Salzburg Mozarteum
       Camerata Academica; Géza Anda. DG 2538 261. (Reissues from
       139447, 139453, 2720 030)
              +Gr 5-75 p1967   /RR 6-75 p44
              +HFN 5-75 p135
1517 Concerto, piano, no. 1, K 37, F major. Concerto, piano, no. 2,
       K 39, B flat major. Concerto, piano, no. 3, K 40, D major.
       Concerto, piano, no. 4, K 41, G major. Ingrid Haebler, forte-
       piano; Vienna Capella Academica; Eduard Melkus. Philips 6500
       773.
              +Gr 12-75 p1037   ++RR 10-75 p45
              +HFN 10-75 p144   ++STL 11-2-75 p38
       Concerto, piano, no. 2, K 39, B flat major. cf Concerto, piano,
       no. 1, K 37, F major.
       Concerto, piano, no. 3, K 40, D major. cf Concerto, piano, no. 1,
       K 37, F major.
       Concerto, piano, no. 4, K 41, G major. cf Concerto, piano, no. 1,
       K 37, F major.
1518 Concerti, piano, nos. 8, 23, 24, 27. Wilehlm Kempff, pno; Bamberg
       Symphony Orchestra; Ferdinand Leitner. DG 2726 024 (2). (Re-
       issues from SLPM 138812, 138645)
              +-Gr 10-75 p621   +RR 12-75 p56
              ++HFN 9-75 p108
1519 Concerto, piano, no. 8, K 246, C major. Concerto, piano, no. 25,
       K 503, C major. Daniel Barenboim, pno; ECO; Daniel Barenboim.
       HMV ASD 3033.
              +Gr 1-75 p1349

1520 Concerto, piano, no. 9, K 271, E major. Concerto, piano, no. 20,
     K 466, D minor. Felicja Blumenthal, pno; Salzburg Mozarteum
     Orchestra; Leopold Hager. Everest SDBR 3381.
          +NR 11-75 p7
     Concerto, piano, no. 9, K 271, E flat major. cf BEETHOVEN: Concerto,
     piano, no. 1, op. 15, C major.
1521 Concerto, piano, no. 11, K 413, F major. Concerto, piano, no. 15,
     K 450, B flat major. Peter Frankl, pno; Würtemburg Chamber
     Orchestra; Jörg Faerber. Turnabout TV 34027. Tape (c) KTVC
     34027.
          +Gr 12-74 p1237 tape   +RR 1-75 p70 tape
1522 Concerto, piano, no. 12, K 414, A major. Concerto, piano, no. 27,
     K 595, B flat major. Jörg Demus, pno; Collegium Aureum. BASF
     BAC 3066.
          -Gr 3-75 p1654   +RR 2-75 p36
1523 Concerto, piano, no. 12, K 414, A major. Concerto, piano, no. 21,
     K 467, C major. Radu Lupu, pno; ECO; Uri Segal. Decca SXL
     6698. (also London CS 6894)
          +Gr 3-75 p1657   ++SFC 7-6-75 p16
          +NR 10-75 p4   +St 11-75 p128
          ++RR 4-75 p29
     Concerto, piano, no. 12, K 414, A major. cf HAYDN: Concerto, harp-
     sichord, op. 21, D major.
1524 Concerti, piano, nos. 14-19. Peter Serkin, pno; ECO; Alexander
     Schneider. RCA ARL 3-0732 (3).
          +Gr 11-75 p819   +RR 11-75 p44
          *HF 5-75 p79   ++SFC 1-26-75 p26
          +HFN 11-75 p157   +SR 2-8-75 p37
          ++NR 2-75 p6   +St 4-75 p71
1525 Concerto, piano, no. 14, K 449, E flat major. Concerto, piano,
     no. 20, K 466, D minor. Dame Myra Hess, pno; Orchestra; Bruno
     Walter. Bruno Walter Society PR 36.
          +NR 3-73 p6   +NR 9-75 p7
1526 Concerto, piano, no. 14, K 449, E flat major. Concerto, piano,
     no. 22, K 482, E flat major. Paul Badura-Skoda, pno; VSO;
     William Steinberg. Ember ECL 9014.
          -HFN 8-75 p79   -RR 9-75 p40
1527 Concerto, piano, no. 14, K 449, E flat major. Concerto, piano,
     no. 18, K 456, B flat major. Walter Klien, pno; Vienna Pro
     Musica, Mainz Chamber Orchestra; Paul Angerer, Gunter Kehr.
     Turnabout TV 34503. (Reissue from Vox PL 11650)
          +-Gr 2-75 p1492   +-RR 1-75 p29
1528 Concerto, piano, no. 15, K 450, B flat major. Concerto, piano,
     no. 20, K 466, D minor. Arturo Benedetti Michelangeli, pno;
     RAI Torino Orchestra; Mario Rossi. Bruno Walter Society RR 422.
          +-NR 12-75 p6
1529 Concerto, piano, no. 15, K 450, B flat major. Concerto, piano,
     no. 25, K 503, C major. Andor Foldes, pno; BPhO; Leopold Lud-
     wig. DG 2548 193. (Reissue from 138796)
          /Gr 11-75 p840   +-RR 12-75 p56
          +HFN 10-75 p152
     Concerto, piano, no. 15, K 450, B flat major. cf Concerto, piano,
     no. 11, K 413, F major.
1530 Concerto, piano, no. 16, K 451, D major. Concerto, piano, no. 23,
     K 488, A major. Walter Klien, pno; Vienna Volksoper Orchestra;
     Paul Angerer, Peter Maag. Turnabout Tape (c) KTVC 34286.
          -RR 11-75 p94 tape

1531 Concerti, piano, nos. 17, 20, 21, 23-24. Rondo, K 511, A minor.
      Artur Rubinstein, pno; RCA Victor Symphony Orchestra; Alfred
      Wallenstein, Josef Krips. RCA SER 5716/8 (3). (Reissues from
      SB 2117, 6532, 6570, 6578)
           ++Gr 5-75 p1968   +RR 7-75 p31
           +HFN 7-75 p85
1532 Concerto, piano, no. 17, K 453, C major. Concerto, piano, no. 24,
      K 491, C minor. André Previn, pno; LSO; Adrian Boult. HMV ASD
      2951. Tape (c) TC ASD 2951. (also Angel S 37002)
           +Gr 1-74 p1378   ++NR 11-74 p5
           +Gr 7-74 p274 tape   +RR 1-74 p43
           +-HF 5-75 p79   +-RR 5-74 p86 tape
1533 Concerto,piano, no. 17, K 453, G major. Concerto, piano, no. 23,
      K 488, A major. Karl Engel, pno; Salzburg Mozarteum Orchestra;
      Leopold Hager. Telefunken AV 6-41888. Tape (c) CX 4-41888.
           /Gr 9-75 p466   +-HFN 8-75 p79
           /Gr 10-75 p721 tape   /RR 8-75 p40
1534 Concerto, piano, no. 18, K 456, B flat major. Concerto, piano,
      no. 27, K 595, B flat major. Alfred Brendel, pno; AMF; Neville
      Marriner. Philips 6500 948. Tape (c) 7300 383.
           +-Gr 4-75 p1811   ++NR 11-75 p6
           +HFN 7-75 p90   ++RR 5-75 p29
           ++MT 10-75 p885   ++SFC 10-5-75 p38
      Concerto, piano, no. 18, K 456, B flat major. cf Concerto, piano,
      no. 14, K 449, E flat major.
1535 Concerto, piano, no. 20, K 466, D minor. Concerto, piano, no. 21,
      K 467, C major. Friedrich Gulda, pno; VPO; Claudio Abbado.
      DG 2530 548. Tape (c) 3300 492.
           +-Gr 11-75 p820   +-HFN 12-75 p173 tape
           +-HFN 10-75 p144   /RR 10-75 p45
1536 Concrto, piano, no. 20, K 466, D minor. Sviatoslav Richter, pno;
      Warsaw Philharmonic Orchestra; Witold Rowicki. DG 2548 106.
      (Reissue from 138075)
           +-HFN 9-75 p108
1537 Concerto, piano, no. 20, K 466, D minor. Concerto, piano, no. 21,
      K 467, C major. Géza Anda, pno; VSO; Géza Anda. Eurodisc (Q)
      86947.
           +-St 11-75 p128
1538 Concerto, piano, no. 20, K 466, D minor. Concerto, piano, no. 23,
      K 488, A major. Daniel Barenboim, pno; ECO. HMV ASD 2318.
      Tape (c) TC ASD 2318.
           +Gr 2-75 p1562 tape   ++RR 5-75 p77 tape
1539 Concerto, piano, no. 20, K 466, D minor.* Concerto, piano, no. 24,
      K 491, C minor. Alfred Brendel, pno; AMF; Neville Marriner.
      Philips 6500 533. (*Reissue from 6833 119)
           +-Gr 8-75 p325   +NR 1-75 p5
           +-HF 5-75 p79   +-RR 7-75 p31
           +HFN 7-75 p85   /SFC 1-26-75 p26
           +MJ 3-75 p26
1540 Concerto, piano, no. 20, K 466, D minor. Concerto, piano, no. 21,
      K 467, C major. Géza Anda, pno; VSO; Géza Anda. RCA ARL 1-0610.
      Tape (c) ARK 1-0610 (ct) ARS 1-0610. (also RCA LRL 1-5020)
           +-Gr 9-74 p505   +-RR 10-74 p50
           -HF 5-75 p79   /SFC 1-5-75 p23
           +-MJ 2-75 p39   +-St 11-75 p128
           +-NR 12-74 p6

1541 Concerto, piano, no. 20, K 466, D minor. Deutsche Tänze, K 605 (3).
     Serenade, no. 13, K 525, G major. Bruno Walter, pno; VPO; Bruno
     Walter. Turnabout THS 65036.
          +—NR 7-75 p6
1542 Concerto, piano, no. 20, K 466, D minor. Concerto, piano, no. 24,
     K 491, C minor. Artur Schnabel, pno; PhO; Walter Susskind.
     Turnabout THS 65046.
          +NR 12-75 p6
     Concerto, piano, no. 20, K 466, D minor. cf Concerto, piano, no.
     9, K 271, E major.
     Concerto, piano, no. 20, K 466, D minor. cf Concerto, piano,
     no. 14, K 449, E flat major.
     Concerto, piano, no. 20, K 466, D minor. cf Concerto, piano, no.
     15, K 450, B flat major.
     Concerto, piano, no. 20, K 466, D minor. cf BEETHOVEN: Rondo,
     WoO6, B flat major.
1543 Concerto, piano, no. 21, K 467, C major. HAYDN: Concerto, harpsi-
     chord, no. 21, D major. Emil Gilels, pno; Moscow Chamber Orch-
     estra; Rudolf Barshai. Columbia/Melodiya M 33098. (Reissue
     from Artia ALP 159)
          -SFC 1-5-75 p23   ++St 5-75 p97
1544 Concerto, piano, no. 21, K 467, C major. Concerto, piano, no. 23,
     K 488, A major. Ilana Vered, pno; LPO; Uri Segal. Decca PFS
     4340.
          +Gr 12-75 p1038   +RR 11-75 p44
          +HFN 12-75 p157
1545 Concerto, piano, no. 21, K 467, C major. Concerto, piano, no. 23,
     K 488, A major. Felicja Blumenthal, pno; Salzburg Mozarteum
     Orchestra; Leopold Hager. Everest SDBR 3374.
          /NR 7-75 p6
1546 Concerto, piano, no. 21, K 467, C major. Concerto, piano, no. 25,
     K 503, C major. Stephen Bishop, pno; LSO; Colin Davis. Philips
     6500 431. Tape (c) 7300 250.
          +—Gr 4-74 p1858   ++NR 1-74 p6
          +Gr 7-75 p254 tape   +NYT 1-27-74 pD24
          +—HF 3-74 p91   +—RR 3-74 p44
          ++HFN 3-74 p111   ++SFC 9-30-73 p26
          +HFN 7-75 p90 tape   ++St 5-74 p105
     Concerto, piano, no. 21, K 467, C major. cf Concerto, piano, no. 12,
     K 414, A major.
     Concerto, piano, no. 21, K 467, C major. cf Concerto, piano, no. 20,
     K 466, D minor (DG 2530 548).
     Concerto, piano, no. 21, K 467, C major. cf Concerto, piano, no. 20,
     K 466, D minor (Eurodisc 86947).
     Concerto, piano, no. 21, K 467, C major. cf Concerto, piano, no. 20,
     K 466, D minor (RCA ARL 1-0610).
     Concerto, piano, no. 21, K 467, C major. cf HAYDN: Concerto, harp-
     sichord, op. 21, D major.
     Concerto, piano, no. 21, K 467, C major: Adagio. cf Works, selec-
     tions (Philips 6833 163).
     Concerto, piano, no. 21, K 467, C major: Andante. cf CBS 77513.
     Concerto, piano, no. 21, K 467, C major: 2nd movement. cf Sound
     Superb SPR 90049.
1547 Concerto, piano, no. 22, K 482, E flat major. Concerto, piano, no.
     23, K 488, A major. Robert Casadesus; Columbia Symphony Orches-
     tra; Georg Szell. CBS 61021. (also Columbia MS 6194)

+–Gr 8-75 p325   +–RR 9-75 p40
++HFN 7-75 p85
Concerto, piano, no. 22, K 482, E flat major. cf Concerto, piano,
    no. 14, K 449, E flat major.
1548 Concerto, piano, no. 23, K 488, A major. Symphony, no. 41, K 551,
    C major. Michael Roll, pno; London Mozart Players; Harry Blech.
    Abbey ABY 746.
        +–RR 3-75 p30
Concerto, piano, no. 23, K 488, A major. cf Concerto, piano, no.
    21, K 467, C major.
Concerto, piano, no. 23, K 488, A major. cf Concerto, piano, no.
    20, K 466, D minor.
Concerto, piano, no. 23, K 488, A major. cf Concerto, piano, no.
    17, K 453, G major.
Concerto, piano, no. 23, K 488, A major. cf Concerto, piano, no.
    16, K 451, D major.
Concerto, piano, no. 23, K 488, A major. cf Concerto, piano, no.
    21, K 467, C major.
Concerto, piano, no. 23, K 488, A major. cf Concerto, piano, no.
    22, K 482, E flat major.
1549 Concerto, piano, no. 24, K 491, C minor. Fantasia, K 475, C minor.
    Sonata, piano, no. 14, K 457, C minor. Walter Klien, pno;
    Vienna Volksoper Orchestra; Peter Maag. Turnabout TV 34178.
    Tape (c) KTVC 34178.
        +–RR 1-75 p70 tape
Concerto, piano, no. 24, K 491, C minor. cf Concerto, piano, no.
    17, K 453, C major.
Concerto, piano, no. 24, K 491, C minor. cf Concerto, piano, no.
    20, K 466, D minor.
Concerto, piano, no. 24, K 491, C minor. cf Concerto, piano, no.
    20, K 466, D minor.
1550 Concerto, piano, no. 25, K 503, C major. Fantasia, K 475, C minor.
    Ivan Moravec, pno; CPhO; Josef Vlach. Vanguard SU 11.
        +NR 11-75 p6
Concerto, piano, no. 25, K 503, C major. cf Concerto, piano, no.
    8, K 246, C major.
Concerto, piano, no. 25, K 503, C major. cf Concerto, piano, no.
    15, K 450, B flat major.
Concerto, piano, no. 25, K 503, C major. cf Concerto, piano, no.
    21, K 467, C major.
Concerto, piano, no. 25, K 503, C major. cf Works, selections
    (Columbia D3M 33261).
1551 Concerto, piano, no. 26, K 537, D major. Concerto, piano, no. 27,
    K 595, B flat major. Robert Casadesus, pno; Columbia Symphony
    Orchestra; Georg Szell. CBS 61597. Tape (c) 40-61597. (Re-
    issue from SBR 72107) (also Columbia MS 6403)
        +–Gr 2-75 p1492   +–RR 8-75 p40
        +–HFN 12-75 p173 tape   –RR 12-75 p99 tape
1552 Concerto, piano, no. 26, K 537, D major. Fantasia, K 397, D minor.
    Sonata, piano, no. 17, K 576, D major. Variations on "Ah, vous
    dirai-je, Maman," K 265, C major. Walter Klien, pno; Vienna
    Volksoper Orchestra; Peter Maag. Turnabout Tape (c) KTVC 134194.
        +RR 7-75 p71 tape
1553 Concerto, piano, no. 27, K 595, B flat major. Sonata, piano, no.
    10, K 330, C major. Sonata, piano, no. 11, K 331, A major.
    Wilhelm Backhaus, pno; VPO; Karl Böhm. Decca ECS 749. (Reissue
    from SXL 2214, 6301)

```
            +-Gr 7-75 p188                    +-RR 8-75 p40
            +-HFN 9-75 p108
```
1554 Concerto, piano, no. 27, K 595, B flat major. Concerto, 2 pianos,
     K 365, E flat major. Emil Gilels, Elena Gilels, pno; VPO;
     Karl Böhm. DG 2530 456. Tape (c) 3300 406.
```
            +Gr 11-74 p903                   +SFC 3-23-75 p22
            +-NR 1-75 p5                     +SR 2-22-75 p47
            +RR 11-74 p45                    ++St 5-75 p101
            +-RR 1-75 p71 tape
```
     Concerto, piano, no. 27, K 595, B flat major. cf Concerto, piano,
     no. 12, K 414, A major.
     Concerto, piano, no. 27, K 595, B flat major. cf Concerto, piano,
     no. 18, K 456, B flat major.
     Concerto, piano, no. 27, K 595, B flat major. cf Concerto, piano,
     no. 26, K 537, D major.
1555 Concerto, 2 pianos, K 365, E flat major. Concerto, 3 pianos, K 242,
     F major. Vladimir Ashkenazy, Daniel Barenboim, Fou Ts'ong, pno;
     ECO; Daniel Barenboim. Decca SXL 6716.
```
            -Gr 6-75 p45                     +-RR 6-75 p43
            +-HFN 6-75 p89
```
1556 Concerto, 2 pianos, K 365, E flat major. Concerto, 3 pianos, K 242,
     F major. Zoltán Kocsis, Dezsö Ránki, András Schiff, pno; HSO;
     János Feréncsik. Hungaroton SLPX 11631.
```
            +-RR 3-74 p41                    ++SFC 1-19-75 p28
```
     Concerto, 2 pianos, K 365, E flat major. cf Concerto, piano, no.
     27, K 595, B major.
     Concerto, 3 pianos, K 242, F major. cf Concerto, 2 pianos, K 365,
     E flat major (Decca SXL 6716).
     Concerto, 3 pianos, K 242, F major. cf Concerto, 2 pianos, K 365,
     E flat major (Hungaroton SLPX 11631).
1557 Concerti, violin, nos. 1-5. Adagio, K 261, E major. Rondo, violin,
     K 373, C major. Rondo, K 269, B flat major. Pinchas Zukerman,
     vln; ECO; Daniel Barenboim. CBS 77381 (3).
```
            +-Gr 1-75 p1350                  +-RR 1-75 p30
```
1558 Concerti, violin, nos. 1-5. Adagio, K 261, E major. Rondo, violin,
     K 269, B flat major. Rondo, violin, K 373, C major. Wolfgang
     Schneiderhan, vln; BPhO; Wolfgang Schneiderhan. DG 2740 116 (3).
     (Reissue from 139350/2)
```
            +Gr 8-75 p326                    +RR 7-75 p32
            +HFN 8-75 p89
```
1559 Concerto, violin, no. 1, K 207, B flat major. Concerto, violin,
     no. 3, K 216, G major. Pinchas Zukerman, vln; ECO; Daniel
     Barenboim. Columbia M 32301. Tape (c) MT 32301 (Q) MQ 32301
     Tape (ct) MAQ 32301.
```
            +HF 3-75 p82                     ++St 12-74 p138
            ++NR 8-74 p5
```
1560 Concerto, violin, no. 1, K 207, B flat major. Concerto, violin,
     no. 2, K 211, D major. Josef Suk, vln; Prague Chamber Orchestra;
     Josef Suk. RCA LRL 1-5084.
```
            +Gr 4-75 p1811                   +RR 3-75 p30
```
     Concerto, violin, no. 1, K 207, B flat major. cf Works, selections
     (Philips 6775 012).
1561 Concerto, violin, no. 2, K 211, D major. Sinfonia concertante,
     K 364, E flat major. David Oistrakh, vln and vla; Igor Oistrakh,
     vln; BPhO; David Oistrakh. Angel S 36892.
```
            ++NR 3-75 p5                     ++St 7-75 p102
```
1562 Concerto, violin, no. 2, K 211, D major. Rondo, violin, K 373,

C major. Rondo, violin, K 269 (261a), B flat major. Adagio,
K 261, E major. Pinchas Zukerman, vln; ECO; Daniel Barenboim.
Columbia M 33206.
      -Audio 11-75 p95   +-St 6-75 p102
      -NR 5-75 p7
Concerto, violin, no. 2, K 211, D major. cf Adagio, K 261, E
  major.
Concerto, violin, no. 2, K 211, D major. cf Concerto, violin, no.
  1, K 207, B flat major.
Concerto, violin, no. 2, K 211, D major. cf Works, selections
  (Philips 6775 012).
1563 Concerto, violin, no. 3, K 216, G major. BEETHOVEN: Concerto,
  piano, no. 4, op. 58, G major. David Oistrakh, vln; Emil
  Gilels, pno; PhO; Leopold Ludwig. HMV Tape (c) TX ECE 156.
    ++Gr 10-75 p721 tape   +-HFN 10-75 p155 tape
1564 Concerto, violin, no. 3, K 216, G major. Concerto, violin, no.
  5, K 219, A major. Arthur Grumiaux, vln; LSO; Colin Davis.
  Philips 835112.
    +HF 3-75 p82   ++SFC 11-17-74 p31
    ++MJ 1-75 p49   ++St 6-75 p102
    ++NR 12-74 p6
1565 Concerto, violin, no. 3, K 216, G major. Concerto, violin, no. 4,
  K 218, D major. Henryk Szeryng, vln; NPhO; Alexander Gibson.
  Philips 6500 036. Tape (c) 7300 054.
    ++HFN 5-75 p142 tape   ++HFN 7-75 p90 tape
Concerto, violin, no. 3, K 216, G major. cf Adagio, K 261, E
  major.
Concerto, violin, no. 3, K 216, G major. cf Concerto, violin,
  no. 1, K 207, B flat major.
Concerto, violin, no. 3, K 216, G major. cf BEEHOVEN: Concerto,
  piano, no. 4, op. 58, G major.
Concerto, violin, no. 4, K 218, D major. cf Concerto, violin,
  no. 3, K 216, G major.
Concerto, violin, no. 4, K 218, D major. cf Pearl GEM 132.
Concerto, violin, no. 4, K 218, D major: Andante cantabile. cf
  Works, selections (Philips 6833 163).
1566 Concerto, violin, no. 5, K 219, A major. TCHAIKOVSKY: Concerto,
  violin, op. 35, D major. David Oistrakh, vln; USSR; Kiril
  Kondrashin. Everest SDBR 3375.
    ++NR 8-75 p6
1567 Concerto, violin, no. 5, K 219, A major. Concerto, violin, no. 6,
  K 268, E flat major. Josef Suk, vln; Prague Chamber Orchestra;
  Josef Suk. RCA LRL L-5089.
    +-Gr 6-75 p45   +RR 6-75 p45
    ++HFN 6-75 p89
Concerto, violin, no. 5, K 219, A major. cf Concerto, violin, no.
  3, K 216, G major.
Concerto, violin, no. 5, K 219, A major. cf RCA ARM 4-0942/7.
Concerto, violin, no. 5, K 219, A major. cf RCA CRL 6-0720.
Concerto, violin, no. 6, K 268, E flat major. cf Concerto, violin,
  no. 5, K 219, A major.
1568 Concertone, 2 violins, K 190 (K 166b), C major. PLEYEL: Sinfonie
  concertante, violin and viola, op. 29, B flat major. Isaac
  Stern, Pinchas Zukerman, vln; Neil Black, ob; Pinchas Zukerman,
  vla; ECO; Daniel Barenboim. CBS 76310. (also Columbia M 32937)
    +Gr 12-74 p1138   +-RR 12-74 p36
    +-HF 3-75 p82   +St 6-75 p143

1569 Così fan tutte, K 588. Elisabeth Schwarzkopf, Hanny Steffek, s;
     Christa Ludwig, ms; Alfredo Kraus, t; Guiseppe Taddei, bar;
     Walter Berry, bs; PhO and Chorus; Karl Böhm. Angel S 3631.
     (also HMV SLS 5028)
          +—Gr 10-75 p679   +RR 11-75 p34
          +—HFN 11-75 p159   +St 4-75 p68
1570 Così fan tutte, K 588. Pilar Lorengar, Jane Berbie, s; Teresa
     Berganza, ms; Ryland Davies, t; Tom Krause, Gabriel Bacquier,
     bar; ROHO Chorus; LPO; Goerg Solti. Decca SET 575/8 (3).
     (also London OSA 1442. Tape (c) Q51442 (r) S 41442)
          +—Gr 6-74 p105   +ON 2-8-75 p33
          +HF 10-74 p81   +Op 9-74 p798
          +HF 4-75 p112 tape   +—RR 6-74 p31
          +—MJ 3-75 p25   ++SFC 9-15-74 p28
          ++NR 1-75 p13   +—St 10-74 p85
          +NYT 12-15-74 pD21
1571 Così fan tutte, K 588. Gundula Janowitz, Reri Grist, s; Brigitte
     Fassbaender, ms; Peter Schreier, t; Hermann Prey, Rolando Pan-
     erai, bar; Vienna State Opera Chorus; VPO; Karl Böhm. DG 2709
     059 (3). (also 2740 118)
          +—Gr 10-75 p679   +ON 12-20-75 p38
          +—HFN 10-75 p145   +—RR 10-75 p29
1572 Così fan tutte, K 588. Montserrat Caballé, Ileana Cotrubas, s;
     Janet Baker, ms; Nicolai Gedda, t; Wladimiro Ganzarolli, bar;
     Richard Van Allan, bs; ROHO; Colin Davis. Philips 6707 025 (4).
          +Gr 2-75 p1543   +ON 2-8-75 p33
          +—HF 3-75 p84   .+—RR 2-75 p24
          ++MJ 3-75 p25   ++SFC 12-15-74 p33
          +—NR 4-75 p8   +St 4-75 p101
1573 Così fan tutte, K 588. Elizabeth Schwarzkopf, Lisa Otto, s; Nan
     Merriman, ms; Leopold Simoneau, t; Rolando Panerai, Sesto
     Bruscatini, bar; PhO and Chorus; Herbert von Karajan. World
     Records SOC 195/7 (3).
          +—HFN 12-75 p157
     Così fan tutte, K 588: Un aura amorosa; In qual fiero contrasto...
        Tradito, schernito. Don Giovanni, K 527: Dalla sua pace, Il
        mio tesoro. Die Entführung aus dem Serail, K 384: Heir soll ich
        dich denn sehen; Constanze, dich wieder zu sehen; Wenn der
        Freude Tränen fliessen; Ich baue ganz. Le nozze di Figaro,
        K 492: In quegl'anni. Die Zauberflöte, K 620: Dies Bildnis
        ist bezaubernd schön; Wie stark ist nicht dein Zauberton. József
        Réti, t; Budapest Philharmonic Orchestra; Adám Medveczky. Hun-
        garoton SLPX 11679.
          /RR 3-75 p18
     Così fan tutte, K 588: Arias. cf Arias (Hungaroton 11679).
     Così fan tutte, K 588: Der Odem der Liebe. cf Telefunken AJ 6-41867.
     Così fan tutte, K 588: Per pieta ben mio. cf HANDEL: Giulio Cesare:
        Hast due mich ganz berauscht, Es blaut die Nacht, Breite aus die
        gnäd'gen Hande, Weine nur klage nur, Heil und sicher kam mein
        Nachen.
     Così fan tutte, K 588: Temerari...Come scoglio. cf Arias (Philips
        6500 544).
     Country dances (Contredances), K 609 (5). cf Nonesuch H 71141.
     Contredances, K 609 (5). cf DG Archive 2533 182.
     Deutsche Tänze, K 605 (3). cf Concerto, piano, no. 20, K 466, D
        minor.

Deutsche Tänze (German dance), no. 3, K 605. cf HMV ASD 3017.
Divertimento, K 63, G major. cf Cassation, no. 2, K 99, B flat
   major.
Divertimento, K Anh 229, B flat major. cf Works, selections (Supraphon 111 1671/2).
Divertimenti, nos. 1-3, K 113, K 131, K 166. cf Adagio and fugue,
   strings, K 546, C minor
Divertimento, no. 1, K 136, D major. cf Serenade, no. 13, K 525,
   G major.
Divertimento, no. 5, K 187, C major. cf Works, selections (Supraphon 111 1671/2).
1575 Divertimento, no. 15, K 287, B flat major. Divertimento, no. 17,
   K 334, D major. Serenade, no. 13, K 525, G major. BPhO; Herbert
   von Karajan. DG 2726 032 (2). (Reissues from SLPM 139004, 139
   008)
       +Gr 3-75 p1657   -RR 2-75 p36
Divertimento, no. 15, K 287, B flat major. cf BACH: Sonata and
   partita, solo violin, S 1003, A minor.
1576 Divertimento, no. 17, K 334, D major. Vienna Mozart Ensemble; Willi
   Boskovsky. Decca SXL 6724.
       +-Gr 11-75 p819   ++RR 10-75 p45
       ++HFN 10-75 p144
1577 Divertimento, no. 17, K 334, D major. March, K 445, D major. New
   York Philomusica Chamber Ensemble. Vox STGBY 678.
       ++Gr 7-75 p188   +RR 6-75 p44
       +HFN 8-75 p78
Divertimento, no. 17, K 334, D major. cf Divertimento, no. 15,
   K 287, B flat major.
Divertimento, no. 17, K 334, D major: Minuet. cf Works, selections
   (Philips 6833 163).
Divertimento, no. 17, K 334, D major: Minuet. cf Philips 6580 098.
Divertimento, no. 17, K 334, D major: Menuetto. cf CBS 77513.
Divertimento, no. 17, K 334, D major: Minuet. cf Discopaedia MB
   1007.
Divertimento, no. 17, K 334, D major: Minuet. cf Philips 6580 066.
Divertimento, no. 17, K 334, D major: Minuet (2). cf RCA ARM
   4-0942/7.
1578 Divertimento, string trio, K 563, E flat major. Isaac Stern, vln;
   Pinchas Zukerman, vla; Leonard Rose, vlc. Columbia M 33266.
   (also CBS 76381)
       ++Gr 8-75 p342   +RR 8-75 p52
       +HF 8-75 p90   +-SFC 5-18-75 p23
       +HFN 8-75 p79   +St 8-75 p101
       +-NR 6-75 p8
1579 Divertimento, string trio, K 563, E flat major. Dénes Kovács, vln;
   Geza Németh, vla; Ede Banda, vlc. Hungaroton SLPX 11590.
       ++RR 5-73 p76   +SFC 5-18-75 p23
       ++SFC 4-14-74 p26
1580 Divertimento, string trio, K 563, E flat major. Bell Arte Trio.
   Turnabout (Q) QTVS 34567.
       +NR 6-75 p8   +St 8-75 p101
       +-SFC 5-18-75 p23
1581 Divertimento, winds K 196, B flat major.* SCHUBERT: Quintet, piano,
   op. 114, D 667, A major. Detmold Wind Sextet; Jörg Demus, pno;
   Schubert Quartet. DG 2548 122. (*Reissue from 136038)
       +Gr 4-75 p1867   /RR 7-75 p43

1582 Don Giovanni, K 527. Antigone Sgourda, Heather Harper, Helen
Donath, s; Luigi Alva, t; Geraint Evans, Alberto Rinaldi, bar;
Roger Soyer, Peter Lagger, bs; Scottish Opera Chorus; ECO;
Daniel Barenboim. Angel SDL 3811 (4). (also HMV SLS 978)
    +-Gr 4-75 p1863   -NYT 1-26-75 pD26
    -HF 4-75 p89   -ON 4-5-75 p40
    -MT 7-75 p631   +-RR 4-75 p17
    -NR 4-75 p8   -St 6-75 p102
1583 Don Giovanni, K 527. Birgit Nilsson, Martina Arroyo, s; Dietrich
Fischer-Dieskau, bar; Ezio Flagello, bs; Prague National Theatre
Orchestra and Chorus; Karl Böhm. DG 2711 006. Tape (c) 3371
014.
    ++Gr 1-75 p1402 tape   +RR 7-75 p68 tape
1584 Don Giovanni, K 527: Suite (arr. J. G. Trienbensee). The abduction
from the Seraglio, K 384: Suite (arr. attr. to J. Wendt). NWE.
Philips 6500 783.
    +Gr 1-75 p1349   ++NR 1-75 p8
    +HF 6-75 p1pp   ++RR 1-75 p43
    +-MJ 2-75 p40   ++St 5-75 p97
    Don Giovanni, K 527: Arias. cf Arias (Hungaroton 11679).
    Don Giovanni, K 527: Arias. cf London OS 26381.
    Don Giovanni, K 527: Arias. cf London OS 26277.
    Don Giovanni, K 527: Batti, batti; Vedrai, carino. cf Arias
    (Philips 6500 544).
    Don Giovanni, K 527: Crudele...Non mir dir; Il quali accesssi...Mi
    tradi. cf Decca GOSC 666/8.
    Don Giovanni, K 527: Dalla sua pace; Il mio tesoro. cf Così fan
    tutte, K 588: Un aura amorosa; In qual fiero contrasto...Tradito,
    schernito.
    Don Giovanni, K 527: Deh, vieni alla finestra. cf Ember GVC 37.
    Don Giovanni, K 527: Finch'han dal vino. cf Columbia/Melodiya
    M 33120.
    Don Giovanni, K 527: In quali accessi...Mi tradi, Ah, fuggi il
    traditor, Crudele...Non mir dir. cf HANDEL: Giulio Cesare: Hast
    du mich ganz berauscht, Es blaut die Nacht, Breite aus die
    gnäd'gen Hande, Weine nur klage nur, Heil und sicher kam mein
    Nachen.
    Don Giovanni, K 527: Minuet. cf CBS 77513.
    Don Giovanni, K 527: Non mi dir, bell'idol mio. cf Arias (EMI
    Electrola C 063-29082).
    Don Giovanni, K 527: Nur ihrem Frieden. cf Telefunken AJ 6-41867.
    Don Giovanni, K 527: Overture. cf Symphonies, nos. 29, 31, 34-36,
    38-41.
    Don Giovanni, K 527: Wenn du fein fromm bist. cf Discophilia KGB 2.
    Duos, 2 horns, K 487 (12). cf Works, selections (Supraphon 111
    1671/2).
1585 Duo, violin and viola, no. 1, K 423, G major. Sinfonia concertante,
K 364, E flat major. Igor Oistrakh, vln; David Oistrakh, vla;
MPO; Kiril Kondrashin. Decca SXL 6088. Tape (c) KSXC 6088.
(Reissues)
    +-RR 3-75 p74 tape
1586 Die Entführung aus dem Serail, K 384. Der Schauspieldirektor, K
486. Otto Mellies, speaker; Arleen Auger, Regina Jeske, Reri
Grist, Helga Piur, s; Peter Schreier, Hans-Jorn Weber, Harald
Neukirch, Kurt Hachlicki, t; Kurt Moll, Wolfgang Dehler, bs;
Leipzig Radio Chorus; Dresden Staatskapelle; Karl Böhm. DG
2740 112 (3). Tape (c) 3371 013. (also DG 2709 051)

```
      +Gr 10-74 p742              +-ON 2-1-75 p29
      +Gr 1-75 p1402 tape         +-RR 10-74 p24
     +-HF 2-75 p97                 +RR 7-75 p68 tape
      +NR 1-75 p14               +-SFC 12-15-74 p33
      +OC 9-75 p48              +-St 2-75 p82
```

1587 Die Entführung aus dem Serail (The abduction from the Seraglio),
     K 384, excerpts. Margaret Price, Danièle Perriers, s; Ryland
     Davies, Kimmo Lappaleinen, t; Noel Mangin, bs; LPO; John Prit-
     chard. Classics for Pleasure CFP 40032. (also Vanguard VSD
     71203)

```
      +Gr 12-72 p1197             +-Op 4-73 p340
     +-HFN 12-72 p2443           +-RR 12-72 p41
      +NR 8-75 p9               ++SFC 10-19-75 p33
```

   Die Entführung aus dem Serail, K 384: Arias. cf Arias (Hungaroton
     11679).
   Die Entführung aus dem Serail, K 384: Arias and scenes. cf London
     OS 26306.
   Die Entführung aus dem Serail, K 384: Hier soll ich dich denn sehen;
     Constanze, dich wieder zu sehen; Wenn der Freude Thränen fliessen;
     Ich baue ganz. cf Così fan tutte, K 588: Un aura amorosa; In
     qual fiero contrasto...Tradito, schernito.
   Die Entführung aus dem Serail, K 384: Constanze, dich wieder zu
     sehen. cf Telefunken AJ 6-41867.
   Die Entführung aus dem Serail, K 384: Martern aller Arten. cf
     Arias (EMI Electrola C 063-29082).
   The abduction from the Seraglio, K 384: Suite. cf Don Giovanni,
     K 527: Suite.
   Exsultate jubilate, K 165. cf BACH: Cantata, no. 51, Jauchzet
     Gott in allen Landen.
   Exsultate jubilate, K 165. cf Philips 6833 105.
   Exsultate jubilate, K 165: Alleluia. cf Works, selections (Philips
     6833 163).
   Exsultate jubilate, K 165: Alleluia. cf London SPC 21100.
1588 Fantasia, K 397, D minor. Fantasia, K 475, C minor. Rondo, piano,
     K 511, A minor. Sonata, piano, no. 14, K 457, C minor. Claudio
     Arrau, pno. Philips 6500 782.

```
      +Gr 11-75 p863              +RR 11-75 p79
     ++HFN 11-75 p157
```

   Fantasia, K 397, D minor. cf Concerto, piano, no. 26, K 537,
     D major.
   Fantasia, K 397, D minor. cf Sonatas, piano, nos. 11-17.
   Fantasia, K 397, D minor. cf DG 2548 137.
1589 Fantasia, K 475, C minor. Sonata, piano, no. 14, K 457, C minor.
     Sonata, piano, no. 16, K 570, B flat major. Sonata, piano, no.
     17, K 576, D major. Glenn Gould, pno. Columbia M 33515.

```
     +-HF 11-75 p108              +NR 11-75 p12
```

1590 Fantasia, K 475, C minor. Rondo, piano, K 485, D major. Sonata,
     piano, no. 11, K 331, A major. BACH (Busoni): Chaconne, D
     minor. Alicia de Larrocha, pno. London CS 6866. Tape (c)
     56866 (r) 86866. (also Decca SXL 6669)

```
      +Gr 4-75 p1830             ++RR 4-75 p52
      +HF 11-74 p114            ++SFC 8-11-74 p31
     ++HF 3-75 p108 tape        ++SFC 1-26-75 p26 tape
     +-NR 4-75 p11              ++St 10-74 p144
```

   Fantasia, K 475, C minor. cf Concerto, piano, no. 24, K 491, C
     minor.

Fantasia, K 475, C minor. cf Concerto, piano, no. 25, K 503,
   C major.
Fantasia, K 475, C minor. cf Fantasia, K 397, D minor.
Fantasia, K 475, C minor. cf Sonatas, piano, nos. 1-18.
Fantasia, K 475, C minor. cf Sonatas, piano, nos. 11-17.
Fantasia, K 594, F minor. cf Decca 5BBA 1013-5.
Fantasia, K 608, F minor. cf Organ works (Philips 6500 598).
1591 Idomeneo, Ré di Creta, K 366: Overture, ballet music. Symphony,
   no. 21, K 134, A major. Miskolc Symphony Orchestra; Péter Mura.
   Hungaroton SLPX 11693.
            +HFN 12-21-75 p39   +RR 3-75 p28
   Idomeneo, Ré di Creta, F 366: Ballet music. cf Marches and dances,
   complete.
   Eine Kleine Gigue, K 574, G major. cf HAYDN: Piano works (Hungaro-
   ton 11638).
   Ländler, K 606 (6). cf DG Archive 2533 182.
   Ländler, K 606 (6). cf Saga 5411.
   Lucia Silla, K 135: Overture. cf Works, selections (Decca ECS 740).
1592 Marches and dances, complete. Idomeneo, Ré di Creta, K 366: Ballet
   music. Les petits riens, K Anh 10. VPO; Willi Boskovsky.
   London STS 15275/9, STS 15280/4 (10).
            ++NR 7-75 p1   ++SFC 7-13-75 p21
   March, K 237, D major. cf DG 2548 148.
   March, K 335, D major. cf HMV ASD 3017.
1593 Marches, K 335, nos. 1 and 2, D major. Serenade, no. 9, K 320,
   D major. Dresden State Orchestra; Edo de Waart. Philips 6500
   627.
            +Gr 4-75 p1808   +RR 4-75 p29
            +-HF 9-75 p87   +SFC 5-18-75 p23
            +NR 7-75 p2   +STL 5-4-75 p37
   Marches, K 335, nos. 10 and 11, D major. cf Serenade, no. 9, K 320,
   D major.
   March, K 385d, D major. cf Philips 6747 199.
   March, K 408 (K385a), no. 2, D major. cf Symphony, no. 35, K 385,
   D major.
   March, K 445, D major. cf Divertimento, no. 17, K 334, D major.
   March for a cassation, K 62, D major. cf Cassation, no. 1, K 63,
   G major.
   Mass: Dies Irae. cf HMV Tape (c) TC MCS 14.
   Mass, no. 13, K 259, C major. cf Missa brevis, K 192, F major.
1594 Mass, no. 16, K 317, C major. Missa brevis, K 220, C major. Ave
   verum corpus, K 618. Edith Mathis, s; Norma Procter, Tatiana
   Troyanos, ms; Donald Grobe, Horst Laubenthal, t; John Shirley-
   Quirk, Keith Engen, bs; Elmar Schloter, org; Regensburg Cathed-
   ral Choir; Bavarian Radio Symphony Orchestra; Rafael Kubelik.
   DG 2530 356. Tape (c) 3300 340.
            +-Gr 3-74 p1729   +-ON 4-5-75 p40
            +-HF 3-74 p91   +RR 3-74 p62
            +HFN 3-74 p111   +RR 1-75 p70 tape
            ++MJ 9-74 p66   +SFC 4-14-74 p26
            +NR 2-74 p7   ++St 5-74 p105
   Mass, no. 16, K 317, C major. cf Missa brevis, K 257, C major.
1595 Mass, no. 18, K 427, C minor. Helen Donath, Heather Harper, s;
   Ryland Davies, t; Stafford Dean, bs; LSO and Chorus; Colin
   Davis. Philips 6500 235. Tape (c) 7300 162 (r) L 45235.
            +HF 7-74 p128 tape   +RR 5-75 p77 tape

```
       +-HFN 11-73 p2321           ++SFC 11-18-73 p32
       +MJ 7-74 p46                ++SFC 4-14-74 p26 tape
       +RR 11-73 p82
```
1596 Mass, no. 18, K 427, C minor. Pro Musica Orchestra and Chorus;
     Ferdinand Grossmann. Turnabout 34174. Tape (c) KTVC 34174.
```
       +-HFN 9-75 p110 tape
```
1597 Mass, no. 19, K 626, D minor (edit. Beyer). Hans Buchhierl, treble;
     Mario Krämer, alto; Werner, Krenn, t; Barry McDaniel, bar; Tölzer
     Boys' Choir; Collegium Aureum; Gerhard Schmidt-Gaden. BASF BAC
     3091.
```
       +Gr 5-75 p2006              +-MT 12-75 p1071
       +-HFN 5-75 p136             +RR 5-75 p61
```
     Minuet, K 355, D major. cf HAYDN: Piano works (Hungaroton 11638).
1598 Missa brevis, K 192, F major. Mass, no. 13, K 259, C major. Celes-
     tina Casapietra, s; Annelies Burmeister, con; Peter Schreier, t;
     Hermann Christian Polster, bs; Walter Heinz Bernstein, org;
     Leipzig Radio Symphony Orchestra and Chorus; Herbert Kegel.
     Philips 6500 867.
```
       +-Gr 11-75 p881             ++RR 10-75 p86
       +HFN 10-75 p145
```
     Missa brevis, K 220, C major. cf Mass, no. 16, K 317, C major.
1599 Missa brevis, K 257, C major. Mass, no. 16, K 317, C major. Helen
     Donath, s; Gillian Knight, con; Ryland Davies, t; Clifford Grant,
     Stafford Dean, bs; John Constable, org; John Alldis Choir; LSO;
     Colin Davis. Philips 6500 234. Tape (c) 7300 161.
```
       +Gr 7-72 p247 tape          /NYT 5-14-74 pD25
       +-HF 6-72 p86               +-RR 4-75 p77 tape
       +MJ 6-72 p34               /SFC 10-8-72 p37
       +NR 6-72 p13               +St 8-72 p76
```
1600 Missa solemnis, K 139, C minor. Celestina Casapietra, s; Annelies
     Burmeister, ms; Peter Schreier, t; Hermann Christian Polster,
     bs; Leipzig Radio Symphony Orchestra and Choir; Herbert Kegel.
     Philips 6500 866.
```
       +HF 12-75 p100              +NYT 9-21-75 pD18
       +HFN 11-75 p157             +ON 12-20-75 p38
       ++NR 11-75 p9              +RR 11-75 p87
```
1601 Musical joke (Ein musikalischer Spass), K 522. Serenade, no. 1,
     K 100, D major. Serenade, no. 13, K 525, G major. Vienna
     Mozart Ensemble; Willi Boskovsky. London STS 15301.
```
       ++Audio 11-75 p96           ++SFC 8-10-75 p26
       +NR 7-75 p2
```
     Nocturne, 4 orchestras, K 286, D major. cf Works, selections
     (Decca ECS 740).
1602 Le nozze di Figaro (The marriage of Figaro), K 492. Elisabeth
     Schwarzkopf, Irmgard Seefried, Hilde Gueden, Sieglinde Wagner,
     s; Peter Klein, Erich Majkut, t; Paul Schöffler, Erich Kunz,
     bar; Endre Koreh, Alois Pernerstorfer, bs; Vienna State Opera
     Chorus; VPO; Wilhelm Furtwängler. Bruno Walter Society IGI
     343 (3).
```
       +NR 8-75 p10
```
1603 Le nozze di Figaro, K 492. Lisa Della Casa, Roberta Peters, Anni
     Felbermayer, s; Rosalind Elias, Sandra Warfield, ms; Gabor
     Carelli, t; George London, bar; Giorgio Tozzi, Fernando Corena,
     Ljubomir Pantscheff, bs; VSOO and Chorus; Erich Leinsdorf.
     Decca ECS 743/5 (3). (Reissue from RCA SER 4508/11)
```
       +-Gr 3-75 p1701            +-RR 3-75 p17
```

Le nozze di Figaro, K 492: Arias. cf Arias (Hungaroton 11679).
Le nozze di Figaro, K 492: Aprite un pi' quegli occhi. cf Ember
GVC 37.
Le nozze di Figaro, K 492: Dove sono. cf HANDEL: Giulio Cesare:
Hast du mich ganz berauscht, Es blaut die Nacht, Breite aus
die gnäd'gen Hande, Weine nur klage nur, Heil und sicher Kam
mein Nachen.
Le nozze di Figaro, K 492: In quegl'anni. cf Così fan tutte, K 588:
Un aura amorosa; In qual fiero contrasto...Tradito, schernito.
Le nozze di Figaro, K 492: Non so più. cf Columbia M 33307.
Le nozze di Figaro, K 492: Non più andrai. cf Decca GOSC 666/8.
Le nozze di Figaro, K 492: Non so più; Voi che sapete; Giunse
alfin il momento...Deh vieni, non tardar. cf Arias (Philips
6500 544).
The marriage of Figaro, K 492: Overture. cf Concerto, horn and
strings, no. 4, K 495, E flat major.
Le nozze di Figaro, K 492: Overture. cf Symphonies, nos. 29, 31,
34-36, 38-41.
Le nozze di Figaro, K 492: Overture. cf Works, selections (Columbia
D3M 33261).
Le nozze di Figaro, K 492: Overture. cf CBS 77513.
Le nozze di Figaro, K 492: Overture. cf Decca SXL 6643.
Le nozze di Figaro, K 492: Overture. cf London CS 6856.
Le nozze di Figaro: Overture; Deh vieni non tardar; Venite, ingin-
occhiatevi. cf LEHAR: Eva: Wär es auch nichts als ein Traum
vom Gluck.
Le nozze di Figaro, K 492: Overture; Non più andrai. cf Works,
selections (Philips 6833 163).
Le nozze di Figaro, K 492: So lang hab ich geschmachtet. cf
Discophilia KGB 2.
Le nozze di Figaro, K 492: Voi che sapete. cf Philips 6747 204.
Le nozze di Figaro, K 492: Voi che sapete. cf Philips 6833 105.
Les petits riens, K Anh 10. cf Marches and dances, complete.
1604 Organ works: Adagio and allegro, K 594, F minor. Andante, mechani-
cal organ, K 616, F major. Fantasia, K 608, F minor. Sonatas,
organ and orchestra, nos. 10, 11, 17. Daniel Chorzempa, org;
German Bach Soloists; Helmut Winschermann. Philips 6500 598.
+Gr 4-75 p1839   +MT 10-75 p886
++HFN 5-75 p135   +RR 4-75 p52
Popoli di Tessaglia...Io non chiedo, eterni dei, K 316. cf Arias
(EMI Electrola C 063-29082).
1605 Preludes and fugues (after J. S. and W. F. Bach), K 404a. Grumiaux
Trio. Philips 6500 605.
+Gr 5-75 p1986   ++NR 1-75 p8
+HF 5-75 p84   ++RR 1-75 p8
1606 Quartets, flute, K 285, D major; K 285a, G major; K 285b, C major;
K 298, A major. Michel Debost, flt; Trio à Cordes Français.
Seraphim S 60246.
++NR 10-75 p5   ++St 12-75 p129
Quartet, flute, K 285, D major: Allegro. cf Works, selections
(Philips 6833 163).
1607 Quartet, oboe, K 370, F major. Quintet, clarinet, K 581, A major.
George Pieterson, clt; Koji Toyoda, Arthur Grumiaux, vln; Pierre
Pierlot, ob; Max Lesueur, vla; János Scholz, vlc. Philips
6500 924. Tape (c) 7300 414.
+-NR 12-75 p9
Quartet, oboe, K 370, F major. cf Works, selections (Supraphon
111 1671/2).

Quartet, oboe, K 370, F major. cf ALBINONI: Concerto, oboe, op. 7,
    no. 6.
1608 Quartet, piano, no. 1, K 478, G minor. Quartet, piano, no. 2,
      K 493, E flat major. Gyula Kiss, pno; Tatrai Quartet, Members.
      Hungaroton SLPX 11668.
          ++NR 6-75 p8   ++SFC 6-13-75 p21
          +RR 8-75 p63
    Quartet, piano, no. 2, K 493, E flat major. cf Quartet, piano, no.
      1, K 478, G minor.
    Quartet, piano, no. 2, K 493, E flat major. cf Classic Record
      Library SQM 80-5731.
1609 Quartets, strings, nos. 9-12 (4). Quartetto Italiano. Philips
      6500 644.
          ++HF 1-75 p84   ++RR 10-74 p71
          ++NR 10-74 p7
    Quartets, strings, nos. 9-13. cf Adagio and fugue, strings, K 546,
      C minor.
1610 Quartets, strings, nos. 14-19. Quartetto Italiano. Philips SC
      71AX 301 (3).
          ++HF 9-75 p87   ++SFC 7-6-75 p16
          ++NR 6-75 p8   ++St 8-75 p101
1611 Quartets, strings, nos. 14-19. Prague Quartet. Supraphon 111 1471/
      73.
          +HFN 12-75 p159   +-RR 10-75 p61
          +NR 11-75 p8
1612 Quartet, strings, no. 14, K 387, G major. Quartet, strings, no.
      15, K 421, D minor. Guarneri Quartet. RCA ARL 1-0760.
          +HF 3-75 p85   +St 4-75 p103
          +NR 1-75 p8
    Quartet, strings, no. 15, K 421, D minor. cf Quartet, strings,
      no. 14, K 387, G major.
    Quartet, strings, no. 15, K 421, D minor: Minuet and trio. cf
      Amon Ra SARB 01.
1613 Quartet, strings, no. 16, K 428, E flat major. Quartet, strings,
      no. 17, K 458, B flat major. RCA ARL 1-0762.
          +HF 3-75 p85   +St 4-75 p103
          ++NR 2-75 p6
    Quartet, strings, no. 16, K 428, E flat major: Menuetto. cf
      Canon CNN 4983.
1614 Quartet, strings, no. 17, K 458, B flat major. Quartet, strings,
      no. 19, K 465, C major. Collegium Aureum Quartet. BASF KHB
      20344.
          +-HF 7-73 p102   ++NR 5-73 p6
          +MQ 4-74 p312   +St 4-75 p103
    Quartet, strings, no. 17, K 458, B flat major. cf Quartet, strings,
      no. 16, K 428, E flat major.
    Quartet, strings, no. 17, K 458, B flat major. cf BEETHOVEN:
      Quartet, strings, no. 11, op. 95, F minor.
1615 Quartet, strings, no. 18, K 464, A major. Quartet, strings, no.
      19, K 465, C major. Guarneri Quartet. RCA ARL 1-1153. Tape
      (c) ARS 1-1153 (ct) ARK 1-1153.
          ++NR 11-75 p8
1616 Quartet, strings, no. 19, K 465, C major. Quartet, strings, no.
      22, K 589, B flat major. Tokyo Quartet. DG 2530 468.
          +Gr 3-75 p1674   ++SFC 12-15-74 p33
          +HF 3-75 p85   ++SR 2-22-75 p47
          ++NR 1-75 p8   ++St 4-75 p103
          +RR 3-75 p39

Quartet, strings, no. 19, K 465, C major. cf Quartet, strings,
    no. 17, K 458, B flat major.
Quartet, strings, no. 19, K 465, C major. cf Quartet, strings,
    no. 18, K 464, A major.
Quartet, strings, no. 22, K 589, B flat major. cf Quartet, strings,
    no. 19, K 465, C major.
Quintet, clarinet, K 581, A major. cf Quartet, oboe, K 370, F
    major.
Quintet, clarinet, K 581, A major. cf KHACHATURIAN: Trio, clarinet,
    violin and piano.
Quintet, clarinet, K 581, A major. cf Camden CCV 5000-12, 5014-24.
Quintet, clarinet, K 581, A major: 4th movement. cf Classics for
    Pleasure CFP 40205.
Quintet, clarinet, K 581, A major: Larghetto. cf Ember GVC 42.
Quintet, horn, K 407, E flat major. cf Works, selections (Supra-
    phon 111 1671/2).

1617 Quintets, strings, nos. 1-6. Amadeus Quartet; Cecil Aronowitz,
    vla. DG 2740 122 (3).
            ++Gr 12-75 p1065   ++RR 11-75 p60
            +HFN 11-75 p173

1618 Quintets, strings, nos. 1-6. Fine Arts Quartet; Francis Tursi, vla.
    Vox SVBX 557 (3).
            +-HF 11-75 p106   +SFC 7-13-75 p21
            ++NR 8-75 p8

1619 Quintet, strings, no. 1, K 174, B flat major. Quintet, strings,
    no. 3, K 515, C major. Grumiaux Trio; Arpad Gérecz, vln; Max
    Lesueur, vla. Philips 6500 619.
            +HF 11-75 p106   +SFC 11-23-75 p26
            ++NR 8-75 p8

1620 Quintet, strings, no. 3, K 515, C major. Quintet, strings, no. 4,
    K 516, G minor. Griller Quartet; William Primrose, vla. Van-
    guard Bach Guild HM 29.
            +-HFN 10-75 p153   +-RR 10-75 p67
Quintet, strings, no. 3, K 515, C major. cf Quintet, strings,
    no. 1, K 174, B flat major.
Quintet, strings, no. 4, K 516, G minor. cf Quintet, strings,
    no. 3, K 515, C major.

1621 Il Re pastore, K 208. Edith Mathis, Arleen Auger, Sona Ghazarian,
    s; Peter Schreier, Werner Krenn, t; Salzburg Mozarteum Orchestra;
    Leopold Hager. BASF KBL 22043 (3). (also BAC 3072/4)
            +Gr 8-75 p359   +ON 4-5-75 p40
            +HF 6-75 p100   +RR 8-75 p22
            +HFN 8-75 p79   ++St 6-75 p102
            +-NR 6-75 p12
Il Re pastore, K 208: L'amerò sarò costante. cf LEHAR: Eva: Wär
    es auch nichts als ein Traum vom Gluck.
Il Re pastore, K 208: L'amerò, serò costante. cf Pearl GEM 121/2.
Rondo, horn, K 371, E flat major. cf Concerti, horn and strings,
    nos. 1-2, 4.
Rondo, horn, K 371, E flat major. cf Concerto, horn and strings,
    no. 1, K 412, D major.
Rondo, piano, K 485, D major. cf Fantasia, K 475, C minor.
Rondo, piano, K 485, D major. cf Sonatas, piano, nos. 11-17.
Rondo, piano, K 511, A minor. cf Concerti, piano, nos. 17, 20-21,
    23-24.
Rondo, piano, K 511, A minor. cf Fantasia, K 397, D minor.

Rondo, piano, K 511, A minor. cf HAYDN: Piano works (Hungaroton
    11638).
Rondo, violin, K 269, B flat major. cf Adagio, K 261, E major.
Rondo, violin, K 269, B flat major. cf Concerti, violin, nos. 1-5.
Rondo, violin, K 269, B flat major. cf Concerti, violin, nos. 1-5
    (DG 2740 116).
Rondo, violin, K 269 (261a), B flat major. cf Concerto, violin,
    no. 2, K 211, D major.
Rondo, violin, K 373, C major. cf Adagio, K 261, E major.
Rondo, violin, K 373, C major. cf Concerti, violin, nos. 1-5 (CBS
    77381).
Rondo, violin, K 373, C major. cf Concerti, violin, nos. 1-5 (DG
    2740 116).
Rondo, violin, K 373, C major. cf Concerto, violin, no. 2, K 211,
    D major.
Rondo, violin, K 373, C major. cf BEETHOVEN: Romance, no. 1, op.
    40, G major.
Der Schauspieldirektor, K 486. cf Die Entführung aus dem Serail,
    K 384.
Serenade, D major: Rondo. cf RCA ARM 4-0942/7.
Serenade, K 62a, D major. cf Cassation, no. 1, K 63, G major.
Serenade, no. 1, K 100, D major. cf Musical joke, K 522.
Serenade, no. 3, K 185, D major: Andante, allegro. cf Works,
    selections (Philips 6775 012).
1622 Serenade, no. 4, K 203, D major. Pinchas Zukerman, vln; ECO;
    Pinchas Zukerman. CBS 76383.
            +Gr 7-75 p188   +RR 7-75 p31
            +HFN 7-75 p84
Serenade, no. 4, K 203, D major: Andante, menuetto, allegro. cf
    Works, selections (Philips 6775 012).
Serenade, no. 5, K 213a (204), D major: Andante moderato, menuetto,
    allegro. cf Works, selections (Philips 6775 012).
Serenade, no. 6, K 239, D major. cf Serenade, no. 13, K 525, G
    major.
Serenade, no. 6, K 239, D major. cf Works, selections (Decca ECS
    740).
1623 Serenade, no. 7, K 250, D major. Franzjosef Maier, vln; Collegium
    Aureum. BASF BHM 19 29310. Tape (c) KBACC 3015.
            +Gr 10-72 p700   +NYT 10-14-73 pD33
            +HFN 2-73 p340   +RR 10-72 p66
            +MQ 4-74 p312   +RR 7-75 p71 tape
            ++NR 12-73 p5
1624 Serenade, no. 9, K 320, D major. Serenade, no. 13, K 525, G major.
    CO; Georg Szell. CBS 61585. (Reissue from 72772) (also
    Columbia MS 7273)
            /Gr 1-75 p1350   +RR 11-74 p46
1625 Serenade, no. 9, K 320, D major. Marches, K 335, nos. 10 and 11,
    D major. Dresden State Orchestra; Edo de Waart. Philips 6500
    627.
            +Gr 4-75 p1808   +RR 4-75 p29
            +HF 9-75 p87   +SFC 5-18-75 p23
            +NR 7-75 p2   +STL 5-4-75 p37
Serenade, no. 9, K 320, D major. cf Marches, K 335, nos. 1 and 2,
    D major.
Serenade, no. 9, K 320, D major. cf Works, selections (Columbia
    D3M 33261).

1626 Serenade, no. 10, 13 wind instruments, K 361, B flat major. Colleg-
      ium Aureum Wind Ensemble. BASF KHB 21414. (also BAC 3100)
                +Gr 6-75 p46   +NYT 11-14-73 pD33
                +HFN 5-75 p135   ++RR 5-75 p36
                ++MQ 4-74 p312   +St 11-73 p122
                ++NR 9-73 p2
1627 Serenade, no. 11, K 375, E flat major. Serenade, no. 2, K 388,
      C minor. New London Wind Ensemble. Classics for Pleasure CFP
      40211.
                ++Gr 11-75 p819   +RR 6-75 p44
      Serenade, no. 11, K 375, E flat major. cf BEETHOVEN: Trio, piano,
      clarinet and violoncello, op. 38, E flat major.
      Serenade, no. 12, K 388, C minor. cf Serenade, no. 11, K 375,
      E flat major.
1628 Serenade, no. 13, K 525, G major (Eine kleine Nachtmusik). Adagio
      and fugue, K 546, C minor. Divertimento, no. 1, K 136, D major.
      Serenade, no. 6, K 239, D major. I Musici. Philips 6580 030.
      Tape (c) 7300 273. (Reissue from Festivo SFM 23010, Philips
      SABL 127)
                ++Gr 9-74 p592 tape   +RR 1-75 p70 tape
                +HFN 4-72 p713
      Serenade, no. 13, K 525, G major. cf Concerto, piano, no. 20,
      K 466, D minor.
      Serenade, no. 13, K 525, G major. cf Divertimento, no. 15, K 287,
      B flat major.
      Serenade, no. 13, K 525, G major. cf Musical joke, K 522.
      Serenade, no. 13, K 525, G major. cf Serenade, no. 9, K 320, D
      major.
      Serenade, no. 13, K 525, G major. cf Works, selections (Columbia
      D3M 33261).
      Serenade, no. 13, K 525, G major. cf BIBER: Serenade, C major.
      Serenade, no. 13, K 525, G major. cf Camden CCV 5000-12, 5014-24.
      Serenade, no. 13, K 525, G major. cf CBS 77513.
      Serenade, no. 13, K 525, G major: Rondo. cf Works, selections
      (Philips 6833 163).
      Serenade, no. 13, K 525, G major: 3rd movement. cf Sound Superb
      SPR 90049.
      Sinfonia concertante, oboe, clarinet, horn, bassoon and orchestra,
      K 297b, E flat major. cf Concerto, flute and harp, K 299, C
      major.
      Sinfonia concertante, K 364, E flat major. cf Adagio, K 261, E
      major.
      Sinfonia concertante, K 364, E flat major. cf Concerto, violin,
      no. 2, K 211, D major.
      Sinfonia concertante, K 364, E flat major. cf Duo, violin and
      viola, no. 1, K 423, G major.
      Sinfonia concertante, K 364, E flat major. cf Works, selections
      (Columbia D3M 33261).
      Sinfonia concertante, K 364, E flat major. cf HAYDN: Sinfonia
      concertante, op. 84, B flat major.
1629 Sonatas, flute and harpsichord, nos. 1-6, K 10-K 15. Wolfgang
      Schulz, flt; Heinz Medjimorec, pno. Decca SDD 449.
                -Gr 4-75 p1826   +RR 3-75 p50
1630 Sonatas, flute and harpsichord, nos. 1-6, K 10-K 15. Thomas Brand-
      dis, vln; Karl-Heinz Zöller, flt; Waldemar Döling, hpd; Wolfgang
      Boettcher, vlc. DG Archive 2533 135.
                ++HF 12-75 p101   ++NR 12-75 p9

1631 Sonatas, flute and harpsichord, nos. 1-6, K 10-K 15.   Jean-Pierre
     Rampal, flt; Robert Reyron-Lacroix, hpd. Odyssey Y 32970.
          +NR 12-74 p8   ++SFC 3-23-75 p22
1632 Sonatas, organ and orchestra (17). HAYDN: Concertos, organ and
     orchestra, nos. 1-3, C major. E. Power Biggs, org. Columbia
     MG 32985 (2).
          ++St 4-75 p102
1633 Sonatas, organ and orchestra (17). Daniel Chorzempa, org; German
     Bach Soloists; Helmut Winschermann. Philips 6700 061 (2).
          +HF 2-75 p98   ++SFC 1-26-75 p27
          +MJ 11-74 p48   ++St 4-75 p102
          +NR 10-74 p15
     Sonatas, organ and orchestra, nos. 10, 11, 17. cf Organ works
     (Philips 6500 598).
1634 Sonatas, piano, complete. Artur Balsam, pno. L'Oiseau-Lyre OLS
     177/81 (5). (Reissues from SOL 252, 253, 254, 258, 259, 60023)
          +-Gr 1-75 p1371   /RR 6-75 p71
1635 Sonatas, piano, nos. 1-10. Lili Kraus, pno. Odyssey Y 33220 (3).
          +HF 8-75 p91   +St 10-75 p112
          ++NR 3-75 p11
1636 Sonatas, piano, nos. 1-18. Fantasia, K 475, C minor. Artur Balsam,
     pno. Musical Heritage Society MHS 3056/3063 (8).
          +St 10-75 p113
1637 Sonata, piano, no. 8, K 310, A minor. Sonata, piano, no. 12, K
     332, F major. Sonata, piano, no. 15, K 545, C major. John
     McCabe, pno. Oryx Tape (c) BRL 27.
          -RR 4-75 p77 tape
1638 Sonata, piano, no. 9, K 311, D major. Sonata, 2 pianos, K 381,
     D major. SCHUBERT: Fantasia, 2 pianos, D 940, F minor. Victoria
     Postnikova, Gennady Rozhdestvensky, pno. HMV SXLP 30189.
          +-Gr 8-75 p348   +-RR 8-75 p58
          +HFN 11-75 p157
     Sonata, piano, no. 9, K 311, D major. cf Adagio, K 540, B minor.
     Sonata, piano, no. 10, K 330, C major. cf Concerto, piano, no. 27,
     K 595, B flat major.
1639 Sonatas, piano, nos. 11-17. Fantasia, K 397, D minor. Fantasia,
     K 475, C minor. Rondo, piano, K 485, D major. Lili Kraus, pno.
     Odyssey Y 33224 (3).
          +HF 8-75 p92   +St 10-75 p112
          ++NR 3-75 p11
1640 Sonata, piano, no. 11, K 331, A major. Sonata, piano, no. 12,
     K 332, F major. Valentina Kamenikova, pno. Supraphon 111 1417.
          +HFN 8-75 p79   +-RR 9-75 p60
     Sonata, piano, no. 11, K 331, A major. cf Concerto, piano, no. 27,
     K 595, B flat major.
     Sonata, piano, no. 11, K 331, A major. cf Fantasia, K 475, C minor.
     Sonata, piano, no. 11, K 331, A major: Rondo alla turca. cf Works,
     selections (Philips 6833 163).
     Sonata, piano, no. 11, K 331, A major: Rondo alla turca. cf CBS
     77513.
     Sonata, piano, no. 11, K 331, A major: Rondo alla turca. cf RCA
     VH 020.
     Sonata, piano, no. 12, K 332, F major. cf Sonata, piano, no. 8,
     K 310, A minor.
     Sonata, piano, no. 12, K 332, F major. cf Sonata, piano, no. 11,
     K 331, A major.

Sonata, piano, no. 13, K 333, B flat major. cf Adagio, K 540, B
  minor.
Sonata, piano, no. 13, K 333, B flat major. cf BEETHOVEN: Sonata,
  piano, no. 29, op. 106, B flat major.
Sonata, piano, no. 14, K 457, C minor. cf Concerto, piano, no. 24,
  K 491, C minor.
Sonata, piano, no. 14, K 457, C minor. cf Fantasia, K 397, D minor.
Sonata, piano, no. 14, K 457, C minor. cf Fantasia, K 475, C minor.
Sonata, piano, no. 15, K 545, C major. cf Sonata, piano, no. 8,
  K 310, A minor.
Sonata, piano, no. 15, K 545, C major. cf Columbia MG 33202.
Sonata, piano, no. 15, K 545, C major. cf Connoisseur Society
  (Q) CSQ 2066.
Sonata, piano, no. 15, K 545, C major: 1st movement. cf CBS 77513.
Sonata, piano, no. 16, K 570, B flat major. cf Fantasia, K 475,
  C minor.
Sonata, piano, no. 17, K 576, D major. cf Concerto, piano, no. 26,
  K 537, D major.
Sonata, piano, no. 17, K 576, D major. cf Fantasia, K 475, C minor.
Sonata, piano, no. 17, K 576, D major. cf BEETHOVEN: Andante favori,
  F major.
Sonata, 2 pianos, K 19d, C major. cf Adagio and allegro, K 594,
  F minor.
Sonata, 2 pianos, K 358, B flat major. cf Adagio and allegro,
  K 594, F minor.
Sonata, 2 pianos, K 381, D major. cf Adagio and allegro, K 594,
  F minor.
Sonata, 2 pianos, K 381, D major. cf Sonata, piano, no. 9, K 311,
  D major.
Sonata, 2 pianos, K 448 (375a), D major. cf BRAHMS: Variations on
  a theme by Haydn, op. 56a.
1641 Sonatas, violin and piano, nos. 17-28, 32-34. Szymon Goldberg, vln;
  Radu Lupu, pno. Decca 13BB 207/12 (6).
          +Gr 11-75 p853   +RR 11-75 p80
          ++HFN 10-75 p145
1642 Sonata, violin and piano, no. 18, K 301, G major. Sonata, violin
  and piano, no. 22, K 305, A major. Sonata, violin and piano,
  no. 24, K 376, F major. Sonata, violin and piano, no. 27, K 379,
  G major. Henryk Szeryng, vln; Ingrid Haebler, pno. Philips
  6500 143.
          +Gr 1-75 p1362   +RR 1-75 p51
1643 Sonatas, violin and piano, nos. 19-20, 23, 28. Sonata, violin and
  piano, K 547, F major. Variations on "La bergère Célimène",
  K 359, G major (12). Variations on "Helas, j'ai perdu mon amant",
  K 360 (6). Henryz Szeryng, vln; Ingrid Haebler, pno. Philips
  6500 144/5 (2).
          ++Gr 3-75 p1674   ++RR 3-75 p51
          ++NR 7-75 p7
Sonata, violin and piano, no. 22, K 305, A major. cf Sonata,
  violin and piano, no. 18, K 301, G major.
Sonatas, violin and piano, nos. 23, 26, 27, K 306, K 378, K 379.
  cf Andante and allegretto, K 404, C major.
Sonata, violin and piano, no. 23, K 306, D major. cf Sonata,
  violin and piano, no. 28, K 380, E flat major.
Sonata, violin and piano, K 372, B flat major. cf Andante and
  allegretto, K 404, C major.

Sonata, violin and piano, no. 24, K 376, F major. cf Sonata, violin
    and piano, no. 18, K 301, G major.
Sonata, violin and piano, no. 27, K 379, G major. cf Sonata, violin
    and piano, no. 18, K 301, G major.
1644 Sonata, violin and piano, no. 28, K 380, E flat major. Sonata,
    violin and piano, no. 23, K 306, D major. Variations on "Helas,
    j'ai perdu mon amant", K 360. Ingrid Haebler, pno; Henryk Szer-
    yng, vln. Philips 6500 144.
        ++Gr 3-75 p1674   ++RR 3-75 p51
        ++NR 7-75 p7
Sonata, violin and piano, K 547, F major. cf Sonatas, violin and
    piano, nos. 1-20, 23, 28.
1645 Songs: Abendempfindung, K 523; Ah, Spiegarti, Oh Dio, K 178; Als
    Luise die Briefe ihres ungetreuen Liebhabers verbrannte, K 520;
    Die Alte, K 517; Im Frühlingsanfange, K 597; Das Kinderspiel,
    K 598; Die kleine Spinnerin, K 531; Die kleine Friedrichs Geburt-
    stag, K 529; Un moto di Gioia, K 579; Oiseaux, si tous les ans,
    K 307; Ridente la Calma, K 152; Sei du mien Trost, K 391; Sehn-
    sucht nach dem Frühlinge, K 596; Die Verschweignung, K 518;
    Das Veilchen, K 476; Der Zauberer, K 472; Die Zufriedenheit,
    K 349. Edith Mathis, s; Bernhard Klee, pno; Takashi Ochi, mand.
    DG 2530 319.
        +Gr 8-73 p368   +-RR 8-73 p74
        +HF 4-74 p100   +SFC 7-7-74 p18
        ++NR 5-74 p10   +St 6-74 p122
        +ON 2-1-75 p29
Songs: Abendempfindung, K 23; Das Veilchen, K 476. cf HAYDN:
    Arianna a Naxos.
1646 Symphonies, nos. 1-24. BPhO; Karl Böhm. DG 2740 109 (8). (Re-
    issue from 643521/35) (also DC 2721 013)
        *Gr 3-75 p1657   +RR 3-75 p29
1647 Symphonies, nos. 1-31. AMF; Neville Marriner. Philips 6747 099 (8).
        +Gr 10-74 p697   +-RR 10-74 p51
        +HF 2-75 p81   +SFC 10-13-74 p26
        ++NR 12-74 p2
1648 Symphony, no. 1, K 16, E flat major. Symphony, no. 4, K 19, D
    major. Symphony, no. 5, K 22, B flat major. Symphony, no. 10,
    K 74, G major. Symphony, K 81, D major. AMF; Neville Marriner.
    Philips 6500 532.
        +HF 10-74 p108   ++SFC 7-14-74 p25
        +MJ 3-75 p26   ++St 8-74 p119
        +NR 9-74 p4
1649 Symphony, no. 4, K 19, D major. Symphony, no. 5, K 22, B flat
    major. Symphony, no. 29, K 201, A major. Delos Chamber Orches-
    tra; James De Preist. Delos DEL 14401.
        +MJ 3-75 p25   /SR 2-22-75 p47
        +NR 3-75 p2
Symphony, no. 4, K 19, D major. cf Symphony, no. 1, K 16, E flat
    major.
Symphony, no. 5, K 22, B flat major. cf Symphony, no. 1, K 16,
    E flat major.
Symphony, no. 5, K 22, B flat major. cf Symphony, no. 4, K 19,
    D major.
Symphony, no. 10, K 74, G major. cf Symphony, no. 1, K 16, E flat
    major.
Symphony, no. 21, K 134, A major. cf Idomeneo, Ré di Creta, K 366:
    Overture, ballet music.

1650 Symphonies, nos. 25-41. BPhO; Karl Böhm. DG 2740 110 (7). (also
     2721 007) (Reissues)
          +HFN 8-75 p78   +RR 5-75 p35
1651 Symphony, no. 25, K 183, G minor. Symphony, no. 40, K 550, G
     minor. VPO; Istvan Kertesz. Decca SCL 6617. (also London CS
     6831)
          ++Gr 3-74 p1705   +-NR 11-74 p4
          +HF 3-75 p85   +-RR 3-74 p44
          +HFN 3-74 p111   +SFC 1-26-75 p27
1652 Symphonies, nos. 29, 31, 34-36, 38-41. Overtures: Don Giovanni,
     K 527. Le nozze di Figaro, K 492. LPO; Thomas Beecham. Vox/
     Turnabout THS 65022/26 (5).
          +St 7-75 p110
     Symphony, no. 29, K 201, A major. cf Symphony, no. 4, K 19, D major.
1653 Symphony, no. 31, K 297, D major. Symphony, no. 35, K 385, D major.
     Bamberg Symphony Orchestra; Hans Schmidt-Isserstedt. BASF 20
     21983-6.
          +Gr 10-75 p621   +RR 10-75 p45
          +HFN 10-75 p144
     Symphony, no. 32, K 318, G major. cf Works, selections (Decca ECS
     740).
1654 Symphony, no. 33, K 319, B flat major. Symphony, no. 40, K 550,
     G minor. HCO; Vilmos Tátrai. Hungaroton SHLX 90012.
          +-RR 1-75 p30
1655 Symphony, no. 35, K 385, D major. Symphony, no. 41, K 551, C major.
     Israel Philharmonic Orchestra; Josef Krips. Decca SPA 336.
     Tape (c) KCSP 336. (Reissue from SXL 2220)
          /Gr 1-75 p1349   -RR 11-74 p48
          /Gr 5-75 p203 tape
1656 Symphony, no. 35, K 385, D major. Symphony, no. 41, K 551, C major.
     COA; Josef Krips. Philips 6500 429. Tape (c) 7300 270.
          +-HFN 9-75 p110 tape   ++NR 3-75 p2
1657 Symphony, no. 35, K 385, D major. Symphony, no. 40, K 550, G minor.
     March, K 408 (K 385a), no. 2, D major. AMF; Neville Marriner.
     Philips 6500 162. (Reissue from Philips 6707 013)
          +HFN 6-72 p1113   ++SFC 7-13-75 p21
          ++NR 7-75 p5
1568 Symphonies, no. 35-36, 38-41. LPO, VSOO, VPO, Vienna Pro Musica,
     NBC Symphony Orchestra; Thomas Beecham, Karl Böhm, Otto Klemp-
     erer, Arturo Toscanini, Bruno Walter, Erich Kleiber. Turnabout
     THS 65033/5 (3). (Reissues)
          +Audio 11-75 p88   +NR 7-75 p4
     Symphony, no. 35, K 385, D major. cf Symphony, no. 31, K 297,
     D major.
     Symphony, no. 35, K 385, D major. cf Symphony, no. 41, K 551,
     C major.
1659 Symphony, no. 39, K 543, E flat major. Symphony, no. 40, K 550,
     G minor. COA; Josef Krips. Philips 6500 430. Tape (c) 7300
     271.
          +-Gr 5-73 p2051   -RR 5-73 p57
          +-HFN 5-73 p986   -RR 1-75 p71 tape
1670 Symphony, no. 40, K 550, G minor. Symphony, no. 41, K 551, C major.
     VPO; Herbert von Karajan. Decca SDD 361. Tape (c) KSCD 361.
     (Reissues from SB 2092, SXL 6067)
          +-Gr 2-73 p1551   -RR 2-73 p61
          +-Gr 10-74 p766   -RR 1-75 p71 tape
          -HFN 2-73 p347

Symphony, no. 40, K 550, G minor. cf Concerto, horn and strings,
    no. 4, K 495, E flat major.
Symphony, no. 40, K 550, G minor. cf Symphony, no. 25, K 183, G
    minor.
Symphony, no. 40, K 550, G minor. cf Symphony, no. 33, K 319, B
    flat major.
Symphony, no. 40, K 550, G minor. cf Symphony, no. 35, K 385, D
    major.
Symphony, no. 40, K 550, G minor. cf Symphony, no. 39, K 543, E
    flat major.
Symphony, no. 40, K 450, G minor. cf Columbia M2X 33014.
Symphony, no. 40, K 550, G minor. cf HMV SLS 5003.
Symphony, no. 40, K 550, G minor: Menuetto. cf Philips 6580 098.
Symphony, no. 40, K 550, G minor: Molto allegro. cf Works, selec-
    tions (Philips 6833 163).
1661 Symphony, no. 41, K 551, C major. Symphony, no. 35, K 385, D major.
    COA; Josef Krips. Philips 6500 429.
        +—Gr 4-73 p1876   +—RR 4-73 p61
        +—HFN 4-73 p783   +SFC 3-2-75 p25
Symphony, no. 41, K 551, C major. cf Concerto, piano, no. 23,
    K 488, A major.
Symphony, no. 41, K 551, C major. cf Symphony, no. 35, K 385,
    D major (Decca SPA 336).
Symphony, no. 41, K 551, C major. cf Symphony, no. 35, K 385, D
    major (Philips 6500 429).
Symphony, no. 41, K 551, C major. cf Symphony, no. 40, K 550, G
    minor.
Symphony, no. 41, K 551, C major. cf SCHUBERT: Symphony, no. 8,
    D 759, B minor.
Symphony, no. 41, K 551, C major. cf Camden CCV 5000-12, 5014-24.
Symphony, K 81, D major. cf Symphony, no. 1, K 16, E flat major.
Thamos, King of Egypt, K 345: Interludes. cf Works, selections
    (Decca ECS 740).
Variations on "Ah, vous dirai-je Maman", K 265, C major. cf
    BEETHOVEN: Sonata, piano, no. 29, op. 106, B flat major.
Variations on "Ah, vous dirai-je Maman", K 265, C major. cf CBS
    77513.
Variations on "Ah, vous dirai-je, Maman", K 265, C major. cf Con-
    certo, piano, no. 26, K 537, D major.
Variations on "La bergère Célimène", K 359, G major. cf Sonatas,
    violin and piano, nos. 19-20, 23, 28.
Variations on "Helas, j'ai perdu mon amant", K 360. cf Sonatas,
    violin and piano, nos. 19-20, 23, 28.
Variations on "Helas, j'ai perdu mon amant", K 360. cf Sonata,
    violin and piano, no. 28, K 380, E flat major.
Variations on "Salve tu Domine", K 398, F major. cf Adagio, K 540,
    B minor.
Vesperae de Dominica, K 321. cf Arias (Crystal S 902).
1662 Works, selections: Concerto, clarinet, K 622, A major. Concerto,
    piano, no. 25, K 503, C major. Le nozze di Figaro, K 492: Over-
    ture. Serenade, no. 9, K 320, D major. Serenade, no. 13, K 525,
    G major. Sinfonia concertante, violin and viola, K 364, E flat
    major. Leon Fleisher, pno; Robert Marcellus, clt; CO; Georg
    Szell. Columbia D3M 33261 (3).
        ++SFC 7-27-75 p22
1663 Works, selections: Lucia Silla, K 135: Overture. Nocturne, 4

orchestras, K 286, D major. Serenade, no. 6, K 239, D major.
Symphony, no. 32, K 318, G major. Thamos, King of Egypt, K 345:
Interludes. Hugh Maguire, Neville Marriner, vln; Simon Streat-
field, vla; Stuart Knusson, double bs; LSO; Peter Maag. Decca
ECS 740. (Reissues from SXL 2135, 2196)
            +Gr 3-75 p1654
1664 Works, selections: Cassation, no. 1, K 63, G major: Adagio. Sere-
     nade, no. 4, K 203, D major: Andante, menuetto, allegro. Sere-
     nade, no. 3, K 185, D major: Andante, allegro. Serenade, no.
     5, K 213a (204), D major: Andante moderato, menuetto, allegro.
     Concerto, violin, no. 1, K 207, B flat major. Concerto, violin,
     no. 2, K 211, D major. Jaap Schröder, vln; Amsterdam Mozart
     Emsenble; Frans Brüggen. Philips 6775 012 (2).
            /Gr 7-75 p191   +RR 7-75 p32
            +HFN 7-75 p85
1665 Works, selections: Concerto, horn, no. 4, K 495, E flat major: Rondo.
     Concerto, piano, no. 21, K 467, C major: Adagio. Concerto,
     violin, no. 4, K 218, D major: Andante cantabile. Divertimento,
     no. 17, K 334, D major: Minuet. Exsultate jubilate, K 165:
     Alleluia. Le nozze di Figaro, K 492: Overture; Non più andrai.
     Quartet, flute, K 285, D major: Allegro. Serenade, no. 13,
     K 525, G major: Rondo. Sonata, piano, no. 11, K 331, A major:
     Rondo alla turca. Symphony, no. 40, K 550, G minor: Molto
     allegro. Various soloists, orchestras and conductors. Philips
     6833 163. (Reissues)
            +Gr 11-75 p915   +-RR 11-75 p44
1666 Works, selections: Divertimento, K Anh 229, B flat major. Diverti-
     mento, no. 5, K 187, C major. Duos, 2 horns, K 487 (12). Quar-
     tet, oboe, K 370, F major. Quintet, horn, K 407, E flat major.
   * Czech Philharmonic Wind Ensemble; CPhO Members. Supraphon
     111 1671/2 (2).
            +NR 11-75 p8
1667 Die Zauberflöte, K 620. Wilma Lipp, Irmgard Seefired, s; Walther
     Ludwig, t; Karl Schmitt-Walter, bar; Josef Griendl, Paul Schöff-
     ler, bs; Vienna State Opera Chorus; VPO; Wilhelm Furtwängler.
     Bruno Walter Society IGI 337 (3).
            +-NR 8-75 p12
1668 Die Zauberflöte, K 620. Cristina Deutekom, Pilar Lorengar, Hanneke
     van Bork, s; Yvonne Minton, ms; Stuart Burrows, Gerhard Stolze,
     René Kollo, t; Herman Prey, bar; Martti Talvela, Hans Sotin, bs;
     VSO and Chorus; Georg Solti. Decca Tape (c) K2A4 (2).
            +Gr 5-75 p2031 tape   +RR 5-75 p76 tape
            ++HFN 6-75 p109 tape
     Die Zauberflöte, K 620: Arias. cf London OS 26381.
     The magic flute, K 680: Arias. cf Arias (Hungaroton 11679).
     Die Zauberflöte, K 620: Arias and scenes. cf London OS 26306.
     Die Zauberflöte, K 620: Dies Bildnis. cf Decca GOSC 666/8.
     Die Zauberflöte, K 620: Dies Bildnis ist bezaubernd schön; Wie
     stark ist nicht dein Zauberton. cf Così fan tutte, K 588: Un
     aura amorosa; In qual fiero contrasto...Tradito, schernito.
     Die Zauberflöte, K 620: O zittre nicht, mein lieber Sohn; Der Hölle
     Rache. cf Arias (EMI Electrola C 063-29082).
     Die Zauberflöte, K 620: Overture. cf Philips 6747 199.
     Die Zauberflöte, K 620: Overture. cf Philips 6747 204.
     The magic flute, K 620: Queen of the night's aria, Act 1 and Act 2.
     cf London SPC 21100.

MUCZYNSKI, Robert
    Suite, op. 13.   cf KABALEVSKY: Sonata, piano, no. 3, op. 46.
MUDARRA, Alonso de
    Diferencias sobre El Conde claros.   cf MILAN: Pavanas, nos. 1-6.
    Diferencias sobre El Conde claros.   cf CBS 72526.
    Dulces exuviae.   cf L'Oiseau-Lyre 12BB 203/6.
    Fantasia.   cf CBS 72526.
    Fantasia.   cf CBS 72950.
    Fantasia que contrahaza la harpa en la manera de Ludovico.   cf
        MILAN: Pavanas, nos. 1-6.
    Gallarda.   cf MILAN: Pavanas, nos. 1-6.
    Gallarda, D major.   cf RCA ARL 1-0485.
    O guárdame las vacas.   cf MILAN: Pavanas, nos. 1-6.
    Pavana de Alexandre.   cf MILAN: Pavanas, nos. 1-6.
    Romanesca, Guárdame las vacas.   cf DG Archive 2533 111.
    Si me llaman a mi.   cf DG 2530 504.
    Songs: Claros y frescos rios; Isabel, perdiste la tua faxa; Triste
        estaba el Rey David.   cf DG 2530 504.
    Songs: Claros y frescos rios; Tiento; Triste estaba el Rey David;
        Si me llaman.   cf Telefunken SAWT 9620/1.
MUFFAT, Georg
    Apparatus Musico-Organisticus: Toccata decima.   cf Pelca PSRK 41013/6.
    Fugue, G minor.   cf Philips 6775 006.
    Nova Cyclopeias harmonica.   cf Telefunken TK 11567/1-2.
MUFFAT, Gottlieb
    Versetti (6).   cf Hungaroton SLPX 11601/2.
    Toccata, no. 6.   cf Hungaroton SLPX 11601/2.
MUHLBERGER
    Mir sein die Kaiserjäger.   cf DG 2721 077.
MULET, Henri
    Carillon-sortie.   cf Decca SDD 463.
MULLER, Marion
    Prelude and fugue, E flat major.   cf Hungaroton SLPX 11601/2.
MUNDY, John
    Sing joyfully.   cf Argo ZRG 789.
MURSCHHAUSER, Franz
    Octi-Tonium novum organicum: Aria pastoralis variata; Variationen
        über das Lied "Lasst uns das Kindelein wiegen", per imitationem
        cuculi.   cf Pelca PSRK 41013/6.
    Praeambulum, fugae, finale tertii toni, A minor.   cf Telefunken
        TK 11567/1-2.
MURTULA, Giovanni
    Tarantella.   cf DG 2530 561.
MUSET, Colin
    Quand je voi.   cf BIS LP 3.
MUSGRAVE, Thea
    Concerto, clarinet.   cf BANKS: Concerto, horn.
1669 Concerto, horn.   Concerto, orchestra.   Barry Tuckwell, hn; Keith
        Pearson, clt; Scottish National Orchestra; Thea Musgrave,
        Alexander Gibson.   Decca HEAD 8.   (also London HEAD 8)
            +Gr 6-75 p46   ++RR 5-75 p36
            +-HFN 6-75 p89   ++STL 6-8-76 p36
            ++MT 10-75 p886
    Concerto, orchestra.   cf Concerto, horn.
1670 Night music, chamber orchestra.   RIEGGER: Dichotomy, chamber orch-
        estra, op. 12.   SESSIONS: Rhapsody, orchestra.   Symphony, no. 8.

London Sinfonietta, NPhO; Frederick Prausnitz.   Argo ZRG 702.
   +Gr 5-73 p2051   +RR 5-73 p25
   +HF 11-74 p97   +SFC 10-6-74 p26
   +HFN 6-73 p1180   ++St 3-75 p106
   +NR 10-74 p3

MUSIC FOR BRASS, 1500-1970. cf Desto DC 6474/7.
MUSIC FOR EVENSONG. cf Polydor 2460 250.
MUSIC FOR FLUTE AND HARP. cf Musical Heritage Society MHS 1345.
MUSIC FOR PIANO, four hands. cf Desto DC 7131.
MUSIC FOR TRUMPET AND ORGAN. cf RCA CRL 2-7001.
MUSIC FOR VOICE AND VIOLIN. cf Musical Heritage Society MHS 1976.
MUSIC IN HONOR OF ST. THOMAS OF CANTERBURY. cf Nonesuch H 71292.
MUSIC OF WELLINGTON's TIME. cf Classics for Pleasure CFP 40230.
MUSIC TO SHAKESPEAR'S PLAYS. cf BBC REB 191.
MUSICA IBERICA, 1100-1600. cf Telefunken SAWT 9620/1.
THE MUSICAL HERITAGE OF AMERICA. cf CMS Records 650/4.
MUSICAL LIFE IN OLD HUNGARY, 13-18th centuries. cf Hungaroton
   SLPX 11491/3.
MUSICKE OF SUNDRE KINDES: Renaissance secular music, 1480-1620.
   cf L'Oiseau-Lyre 12BB 203/6.

MUSSORGSKY, Modest
1671 Boris Godunov: Elisaveta Shumskaya, Irina Arkhipova, s; Maria
   Mitukova, Veronika Borisenko, ms; Eugenia Verbitsky, con; Vlad-
   imir Ivanovsky, Georgi Shulpin, Nikolai Zahkarov, Anton Grigoryev,
   t; Alexie Ivanov, George London, bar; Mark Reshetin, Eugene
   Kibkalo, Alexei Gueleva, Leonid Ktitorov, Vladimir Valaitis,
   Yuri Dementiev, bs; Bolshoi Theatre Orchestra and Chorus;
   Alexander Melik-Pashayev. CBS 77396 (3). (Reissue)
      +Gr 10-75 p680   +-RR 9-75 p23
      +HFN 9-75 p101

1672 Boris Godunov, symphonic synthesis. A night on the bare mountain.
   Pictures at an exhibition (transcriptions by Stokowski). OSR,
   LSO, NPhO; Leopold Stokowski. London SPC 21110. Tape (c) M5
   2110 (ct) M8 21110 (r) L 475110. (Reissues from SPC 21006,
   21026, 21032) (also Decca SDD 456)
      +Gr 5-75 p2023   +RR 5-75 p37
      +HF 11-74 p91   +SFC 4-28-74 p29
      +HFN 5-75 p136

1673 Boris Godunov, excerpts. Ezio Pinza, bs; Metropolitan Opera Orchestra and Chorus; Emil Cooper. Odyssey Y 33129.
      +NR 2-75 p9

Boris Godunov: Chorus from scene at Kromy. cf HMV Melodiya SXL
   30190.
Boris Godunov: Coronation scene. cf Decca PFS 4323.
Boris Godunov: Death of Boris. cf Decca GOSC 666/8.
Boris Godunov: Introduction and polonaise. cf Works, selections
   (HMV Melodiya 3101).
By the water. cf RCA VH 020.
The capture of Kars, trimphal march. cf Works, selections (BASF
   22128).
Fair at Sorochinsk: Gopak. cf Connoisseur Society CS 2070.
Fair at Sorochinsk: Gopak (trans. Rachmaninoff). cf HMV SXLP 30193.
Fair at Sorochinsk: Introduction, Gopak. cf Works, selections
   (HMV Melodiya 3101).
Intermezzo, piano. cf Khovanschina: Prelude.
Intermezzo, B minor. cf Works, selections (HMV Melodiya 3101).

Intermezzo in modo classico. cf Works, selections (BASF 22128).
1674 Khovanschina (edit. Rimsky-Korsakov). Maria Dimchevska, Nadya
     Dobriyanova, s; Alexandrina Milcheva-Nonova, ms; Lyubomir
     Bodurov, Lyuben Mikhailov, Milen Payunov, Dimiter Dimitrov,
     Verter Vrachovsky, t; Stoyan Popov, bar; Dimiter Petkov, Nikola
     Gyuselev, bs; Svetoslav Obretenov Chorus; Sofia National Opera
     Orchestra; Athanas Margaritov. CRD Harmonia Mundi HMB 4-124 (4).
          +-Gr 11-75 p899   +-RR 12-75 p28
          +-HFN 12-75 p159
1675 Khovanschina. Tamara Sorokina, s; Irina Arkhipova, ms; Vladislav
     Pyavko, Aleksei Maslennikov, Gennady Yefimov, Yuri Grigoriev, t;
     Viktor Nechipailo, bar; Aleksei Krivchenya, Aleksander Ogniv-
     tsev, Yuri Korolev, bs; Bolshoi Theatre Orchestra and Chorus;
     Boris Khaikin. Melodiya/Angel SRDL 4125 (4). (also HMV Melodiya
     SLS 5023)
          +-Gr 11-75 p899   +-NYT 12-1-74 pD18
          +-HF 12-74 p89   +-ON 12-14-74 p46
          +-HFN 10-75 p146   +-RR 10-75 p29
          +NR 12-74 p12   +-St 10-74 p140
1676 Khovanschina: Prelude. Intermezzo, piano. A night on the bare
     mountain. Pictures at an exhibition. André Previn, pno; NYP,
     CO; Thomas Schippers, Georg Szell, Leonard Bernstein. CBS Tape
     (c) 40-30050.
          +Gr 2-75 p1561 tape
     Khovanschina: Dawn on the Moscow River; Dance of the Persian slaves;
     Galitzin's journey. cf Works, selections (HMV Melodiya 3101).
     Khovanschina: Father, come out; The fair swan swims. cf HMV Melo-
     diya SXL 30190.
     Khovanschina: Prelude and Persian dance. cf BORODIN: Prince Igor:
     Overture.
     Khovanschina: Prelude; Dance of the Persian slaves. cf BORODIN:
     Prince Igor: Dance of the Polovtsian maidens; Polovtsian dances.
     Mlada: Triumphal march. cr Works, selections (HMV Melodiya 3101).
1677 A night on the bare mountain. Pictures from an exhibition (orch.
     Ravel). COA; Bernard Haitink, Willem van Otterloo. Philips
     Tape (c) 7300 022.
          -Gr 8-75 p376 tape
1678 A night on the bare mountain (arr. Rimsky-Korsakov). Pictures at
     an exhibition (orch. Ravel). LSO; Malcolm Sargent. Saga 5383.
     Tape (c) CA 5383. (Reissue from Vox GBYE 15020)
          +-Gr 4-75 p1811   +-RR 9-75 p78
     A night on the bare mountain. cf Boris Godunov, symphonic synthesis.
     A night on the bare mountain. cf Khovanschina: Prelude.
     A night on the bare mountain. cf Works, selections (HMV Melodiya
     3101).
     A night on the bare mountain. cf BORODIN: Prince Igor: Dance of
     the Polovtsian maidens; Polovtsian dances.
     A night on the bare mountain. cf BORODIN: Prince Igor: Overture.
     A night on the bare mountain. cf Decca DPA 519/20.
1679 Pictures at an exhibition. János Solyom, pno. BIS LP 16.
          +-Gr 11-75 p863   +RR 11-75 p80
1680 Pictures at an exhibition (orch. Ravel). TCHAIKOVSKY: Romeo and
     Juliet: Overture. LPO; John Pritchard. Classics for Pleasure
     CFP 106.
          +-RR 2-75 p40
1681 Pictures at an exhibition (orch. Ravel). RAVEL: Bolero. Rotterdam

Philharmonic Orchestra; Edo de Waart. Philips 6500 882. Tape
(c) 7300 363.

+-Gr 11-75 p820   +NR 6-75 p1
+HF 9-75 p95   ++RR 10-75 p48
+HFN 10-75 p145   ++SFC 5-18-75 p23

1682 Pictures at an exhibition. RAVEL: Bolero. PO; Eugene Ormandy.
RCA ARL 1-0451. Tape (c) ARK 1-0451 (ct) ARS 1-0451.

+-Gr 5-75 p1968   ++NR 7-74 p1
++HF 8-74 p100   ++RR 5-75 p36
+HFN 7-75 p87   ++SFC 5-19-74 p29
++MJ 3-75 p33   ++St 9-74 p127

1683 Pictures at an exhibition. Isao Tomita, Moog synthesizer. RCA
ARD 1-0838. (Q) CD 4.

+Audio 10-75 p117   +-NYT 6-15-75 pD18
-NR 9-75 p14   -SFC 5-18-75 p23

1684 Pictures at an exhibition (arr. Chapman). Keith Chapman, org.
Stentorian SC 1710.

++NR 12-75 p13

Pictures at an exhibition. cf Boris Godunov, symphonic synthesis.
Pictures at an exhibition. cf Khovanschina: Prelude.
Pictures at an exhibition. cf A night on the bare mountain (Phil-
ips 7300 022).
Pictures at an exhibition. cf A night on the bare mountain (Saga
5383).
Pictures at an exhibition. cf Works, selections (BASF 22128).
Pictures at an exhibition. cf FISER: Sonata, piano, no. 3.
Pictures at an exhibition. cf HMV SLS 5019.
Pictures at an exhibition: Bydlo and ballet. cf Ember GVC 40.
Scherzo, B flat major. cf Works, selections (BASF 22128).
Scherzo, B flat major. cf Works, selections (HMV Melodiya 3101).
The song of the flea. cf RCA ARL 1-5007.
Songs and dances of death. cf Odyssey Y2 32880.

1685 Works, selections: The capture of Kars, triumphal march. Pictures
at an exhibition (orch. Tushmalov, Rimsky-Korsakov). Scherzo,
B flat major. Intermezzo in modo classico. Munich Philharmonic
Orchestra; Marc Andreae. BASF BC 21128.

+HF 11-75 p108   +SFC 8-24-75 p28
+-NYT 8-10-75 pD14

1686 Works, seledtions: Boris Godunov: Introduction and polonaise (arr.
Rimsky-Korsakov). The fair at Sorochinsk: Introduction, Gopak.
Intermezzo, B minor (arr. Rimsky-Korsakov). Khovanschina: Dawn
on the Moscow River; Dance of the Persian slaves; Galitzin's
journey (arr. Rimsky-Korsakov). Mlada: Triumphal march (arr.
(Rimsky Korsakov). A night on the bare mountain (arr. Rimsky-
Korsakov). Scherzo, B flat major. USSR; Yevgeny Svetlanov.
HMV Melodiya ASD 3101.

+-Gr 8-75 p326   +RR 8-75 p40
+-HFN 8-75 p80

MYSLIVECEK, Josef

1687 Abramo ed Isacco. Gianfranca Ostini, Jana Jonásová, s; Anna Viga-
noni, alto; Choichiro Tahara, t; Gianni Matteo, bar; CPhO;
Prague Chamber Orchestra; Peter Maag. Supraphon 112 1021/2 (2).

+Gr 9-74 p571   +RR 2-74 p56
-HF 5-74 p91   +SFC 10-19-75 p33
/HFN 2-74 p341   ++St 11-74 p131
+NR 4-74 p8

1688 Kasace. B flat major. Octet, B flat major. Quod est inigne calor.
      Symphony, C major. Marta Boháčová, s; Musici de Prague; Franti-
      sek Vajnar; Prague Collegium Musicum. Panton 110 229.
           +RR 12-75 p57
      Octet, B flat major. cf Kasace, B flat major.
      Octets, no. 1, E flat major; no. 2, E flat major. cf KRAMAR-
         KROMMER: Harmonie, op. 57, F major.
      Quod est inigne calor. cf Kasace, B flat major.
      Symphony, C major. cf Kasace, B flat major.
NARDELLA
      Chiove. cf Musical Heritage Society MHS 1951.
NARVAEZ, Luis de
      Baxa de contrapunto. cf MILAN: Pavanas, nos. 1-6.
      Con qué la lavaré. cf DG 2530 504.
      Diferencias sobre Guárdame las vacas. cf MILAN: Pavanas, nos. 1-6.
      Diferencias sobre Guárdame las vacas. cf Everest SDBR 3380.
      Fantasia. cf MILAN: Pavanas, nos. 1-6.
      Fantasia. cf L'Oiseau-Lyre 12BB 203/6.
      Guárdame las vacas. cf Swedish Society SLT 33189.
      Mille regretz. cf MILAN: Pavanas, nos. 1-6.
      Mille regretz. cf L'Oiseau-Lyre 12BB 203/6.
      Variations on a Spanish folk tune. cf RCA ARL 1-0485.
NEDBAL, Oskar
      The simple Johnny: Valse triste. cf Supraphon 110 1429.
NEEFE, Christian
      Serenate. cf DG Archive 2533 149.
NEGRI, Marc Antonio
      Balletto. cf DG Archive 2533 184.
      Il bianco fiore. cf DG Archive 2533 173.
      Lo spagnoletto. cf DG Archive 2533 173.
NEJEDLY, Vit
      Vitězství bude nase. cf Panton 110 361.
NERUDA
      Berceuse slav d'apres un chant polonais, op. 11. cf Discopaedia
         MB 1005.
      Concerto, trumpet. cf BACH: Brandenburg concerto, no. 2, S 1047,
         F major.
NEUSIDLER, Hans
      Der Judentanz. cf DG Archive 2533 111.
      Der Judentanz. cf Vanguard SRV 316.
      Welscher Tanz, Wascha mesa: Hupfauff. cf DG Archive 2533 111.
      Welscher Tanz, Wascha mesa: Hupfauff. cf DG Archive 2533 184.
NEWARK, William
      The farther I go, the more behind. cf BASF 25 22286-1.
NEWMAN (16th century England)
      Pavan. cf Philips 6775 006.
NICOLAI, Otto
      The merry wives of Windsor: Overture. cf Philips 6747 199.
NIELSEN, Carl
      At a young artist's bier. cf Works, selections (HMV 5027).
      Bohemian-Danish folk tunes. cf Works, selections (HMV 5027).
      Chaconne, op. 32. cf Piano works (Decca SDD 476).
      Concerto, clarinet. cf MOZART: Concerto, clarinet, K 622, A major.
      Concerto, clarinet, op. 57. cf Works, selections (HMV 5027).
      Concerto, flute. cf Works, selections (HMV 5027).
      Concerto, violin, op. 33. cf Works, selections (HMV 5027).

Dance of the lady's maids. cf Piano works (Decca SDD 476).
Festival prelude. cf Piano works (Decca SDD 475).
Helios overture, op. 17. cf Works, selections (HMV 5027).
Helios overture, op. 17. cf GADE: Echoes of Ossian.
Humoresque-bagatelles, op. 11. cf Piano works (Decca SDD 476).
Little suite, op. 1, A minor. cf Serenato in vano.
Pan and syrinx, op. 49. cf Works, selections (HMV 5027).
Piano music for young and old, op. 53. cf Piano works (Decca SDD
  475).
1689 Piano works: Festival prelude. Piano music for young and old, op.
  53. Pieces, piano, op. 3 (5). Pieces, piano, op. 59 (3). Sym-
  phonic suite, op. 8. John McCabe, pno. Decca SDD 475.
        +HFN 11-75 p161   +-RR 11-75 p81
1690 Piano works: Chaconne, op. 32. Dance of the lady's maids. Humoresque-
  bagatelles, op. 11. Suite, op. 45. Theme and variations, op. 40.
  John McCabe, pno. Decca SDD 476.
        +HFN 11-75 p161   +-RR 11-75 p81
Pieces, piano, op. 3 (5). cf Piano works (Decca SDD 475).
Pieces, piano, op. 59 (3). cf Piano works (Decca SDD 475).
Rhapsodie overture (An imaginary journey to the Faroe Islands). cf
  Works, selections (HMV 5027).
Saga-Drøm, op. 39. cf Symphony, no. 5, op. 50.
Saga-Drøm, op. 39. cf Works, selections (HMV 5027).
Saga-dream, op. 39. cf GADE: Echoes of Ossian.
1691 Serenata in vano. Little suite, op. 1, A minor. Symphony, no. 2,
  op. 16. H. Nielsen, clt; J. Nilsson, bsn; H. C. Sørensen, hn;
  A. L. Christiansen, vlc; John Poulsen, double bs; Tivoli Symph-
  ony Orchestra; Carl Garaguly. Turnabout 34049. Tape (c) KTVC
  34049.
        -RR 1-73 p71 tape
Suite, op. 45. cf Piano works (Decca SDD 476).
Symphonic rhapsody. cf Works, selections (HMV 5027).
Symphonic suite, op. 8. cf Piano works (Decca SDD 475).
1692 Symphonies (6). Jill Gomez, s; Brian Royner Cook, bar; LSO; Ole
  Schmidt. Unicorn RHS 324/330 (6).
        +Gr 1-75 p1350   +RR 12-74 p16
        ++HF 4-75 p73   ++SFC 2-16-75 p23
        ++NR 4-75 p2   +-St 6-75 p100
Symphonies, nos. 1-6. cf Works, selections (HMV 5027).
1693 Symphony, no. 2, op. 16. NYP; Leonard Bernstein. Columbia M 32779.
  (Q) MQ 32779. Tape (ct) 32779.
        +HF 7-74 p99   ++SFC 4-21-74 p27
        +NR 7-74 p2   +-St 6-75 p100
Symphony, no. 2, op. 16. cf Serenata in vano.
1694 Symphony, no. 3, op. 27. Felicity Palmer, s; Thomas Allen, bar;
  LSO; François Huybrechts. Decca SXL 6695.
        +-Gr 3-75 p1657   +-RR 3-75 p30
1695 Symphony, no. 4, op. 29. LAPO; Zubin Mehta. London CS 6848.
  (also Decca SXL 6633)
        +-Gr 12-74 p1143   +-RR 12-74 p16
        +-HF 11-74 p117   ++SFC 9-22-74 p22
        +NR 11-74 p2   ++St 2-75 p110
1696 Symphony, no. 5, op. 50. Bournemouth Symphony Orchestra; Paavo
  Berglund. HMV ASD 3063.
        +Gr 6-75 p46   +-RR 6-75 p45
        +-HFN 6-75 p91

1697 Symphony, no. 5, op. 50. Saga-Drøm, op. 39. NPhO; Jascha Horen-
        stein. Unicorn RHS 300. Tape (c) ZCUN 300.
                +Gr 7-72 p247 tape   ++NR 8-75 p5
        Theme and variations, op. 40. cf Piano works (Decca SDD 476).
1698 Works, selections: At a young artist's bier. Concerto, clarinet,
        op. 57. Concerto, flute. Concerto, violin, op. 33. Bohemian-
        Danish folk tunes. Helios overture, op. 17. Saga-Drøm, op. 39.
        Pan and syrinx, op. 49. Rhapsodie overture (An imaginary jour-
        ney to the Faroe Islands) Symphonic rhapsody. Symphonies, nos.
        1-6. Kjell-Inge Stevensson, clt; Frantz Lemsser, flt; Arve
        Tellefsen, vln; Danish State Radio Symphony Orchestra; Herbert
        Blomstedt. HMV SLS 5027 (8).
                +Gr 10-75 p621   +RR 11-75 p45
                ++HFN 11-75 p161

NIGER
        Preludes, G major, C major, F major. cf Hungaroton SLPX 11601/2.
NILSSON, Bo
1699 Quantitaten. PAULSON: Mode, op. 108b. SCHUMANN: Kinderscenen,
        op. 15. STRAVINSKY: Serenade, A major. Hans Pålsson, pno.
        Caprice RIKS LP 79.
                +RR 10-75 p80
NIN, Joaquin
        Cantilena asturiana. cf RCA ARM 4-0942/7.
        Paño murciano. cf Odyssey Y3 32880.
        Songs: Jesús de Nazareth; Villancico andaluz; Villancico asturiano;
        Villancico castellano. cf BIZET: Songs (Decca SXL 6577).
        Variations on a theme by Milan. cf Nonesuch H 71233.
NOLA, Gian Domenico de
        Chi chi li chi. cf Telefunken TK 11569/1-2.
NONO, Luigi
1700 Como una ola de fuerza y luz. Y entonces comprendió. Slavka
        Taskova, Mary Lindsay, Liliana Poli, Gabriela Ravazzi, s;
        Kadigia Bove, Miriam Acevedo, Elena Vicini, speakers; Maurizio
        Pollini, pno; Bavarian Radio Symphony Orchestra; RAI Rome Chamber
        Choir; Claudio Abbado, Nino Antonellini; Magnetic tape, Luigi
        Nono, sound direction. DG 2530 436.
                +-Gr 7-74 p253   +-NR 8-74 p13
                +HF 9-74 p98   +RR 7-74 p77
                +MJ 10-75 p39
        Y entonces comprendió. cf Como una ola de fuerza y luz.
NORDQUIST, Gustav
        Till havs. cf RCA SER 5704/6.
NORGARD, Per
        Constellations, 12 strings, op. 22. cf HOLMBOE: Symphony, no. 8,
        op. 56.
NOTRE DAME SCHOOL
        Flos filies and motet. cf BIS LP 2.
NOVAK, Vitĕzslav
        Youth, op. 55, no. 21: The devil's polka. cf Supraphon 110 1429.
NOVIKOV, Anatolii
        Vassya-Vassilyok. cf HMV ASD 3116.
NOVOTNY
        Prelude, G major. cf Hungaroton SLPX 11601/2.
        Prelude, D major. cf Hungaroton SLPX 11601/2.
        Prelude and fugue from the Stark Tablature Book. cf Hungaroton
        SLPX 11601/2.

NUNES GARCIA, José Mauricio
1701 Requiem (ed. Lerma). Doralene Davis, s; Betty Allen, ms; William
     Brown, t; Matti Tuloisela, bs-bar; Morgan State College Choir;
     Helsinki Philharmonic Orchestra; Paul Freeman. Columbia M 33431.
         +HF 10-75 p65   ++NYT 8-10-75 pD14
         +NR 11-75 p9
NYSTROEM, Gösta
1702 Sinfonia concertante, violoncello and orchestra. KOCH: Nordic
     capriccio, op. 26. Erling Blöndal Bengtsson, vlc; Swedish Radio
     Symphony Orchestra; Stig Westerberg. Swedish Discofil SLT 33136.
         +St 6-75 p51
OBRADORS
     Coplas de curro dulce. cf Odyssey Y2 32880.
OBRECHT, Jacob
     Ic draghe de mutze clutze. cf Telefunken TK 11569/1-2.
     Ic draghe de mutze clutze; Mijn morken gaf; Pater noster. cf
     L'Oiseau-Lyre 12BB 203/6.
     Magnificat. cf Turnabout TV 34569.
1703 Missa caput. Salve crux, arbor vitae. Capella Cordina; Alejandro
     Planchart. Lyrichord LLST 7273.
         +NR 11-75 p9
     Salve crux, arbor vitae. cf Missa caput.
OCHS, Siegfried
     's kommt ein Vogerl geflogen. cf HUMPERDINCK: Hänsel und Gretel,
     excerpts.
OCKEGHEM, Johannes
1704 Marienmotetten. Prague Madrigal Singers; Miroslav Venhoda. Tele-
     funken AW 6-41878.
         -Gr 6-75 p79   *NR 10-75 p6
         +-HFN 8-75 p80   +RR 6-75 p81
         +-MT 10-75 p886
OFFENBACH, Jacques
     Gaité parisiénne, excerpts. cf HMV SLS 5019.
     Genevieve de Brabant: Gendarmes' duet. cf Decca SXL 6719.
     Orphée aux enfers: Overture. cf HEROLD: Zampa: Overture.
     Orphée aux enfers: Overture. cf Telefunken DX 6-35262.
     La Périchole: Laughing song. cf Desto DC 7118/9.
     La Périchole: O mon cher amant; Ah, quel diner. cf Columbia M 32231.
     La Périchole: Tu n'est pas beau. cf RCA ARL 1-5007.
     Que voulez-vous faire. cf RCA ARL 1-5007.
1705 The tales of Hoffmann (Les contes de Hoffmann). Dorothy Bond,
     Margherita Grandi, Ann Ayars, s; Monica Sinclair, ms; Robert
     Tounseville, t; Bruce Dargavel, bar; Sadler's Wells Chorus;
     RPO; Thomas Beecham. Vox/Turnabout THS 65012/14 (3).
         +St 7-75 p110
     Les contes de Hoffman: Barcarolle. cf Discophilia KGP 4.
     Les contes de Hoffman: Ha wie in meiner Seele. cf Bel Canto Club
     No. 3.
     The tales of Hoffmann: C'est une chanson d'amour. cf ABC ATS 20018.
1706 La vie Parisienne, excerpts. Nicole Broissin, Madeleine Vernon, s;
     Christiane Harbell, Danièlle Millet, ms; Michel Caron, Jean-
     Marc Recchia, t; Henri Gui, Jacques Marfeuil, Luc Barney, Phil-
     ippe Ariotti, bar; Orchestra and Chorus; François Rauber. French
     Decca 117 011.
         +-Gr 10-75 p688   +-HFN 9-75 p101

O'GALLAGHER, Liam
     Border dissolve in audiospace.   cf 1750 Arch 1752.
O'KOEVER, John
     Fantasie.   cf Turnabout TV 37071.
OLDFIELD, Mike
1707 Tubular bells (arr. Bedford).   RPO; David Bedford.   Virgin V 2026.
          Tape (c) TCV 2026 (ct) 8XV 2026.
               -Gr 4-75 p1811   +St 8-75 p103
               -HFN 5-75 p136
OLSSON, Otto
     Prelude and fugue, op. 56, D sharp minor.   cf BIS LP 7.
     OPERATIC SONGS AND ARIAS, Jussi Bjorling.   cf RCA SER 5704/6.
ORBON, Julian
     Preludio y tocata.   cf Nonesuch H 71233.
O'REILLY
     Metropolitan quintet.   cf Kendor KE 22174.
ORFF, Carl
1708 Carmina burana.   Judith Blegen, s; Kenneth Riegel, t; Peter Binder,
          bar; Cleveland Orchestra Chorus and Boys' Choir; CO; Michael
          Tilson Thomas.   Columbia MX 33172.   Tape (c) MT 33172 (ct) MAX
          33172 (Q) MQ 33172, M 33172 Tape (ct) MAQ 33172.   (also CBS
          76372.   Tape (c) 40-76372)
               +-Audio 10-75 p119   +-ON 11-75 p70
               -Gr 5-75 p2006   +-RR 5-75 p62
               -Gr 12-75 p1121 tape   +-RR 12-75 p99 tape
               +HF 5-75 p84   -SFC 5-11-75 p23
               ++HFN 5-75 p136   ++St 5-75 p74
               ++NR 9-75 p10
1709 Carmina burana.   Jutta Vulpius, s; Hans Joachim Rotzsch, t; Kurt
          Rehm, Kurt Hübenthal, bar; Leipzig Radio Childrens' Choruses;
          Leipzig Radio Symphony Orchestra; Herbert Kegel. DG 2548 194.
          Tape (c) 3318 051.   (Reissue from 89525)
               -Gr 9-75 p500   -RR 9-75 p70
               +HFN 9-75 p109   -RR 11-75 p94 tape
1710 Carmina burana.   Sheila Armstrong, s; Gerald English, t; Thomas
          Allen, bar; St. Clement Danes Grammar School Boys' Choir; LSO
          and Chorus; André Previn.   HMV (Q)ASD 3117.   Tape (c) TC ASD
          3117.
               +Gr 10-75 p670   +HFN 12-75 p160
1711 Carmina burana.   Gerda Hartman, s; Richard Brunner, t; Rudolf Knoll,
          bar; Salzburg Mozarteum Orchestra and Choir; Kurt Prestel.   Pye
          GSGC 15001.   Tape (c) ZCCCB 15001.
               +-Gr 8-75 p355   +-HFN 8-75 p81
               /Gr 12-75 p1121 tape   -RR 8-75 p63
1712 Carmina burana.   Catulla carmina.   Trionfo di Afrodite.   Milada
          Subrtová, Helena Tattermuschova, s; Marta Boháčová, ms; Jaros-
          lav Tománek, Ivo Zidek, Oldrich Lindauer, t; Teodor Srubař, bar;
          Karel Berman, bs; Ludmila Trzcka, Vladimir Topinka, Vladimir
          Mencl, Oldřich Kredba, pno; CPhO and Chorus; PSO; Václav Smetá-
          ček.   Supraphon 112 1461/3 (3).
               +-HFN 10-75 p153   ++RR 10-75 p88
1713 Catulla carmina.   Ute Mai, s; Eberhard Büchner, t; Leipzig Radio
          Symphony Orchestra and Chorus; Herbert Kegel.   Philips 6500 815.
               +HF 11-75 p94   ++St 11-75 p130
               +NR 9-75 p10
     Catulla carmina.   cf Carmina burana.

1714 De Temporum Fine Comoedia (Play about the end of time). Sylvia
        Anderson, Colette Lorand, Jane Marsh, Kay Griffel, Gwendolyn
        Killebrew, Kari Lövaas, Anna Tomova-Sintow, Heljä Angervo,
        Glenys Louis, Erik Geisen, Hans Wegmann, Hans Helm, Wolfgang
        Annheisser, Siegfried Rudolf Frese, Hermann Patzalt, Hannes
        Jokel, Anton Diakov, Boris Carmeli, Christa Ludwig, Peter Schrei-
        er, Josef Greindl, vocalists; Rolf Boysen, narrator; Kölner
        Rundfunkchor, RIAS Kammerchor, Tölzer Knabenchor; Kölner Rundfunk
        Symphony Orchestra; Herbert von Karajan. DG 2530 432.
                    -Gr 10-74 p746   +ON 9-74 p64
                    +HF 2-75 p98   +-RR 11-74 p89
                    +-MJ 10-75 p39   -SR 2-8-75 p37
                    +NR 2-75 p8   ++St 11-74 p131
                    *NYT 10-10-74 pD1
1715 Der Mond. Eberhard Büchner, Helmut Klotz, t; Horst Lunow, Wilfred
        Schaal, bar; Reiner Süss, Fred Teschler, Armin Terzibaschian,
        bs; Leipzig Radio Symphony Orchestra and Chorus; Herbert Kegel.
        Philips 6700 083 (2).
                    ++Audio 11-75 p95   +ON 11-75 p70
                    +Gr 11-75 p900   ++RR 10-75 p30
                    +HF 11-75 p94   ++SFC 8-17-75 p22
                    +HFN 10-75 p146   ++St 10-75 p71
                    +NR 9-75 p10
1716 Schulwerk: Street songs (Gassenhauer). Tölzer Boys' Choir; Instru-
        mental Ensemble; Carl Orff. BASF HC 25122.
                    +HF 11-75 p94   +St 10-75 p111
                    +ON 11-75 p70
      Trionfo di Afrodite. cf Carmina burana.
      ORGAN CONCERT AR INSELKIRCHE, St. Nicolai, Helgoland. cf Pelca
        PSR 40598.
      ORGAN MUSIC OF THE 16th AND 17th CENTURIES. cf Saga 5374.
      ORGANS OF THE SWISS COUNTRYSIDE. cf Pelca PSRK 41013/6.
OROLOGIO, Alessandro
      Intradas a 5 (2). cf Erato STU 70847.
      Occhi miei. cf Panton 010335.
ORR, Robin
      Preludes on a Scottish psalm tune. cf Wealden WS 145.
ORTEGA
      Pues aun me tienes, Miguel. cf HMV SLS 5012.
ORTIZ, Diego
      Dulce memoire; Recercada. cf L'Oiseau-Lyre 12BB 203/6.
      Recercada (2). cf Telefunken SAWT 9620/1.
OVALLE-BANDEIRO
      Modinha. cf Angel S 36076.
OWEN, Morfydd
      Madonna songs. cf Argo ZRG 769.
PACHELBEL, Johann
      Aria sebaldina, variations, F minor. cf BACH: Toccata and fugue,
        S 916, G major.
      Aria sebaldina, variations, F minor. cf Organ works (Arion ARN
        38273).
      Canon. cf Decca SDD 411.
      Canon, D minor. cf CORELLI: Concerto grosso, op. 6, no. 8, C minor.
      Canon a 3 on a ground, D major. cf HMV ASD 3017.
      Chaconne, F minor. cf Organ works (Arion ARN 38273).
      Chaconne, F minor. cf Pelca PSR 40598.

Chorale partita, Alle Menschen müssen sterben. cf Telefunken TK
    11567/1-2.
Chorale preludes: Da Jesus an dem Kreuze stund; Christus, der ist
    mein Leben; Vom Himmel hoch; Wir glauben all an einen Gott;
    Wie schön leuchtet der Morgernstern; Vater unser im Himmelreich;
    O Lamm Gottes unschuldig. cf Organ works (Arion ARN 38273).
Chorale prelude, Alle Menschen müssen sterben. cf Philips 6775 006.
Chorale preludes, Ein fest Burg; Komm, Gott, Schöpfer, Heiliger
    Geist. cf Telefunken TK 11567/1-2.
Fantasia, G minor. cf Pelca PSR 40598.
Fugue, C major. cf Telefunken TK 11567/1-2.
Magnificat fugues, nos. 4-5, 10, 13. cf Philips 6775 006.
1717 Organ works: Aria sebaldina, variations, F minor. Chaconne, F
    minor. Chorale preludes: Da Jesus an dem Kreuze stund; Christus,
    der ist mein Leben; Vom Himmel hoch; Wir glauben all an einen
    Gott; Wie schön leuchtet der Morgernstern; Vater unser im Himmel-
    reich; O Lamm Gottes unschuldig. Ricercar. Toccata, E minor.
    Bernard Lagacé, org. Arion ARN 38273.
        +RR 11-75 p82
Ricercar. cf Organ works (Arion ARN 38273).
Toccata, E minor. cf Organ works (Arion ARN 38273).
Toccata, E minor. cf Telefunken TK 11567/1-2.
Toccata, F major. cf Pelca PSR 40598.
Toccata and fugue, B flat major. cf Philips 6775 006.
Vom Himmel hoch. cf University of Colorado, unnumbered.
Vom Himmel hoch. cf University of Iowa Press, unnumbered.
PACOLONI
    Padoana commun; Passamezzo commun. cf L'Oiseau-Lyre 12BB 203/6.
PADEREWSKI, Ignace Jan
    Caprice. cf International Piano Library IPL 5005/6.
    Caprice, G major. cf CHOPIN: Polish songs, op. 74: Maiden's wish.
    Fantaisie polonaise, op. 19. cf ARENSKY: Concerto, piano, no. 2.
    Genre Scarlatti. cf International Piano Library IPL 5005/6.
    Légende, op. 16, no. 1. cf CHOPIN: Chants polonaise, op. 74, no. 1:
        A maiden's wish.
    Melodie. cf CHOPIN: Polish songs, op. 74, no. 1: A maiden's wish.
    Minuet, G major. cf CHOPIN: Polish songs, op. 74, no. 1: A maiden's
        wish.
    Minuet, op. 14, no. 1, G major. cf Discopaedia MB 1007.
    Moja piesczotka. cf Discophilia KGS 3.
    Nocturne, op. 16, no. 4, B flat major. cf CHOPIN: Chants polonaise,
        op. 74, no. 1: A maiden's wish.
    Sonata, violin and piano, op. 13, A minor. cf BUSONI: Sonata,
        violin and piano, no. 2, op. 36a, E minor.
PAER, Ferdinando
    Il bacio della partenza. cf Rococo 5368.
PAGANINI, Niccolo
1718 Cantabile, violin and guitar, op. 17. Centones de Sonate, nos. 1,
    3-4, 6. Sonata, violin and guitar, op. 2, no. 6, A minor.
    Sonata, violin and guitar, op. 3, no. 4, A minor. György Tere-
    besi, vln; Sonja Prunnbauer, gtr. Telefunken SAT 22548.
        +HF 2-75 p99   ++SFC 1-12-75 p26
        ++NR 12-74 p5
    Caprice, op. 1, no. 2, B minor. cf Discopaedia MB 1002.
    Caprice, op. 1, no. 2, B minor. cf Rococo 2072.
    Caprices, op. 1, nos. 9, 14, 20, 24. cf Classics for Pleasure CFP
        40205.

Caprices, op. 1, nos. 9, 21. cf Hungaroton SLPX 11677.
Caprices, op. 1, nos. 13, 20. cf RCA ARM 4-0942/7.
1719 Caprices, op. 1, nos. 17, E major; no. 24, A minor. Variations
on "God save the Queen". Variations on "Nel cor più non mi
sento". VARGA: Prelude and caprices (4). Sonata, violin, G
minor. Ruben Varga, vln. Audio Fidelity FCS 50061.
        +-NR 10-75 p11
Caprices, op. 1, nos. 17, 24. cf RCA ARL 1-0735.
1720 Caprice, op. 1, no. 24, A minor (arr. Callimahos). PROKOFIEV:
Sonata, flute and piano, op. 94, D major. SANCAN: Sonatine
pour flute et piano. Louise di Tullio, flt; Virginia di Tullio,
pno. Crystal S 311.
        +NR 12-75 p8
Centones di Sonate, nos. 1, 3-4, 6. cf Cantabile, violin and
guitar, op. 17.
1721 Concerti, violin, nos. 1-6. Salvatore Accardo, vln; LPO; Charles
Dutoit. DG 2740 121 (5). (No. 6 available on 2530 467)
        ++Gr 11-75 p827   ++RR 11-75 p46
        +HFN 12-75 p160   ++SFC 12-7-75 p31
1722 Concerto, violin, no. 1, op. 6, D major. WIENIAWSKI: Concerto,
violin, no. 2, op. 22, D minor. Michael Rabin, vln; PhO;
Eugene Goossens. Seraphim S 60222.
        ++St 1-75 p116
1723 Concerto, violin, no. 6, op. posth., E minor (orch. Mompellio).
Salvatore Accardo, vln; LPO; Charles Dutoit. DG 2530 467.
        +Gr 11-74 p904   +RR 11-74 p48
        ++HF 10-74 p110   ++SFC 4-20-75 p23
        ++MJ 1-75 p48   +SR 1-11-75 p51
        +NR 12-74 p5   ++St 11-74 p132
Fantaisie on "Dal tuo stellato sogio" (Rossini). cf LALO: Symphonie
espagnole, op. 21.
Moto perpetuo, op. 11 (arr. and orch. Gerhardt). cf RCA LRL 1-5094.
Moto perpetuo, op. 11. cf RCA ARM 4-0942/7.
Sonata, violin, op. 25, C major. cf DG 2530 561.
Sonata, violin and guitar, op. 2, no. 6, A minor. cf Cantabile,
violin and guitar, op. 17.
Sonata, violin and guitar, op. 3, no. 4, A minor. cf Cantabile,
violin and guitar, op. 17.
Valse. cf Decca SPA 405.
Variations on "God save the Queen". cf Caprices, op. 1, no. 17,
E major; no. 24, A minor.
Variations on "Nel cor più non mi sento". cf Caprices, op. 1, no.
17, E major; no. 24, A minor.
Variations on "Nel cor più non mi sento". cf LALO: Symphonie
espagnole, op. 21.
Variations on a theme by Rossini, D minor. cf BOCCHERINI: Concerto,
violoncello, B flat major.
Variations on a theme by Rossini, D minor. cf HMV SEOM 19.
PAISIELLO, Giovanni
Marche de la Garde Consulaire à Marengo. cf Classics for Pleasure
CFP 40230.
Marche du premier Consul. cf Classics for Pleasure CFP 40230.
La molinara: Nel cor più non mi sento. cf Decca SXL 6629.
I Zingari in Fiera: Chi vuol la zingarella. cf Decca SXL 6629.
PAIX, Jacob
Schirazula Marazula. cf DG Archive 2533 111.

Ungarescha: Saltarello. cf DG Archive 2533 111.
PALADILHE, Emile
    Psyché. cf Rococo 5365.
PALESTRINA, Giovanni de
    Missa Sine Nomine: Kyrie, Sanctus. cf Gemini GM 2022.
    Ricercar sopra il primo tuono. cf Desto DC 6474/7.
1724 The song of songs, nos. 1-29 (Moctecta quinque vocum ex Cantico
       Canticorum). Cantores in Ecclesia; Michael Howard. L'Oiseau-
       Lyre SOL 338/9 (2).
           ++Gr 10-74 p741   +RR 10-74 p87
           +HF 6-75 p100
PANELLA
    On the square. cf Michigan University SM 0002.
PANUFNIK, Andrzej
1725 Autumn music. Heroic overture. Nocturne. Tragic overture. LSO;
       Jascha Horenstein. Unicorn RHS 306.
           ++NR 8-75 p5
1726 Concerto, violin. Sinfonia concertante, flute, harp and strings.
       Yehudi Menuhin, vln; Aurèle Nicolet, flt; Osian Ellis, hp;
       Menuhin Festival Orchestra; Andrzej Panufnik. HMV EMD 5525.
           +Gr 12-75 p1038   +RR 12-75 p57
    Heroic overture. cf Autumn music.
    Nocturne. cf Autumn music.
    Sinfonia concertante, flute, harp and strings. cf Concerto, violin.
1727 Symphony, no. 1. Symphony, no. 3. Monte Carlo Opera Orchestra;
       Andrzej Panufnik. Unicorn RHS 315. (Reissue from HMV ASD 2298)
           +Gr 9-74 p511   +SFC 12-8-74 p39
           ++NR 4-75 p3   ++St 6-75 p103
           +RR 9-74 p57
    Symphony, no. 3. cf Symphony, no. 1.
    Tragic overture. cf Autumn music.
PAPP, Lajos
    Dialogue, piano and orchestra. cf KOCSAR: Lonely song on poems
      by Attila József.
    Meditations in memory of Milán Füst. cf KOCSAR: Lonely song on
      poems by Attila József.
PARCHAM, Andrew
    Solo, G major. cf Argo ZRG 746.
    Solo, recorder and harpsichord, G major. cf Telefunken Tape (c)
      CX 4-41203.
PARISH-ALVARS, Elias
    Concerto, harp, op. 81, G minor. cf RODRIGO: Concierto de Aranjuez,
      guitar and orchestra.
PARKHURST, E. A.
    Father's a drunkard. cf RCA ARL 1-5007.
PARMA, Nicola
    Aria del Gran Duca. cf DG Archive 2533 173.
    Ballo del serenissimo Duca di Parma. cf DG Archive 2533 173.
    La Cesarina. cf DG Archive 2533 173.
    Corenta. cf DG Archive 2533 173.
    Gagliarda Manfredina. cf DG Archive 2533 173.
    La Mutia. cf DG Archive 2533 173.
    La ne mente per la gola. cf DG Archive 2533 173.
PARRIS, Robert
    The book of imaginary beings. cf EVETT: Quintet, piano and strings.
1728 Concerto, trombone. ROCHBERG: Contra mortem et tempus. WUORINEN:

PARRIS (cont.)   304

   Janissary music. Roman Siwek, trom; Raymond DesRoches, perc;
      Polish Radio Symphony Orchestra; Aeolian Quartet of Sarah Lawrence
      College; Zdislav Szostak. CRI SD 231.
         +Gr 8-75 p326   *RR 6-75 p44
         +-HFN 12-75 p160
PARRY, Hubert
   Chorale prelude on Rockingham. cf Wealden WS 139.
   Jerusalem, op. 208. cf Sound Superb SPR 90049.
   O how amiable. cf HMV Tape (c) TC MCS 13.
   Sonata, harp, D major. cf RCA LRL 1-5087.
PASQUINI, Bernardo
   Canzone francese, no. 7. cf Philips 6775 006.
   Ricercare, no. 4. cf Philips 6775 006.
PATTERSON, Paul
   Time piece. cf HMV EMD 5521.
PAUER, Jiri
   Hrdinum práce. cf Panton 110 361.
PAULSON
   Modi, op. 108b. cf NILSSON: Quantitaten.
PAUMANN
   Ellend du hast. cf BIS LP 3.
PAVER, Jiri
   Panychida. cf CHAUN: Proces.
PEARSON, Leslie
   An Elizabethan fantasy: Now is the month of Maying; The willow
      song; The night watch. cf Argo ZDA 203.
   A medieval pageant: Agincourt song; Greensleeves; Summer is icumen
      in. cf Argo ZDA 203.
PEASLEE
   Alice. cf Argo ZRG 765.
PEERSON, Martin
   The fall of the leafe. cf Angel SB 3816.
   The primrose. cf Angel SB 3816.
   Sing, love is blind. cf Turnabout TV 37079.
PEETERS, Flor
   Chorale preludes, op. 68: Ach bleib mit Deiner Gnade; O Gott, du
      frommer Gott; Wachet auf. cf ALCOCK: Introduction and passa-
      caglia.
   Jubilate Deo Omnis Terra. cf Wayne State University, unnumbered.
   Suite modale, op. 43. cf ALCOCK: Introduction and passacaglia.
PENDERECKI, Krzysztof
1729 Anaklasis. De natura sonoris II. Fluorescences. Kosmogonia.
      Stefania Woytowicz, s; Kazimierz Pustelak, t; Bernard Ladysz,
      bs; Warsaw, National Philharmonic Orchestra and Chorus; Andrzej
      Markowski. Philips 6500 683.
         ++HF 1-75 p84   ++SFC 12-8-74 p36
          *NR 11-74 p7   +St 5-75 p98
   Capriccio per Siegfried Palm. cf DG 2530 562.
   De natura sonoris II. cf Anaklasis.
   Ecloga VIII. cf HMV EMD 5521.
   Flourescences. cf Anaklasis.
   Kosmogonia. cf Anaklasis.
1730 Magnificat. Peter Lagger, bs; Krakow Radio and Children's Choir;
      Polish National Symphony Orchestra; Krzysztof Penderecki. Angel
      S 37141. (also HMV EMD 5524)
         ++NR 11-75 p9   ++RR 12-75 p89
          *NYT 12-21-75 pD18   +SFC 9-28-75 p30

PENN, William
1731 Ultra mensuram. ROSS: Concerto, trombone and orchestra. Prelude,
      fugue and big apple, bass trombone and tape. SHCWANTNER: Modus
      caelestis. Western Michigan University Wind Ensemble; Per Brevig,
      trom; Bergen Symphony Orchestra; Gregg Shearer, flt; New England
      Conservatory Repertory Orchestra; Karsten Andersen, Richard
      Pittman. CRI SD 340.
            +HF 12-75 p110   +-NR 9-75 p8
PENNINO
      Pecchê. cf Musical Heritage Society MHS 1951.
PEPUSCH, John
      Preludes, cf Argo ZRG 746.
      Trio sonata, G minor. cf Orion ORS 73114.
      Trio sonata, flute and harpsichord, G major. cf Ember ECL 9040.
PERGOLESI, Giovanni
      Concerto, flute, no. 1, G major. cf Decca SPA 394.
1732 La serva padrona. Annette Celine, s; Sesto Bruscantini, bs; Rome
      Radio Italiana Symphony Orchestra; Alberto Zedda. Everest SDBR
      3373.
            +NR 4-75 p8
1733 La serva padrona. Carmen Bustamente, s; Renato Capecchi, bar;
      ECO; Antonio Ros-Marbá. Pye EnSayo NEL 2014.
            +-Gr 1-75 p1384   +RR 12-74 p72
1734 La serva padrona. Virginia Zeani, s; Nicola Rossi-Lemeni, bs;
      Hamburg Radio Orchestra, Members; Gerhard Gregor, hpd; George
      Singer. Saga 5360. (Reissue from Vox SLPX 50380)
            +-Gr 1-75 p1384   +STL 1-12-75 p36
            +-RR 12-74 p72
1735 La serva padrona. Reiner Süss, bs; Olivera Miljakovic, s; Berlin
      Staatskapelle Orchestra; Helmut Koch. Telefunken SLT 43126.
            -HF 6-73 p89   +-RR 5-75 p62
            /HFN 8-75 p81   /St 4-73 p118
            /NR 3-73 p9
      La serva padrona: Stizzoso, mio stizzoso. cf Decca SXL 6629.
      Songs: Tre giorni son che Nina. cf Decca SXL 6629.
PERNAMBUCO, João
      Sons de carrilhoes. cf Amberlee ACL 501X.
PERLE, George
1736 Toccata. WOLPE: Form. Form IV. WUORINEN: Sonata, piano. WYNER:
      Short fantasies (3). Robert Miller, pno. CRI SD 306.
            +HF 9-74 p107   +-NR 5-74 p12
            +LJ 9-75 p37   +-SFC 12-8-74 p36
            +MQ 4-75 p326
PERSICHETTI, Vincent
      Gloria. cf Wayne State University, unnumbered.
      Songs: The death of a soldier; The grass; Of the surface of things;
      The snow man; Thou child so wise. cf Duke University Press DWR
      7306.
PERUSIO, Matheus de
      Andray soulet. cf 1750 Arch S 1753.
PESCETTI, Giovanni
      Allegretto, C major. cf Pandora PAN 101.
      Presto, C minor. cf Pandora PAN 101.
      Sonata, organ, C minor. cf Argo ZRG 783.
PESSARD, Emile
      Songs: L'Adieu du matin; Bergère legère. cf Discophilia KGC 1.

PETERSON-BERGER
    Songs: When I walk alone; Among the fir trees.   cf RCA SER 5719.
PETRASSI, Goffredo
    Suoni notturni.   cf Hungaroton SLPX 11629.
PETROVICS, Emil
    Quintet, winds.   cf BOZAY: Quintet, winds, op. 6.
PETTERSSON, Gustaf Allan
1737 The barefoot songs.   Sonatas, violin, nos. 3, 7.   Karl Sjunnesson,
        bar; Stockholm University Chorus; Carl Rune Larsson, pno; Josef
        Grünfarb, Karl-Ove Mannberg, vln; Eskil Hemberg, cond.   Caprice
        RIKSLP 28.
            +Gr 6-75 p79
    Sonatas, 2 violins, nos. 3, 7.   cf The barefoot songs.
1738 Symphony, no. 2.   Swedish Radio Symphony Orchestra; Stig Westerberg.
        Swedish Society SLT 33219.
            +Gr 3-75 p1658
1739 Symphony, no. 7.   Stockholm Philharmonic Orchestra; Antal Dorati.
        Decca SXL 6538.   (also London CS 6740)
            +Gr 5-72 p1894   ++NR 3-73 p2
            +HF 5-73 p88   ++SR 3-73 p47
            +HFN 6-72 p1116   ++St 5-73 p118
            ++MJ 5-73 p35   +St 6-75 p51
    Symphony, no. 10.   cf BLOMDAHL: Symphony, no. 2.
PEUERL, Paul
    Dance.   cf Pelca PSR 40589.
PEZEL, Johann
    Seventeenth century dances (6).   cf Desto DC 6474/7.
    Sonatinas, nos. 61-62, 65-66.   cf Nonesuch H 71301.
    Suite, C major.   cf Saga 5417.
PFAUTSCH, Lloyd
    Songs: A day for dancing: What shall I bring.   cf University of
        Colorado, unnumbered.
PFITZNER, Hans
1740 Palestrina.   Helen Donath, s; Brigitte Fassbaender, ms; Nicolai
        Gedda, Heribert Steinbach, Friedrich Lenz, t; Dietrich Fischer-
        Dieskau, Hermann Prey, Bernd Weikl, bar; Karl Ridderbusch, bs;
        Tölzer Boys' Choir; Bavarian Radio Symphony Orchestra and Chorus;
        Rafael Kubelik.   DG 2711 013 (4).
            ++Gr 2-74 p1596   +-NYT 11-11-73 pD28
            ++HF 1-74 p56   +-Op 4-74 p324
            ++LJ 3-75 p35   ++RR 3-74 p28
            ++MJ 3-74 p47   ++St 12-73 p134
            ++NR 4-74 p6
    Songs: Mailied; Trauerstille.   cf WAGNER: Lohengrin.
    Voll jener süsse.   cf DG 2530 332.
PHALESE, Pierre
    Dances.   cf BIS LP 3.
    Passamezzo: Saltarello.   cf DG Archive 2533 111.
    Passamezzo d'Italie: Reprise, Gaillarde.   cf DG Archive 2533 111.
    Reprise, galliard.   cf DG Archive 2533 184.
PHILE
    Hail, Columbia.   cf Michigan University SM 0002.
    PHILIPPE ENTREMONT, A la Française.   cf Columbia M 32070.
PHILIPS, Peter
    Amarilli di Julio Romano.   cf Saga 5402.
    Ave, Jesu Christe.   cf Abbey LPB 734.

Philips' pavan and galliard. cf L'Oiseau-Lyre SOL 336.
PICK (attrib.)
     March and troop. cf Saga 5417.
     Suite, B flat major. cf Saga 5417.
     PICTURES FROM ISRAEL. cf Da Camera Magna SM 93399.
PIEFKE, G.
     Königgratzer März̈ch. cf BASF 292 2116-4.
     Königgratzer Märsch. cf DG 2721 077.
     Preussens Gloria. cf DG 2721 077.
PIERNE, Gabriel
     Giration. cf HAHN: Concerto, piano, E major.
     Serenade, op. 7, A major. cf Discopaedia MB 1008.
PIJPER, Willem
     Concerto, piano. cf DRESDEN: Concerto, oboe.
     PILAR LORENGAR, aria recital. cf London OS 26381.
PILKINGTON, Francis
     Sweet Phillida. cf Harmonia Mundi 593.
PINKHAM, Daniel
     For evening draws on. cf MARTINO: Paradiso choruses.
     Liturgies. cf MARTINO: Paradiso choruses.
     Toccatas for the vault of heaven. cf MARTINO: Paradiso choruses.
PIPO, Ruiz
     Canción y danza. cf Everest SDBR 3380.
PISADOR, Diego
     Pavana, E minor. cf RCA ARL 1-0485.
PISTON, Walter
     Incredible flutist. cf CHADWICK: Symphonic sketches.
     Serenata. cf BENTZON: Pezzi sinfonici, op. 109.
PIXIS, Johann
     Concerto, violin, piano and strings. cf MOSCHELES: Grande sonate
          symphonique, op. 112.
PLATTI, Giovanni
     Concerto, flute and orchestra, G major. cf RCA CRL 2-7003.
PLAYFORD, John
     The English dancing master: Country dances. cf DG Archive 2533 172.
PLESKOW, Raoul
     Motet and madrigal. cf CRI SD 342.
PLEYEL, Ignaz
     Sinfonie concertante, violin and viola, op. 29, B flat major. cf
          MOZART: Concertone, 2 violins, K 190 (K 166b), C major.
POGLIETTI, Alessandro
     Balletto. cf DG Archive 2533 172.
POLAK, Jakub
     Praeludium. cf DG Archive 2533 294.
POLDINI, Eduard
     Caprice viennois. cf Argo ZRG 805.
     Dancing doll. cf Connoisseur Society (Q) CSQ 2065.
     Poupée valsante. cf BIZET: Jeux d'enfants, op. 22.
     Poupée valsante. cf Argo ZRG 805.
     Romance. cf Argo ZRG 805.
POLDOWSKI, Irene Regine Wieniawska
     Tango. cf RCA ARM 4-0942/7.
PONCE, Manuel
     Campo. cf LAURO: Danza negra.
     Estrellita. cf RCA ARM 4-0942/7.
     Prelude, E major. cf RCA ARL 1-0864.

Preludes, nos. 1-2, 4-6, 12. cf CBS 61654.
Suite, A minor (Suite antica). cf GIULIANI: Sonata eroica, op.
    150, A major.
Thème varié et finale. cf CBS 61654.
Variations and fugue on "La folias". cf Hungaroton SLPX 11629.

PONCHIELLI, Amilcare
1741 La gioconda. Renata Tebaldi, s; Marilyn Horne, ms; Carlo Bergonzi,
    t; Robert Merrill, bar; Rome, Saint Cecilia Orchestra and Chorus;
    Lamberto Gardelli. London 1388. Tape (r) LOR 90139.
        ++LJ 9-75 p38
La gioconda: Arias. cf London OS 26277.
La gioconda: Cielo e mar. cf Columbia D3M 33448.
La gioconda: Cielo e mar. cf Ember GVC 41.
La gioconda: Cielo e mar. cf Supraphon 10924.
La gioconda: Dance of the hours, Act 3. cf HMV SLS 5019.
La gioconda: Ma chi vien...Oh, la sinistra voce. cf London OS 26315.
I promessi sposi: Overture. cf Erato STU 70880.

PONIRIDY, George
Rythemes Grecs. cf HADZIDAKIS: For a little white seashell, op. 1.

PONS
Laura: Serenade Napolitaine. cf Rubini RS 300.

POPPER, David
1742 Hungarian rhapsodie, op. 68. In Walde suite, op. 50. Virtuoso
    pieces (5). Gayle Smith, vlc; John Ritter, pno. Genesis GS
    1050.
        ++Gr 4-75 p1826   +NR 8-75 p15
        +MJ 12-74 p45   ++SFC 10-27-74 p10
Im Walde suite. cf Hungarian rhapsodie.
Mazurka, op. 11, no. 3, G major. cf Ember GVC 42.
Virtuoso pieces (5). cf Hungarian rhapsodie.
POPULAR RUSSIAN SONGS. cf Anonymous works.

PORCELIJN, David
Continuations. cf KEURIS: Concerto, saxophone.
10-5-6-5 (a). cf KEURIS: Concerto, saxophone.

PORPORA, Nicola
1743 Concerto, violoncello, G major. SAMMARTINI: Concerto, viola pom-
    posa, C major. VIVALDI: Concerto, viola d'amore, A major.
    Thomas Blees, vlc; Ulrich Koch, vla d'amore, vla pomposa; South
    West German Orchestra; Paul Angerer. Turnabout TV 34574.
        ++NR 6-75 p6

PORTER, Walter
Thus sang Orpheus. cf L'Oiseau-Lyre 12BB 203/6.

POSTON, Elizabeth
Jesus Christ the apple tree. cf Argo ZRG 785.
Jesus Christ the apple tree. cf HMV Tape (c) TC MCS 13.
Sugar plums. cf HMV SLA 870.

POULENC, Francis
Aubade, piano and 18 instruments. cf d'INDY: Symphony on a French
    mountain air, op. 25.
Les biches, Ballet suite. cf HONEGGER: Pacific 231.
Gloria. cf HMV Tape (c) TC MCS 14.
Marches et un intermède. cf IBERT: Paris 32.
Mouvements perpetuels, no. 1. cf RCA ARM 4-0942/7.
Pieces: Toccata. cf Columbia M 32070.
Sonata, clarinet and piano. cf Crystal S 331.
Sonata, trumpet, horn and trombone. cf Desto DC 6474/7.

1744 Songs: Chansons de F. Garcia Lorca (3); La courte paille; Fian-
     çailles pour rire; Métamorphoses; Poèmes de Louise de Vilmorin
     (3); Poèmes de Max Jacob (5). Felicity Palmer, s; John Constable,
     pno. Argo ZRG 804.
          +HFN 8-75 p81   +STL 10-5-75 p36
          +RR 7-75 p59
     Songs: L'Anguille; La belle jeunesse; Priez pour paix; Serenade.
     cf FAURE: Songs (1750 Arch S 1754).
1745 The story of Babar, the little elephant. SATIE: Sports and diver-
     tissements. Mildred Natwick, narrator; Grant Johannesen, pno.
     Golden Crest CRS 4133.
          +St 5-75 p98
POWER, Lionel
     Opem nobis, Credo. cf Nonesuch H 71292.
PRAETORIUS, Michael
     Ballet. cf Swedish Society SLT 33189.
     La bourrée. cf DG Archive 2533 184.
     La bourrée and Der Schützkonig. cf Oryx BRL 1.
     Galliarde de la guerre. cf DG Archvie 2533 184.
     Galliarde de Monsieur Wustron. cf DG Archive 2533 184.
     Gavotte. cf DG Archive 2533 184.
1746 Hosianna dem Sohne Davids. Ein Kind geborn zu Bethlehem. Nun
     Komm, der Heiden Heiland. Psallite unigenito Christo. Terpsi-
     chore: Ballet des sorciers; Bransle double; Gaillarde; Sarabande;
     Ballet des feus; Pavane Spaigne; La rosette; Bransle gentil;
     Volte; Courante. Von Himmel hoch. SCHEIN: Banchetto musicale:
     Suite, no. 1, G major; No. 2, D minor. Hannover Niedersäch-
     sischer Choir; Ferdinand Conrad Instrumental Ensemble; Willi
     Träder. Nonesuch H 71128.
          /RR 12-75 p91
     Ein Kind geborn zu Bethlehem. cf Hosianna dem Sohne Davids.
     Lo, how a rose e'er blooming. cf University of Colorado, unnumbered.
     Motets: Allein Gott in der Höh sei Ehr; Aus tiefer Not schrei ich
          zu dir; Christus, der uns selig macht; Erhalt uns, Herr, bei
          deinem Wort; Gott der Vater wohn uns bei; Resonet in laudibus.
          cf Terpsichore dances.
     Nun Komm, der Heiden Heiland. cf Hosianna dem Sohne Davids.
     O vos omnes. cf BASF BAC 3087.
     Peasant dances. cf CRD CRD 1019.
     Psallite unigenito Christo. cf Hosianna dem Sohne Davids.
     Quem pastores laudavere; Wie schön leuchtet der Morgernstern.
          cf Turnabout TV 34569.
     Reprinse. cf DG Archive 2533 184.
     La rosette. cf Vanguard SRV 316.
     Spagnoletta. cf DG Archive 2533 184.
1746 Terpsichore dances (1612): La bourrée; Courante M. M. Wustrow;
     Galliard; Reprinse secundam inferiorem; Passameze; Pavane de
     Spaigne; La sarabande; Spagnoletta; Suite de ballets; Suite de
     voltes. Motets: Allein Gott in her Höh sei Ehr; Aus tiefer
     Not schrei ich zu dir; Christus, der uns selig macht; Erhalt
     uns, Herr, bei deinem Wort; Gott der Vater wohn uns bei; Resonet
     in laudibus. London Early Music Consort; St. Alban's Abbey Choir
     Boys' Voices; David Munrow. HMV CSD 3761. (also Angel S 37091)
          +Gr 11-74 p904   +-RR 12-74 p46
          ++HF 6-75 p101   +St 11-75 p142
          -NR 4-75 p6

Terpsichore: Ballet. cf CBS 72526.

Terpsichore: Ballet des sorciers; Bransle double; Gaillarde; Sarabande; Ballet des feus; Pavane Spaigne; La rosette; Bransle gentil, Volte; Courante. cf Hosianna dem Sohne Davids.

Terpsichore: Fire dance; Stepping dance; Windmills; Village dance; Sailor's dance; Fisherman's dance; Festive march. cf CRD CRD 1019.

Volta. cf Swedish Society SLT 33189.

Von Himmel hoch. cf Hosianna dem Sohne Davids.

Winter is an icy guest. cf Oryx BRL 1.

PREIS

O Du mein Oesterreich. cf DG 2721 077.

PREVIN, André

André Previn's music night: Signature tune. cf HMV (Q) ASD 3131.

PROKOFIEV, Serge

1748 Alexander Nevsky, op. 78. Rosalind Elias, ms; CSO; Fritz Reiner. RCA LSC 2395. Tape (c) ERPA 23950.
　　　　+HF 4-75 p112 tape

1749 Alexander Nevsky, op. 78. Betty Allen, ms; Mendelsohn Club Chorus; PO; Eugene Ormandy. RCA ARL 1-1151. Tape (c) ARK 1-1151 (ct) ARS 1-1151.
　　　　+-NR 11-75 p9　　　　　　　　++SFC 11-9-75 p22

1750 Alexander Nevsky, op. 78. Věra Soukupová, con; CPhO and Chorus; Karel Ančerl. Supraphon SUAST 50429. Tape (c) 045 0429.
　　　　+Gr 5-75 p2032 tape　　　　　-RR 8-74 p85 tape
　　　　+HFN 5-75 p142 tape　　　　 +RR 6-75 p92 tape

Alexander Nevsky, op. 78. cf BORODIN: Prince Igor: Polovtsian march.

Alexander Nevsky, op. 78: Alexander's entry into Pskov. cf HMV SEOM 14.

1751 Betrothal in the monastery, op. 86: Summer night suite. RIMSKY-KORSAKOV: Le coq d'or: Suite. Bournemouth Symphony Orchestra; Paavo Berglund. HMV (Q) ASD 3141.
　　　　+Gr 12-75 p1045

Chose en soi, op. 45. cf Piano works (Turnabout TV 37065/70).

Cinderella, op. 87: Gavotte. cf CHOPIN: Scherzo, no. 4, op. 54, E major.

Cinderella, op. 87: Introduction; Quarrel; The dancing lesson; Spring fairy; Summer fairy; Grasshoppers dance; Winter fairy; The interrupted departure; Clock scene; Cinderella's arrival at the ball; Grande valse; Cinderella's waltz; Midnight; Apotheosis. cf BRITTEN: Young person's guide to the orchestra, op. 34.

1752 Cinderella, op. 87: Suites, nos. 1 and 2. ROHO: Hugo Rignold. London STS 15193.
　　　　+NR 7-75 p4

1753 Concerti, piano, nos. 1-5. In autumn, op. 8. Overture on Hebrew themes, op. 34. Symphony, no. 1, op. 25, D major. Vladimir Ashkenazy, pno; Keith Puddy, clt; LSO, Gabrieli Quartet; Vladimir Ashkenazy. Decca 15BB 218/20 (3).
　　　　++Gr 10-75 p622　　　　　　　+RR 9-75 p45
　　　　++HFN 9-75 p102

1754 Concerto, piano, no. 2, op. 16, G minor. TCHAIKOVSKY: Concerto, piano, no. 1, op. 23, B flat minor. Tedd Joselson, pno; PO; Eugene Ormandy. RCA ARL 1-0751.
　　　　-HF 4-75 p90　　　　　　　　 +-SR 1-25-75 p50

         +MJ 2-75 p40   ++St 3-75 p70
         +-NR 3-75 p4
1755 Concerto, piano, no. 2, op. 16, G minor. Concerto, piano, no. 5,
      op. 55, G major. Jorge Bolet, Alfred Brendel, pno; Cincinnati
      Symphony Orchestra, VSOO; Thor Johnson, Jonathan Sternberg.
      Turnabout TV 34543.
         +HFN 11-75 p173   +-RR 11-75 p47
1756 Concerto, piano, no. 3, op. 26, C major. Concerto, piano, no. 5,
      op. 55, G major. Michel Beroff, pno; Leipzig Gewandhaus Orches-
      tra; Kurt Masur. Angel S 37084.
         +HF 6-75 p102   +SFC 3-9-75 p27
         +NR 5-75 p5   +-St 10-75 p111
1757 Concerto, piano, no. 3, op. 26, C major.* RAVEL: Concerto, piano,
      G major.** Marguerite Long, Serge Prokofiev, pno; LSO; Symphony
      Orchestra; Piero Coppola, Maurice Ravel. World Records SH 209.
      (*Reissue from HMV DB 1725/6; **from Columbia LX 194/6)
         +Gr 2-75 p1492   +-RR 12-74 p39
   Concerto, piano, no. 3, op. 26, C major. cf GERSHWIN: Rhapsody
      in blue.
   Concerto, piano, no. 3, op. 26, C major. cf RAVEL: Concerto, piano,
      for the left hand, D major.
1758 Concerto, piano, no. 5, op. 55, G major. RAVEL: Concerto, piano,
      G major. Monique Haas, Sviatoslav Richter, pno; Paris National
      Orchestra; Warsaw National Philharmonic Orchestra; Paul Paray,
      Witold Rowicki. DG 2548 109. (Reissues from 138988, 138075)
         +Gr 11-75 p840   +RR 10-75 p48
         +-HFN 10-75 p152
1759 Concerto, piano, no. 5, op. 55, G major. Sonata, piano, no. 5,
      op. 38, C major. Alfred Brendel, pno; VSO; Jonathan Sternberg.
      Everest SDBR 3385.
            *NR 11-75 p6
   Concerto, piano, no. 5, op. 55, G major. cf Concerto, piano, no.
      2, op. 16, G minor.
   Concerto, piano, no. 5, op. 55, G major. cf Concerto, piano, no.
      3, op. 26, C major.
1760 Concerto, violin, no. 1, op. 19, D major. Concerto, violin, no.
      2, op. 63, G minor. Ruggiero Ricci, vln; Luxembourg Radio
      Orchestra; Louis de Froment. Candide (Q) QCE 31081.
         +NR 2-75 p5
1761 Concerto, violin, no. 1, op. 19, D major. Concerto, violin, no. 2,
      op. 63, G minor. Pierre Amoyal, vln; Strasbourg Philharmonic
      Orchestra; Alain Lombard. Erato STU 70866.
         +Gr 12-75 p1038   ++RR 12-75 p57
   Concerto, violin, no. 1, op. 19, D major. cf KHACHATURIAN: Concerto,
      violin.
   Concerto, violin, no. 1, op. 19, D major. cf HMV SLS 5004.
1762 Concerto, violin, no. 2, op. 63, G minor. MIASKOVSKY: Concerto,
      violoncello, op. 66, C minor. David Oistrakh, vln; Mstislav
      Rostropovich, vlc; PhO; Alceo Galliera, Malcolm Sargent.
      Seraphim S 60223.
            ++St 2-75 p111
   Concerto, violin, no. 2, op. 63, G minor. cf Concerto, violin,
      no. 1, op. 19, D major (Candide QCE 31081).
   Concerto, violin, no. 2, op. 63, G minor. cf Concerto, violin,
      no. 1, op. 19, D major (Erato 70866).
   Concerto, violin, no. 2, op. 63, G minor. cf RCA ARM 4-0942/7.

Concerto, violin, no. 2, op. 63, G minor. cf RCA CRL 6-0720.
Episodes, op. 12, nos. 1-10. cf Piano works (Turnabout TV 37065/70).
Episodes, op. 12, nos. 1-3, 7, 10. cf Ember GVC 40.
Etudes, op. 2, nos. 1-4. cf Piano works (Turnabout TV 37065/70).
Gavotte. cf RCA ARM 4-0942/7.
In autumn, op. 8. cf Concerti, piano, nos. 1-5.
Lieutenant Kijé suite, op. 60. cf KODALY: Háry János: Suite.
Lieutenant Kijé suite, op. 60. cf SHOSTAKOVICH: Symphony, no. 6,
    op. 54, B minor.
Love for three oranges, op. 33: Intermezzo. cf Ember GVC 40.
Love for three oranges, op. 33: March. cf RCA ARL 1-0735.
Love for three oranges, op. 33: March. cf Westminster WGM 8309.
March, F minor. cf RCA ARM 4-0942/7.
1763 Marches, nos. 1-3, op. 69. March, op. 99, B flat major. RIMSKY-
    KORSAKOV: Concerto, clarinet, E flat major. Concerto, trombone.
    Variations, oboe and brass band. TCHAIKOVSKY: March, B flat
    major. Victor Batashov, trom; Lev Mikhailov, clt; Yevgeny Lyak-
    hovitsky, ob; USSR Ministry of Defense Symphonic Band; Nicolai
    Navarov, Nicolai Sergeyev, Yuri Pitirov, Leonid Dunsev. HMV
    Melodiya ASD 3107.
        +-Gr 8-75 p332   +RR 8-75 p42
        +-HFN 9-75 p104
March, op. 99, B flat major. cf Marches, nos. 1-3, op. 69.
March, op. 99, B flat major. cf DG 2548 148.
March, op. 99, B flat major. cf HMV SEOM 20.
Music for children, op. 65. cf Piano works (Turnabout TV 37065/70).
Overture on Hebrew themes, op. 34. cf Concerti, piano, nos. 1-5.
Pensées, op. 62. cf Piano works (Turnabout TV 37065/70).
1764 Peter and the wolf, op. 67. SAINT-SAENS: Le carnaval des animaux.
    Eric Shilling, narrator; Pavel Stepán, Olja Hurnik, pno; CPhO;
    Prague Symphony Orchestra; Karel Ancerl, Martin Turnovský.
    Supraphon 50865. Tape (c) 0450865.
        +Gr 6-75 p106   *HFN 5-75 p142 tape
        +Gr 4-75 p1872 tape   -RR 5-75 p37
        +-HFN 6-75 p91   -RR 6-75 p92 tape
Peter and the wolf, op. 67. cf BRITTEN: The young person's guide to
    the orchestra, op. 34 (Philips Tape (c) 7317 093).
Peter and the wolf, op. 67. cf BRITTEN: The young persons's guide
    to the orchestra, op. 34 (Vanguard 71189).
1765 Piano works: Chose en soi, op. 45. Episodes, op. 12, nos. 1-10.
    Etudes, op. 2, nos. 1-4. Music for children, op. 65. Pensées,
    op. 62. Pieces, op. 3, nos. 1-4. Pieces, op. 4, nos. 1-4.
    Pieces, op. 32, nos. 1-4. Pieces, op. 59, nos. 1-3. Scarcasms,
    op. 17. Sonatas, piano, nos. 1-9. Sonatines, op. 54. Tales
    of an old grandmother, op. 31, no. 3. Toccata, op. 11. Visions
    fugitives, op. 22. György Sandor, pno. Turnabout TV 37065/70
    (6). (Reissues from STGBY 601, 609, 617, 621, 627, 631)
        /Gr 6-75 p69   +HFN 8-75 p81
Pieces, piano, op. 3, nos. 1-4. cf Piano works (Turnabout TV
    37065/70).
Pieces, piano, op. 4, nos. 1-4. cf Piano works (Turnabout TV
    37065/70).
Pieces, piano, op. 32, nos. 1-4. cf Piano works (Turnabout TV
    37065/70).
Pieces, piano, op. 59, nos. 1-3. cf Piano works (Turnabout TV
    37065/70).

Pieces, piano, op. 96. cf KABALEVSKY: Sonata, piano, no. 3, op. 46.
Prelude, op. 12, no. 7, C major. cf Saga 5356.
1766 Romeo and Juliet, op. 64: Act 1, Introduction, The street awakens,
The Duke's command, Interlude, Juliet the girl, Masks, Dance
of the Knights, Balcony scene, Romeo's variation, Love duet;
Act 2, Dance of the five couples, Dance with mandolins, Romeo
at Friar Lawrence's, Romeo decides to avenge Mercutio's death,
Death of Tybalt; Act 3, Aubade, Dances of the girls with lilies;
Act 4, Juliet's funeral, Juliet's death. CO; Lorin Maazel.
Decca SXL 6668. Tape (c) KSXC 6668. (Reissue from SXL 6620/2)
        +Gr 1-75 p1353   +RR 1-75 p33
        +Gr 2-75 p1562 tape   +RR 1-75 p71 tape
1767 Romeo and Juliet, op. 64: Act 1, Introduction, Juliet as a child,
Masks, Scene at the balcony, Romeo's variation, Love dance; Act
2, Dance of the five pairs, Dance with mandolines, Romeo decides
to avenge Mercutio's death, Finale; Act 3, Introduction, Romeo
and Juliet, Juliet's bedroom, The last farewell, Morning serenade, Dance of the girls with lilies, Epilogue, Juliet's funeral,
Death of Juliet. LSO; André Previn. HMV ASD 3054. (Reissue
from SLS 864)
        +-Gr 3-75 p1658   +-RR 3-75 p33
1768 Romeo and Juliet, op. 64: Montagues and Capulets; Juliet, the maid-
en; Masques; Minuet; Friar Lawrence; Dance; Dance of the West
Indian slave girls; Scene; The death of Tybalt; Romeo and Juliet
before parting; Romeo at Juliet's grave. Rotterdam Philharmonic
Orchestra; Edo de Waart. Philips 6500 640. Tape (c) 7300 305.
        +Gr 3-74 p1706   +NR 5-74 p6
        +HF 5-74 p93   +RR 3-74 p18
        +HF 3-75 p108 tape   -RR 2-75 p75 tape
        +HFN 3-74 p112   +SFC 6-9-74 p28
        /MJ 5-74 p47
Romeo and Juliet, op. 64: 3 pieces. cf Hungaroton SLPX 11677.
Romeo and Juliet: Maska; Balcony scene; Death of Tybalt. cf Decca
SDDJ 393/95.
Scarcasms, op. 17. cf Piano works (Turnabout TV 37065/70).
Scarcasms, op. 17, nos. 1, 2. cf Ember GVC 40.
1769 Scythian suite, op. 20. Seven, they are seven, op. 30.* SHOSTA-
KOVICH: Symphony, no. 2, op. 14, C major. Yuri Elnikov, t;
MPO, MRSO and Chorus, RSFSR Russian Chorus; Gennady Rozhdest-
vensky, Kyril Kondrashin. HMV Melodiya ASD 3060. (*Reissue
from ASD 2669)
        +Gr 5-75 p1968   +RR 5-75 p40
        +-HFN 5-75 p137
Scythian suite, op. 20. cf BARTOK: Dance suite.
Seven, they are seven, op. 30. cf Scythian suite, op. 20.
Sonata, flute and piano, op. 94, D major. cf FRANCK: Sonata, flute,
A major.
Sonata, flute and piano, op. 94, D major. cf PAGANINI: Caprice,
op. 1, no. 24, A minor.
Sonatas, piano, nos. 1-9. cf Piano works (Turnabout TV 37065/70).
Sonata, piano, no. 5, op. 38, C major. cf Concerto, piano, no. 5,
op. 55, G major.
1770 Sonata, violin and piano, no. 1, op. 80, F minor. JANACEK: Sonata,
violin and piano, op. 21. David Oistrakh, vln; Frieda Bauer,
pno. Westminster WGS 8292.
        ++St 4-75 p105

Sonata, violin and piano, no. 1, op. 80, F minor. cf BARTOK:
   Sonata, violin and piano, no. 1.
Sonata, violin and piano, no. 2, op. 94bis, D major. cf DEBUSSY:
   Sonata, violin and piano, no. 3, G minor.
Sonata, violoncello and piano, op. 119. cf DEBUSSY: Sonata, violon-
   cello and piano, no. 1, D minor.
Sonatines, op. 54, G major. cf Piano works (Turnabout TV 37065/70).
1771 The stone flower. Bolshoi Theatre Orchestra; Gennady Rozhdestvensky.
   Columbia/Melodiya M3 33215 (3). (also HMV Melodiya SLS 5024)
        +-Gr 9-75 p466   +NYT 3-9-75 pD23
        +HF 5-75 p86   +RR 8-75 p41
        +HFN 9-75 p102   +Te 12-75 p42
        +NR 4-75 p4
1772 The story of a real man, op. 117. Glafira Deomidova, s; Kira Leonova, ms; Gyorgy Shulpin, Aleksei Maslennikov, t; Evgeni Kibkalo,
   bar; Gennadi Pankov, Mark Reshetin, Artur Eizen, bs; Bolshoi
   Theatre Orchestra and Chorus; Mark Ermler. Westminster WGSO
   8317/2 (2).
        +ON 7-75 p30   +St 10-75 p111
A summer day, suite for small orchestra, op. 65b. cf Winter bonfire suite, op. 122.
1773 Symphony, no. 1, op. 25, D major. Symphony, no. 7, op. 131, C
   sharp minor. LSO; Walter Weller. Decca SXL 6702.
        ++Gr 5-75 p1968   +-RR 5-75 p38
        +HFN 5-75 p137
Symphony, no. 1, op. 25, D major. cf Concerti, piano, nos. 1-5.
1774 Symphony, no. 5, op. 100, B flat major. LSO; André Previn. HMV
   ASD 3115. Tape (c) TC ASD 3115. (also Angel S 37100)
        +-HFN 9-75 p102   +RR 9-75 p46
        +-HFN 12-75 p173 tape   +-RR 12-75 p99 tape
Symphony, no. 7, op. 131, C sharp minor. cf Symphony, no. 1, op.
   25, D major.
Tales of an old grandmother, op. 31, no. 3. cf Piano works
   (Turnabout TV 37065/70).
Tales of an old grandmother, op. 31, no. 3. cf Ember GVC 40.
Toccata. cf RCA VH 020.
Toccata, op. 11. cf Piano works (Turnabout TV 37065/70).
Toccata, op. 11. cf Ember GVC 40.
Visions fugitives, op. 22. cf Piano works (Turnabout TV 37065/70).
Visions fugitives, op. 22. cf CHOPIN: Scherzo, no. 4, op. 54, E
   major.
1775 War and peace, op. 91. Galina Vishnevskaya, Nedezhda Kossitsyna,
   Margarita Miglayu, s; Valentina Klepatskaya, Yevgenia Verbit-
   skaya, Irina Arkhipova, Kira Leonova, ms; Yelena Grivoba, Valen-
   tina Petrova, con; Vitali Vlassov, Aleksei Maslennikov, Vladimir
   Petrov, Gyorgy Shulpin, Nikolai Zakharov, Nikolai Timchanko, t;
   Yevgeny Kibkalo, Boris Shapenko, Pavel Lisitsian, Vladimir Val-
   aitis, bar; Leonid Ktitorov, Yevgeny Belov, Mark Reshetin, bs-
   bar; Nikolai Shchegolkov, Gennady Pankov, Alexander Vedernikov,
   Viktor Nechinailo, Alexei Krivchenya, Leonid Masslov, Yuri
   Galkin, Artur Eizen, Ivan Sipaev, Alexei Geleva, bs; Bolshoi
   Theatre Orchestra and Chorus; Aleksander Melik-Pashayev. HMV
   Melodiya SLS 837 (4). (Reissue from Artia/MK 218D) (also Col-
   umbia M4 33111; Angel/Melodiya SR 40053)
        ++Gr 2-73 p1545   +Op 6-73 p532
        +-HF 1-75 p67   +RR 2-73 p48
        +HFN 3-73 p567   ++St 2-75 p108
        +NYT 12-1-74 pD18

1776 War and peace, op. 91: Duets, Scenes 1 and 3. PUCCINI: Madama
     Butterfly: Un bel dì vedremo; Final scene. TCHAIKOVSKY: Eugene
     Onegin: Final scene. VERDI: Aida: Ritorna vincitor; O patria
     mia. Galina Vishnevskaya, s; Valentina Klepatskaya, ms; Georg
     Otts, bar; Orchestra. Westminster WGS 8267.
        ++St 1-75 p120
1777 Winter bonfire suite, op. 122. A summer day, suite for small orch-
     estra, op. 65b. Prague Radio Symphony Orchestra; Prague Radio
     Children's Chorus; PCO. Supraphon SUAST 50773. Tape (c) 04
     50773.
        +Gr 4-72 p1721   +-HFN 3-72 p509
        +Gr 5-75 p2032 tape   +HFN 5-75 p142 tape
PROVAZNIK
     Valse joyeuse, op. 137. cf Discopaedia MB 1004.
PROVOST
     Intermezzo. cf Discopaedia MB 1007.
     PRUSSIAN AND AUSTRIAN MARCHES. cf DG 2721 077.
PUCCINI, Giocomo
1778 Arias: La bohème: O soave fanciulla. Gianni Schicchi: Lauretta
     mia. Madama Butterfly: Bimba, dagli occhi. Manon Lescaut: Tu,
     tu, amore tu. VERDI: Otello: Già nella notte densa. La travia-
     ta: Parigi o cara. Mirella Freni, s; Franco Bonisolli, t;
     Hamburg State Philharmonic Orchestra; Berlin State Opera Orches-
     tra; Leone Magiera, Lamberto Gardelli. BASF KBC 22007.
        +-ON 3-8-75 p34   +St 5-75 p104
1779 Arias: La bohème: Mi chiamano Mimi; Musetta's waltz; Addio di Mimi.
     Gianni Schicchi: O mio babbino caro. Madama Butterfly: Un bel
     di; Death of Butterfly. Manon Lescaut: In quelle trine morbide.
     Suor Angelica: Senza mamma. La rondine: Chi il bel sogno di
     Doretta. Tosca: Vissi d'arte. Turandot: Signore, ascolta; Tu
     chi di gel sei cinta. Le Villi: Se come voi piccino io fossi.
     Licia Albanese, s; Orchestra. RCA AVM 1-0715.
        +NR 12-74 p12   +ON 3-8-75 p34
1780 La bohème. Victoria de los Angeles, Lucine Amara, s; Jussi Björ-
     ling, William Nahr, t; Robert Merrill, John Reardon, Thomas
     Powell, George del Monte, bar; Giorgio Tozzi, Fernando Corena,
     bs; Columbus Boys' Choir; RCA Victor Orchestra and Chorus; Thomas
     Beecham. HMV SLS 896 (2). Tape (c) TC SLS 896. (Reissue from
     ALP 1409/10) (also Seraphim SIB 6099)
        ++Gr 11-74 p963   +RR 11-74 p28
        +Gr 10-75 p721 tape   +St 7-75 p110
1781 La bohème. Mirella Freni, Elizabeth Harwood, s; Luciano Pavarotti,
     Michel Sénéchal, Gernot Pietsch, t; Rolando Panerai, Gianni
     Maffeo, Hans Dietrich Pohl, Hans-Dieter Appelt, bar; Nicolai
     Ghiaurov, bs; Berlin German Opera Chorus; BPhO; Herbert von
     Karajan. London OSA 1299 (2). Tape (c) D 31235. (also Decca
     SET 565/6. Tape (c) K2B2)
        ++Gr 8-73 p373   +-Op 10-73 p906
        +Gr 5-75 p2031 tape   +-RR 8-72 p20
        +-HF 11-73 p112   +RR 5-75 p76 tape
        ++HFN 6-75 p109 tape   ++SFC 7-22-73 p28
        +-NR 11-73 p7   +SR 9-11-73 p80
        +NYT 9-30-73 pD25   +St 10-73 p146
        +-ON 9-73 p62
1782 La bohème. Monteserrat Caballé, Judith Blegen, s; Placido Domingo,
     Nico Castel, Alan Byers, t; Vicenzo Sardinero, Sherrill Milnes,

bar; Noël Mangin, Ruggero Raimondi, William Mason, Franklyn
Whitely, bs; Wandsworth School Boys' Choir; John Alldis Choir;
LPO; Georg Solti. RCA ARL 2-0371 (2). (Q) ARD 2-0371.

 +Gr 10-74 p746   +-ON 12-7-74 p42
 +-HF 1-75 p84   +-RR 10-74 p25
 ++MJ 2-75 p30   ++SFC 11-10-74 p27
 +NR 11-74 p8   +St 12-74 p141
 +NYT 12-15-74 pD21

1783 La bohème for orchestra. Orchestra; André Kostelanetz. CBS 61610.
 +Gr 2-75 p1555   /RR 2-75 p37

1784 La bohème: Che gelida manina; Si, mi chiamano Mimì...O soave fanci-
ulla; In un coupé...O Mimì tu più non torni. Madama Butterfly:
Bimba dagli occhi; Un bel di, Humming chorus. Tosca: Tre sbirri
...Te deum; Vissi d'arte; E lucevan le stelle. Turandot: Sig-
nore, ascolta...Non piangere, Liù...Ah, per l'ultima volta;
Nessun dorma. Renata Tebaldi, s; Carlo Bergonzi, Giuseppe di
Stefano, Mario del Monaco, Piero de Palma, Mario Carlin, Renato
Ercolani, t; Ettore Bastianini, George London, Renato Cesari,
bar; Cesare Siepi, Fernando Corena, Nicolas Zaccaria, bs; Rome,
Santa Cecilia Orchestra and Chorus; Tullio Serafin, Francesco
Molinari-Pradelli, Gianandrea Gavazzeni, Alberto Erede, Franco
Patané. Decca SPA 365. (Reissues)
 ++Gr 4-75 p1872   +RR 3-75 p18
 +Gr 2-75 p1556

La bohème: Arias. cf London OS 26373.
La bohème: Arias. cf London OS 26381.
La bohème: Addio senza rancor. cf Rubini RS 300.
La bohème: Che gelida manina. cf BASF BC 21723.
La bohème: Che gelida manina. cf Columbia D3M 33448.
La bohème: Che gelida manina. cf Decca SXL 6649.
La bohème: Che gelida manina. cf Supraphon 10924.
La bohème: Che gelida manina. cf Supraphon 012 0789.
La bohème: Che gelida manina; O Mimi, tu più non torni. cf RCA
SER 5704/6.
La bohème: Donde lieta uscì. cf HMV SLS 5012.
La bohème: Man nennt mich jetzt nur Mimi. cf Disophilia KGP 4.
La bohème: O soave fanciulla. cf Arias (BASF KBC 22007).
La bohème: O soave fanciulla; O Mimì, tu più non torni. cf Decca
DPA 517/8.
La bohème: Mi chiamano Mimi; Musetta's waltz; Addio di Mimi. cf
Arias (RCA AVM 1-0715).
La bohème: O soave fanciulla. cf Rubini RS 300.
La bohème: Quando m'en vo' soletta. cf CATALANI: La wally: Ebben,
Ne andrò lontana.
La bohème: Si mi chiamano Mimi; O soave fanciulla; Donde lieta
uscì; Addio, dolce svegliare; Sono andate; Finale. cf Pearl
GEM 121/2.
La bohème: Vecchia zimarra senti. cf Discophilia KGC 2.
I crisantemi. cf CBS 76267.
La fanciulla del West: Ch'ella mi creda. cf Columbia D3M 33448.
La fanciulla del West: Ch'ella mi creda. cf Decca GOSC 666/8.
La fanciulla del West: Lasset sie glauben. cf Bel Canto Club No. 3.
La fanciulla del West: Minnie dalla mia casa. cf Ember GVC 37.
Gianni Schicchi: Firenze è come un albero. cf BASF BC 21723.
Gianni Schicchi: Lauretta mia. cf Arias (BASF KBC 22007).
Gianni Schicchi: O mio babbino caro. cf Arias (RCA AVM 1-0715).

Gianni Schicchi: O mio babbino caro. cf CATALANI: La wally: Ebben,
    Ne andrò lontana.
1785 Madama Butterfly. Maria Chiara, s; Trudeliese Schmidt, ms; James
    King, Perry Gruber, t; Hermann Prey, Anton Rosner, bar; Richard
    Kogel, bs; Bavarian Radio Orchestra and Chorus; Giuseppe Patané.
    Eurodisc 86 515 XR (3).
        +-HF 1-74 p78   +St 4-74 p115
        +-ON 1-11-74 p35
1786 Madama Butterfly. Mirella Freni, s; Christa Ludwig, ms; Luciano
    Pavarotti, t; Robert Kerns, bar; Vienna State Opera Chorus;
    VPO; Herbert von Karajan. London OSA 13110 (3). (also Decca
    SET 584/6. Tape (c) K2A1)
        +Gr 2-75 p1544   +ON 1-11-75 p35
        -Gr 5-75 p2031   +-RR 2-75 p26
        +-HF 2-75 p99   +-RR 5-75 p76 tape
        ++HFN 6-75 p109 tape   ++SFC 11-24-74 p31
        +MJ 9-75 p51   +-SR 2-8-75 p37
        +MT 10-75 p887   ++St 2-75 p111
        +NR 6-75 p11   ++STL 2-9-75 p37
1787 Madama Butterfly. Victoria de los Angeles, s; Anna Maria Canali,
    ms; Giuseppe di Stefano, Renato Ercolani, t; Tito Gobbi, bar;
    Rome Opera Orchestra and Chorus; Gianandrea Gavazzeni. Seraphim
    IC 6090 (3). (Reissue from RCA LM 6121, Captiol GCR 7137)
        +-HF 7-75 p80
1788 Madama Butterfly. Licia Albanese, s; James Melton, t; John Brown-
    lee, bar; Metropolitan Opera Orchestra; Pietro Cimara. Metro-
    politan Opera, unnumbered.
        +NYT 7-6-75 pD 11   +SR 11-29-75 p50
1789 Madama Butterfly: Act 1, Amore o grillo (Enrico Caruso, t; Antonio
    Scotti, bar); Ah, quanto cielo...Ancora un passo (Giannina Russ,
    s); Iero son salita (Geraldine Farrar, s); Vogliateme bene (Linda
    Cannetti, s; Giovanni Zenatello, t); Act 2, Piani, Perchè...Un
    bel dì, vedremo (Emmy Destinn, s); Ora a noi (Lina Pasini-
    Vitale, s; Ferruccio Corradetti, bar); Saicos' ebbe cuore...Che
    tue madre (Emmy Destinn, s); Tutti i fior (Geraldine Farrar, s;
    Louise Homer, con); Non v'è l'avevo detto...Addio, fiorito asil
    (Enrico Caruso, t; Antonio Scotti, bar); Tu, tu, piccolo iddio
    (Emmy Destinn, s). Various orchestral accompaniments. Saga
    7021. (Reissues)
        +-Gr 3-75 p1701   +-RR 1-75 p24
    Madama Butterfly: Arias. cf London OS 26381.
    Madama Butterfly: Bimba, dagli occhi. cf Arias (BASF KBC 22007).
    Madama Butterfly: Bimba dagli occhi; Un bel di, Humming chorus.
        cf La bohème: Che gelida manina (Decca SPA 365).
    Madama Butterfly: Bimba dagli occhi pieni di malia; Una nava da
        guerra...Scouti quella fronda. cf Decca DPA 517/8.
    Madama Butterfly: Con onor muore. cf Gecca GOSC 666/8.
    Madama Butterfly: Eines Tages sehn wir. cf Discophilia KGB 2.
    Madama Butterfly: Give me your darling hands. cf HMV RLS 714.
    Madama Butterfly: Humming chorus. cf Decca ECS 2159.
    Madama Butterfly: Humming chorus, Act 2. cf Decca PFS 4323.
    Madama Butterfly: Un bel dì, vedremo. cf HMV SLS 5012.
    Madama Butterfly: Un bel dì, vedremo. cf Pearl GEM 121/2.
    Madama Butterfly: Un bel dì; Death of Butterfly. cf Arias (RCA
        AVM 1-0715).
    Madama Butterfly: Un bel dì, vedremo; Final scene. cf PROKOFIEV:

War and peace, op. 91: Duets, Scenes 1 and 3.
Manon Lescaut: Donna non vidi mai. cf BASF BC 21723.
Manon Lescaut: Donna non vidi mai. cf Saga 7206.
Manon Lescaut: Donna non vidi mai; Guardate passo son. cf Columbia
     D3M 33448.
Manon Lescaut: Donna non vidi mai; Ah, Manon, mi tradice; Presto
     in fila; Ah, non v'avvicinata; No, pezzo son; Guardate. cf
     RCA SER 5704/6.
Manon Lescaut: In quelle trine morbide. cf Arias (RCA AVM 1-0715).
Manon Lescaut: In quella trine morbide. cf Rubini RS 300.
Manon Lescaut: In quelle trine morbide; Sola, perduta, abbandonata.
     cf CATALANI: La wally: Ebben, Ne andrò lontana.
Manon Lescaut: Intermezzo, Act 3. cf HMV SLS 5019.
Manon Lescaut: Tra voi belle. cf Supraphon 012 0789.
Manon Lescaut: Tu, tu, amore tu. cf Arias (BASF KBC 22007).
1790 Messa di Gloria, A major. William John, t; Philippe Huttenlocher,
     bs; Gulbenkian Foundation Orchestra and Choir; Michel Corbóz.
     Erato STU 70890.
          +-Gr 2-75 p1535
La rondine: Arias. cf London OS 26381.
La rondine: Chi il bel sogno di Doretta. cf Arias (RCA AVM 1-0715).
La rondine: Chi il bel sogno di Doretta. cf CATALANI: La wally:
     Ebben, Ne andrò lontana.
Suor Angelica: Arias. cf BELLINI: La sonnambula: Arias.
Suor Angelica: Senza mamma. cf Arias (RCA AVM 1-0715).
Suor Angelica: Senza mamma. cf CILEA: Adriana Lecouvreur: Io sono
     l'umile ancella.
Suor Angelica: Senza mamma. cf RCA ARL 1-0702.
Suor Angelica: Senza mamma, o bimbo. cf CATALANI: La wally: Ebben,
     Ne andrò lontana.
1791 Tosca. Maria Callas, s; Giuseppe di Stefano, Angelo Mercuriali, t;
     Tito Gobbi, bar; Alvaro Cordova, treble; Franco Calabrese, Dario
     Caselli, Melchiorre Luise, bs; La Scala Orchestra and Chorus;
     Victor de Sabata. HMV SLS 825. Tape (c) RC SLS 825. (Reissue
     from Columbia CX 1094/5)
          +-Gr 3-73 p1729   +Op 5-73 p438
          +HFN 3-73 p567   +RR 3-73 p38
          +-HFN 12-75 p173 tape
Tosca: E lucevan le stelle. cf Columbia D3M 33448.
Tosca: E lucevan le stelle. cf Decca SXL 6649.
Tosca: E lucevan le stelle. cf RCA SER 5704/6.
Tosca: Mario, Mario. cf Decca DPA 517/8.
Tosca: Nur der Schonheit (2). cf Discophilia KGP 4.
Tosca: Recondita armonia. cf Decca GOSC 666/8.
Tosca: Recondita armonia. cf Ember GVC 41.
Tosca: Recondita armonia. cf Saga 7206.
Tosca: Recondita armonia; E lucevan le stelle. cf Decca SDD 390.
Tosca: Tre sbirri...Te deum; Vissi d'arte; E lucevan le stelle.
     cf La bohème: Che gelida manina (Decca SPA 365).
Tosca: Vissi d'arte. cf Arias (RCA AVM 1-0715).
Tosca: Vissi d'arte. cf RCA ARL 1-0702.
1792 Turandot. Joan Sutherland, Montserrat Caballé, s; Luciano Pavarotti,
     Peter Pears, Piero de Palma, Pier Francesco Poli, t; Tom Krause,
     Sabin Markov, bar; Nicolai Ghiaurov, bs; John Alldis Choir;
     Wandsworth School Boys' Choir; LPO; Zubin Mehta. London OSA
     13108 (3). Tape (c) D 31244 (r) 490244. (also Decca SET 561/3

Tape (c) K2A3)

+Gr 9-73 p524	+ON 12-15-73 p44
++Gr 5-75 p2031 tape	+Op 11-73 p1006
+HF 1-74 p76	+RR 9-73 p47
+HF 12-74 p146 tape	+RR 5-75 p76 tape
++HFN 6-75 p109 tape	++SFC 10-7-73 p28
+MJ 1-74 p41	+-SR 12-4-73 p33
+NR 1-74 p11	++St 11-73 p74
+-NYT 10-21-73 pD30	

1793 Turandot: Gira la cote...Perche tarda la luna; Signore, ascolta...
    Non piangere, Liù...Ah, per l'ultima volta, Act 1; Ola, Pang,
    Ola, Pong...Ho una casa; In questa reggia; Gelo che ti da foco...
    Figlio del cielo, Act 2; Nessun dorma; Quel nome...Tanto amore
    segreto...Tu che di gel sei cinta; C'era negli occhi tuoi...
    Diecimila anni. Joan Sutherland, Montserrat Caballé, s; Peter
    Pears, Luciano Pavarotti, Pier Francesco Poli, Piero de Palma,
    t; Tom Krause, bar; Nicolai Ghiaurov, bs; Wandsworth School Boys'
    Choir, John Alldis Choir; LPO; Zubin Mehta. Decca SET 573.
    Tape (c) KCET 573. (Reissue from SET 561/3)
        +Gr 8-75 p360   ++HFN 9-75 p110 tape
        +HFN 8-75 p91   +RR 7-75 p16
    Turandot: Nessun dorma. cf BASF BC 21723.
    Turandot: Nessun dorma. cf Decca SXL 6649.
    Turandot: Nessun dorma; Non piangere Liù. cf Columbia D3M 33448.
    Turandot: Non piangere Liù; Nessun dorma. cf Decca SDD 390.
    Turandot: Signore, ascolta...Non piangere, Liù...Ah, per l'ultima
        volta; Nessun dorma. cf La bohème: Che gelida manina.
    Turandot: Signore, ascolta; Tu chi di gel sei cinta. cf Arias
        (RCA AVM 1-0715).
    Turandot: Signore, ascolta; Tu che di gel sei cinta. cf RCA ARL
        1-0702.
    Le Villi: Non ti scordar di me. cf CATALANI: La wally: Ebben, Ne
        andrò lontana.
    Le Villi: Se come voi piccino io fossi. cf Arias (RCA AVM 1-0715).
PUJOL VILARRUBI, Emilio
    El abejorro. cf Swedish Society SLT 33189.
PURCELL, Daniel
    Sonata, flute, D minor. cf Argo ZRG 746.
PURCELL, Henry
    Abdelazer: Suite. cf Works, selections (Classics for Pleasure
        40208).
    Ayre, G major. cf Columbia M 33514.
    Bonduca. cf Columbia M 33514.
1794 Ceremonial music: Funeral sentences, Z 27: Man that is born of woman.
    March and canzona, 4 trombones, Z 860. Remember not Lord our
    offences, Z 50. Te Deum Laudamus and Jubilate Deo, Z 232, D
    major. Thou knowest Lord, Z 58c. James Bowman, Charles Brett,
    c-t; Ian Partridge, t; Forbes Robinson, bs; St. John's College
    Chapel Choir, Cambridge; ECO, Symphoniae Sacrae Chamber Ensemb-
    le; George Guest. Argo ZRG 724.
        +-Gr 11-72 p946   +-RR 11-72 p100
        +HF 7-73 p74   ++SFC 1-19-75 p27
        +MQ 1-74 p151   +St 9-73 p122
        +NR 7-73 p6
    Chacony, G minor. cf Works, selections (Classics for Pleasure
        40208).

Come ye sons of art: Overture. cf Works, selections (Classics for
Pleasure 40208).
1795 The fairy queen. The Deller Consort, Stour Music Festival Orches-
tra and Chorus; Alfred Deller. Vanguard SRV 311/12 (2).
+NR 8-75 p11   ++St 11-75 p88
Fanfare, C major. cf Columbia M 33514.
The Indian Queen: Trumpet overture. cf Classics for Pleasure 40205.
The Indian Queen: Trumpet overture, D major. cf Works, selections
(Classics for Pleasure 40208).
Musick and ayres: Rondeau, Gavotte, Minuet, Trumpet tune. cf
Argo ZDA 203.
Musick's handmaid, Z 648, 653, 655, 656. cf Angel SB 3816.
Nymphs and shepherds. cf RCA ARL 1-5007.
1796 Ode for St. Cecilia's day. April Cantelo, s; Alfred Deller, Peter
Salmon, c-t; Wilfred Brown, t; Maurice Bevan, bar; John Frost,
bs; Ambrosian Singers; Kalmar Chamber Orchestra; Michael Tippett;
George Eskdale, tpt; Walter Bergmann, hpd. Pye Vanguard Bach
Guild HM 33. (Reissue from NCL 16021)
+Gr 1-75 p1379   +RR 12-74 p73
Prelude. cf Argo ZRG 746.
Rigaudon. cf Columbia M 33514.
Rondeau. cf Philips 6580 066.
Sonata, trumpet, C major. cf CRD CRD 1008.
Sonatas, trumpet, nos. 1 and 2, D major. cf Erato STU 70871.
Sonata, 2 violins and bass, no. 9, F major. cf Works, selections
(Classics for Pleasure 40208).
1797 Songs: Come ye sons of art; My beloved spake; Rejoice in the Lord
alway. Mary Thomas, Honor Sheppard, April Cantelo, s; Alfred
Deller, Mark Deller, c-t; Robert Tear, Max Worthley, Gerald
English, t; Maurice Bevan, bar; Oriana Concerto Orchestra and
Choir, Kalmar Orchestra; Alfred Deller. Vanguard Bach Guild HM
14.
+-RR 9-75 p71
Songs: As Roger last night; He that drinks is immortal; I gave her
cakes and ale; My Lady's coachman John; Sir Walter; Tom the
taylor; Upon Christ Church bells in Oxford; When the cock begins
to crow. cf BASF BAC 3081.
Songs: Let us wander; Sound the trumpet; Two daughters of this
aged stream. cf BIS LP 17.
Songs: Stript of their green. cf HMV RLS 714.
Suite, no. 8, F major. cf Angel SB 3816.
Three parts upon a ground: Fantasia. cf HMV SLS 5022.
Trumpet tune (arr. Trevor). cf Decca 5BBA 1013-5.
Trumpet tunes (4). cf London SPC 21100.
Trumpet tunes and air (2). cf CRD CRD 1008.
Voluntary, C major. cf Columbia M 33514.
Voluntary on the old 100th. cf Klavier KS 536.
1798 Works, selections: Abdelazer: Suite. Chacony, G minor. Come ye
sons of art: Overture. The Indian Queen: Trumpet overture, D
major. Sonata, 2 violins and bass, no. 9, F major. Virtuosi
of England; Arthur Davison. Classics for Pleasure CFP 40208.
/Gr 11-75 p820   +-RR 12-75 p58
+HFN 11-75 p163
PURSWELL, Patrick
It grew and grew. cf CRI SD 324.

PURVIS
    Les petits cloches. cf Stentorian SC 1685.
    QUADRAPHONIC DEMONSTRATION/sampler records. cf Vanguard (Q) VSS
        1, 22.
QUANTS, Johann
    Sonata, flute, D major. cf Pandora PAN 103.
RACHMANINOFF, Sergei
1799 Aleko. Blagovesta Karnobatlova-Dobreva, s; Tony Khristova, con;
        Pavel Kurshumov, t; Nikola Gyuzelev, Dimitre Petkov, bs; Bulgar-
        ian Radio and Television Vocal Ensemble; Plovdiv Symphony Orch-
        estra; Ruslan Raychev. Balkaton BOA 1530 (2). (also Monitor
        HS 90102/3, also Harmonia Mundi HMV 135)
            +Gr 9-74 p578   +SFC 11-9-75 p22
            +-NR 8-75 p11   /St 10-75 p116
            +RR 11-75 p34
    Aleko: Men's dance. cf HMV SEOM 20.
    Aleko: The moon is high. cf Columbia/Melodiya M 33120.
    Barcarolle, op. 10, no. 3, G minor. cf CHOPIN: Piano works
        (Everest SDBR 3377).
1800 Concerti, piano, nos. 1-4. Rhapsody on a theme by Paganini, op.
        43. Agustin Anievas, pno; NPhO; Rafael Frühbeck de Burgos,
        Moshe Atzmon, Aldo Ceccato. HMV SLS 855 (3). (also Angel S
        3801) (Reissue from HMV ASD 2361; Seraphim S 60091)
            +Gr 8-73 p339   +-NYT 11-4-73 pD23
            +-HF 5-74 p93   +RR 8-73 p51
            +LJ 9-75 p37   +St 4-74 p115
            +-NR 2-74 p5
1801 Concerti, piano, nos. 1-4. Rhapsody on a theme by Paganini, op.
        43. Sergei Rachmaninoff, pno; PO; Eugene Ormandy, Leopold
        Stokowski. RCA ARM 3-0296 (3).
            +Gr 7-75 p191   +NYT 11-4-73 pD23
            +-HF 2-74 p110   +RR 7-75 p32
            +HFN 8-75 p82   +SR 12-4-73 p33
            +NR 1-74 p13   +-St 2-74 p124
            +NR 2-74 p12
1802 Concerto, piano, no. 2, op. 18, C minor. Prelude, op. 3, no. 2,
        C sharp minor. Rhapsody on a theme by Paganini, op. 43. Ilana
        Vered, pno; LSO; NPhO; Andrew Davis, Hans Vonk. London SPC
        21099. Tape (c) 521099 (ct) 821099. (also Decca PFS 4327.
        Tape (c) KPRC 14327)
            -Gr 3-75 p1658   +NR 7-75 p6
            +-Gr 5-75 p2032 tape   +RR 3-75 p33
            +HF 7-75 p81   +-St 7-75 p105
            +HFN 5-75 p142 tape
1803 Concerto, piano, no. 2, op. 18, C minor. TCHAIKOVSKY: Concerto,
        piano, no. 1, op. 23, B flat minor. Byron Janis, pno; Minnea-
        polis Symphony Orchestra, LSO; Antal Dorati, Herbert Menges.
        Mercury SRI 75032.
            +St 7-75 p105
1804 Concerto, piano, no. 2, op. 18, C minor. Prelude, op. 3, no. 2,
        C sharp minor. Prelude, op. 23, no. 5, G minor. Prelude, op.
        32, no. 10, B minor. Gina Bachauer, pno; Strasbourg Philharmon-
        ic Orchestra; Alain Lombard. Musical Heritage Society MHS 1924.
            ++St 7-75 p105
1805 Concerto, piano, no. 2, op. 18, C minor. TCHAIKOVSKY: Concerto,
        piano, no. 1, op. 23, B flat minor. Yuri Boukoff, pno; VSO;

Jean Fournet. Philips 6833 156. (Reissue from ABL 3278)
        -Gr 8-75 p326   -RR 6-75 p54
     +-HFN 8-75 p87

1806 Concerto, piano, no. 2, op. 18, C minor. Etudes tableaux, op. 33,
     nos. 3, 4; op. 39, no. 5, E flat minor. Rafael Orozco, pno;
     RPO; Edo de Waart. Philips Tape (c) 7300 201.
        +RR 5-75 p77 tape

1807 Concerto, piano, no. 2, op. 18, C minor. Artur Rubinstein, pno;
     PO; Eugene Ormandy. RCA ARD 1-0031. Tape (c) ARK 1-0031 (ct)
     ARS 1-0031 (Q) ARD 1-0031.
        -Gr 6-73 p56   ++RR 7-73 p56
        +HF 6-73 p94   ++SFC 3-9-75 p26
        -HF 1-74 p84 Quad   ++SFC 3-4-73 p33 tape
     +-HFN 6-73 p1181   +St 6-73 p121
        +MJ 1973 annual p18   +St 7-75 p105
        ++NR 4-73 p6

1808 Concerto, piano, no. 2, op. 18, C minor. Concerto, piano, no. 3,
     op. 30, D minor. Rhapsody on a theme by Paganini, op. 43. Van
     Cliburn, pno; CSO, Symphony of the Air, PO; Fritz Reiner, Kiril
     Kondrashin, Eugene Ormandy. RCA ARL 2-0318 (2). Tape (c) CRK
     2-0318 (ct) CRS 2-0318.
        -LJ 9-75 p37   +-SFC 3-31-74 p29
        +NR 3-74 p10

     Concerto, piano, no. 2, op. 18, C minor. cf GRIEG: Concerto,
     piano, op. 16, A minor.
     Concerto, piano, no. 2, op. 18, C minor. cf HMV SLS 5033.
1809 Concerto, piano, no. 3, op. 30, D minor. Vladimir Ashkenazy, pno;
     LSO; André Previn. London CS 6775.
        +NR 1-75 p5

1810 Concerto, piano, no. 3, op. 30, D minor. Rafael Orozco, pno; RPO;
     Edo de Waart. Philips 6500 540. Tape (r) E45540.
        -Gr 7-74 p214   ++SFC 3-31-74 p29
     +-HF 5-74 p93   ++SFC 6-29-75 p26
        +LJ 5-75 p44   ++SFC 8-24-75 p32 tape
        +NR 2-74 p5   +-St 4-74 p115
        +RR 6-74 p46

     Concerto, piano, no. 3, op. 30, D minor. cf Concerto, piano, no.
     2, op. 18, C minor.
     Concerto, piano, no. 3, op. 30, D minor. cf BEETHOVEN: Concerto,
     piano, no. 5, op. 73, E flat major.
1811 Concerto, piano, no. 4, op. 40, G minor. Rhapsody on a theme by
     Paganini, op. 43. Vladimir Ashkenazy, pno; LSO; André Previn.
     London CS 6776. Tape (c) 056776 (ct) 086776.
        +-NR 6-75 p7
     Daisies, op. 38, no. 3. cf Lilacs, op. 21, no. 5.
     Daisies, op. 38, no. 3. cf RCA ARM 4-0942/7.
     Elegie, op. 3, no. 1. cf CHOPIN: Piano works (Everest SDBR 3377).
1812 Etudes tableaux, op. 33 and op. 39. Jean-Philippe Collard, pno.
     French EMI C 063 11326. (also Connoisseur Society CS 2075)
        +HF 6-75 p94   ++SFC 1-7-73 p23
        ++NR 10-75 p9   ++SFC 3-23-75 p22
        ++NR 8-75 p104
     Etudes tableaux, op. 33, nos. 3, 4. cf Concerto, piano, no. 2,
     op. 18, C minor.
     Etude tableaux, op. 33, no. 4, B minor. cf CBS 76420.
     Etude tableaux, op. 37, no. 2. cf RCA ARM 4-0942/7.

Etude tableaux, op. 39, no. 5, E flat minor. cf Concerto, piano,
   no. 2, op. 18, C minor.
Etude tableaux, op. 39, no. 5, E flat minor. cf Piano works
   (RCA ARL 1-0352).
The isle of the dead, op. 29. cf BEETHOVEN: Sonata, violin and
   piano, no. 8, op. 30, no. 3, G major.
1813 Lilacs, op. 21, no. 5. Daisies, op. 38, no. 3. SCRIABIN: Etudes,
   op. 8. Morton Estrin, pno. Connoisseur Society CS 2009.
      +Gr 2-75 p1519   +RR 3-75 p52
Melodie, op. 3, no. 3, E major. cf CHOPIN: Piano works (Everest
   SDBR 3377).
O cease they singing, maiden fair, op. 4, no. 4. cf Columbia M
   32231.
Oriental sketch. cf RCA ARM 4-0942/7.
1814 Piano works: Etude tableaux, op. 39, no. 5, E flat minor. Prelude,
   op. 23, no. 4, D major. Prelude, op. 23, no. 5, G minor. Prelude, op. 26, no. 6, E flat major. Prelude, op. 23, no. 7,
   C minor. Prelude, op. 32, no. 5, G major. Sonata, piano, no.
   2, op. 36, B flat minor. Van Cliburn, pno. RCA ARL 1-0352.
   Tape (c) ARK 1-0352 (ct) ARS 1-0352.
      ++HF 4-74 p103   ++SFC 3-17-74 p26
      -LJ 2-75 p36   ++St 5-74 p74
      ++NR 3-74 p10
1815 Pieces, op. 3 (5). Preludes, op. 23. Ruth Laredo, pno. Columbia
   M 32938.
      ++HF 3-75 p86   ++SFC 10-6-74 p26
      ++NR 11-74 p10   ++St 10-74 p135
Polichinelle, op. 3, no. 4, F sharp minor. cf CHOPIN: Piano works
   (Everest SDBR 3377).
Polka de W. R. cf CHOPIN: Piano works (Everest SDBR 3377).
Preludes. cf CHOPIN: Scherzo, no. 4, op. 54, E major.
1816 Preludes (24). Constance Keene, pno. Laurel Protone LP 11 (2).
      ++NR 6-75 p13
Prelude, op. 3, no. 2, C sharp minor. cf Concerto, piano, no. 2,
   op. 18, C minor (London SPC 21099).
Prelude, op. 3, no. 2, C sharp minor. cf Concerto, piano, no. 2,
   op. 18, C minor (Musical Heritage Society MHS 1924).
Prelude, op. 3, no. 2, C sharp minor. cf CHOPIN: Piano works
   (Everest SDBR 3377).
Preludes, op. 23. cf Pieces, op. 3.
1817 Preludes, op. 23, nos. 1-2, 4-5, 7-8. Preludes, op. 32, nos. 1-2,
   6-7, 9-10, 12. Sviatoslav Richter, pno. Melodiya/Angel SR
   40235.
      ++Gr 9-74 p553   ++NR 1-74 p13
      ++HF 3-74 p92   +St 6-74 p123
      ++LJ 10-75 p48
Prelude, op. 23, no. 4, D major. cf Piano works (RCA ARL 1-0352).
Prelude, op. 23, no. 5, G minor. cf Concerto, piano, no. 2, op.
   18, C minor.
Prelude, op. 23, no. 5, G minor. cf Piano works (RCA ARL 1-0352).
Prelude, op. 23, no. 5, G minor. cf CHOPIN: Piano works (Everest
   SDBR 3377).
Prelude, op. 23, no. 6, E flat major. cf Piano works (RCA ARL
   1-0352).
Prelude, op. 23, no. 7, C minor. cf Piano works (RCA ARL 1-0352).
1818 Preludes, op. 32 (13). Prelude, op. posth., D minor. Ruth Laredo,

pno. Columbia M 33430.
    +-HF 10-75 p81   ++NR 8-75 p12
Preludes, op. 32, nos. 1-2, 6-7, 9-10, 12. cf Preludes, op. 23,
    no. 1-2, 4-5, 7-8.
Prelude, op. 32, no. 5, G major. cf Piano works (RCA ARL 1-0352).
Prelude, op. 32, no. 10, B minor. cf Concerto, piano, no. 2, op.
    18, C minor.
Prelude, op. 32, no. 12, G sharp minor. cf Columbia Special Pro-
    ducts AP 12411.
Prelude, op. posth., D minor. cf Preludes, op. 32.
1819 Prince Rostislav. The rock, fantasia for orchestra, op. 7. Vocal-
    ise, op. 34, no. 14. USSR Symphony Orchestra; Yevgeny Svetlanov.
    Melodiya/Angel SR 40252.
        +HF 8-75 p95   *   +St 10-75 p114
        +NR 8-75 p2
Rhapsody on a theme by Paganini, op. 43. cf Concerti, piano,
    nos. 1-4 (HMV SLS 855).
Rhapsody on a theme by Paganini, op. 43. cf Concerti, piano, nos.
    1-4 (RCA ARM 3-0296).
Rhapsody on a theme by Paganini, op. 43. cf Concerto, piano, no.
    2, op. 18, C minor (London SPC 21099).
Rhapsody on a theme by Paganini, op. 43. cf Concerto, piano, no.
    2, op. 18, C minor (RCA ARL 2-0318).
Rhapsody on a theme by Paganini, op. 43. cf Concerto, piano, no.
    4, op. 40, G minor.
1820 The rock, fantasia for orchestra, op. 7. Symphony, no. 3, op. 44,
    A minor. LPO; Walter Weller. Decca SXL 6720.
        +Gr 10-75 p625   -RR 10-75 p48
        +HFN 10-75 p148
The rock, fantasia for orchestra, op. 7. cf Prince Rostislav.
1821 Sonata, piano, no. 2, op. 36, B flat minor (original version).
    Variations on a theme by Corelli, op. 42. Jean-Philippe Collard,
    pno. Connoisseur Society CS 2082.
        ++HF 12-75 p102   ++SFC 9-28-75 p30
        ++NR 11-75 p13
Sonata, piano, no. 2, op. 36, B flat minor. cf Piano works (RCA
    ARL 1-0352).
Sonata, violoncello, op. 19, G minor: 2d movement. cf HMV SEOM 19.
1822 Songs: A-oo, op. 38, no. 6; Daisies, op. 38, no. 3; Dissonance,
    op. 34, no. 13; Dreams, op. 38, no. 5; The harvest of sorrow,
    op. 4, no. 5; How fair this spot, op. 21, no. 7; In my garden
    at night, op. 38, no. 1; The morn of life, op. 34, no. 10; The
    muse, op. 34, no. 1; Oh, never sing to me again, op. 4, no. 4;
    The pied piper, op. 38, no. 4; The poet, op. 34, no. 9; The
    storm, op. 34, no. 3; To her, op. 38, no. 2; Vocalise, op. 34,
    no. 14; What wealth of rapture, op. 34, no. 12. Elisabeth Söder-
    strom, s; Valdimir Ashkenazy, pno. Decca SXL 6718.
        ++Gr 7-75 p228   ++RR 7-75 p59
        ++HFN 7-75 p87
1823 Songs: All love you so, op. 14, no. 6; A-oo, op. 38, no. 6; At
    night in my garden, op. 38, no. 1; Child, like a flower, op. 8,
    no. 2; Daisies, op. 38, no. 3; Dream, op. 8, no. 5 (2); Floods
    of spring, op. 14, no. 11; He took it all away from me, op. 26,
    no. 2; Love has lost its joy, op. 14, no. 3; Night is sad, op.
    26, no. 12; The rat catcher, op. 38, no. 4; Time, op. 14, no.
    12; To the children, op. 26, no. 7; To her, op. 38, no. 2; Water

lily, op. 8, no. 1. Peter del Grande, bar; Vladimir Pleshakov,
  pno. Orion ORS 75180.
        +NR 7-75 p12   -NYT 6-8-75 pD19
  Songs: When night descends. cf Ember GVC 46.
  Songs: In the silent night, op. 43, no. 3. cf RCA SER 5704/6.
1824 Suite, 2 pianos, no. 1, op. 5. Suite, 2 pianos, no. 2, op. 17.
  Vladimir Ashkenazy, André Previn, pno. Decca SXL 6697. (also
  London CS 6893)
        +Gr 6-75 p62   +NR 11-75 p13
        +HFN 8-75 p82   ++RR 6-75 p72
        +MT 10-75 p887
1825 Suite, 2 pianos, no. 1, op. 5. Suite, 2 pianos, no. 2, op. 17.
  John Ogdon, Brenda Lucas, pno. HMV HQS 1340.
        +Gr 2-75 p1510   +-RR 3-75 p51
  Suite, 2 pianos, no. 2, op. 17. cf Suite, 2 pianos, no. 1, op. 5
  (Decca 6697)
  Suite, 2 pianos, no. 2, op. 17. cf Suite, 2 pianos, no. 1, op. 5
  (HMV 1340).
  Symphonic dances, op. 45. cf DEBUSSY: Nocturnes.
1826 Symphony, no. 1, op. 13, D minor. LSO; André Previn. HMV (Q) ASD
  3137. (also Angel (Q) S 37120)
        +Gr 11-75 p827   +NR 11-75 p3
        +HF 12-75 p103   -RR 11-75 p53
        ++HFN 11-75 p163
1827 Symphony, no. 2, op. 27, E minor. PO; Eugene Ormandy. CBS 73042.
  (Reissue from 77345) (also Columbia MS 6110)
        +Gr 6-75 p49   -RR 5-75 p38
        -HFN 8-75 p82
1828 Symphony, no. 2, op. 27, E minor. Bolshoi Theatre Orchestra;
  Yevgeny Svetlanov. Columbia/Melodiya M 33121.
        +HF 7-75 p82   +-NR 2-75 p4
1829 Symphony, no. 2, op. 27, E minor. PO; Eugene Ormandy. RCA ARL
  1-1150. Tape (c) ARK 1-1150 (ct) ARS 1-1150.
        -HF 12-75 p103   ++NR 11-75 p3
  Symphony, no. 3, op. 44, A minor. cf The rock, fantasia for
  orchestra, op. 7.
  Symphony, no. 3, op. 44, A minor. cf BEETHOVEN: Sonata, violin
  and piano, no. 3, op. 30, G major.
1830 Trio, piano, no. 2, op. 9, D major. Leonid Kogan, vln; Fedor
  Luzanov, vlc; Yevgeny Svetlanov, pno. HMV Melodiya ASD 3061.
        +Gr 5-75 p1986   +RR 5-75 p50
        ++HFN 5-75 p137
  Variations on a theme by Chopin, op. 22. cf KABALEVSKY: Sonata,
  piano, no. 3, op. 46.
  Variations on a theme by Corelli, op. 42. cf Sonata, piano, no.
  2, op. 36, B flat minor.
  Vocalise, op. 34, no. 14. cf Prince Rostislav.
  Vocalise, op. 34, no. 14. cf BEETHOVEN: Sonata, violin and piano,
  no. 8, op. 30, no. 3, G major.
  Vocalise, op. 34, no. 14. cf CANTELOUBE: Songs of the Auvergne.
  Vocalise, op. 34, no. 14. cf Command COMS 9006.
  Vocalise, op. 34, no. 14. cf HMV SXLP 30193.
RADECK
  Fridericus-Rex-Grenadiermarsch. cf DG 2721 077.
RAFF, Joachim
  Cavatina, D major. cf Discopaedia MB 1005.

1831 Symphony, no. 5, E major. LPO; Bernard Herrmann. Nonesuch H 71287.
        +HF 5-74 p95   +-SFC 1-13-74 p20
        +LJ 3-75 p34   +St 3-74 p117
        +NR 3-74 p2
    RAGAS, MUSIC FOR MEDITATION. cf Connoisseur Society CS 2063.
    RAGS AND TANGOS. cf Golden Crest CRS 4132.
RAMEAU, Jean
    Castor et Pollux: Gavotte. cf Discopaedia MB 1003.
    Les cyclopes, D minor. cf Pièces de clavecin.
    La Dauphine, G minor. cf Pièces de clavecin.
    L'Enharmonique, G minor. cf Pièces de clavecin.
    L'Enharmonique, G minor. cf Harmonia Mundi HMU 334.
    L'Entretien des Muses, D minor. cf Pièces de clavecin.
1832 Les fetes d'Hebe: Ballet. music. Ursula Connors, s; Ambrosian
    Singers; ECO; Raymond Leppard. Angel S 37105.
        ++NR 8-75 p4   +St 12-75 p129
        +-ON 10-75 p56
1833 Les Indes galantes. Anne-Marie Rodde, Rachel Yakar, Sonia Nigog-
    hossian, Janine Micheau, s; Bruce Brewer, Jean-Marie Gouelou,
    t; Jean-Christophe Benoit, bar; Christian Treguier, Pierre-Ives
    le Maigat, bs; Raphael Passaquet Vocal Ensemble; Le Grande
    Ecurie et Le Chambre du Roy; Jean-Claude Malgoire. CBS 77365
    (3). (also Columbia M3 32973)
        +Gr 7-74 p262   +-ON 12-14-74 p46
        +-HF 5-75 p88   +-Op 11-74 p988
        +-MQ 7-75 p499   +RR 6-74 p33
        -NR 1-75 p11
1834 Les Indes galantes. Gerda Hartman, Jennifer Smith, s; Louis Devos,
    John Elwes, t; Philippe Huttenlocher, bs; Jean-François Paillard
    Orchestra; Jean-François Paillard. Erato STU 70850/3 (4).
    (also Musical Heritage Society MHS 3114/7)
        +Gr 2-75 p1547   +RR 2-75 p28
        /ON 10-75 p56
1835 Les Indes galantes: Prologue; First entrée: The generous Turk,
    Storm and air of Emilie; Second entrée: The Incas of Peru, Scene
    between the evil Huascar and the virtuous Phani; Dance of the
    Incas; Earthquake and volcano eruption; Third entrée: The flow-
    ers of the Persian festival; Italian aria, Scene between the
    adoring Tacmas, the lovely Zaire, and a chorus of flowers; Fourth
    entrée: The savages of Illinois, Air of adagio; Dances and
    ballet of French warriors and savages, of French women and Ama-
    zons and of shepherds and shepherdesses. Le Grande Ecurie et
    Le Chambre du Roy; Jean-Claude Malgoire.
        +-Gr 12-75 p1093   +-RR 10-75 p30
        +HFN 10-75 p153
    Menuets (2). cf Pièces de clavecin.
    Menuets. cf Angel S 37015.
1836 Les paladins, excerpts. Anne-Marie Rodde, s; Henri Farge, c-t;
    Jean-Christophe Benoit, bar; Le Grande Ecurie et Le Chambre du
    Roy; Jean-Claude Malgoire. Vanguard Everyman SRV 318 SD.
        -NR 4-75 p9   +St 7-75 p105
        +SFC 3-30-75 p16
1827 Pièces de clavecin: Les cyclopes, D minor. La Dauphine, G minor.
    L'Enharmonique, G minor. L'Entretien des Muses, D minor. Men-
    uets (2). La Poule, G minor. Suite, E minor. Albert Fuller,
    hpd. Nonesuch H 71278.

++HF 8-73 p86   ++NR 9-73 p13
+LJ 9-74 p30   +-RR 3-75 p51
-MJ 9-73 p44

Le Poule, G minor. cf Pièces de clavecin.
Renaissance suite: Sarabande. cf International Piano Library IPL
    5005/6.
Le Reppel des oiseaux. cf Westminster WGM 8309.
Suite, E minor. cf Pièces de clavecin.
Suite, E minor. cf Harmonia Mundi HMU 334.
1838 Suites, harpsichord, A minor, E minor. Trevor Pinnock, hpd. CRD
    CRD 1010.
        ++Gr 2-75 p1519   ++RR 2-75 p56
Tambourin. cf L'Oiseau-Lyre DSLO 7.
Tambourin chinois (arr. Kreisler). cf Decca SLA 405.
RANDALL, J. K.
    Improvisation. cf ERICKSON: End of the mime.
RANDALL, John   .
    Music for the film "Eakins". cf CEELY: Elegia.
RANGSTROM, Ture
    Songs: Flickan under Nymanen; Pan; Semele; Sköldmön; Villemo. cf
        GRIEG: Af Haugtussa, op. 67.
    Songs: Vingar i natten (Wings in the night); Vinden och trädet
        (The wind and the tree); Sköldmön (Amazon). cf LARSSON: Förk-
        lädd Gud (A God in disguise), op. 24.
    Tristan's död. cf RCA SER 5704/6.
RANKI, Ryorgy
    Cantus Urbis. cf DAVID: Concerto, horn.
RASELIUS, Andreas
    Nun komm der heiden Heiland. cf Turnabout TV 34569.
RATHAUS, Karol
    Tower music. cf Crystal S 201.
RATHGEBER, Johann Valentin
    Aria pastorella, C major. cf Hungaroton SLPX 11601/2.
    Aria pastorella, G major. cf Hungaroton SLPX 11601/2.
    Concerto, trumpet, op. 6, no. 15, E flat major. cf Nonesuch H 71301.
    Schlag-arie, E flat major, F major, G major, D minor. cf Hungaro-
        ton SLPX 11601/2.
1839 Songs and quodlibets, from the Augsburg Tafel-Confect, 1733-46.
        Von der Edlen Music; Von der Music und Jägerey; Von allerhand
        Nasen; Von der Weibsbildern wird aus der Schrifft probirt; Von
        der Bedierd zum Geld; Von einem Politico; Von der Gedult.
        SEYFERT: Die frohe Compagnie; Die Beschwerlichkeiter des Ehes-
        tandes; Die lustige Tyrolerin; Wir haben drey Katzen; Der
        verachtete Liebhaber. Rita Streich, s; Udo Wildenblanck, boy
        soprano; Willi Brokmeier, t; Wolfgang Annheisser, bar; Gottlob
        Frick, bs; Munich Instrumental Ensemble; Fritz Neumeyer. Angel
        D 37107.
        +HF 9-75 p101   +NR 10-75 p8
RAUTAVAARA, Rinojuhani
1840 Pelimannit, op. 1. SEGERSTAM: Divertimento. SIBELIUS: Canzonetta,
        op. 62, no. 1. Rakastava, op. 14. Suite mignonne, op. 98.
        Helsinki Chamber Orchestra; Leif Segerstam. BIS LP 9.
        +Gr 11-75 p827   ++HFN 11-75 p163
RAVEL, Maurice
    A la manière de Borodine. cf Piano works (CBS 77380).
    A la manière de Borodine. cf Piano works (Decca SXL 6700).

A la manière de Borodine. cf Piano works (London 6985).
A la manière de Borodine. cf Piano works (Vox SVBX 5473).
A la manière de Borodine. cf Works, selections (CRD 1821-6).
A la manière de Chabrier. cf Piano works (CBS 77380).
A la manière de Chabrier. cf Piano works (Decca SXL 6700).
A la manière de Chabrier. cf Piano works (London 6985).
A la manière de Chabrier. cf Piano works (Vox SVBX 5473).
A la manière de Chabrier. cf Works, selections (CRD 1821-6).
Berceuse sur le nom de Gábriel Fauré. cf Works, selections (CRD
      1821-6).
1841 Boléro. Rapsodie espagnole. Shéhérazade: Ouverture de Feerie.
      La valse. Orchestra de Paris; Jean Martinon. Angel S 37147.
        +-NR 12-75 p4   ++SFC 11-9-75 p22
1842 Boléro. Daphnis and Chloe: Suite, no. 2. Ma mère l'oye. La valse.
      LAPO: Zubin Mehta. Decca SXL 6488. Tape (c) KXSC 16488.
        ++HFN 7-75 p91 tape
1843 Boléro. Rapsodie espagnole. La valse. BSO; Seiji Ozawa. DG
      2530 475.
        +-Audio 8-75 p80   +RR 3-75 p33
        -Gr 4-75 p1812   ++SFC 4-27-75 p23
        /HF 6-75 p96   +St 9-75 p111
        ++NR 6-75 p4
1844 Boléro. Ma mère l'oye. Pavane pour une infante défunte. Rapsodie
      espagnole. Budapest Philharmonic Orchestra; András Kórody.
      Hungaroton SLPX 11644.
        +-NR 4-75 p3   +-St 9-75 p111
1845 Boléro. Miroirs: Alborada del gracioso. Pavane pour une infante
      défunte. Rapsodie espagnole. La valse. Detroit Symphony Orch-
      estra; Paul Paray. Mercury 75033.
        +HF 3-75 p64
1846 Boléro. Ma mère l'oye, ballet. La valse. LSO; Pierre Monteux.
      Philips 6580 106. (Reissue from SAL 3500) (also Philips 835
      258)
        +Gr 6-75 p49   +RR 5-75 p38
        +HFN 8-75 p82
1847 Boléro. Pavane pour une infante défunte. La valse. BSO; Charles
      Munch. RCA LSC 2664.
        +HF 3-75 p65
    Boléro. cf Works, selections (Decca SPA 392).
    Boléro. cf Works, selections (HMV 5016).
    Boléro. cf Works, selections (DG 2740 120).
    Boléro. cf Works, selections (Vox SVBX 5133).
    Boléro. cf MUSSORGSKY: Pictures at an exhibition (Philips 6500 882).
    Boléro. cf MUSSORGSKY: Pictures at an exhition (RCA ARL 1-0451).
    Boléro. cf HMV ASD 3008.
    Boléro. cf Toscanini Society ATS GC 1201/6.
1848 Chansons madécasses. Introduction and allegro, harp, flute, clari-
      net and string quartet. Quartet, strings. Marie-Clair Jamet
      Ensemble, Debussy Quartet; Jacques Herbillon, bar. Calliope
      CAL 1823.
        +HFN 11-75 p163
    Chansons madécasses: Nahandove; Méfiez-vous de blancs; Il est doux
      de se coucher. cf Works, selections (CRD 1821-6).
1849 Concerto, piano, G major. Concerto, piano, for the left hand, D
      major. Julius Katchen, pno; LSO; István Kertész. Decca SDD
      486. (Reissue)

+HFN 12-75 p171   /RR 12-75 p58
1850 Concerto, piano, G major. Concerto, piano, for the left hand, D
     major. Werner Haas, pno; Monte Carlo Opera Orchestra; Alceo
     Galliera. Philips 839 755.
     +HF 3-75 p66
     Concerto, piano, G major. cf Works, selections (HMV 5016).
     Concerto, piano, G major. cf FAURE: Fantaisie, piano and orchestra,
        op. 111.
     Concerto, piano, G major. cf GRIEG: Concerto, piano, op. 16, A
        minor.
     Concerto, piano, G major. cf PROKOFIEV: Concerto, piano, no. 3,
        op. 26, C major.
     Concerto, piano, G major. cf PROKOFIEV: Concerto, piano, no. 5,
        op. 55, G major.
     Concerto, piano, G major: Presto. cf Works, selections (Decca
        SPA 392).
1851 Concerto, piano, for the left hand, D major. GERSHWIN: Rhapsody
     in blue. PROKOFIEV: Concerto, piano, no. 3, op. 26, C major.
     Julius Katchen, pno; LSO; István Kertész. London 6633.
     +HF 3-75 p66
     Concerto, piano, for the left hand, D major. cf Concerto, piano,
        G major (Decca SDD 486).
     Concerto, piano, for the left hand, D major. cf Concerto, piano,
        G major (Philips 839755).
     Concerto, piano, for the left hand, D major. cf Works, selections
        (HMV 5016).
     Concerto, piano, for the left hand, D major. cf FAURE: Fantaisie,
        piano and orchestra, op. 111.
     Concerto, piano, for the left hand, D major. cf GERSHWIN: Rhapsody
        in blue.
1852 Daphnis et Chloé. NPhO; Rafael Frühbeck de Burgos. Angel S 36471.
     +HF 3-75 p65
1853 Daphnis et Chloé. Orchestra; Jean Martinon. Angel S 37148.
     ++SFC 12-7-75 p31
1854 Daphnis et Chloé. Camerata Singers; NYP; Pierre Boulez. Columbia
     M 33523. (Q) MQ 33523. (also CBS 76425. Tape (c) 40-76425)
          +-Gr 12-75 p1045   +-MJ 11-75 p41
          +HF 12-75 p104   ++RR 12-75 p58
          +HFN 12-75 p163   ++St 11-75 p132
1855 Daphnis et Chloé. CO and Chorus; Lorin Maazel. Decca SXL 6703.
     (also London CS 6898)
          +Gr 4-75 p1812   +RR 3-75 p33
          +-HF 9-75 p88   ++SFC 7-20-75 p27
          +HFN 8-75 p82   ++St 11-75 p132
          +NR 8-75 p5
1856 Daphnis et Chloé. Tanglewood Festival Chorus; BSO; Seiji Ozawa.
     DG 2530 563.
          +-HF 12-75 p104   ++SFC 10-12-75 p22
          +NR 12-75 p4
     Daphnis et Chloé. cf Works, selections (DG 2740 120).
     Daphnis et Chloé. cf Works, selections (HMV 5016).
     Daphnis et Chloé: Lever du jour. cf Works, selections (Decca SPA
        392).
1857 Daphnis et Chloé: Suites, nos. 1 and 2. Mother Goose ballet. COA:
     Bernard Haitink. Philips 6500 311. Tape (c) 7300 166.
          ++Gr 10-72 p707   +HFN 8-75 p110 tape

```
        +-HF 12-71 p118                    +NR 11-72 p3
        +HF 3-75 p66                      ++RR 10-72 p72
        ++HFN 10-72 p1917                 ++SFC 11-19-72 p31
```
Daphnis et Chloé: Suites, nos. 1 and 2. cf Works, selections
    (Vox SVBX 5133).
Daphnis et Chloé: Suite, no. 1. cf Decca DPA 519/20.
1858 Daphnis et Chloé: Suite, no. 2. Pavane pour une infante défunte.
    DEBUSSY: Nocturnes. BSO; New England Conservatory Chorus;
    Claudio Abbado. DG 2530 038. Tape (c) 3300 016.
        ++HF 3-75 p65
Daphnis and Chloé: Suite, no. 2. cf Boléro.
Daphnis et Chloé: Suite, no. 2. cf DEBUSSY: La mer (Pye 15013).
Daphnis and Chloé: Suite, no. 2. cf DEBUSSY: La mer (CBS 61075).
Daphnis and Chloé: Suite, no. 2. cf DEBUSSY: La mer (RCA (Q) ARD
    1-0029).
Fanfare for "L'Eventail de Jeanne." cf Works, selections (Vox
    SVBX 5133).
1859 Gaspard de la nuit. Sonatine. Valses nobles et sentimentales.
    Martha Argerich, pno. DG 2530 540.
        +-Gr 8-75 p348   +HFN 9-75 p103
        ++HF 12-75 p104   +RR 8-75 p59
1860 Gaspard de la nuit. Sonatine. Le tombeau de Couperin. John
    Browning, pno. RCA LSB 4096. (Reissue) (also RCA LSC 3028.
    Tape (r) ERPA 3028)
        ++HF 9-74 p126 tape   +St 5-75 p92
        +RR 9-73 p96
Gaspard de la nuit. cf Miroirs: Alborado del gracioso.
Gaspard de la nuit. cf Piano works (CBS 77380).
Gaspard de la nuit. cf Piano works (Decca SXL 6700).
Gaspard de la nuit. cf Piano works (London 6985).
Gaspard de la nuit. cf Piano works (Philips 6580 046).
Gaspard de la nuit. cf Piano works (Telefunken 6 DX 48068).
Gaspard de la nuit. cf Piano works (Vox SVBX 5473).
Gaspard de la nuit. cf Works, selections (CRD 1821-6).
Gaspard de la nuit. cf DEBUSSY: Images, Bks 1 and 2.
Gaspard de la nuit. cf DEBUSSY: Images, Bk 1.
Gaspard de la nuit, no. 2: Le gibet. cf Piano works (Ember GVC 39).
Introduction and allegro, harp, flute, clarinet and string quartet.
    cf Chansons madécasses.
Introduction and allegro, harp, flute, clarinet and string quartet.
    cf Works, selections (CRD 1821-6).
Introduction and allegro, harp, flute, clarinet and string quartet.
    cf Works, selections (Decca SPA 392).
Introduction and allegro, harp, flute, clarinet and string quartet.
    cf DEBUSSY: Works, selections (DG 2533 007).
1861 Jeux d'eau. Ma mère l'oye. Miroirs. Pascal Rogé, Denise Françoise
    Rogé, pno. Decca SXL 6715.
        +Gr 11-75 p864   +RR 11-75 p82
        +HFN 11-75 p163
1862 Jeux d'eau. Miroirs. Pavane pour une infante défunte. Sonatine.
    Daniel Adni, pno. HMV HQS 1336.
        -Gr 3-75 p1681   +RR 3-75 p52
Jeux d'eau. cf Piano works (CBS 77380).
Jeux d'eau. cf Piano works (Philips 6580 046).
Jeux d'eau. cf Piano works (Telefunken 6 DX 48068).
Jeux d'eau. cf Piano works (Vox SVBX 5473).
```
```

Jeux d'eau. cf Works, selections (CRD 1821-6).
Jeux d'eau. cf CHOPIN: Scherzo, no. 4, op. 54, E major.
Jeux d'eau. cf Monitor MCS 2143.
1863 Ma mère l'oye. Menuet antique. La valse. NYP; Pierre Bouelez.
     Columbia M 32838. (Q) MQ 32838 Tape (ct) MAQ 32838. (also CBS
     76306)
                +Gr 5-75 p1968   +MT 10-75 p888
                +HF 6-75 p96   +NR 6-75 p1
                +HFN 5-75 p137   +-RR 4-75 p30
                -MJ 12-75 p38   ++St 6-75 p104
Ma mère l'oye. cf Boléro (Decca 6488).
Ma mère l'oye. cf Boléro (Hungaroton 11644).
Ma mère l'oye, ballet. cf Bolero (Philips 5680 106).
Mother Goose ballet. cf Daphnis et Chloé: Suites, nos. 1 and 2.
Ma mère l'oye. cf Jeux d'eau.
Ma mère l'oye. cf Piano works (CBS 77380).
Ma mère l'oye. cf Works, selections (CRD 1821-6).
Ma mère l'oye. cf Works, selections (DG 2740 120).
Ma mère l'oye. cf Works, selections (HMV 5016).
Ma mère l'oye. cf Works, selections (Vox SVBX 5133).
Ma mère l'oye. cf BIZET: Jeux d'enfants, op. 22.
Ma mère l'oye. cf BRAHMS: Variations on a theme by Haydn, op. 56a.
Ma mère l'oye, excerpts. cf BRITTEN: Young person's guide to the
     orchestra, op. 34.
1864 Menuet antique. Miroirs: Une barque sur l'océan. Le tombeau de
     Couperin. Valses nobles et sentimentales. OSCCP; André Cluy-
     tens. Classics for Pleasure CFP 40093. (Reissue from Columbia
     SAX 2479, 2478)
                +Gr 12-75 p1045   +RR 11-75 p54
Menuet antique. cf Ma mère l'oye.
Menuet antique. cf Miroirs: Alborada del gracioso; Une barque sur
     l'océan.
Menuet antique. cf Piano works (CBS 77380).
Menuet antique. cf Piano works (Decca SXL 6700).
Menuet antique. cf Piano works (London 6985).
Menuet antique. cf Piano works (Vox SVBX 5473).
Menuet antique. cf Works, selections (CRD 1821-6).
Menuet antique. cf Works, selections (DG 2740 120).
Menuet antique. cf Works, selections (HMV 5016).
Menuet antique. cf Works, selections (Vox SVBX 5133).
Menuet sur le nom de Haydn. cf Piano works (CBS 77380).
Menuet sur le nom de Haydn. cf Piano works (Decca SXL 6700).
Menuet sur le nom de Haydn. cf Piano works (London 6985).
Menuet sur le nom de Haydn. cf Piano works (Vox SVBX 5473).
Menuet sur le nom de Haydn. cf Works, selections (CRD 1821-6).
Miroirs. cf Jeux d'eau (Decca 6715).
Miroirs. cf Jeux d'eau (HMV 1336).
Miroirs. cf Piano works (CBS 77380).
Miroirs. cf Piano works (Telefunken 6 DX 48068).
Miroirs. cf Piano works (Vox SVBX 5473).
Miroirs. cf Works, selections (CRD 1821-6).
1865 Miroirs: Alborada del gracioso; Une barque sur l'océan. Menuet
     antique. Pavane pour une infante défunte. Le tombeau de Coup-
     erin. OSCCP; André Cluytens. Angel S 36111.
                +HF 3-75 p65
1866 Miroirs: Alborado del gracioso. Gaspard de la nuit. Valses nobles

et sentimentales. Alicia de Larrocha, pno. Columbia M 30115.

/ARG 3-71 p421	+NR 2-71 p13
/Gr 5-71 p1796	+SR 4-24-71 p54
/HF 5-71 p81	++St 5-71 p87
+MJ 2-71 p78	+St 5-75 p92

Miroirs: Alborada del gracioso. cf Boléro.
Miroirs: Alborada del gracioso. cf Columbia M 32070.
Miroirs: Alborada del gracioso: Une barque sur l'océan. cf Works, selections (DG 2740 120).
Miroirs: Alborada del gracioso; Une barque sur l'océan. cf Works, selections (HMV 5016).
Miroirs: Alborada del gracioso; Une barque sur l'océan. cf Works, selections (Vox SVBX 5133).
Miroirs: Une barque sur l'océan. cf Menuet antique.
Miroirs: Oiseau tristes; La vallée des cloches. cf Piano works (Ember GVC 39).
Miroirs: La vallée des cloches. cf CHOPIN: Scherzo, no. 4, op. 54, E major.
Pavane pour une infante défunte. cf Boléro (Hungaroton 11644).
Pavane pour une infante défunte. cf Boléro (Mercury 75033).
Pavane pour une infante défunte. cf Boléro (RCA 2664).
Pavane pour une infante défunte. cf Daphnis et Chloé: Suite, no. 2.
Pavane pour une infante défunte. cf Jeux d'eau.
Pavane pour une infante défunte. cf Miroirs: Alborada del gracioso; Une barque sur l'océan.
Pavane pour une infante défunte. cf Piano works (CBS 77380).
Pavane pour une infante défunte. cf Piano works (Ember GVC 39).
Pavane pour une infante défunte. cf Piano works (Decca SXL 6700).
Pavane pour une infante défunte. cf Piano works (London 6985).
Pavane pour une infante défunte. cf Piano works (Telefunken 6 DX 48068).
Pavane pour une infante défunte. cf Piano works (Vox SVBX 5473).
Pavane pour une infante défunte. cf Works, selections (CRD 1821-6).
Pavane pour une infante défunte. cf Works, selections (Decca SPA 392).
Pavane pour une infante défunte. cf Works, selections (DG 2740 120).
Pavane pour une infante défunte. cf Works, selections (HMV 5016).
Pavane pour une infante défunte. cf Works, selections (Vox SVBX 5133).
Pavane pour une infante défunte. cf DEBUSSY: Danse sacrée et profane.
Pavane pour une infante défunte. cf DEBUSSY: La mer.
Pavane pour une infante défunte. cf Angel S 36076.
Pavane pour une infante défunte. cf CBS 73396.
Pavane for a dead princess. cf Columbia M 32070.
1867 Piano works: A la manière de Borodine. A la manière de Chabrier. Gaspard de la nuit. Rapsodie espagnole: Habañera. Jeux d'eau. Ma mère l'oye. Menuet antique. Menuet sur le nom de Haydn. Miroirs. Pavane pour une infante défunte. Prelude. Sonatine. Le tombeau de Couperin. Valses nobles et sentimentales. Philippe Entremont, pno. CBS 77380 (3). (also Columbia D3M 33311)

+-Gr 12-74 p1181	+-RR 12-74 p63
+-HF 4-75 p90	+-SFC 6-8-75 p23
-MT 10-75 p888	+-St 5-75 p92
+NR 7-75 p15	

1868 Piano works: A la manière de Borodine. A la manière de Chabrier.
     Gaspard de la nuit. Menuet antique. Menuet sur le nom de Haydn.
     Pavane pour une infante défunte. Prélude. Pascal Rogé, pno.
     Decca SXL 6700. (also London CS 6873)
              +Audio 8-75 p80   ++RR 3-75 p51
              +Gr 3-75 p1682
1869 Piano works: Gaspard de la nuit, no. 2: Le gibet. Miroirs: Oiseau
     tristes; La vallée des cloches. Pavane pour une infante dé-
     funte. Le tombeau de Couperin: Toccata. Maurice Ravel, pno.
     Ember GVC 39.
              +-Gr 6-75 p69   +-RR 6-75 p72
              +-HFN 8-75 p82
1870 Piano works: A la manière de Borodine. A la manière de Chabrier.
     Gaspard de la nuit. Menuet antique. Menuet sur le nom de
     Haydn. Pavane pour une infante défunte. Prélude. Pascal Rogé,
     pno. London CS 6985.
              +-HF 12-75 p104   +-St 11-75 p130
              +NR 11-75 p13
1871 Piano works: Gaspard de la nuit. Jeux d'eau. Prélude. Sonatine.
     Le tombeau de Couperin. Werner Haas, pno. Philips 6580 046.
              -Gr 5-75 p1996   +RR 6-75 p75
              +HFN 6-75 p91
1872 Piano works: Gaspard de la nuit. Jeux d'eau. Miroirs. Pavane
     pour une infante défunte. Sonatine. Le tombeau de Couperin.
     Valses nobles et sentimentales. Noël Lee, pno. Telefunken
     6 DX 48068.
              +-Gr 7-75 p220   +-RR 5-75 p56
1873 Piano works: A la manière de Borodine. A la manière de Chabrier.
     Gaspard de la nuit. Jeux d'eau. Menuet antique. Menuet sur
     le nom de Haydn. Miroirs. Pavane pour une infante défunte.
     Prélude. Sonatine. Le tombeau de Couperin. La valse. Valses
     nobles et sentimentales. Abbey Simon, pno. Vox SVBX 5473 (3).
              ++HF 4-75 p90   -SFC 3-23-75 p22
              ++NR 12-74 p14   ++St 5-75 p92
     Pièce en forme de habañera. cf Orion ORS 75181.
     Prélude. cf Piano works (CBS 77380).
     Prélude. cf Piano works (Decca SXL 6700).
     Prélude. cf Piano works (London 6985).
     Prélude. cf Piano works (Philips 6580 046).
     Prélude. cf Piano works (Vox SVBX 5473).
     Prélude. cf Works, selections (CRD 1821-6).
     Quartet, strings. cf Chansons madécasses.
     Quartet, strings, F major. cf Works, selections (CRD 1821-6).
     Quartet, strings, F major. cf DEBUSSY: Quartet, strings, op. 10,
        G minor (RCA ARL 1-0187).
     Quartet, strings, F major. cf DEBUSSY: Quartet, strings, op. 10,
        G minor (Telefunken SAT 22541).
     Quartet, string, F major. cf DEBUSSY: Works, selections (DG 2533
        007).
1874 Rapsodie espagnole. DEBUSSY: La mer. BSO; Charles Munch. RCA
     VICS 1041. Tape (ct) V8S 1040.
              +HF 3-75 p65
     Rapsodie espagnole. cf Boléro (Angel 37147).
     Rapsodie espagnole. cf Boléro (DG 2530 475).
     Rapsodie espagnole. cf Boléro (Hungaroton 11644).
     Rapsodie espagnole. cf Boléro (Mercury 75033).

Rapsodie espagnole. cf Works, selections (DG 2740 120).
Rapsodie espagnole. cf Works, selections (HMV 5016).
Rapsodie espagnole. cf Works, selections (Vox SVBX 5133).
Rapsodie espagnole: Feria. cf Columbia M2X 33014.
Rapsodie espagnole: Habañera. cf Piano works (CBS 77380).
Rapsodie espagnole: Habañera. cf Works, selections (CRD 1821-6).
Rapsodie espagnole: Habañera. cf Works, selections (Decca SPA 392).
Shéhérazade. cf BERLIOZ: Les nuits d'été, op. 7.
Shéhérazade: La flute enchantée. cf Works, selections (Decca SPA 392).
Shéhérazade: Ouverture de Feerie. cf Bolero (Angel 37147).
Shéhérazade: Ouverture de Feerie. cf Works, selections (HMV 5016).
Sonata, violin. cf Works, selections (CRD 1821-6).
Sonata, violin and piano. cf DEBUSSY: Works, selections (DG 2533 007).
1875 Sonata, violin and violoncello. Trio, violin, violoncello and piano. Jaime Laredo, vln; Jeffrey Solow, Leslie Parnas, vlc; Ruth Laredo, pno. Columbia M 33529.
    +HF 10-75 p82   ++SFC 8-31-75 p20
    ++NR 8-75 p7   ++St 11-75 p134
1876 Sonata, violin and violoncello. Trio, violin, violoncello and piano. Jean-Jacques Kantorow, vln; Philippe Muller, vlc; Jacques Rouvier, pno. Erato STU 70861.
    +Gr 4-75 p1826   +RR 5-75 p56
Sonata, violin and violoncello. cf Works, selections (CRD 1821-6).
1877 Sonatine. Le tombeau de Couperin. Valses nobles et sentimentales. Pascal Rogé, pno. Decca SXL 6674. (also London CS 6873)
    +Gr 11-74 p922   +RR 11-74 p83
    +-HF 4-75 p90   ++SFC 5-4-75 p35
    +NR 7-75 p15   +-SR 2-8-75 p37
Sonatine. cf Gaspard de la nuit (DG 2530 540).
Sonatine. cf Gaspard de la nuit (RCA LSB 4096).
Sonatine. cf Jeux d'eau.
Sonatine. cf Piano works (CBS 77380).
Sonatine. cf Piano works (Philips 6580 046).
Sonatine. cf Piano works (Telefunken 6DX 48068).
Sonatine. cf Piano work (Vox SVBX 5473).
Sonatine. cf Works, selections (CRD 1821-6).
Sonatine. cf DEBUSSY: Estampes, no. 3: Jardins sous la pluie.
Sonatine. cf Draco DR 1333.
1878 Songs: Don Quichotte à Dulcinée; Un grand sommeil noir; Histoires naturelles; Mélodies hébraïques: Kaddisch, L'Enigme éternelle; Mélodies populaires grecques (5); Rêves; Ronsard à son âme; Sainte. Bernard Kruysen, bar; Noël Lee, pno. Telefunken AW 6-41873.
    ++HF 10-75 p81   ++RR 5-75 p62
    ++HFN 5-75 p137   ++STL 5-4-75 p37
Songs: Chansons madécasses; Vocalise. cf Odyssey Y2 32880.
1879 Le tombeau de Couperin. STRAVINSKY: Pulcinella: Suite. London Mozart Players; Yuval Zaliouk. Unicorn UNS 253. Tape (c) ZCUN 253.
    +Gr 4-73 p1879   +RR 4-73 p64
    +Gr 2-74 p1611 tape   +-RR 4-74 p93 tape
    +-HFN 4-73 p184   ++SFC 3-23-75 p22
Le tombeau de Couperin. cf Menuet antique.
Le tombeau de Couperin. cf Miroirs: Alborada del gracioso; Une barque sur l'océan.

Le tombeau de Couperin. cf Gaspard de la nuit.
Le tombeau de Couperin. cf Piano works (CBS 77380).
Le tombeau de Couperin. cf Piano works (Philips 6580 046).
Le tombeau de Couperin. cf Piano works (Telefunken 6 DX 48068).
Le tombeau de Couperin. cf Piano works (Vox SVBX 5473).
Le tombeau de Couperin. cf Sonatine.
Le tombeau de Couperin. cf Works, selections (CRD 1821-6).
Le tombeau de Couperin. cf Works, selections (DG 2740 120).
Le tombeau de Couperin. cf Works, selections (Vox SVBX 5133).
Le tombeau de Couperin: Menuet. cf Angel S 36076.
Le tombeau de Couperin: Rigaudon. cf Columbia M 32070.
Le tombeau de Couperin: Rigaudon. cf Draco DR 1333.
Le tombeau de Couperin: Suite. cf Works, selections (HMV 5016).
Le tombeau de Couperin: Toccata. cf Piano works (Ember GVC 39).
Trio, piano, A minor. cf Works, selections (CRD 1821-6).
Trio, piano, A minor. cf DEBUSSY: Works, selections (DG 2533 007).
Trio, violin, violoncello and piano. cf Sonata, violin and violon-
     cello (Columbia 33529).
Trio, violin, violoncello and piano. cf Sonata, violin and violon-
     cello (Erato STU 70861).
1880 Tzigane. DVORAK: Concerto, violin and piano, op. 53, A minor.
     Edith Peinemann, vln; CPhO; Peter Maag. DG 193 120.
          +HF 3-75 p66
     Tzigane. cf Works, selections (CRD 1821-6).
     Tzigane. cf Works, selections (HMV 5016).
     Tzigane. cf CHAUSSON: Poème, op. 25.
     Tzigane. cf DEBUSSY: Danse sacrée et profane.
     Tzigane. cf DEBUSSY: Sonata, violin and piano, no. 3, G minor.
     Tzigane. cf CBS 76420.
     Tzigane. cf Hungaroton SLPX 11677.
     Tzigane. cf RCA ARM 4-0942/7.
     La valse. cf Boléro (Angel 37147).
     La valse. cf Boléro (Decca 6488).
     La valse. cf Boléro (DG 2530 475).
     La valse. cf Boléro (Mercury 75033).
     La valse. cf Boléro (Philips 6580 106).
     La valse. cf Boléro (RCA 2664).
     La valse. cf Ma mère l'oye.
     La valse. cf Piano works (Vox SVBX 5473).
     La valse. cf Works, selections (Decca SPA 392).
     La valse. cf Works, selections (DG 2740 120).
     La valse. cf Works, selections (HMV 5016).
     La valse. cf Works, selections (Vox SVBX 5133).
     La valse. cf DEBUSSY: La mer.
     La valse. cf HMV (Q) ASD 3131.
     La valse. cf Toscanini Society ATS GC 1201/6.
1881 Valses nobles et sentimentales. BERLIOZ: The damnation of Faust,
     selections. FAURE: Pelléas et Mélisande, op. 80. PO; Charles
     Munch. Odyssey Y 31017.
          +HF 3-75 p65
     Valses nobles et sentimentales. cf Gaspard de la nuit.
     Valses nobles et sentimentales. cf Menuet antique.
     Valses nobles et sentimentales. cf Miroirs: Alborado del gracioso.
     Valses nobles et sentimentales. cf Piano works (CBS 77380).
     Valses nobeles et sentimentales. cf Piano works (Telefunken 6 DX
     48068).

Valses nobles et sentimentales. cf Piano works (Vox SVBX 5473).
Valses nobles et sentimentales. cf Sonatine.
Valses nobles et sentimentales. cf Works, selections (CRD 1821-6).
Valses nobles et sentimentales. cf Works, selections (DG 2740 120).
Valses nobles et sentimentales. cf Works, selections (HMV 5016).
Valses nobles et sentimentales. cf Works, selections (Vox SVBX
    5133).
Valses nobles et sentimentales, nos. 6 and 7. cf RCA ARM 4-0942/7.
1882 Works, selections: A la manière de Borodine. A la manière de
    Chabrier. Berceuse sur le nom de Gábriel Fauré. Chansons
    madécasses: Nahandove; Méfiez-vous de blancs; Il est doux de
    se coucher. Gaspard de la nuit. Rapsodie espagnole: Habañera.
    Introduction and allegro, harp, flute, clarinet and string
    quartet. Jeux d'eau. Ma mère l'oye. Menuet antique. Menuet
    sur le nom de Haydn. Miroirs. Pavane pour une infante défunte.
    Prélude. Quartet, strings, F major. Sonata, violin. Sonata,
    violon and violoncello. Sonatine. Le tombeau de Couperin.
    Trio, piano, A minor. Tzigane. Valses nobles et sentimentales.
    Jacques Rouvier, Théodore Paraskivesco, Henri Bard, pno; Yvon
    Carracilly, Karl Heinrich von Stumpff, Hervé Le Floch, Pierre
    Hofer, vln; Christoph Killian, Klaus Heitz, Pierre Degenne, vlc;
    Marie-Claire Jamet, hp; Christian Larde, flt; Guy Deplus, clt;
    Colette Lequien, vla; Jacques Herbillon, bar; Debussy Quartet.
    CRD CAL 1821-6 (6).
         +-Gr 10-75 p646   +RR 11-75 p24
         +HFN 9-75 p103
1883 Works, selections: Boléro. Concerto, piano, G major: Presto.
    Daphnis et Chloé: Lever du jour. Introduction and allegro,
    harp, flute, clarinet and string quartet. Pavane pour une
    infante défunte. Rapsodie espagnole: Habañera. Shéhérezade:
    L aflute enchantée. La valse. Regine Crespin, s; Osian Ellis,
    hp; Julius Katchen, pno; ROHO Chorus; LSO, OSR, Melos Ensemble;
    Pierre Monteux, Ernest Ansermet, István Kertész. Decca SPA 392.
    Tape (c) KCSP 392.
         +Gr 7-75 p254 tape   +RR 5-75 p38
         +HFN 8-75 p91
1884 Works, selections: Boléro.* Daphnis et Chloé. Ma mère l'oye.
    Menuet antique. Pavane pour une infante défunte. Miroirs:
    Alborada del gracioso: Une barque sur l'océan. Rapsodie espag-
    nole.* Le tombeau de Couperin. La valse.* Valses nobles et
    sentimentales. Tanglewood Festival Chorus; BSO; Seiji Ozawa.
    DG 2740 120 (4). (*available on 2530 475)
         +-Gr 11-75 p828   ++RR 11-75 p53
1885 Works, selections: Boléro. Concerto, piano, G major. Concerto,
    piano, for the left hand, D major. Daphnis et Chloé. Ma mère
    l'oye. Menuet antique. Miroirs: Alborada del gracioso: Une
    barque sur l'océan. Pavane pour une infante défunte. Rapsodie
    espagnole. Shéhérazade: Ouverture de feerie. Le tombeau de
    Couperin: Suite. Tzigane. La valse. Valses nobles et senti-
    mentales. Aldo Ciccolini; pno; Orchestre de Paris; Jean Marti-
    non. HMV (Q) SLS 5016 (5).
         +Gr 10-75 p616   +RR 11-75 p54
1886 Works, selections: Boléro. Daphnis et Chloé: Suites, nos. 1 and 2.
    Fanfare for "L'Eventail de Jeanne." Ma mère l'oye. Menuet
    antique. Miroirs: Alborada del gracioso; Une barque sur l'océan.
    Pavane pour une infante défunte. Rapsodie espagnole. Le tombeau

de Couperin. La valse. Valses nobles et sentimentales. Minnesota Orchestra; Stanislaw Skrowaczewski. Vox SVBX 5133 (4).
(Q) QSVBX 5133.
    ++HF 7-75 p66   +St 8-75 p101 Quad
    ++NR 6-75 p1

RAVENSCROFT, Thomas
    Rustic lovers. cf Turnabout TV 37079.
    Songs: By a bank; The marriage of the frog and the mouse; Of all the birds; Remember O thou man; A round of three country dances in one; The three ravens; Trudge away quickly; We be soldiers three; We be three poor mariners. cf BASF BAC 3081.

RAWSTHORNE, Alan
    Ballade. cf HMV HQS 1337.

RAXACH, Enrique
    Imaginary landscape. cf EISMA: Le gibet.
    Paraphrase. cf EISMA: Le gibet.
    Quartet, strings, no. 2. cf CRUMB: Black angels, electric string quartet.

READ, Daniel
    Newport. cf Nonesuch H 71276.

REDFORD, John
    Eterne rex altissime. cf Angel SB 3816.

REEVES, D. W.
    Second Connecticut Regiment. cf Columbia 33513.

REGER, Max
    Allegro, 2 violins, op. posth., A major. cf Works, selections (Turnabout TV 37056/61).
    Canons and fugues, 2 violins, op. 131b, nos. 1-3. cf Works, selections (Turnabout TV 37056/61).
    Dankpsalm, op. 145, no. 2. cf University of Iowa Press, unnumbered.
1887 Introduction, passacaglia and fugue, op. 96, C minor. Variations and fugue on a theme by Beethoven, op. 86, B flat major. Frank Merrick, Michael Round, pno. Cabaletta HRS 2003.
    /HFN 5-75 p137
    Maria Wiegenlied, op. 76, no. 52. cf Pearl GEM 121/2.
    Pieces, op. 59: Gloria in Excelsis. cf Hungaroton SLPX 11548.
1888 Quartet, piano, op. 133, A minor. STRAUSS, R.: Quartet, piano, op. 13, C minor. Cardiff Festival Ensemble. Argo ZRG 809.
    +-Gr 9-75 p485   +RR 7-75 p44
    +-HFN 7-75 p87
    Quartets, strings, nos. 1-5. cf Works, selections (Turnabout TV 37056/61).
1889 Quintet, clarinet and strings, op. 146, A major. Karl Leister, clt; Drolc Quartet. DG 2530 303.
    +Gr 6-73 p68   +RR 6-73 p66
    +HFN 6-73 p1181   +STL 7-8-73 p36
    ++NR 12-75 p8
    Quintet, clarinet and strings, op. 146, A major. cf Works, selections (Turnabout TV 37056/61).
    Serenade, flute, violin and viola, op. 77a, D major. cf Works, selections (Turnabout TV 37056/61).
    Serenade, flute, violin and viola, op. 141a, G major. cf Works, selections (Turnabout TV 37056/61).
    Sonata, violin, no. 2, A major: Andantino. cf Discopaedia MB 1008.
    Toccata and fugue, op. 59. cf Decca 5BBA 1013-5.
    Trio, strings, op. 77b, A minor. cf Works, selectons (Turnabout TV 37056/61).

Trio, strings, op. 141b, D minor. cf Works, selections (Turnabout
    TV 37056/61).
Variations and fugue on a theme by Beethoven, op. 86, B flat major.
    cf Introduction, passacaglia and fugue, op. 96, B minor.
1890 Works, selections: Allegro, 2 violins, op. posth., A major. Canons
    and fugues, 2 violins, op. 131b, nos. 1-3. Quartets, strings,
    nos. 1-5. Quintet, clarinet and strings, op. 146, A major.
    Serenade, flute, violin and viola, op. 77a, D major. Serenade,
    flute, violin and viola, op. 141a, G major. Trio, strings, op.
    77b, A minor. Trio, strings, op. 141b, D minor. Susanne Lauten-
    bacher, Georg Egger, vln; Bell'Arte Trio, Reger String Quartet.
    Turnabout TV 37056/61 (6). (also Vox SVBX 586)
        +Gr 12-74 p1167   +RR 9-75 p52
REGNART, Jacob
    Ardo si, ma non t'amo. cf Panton 010335.
REICH, Steve
1891 Drumming. Music for mallet instruments, voices and organ. Six
    pianos. Steve Reich and Musicians. DG 2740 106 (3).
        +Gr 1-75 p1353   +-St 6-75 p110
    Music for mallet instruments, voices and organ. cf Drumming.
    Six pianos. cf Drumming.
REICHA (Rejcha), Antonin
1892 Quartet, flute and strings, op. 98, no. 1, G minor. KRAMAR: Quar-
    tet, flute and strings, op. 75, D major. Peter Brock, flt;
    Josef Vlach, vln; Josef Kodoušek, vla; Viktor Moučka, vlc.
    Supraphon 111 1450.
        +NR 3-75 p6   ++RR 12-75 p73
        ++RR 10-74 p71
REICHARDT, Johann Friedrich
    Songs: Canzon, s'al dolce loco; Di tempo in tempo; Erano i capei
        d'oro; O poggi, O valli, O fiumi; Or ch'il ciel; Più volte già
        dal bel sembiante. cf DG 2530 332.
    Songs: Gott; Feiger Gedanken; Die schöne Nacht; Einzeiger Augen-
        blick; Einschränkung; Mut; Rhapsodie; An Lotte; Tiefer liegt
        die Nacht um mich her. cf DG Archive 2533 149.
REICHE
    Baroque suite. cf Desto DC 6474/7.
REINECKE, Carl
    Children's symphony: Slow movement. cf International Piano
        Library IPL 5005/6.
1893 Concerto, piano, no. 1, op. 72, F sharp minor. Concerto, piano,
    no. 2, op. 120, E minor. Gerald Robbins, pno; Monte Carlo
    Opera Orchestra; Edouard van Remoortel. Genesis GS 1034.
        +Gr 10-74 p707   +RR 6-74 p51
        +HF 6-74 p92   +SFC 2-2-75 p25
        ++MJ 3-74 p6   ++St 2-74 p119
        +NR 10-73 p4
    Concerto, piano, no. 1, op. 72, F sharp minor. cf d'ALBERT:
        Concerto, piano, no. 2, op. 12, E major.
    Concerto, piano, no. 2, op. 120, E minor. cf Concerto, piano, no.
        1, op. 72, F sharp minor.
    Sonata, flute and piano, op. 167, E minor. cf KUHLAU: Grand solo,
        piano and flute, op. 57, no. 3.
    Toy symphony, C major. cf Angel S 36080.
REINHOLD
    Impromptu, C sharp minor. cf Connoisseur Society (Q) CSQ 2066.

REINKEN, Johann
    Toccata. cf Telefunken DX 6-35265.
REIZENSTEIN, Franz
    Concerto popolare. cf HMV SLA 870.
1894 Impromptu, op. 14. Legend, op. 24. Scherzo, op. 21, A major.
        Scherzo fantastique, op. 26. Sonata, piano, op. 19, B major.
        Franz Reizenstein, pno. Lyrita RCS 19. (Reissue from 1960)
            +Gr 9-75 p492   +RR 7-75 p50
            +HFN 12-75 p156
    Legend, op. 24. cf Impromptu, op. 14.
1895 Partita, treble recorder and piano, op. 13. Quintet, piano and
        strings, op. 23, D major. Sonatina, oboe and piano, op. 11.
        Melos Ensemble. L'Oiseau-Lyre SOL 344.
            +Gr 7-75 p206   +MT 12-75 p1071
            +HF 11-75 p110   ++NR 10-75 p4
            +HFN 12-75 p163   +-RR 7-75 p43
    Quintet, piano and strings, op. 23, D major. cf Partita, treble
        recorder and piano, op. 13.
    Scherzo, op. 21, A major. cf Impromptu, op. 14.
    Scherzo fantastique, op. 26. cf Impromptu, op. 14.
    Sonata, piano, op. 19, B major. cf Impromptu, op. 14.
    Sonatina, oboe and piano, op. 11. cf Partita, treble recorder
        and piano, op. 13.
    RENAISSANCE BRASS MUSIC. cf Turnabout TV 37071.
    A RENAISSANCE CHRISTMAS. cf Turnabout TV 34569.
    RENAISSANCE HITS. cf Oryx BRL 1.
RENIE, Henriette
    Contemplation. cf GLINKA: Nocturne.
    Legende. cf GLINKA: Nocturne.
    Pièce symphonique. cf GLINKA: Nocturne.
RESINARIUS, Balthasar
    Nun komm der Heiden Heiland. cf Turnabout TV 34569.
RESPIGHI, Ottorino
1896 La boutique fantasque: Ballet suite. The pines of Rome. CPhO;
        Antonio Pedrotti. Supraphon 110 1204. Tape (c) 04 1204.
            +-Gr 4-75 p1872   /NR 1-74 p4
            ++HF 3-74 p104   -RR 7-75 p71
            /HFN 5-75 p142 tape   +SFC 1-12-75 p29
    La boutique fantasque. cf Fountains of Rome.
    La boutique fantasque. cf DUKAS: The sorcerer's apprentice.
1897 Feste Romane (Roman festivals). STRAUSS, R.: Don Juan, op. 20.
        LAPO; Zubin Mehta. RCA LSB 4109. Tape (c) MCK 579.
            +Gr 9-74 p512   /RR 9-74 p59
            +-Gr 6-75 p111 tape   +-RR 9-75 p79 tape
1898 Fountains of Rome. La boutique fantasque. PhO, RPO; Eugene Goos-
        sens. Classics for Pleasure CFP 40204. (Reissue from HMV ASD
        366, BSD 752)
            +Gr 4-75 p1812   +-RR 2-75 p38
    The pines of Rome. cf La boutique fantasque: Ballet suite.
    The pines of Rome. cf HMV SLS 5019.
    Songs: Nevicata; Nebbie; Pioggia. cf London OS 26391.
    Il tramonto. cf DEBUSSY: Le promenoir des deux amants.
REUSNER, Esaias
    German dances. cf DG Archive 2533 172.
    Suite, E minor. cf Pelca PSR 40589.
    Suite, no. 2, C minor: Paduana. cf CBS 72526.

REYNALDO HAHN AND HIS SONGS. cf Rococo 5365.
REYNOLDS, Roger
    Ambages. cf Nonesuch HB 73028.
REZAC, Ivan
    Sisyfova nedele (The Sunday of Sisyfos). cf DEBUSSY: Images, Bk 1.
RHEINBERGER, Josef
1899 Sonata, organ, no. 1, op. 27, C minor. Sonata, organ, no. 20,
        op. 196, F major. Conrad Eden, Timothy Farrell, org. Vista
        VPS 1011.
            +-Gr 5-75 p1996   +-RR 4-75 p53
            +HFN 5-75 p138
1900 Sonata, organ, no. 2, op. 63, A flat major. Sonata, organ, no. 9,
        op. 42, B flat minor. Conrad Eden, Robert Munns, org. Vista
        VPS 1013.
            /Gr 8-75 p348   +-RR 9-75 p61
            +-HFN 8-75 p82
1901 Sonata, organ, no. 6, op. 119, E flat minor. Sonata, organ, no. 11,
        op. 148, D minor. Robert Munns, Roger Fisher, org. Vista VPS
        1012.
            +Gr 6-75 p69   +RR 7-75 p53
            ++HFN 6-75 p91
    Sonata, organ, no. 6, op. 119, E flat minor. cf Wealden WS 110.
    Sonata, organ, no. 9, op. 142, B flat minor. cf Sonata, organ,
        no. 2, op. 63, A flat major.
    Sonata, organ, no. 11, op. 148, D minor. cf Sonata, organ, no. 6,
        op. 119, E flat minor.
1902 Sonata, organ, no. 12, op. 154, D flat major. Sonata, organ, no.
        17, op. 181, B major. Roger Fisher, Timothy Farrell, org. Vista
        VPS 1014.
            +Gr 10-75 p658   +RR 12-75 p84
            +HFN 12-75 p163
    Sonata, organ, no. 17, op. 181, B major. cf Sonata, organ, no. 12,
        op. 154, D flat major.
    Sonata, organ, no. 20, op. 196, A major. cf Sonata, organ, no. 1,
        op. 27, C minor.
    Sonata, violoncello, op. 92, C major. cf GRIEG: Sonata, violoncello
        op. 36, A minor.
RHODES, Phillip
    Duo, violin and violoncello. cf WOLPE: Piece in two parts for solo
        violin.
RICCIOTTI, Carlo
    Concertino, no. 2, G major. cf Decca SDD 411.
    Concerto, no. 2, G major. cf CORELLI: Concerto grosso, op. 6, no.
        8, C minor.
    RICHARD TUCKER: In memoriam. cf Columbia D3M 33448.
RICHTER, Franz Xaver
    Sonata, violin and harpsichord, G major: Andante. cf Oryx ORYX 1720
1903 Super Flumina Babylonis (Psalm 137). Maria Venuti, s; Marie Thér-
        èse Mercanton, con; Dieter Ellenbeck, t; Friedhelm Hessenbruch,
        bs; Heidelberg Bach Choir; South West German Chamber Orchestra;
        Erich Hubner; Ursula Trede-Boettcher, org and hpd. Pelca PSR
        40604.
            +NR 11-75 p9
RIDKY
    March. cf Panton 110 361.

RIDOUT, Alan
    Scherzo. cf Vista VPS 1021.
RIEGGER, Wallingford
    Dichotomy, chamber orchestra, op. 12. cf MUSGRAVE: Night music,
        chamber orchestra.
RIES, Ferdinand
    Sonata, flute and piano, op. 169, E flat major. cf KUHLAU: Grand
        solo, piano and flute, op. 57, no. 3.
RIGLER
    Fugue, A major. cf Hungaroton SLPX 11601/2.
RIISAGER, Knudåge
    Etude. cf GADE: Echoes of Ossian.
    Qarrtsiluni. cf GADE: Echoes of Ossian.
RILEY, Dennis
    Variations II: Trio. cf CRI SD 324.
RIMMER, Frederick
    Singe we merrily. cf Gemini GM 2022.
RIMSKY-KORSAKOV, Nikolai
1904 Capriccio espagnol, op. 34. Mlada: Procession of the nobles.
        TCHAIKOVSKY: Capricio italien, op. 45. Marche slav, op. 31.
        Mazeppa: Gopak. LPO; Adrian Boult. HMV ASD 3093.
            +Gr 7-75 p245   +RR 8-75 p42
            ++HFN 7-75 p87
    Capriccio espagnol, op. 34. cf DVORAK: Slavonic dance, op. 72,
        no. 2, E minor.
    Concerto, clarinet, E flat major. cf PROKOFIEV: Marches, nos. 1-3,
        op. 69.
    Concerto, trombone. cf PROKOFIEV: Marches, nos. 1-3, op. 69.
    Le coq d'or: Suite. cf PROKOFIEV: Betrothal in the monastery, op.
        86: Summer night suite.
    Le coq d'or: Wedding march. cf BERLIOZ: Les Troyens: Overture,
        March Troyenne.
    Enslaved the rose, the nightingale. cf RCA ARL 1-5007.
    Fantasy on Russian themes, op. 33. cf HMV SXLP 30193.
    The legend of Sadko: Chant Hindou. cf Decca SPA 405.
    The legend of Sadko: Song of the Venetian guest. cf Columbia/
        Melodiya M 33120.
    The legend of the invisible city of Kitezh and the maiden Fevronia:
        Fevronia's wedding train; Prayer. cf HMV Melodiya SXL 30190.
    Mlada: Procession of the nobles. cf Capriccio espagnol, op. 34.
1905 Quintet, piano and winds, op. posth., B flat major. RUBINSTEIN:
        Quintet, piano and winds, op. 55, F major. New Philharmonic
        Wind Ensemble, London; Felicja Blumenthal, pno. Turnabout TV
        34477.
            +NR 10-74 p7   +SFC 6-29-75 p26
    Quintet, piano and winds, op. posth., B flat major. cf MENDELSSOHN:
        Octet, strings, op. 20, E flat major.
    Russian Easter festival overture, op. 36. cf Philips 6747 204.
1906 Scheherazade, op. 35. Ruben Yordanoff, vln; Orchestre de Paris;
        Mstislav Rostropovich. Angel S 37061. (also HMV ASD 3047 (Q)
        Q4 ASD 3047)
            +-Gr 2-75 p1499   +-RR 2-75 p37
            +-HF 5-75 p89   +-SFC 4-13-75 p23
            -NR 3-75 p2   ++St 6-75 p104
1907 Scheherazade, op. 35. LAPO; Sydney Harth, vln; Zubin Mehta. Decca
        SXL 6731. Tape (c) KSXC 6731. (also London 6950. Tape (c)

56590 (ct) 6950)
    -Gr 11-75 p828   +-RR 10-75 p49
    +HFN 10-75 p148   +SFC 9-14-75 p28
Scheherazade, op. 35. cf Camden CCV 5000-12, 5014-24.
Scheherazade, op. 35: Fantasy. cf Ember GVC 40.
Scheherazade, op. 35: The sea and Sinbad's ship. cf Decca 396.
Scheherazade, op. 35: 3rd movement. cf Discopaedia MB 1007.
The snow maiden: Mizghir's arioso. cf Columbia/Melodiya M 33120.
Symphony, no. 2, op. 9. cf HMV SEOM 20.
The tale of the Tsar Sultan: Flight of the bumblebee. cf Monitor
    MCS 2143.
The tale of the Tsar Sultan: Flight of the bumblebee (2). cf RCA
    ARM 4-0942/7.
The tale of the Tsar Sultan: The flight of the bumblebee (arr. and
    orch. Gerhardt). cf RCA LRL 1-5094.
The tale of the Tsar Sultan: May you grow to be like a mighty oak.
    cf HMV Melodiya SXL 30190.
The Tsar's bride: Griaznoy's aria. cf Columbia/Melodiya M 33120.
The Tsar's bride: Scene from Act 2. cf HMV Melodiya SXL 30190.
Variations, oboe and brass band. cf PROKOFIEV: Marches, nos. 1-3,
    op. 69.
RITTER, Christian
    Sonatina, D minor. cf Argo ZRG 783.
ROBINSON, Thomas
    Fantasia. cf L'Oiseau-Lyre SOL 336.
    A plainsong. cf L'Oiseau-Lyre SOL 336.
    A toy: Bo Peep. cf L'Oiseau-Lyre SOL 336.
ROBISON
    Fiducia. cf Nonesuch H 71276.
ROCHBERG, George
    Blake songs. cf JONES: Ambiance.
    Blake songs. cf Nonesuch H 71302/3.
    Contra mortem et tempus. cf PARRIS: Concerto, trombone.
    Night music. cf SAEVERUD: Peer Gynt, op. 28.
1908 Quartet, strings, no. 2. SUDERBURG: Chamber music II. Phyllis
    Bryn-Julson, s; Concord String Quartet, Philadelphia String
    Quartet. Turnabout TVD 34524.
      +-HF 11-74 p117   +St 1-75 p106
      +NR 10-74 p8
RODGERS
    You'll never walk alone. cf Columbia D3M 33448.
RODNEY
    Calvary. cf RCA ARL 1-0561.
RODRIGO, Joaquin
1909 Concierto andaluz, 4 guitars and orchestra. Concierto de Aranjuez,
    guitar and orchestra. Angel, Celedonio, Celin, Pepe Romero,
    guitar; San Antonio Symphony Orchestra; Victor Alessandro.
    Mercury SRI 75021.
      ++SFC 7-21-74 p26   +-St 7-75 p104
1910 Concierto de Aranjuez, guitar and orchestra. VILLA-LOBOS: Concerto,
    guitar and small orchestra. John Williams, gtr; James Brown,
    cor anglais; ECO; Daniel Barenboim. CBS 76369. Tape (c) 40-
    76369. (also Columbia M 33208. Tape (c) MT 33208 (Q) MQ 33208
    Tape (ct) MAQ 33208)
      ++Gr 1-75 p1354   ++RR 1-75 p33
      +Gr 5-75 p2032   ++RR 6-75 p93 tape
      +NR 4-75 p5   ++St 7-75 p104

1911 Concierto de Aranjuez, guitar and orchestra. PARISH-ALVARS: Con-
     certo, harp, op. 81, G minor. Nicanor Zabaleta, hp; Spanish
     National Orchestra; Rafael Frühbeck de Burgos. Angel S 37042.
          ++NR 3-75 p5   +St 7-75 p104
          +SR 1-25-75 p50
1912 Concierto de Aranjuez, guitar and orchestra. SOR: Estudio, no. 9,
     en si menor. Minuetto, op. 22. TARREGA: Recuerdos de la
     Alhambra. Mazurka. TORROBA: Zapateado. Rumor de copla. Ren-
     ata Tarragó, gtr; Madrid, Orquesta de Conciertos; Odón Alonso.
     Olympic OL 8100.
          +NR 8-74 p13   +-St 7-75 p104
1913 Concierto de Aranjuez, guitar and orchestra. Fantasia para un
     gentilhombre, guitar. Alexandre Lagoya, gtr; Monte Carlo
     National Orchestra; Antonio de Almeida. Philips 6500 454. Tape
     (c) 7300 241.
          +-Gr 4-74 p1858   +RR 4-74 p48
          +Gr 7-73 p247 tape   +RR 1-75 p72 tape
          +NR 3-74 p5   +St 7-75 p104
     Concierto de Aranjuez, guitar and orchestra. cf Concierto andaluz,
          4 guitars and orchestra.
     Concierto de Aranjuez, guitar and orchestra. cf BERKELEY: Concerto,
          guitar.
     Concierto de Aranjuez, guitar and orchestra. cf Camden CCV 5000-12,
          5014-24.
     Concierto madrigal, 2 guitars and orchestra. cf GIULIANI: Concerto,
          guitar, op. 30, A major.
     Concierto serenata. cf BOIELDIEU: Concerto, harp, C major.
     De los álamos vengo, madre. cf HMV SLS 5012.
     Fantasia para un gentilhombre, guitar. cf Concierto de Aranjuez,
          guitar and orchestra (Philips 6500 454).
     Fantasia para un gentilhombre. cf CASTELNUOVO-TEDESCO: Concertino,
          harp and chamber orchestra, op. 93.
     Fantasia para un gentilhombre. cf Camden CCV 5000-12, 5014-24.
     Zarabanda lejana. cf Nonesuch H 71233.
ROGERS, Bernard
     Soliloquy, flute and strings. cf Era 1001.
ROGIER
     D'amours me plains. cf Hungaroton SLPX 11549.
ROMAN, Johann Helmich
     Music for the royal nuptials. cf BRYCE: Promenade.
     Sonata, harpsichord, C major. cf AGRELL: Sonata, harpsichord,
          no. 1, B flat major.
     Sonata IX. cf AGRELL: Sonata, harpsichord, no. 1, B flat major.
     Suite, no. 2. cf CRD CRD 1008.
ROMANINI
     Toccata in Mixolydian mode. cf Hungaroton SLPX 11601/2.
RONCALLI
     Passacaglia. cf Angel S 37015.
     Gigue. cf Angel S 37015.
     Suite, G major. cf DG 2530 561.
     Suite, G major (Capriccio armonici). cf DG 2530 561.
RONTANI, Raffaello
     Nerinda bella. cf L'Oiseau-Lyre 12BB 203/6.
RORE, Cyprien
     De la belle contrade. cf L'Oiseau-Lyre 12BB 203/6.

ROREM, Ned
    Concerto, piano, in six movements. cf BRICCETTI: The fountain
        of youth overture.
    Songs: A Christmas carol; For Susan; Clouds; Guilt; What sparks
        and wiry cries. cf Duke University Press DWR 7306.
ROSE, Bernard
    Lord, I have loved the habitation of Thy house. cf Abbey LPB 734.
ROSENBERG, Hilding
    Resan till America (Voyage to America), excerpts. cf BLOMDAHL:
        Sisyphos.
    Symphony, no. 2. cf WIREN: Symphony, no. 4, op. 27.
1914 Symphony, no. 3. Stockholm Philharmonic Orchestra; Herbert Blom-
        stedt. Odeon SCLP 1071.
            *St 6-75 p52
    Symphony, no. 6. cf BLOMDAHL: Symphony, no. 3.
ROSENMULLER, Johann
    Pavane. cf Pye GSGC 14154.
ROSETTI, Francesco Antonio (Franz Anton Rössler)
1915 Concerto, horn, D minor. TELEMANN: Suite, 4 horns and strings,
        F major. VIVALDI: Concerto, 2 horns, P 320, F major. Erich
        Penzel, Alois Spach, Gottfried Roth, Joachim Schollmeyer, Alfred
        Balsar, hn; Württemberg Chamber Orchestra; Mainz Chamber Orches-
        tra; Jörg Faerber, Günter Kehr. Turnabout TV 34078. Tape (c)
        KTVC 34078.
            +-RR 2-75 p76 tape   +RR 4-75 p78 tape
    Sinfonia, G minor. cf CANNABICH: Sinfonia concertante, F major.
ROSS, Walter
    Concerto, trombone and orchestra. cf PENN: Ultra mensuram.
    Prelude, fugue and big apple, bass trombone and tape. cf PENN:
        Ultra mensuram.
ROSSI, Luigi
    Toccata, no. 7, D minor. cf Telefunken AW 6-41890.
ROSSINI, Gioacchino
1916 The barber of Seville. Beverly Sills, s; Fedora Barbieri, ms;
        Nicolai Gedda, t; Joseph Galiano, Sherrill Milnes, Renato
        Capecchi, Michael Rippon, bar; Ruggero Raimondi, bs; John Alldis
        Choir; LSO; James Levine. Angel SCLX 3761 (3). Tape (c) 4X3S
        3761. (also HMV (Q) SLS 985)
            +Gr 11-75 p903   +-ON 11-75 p70
            +-HF 11-75 p111   +-RR 10-75 p28
            +HFN 12-75 p163   +-SR 10-4-75 p50
            -NR 12-75 p10   ++St 10-75 p117
            +OC 12-75 p49
1917 Il barbiere di Siviglia. Gianna d'Angelo, Gabriella Carturan, s;
        Nicola Monti, t; Renato Capecchi, bar; Giorgio Tadeo, Carlo
        Cava, Giorgio Giorgetti, bs; Bavarian Radio Orchestra and Chorus;
        Renato Sabbioni, hpd; Fred Artmeier, gtr; Bruno Bartoletti.
        DG 2728 005 (3). (Reissue from SLPM 138665/7)
            +Gr 2-75 p1547   +-RR 2-75 p28
1918 Il barbiere di Siviglia. Teresa Berganza, Stefania Malagù, ms; Ugo
        Benelli, t; Fernando Corena, Nicolai Ghiaurov, bs-bar; Manuel
        Ausensi, bs; Naples, Rossini Orchestra and Chorus; Silvio Var-
        viso. London 1381. Tape (r) LON 90105R (2).
            ++LJ 9-75 p38
    Il barbiere di Siviglia: A un dottor della mia sorte. cf Decca 666/8.
    Il barbiere di Siviglia: Overture. cf Overtures (Decca SDD 392).

Il barbiere di Siviglia: Overture. cf Overtures (Philips 6500 878).
Il barbiere di Siviglia: Overture. cf BEETHOVEN: Egmont overture,
op. 84.
Il barbiere di Siviglia: Overture. cf HMV SLS 5019.
Il barbiere di Siviglia: Overture and storm music. cf Overtures
(DG 2548 171).
Il barbiere di Siviglia: Sieh schon die Morgenröte. cf Telefunken
AJ 6-41867.
Il barbiere di Siviglia: Una voce poco fa. cf BELLINI: Capuletti
ed i Montecchi: O, quante volte, O quante.
Il barbiere di Siviglia: Una voce poco fa. cf HMV SLS 5012.
Il barbiere di Siviglia: Una voce poco fa. cf Pearl GEM 121/2.
Il barbiere di Siviglia: Una voce poco fa. cf Odyssey Y2 32880.
La boutique fantasque. cf DUKAS: L'apprenti sorcier.
La boutique fantasque: Overture, Tarantella, Mazurka, Can-Can,
Galop, Finale. cf Decca SDDJ 393/5.
La boutique fantasque: Overture, Tarantella, Mazurka, Can-Can,
Galop, Finale. cf Decca DPA 515/6.
La cambiale di matrimonio: Arias and scenes. cf London OS 26306.
La cambiale di matrimonio: Overture. cf Overtures (Philips 6500
878).
La cenerentola: Nacqui all'affano. cf Odyssey Y2 32880.
La cenerentola: Overture. cf DANKWORTH: Tom Sawyer's Saturday.
La danza. cf Columbia D3M 33448.
La danza. cf London OS 26391.
La danza. cf Polydor 2489 519.
La donna del lago: Mura felici; Tanti affetti. cf L'Assedio di
Corinto (The siege of Corinth): Svanziam...Non temer d'un basso
affetto.
La donna del lago: Tanti affetti. cf Decca GOSC 666/8.
La fioraria fiorentian. cf DONIZETTI: Lucia di Lammermoor: Sulla
tomba to end of Act 1.
La gazza ladra: Di piacer mi balza il cor. cf BELLINI: Capuletti
ed i Montecchi: O, quante volte, O quante.
La gazza ladra: Overture. cf Overtures (Decca SDD 392).
La gazza ladra: Overture. cf Overtures (DG 2548 171).
La gazza ladra: Overture. cf Decca SXL 6643.
La gazza ladra: Overture. cf HMV SLS 5019.
La gazza ladra: Overture. cf London CS 6856.
Guillaume Tell: Ah, Mathilde. cf Rubini RS 300.
William Tell: Arias. cf London OS 26373.
William Tell: Ballet music. cf DELIBES: Coppelia: Ballet music.
Guillaume Tell: Overture. cf Overtures (Decca SDD 392).
Guillaume Tell: Overture. cf Overtures (DG 2548 171).
Guglielmo Tell: Overture. cf HEROLD: Zampa: Overture.
William Tell: Overture. cf Columbia 31726.
Guglielmo Tell: Resta immobile. cf Ember GVC 37.
L'Inganno felice: Overture. cf Overtures (Philips 6500 878).
Introduction and variations. cf Musical Heritage Society MHS 1345.
L'Italiana in Algeri: Amici in ogni evento...Pensa all patria.
cf Rococo 5368.
L'Italiana in Algeri: Cruda sorte. cf Odyssey Y2 32880.
L'Italiana in Algeri: Overture. cf Overtures (Philips 6500 878).
1919 Messa di gloria. Margherita Rinaldi, s; Ameral Gunson, con; Ugo
Benelli, John Mitchinson, t; Jules Bastin, bs; BBC Singers; ECO;
Herbert Handt. Philips 6500 612.

             +Gr 6-74 p95   ++ON 12-28-74 p48
           +-HF 12-74 p112   +RR 5-74 p66
            -MJ 3-75 p24   ++SFC 6-16-74 p29
            +NR 8-74 p7   +SR 11-30-74 p41
            +NYT 8-18-74 pD20   +St 8-74 p73
      Moderato. cf Philips 6580 066.
1920 Overtures: Il barbiere di Siviglia. La gazza ladra. Guillaume
       Tell. La scala di seta. Le siège de Corinth. Il Signor Brus-
       chino. NPhO; Lamberto Gardelli. Decca SDD 392. Tape (c) KSDC
       392. (also London STS 15307)
             +Gr 2-75 p1561 tape   +RR 10-73 p78
            ++NR 8-75 p3   +RR 3-75 p74 tape
             -Op 5-74 p422
1921 Overtures: Il barbiere di Siviglia: Overture and storm music. La
       gazza ladra. Guillaume Tell. La scala di seta. Semiramide.
       Rome Opera Orchestra; Tullio Serafin. DG 2548 171. (Reissue
       from 136395)
            ++Gr 11-75 p840   +HFN 10-75 p152
1922 Overtures: Il barbiere di Siviglia. La cambiale di matrimonio.
       L'Inganno felice. L'Italiana in Algeri. La scala di seta.
       Il Signor Bruschino. Tancredi. Il Turco in Italia. AMF;
       Neville Marriner. Philips 6500 878.
             +HFN 10-75 p148   ++SFC 12-28-75 p30
             +RR 11-75 p54
      Overtures (6). cf Camden CCV 5000-12, 5014-24.
      Petite caprice à la Offenbach (piano solo). cf RCA ARL 1-5007.
1923 Petite messe solennelle. Grace de la Cruz, s; Marie Louise Gilles,
       con; Hans Dieter Saretzki, t; Hans Günther Grimm, bs; Paderborn
       State Music Society, Northwest German Philharmonic Orchestra
       and Choir; Hans Josef Roth, org; Werner A. Albert. Peerless
       Oryx ORYX 1826/7 (2).
             -Gr 11-75 p881   +-HFN 8-75 p82
      La scala di seta: Overture. cf Overtures (Decca SDD 392).
      La scala di seta: Overture. cf Overtures (DG 2548 171).
      La scala di seta: Overture. cf Overtures (Philips 6500 878).
      Semiramide: Arias. cf London OS 26277.
      Semiramide: Bel raggio lusinghier. cf Odyssey Y2 32880.
      Semiramide: Bel raggio lusinghier. cf RCA ARL 1-0702.
      Semiramide: Overture. cf Overtures (DG 2548 171).
      Semiramide: Overture. cf Columbia 31726.
      Semiramide: Overture. cf Erato STU 70880.
      Semiramide: Serbami ognor. cf Decca DPA 517/8.
1924 The siège of Corinth. Beverly Sills, Delia Wallis, s; Shirley
       Verrett, ms; Harry Theyard, Gaetano Scano, t; Justino Diaz,
       Gwynne Howell, Robert Lloyd, bs; Ambrosian Opera Chorus; LSO;
       Thomas Schippers. Angel SLCX 3819 (3). (also HMV SLS 981)
             +-Gr 6-75 p83   +ON 4-19-75 p54
             +-HF 6-75 p75   +RR 6-75 p24
             +HFN 7-75 p87   ++SFC 3-30-75 p16
             +-MQ 10-75 p626   +-SR 5-3-75 p33
             +NR 5-75 p11   +St 5-75 p98
             +NYT 4-6-75 pD18
1925 L'Assedio di Corinto (Le siège de Corinth): Avanziam...Non temer
       d'un basso effetto...Sei tu che stendi, O Dio; L'ora fatal
       s'appressa...Giusto ciel. La donna del lago: Mura felici; Tanti
       affetti. Marilyn Horne, ms; Ambrosian Opera Chorus; RPO; Henry

Lewis. London OS 26305. Tape (r) 490 229. (also Decca SXL
        6584)
    +-Gr 10-73 p728   +NYT 8-12-73 pD24
    ++HF 7-73 p92   +OC 12-75 p49
    ++LJ 4-75 p70 tape   +Op 12-73 p1104
    ++MJ 9-73 p46   +RR 10-73 p40
    +NR 11-73 p9   ++St 8-73 p108
Le siège de Corinth: Overture. cf Overtures (Decca SDD 392).
Il Signor Bruschino: Overture. cf Overtures (Decca SDD 392).
Il Signor Bruschino: Overture. cf Overtures (Philips 6500 878).
1926 Sonata, strings, no. 1, G major. VERDI: Quartet, strings, E minor
     (arranged for string orchestra). ECO; Pinchas Zukerman. Col-
     umbia M 33415. (also CBS 76382)
    +-Gr 8-75 p342   ++NR 7-75 p5
    -HF 9-75 p95   +RR 8-75 p45
    +HFN 8-75 p83   ++St 9-75 p115
Songs: Duetto buffo di due gatti; La pesca; La regata Veneziana.
     cf BIS LP 17.
1927 Stabat Mater. Ilona Steingruber, s; Dagmar Hermann, ms; Anton
     Dermota, t; Paul Schöffler, bs; Vienna Akademiekammerchor;
     VSOO; Jonathan Sternberg. Olympic 9108.
    +-St 1-75 p116
1928 Stabat Mater. Isabella Aidinian, s; Goar Calachian, ms; Michael
     Dovenman, t; Migran Erkat, bs; Studio Orchestra State Cinema;
     Armenia, State Academic Choir, Oganes Chekidjian. Westminster
     WGS 8268.
    ++MJ 11-74 p48   -St 1-75 p116
Stabat Mater: Pro peccatis. cf Rubini RS 300.
Tancredi: O patria, dolce, ingrata...Di tanti palpita. cf Rococo
     5368.
Tancredi: Overture. cf Overtures (Philips 6500 878).
Il Turco in Italia: Overture. cf Overtures (Philips 6500 878).
ROTA, Nino
     Romeo and Juliet: What is a youth. cf BBC REB 191.
ROUSSAKIS, Nicolas
     Short pieces, 2 flutes (6). cf Nonesuch HB 73028.
ROUSSEL, Albert
     Le festin de l'araignée, op. 17. cf HONEGGER: Pacific 231.
     Petite suite, op. 39. cf HONEGGER: Pacific 231.
1929 Songs: A flower given to my daughter; A un jeune gentilhomme;
     Adieux; Amoureux separés; Le bachelier de Salamanque; Chansons
     anacréontiques; Des fleurs font une broderie; Jazz dans la
     nuit; Light; Odelette; Reponse d'une epouse sage; Sarabande.
     ENESCO: Songs: Aux demoiselles pareseuses; Changeons propos
     c'est trop; Chanté d'amour; Du confict en douleur; d'Ecrire à
     leurs amis; Estrene à Anne; Estrene de a rose; Languir me
     fais; Présent de couleur blanche. Yolanda Marcoulescou, s;
     Katja Phillabaum, pno. Orion ORS 75184.
    ++HF 10-75 p78   +NYT 6-8-75 pD19
    +NR 7-75 p12
Songs: Adieu; A flower given to my daughter; Jazz dans la nuit;
     Light; Mélodies, op. 20; Poèmes chinois, op. 12; Poèmes chinois,
     op. 35; Odes anacréontiques, nos. 1, 5; Odelette. cf ENESCO:
     Songs, op. 15.
Suite, op. 33, F major. cf DUTILLEUX: Symphony, no. 2.
1930 Symphony, no. 3, op. 42, G minor. Symphony, no. 4, op. 53, A major.

Lamoureux Orchestra; Charles Munch. Musical Heritage Society
MHS 1879.
+HF 2-75 p100
Symphony, no. 4, op. 53, A major. cf Symphony, no. 3, op. 42, G
minor.
ROZSA, Miklos
1931 Film music to Julius Caesar: Suite. SHOSTAKOVICH: Hamlet: Suite.
WALTON: Richard III: Prelude. National Philharmonic Orchestra;
Bernard Herrmann. London SPC 21132. Tape (c) 0521132 (ct)
0821132.
+HFN 8-75 p80   +NR 10-75 p2
Julius Caesar. cf SHOSTAKOVICH: Hamlet suite.
RUBBRA, Edmund
Songs: Psalm, no. 6, O Lord, rebuke me not; Psalm, no. 23, The Lord
is my Shepherd; Psalm, no. 150, Praise ye the Lord. cf Decca
6BB 197/8.
RUBINSTEIN, Anton
Concerto, piano, no. 3, op. 45, G major. cf KABALEVSKY: Concerto,
piano, no. 3, op. 50, D major.
Es blinkt der Tan. cf Rubini RS 300.
Etude, op. 23, no. 2, C major. cf RCA ARL 2-0512.
Melody, op. 3, no. 1, F major. cf L'Oiseau-Lyre DSLO 7.
Quintet, piano and winds, op. 55, F major. cf RIMSKY-KORSAKOV:
Quintet, piano and winds, op. posth., B flat major.
Romance, E flat major. cf Discopaedia MB 1006.
Songs: Melody. cf London OS 26249.
Songs: Now shines the dew; Spring song. cf Discophilia KGS 3.
RUBIRA, Antonio
Romance. cf Amberlee ACL 501X.
RUE, Pierre de la
Mijn hert. cf Telefunken TK 11569/1-2.
Pour ung jamais. cf L'Oiseau-Lyre 12BB 203/6.
RUFFO, Vincenzo
Dormendo un giorno: Capriccio. cf Erato STU 70847.
La gamba in basso e soprano. cf Erato STU 70847.
RUIZ, Lucas
Galliards. cf Amberlee ACL 501X.
RUSSIAN AND BULGARIAN ORTHODOX CHANTS. cf Harmonia Mundi HMU 133.
RUSSIAN CHORAL WORKS OF THE SEVENTEENTH AND EIGHTEENTH CENTURIES.
cf HMV Melodiya ASD 3102.
RUSSIAN ORTHODOX CHURCH MUSIC. cf Anonymous works.
RUSSIAN SONGS. cf London OS 26249.
RUTTER, John
Praise ye the Lord. cf Gemini GM 2022.
RYBA, Jan Yakub
1932 Czech Christmas mass. Pastorella. Jaroslava Vymazalová, Helena
Tattermuschová, s; Marie Mrázova, con; Beno Blachut, t; Zdeněk
Kroupa, bs; Czech Philharmonic Chorus; PSO; Václav Smetáček.
Supraphon 50768.
++RR 12-75 p91
Pastorella. cf Czech Christmas mass.
RYSBYE (16th century England)
Who so that will himself apply. cf L'Oiseau-Lyre SOL 329.
RYTERBAND, Roman
1933 Ballades hébraïques (3). Pièce sans Titre, 2 flutes. Sonata brève,
violin and harp. Sonnets, contralto, flute and harp. Suite

polonaise, excerpts. Lyn Vernon, ms; Elemér Glanz, vln; Eva
Kauffunger, hp; Alexandre Magnin, Georges Gueneux, flt; Boris
Mersson, pno. Orion ORS 74167.
        +NR 3-75 p6
Pièce sans Titre, 2 flutes. cf Ballades hébraïques.
Sonata brève, violin and harp. cf Ballades hébraïques.
Sonnets, contralto, flute and harp. cf Ballades hébraïques.
Suite polonaise, excerpts. cf Ballades hébraïques.
SABATA, Victor de
1934 Juventus. Mille e una notte: Prima quadro. La notte di platon.
        DEBUSSY: Jeux: Poeme danse. MOSSOLOV: Music of the machines.
        STRAVINSKY: Fireworks, op. 5. EIAR Symphony Orchestra, Rome,
        Santa Cecilia Symphony Orchestra, RAI Rome Orchestra, RAI Torino
        Orchestra, RAI Milan Orchestra; Victor de Sabata, Armando Rosa
        Parodi, Lorin Maazel, Aldo Ceccato. Rococo 2075.
            +NR 3-75 p3
Mille e una notte: Prima quadro. cf Juventus.
La notte di platon. cf Juventus.
SACHS
    Nachdem David. cf Telefunken TK 11569/1-2.
SACHSEN-WEIMAR, Anna Amalie of
    Songs: Auf dem Land und in der Stadt; Sie scheinen zu spielen.
        cf DG Archive 2533 149.
SAEVERUD, Harald
1935 Peer Gynt, op. 28. ROCHBERG: Night music. Louisville Orchestra;
        Robert Whitney. Louisville LOU 623.
            *St 6-75 p52
SAINT-GEORGES, Joseph Boulogne
    Ernestine: Scena. cf Quartet, strings, no. 1, op. 1, no. 1, C major.
1936 Quartet, strings, no. 1, op. 1, no. 1, C major. Symphonie concer-
        tante, 2 violins and orchestra, op. 13, G major. Symphony, no.
        1, op. 11, no. 1, G major. Ernestine: Scena. Faye Robinson,
        s; Miriam Fried, Jaime Laredo, vln; Juilliard Quartet, LSO;
        Paul Freeman. Columbia M 32781.
            +HF 6-74 p71   +NYT 5-12-74 pD26
            ++MQ 10-75 p645   *St 7-74 p105
            +-NR 7-74 p13
    Symphonie concertante, 2 violins and orchestra, op. 13, G major.
        cf Quartet, strings, no. 1, op. 1, no. 1, C major.
    Symphony, no. 1, op. 11, no. 1, G major. cf Quartet, strings, no.
        1, op. 1, no. 1, C major.
SAINT-SAENS, Camille
    Africa, op. 89. cf GOUNOD: Fantasy on the Russian national hymn.
    Air de ballet. cf Armstrong 721-4.
    Album, op. 72. cf Piano works (Vox 5476/7).
    Allegro appassionato, op. 43. cf Concerto, violoncello, no. 1,
        op. 33, A minor.
    Allegro appassionato, op. 43. cf Works, selections (HMV ASD 3058).
    Allegro appassionato, op. 43. cf Works, selections (Vox 5134).
    Allegro appassionato, op. 70. cf Piano works (Vox 5476/7).
    Ascanio: Ballet music, Adagio and variation. cf RCA LRL 1-5094.
    Bagatelles, op. 3. cf Piano works (Vox 5476/7).
    Le bonheur est chose légère. cf Columbia M 33307.
    Caprice, violin, op. 122. cf Works, selections (HMV ASD 3058).
    Caprice andalous, op. 122. cf Works, selections (Vox 5134).
    Caprice arabe, op. 96. cf Piano works (Vox 5476/7).

Caprice héroique, op. 106. cf Piano works (Vox 5476/7).
1937 Caprice on Danish and Russian airs, op. 79. Sonata, oboe and piano,
op. 166, D major. Sonata, clarinet and piano, op. 167, E flat
major. Samuel Baron, flt; Joseph Rabbai, clt; Ronald Roseman,
ob; Gilbert Kalish, pno. Desto DC 7146. (also Peerless PRCM
206)

  +Gr 4-75 p1826   ++MJ 3-75 p34
  +HF 10-74 p116   ++NR 1-75 p6
  +-HFN 5-75 p138   ++St 12-74 p141

Caprice on Danish and Russian airs, op. 79. cf Classic Record
Library SQM 80-5731.
Caprice sur des airs de ballet d'Alceste. cf Piano works (Vox
5476/7).
Le carnaval des animaux. cf BIZET: Jeux d'enfants, op. 22.
Le carnaval des animaux. cf PROKOFIEV: Peter and the wolf, op. 67.
Le carnaval des animaux. cf HMV SLS 5033.
Le carnaval des animaux: Le cynge. cf Works, selections (HMV ASD
3058).
Le carnaval des animaux: Le cynge. cf Discopaedia MB 1008.
Le carnaval des animaux: The swan. cf L'Oiseau-Lyre DSLO 7.
Les cloches du soir, op. 85. cf Piano works (Vox 5476/7).
1938 Concerto, piano, no. 2, op. 22, G minor. Concerto, piano, no. 5,
op. 103, F major. Gabriel Tacchino, pno; Luxembourg Radio
Orchestra; Louis de Froment. Candide (Q) QCE 31080.

  ++HF 7-75 p82   +SFC 12-22-74 p20
  ++NR 1-75 p5

Concerto, piano, no. 3, op. 29, E flat major: Allegro. cf Piano
works (Vox 5476/7).
Concerto, piano, no. 5, op. 103, F major. cf Concerto, piano,
no. 2, op. 22, G minor.
Concerti, violin (3). cf Works, selections (Vox 5134).
1939 Concerto, violin, no. 3, op. 61, B minor. Havanaise, op. 83.
Louis Kaufman, vln; Netherlands Philharmonic Orchestra; Maurits
Van Den Berg. Orion ORS 75177.

  +-HF 9-75 p89   +-St 9-75 p111
  -NR 6-75 p7

Concerti, violoncello (2). cf Works, selections (Vox 5134).
1940 Concerto, violincello, no. 1, op. 33, A minor. LALO: Concerto,
violoncello, D minor. André Navarra, vlc; Lamoureux Orchestra;
Charles Munch. Musical Heritage Society MHS 3023.

  +St 6-75 p104

1941 Concerto, violoncello, no. 1, op. 33, A minor. Concerto, violoncello, no. 2, op. 119, D minor. Suite, violoncello, op. 16.
Allegro appassionato, op. 43. Christine Walevska, vlc; Monte
Carlo Opera Orchestra; Eliahu Inbal. Philips 6500 459. Tape
(c) 7300 343.

  +-HF 9-75 p89   +NR 3-75 p4
  +HFN 7-75 p91   ++St 6-75 p106

Concerto, violoncello, no. 1, op. 33, A minor. cf Works, selections
(HMV ASD 3058).
Concerto, violoncello, no. 2, op. 119, D minor. cf Concerto, violon-
cello, no. 1, op. 33, A minor.
Danse macabre, op. 40. cf Decca DPA 519/20.
Danse macabre, op. 40. cf HMV ASD 3008.
Danse macabre, op. 40. cf RCA VH 020.
Danse macabre, op. 40. cf RCA ARL 1-5007.

Le déluge, op. 45: Prelude.   cf Works, selections (HMV ASD 3058).
Le déluge, op. 45: Prelude.   cf Discopaedia MB 1007.
Le déluge, op. 45: Prelude.   cf Discopaedia MB 1008.
Duettino, op. 11. cf PIano works (Vox 5476/7).
Etudes, op. 52. cf Piano works (Vox 5476/7).
Etudes, op. 111. cf Piano works (Vox 5476/7).
Etudes, left hand alone, op. 135. cf Piano works (Vox 5476/7).
Feuillet d'album, op. 81. cf Piano works (Vox 5476/7).
Feuillet d'album, op. 169. cf Piano works (Vox 5476/7).
Fugues, op. 161. cf Piano works (Vox 5476/7).
Gavotte, op. 23. cf Piano works (Vox 5476/7).
Havanaise, op. 83. cf Concerto, violin, no. 3, op. 61, B minor.
Havanaise, op. 83. cf Works, selections (Vox 5134).
Havanaise, op. 83. cf BIZET: Carmen fantasy, op. 25.
Havanaise, op. 83. cf BRUCH: Concerto, violin, no. 2, op. 44, D
    minor: 2d movement.
Havanaise, op. 83. cf CHAUSSON: Poème, op. 25.
Havanaise, op. 83. cf DEBUSSY: Danse sacrée et profane.
Havanaise, op. 83. cf Discopaedia MB 1010.
Havanaise, op. 83. cf Hungaroton SLPX 11677.
Havaniase, op. 83. cf RCA ARM 4-0942/7.
Henry VIII: Donc, le pape est hostile...Qui donc commande. cf RCA
    ARL 1-0851.
1942 Introduction and rondo capriccioso, op. 28. TCHAIKOVSKY: Concerto,
    violin, op. 35, D major. Eugene Fodor, vln; NYP; Erich Leins-
    dorf. RCA ARL 1-0781. Tape (c) ARK 1-0781 (ct) ARS 1-0781 (Q)
    ARD 1-0781 Tape (ct) ART 1-0781.
        +-Gr 5-75 p1973   +SFC 11-17-74 p32
        +HF 3-75 p88   +St 3-75 p109
        +-NR 12-74 p6
Introduction and rondo capriccioso, op. 28. cf Works, selections
    (Vox 5134).
Introduction and rondo capriccioso, op. 28. cf BIZET: Carmen
    fantasy, op. 25.
Introduction and rondo capriccioso, op. 28. cf CHAUSSON: Poème,
    op. 25.
Introduction and rondo capriccioso, op. 28. cf DEBUSSY: Danse
    sacrée et profane.
Introduction and rondo capriccioso, op. 28. cf TCHAIKOVSKY: Con-
    certo, violin, op. 35, D major.
Introduction and rondo capriccioso, op. 28. cf RCA ARM 4-0942/7.
Marche interalliée, op. 155. cf Piano works (Vox 5476/7).
Marche militaire française: Africa, Valse mignonne, Reverie a
    Blidah, Suite Algérienne. cf Rococo 2049.
Mazurkas, opp. 21, 24, 66. cf Piano works (Vox 5476/7).
Menuet et valse, op. 56. cf Piano works (Vox 5476/7).
Morceau de concerto, op. 62. cf Works, selections (Vox 5134).
Une nuit à Lisbonne, op. 63. cf Piano works (Vox 5476/7).
Pas redoublé, op. 86. cf Piano works (Vox 5476/7).
1943 Piano works : Album, op. 72. Concerto, piano, no. 3, op. 29, E
    flat major: Allegro. Allegro appassionato, op. 70. Bagatelles,
    op. 3 (6). Caprice arabe, op. 96, Caprice héroïque, op. 106.
    Caprice sur des airs de ballet d'Alceste. Les cloches du soir,
    op. 85. Duettino, op. 11. Etudes, op. 52 (6). Etudes, op.
    111 (6). Etudes, left hand alone, op. 135 (6). Feuillet d'
    album, op. 81. Feuillet d'album, op. 169. Fugues, op. 161 (6).
    Gavotte, op. 23. Marche interalliée, op. 155. Mazurkas, opp.

21, 24, 66 (3). Menuet et valse, op. 56. Une nuit à Lisbonne,
op. 63. Pas redoublé, op. 86. Polonaise, op. 77. Scherzo,
op. 87. Souvenir d'Ismailla, op. 100. Souvenir d'Italie,
op. 80. Suite, op. 90. Thème varié, op. 97. Valse canariote,
op. 88. Valse gaie, op. 139. Valse langoureuse, op. 120.
Valse mignonne, op. 104. Valse nonchalante, op. 110. Variations on a theme by Beethoven, op. 35. Marylène Dosse, Annie
Petit, pno. Vox SVBX 5476/7 (6).
    +-HF 9-75 p79   ++NR 6-75 p14
Polonaise, op. 77. cf Pianow works (Vox 5476/7).
Romance, op. 67. cf BEETHOVEN: Sonata, horn, op. 17, F major.
Romance, violin and orchestra, op. 48. cf Works, selections (Vox
5134).
1944 Samson et Dalila. Christa Ludwig, ms; Albert Gassner, James King,
Heinrich Weber, t; Bernd Weikl, bar; Richard Kogel, Alexander
Malta, Peter Schranner, bs; Bavarian Radio Orchestra and Chorus;
Giuseppe Patané. Eurodisc 86 977 XR (3). (Q) 86 977 XR. (also
RCA LRL 3-5017, ARL 3-0662)
        -Gr 6-74 p111   +-ON 4-12-75 p36
    +-HF 7-74 p100   +-Op 7н74 p612
    +-HF 7-75 p82   +-RR 6-74 p22
    +-HF 9-74 p113 Quad   +-SFC 2-23-75 p22
    ++MJ 3-75 p25   ++St 7-74 p109
    +NR 3-75 p9
Samson et Dalila: Je viens celebrer. cf Discophilia KGC 2.
Samson et Dalila: Mon coeur s'ouvre à ta voix. cf Rubini RS 300.
Scherzo, op. 87. cf Piano works (Vox 5476/7).
Sonata, clarinet and piano, op. 167, E flat major. cf Caprice on
Danish and Russian airs, op. 79.
Sonata, oboe and piano, op. 166, D major. cf Caprice on Danish
and Russian airs, op. 79.
Sonata, violin, no. 1, op. 75, D minor. cf RCA ARM 4-0942/7.
Souvenir d'Ismailla, op. 100. cf Piano works (Vox 5476/7).
Souvenir d'Italie, op. 80. cf Piano works (Vox 5476/7).
Suite, op. 90. cf Piano works (Vox 5476/7).
Suite, violoncello, op. 16. cf Concerto, violoncello, no. 1,
op. 33, A minor.
1945 Symphonies, A major, F major. ORTF; Jean Martinon. Angel S 37089.
(also HMV (Q) ASD 3138)
        ++Gr 11-75 p828 Quad   +NR 8-75 p3
        +HF 8-75 p75   +RR 12-75 p61
        +HFN 12-75 p163   ++St 11-75 p135
1946 Symphony, no. 1, op. 2, E flat major. Symphony, no. 2, op. 55,
A minor. ORTF; Jean Martinon. HMV ASD 2946. Tape (c) TC ASD
2946. (also Angel S 36995)
        ++Gr 1-74 p1387   +NR 5-74 p5
        +Gr 10-75 p721 tape   ++NYT 5-19-74 pD27
        +HF 7-74 p100   +RR 12-73 p74
        +HFN 10-75 p155 tape   ++St 9-74 p128
Symphony, no. 2, op. 55, A minor. cf Symphony, no. 1, op. 2, E
flat major.
1947 Symphony, no. 3, op. 78, C minor. Virgil Fox, org; PO; Eugene
Ormandy. RCA ARL 1-0484. (Q) ARD 1-0484 Tape (ct) ART 1-0484.
        +-HF 9-74 p101   ++NR 8-74 p3
        ++MJ 3-75 p26   ++St 9-74 p128
Thème varié, op. 97. cf Piano works (Vox 5476/7).

Valse canariote, op. 88. cf Piano works (Vox 5476/7).
Valse gaie, op. 139. cf Piano works (Vox 5476/7).
Valse langoureuse, op. 120. cf Piano works (Vox 5476/7.
Valse mignonne, op. 104. cf Piano works (Vox 5476/7).
Valse nonchalante, op. 110. cf Piano works (Vox 5476/7).
Variations on a theme by Beethoven, op. 35. cf Piano works
     (Vox 5476/7).
Wedding cake, piano and strings, op. 76. cf Works, selections
     (HMV ASD 3058).
1948 Works, selections: Allegro appassionato, op. 43. Caprice, violin,
     op. 122 (arr. Ysaye). Le carnaval des animaux: Le cynge. Con-
     certo, violoncello, no. 1, op. 33, A minor. Le déluge, op. 45:
     Prelude. Wedding cake, piano and strings, op. 76. Paul Torte-
     lier, vlc; Yan Pascal Tortelier, vln; Maria de la Pau, pno;
     Robert Johnston, hp; Birmingham City Symphony Orchestra; Louis
     Frémaux. HMV ASD 3058. Tape (c) TC ASD 3058.
          +Gr 7-75 p192   +-RR 6-75 p46
          -Gr 6-75 p106 tape   +-RR 7-75 p72
          +HFN 6-75 p91
1949 Works, selections: Allegro appassionato, op. 43. Caprice andalous,
     op. 122. Concerti, violoncello (2). Concerti, violin (3).
     Havanaise, op. 83. Introduction and rondo capriccioso, op. 28.
     Morceau de concert, op. 62. Romance, violin and orchestra,
     op. 48. Ruggiero Ricci, vln; Laszlo Varga, vlc; Luxembourg
     Radio Orchestra, Westphalian Symphony Orchestra, PH; Pierre
     Cao, Reinhard Peters, Siegfried Landau. Vox (Q) SVBX 5134 (3).
          ++NR 10-75 p3   ++SFC 9-28-75 p30
SAINTE-COLOMBE, Sieur de
     Concerts à deux violes esgales (3). cf MARAIS: Suite, viola da
          gamba and harpsichord, E minor.
SAINZ DE LA MAZA, Eduardo
     Homenaje à la guitarra. cf LAURO: Danza negra.
SALIERI, Antonio
     Minuet. cf DG Archive 2533 182.
SALLINEN, Aulis
     Cadenze. cf BIS LP 18.
SALZEDO, Carlos
     Jeux d'eau. cf Saga 5356.
SAMMARTINI, Giuseppe
     Concerto, recorder and strings, F major. cf HANDEL: Concerto,
          recorder and strings, B flat major.
     Concerto, recorder and strings, F major. cf Telefunken SMA 25121/
          1-3 (3).
     Concerto, soprano recorder and strings. cf MARCELLO, A.: Concerto,
          oboe and strings.
     Concerto, viola pomposa, C major. cf PORPORA: Concerto, violon-
          cello, G major.
     Sonata, violin, no. 4, A major: Andante. cf Discopaedia MB 1006.
SANCAN, Pierre
     Sonatine pour flute et piano. cf PAGANINI: Caprice, op. 1, no.
          24, A minor.
SANDRIN, Pierre
     Doulce memoire. cf L'Oiseau-Lyre 12BB 203/6.
     O combien. cf Hungaroton SLPX 11549.
SANTOLIQUIDO, Francesco
     Poesi persiane (3). cf DEBUSSY: Le promenoir des deux amants.

SANZ, Gaspar
    Canarios. cf DG Archive 2533 172.
    Castillian dances (4). cf Swedish Society SLT 33189.
    Españoleta. cf DG Archive 2533 172.
    Gallard y Villano. cf DG Archive 2533 172.
    Passacalle de la Cavalleria de Napoles. cf DG Archive 2533 172.
    Suite española. cf BACH: Bourrée, E minor.
    Suite española. cf Everest SDBR 3380.
SAPOJNIKOV
    Tebe poem. cf Harmonia Mundi HMU 133.
SARAI, Tibor
1950 Diagnosis '69. Quartet, strings, no. 2. Serenade, strings.
        Symphony, no. 1. Sándor Palcsó, t; Kodály Quartet, HCO, HRT
        Orchestra; György Lehel, Miklós Erdélyi. Hungaroton SLPX 11636.
          +HFN 5-75 p138   +RR 5-75 p39
          +NR 4-75 p3
    Quartet, strings, no. 2. cf Diagnosis '69.
    Serenade, strings. cf Diagnosis '69.
    Symphony, no. 1. cf Diagnosis '69.
SARASATE, Pablo
    Caprice basque, op. 24. cf Discopaedia MB 1003.
    Carmen fantasy, op. 25. cf Discopaedia MB 1010.
    Carmen fantasy, op. 25. cf RCA ARM 4-0942/7.
    Danzas españolas, nos. 2, 6. cf Discopaedia MB 1003.
    Danzas españolas, nos. 2, 6. cf Discopaedia MB 1010.
    Danzas españolas, op. 21, no. 1: Malagueña; no. 2: Habañera;
        op. 23, no. 2: Zapateado. cf RCA ARM 4-0942/7.
    Danzas españolas, op. 22, no. 1: Romanza andaluza. cf Connoisseur
        Society CS 2070.
    Danzas españolas, op. 23, no. 2: Zapateado. cf RCA ARM 4-0942/7.
    Danza española, op. 26, no. 2, C major. cf Discopaedia MB 1005.
    Introduction and caprice: Jota. cf Discopaedia MB 1003.
    Introduction and tarantelle. cf RCA ARM 4-0942/7.
    Introduction and tarantelle, op. 43. cf Discopaedia MB 1009.
    Introduction and tarantelle, op. 43. cf Orion ORS 74160.
    Introduction and tarantelle: Tarantelle. cf Discopaedia MB 1003.
    Jota navarra, op. 20, no. 2. cf Discopaedia MB 1004.
    Miramar Zortzico, op. 42. cf Discopaedia MB 1003.
    Zigeunerweisen, op. 20. cf Discopaedia MB 1003.
    Zigeunerweisen, op. 20, no. 1. cf BIZET: Carmen fantasy, op. 25.
    Zigeunerweisen, op. 20, no. 1. cf Discopaedia MB 1004.
    Zigeunerweisen, op. 20, no. 1. cf RCA ARM 4-0942/7.
SARKOZY, István
1951 Concerto grosso (Ricordanze, I). Shepherd's ballad. Sinfonia con-
        certante, clarinet and 24 strings. Songs on poems by András
        Mezei (3). Erika Sziklay, s; Lorand Szücs, pno; Béla Kovács,
        clt; HCO, HRT Orchestra; György Lehel. Hungaroton SLPX 11667.
          +-HFN 5-75 p138   +RR 6-75 p47
          +NR 4-75 p14
    Shepherd's ballad. cf Concerto grosso.
    Sinfonia concertante, clarinet and 24 strings. cf Concerto grosso.
    Songs on poems by András Mezei. cf Concerto grosso.
SAROYAN
    Crickets. cf 1750 Arch 1752.
SARTI, Giuseppe
    Gospodiin pomiluj Ny. cf Russian oratorio.

Lungi dal caro bene. cf Decca SXL 6629.
1952 Russian oratorio. Gospodiin pomiluj Ny. Alena Miková, s; Věra
        Hubáčková, ms; Marie Mrazová, con; Czech Philharmonic Chorus;
        Bratislava Radio Orchestra; Václav Smetáček. Musical Heritage
        Society MHS 1735. Tape (c) MHC 2084.
              +HF 8-75 p116 tape   ++St 8-74 p71
SARY, László
        Catocoustics, 2 pianos. cf JENEY: Alef: Homage á Schonberg.
        Immaginario, no. 1. cf JENEY: Alef: Homage á Schonberg.
        Incanto. cf JENEY: Alef: Homage á Schonberg.
        Sounds for piano. cf Hungaroton SLPX 11692.
SATIE, Erik
1953 Airs à faire fuir (3). Gnossiennes (6). Gymnopédies, nos. 1-3.
        Nouvelles pièces froides (3). Ogives (2). Rêveries nocturnes
        (2). Sarabandes, nos. 1-3. Songe creux. Peter Kraus, Mark
        Bird, gtr. Orion ORS 74163.
              +Audio 12-75 p105   +NR 11-74 p12
        Les aventures de Mercure. cf MILHAUD: Saudades do Brazil.
        La belle excentrique. cf MILHAUD: Saudades do Brazil.
        Le chapelier. cf Columbia M 32231.
        Gnossiennes (6). cf Airs à faire fuir (3).
        Gymnopédies, nos. 1-3. cf Airs à faire fuir (3).
        Gymnopédies, nos. 1-3. cf Columbia M 32070.
        Gymnopédies: no. 1, Lent et grave, no. 3; Lent et douloureux. cf
            HONEGGER: Pacific 231.
        Gymnopédie, no. 1. cf Connoisseur Society (Q) CSQ 2065.
        Jack in the box (orch. Milhaud). cf MILHAUD: Saudades do Brazil.
        Nouvelles pièces froides (3). cf Airs à faire fuir (3).
        Ogives (2). cf Airs à faire fuir (3).
        Petites pièces monées. cf IBERT: Paris 32.
        Rêveries nocturnes (2). cf Airs à faire fuir (3).
        Sarabandes, nos. 1-3. cf Airs à faire fuir (3).
        Songe creux. cf Airs à faire fuir (3).
        Sports and divertissements. cf POULENC: The story of Babar, the
            little elephant.
SAURET, Emile
        Will-o-the-wisp. cf Discopaedia MB 1005.
SAYVE,   de
        Kryie. cf Panton 010335.
SCARLATTI, Alessandro
        Concerto, alto recorder and 2 violins. cf MARCELLO, A.: Concerto,
            oboe and strings.
        Domine, refugium factus es nobis. cf O Magnum mysterium.
        Endimione and Cintia, aria. cf BACH: Brandenburg concerto, no. 2,
            S 1047, F major.
        Endimione e Cinta: Se geloso e il mio core. cf Columbia M 33307.
1954 O magnum mysterium. Domine, refugium factus est nobis. SCARLATTI,
        D.: Stabat Mater. Schütz Choir; Roger Norrington. Argo ZRG 768.
              +Gr 7-74 p253   ++MQ 10-75 p641
              ++HF 3-75 p86   +NR 12-74 p9
              +MJ 2-75 p39   +RR 7-74 p78
        Pompeo: Gia il sole dal Gange. cf Pearl SHE 511.
        Pompeo: Gia il sole dal Gange. cf London OS 26391.
        La rosaura. cf Telefunken AJ 6-41867.
        Sonata, recorder and 2 violins, A minor. cf Telefunken SMA 25121/
            1-3.

Le violette. cf Decca SXL 6629.
SCARLATTI, Domenico
Allegretto, D minor. cf Pandora PAN 101.
Presto, E major. cf Pandora PAN 101.
1955 Sonatas, guitar, L 387, G major; L 238, A major; L 162, D major;
L 257, E major; L 418, D major; L 103, G major; L 349, G
major; L 203, A major; L 497, D major; L 366, D minor; L 23,
E major; L 383, E minor. Leo Brouwer, gtr. Erato STU 70876.
+Gr 5-75 p1999   +RR 4-75 p53
Sonata, guitar, E minor. cf BACH: Bourrée, E minor.
Sonatas, guitar (2). cf RCA ARL 1-0864.
Sonata, guitar, L 23 (K 380), E major. cf Swedith Society SLT
33189.
Sonatas, guitar, L 79, G major; L 352, E minor; L 187, A minor;
L 483, A major. cf Angel S 37015.
Sonata, harp, G major. cf RCA LRL 1-5087.
1956 Sonatas, harpsichord, E major, K 215, K 216, K 264; E minor, K 263;
D major, K 490, K 491, K 492; D minor, K 52; C major, K 308, K
309. Gustav Leonhardt, hpd. BASF BAC 3068.
+Gr 1-75 p1371   ++RR 1-75 p52
1957 Sonatas, harpsichord (36). Fernando Valenti, hpd. Westminster
WGN 8208-3 (3).
++SFC 6-1-75 p21
1958 Sonatas, harpsichord, K 2, 20, 24, 29, 33, 44, 46, 52, 96, 113-116,
119-120, 140-141, 175, 208-209, 215-216, 238-239, 242-243, 252-
253, 259-264, 402-403, 420-421, 426-427, 460-461, 470-471, 490-
492, 513, 518-519, 358-359, 422-423, 428-429, 434-436, 441-442,
478-479, 481, 524-525, 544-545, 552-553. Huguette Dreyfus, hpd.
Telefunken EK 6-35086 (4).
+Gr 4-75 p1840
1959 Sonatas, harpsichord, L 3 (K 502), C major; L 10 (K 84), C minor;
L 14 (K 492), D major; L 189 (K 184), F minor; L 198 (K 296),
F major; L 204 (K 105), G major; L 209 (K 455), G major; L 223
(K 532), A minor; L 238 (K 208), A major; L 281 (K 239), F minor;
L 389 (K 375), G major; L 422 (K 141), D minor; L 483 (K 322),
A major; L 23 (K 380), E major; L 31 (K 318), F sharp major;
L 35 (K 319), F sharp major; L 93 (K 149), A minor; L 148 (K
261), B major; L 205 (K 487), C major; L 225 (K 381), E major;
L 256 (K 247), C sharp minor; L 260 (K 246), C sharp minor; L 446
(K 262), B major; L 457 (K 132), C major; L supp. 31 (K 83), A
major. Kenneth Cooper, hpd. Vanguard VSD 71201/2.
+Audio 12-75 p104   +NR 8-75 p14
+HF 11-75 p115   ++SFC 11-30-75 p34
1960 Sonatas, harpsichord, L 314 (K 511), L 339 (K 512), L 42 (K 217),
L 392 (K 218), L 451 (K 422), L 102 (K 423), L 348 (K 244), L
450 (K 245), L 193 (K 499), L 492 (K 500). Gilbert Rowland,
hpd. Keyboard KGR 1001.
+HFN 8-75 p83   +RR 6-75 p75
1961 Sonatas, harpsichord, K 371 (L 17), E flat major; K 430 (L 463),
D major; K 444 (L 420), D minor; K 28 (L 373), E major; K 446
(L 433), F major; K 6 (L 479), F major; K 141 (L 422), D minor;
K 145 (L 369), D major; K 268 (L 41), A major; K 203 (L 380),
E minor; K 445 (L 385), F major; K 517 (L 266), D minor; K 125
(L 487), G major; K 436 (L 109), D major; K 87 (L 33), B minor;
K 2 (L 388), G major; K 417 (L 462), D minor; K 12 (L 486), G
major; K 17 (L 384), F major; K 10 (L 370), D minor; K 9 (L 413),
D minor; K 159 (L 104), C major; K 533 (L 395), A major. Anthony

di Bonaventura, pno.   Connoisseur CS 2044.
    +HF 8-73 p96   +RR 4-75 p57
    +HFN 9-75 p103   ++SFC 3-4-73 p33
    +NR 8-73 p15

1962 Sonatas, harpsichord, K 2 (L 388), G major; K 20, (L 375), E major;
    K 24 (L 495), A major; K 29 (L 461), D major; K 33 (L 424), D
    major; K 44 (L 432), F major; K 46 (L 25), E major; K 52 (L 267),
    D minor; K 96 (L 465), D major; K 113 (L 345), A major; K 114
    (L 344), A major; K 115 (L 407), C minor; K 116 (L 452), C minor;
    K 119(L 415), D major; K 120 (L 215), D minor; K 140 (L 107),
    D major; K 141 (L 422), D minor; K 175 (L 429), A minor; K 208
    (L 238), A major; K 209 (L 428), A major; K 215 (L 323), E major;
    K 216 (L 273), E major; K 238 (L 27), F minor; K 239 (L 281),
    F minor; K 242 (L 202), C major; K 243 (L 253), C major; K 252
    (L 159), E flat major; K 253 (L 320), E flat major; K 259 (L 103);
    K 260 (L 124), G major; K 261 (L 148), B major; K 262 (L 446),
    B major; K 263 (L 321), E minor; K 264 (L 466), E major; K 402
    (L 427), E minor; K 403 (L 470), E major; K 420 (L 2), C major;
    K 421 (L 252), C major; K 426 (L 128), G minor; K 427 (L 286),
    G major; K 460 (L 324), C major; K 461 (L 8), C major; K 470
    (L 304), G major; K 471 (L 82), G major; K 490 (L 206), D major;
    K 491 (L 164), D major; K 492 (L 14), D major; K 513 (L 3), C
    major; K 518 (L 116), F major; K 519 (L 475), F minor; K 358
    (L 412), D major; K 359 (L 448), D major; K 422 (L 451), C major;
    K 423 (L 2), C major; K 428 (L 131), A major; K 429 (L 132),
    A major; K 434 (L 343), D minor; K 435 (L 361), D major; K 436
    (L 109), D major; K 441 (L 39), B flat major; K 442 (L 319), B
    flat major; K 478 (L 12), D major; K 479 (L 16), D major; K 481
    (L 187), F minor; K 524 (L 283), F major; K 525 (L 188), F major;
    K 544 (L 497), B flat major; K 545 (L 500), B flat major; K 552
    (L 421), D minor; K 553 (L 425), D minor. Huguette Dreyfuss,
    hpd. Telefunken SBA 25127-5/1-4 (4).
       +RR 3-75 p52
  Sonatas, harpsichord, K 87, B minor; K 201, G major; K 370, E flat
    major; K 371, E flat major. cf Saga 5402.
  Sonata, violin, L 413, E minor: Pastorale. cf Discopaedia MB 1006.
  Stabat Mater. cf SCARLATTI, A.: O magnum mysterium.

SCHARWENKA, Philipp
1963 Sonata, violin and piano, op. 110, B minor. SCHARWENKA, X.: Sonata,
    violin and piano, no. 1, op. 2, D minor. Robert Zimansky, vln;
    Leonore Klinckerfuss, Gordon Steel, pno. Genesis GS 1056.
       +NR 8-75 p7

SCHARWENKA, Xaver
  Polish dance, op. 31, no. 1. cf Rococo 2049.
  Sonata, violin and piano, no. 1, op. 2, D minor. cf SCHARWENKA, P.:
    Sonata, violin and piano, op. 110, B minor.

SCHAT, Peter
1964 Canto general. To you. Lucia Kerstens, ms; Vera Beths, vln;
    Reinbert de Leeuw, Maarten Bon, Stanley Hoogland, Bart Berman,
    pno; Bert van Dijk, Christian Engelse, org; Hans Bredenbeek,
    Ton Burmanje, Louis Ignatius Gall, Harmoed Greef, Dick Hoogeveen,
    Franck Noya, Jorge Oraison, Frank Wesstein, Bob Zimmermann, gtr;
    Peter Schat. Donemus Audio-Visual DAVS 7475/1.
      +-RR 10-75 p88
  Thema. cf KEURIS: Concerto, saxophone.
  To you. cf Canto general.

SCHEIDEMANN, Heinrich
  Jesu wolt'st uns weisen. cf Telefunken DX 6-35265.
  Preambulum, D minor. cf Pelca PSRK 41013/6.
  Praeambulum and canzona, F major. cf Saga 5374.
SCHEIDT, Samuel
  Benedicamus Domino. cf Turnabout TV 37071.
  Canzon cornetto. cf Nonesuch H 71301.
  Canzona Aechiopicam. cf Turnabout TV 37071.
  Canzona bergamasca. cf Turnabout TV 37071.
  Canzona gallicam. cf Turanbout TV 37071.
  Ei, du feiner Reiter, variations. cf Saga 5374.
  Galliard battaglia. cf Turnabout TV 37071.
  Tabulatura nova: Veni creator. cf Pelca PSRK 41013/6.
  Wendet euch um ihr Anderlein. cf Turnabout TV 37071.
SCHEIN, Johann
  Allemande-Tripla. cf DG Archive 2533 184.
  Banchetto musicale: Suite, no. 1, G major; no. 2, D minor. cf
    PRAETORIUS: Hosianna dem Sohne Davids.
  Banchetto musicale: Suite, no. 3. Crystal S 202.
  Die mit Tränen Säen. cf Klavier KS 536.
  Paduana. cf Crystal S 201.
SCHERZER
  Bayerischer Defiliermarsch. cf BASF 292 2116-4.
SCHICKHARDT, Johann Christian
  Trio sonata, F major. cf Pelca PSR 40589.
SCHINDLER
  Souvenir poetique. cf Discopaedia MB 1006.
SCHMELZER, Johann
  Sonata a 7 flauti. cf HMV SLS 5022.
SCHMID d. "A", G.
  Englischer Tanz. cf DG Archive 2533 111.
  Tanz Du has mich wollen nemmen. cf DG Archive 2533 111.
SCHMID, Bernhard
  Passamezzo and saltarello. cf Saga 5374.
SCHMIDT, Franz
  Notre Dame: Intermezzo. cf HMV SLS 5019.
SCHMITT, Florent
  Dionysiaques, op. 62, no. 1. cf FAURE: Chant funéraire, op. 117.
SCHNABEL
  Four old Vienna waltzes. cf Musical Heritage Society MHS 1959.
SCHOLZ
  Torgauer Marsch. cf DG 2721 077.
SCHONBERG, Arnold
1965 Chamber symphony, no. 1, op. 9, E major (arrangement for orchestra,
       op. 9b). Chamber symphony, no. 2, op. 38, E flat major. Frank-
       furt Radio Symphony Orchestra; Eliahu Inbal. Philips 6500 923.
         +--Gr 12-75 p1045   +RR 11-75 p55
         +HFN 12-75 p163
     Chamber symphony, no. 1, op. 9, E major. cf Pierrot Lunaire, op. 21.
     Chamber symphony, no. 1, op. 9, E major. cf Works, selections
       (Decca SXLK 6660/4).
     Chamber symphony, no. 1, op. 9, E major. cf BERG: Sonata, piano,
       op. 1.
     Chamber symphony, no. 2, op. 38, E flat major. cf Chamber symphony,
       no. 1, op. 9, E major.
1966 Concerto, string quartet and orchestra. Trio, strings, op. 45.

Lenox Quartet, LSO; Harold Farberman. Desto DC 7170. (also
Peerless PRCM 205)
+—Gr 7-75 p192   +LJ 3-75 p34
+HFN 7-75 p88   +NR 1-75 p6
Die eiserne Brigade. cf Works, selections (Decca SXLK 6660/4).
Fantasia, violin and piano, op. 47. cf Works, selections (Decca
SXLK 6660/4).
Fantasia, violin and piano, op. 47. cf DALLAPICCOLA: Studi, violin
and piano.
1967 Gurrelieder. Marita Napier, s; Yvonne Minton, con; Jess Thomas,
Kenneth Bowen, t; Siegmund Nimsgern, Günter Reich, bs; BBC Sin-
gers, BBC Chorus and Choral Society, Goldsmiths' Choral Union,
London Philharmonic Chorus, Mens' Voices; BBC Symphony Orchestra;
Pierre Boulez. CBS 78264 (2). (also Columbia M2 33303 (2) (Q)
M2Q 33303)
+Gr 4-75 p1855   +ON 9-75 p60
+—HF 8-75 p96   +—RR 4-75 p64
+MT 7-75 p631   +SFC 11-9-75 p22
+NR 7-75 p7   +St 9-75 p120
+—NYT 9-21-75 pD18
1968 Gurrelieder. Martina Arroyo, s; Janet Baker, ms; Alexander Young,
Niels Møller, t; Odd Wolstad, bs; Julius Patzak, speaker; Dan-
ish State Radio Symphony and Concert Orchestras and Chorus;
János Ferencsik. HMV SLS 884 (2). (also EMI Odeon SLS 884)
+—Gr 11-74 p945   +—NYT 9-21-75 pD18
+—HF 8-75 p96   +—RR 11-74 p90
Herzgewächse, op. 20. cf Works, selections (Decca SXLK 6660/4).
Impressions, op. 38. cf BRAHMS: Variations on a theme by Paganini,
op. 35, A minor.
Klavierstück, op. 33a and b. cf BARTOK: Out of doors.
Lied der Waldtaube. cf Works, selections (Decca SXLK 6660/4).
Little pieces, piano, op. 19 (6). cf Piano works (DG 2530 531).
Little pieces, piano, op. 19 (6). cf Piano works (Nonesuch H 71309).
1969 Moses und Aron. Felicity Palmer, Jane Manning,s; Gillian Knight,
ms; Richard Cassilly, John Winfield, t; John Noble, Roland
Hermann, bar; Richard Angas, Michael Rippon, bs; Günter Reich,
speaker; BBC Singers, Orpheus Boys' Choir; BBC Symphony Orches-
tra; Pierre Boulez. CBS 79201 (2).
+Gr 11-75 p900   +—RR 11-75 p35
+HFN 11-75 p167.
1970 Moses und Aron. Eva Csapó, s; Louis Devos, Roger Lucas, t; Werner
Mann, bs; Günter Reich, speaker, Vienna Boys' Choir, Members;
Austrian Radio Symphony Orchestra and Chorus; Micahel Gielen.
Philips 6700 084 (2).
+Gr 1-75 p1383   +ON 2-22-75 p28
+HF 2-75 p82   +RR 12-74 p23
+LJ 11-75 p43   +SFC 2-23-75 p22
+MJ 2-75 p31   +SR 1-25-75 p50
+NR 1-75 p13   +St 2-75 p114
+NYT 1-5-75 pD8   +STL 1-12-75 p36
Nachtwandler. cf Works, selections (Decca SXLK 6660/4).
Der neue Klassizimus, op. 28, no. 3. cf Works, selections (Decca
SXLK 6660/4).
Ode to Napoleon, op. 41. cf Works, selections (Decca SXLK 6660/4).
1971 Pelleas und Melisande, op. 5. BPhO; Herbert von Karajan. DG 2530
485.

                +Audio 9-75 p70   +SR 5-3-75 p33
                +NR 5-75 p3
1972 Pelleas und Melisande, op. 5. WEBERN: Passacaglia, op. 1. CPhO;
      Hans Swarowsky. Supraphon 110 1505.
                +-Gr 11-75 p833   ++NR 9-75 p4
                +HFN 9-75 p103
      Pelleas und Melisande, op. 5. cf BERG: Lyric suite: Pieces.
1973 Piano works: Pieces, op. 11 (3). Pieces, op. 23 (5). Pieces,
      opp. 33a and 33b. Little pieces, op. 19 (6). Suite, piano,
      op. 25. Maurizio Pollini, pno. DG 2530 531.
                ++Gr 5-75 p1999   +NR 12-75 p13
                +HF 11-75 p116   ++RR 6-75 p76
                ++HFN 6-75 p92
1974 Piano works: Pieces, op. 11 (3). Little pieces, op. 19 (6).
      Pieces, op. 23 (5). Pieces, op. 33a and 33b. Suite, op. 25.
      Paul Jacobs, pno. Noneusch H 71309.
                +HF 6-75 p86   +NYT 4-27-75 pD19
                +HFN 7-75 p88   ++SFC 6-15-75 p24
                ++NR 6-75 p13   +St 9-75 p120
      Pieces, chamber orchestra (3). cf Works, selections (Decca SXLK
      6660/4).
      Pieces, piano, op. 11 (3). cf Piano works (DG 2530 531).
      Pieces, piano, op. 11 (3). cf Piano works (Nonesuch H 71309).
      Pieces, pinao, op. 23 (5). cf Piano works (DG 2530 531).
      Pieces, piano, op. 23 (5). cf Piano works (Nonesuch H 71309).
      Pieces, piano, opp. 33a and 33b. cf Piano works (DG 2530 531).
      Pieces, piano, opp. 33a and 33b. cf Piano works (Nonesuch H 71309).
1975 Pierrot Lunaire, op. 21. Chamber symphony, no. 1, op. 9, E major.
      Mary Thomas, s; The Fires of London; Peter Maxwell Davies. Uni-
      corn RHS 319.
                +Gr 9-74 p535   +RR 9-74 p34
                +HF 5-75 p77   ++SFC 6-8-75 p23
                ++NR 4-75 p7   +-St 4-75 p105
      Pierrot Lunaire, op. 21. cf Works, selections (Decca SXLK 6660/4).
      Pierrot Lunaire, op. 21. cf IVES: Songs (RCA LRL 1-5058).
      Quintet, wind instruments, op. 26. cf Works, selections (Decca
      SXLK 6660/4).
      Rondo. cf Works, selections (Decca SXLK 6660/4).
      Serenade, op. 24. cf Works, selections (Decca SXLK 6660/4).
1976 Songs: Cabaret songs (8); Early songs (9). Marni Nixon, s; Leonard
      Stein, pno. RCA ARL 1-1231.
                ++SFC 12-14-75 p39
      Ein Stelldichein. cf Works, selections (Decca SXLK 6660/4).
      Suite, op. 25. cf Piano works (DG 2530 531).
      Suite, op. 25. cf Piano works (Nonesuch H 71309).
      Suite, op. 29. cf Works, selections (Decca SXLK 6660/4).
      Trio, strings, op. 45. cf Concerto, string quartet and orchestra.
      Variations, orchestra, op. 31. cf BERG: Lyric suite: Pieces.
      Verklärte Nacht, op. 4. cf Works, selections (Decca SXLK 6660/4).
      Verklärte Nacht, op. 4. cf BERG: Lyric suite: Pieces.
      Verklärte Nacht, op. 4. cf HINDEMITH: Pieces, string orchestra,
      op. 44, no. 4.
      Weihnachtsmusik. cf Works, selections (Decca SXLK 6660/4).
1977 Works, selections: Chamber symphony, no. 1, op. 9, E major. Die
      eiserne Brigade. Fantasia, violin and piano, op. 47. Herzge-
      wächse, op. 20. Lied der Waldtaube. Nachtwandler. Der neue

Klassizimus, op. 28, no. 3. Ode to Napoleon, op. 41. Pieces, chamber orchestra (3). Pierrot Lunaire, op. 21. Quintet, wind instruments, op. 26. Rondo. Serenade, op. 24. Ein Stelldichein. Suite, op. 29. Verklärte Nacht, op. 4. Weihnachtsmusik. Der wunsch des Liebhabers, op. 27, no. 4. Nona Liddell, vln; John Constable, pno; Mary Thomas, June Barton, s; Anna Reynolds, ms; John Shirley-Quirk, bar; Gerald English, speaker; London Sinfonietta and Chorus; David Atherton. Decca SXLK 6660/4).
  +Gr 9-74 p535   ++SFC 12-14-75 p39
  +RR 9-74 p34   ++Te 9-75 p42
 Der wunsch des Liebhabers, op. 27, no. 4. cf Works, selections (Decca SXLK 6660/4).

SCHONDORFF
 Gloria. cf Panton 010335.
SCHORGE, J.
 Wayward waltz. cf Kendor KE 22174.
SCHRAMMEL
 Wien bleibt Wien. cf DG 2721 077.
SCHREINER
 General Lee's grand march. cf Michigan University SM 0002.
SCHUBEL, Max
 Fracture. cf FUSSELL: Processionals, orchestra.
SCHUBERT, Franz
 L'Abeille. cf Pearl GEM 132.
 Alfonso und Estrella, D 732: Overture. cf Symphonies, nos. 1-6, 8-9.
 Allegretto, D 915, C minor. cf Piano works (Philips 6500 928/9).
 Allegretto, D 915, C minor. cf Piano works (Philips 6747 175).
 Ave Maria, D 839. cf Songs.
 Deutsche Messe, D 872. cf BRUCKNER: Mass, no. 2, E minor.
 Deutsche Tanze, nos. 1-16, D 783. cf Piano works (Philips 6747 175).
 German dances, nos. 1-16, D 783. cf Sonata, piano, no. 17, op. 53, D 850, D major.
 German dances, D 973 (3); D 769 (2); D 820 (6). cf Piano works (Saga 5407).
 German dances with coda and 7 trios. cf Saga 5411.
 Ecossaises, D 781. cf Piano works (Philips 6500 928/9).
 Ecossaises, D 781. cf Piano works (Philips 6747 175).
 Fantasia, D 934, C major. cf BRAHMS: Trio, piano, no. 2, op. 87, C major.
 Fantasia, D 993, C minor. cf Piano works (Saga 5407).
1978 Fantasia, op. 15, D 760, C major. Sonata, piano, no. 14, op. 143, D 784, A minor. Waltzes, D 145 (12). André Watts, pno. Columbia M 33073.
  +-HF 4-75 p84   +-SFC 1-5-75 p23
  /NR 2-75 p11   +St 3-75 p108
1979 Fantasia, op. 15, D 760, C major. SCHUMANN: Sonata, piano, no. 2, op. 22, G minor. Bruno-Leonardo Gelber, pno. Connoisseur Society (Q) CSQ 2085.
  ++SFC 11-16-75 p32
1980 Fantasia, op. 15, D 760, C major. Sonata, piano, no. 16, op. 42, D 845, A minor. Maurizio Pollini, pno. DG 2530 473.
  ++Gr 1-75 p1372   -RR 12-74 p64
  ++HF 5-75 p86   ++SFC 3-2-75 p24
  ++NR 4-75 p12

1981 Fantasia, op. 15, D 760, C major. Sonata, piano, op. posth., D
     894, G major. Impromptu, D 946, E flat minor.   Dezsö Ránki,
     pno. Hungaroton SLPX 11664.
         +HF 12-75 p105   +-RR 9-75 p62
         +NR 9-75 p13
     Fantasia, op. 15, D 760, C major. cf Piano works (Philips 6747 175).
     Fantasia, op. 15, D 760, C major (arr. Liszt). cf LISZT: Toten-
         tanz, G 126.
1982 Fantasia, op. 103, D 940, F minor. Grand duo, D 813, C major.
     Alfred Brendel, Evelyne Crochet, pno. Turnabout TV 34144.
     Tape (c) KTVC 34144.
         +Gr 5-75 p2032 tape   -HFN 5-75 p142 tape
     Fantasia, op. 108, F minor (arr. Kabalevsky). cf BRAHMS: Capriccio,
         op. 116.
     Fantasia, 2 pianos, D 940, F minor. cf MOZART: Sonata, piano, no.
         9, K 311, D major.
     Fugue, D 952, E minor. cf Pelca PSRK 41013/6.
1983 Gesand er Geister über den Wassern, male chorus and strings. Mir-
     jam's Siegesgesang, soprano, chorus and piano. Nachtgesang im
     Walde, D 913. Ursula Buckel, s; Gerd Lohmeyer, pno; South
     German Madrigal Choir; Wolfgang Gönnenwein. Candide (Q) QCE
     31087.
         /HF 11-75 p119   ++SFC 7-6-75 p16
     Grand duo, D 813, C major. cf Fantasia, op. 103, D 940, F minor.
     Hark, hark, the lark. cf CHOPIN: Ballade, no. 1, op. 23, G minor.
     Hungarian melody, D 817, B minor. cf Piano works (Philips 6500
         928/9).
     Hungarian melody, D 817, B minor. cf Piano works (Philips 6747 175).
1984 Impromptus, op. 90 and op. 142, D 899, D 935. Moments musicaux,
     op. 94, D 780. Jörg Demus, pno. BASF KHF 22062.
         ++NR 2-75 p11   +SFC 1-26-75 p27
     Impromptus, opp. 90, 142, D 899, D 935, D 946. cf Piano works
         (Philips 6747 175).
     Impromptus, op. 90, D 935, D 946. cf Piano works (Philips 6500
         928/9).
     Impromptus, op. 90, D 899, nos. 1-4. cf Sonata, piano, no. 19,
         op. posth., D 958, C minor.
     Impromptu, op. 90, no. 2, D 899, E flat major. cf Decca SPA 372.
     Impromptu, op. 90, no. 3, D 899, G flat major. cf Connoisseur
         Society (Q) CSQ 2065.
     Impromptu, op. 90, no. 3, D 899, G flat major. cf RCA ARM 4-0942/7.
     Impromptu, op. 90, no. 4, D 899, A flat major. cf HMV HQS 1353.
     Impromptu, op. 142, no. 2, D 935, A flat major. cf Sonata, piano,
         no. 19, op. posth., D 958, C minor.
     Impromptu, op. 142, no. 3, D 935, B flat major. cf CHOPIN: Ballade,
         no. 1, op. 23, G minor.
     Impromptus, op. 142, nos. 3 and 4, D 935, B flat major and F minor.
         cf BACH: Prelude and fugue, B flat minor.
     Impromptu, D 946, E flat minor. cf Fantasia, op. 15, D 760, C major.
     In the Italian style, D 590/1, C major and D major. cf Symphonies,
         nos. 1-6, 8-9.
     Ländler, D 790. cf Piano works (Philips 6747 175).
     Ländler. cf International Piano Library IPL 5005/6.
     Marche militaire, no. 1, op. 51, D 733, D major. cf Philips 6580
         107.
1985 Mass, no. 6, D 950, E flat major. Felicity Palmer, s; Helen Watts,

con; Kenneth Bowen, Wynford Evans, t; Christopher Keyte, bs;
St. John's College Choir; AMF; George Guest. Argo ZRG 825.
    +Gr 11-75 p881   +—RR 12-75 p92
    +HFN 11-75 p165
1986 Mass, no. 6, D 950, E flat major. Pilar Lorengar, s; Betty Allen,
    ms; Fritz Wunderlich, Manfred Schmidt, t; Josef Greindl, bs;
    St. Hedwig's Cathedral Choir; BPhO; Erich Leinsdorf. Seraphim
    S 60243. (Reissue from Capital SP 8579)
        -HF 11-75 p119
Minuet, D 995, F major. cf Piano works (Saga 5407).
Minuets with 6 trios, D 89 (5). cf Nonesuch H 71141.
Mirjam's Siegesgesang, soprano, chorus and piano. cf Gesang der
    Geister über den Wassern, male chorus and strings.
Moments musicaux, op. 94, D 780. cf Impromptus, op. 90 and op.
    142, D 899, D 935.
Moments musicaux, op. 94, D 780. cf Piano works (Philips 6747 175).
Moments musicaux, op. 94, D 780. cf DG 2548 137.
Moment musicaux, op. 94, D 780, F major. cf HMV HQS 1353.
Moment musical, op. 94, D 780, F major. cf International Piano
    Archives IPA 5007/8.
Moment musicaux, op. 94, D 780, F major. cf L'Oiseau-Lyre DSLO 7.
Motets: Christ ist erstanden, D 440. Gebet, D 815. Gott im
    Ungewitter, D 985. cf BRAHMS: Motets (HMV ASD 3091).
Nachtgesang im Walde, D 913. cf Gesang der Geister über den Wassern,
    male chorus and strings.
1987 Nocturne, piano, violin and violoncello, op. 148, D 897, E flat
    major. Trio, piano, no. 1, op. 99, D 898, B flat major. Suk
    Trio. Vanguard/Supraphon SU 6. (Reissue)
    ++MJ 3-75 p25   ++SFC 1-26-75 p27
    +NR 3-75 p5   ++St 5-75 p102
Nocturne, piano, violin and violoncello, op. 148, D 897, E flat
    major. cf Quintet, op. 144, D 667, A major (BASF KHB 20314).
Nocturne, piano, violin and violoncello, op. 148, D 897, E flat
    major. cf Quintet, piano, op. 114, D 667, A major (Unicorn 311).
1988 Octet, op. 166, D 803, F major. Consortium Classicum. BASF BAC
    3099.
        +Gr 7-75 p206   +RR 7-75 p43
        ++HFN 6-75 p92
1989 Octet, op. 166, D 803, F major. Berlin Octet. Philips 6580 110.
    +HFN 12-75 p164   +—RR 11-75 p60
1990 Octet, op. 166, D 803, F major. Cleveland Quartet; Thomas Martin,
    double bs; Jack Brymer, clt; Barry Tuckwell, hn. RCA ARL 1-1047.
    +HF 11-75 p120   ++SR 9-7-75 p40
    ++NR 11-75 p7
1991 Piano works: Allegretto, D 915, C minor. Ecossaises, D 781 (11).
    Hungarian melody, D 817, B minor. Impromptus, op. 90, D 935,
    D 946. Sonata, piano, no. 16, op. 42, D 845, A minor. Alfred
    Brendel, pno. Philips 6500 298/9 (2).
        +Gr 9-75 p491   +RR 9-75 p62
        ++HFN 9-75 p104
1992 Piano works: Allegretto, D 915, C minor. Deutsche Tanze, nos. 1-
    16, D 783. Ecossaises, D 781 (11). Fantasia, op. 15, D 760,
    C major. Hungarian melody, D 817, B minor. Impromptus, opp.
    90, 142, D 899, D 935, D 946. Ländler, D 790 (12). Moments
    musicaux, op. 94, D 780. Sonatas, piano, nos. 14-21. Alfred
    Brendel, pno. Philips 6747 175 (8). (Reissues from 6500 418,

6500 929, 6500 763, 6500 416, 6500 415, 6500 284, 6500 285,
6500 928)
+Gr 9-75 p491   +RR 11-75 p82
++HFN 9-75 p109

1993 Piano works: Fantasia, D 993, C minor. German dances, D 973 (3);
D 769 (2); D 820 (6). Minuet, D 995, F major. Sonata, piano,
no. 5, D 557, A flat major. Variation on a waltz by Anton
Diabelli, D 718, C minor. Waltz, G flat major (Kupelwieser).
Rosario Marciano, pno. Saga 5407.
+HFN 11-75 p165   +RR 11-75 p82

Psalm no. 23, The Lord is my Shepherd, D 706. cf Argo ZRG 785.

Psalm no. 23, Gott ist mein Hirt, D 706. cf BRAHMS: Motets (HMV
ASD 3091).

1994 Quartets, strings, nos. 1-3, D 18, D 32, D 36. Melos Quartet.
DG 2530 322.
+Gr 8-73 p344   +SFC 12-28-75 p30
+RR 8-73 p57

1995 Quartets, strings, nos. 1-15. Melos Quartet. DG 2740 123 (7).
(Reissues from 2530 322, 2530 533)
+Gr 10-75 p651   +RR 10-75 p67
+HFN 10-75 p148

1996 Quartet, strings, no. 9, D 173, G minor. Quartet, strings, no.
13, op. 29, D 804, A minor. Alban Berg Quartet. Telefunken
AW 6-41882. Tape (c) 4-41882.
+Gr 6-75 p62   +RR 5-75 p50
++HFN 6-75 p92   ++RR 10-75 p97 tape
+HFN 9-75 p110 tape

Quartets, strings, no. 10, D 87, E flat major: Scherzo. cf Amon
Ra SARB 01.

1997 Quartets, strings, nos. 12-15, D 703, D 804, D 810, D 887. New
Hungarian Quartet. Vox SVBX 601 (3).
+HF 1-75 p88   +St 12-74 p142
+NR 10-74 p9

1998 Quartet, strings, no. 12, D 703, C minor. Quintet, strings, op.
163, D 956, C major. Weller Quartet. Decca SDD 441. (Reissue
from SXL 6481) (also London STS 15300)
-Gr 4-75 p1829   ++SFC 7-27-75 p22
++NR 8-75 p8   +St 11-75 p135
+RR 3-75 p39

1999 Quartet, strings, no. 12, D 703, C minor. Quintet, piano, op.
114, D 667, A major. Jörg Demus, pno; Schubert Quartet; Amadeus
Quartet. DG 135 062. Tape (c) 921020, 3318 044.
++HFN 1-72 p117 tape   +-RR 9-75 p78 tape

2000 Quartet, strings, no. 12, D 703, C minor. Quartet, strings, no.
14, D 810, D minor. Melos Quartet. DG 2530 533.
+Gr 5-75 p1991   +-NR 12-75 p9
+-HFN 6-75 p91   +-RR 5-75 p51

Quartet, strings, no. 12, D 703, C minor. cf BEETHOVEN: Quartet,
strings, no. 11, op. 95, F minor.

Quartet, strings, no. 12, D 703, C minor. cf CBS 76267.

2001 Quartets, strings, nos. 13-15. Budapest Quartet. Odyssey Y3 33320
(3). (Reissue from Columbia SL 194)
+-HF 12-75 p92

Quartet, strings, no. 13, op. 29, D 804, A minor. cf Quartet,
strings, no. 9, D 173, G minor.

Quartet, strings, no. 13, op. 29, D 804, A minor: Andante. cf Amon
Ra SARB 01.

2002 Quartet, strings, no. 14, D 810, D minor. Collegium Aureum Quartet.
     BASF KHC 22059.
              +Gr 10-75 p651   ++NR 3-75 p6
              +HF 11-75 p120   ++RR 10-75 p68
              +HFN 10-75 p148
2003 Quartet, strings, no. 14, D 810, D minor. Busch Quartet. DaCapo
     C 047 01374. (Reissue from HMV 78 rpm originals)
              +HF 1-75 p88
2004 Quartet, strings, no. 14, D 810, D minor. MOZART: Adagio and fugue,
     strings, K 546, C minor. Cleveland Quartet. RCA ARL 1-0483.
              +Gr 9-74 p530   +RR 9-74 p69
              +HF 1-75 p88   ++SFC 7-7-74 p18
              ++NR 8-74 p6   +-St 12-74 p142
     Quartet, strings, no. 14, D 810, D minor. cf Quartet, strings,
         no. 12, D 703, C minor.
2005 Quintet, piano, op. 114, D 667, A major. Nocturne, piano, violin
     and violoncello, op. 148, D 897, E flat major. Jörg Demus,
     pno; Franzjosef Maier, vln; Heinz-Otto Graf, vla; Rudolf Mandalka,
     vlc; Paul Breuer, bs. BASF KHB 20314. (also BAC 3004. Tape
     (c) KBACC 3004)
              +Gr 6-74 p66   ++NYT 10-14-73 pD33
              +-Gr 5-75 p2032 tape   +-RR 6-74 p61
              +-HFN 5-75 p142 tape   +RR 6-75 p93 tape
              +MQ 4-74 p312   +SFC 9-30-73 p27
              +NR 9-73 p7   ++St 10-73 p89
2006 Quintet, piano, op. 114, D 667, A major. Clifford Curzon, pno;
     Vienna Octet. London 6090. (also Decca SDD 185. Tape (c)
     KSDC 185)
              +-HFN 7-75 p91 tape   +RR 9-75 p79 tape
2007 Quintet, piano, op. 114, D 667, A major. Nocturne, piano, violin
     and violoncello, op. 148, D 897, E flat major. London Music
     Group. Unicorn RHS 311. Tape (c) ZCUN 311.
              +-Gr 12-73 p1221   ++RR 7-73 p65
              +-Gr 2-74 p1611 tape   +RR 2-74 p71 tape
              +HF 5-75 p89   -RR 2-74 p73 tape
     Quintet, piano, op. 114, D 667, A major. cf Quartet, strings,
         no. 12, D 703, C minor.
     Quintet, piano, op. 114, D 667, A minor. cf MOZART: Divertimento,
         winds, K 196, B flat major.
2008 Quintet, strings, op. 163, D 956, C major. Juilliard Quartet.
     CBS 76268. (also Columbia M 32808)
              +-Gr 1-75 p1362   +RR 1-75 p44
2009 Quintet, strings, op. 163, D 956, C major. Alberni Quartet; Thomas
     Igloi, vlc. CRD CRD 1018.
              +Gr 11-75 p854   ++RR 11-75 p65
              +HFN 12-75 p164
2010 Quintet, strings, op. 163, D 956, C major. Guarneri Quartet;
     Leonard Rose, vlc. RCA ARL 1-1154. Tape (c) ARS 1-1154 (ct)
     ARK 1-1154.
              +NR 11-75 p7
2011 Quintet, strings, op. 163, D 956, C major. Taneyev Quartet; Mstis-
     lav Rostropovich, vlc. Westminster WGS 8299.
              +HF 6-75 p103   +-St 11-75 p135
     Quintet, strings, op. 163, D 956, C major. cf Quartet, strings,
         no. 12, D 703, C minor.
     Rondo, violin and string quartet, D 438, A major. cf BEETHOVEN:

Romance, no. 1, op. 40, G major.

Rondo brillant, op. 70, D 895, B minor. cf BRAHMS: Trio, piano,
no. 2, op. 87, C major.

Rosamunde, op. 26, D 797: Ballet music. cf Philips 6580 066.

Rosamunde, op. 26, D 797: Ballet music, no. 2. cf Philips 6747 199.

Rosamunde, op. 26, D 797: Entr'acte III. cf Discopaedia MB 1005.

Rosamunde, op. 26, D 797: Incidental music. cf Symphony, no. 6,
D 589, C major.

Rosamunde, op. 26, D 797: Overture; Ballet music, nos. 1 and 2.
cf Symphonies, nos. 1-6, 8-9.

Rosamunde, op. 26, D 797: Overture; Entr'acte, Act 3; Ballet music,
no. 2. cf BEETHOVEN: Symphony, no. 1, op. 21, C major.

2012 Die schöne Müllerin, D 795. Dietrich Fischer-Dieskau, bar; Gerald
Moore, pno. DG 2530 544. (Reissue from 2720 059)
++Gr 8-75 p355   ++RR 10-75 p88
+HFN 9-75 p109

2013 Die schöne Müllerin, op. 25, D 975. Fritz Wunderlich, t; Hubert
Giesen, pno. DG 2538 347. (Reissue from SLPM 139219/20)
-Gr 7-75 p228   +-RR 8-75 p63
+-HFN 8-75 p91

2014 Die schöne Müllerin, op. 25, D 795. Nigel Rogers, bar; Richard
Burnett, pno. Telefunken Tape (c) CX 4-41892.
+Gr 8-75 p376 tape   +-HFN 9-75p110 tape

2015 Schwanengesang, D 957. Peter Schreier, t; Walter Olbertz, pno.
DG 2530 469.
+-Gr 3-75 p1691   +-RR 3-75 p65
+-MT 10-75 p888

2016 Schwanengesang, D 957. Tom Krause, bar; Irwin Gage, pno. London
26328. (also Decca SXL 6590)
+Gr 7-73 p221   +-RR 7-73 p79
-HF 1-74 p79   +SFC 6-16-74 p29
+-NR 8-73 p12   +St 10-73 p152
+ON 5-73 p36

Schwanengesang, D 957, excerpt (4). cf Seraphim S 60251.

Schwanengesang, D 957: Die Taubenpost. cf WAGNER: Lohengrin.

2017 Sonata, arpeggione and piano, D 821, A minor. Variations, flute
and piano, K 802, E minor. Klaus Storck, arpeggione; Alfons
Kontarsky, pno; Hans-Martin Linde, flt. DG Archive 2533 175.
+-Gr 2-75 p1510   /+RR 2-75 p57
+-HF 4-75 p91   +-St 5-75 p106
++NR 2-75 p7

2018 Sonatas, piano, complete. Wilhelm Kempff, pno. DG 2740 132 (9).
+-HFN 12-75 p171

Sonata, piano, no. 5, D 557, A flat major. cf Piano works (Saga
5407).

Sonatas, piano, nos. 14-21. cf Piano works (Philips 6747 175).

Sonata, piano, no. 14, op. 143, D 784, A minor. cf Fantasia,
op. 15, D 760, C major.

Sonata, piano, no. 16, op. 42, D 845, A minor. cf Fantasia, op.
15, D 760, C major.

Sonata, piano, no. 16, op. 42, D 845, A minor. cf Piano works
(Philips 6500 928/9).

2019 Sonata, piano, no. 17, op. 53, D 850, D major. German dances,
nos. 1-16, D 783. Alfred Brendel, pno. Philips 6500 763.
++Gr 7-75 p219   +RR 7-75 p53
++HFN 7-75 p88

2020 Sonata, piano, no. 19, op. posth., D 958, C minor. Impromptu, op.
      142, no. 2, D 935, A flat major. Sviatoslav Richter, pno.
      Melodiya/Angel SR 40254.
      +HF 12-74 p112   +St 1-75 p116
      +NR 11-74 p10
2021 Sonata, piano, no. 19, op. posth., D 958, C minor. Impromptus,
      op. 90, D 899, nos. 1-4. Alfred Brendel, pno. Philips 6500 415.
      ++SFC 1-12-75 p29
2022 Sonata, piano, no. 21, op. posth., D 960, B flat major. Christoph
      Eschenbach, pno. DG 2530 477.
      +-Gr 3-75 p1682   +RR 4-75 p54
      /HF 8-75 p99   ++SFC 6-29-75 p26
      +NR 7-75 p14
2023 Sonata, piano, no. 21, op. posth., D 960, B flat major. SCHUMANN:
      Kinderscenen, op. 15. Vladimir Horowitz, pno. RCA VH 016.
      (Reissues from HMV ALP 1430, 1469)
      +-Gr 12-75 p1081   +-RR 12-75 p84
      Sonata, piano, op. posth., D 894, G major. cf Fantasia, op. 15,
      D 760, C major.
      Sonata, violin and piano, op. 162, D 574, A major. cf BRAHMS:
      Trio, piano, no. 2, op. 87, C major.
      Sonata, violin and piano, op. 162, D 574, A major. cf BEETHOVEN:
      Sonata, violin and piano, no. 8, op. 30, no. 3, G major.
      Sonatina, violin and piano, no. 1, op. 137, D 384, D major. cf
      BRAHMS: Trio, piano, no. 2, op. 87, C major.
      Sonatina, violin and piano, no. 1, op. 137, D 384, D major: Rondo.
      cf RCA ARM 4-0942/7.
      Sonatina, violin and piano, no. 3, op. 137, D 408, G minor. cf RCA
      ARM 4-0942/7.
2024 Songs: An die Musik, D 547; An Sylvia, D 891; Auf dem Wasser zu
      singen, D 774; Ganymed, D 544; Gretchen am Spinnrade, D 118;
      Im Frühling, D 882; Die junge Nonne, D 828; Das Lied im Grünen,
      Der Musensohn, D 764; Nachtviolen, D 752; Nahe des Geliebten,
      D 162b; Wehmut, D 772. Elisabeth Schwarzkopf, s; Edwin Fischer,
      pno. Angel 35022.
      +St 4-75 p67
2025 Songs: Du liebst mich nicht, D 756; Gretchen am Spinnrade, D 118;
      Heimliches lieben, D 922; Der Hirt auf dem Felsen, D 965; Im
      Frühling, D 882; Der Jüngling an der Quelle, D 300; Der Musen-
      sohn, D 764; Seligkeit, D 433; Der Vogel, D 691. Ländler, D
      790 (12). Elly Ameling, s; Jörg Demus, pno; Hans Deinzer, clt.
      BASF BAC 3088.
      +-Gr 2-75 p1536   +HFN 5-75 p139
2026 Songs: Fischerweise, D 881; Im Frühling, D 882; Die junge Nonne,
      D 828; Nacht und Träume, D 827. SCHUMANN: Songs: Abendlied,
      op. 107, no. 6; Aus den Hebräischen Gesangen, op. 25, no. 15;
      Der Kartenlegerin, op. 31, no. 2; Die Lotusblume, op. 25, no.
      7. STRAUSS, R.: Songs: Befreit, op. 39, no. 4; Für fünfzehn
      Pfennige, op. 36, no. 2; Schon sind, doch kalt, op. 19, no. 3.
      WOLF: Songs: Auf einer Wanderung; Der Genesene an die Hoffnung;
      Mein Liebster; Kennst du das Land. Marilyn Horne, ms; Martin
      Katz, pno. Decca SXL 6578. (also London OS 26302)
      +-Gr 4-75 p1860   -NYT 6-8-75 pD19
      -HF 8-75 p106   ++ON 5-75 p36
      +-MJ 10-75 p39   +-RR 4-75 p67
      -NR 5-75 p11   ++St 5-75 p105

2027 Songs: Abendstern, D 806; An die Entfernte, D 765; Atys, D 585;
     Auf dem Wasser zu singen, D 774; Auflösung, D 807; Der Einsame,
     D 800; Das Fischermädchen, D 957; Der Geistertanz, D 116; Im
     Frühling, D 882; Lachen und Weinen, D 777; Nacht und Träume,
     D 827; Nachtstück, D 672; Sprache der Liebe, D 410. Peter Pears,
     t; Benjamin Britten, pno. Decca SXL 6722.
          ++Gr 6-75 p79   ++RR 6-75 p82
          ++HFN 6-75 p92

2028 Songs: Abendstern, D 806; Am See, D 746; Auflösung, D 807; Der
     blinde Knabe, D 833b; Grablied, D 218; Herrn Joseph von Spaun,
     D 749; Im Haine, D 738; Der Jüngling auf dem Hügel, D 702;
     Der Jüngling und der Tod, D 545b; Der Knabe in der Wiege, D
     579b; Leiden der Trennung, D 509; Der Strom, D 565; Totengräbers
     Heimweh, D 842; Der Vater mit dem Kind, D 906; Wehmut, D 771;
     Der zürnende Barde, D 785; Der Zwerg, D 771. Dietrich Fischer-
     Dieskau, bar; Gerald Moore, pno. DG 2530 347. (Reissue from
     2270 022/1-13)
          +Gr 11-73 p1003   *ON 5-75 p36
          +HFN 12-73 p2616   +RR 12-73 p96
          ++NR 5-74 p11   +-St 7-74 p111

2029 Songs: Die Advokaten, D 37; Cantate zur fünfzigjahrigen Jubelfeier
     Salieri's, D 441; Cantate zum Geburtstag das Sängers Johann
     Michael Vogl, D 666; Der Hochzeitsbraten, D 930; Punschlied,
     D 277; Trinklied, D 148; Verschwunden sind die Schmerzen, D 88.
     Elly Ameling, s; Peter Schreier, Horst R. Laubenthal, t; Diet-
     rich Fischer-Dieskau, bar; Gerald Moore, pno. DG 2530 361.
          +Gr 11-74 p945   -NYT 4-28-74 pD24
          /HF 5-74 p95   ++RR 11-74 p90
          +-MT 8-75 p715   -St 7-74 p111
          ++NR 5-74 p11

2030 Songs: Am Bach im Frühling, D 361; An die Nachtigall, D 196; Ave
     Maria, D 839; Auf der Donau, D 553; Frühlingsglaube, D 686;
     Gretchen am Spinnrade, D 118; Im Abendrot, D 799; Die junge
     Nonne, D 828; Der König in Thule, D 367; Lachen und Weinen,
     D 777; Des Mädchens Klage, D 6; Mignon's song; D 191b; Romanze,
     D 797; Die Rose, D 745a; Der Tod und das Mädchen, D 531. Chris-
     ta Ludwig, ms; Irwin Gage, pno. DG 2530 404.
          +-Gr 7-74 p254   +ON 5-75 p36
          +-HF 11-74 p121   +-RR 6-74 p74
          +MJ 9-74 p65   ++SFC 6-16-74 p29
          +NR 7-74 p9   +-St 8-74 p121
          +NYT 4-28-74 pD24

2031 Songs: An die Sonne, D 439; Begräbnislied, D 168; Gebet, D 815;
     Gott der Weltschöpfer, D 986; Gott im Ungewitter, D 985; Hymne
     an den Unendlichen, D 232; Lebenslust, D 609; Des Tages Weihe,
     736; Der Tanz, D 826. Elly Ameling, s; Janet Baker, con; Peter
     Schreier, t; Dietrich Fischer-Dieskau, bar; Gerald Moore, pno.
     DG 2530 409.
          +Gr 11-74 p945   ++RR 11-74 p90
          +-HF 11-74 p118   ++SFC 7-7-74 p18
          +-MT 8-75 p715   ++St 8-74 p121
          +NR 8-74 p9

2032 Songs: An den Mond, D 296; Bertha's Lied in der Nacht, D 653; Dass
     sie hier gewesen, D 775; Klärchens Lied, D 210; Lied der Anna
     Lyle, D 830; Lied der Mignon I and II, D 877/2-3; Das Mädchen,
     D 652; Mignons Gesang, D 321; Lilla an die Morgenröte, D 273;

Sehnsucht, D 636b; Ständchen, D 921; Wehmut, D 772; Der Zwerg,
D 771. Christa Ludwig, ms; Irwin Gage, pno. DG 2530 528.
+Gr 6-75 p80   +HFN 7-75 p88
+HF 12-75 p105

2033 Songs: An die Musik, D 547; An Sylvia, D 891; Auf dem See; Auf dem
Wasser zu singen, D 774; Der Einsame, D 800; Erster Verlust;
Frühlingsglaube, D 686; Gretchen am Spinnrade, D 118; Der König
in Thule, D 367; Lachen und Weinen, D 777; Der Musensohn, D 764;
Nähe des Geliebten, D 162; Rastlose Liebe, D 138; Seligkeit,
D 433; Ständchen, D 921; Der Sterne. Brigitte Fassbaender,
ms; Erik Werba, pno. EMI Odeon IC 065 28969.
+HF 12-75 p105

2034 Songs: Gott der Weltschöpfer, D 986; Gott im Ungewitter, D 985;
Hymne an den Unendlichen, D 232; Miriams Siegesgesand, D 942.
SCHUMANN: Spanisches Liederspiel, op. 74. Gabriella Déry,
Margit László, s; József Réti, t; Zsolt Bende, bar; István
Antal, pno; HRT Chorus; Zoltán Vásarhelyi. Hungaroton SHLX
90050. (Reissue)
+-RR 1-75 p58

2035 Songs on Goethe texts: Erlkönig, op. 1, D 328; Ganymed, op. 19,
no. 3, D 544; Gesänge des Harfners, D 478, 479, 480; Heiden-
röslein, op. 3, no. 3, D 257; Die Liebende schreibt, op. 165,
no. 1, D 673; Liebhaber in allen Gestalten, D 558; Mignon Lieder,
D 321, D 877, nos. 2, 3, 4; Nähe des Geliebten, op. 5, no. 2,
D 162; Der Sänger, op. 117, D 149. Elly Ameling, s; Hermann
Prey, bar; Dalton Baldwin, Karl Engel, pno. Philips 6500 515.
+Gr 7-74 p254   +-ON 5-75 p36
+HF 8-74 p102   +RR 6-74 p74
+MJ 3-75 p34   ++SFC 7-7-74 p18
++NR 7-74 p9   ++St 7-74 p110
+NYT 4-28-74 pD24   ++St 8-74 p72

2036 Songs: An die Nachtigall, D 196; Fischerweise, D 869; Die Gebüsche,
D 646; Im Freien, D 880; Im Haine, D 738; Das Lied im Grünen,
D 917; Der Schmetterling, D 633; Die Vögel; D 691; Der Wachtel-
schlag, D 742. SCHUMANN: Frauenliebe und Leben, op. 42. Elly
Ameling, s; Dalton Baldwin, pno. Philips 6500 706.
+-Gr 10-75 p670   +RR 10-75 p88
++HF 12-75 p106   ++SFC 9-14-75 p28
+HFN 10-75 p149   +St 12-75 p83
++NR 11-75 p11

2037 Songs: An die Laute, D 905; An die Leier, D 737; An die Musik, D
547; An Silvia, D 891; Fischerweise, D 881; Die Forelle, D 550;
Frühlingsglaube, D 686; Heidenröslein, D 257; Horch, horch, die
Lerch, D 889; Im Frühling, D 882; Lachen und Weinen, D 777;
Litanei auf das Fest Aller Seelen, D 343; Lob der Tränen, D 711;
Der Musensohn, D 764; Nacht und Träume, D 827; Schwanengesang,
no. 4: Ständchen, D 957; Die Winterreise, no. 5: Du Lindenbaum,
D 911. Martyn Hill, t; Nina Walter, pno. Saga 5404.
+-Gr 11-75 p881   -RR 11-75 p88
+HFN 12-75 p164

2038 Songs: An die Nachtigall, D 196; An die untergehende Sonne, D 457;
Berthas Lied in der Nacht, D 653; Claudine von Villabella: Hin
und wieder, Liebe schwärmt; Delphine, D 857; Ellens Gesange:
Raste, Krieger, D 837, Jäger, ruhe von der Jadg, 838, Ave Maria,
D 839; Epistel an Herrn Josef von Spaun, D 749; Gretchen am
Spinnrade, D 118; Iphigenia, D 573; Die junge Nonne, D 828;

Kennst du das Land, D 321; Das Mädchen, D 652; Des Mädchens
Klage, D 6; Die Männer sind mechant, D 866; Mignon Lieder,
D 877: Heiss' mich nicht reden, Nur wer die Sehnsucht kennt,
So lasst mich scheinen; Schlummerlied, D 527; Schwestergruss,
762; Suleika I and II; Wiegenlied, D 498; Wiegenlied, D 867.
Janet Baker, ms; Gerald Moore, pno. Seraphin SIB 6083 (2).

+HF 4-74 p104   +ON 5-75 p36
++NR 1-74 p11   ++St 2-72 p83

Songs: An die Musik, D 547; Die Allmacht, D 852; Du bist die Ruh,
D 776; Der Jüngling an der Quelle, D 300; Lachen und Weinen,
D 777; Nacht und Träume, D 827; Nachtviolen, D 752; Rastlose
Liebe, D 138. cf BEETHOVEN: Songs (BBC REB 170).

Songs: Dithyrambe, op. 60, no. 2; Ellens zweiter Gesang, op. 52;
Schwanengesang, no. 11: Die Stadt; Wiegenlied. cf BRAHMS:
Songs (Discophilia KGG 4).

Songs: Christ ist erstanden, S 440; Gebet, D 815; Gott im Ungewitter, D 985; Psalm 23, Gott ist mein Hirt, D 706. cf BRAHMS:
Songs (HMV KASD 3091).

Songs: An die Musik, D 547; Auf dem Wasser zu singen, D 774; Ellens
Gesang III, Ave Maria, D 839; Die Forelle, D 550; Der Hirt auf
dem Felsen, D 965; Der Musensohn, D 764. cf MAHLER: Das Knaben
Wunderhorn (HMV SXLP 30182).

Songs: Ave Maria, D 839; Jäger, ruhe von der Jagd, D 838; Raste,
Krieger, D 837; Schwestergruss, D 762; Der Zwerg, D 771. cf
MAHLER: Des Knaben Wunderhorn (Philips 6500 412).

Songs: An die Musik, D 547; An Sylvia, D 891; Auf dem Wasser zu
singen, D 774; Die Forelle, D 550; Lachen und Weinen, D 777;
Lied eines Schiffers an die Dioskuren, D 360; Der Wanderer, D
649; Der Wanderer an den Mond, D 870. cf SCHUMANN: Dichterliebe,
op. 48.

Songs: Du bist die Ruh, D 776; Du liebst mich nicht, D 756; Die
junge Nonne, D 828; Rosamunde, no. 5: Romance, D 797; Suleika,
D 717; Der Tod und das Mädchen, D 531. cf Decca 6BB 197/8.

Songs: Cradle song (arr. Elman). cf Decca SPA 405.

Songs: Allein, nachdenklich, gelähmt, D 629; Apollo, lebet noch
dein hold Verlangen, D 628; Nunmehr, da Himmel Erde schweigt,
D 630. cf DG 2530 332.

Songs: Ave Maria, D 839. cf Discopaedia MB 1002.

Songs: Ständchen, op. 135, D 920. cf Discopaedia MB 1007.

Songs: Ave Maria, D 839. cf Discopaedia MB 1009.

Songs: To music, D 547; The brook. cf HMV RLS 714.

Songs: Ave Maria, D 839. cf London SPC 21100.

Songs: Ave Maria, D 839; Die Forelle, D 550. cf Philips 6747 204.

Songs: Heidenröslein, D 257. cf Philips 6833 105.

Songs: Die Forelle, D 550; Ständchen, D 889; Die böse Farbe; An
die Leier, D 737. cf RCA SER 5704/6.

Songs: Ave Maria, D 839 (2). cf RCA ARM 4-0942/7.

Songs: Ave Maria, D 839. cf Rococo 2035.

Songs: Ave Maria, D 839. cf Rococo 2072.

Songs: Frühlingsglaube, D 686; Ganymed, op. 19, no. 3, D 544;
Liebesbotschaft; Lied eines Schiffers an die Dioskuren, D 360;
Der Musensohn, D 764; Nacht und Träume, D 827; Wohin. cf Tele-
funken DLP 6-48042.

2039 Symphonies, nos. 1-6, 8-9. Rosamunde, op. 26, D 797: Overture,
Ballet music, nos. 1 and 2. BPhO; Karl Böhm. DG 2740 127 (5).

+-HFN 10-75 p152   +-RR 10-75 p49

2040 Symphonies, nos. 1-6, 8-9. Overtures: Alfonso und Estrella, D 732.
       In the Italian style, D 590/1, C major and D major. Die Zwill-
       ingsbruder. Bath Festival Orchestra; Yehudi Menuhin. HMV SLS
       5007 (5).
               +-RR 7-75 p33
2041 Symphonies, nos. 2-6, 8. VPO; Karl Münchinger. Decca ECS 7613 (3).
               -Gr 5-75 p1973
2042 Symphony, no. 2, D 125, B flat major. Symphony, no. 8, D 759, B
       minor. VPO; Karl Münchinger. Decca ECS 761. (also London STS
       15061. Tape (c) A 30661)
               +HFN 5-75 p138   +RR 4-75 p31
2043 Symphony, no. 3, D 200, D major. Symphony, no. 6, D 589, C major.
       VPO; Karl Münchinger. Decca ECS 763. (also London 6453)
               +HFN 5-75 p138   +RR 4-75 p31
2044 Symphony, no. 3, D 200, D major. Symphony, no. 4, D 417, C minor.
       BPhO; Karl Böhm. DG 2530 526. (Reissue from 2720 062)
               ++Gr 7-75 p195   +-NR 12-75 p5
               +HFN 8-75 p89   +RR 7-75 p34
2045 Symphony, no. 3, D 200, D major. Symphony, no. 8, D 759, B minor.
       BPhO; Karl Böhm. DG Tape (c) 3300 475.
               +HFN 12-75 p173 tape   ++RR 9-75 p79 tape
2046 Symphony, no. 3, D 200, D major. WEBER: Concerto, clarinet, no.
       2, op. 74, E flat major. Josef Pacewicz, clt; Lambeth Orches-
       tra. Hello LOR 1001.
               +-RR 2-75 p38
2047 Symphony, no. 4, D 417, C minor. Symphony, no. 5, D 485, B flat
       major. VPO; Karl Münchinger. Decca ECS 762. (also London
       STS 15095)
               +HFN 5-75 p138   +RR 4-75 p31
2048 Symphony, no. 4, D 417, C minor. Symphony, no. 5, D 485, B flat
       major. Orchestra; Karl Böhm. DG Tape (c) 3300 484.
               +HFN 12-75 p173 tape
2049 Symphony, no. 4, D 417, C minor. Symphony, no. 8, D 759, B minor.
       Budapest Philharmonic Orchestra; Géza Oberfrank. Hungaroton
       SLPX 11570.
               +-NR 6-75 p4
    Symphony, no. 4, D 417, C minor. cf Symphony, no. 3, D 200, D
       major.
    Symphony, no. 5, D 485, B flat major. cf Symphony, no. 4, D 417,
       C minor.
    Symphony, no. 5, D 485, B flat major. cf Symphony, no. 4, D 417,
       C minor.
    Symphony, no. 5, D 485, B flat major. cf Camden CCV 5000-12,
       5014-24.
2050 Symphony, no. 6, D 589, C major. Rosamunde, op. 26, D 797: Inci-
       dental music. BPhO; Karl Böhm. DG 2530 422. (Reissue from
       2720 062)
               ++Gr 12-74 p1144   ++RR 11-74 p51
               +NR 1-75 p3
    Symphony, no. 6, D 589, C major. cf Symphony, no. 3, D 200, D
       major.
2051 Symphony, no. 8, D 759, B minor. MOZART: Symphony, no. 41, K 551,
       C major. BSO; Eugen Jochum. DG 2530 357. Tape (c) 3300 318
       (ct) 89 468.
               ++Gr 4-74 p1857   +RR 4-74 p55
               +HF 3-74 p75   +RR 9-75 p71
               ++NR 4-74 p2   ++St 5-74 p107

Symphony, no. 8, D 759, B minor. cf Symphony, no. 2, D 125,
    B flat major.
Symphony, no. 8, D 759, B minor. cf Symphony, no. 3, D 200, D
    major.
Symphony, no. 8, D 759, B minor. cf Symphony, no. 4, D 417, C
    minor.
Symphony, no. 8, D 759, B minor. cf MENDELSSOHN: Symphony, no. 4,
    op. 90, A major.
Symphony, no. 8, D 759, B minor. cf Camden CCV 5000-12, 5014-24.
Symphony, no. 8, D 759, B minor. cf HMV SLS 5003.
2052 Symphony, no. 9, D 944, C major. Utrecht Symphony Orchestra;
    Ignace Neumark. Audio Fidelity FCS 50058.
        /NR 2-75 p4
2053 Symphony, no. 9, D 944, C major. RPO; Hans Vonk. Decca PFS 4335.
        +-Gr 6-75 p49   +-RR 6-75 p47
        +HFN 7-75 p88
2054 Symphony, no. 9, D 944, C major. Bavarian Radio Symphony Orchestra;
    Eugen Jochum. DG Tape (c) 3318 050.
        +-RR 7-75 p72 tape
2055 Symphony, no. 9, D 944, C major. Stuttgart Klassische Philharmonie;
    Karl Münchinger. London STS 15299.
        +NR 8-75 p2
2056 Symphony, no. 9, D 944, C major. Stockholm Philharmonic Orchestra;
    Bruno Walter. Olympic 8123.
        +-NR 2-75 p4
Symphony, no. 9, D 944, C major. cf Toscanini Society ATS GC 1201/6.
2057 Trio, piano, no. 1, op. 99, D 898, B flat major. Trio, piano, no.
    2, op. 100, D 929, E flat major. Henryk Szeryng, vln; Pierre
    Fournier, vlc; Artur Rubinstein, pno. RCA ARL 2-0731 (2).
        +Gr 7-75 p205   +NR 5-75 p8
        ++HF 6-75 p90   +-RR 8-75 p51
        ++HFN 8-75 p83   ++St 7-75 p106
        ++MJ 9-75 p51
Trio, piano, no. 1, op. 99, D 898, B flat major. cf Nocturne,
    piano, violin and violoncello, op. 148, D 897, E flat major.
Trio, piano, no. 1, op. 99, D 898, B flat major. cf BEETHOVEN:
    Trio, piano, no. 6, op. 97, B flat major.
Trio, piano, no. 2, op. 100, D 929, E flat major. cf Trio, piano,
    no. 1, op. 99, D 898, B flat major.
Valse sentimentale. cf Decca SPA 405.
Variation on a waltz by Anton Diabelli, D 718, C minor. cf Piano
    works (Saga 5407).
Variations, flute and piano, D 802, E minor. cf Sonata, arpeggione
    and piano, D 821, A minor.
Die Verschworenen. cf Columbia M 33307.
Die Verschworenen: I must behold her. cf Pye NSPH 6.
Waltz, G flat major (Kupelwieser). cf Piano works (Saga 5407).
Waltzes, D 145 (12). cf Fantasia, op. 15, D 760, C major.
2058 Die Winterreise, op. 89, D 911. Hans Hotter, bar; Erik Werba, pno.
    DG 2726 030 (2). (Reissue from SLPM 138778/9)
        +Gr 2-75 p1536   +RR 2-75 p60
2059 Die Winterreise, op. 89, D 911. Hermann Prey, bar; Wolfgang Sawal-
    lisch, pno. Philips 6747 033 (2).
        +-HF 7-74 p100   ++SFC 6-2-74 p22
        ++NR 5-74 p11   ++St 6-74 p124
        +ON 5-75 p36

Der Zwillingsbruder. cf Symphonies, nos. 1-6, 8-9.
SCHULHOFF, Julius
    Pizzicato polka. cf Musical Heritage Society MHS 1959.
SCHULLER, Gunther
    Dramatic overture. cf Century/Advent OU 97 734.
SCHULTZ
    O hilf, Christe. cf Wayne State University, unnumbered.
SCHULZ-ELVER, Adolf
    Arabesque on "The blue Danube". cf Musical Heritage Society MHS
        1959.
SCHUMAN, William
    Credendum (Article of faith). cf IMBRIE: Symphony, no. 3.
SCHUMANN, Robert
    Abendlied, op. 85, no. 12. cf Ember GVC 42.
    Abendlied, op. 107, no. 6. cf Discopaedia MB 1009.
    Adagio and allegro, op. 70, A flat major. cf BEETHOVEN: Sonata,
        horn, op. 17, F major.
    Adagio and allegro, op. 70, A flat major. cf BRAHMS: Trio, horn,
        violin and piano, op. 40, E flat major.
    Album für die Jugend, op. 68. cf Piano works (Telefunken SKA
        25112 T/1-4).
    Album für die Jugend, op. 68. cf Piano works (Vox SVBX 5468).
2060 Album für die Jugend, op. 68: No. 37, Sailor's songs; Nos. 38 and
        39, Wintertime. Carnaval, op. 9. Arturo Benedetti Michelangeli,
        pno. HMV ASD 3129. (also Angel S 37137)
            ++Gr 10-75 p658   +RR 10-75 p80
            +-HF 12-75 p106   -SFC 9-21-75 p34
            +-HFN 11-75 p165   +-SR 11-29-75 p50
            +-NR 11-75 p5
    Album for the young, op. 68: Knight Rupert. cf Connoisseur Society
        (Q) CSQ 2065.
    Albumblätter, op. 124. cf Piano works (Vox 5469).
    Andante and variatons, op. 46, B flat major. cf Classic Record
        Library SQM 80-5731.
    Arabeske, op. 18, C major. cf Piano works (DG 2740 133).
    Arabeske, op. 18, C major. cf Piano works (Telefunken SKA 25085
        T/1-4).
    Arabeske, op. 18, C major. cf Piano works (Vox SVBX 5468).
    Arabeske, op. 18, C major. cf BRAHMS: Ballades, op. 10, nos. 1-4.
    Arabesque, op. 18, C major. cf HMV HQS 1353.
    Blumenstück, op. 19, D major. cf Piano works (Telefunken SKA
        25085 T/1-4).
    Blumenstück, op. 19, D major. cf Piano works (Vox SVBX 5468).
    Bunte Blätter, op. 99, no. 9. cf BRAHMS: Ballades, op. 10, nos.
        1-4.
2061 Carnaval, op. 9. Papillons, op. 2. Gwyneth Pryor, pno. Classics
        for Pleasure CFP 40212.
            +-HFN 5-75 p139   +-RR 7-75 p53
2062 Carnaval, op. 9. Waldscenen, op. 82. Dezsö Ránki, pno. Hungaro-
        ton SLPX 11659.
            -HF 11-75 p122   +-RR 12-75 p85
            ++NR 7-75 p14   ++SFC 11-16-75 p32
    Carnaval, op. 9. cf Album für die Jugend, op. 68: No. 37, Sailor's
        songs; Nos. 38 and 39, Wintertime.
    Carnaval, op. 9. cf Piano works (Da Capo 1C 147 01544/5).
    Carnaval, op. 9. cf Piano works (DG 2740 133).

Carnaval, op. 9. cf Piano works (Vox SVBX 5468).

Carnaval, op. 9. cf BRAHMS: Variations on a theme by Paganini, op. 35, A minor.

2063 Concerto, piano, op. 54, A minor. Introduction and allegro, op. 92, G major. Wilhelm Kempff, pno; Bavarian Radio Symphony Orchestra; Rafael Kubelik. DG 2530 484.
   -HF 7-75 p83   -RR 4-75 p32
   -NR 5-75 p6   -SFC 6-29-75 p26

2064 Concerto, piano, op. 54, A minor. Introduction and allegro, op. 92, G major. Daniel Barenboim, pno; LPO; Dietrich Fischer-Dieskau. HMV ASD 3053.
   +-Gr 3-75 p1658   +-RR 4-75 p31

2065 Concerto, piano, op. 54, A minor. Symphonic etudes, op. 13. Sviatoslav Richter, pno; Symphony Orchestra; Riccardo Muti. Rococo 2084.
   ++NR 10-75 p3

2066 Concerto, piano, op. 54, A minor. Arturo Benedetti Michelangeli, pno; La Scala Orchestra; Antonio Pedrotti. Telefunken AJ 6-41903. (Reissue from SKB 3260/3)
   +-Gr 12-75 p1045

Concerto, piano, op. 54, A minor. cf BEETHOVEN: Egmont overture, op. 84.

Concerto, piano, op. 54, A minor. cf Introduction and allegro, op. 134, D minor.

Concerto, piano, op. 54, A minor. cf GRIEG: Concerto, piano, op. 16, A minor (Bruno Walter Socity 348).

Concerto, piano, op. 54, A minor. cf GRIEG: Concerto, piano, op. 16, A minor (Decca SXL 6624).

Concerto, piano, op. 54, A minor. cf GRIEG: Concerto, piano, op. 16, A minor (HMV 3133).

Concerto, piano, op. 54, A minor. cf GRIEG: Concerto, piano, op. 16, A minor (Philips 6580 108).

Concerto, piano, op. 6. cf GRIEG: Concerto, piano, op. 16, A minor (Philips 6833 020).

Concerto. piano, op. 54, A minor. cf HMV SLS 5033.

Der contrabandiste. cf Westminster WGM 8309.

Davidsbündlertänze, op. 6. cf Piano works (CaCapo 1C 147 01544/5).

Davidsbündlertänze, op. 6. cf Piano works (DG 2740 133).

2067 Dicterliebe, op. 48. Liederkreis, op. 39. Ian Partridge, t; Jennifer Partridge, pno. Classics for Pleasure CFP 40099.
   ++Gr 2-75 p1536   +RR 1-75 p58
   ++STL 2-9-75 p37

2068 Dichterliebe, op. 48. SCHUBERT: Songs: An die Musik, D 547; An Sylvia, D 891; Auf dem Wasser zu singen, D 774; Die Forelle, D 550; Lachen und Weinen, D 777; Lied eines Schiffers an die Dioskuren, D 360; Der Wanderer, D 649; Der Wanderer an den Mond, D 870. Hermann Prey, bar; Leonard Hokanson, pno. Philips 6520 002.
   +HF 8-74 p102   -ON 5-75 p36
   +NR 7-74 p8   ++St 8-74 p72

2069 Fantasia, op. 17, C major. Kreisleriana, op. 16. Adelina de Lara, pno. Rare Recorded Editions ALP 18. (Reissue from ADLP 4, ADLP 6)
   +-Gr 6-75 p69

Fantasia, op. 17, C major. cf Piano works (DG 2740 133).

Fantasia, op. 17, C major. cf Piano works (Telefunken SKA 25085/ T/1-4).

Fantasia, op. 17, C major. cf Symphonic etudes, op. 13.

2070 Fantasiestücke, op. 12. Waldscenen, op. 82, no. 6. Claudio Arrau,
        pno. Philips 6500 423.
            +Gr 7-75 p220   +-RR 7-75 p54
            +-HF 6-74 p96   +SFC 12-23-73 p18
            +HFN 7-75 p88   ++STL 9-7-75 p37
            +NR 3-74 p11

Fantasiestücke, opp. 12 and 111. cf Piano works (Vox 5469).

Fantasiestücke, op. 12. cf Piano works (Telefunken SKA 25085 T/1-4).

Fantasiestücke, op. 12, no. 3: Warum. cf CBS 73396.

Fantasiestücke, op. 12, no. 3: Warum. cf Connoisseur Society (Q)
    CSQ 2066.

Fantasiestücke, op. 12, no. 7: Traumeswirren. cf Westminster WGM
    8309.

Fantasiestücke, op. 73. cf BRAHMS: Sonata, violoncello, no. 1,
    op. 38, E minor.

Fantasiestücke, op. 73. cf Classic Record Library SQM 80-5731.

Fantasiestücke, op. 73. cf Crystal S 331.

Fantasiestücke, op. 111. cf Piano works (Telefunken SKA 25085 T/1-4).

Fantasiestücke, op. 111. cf Piano works (Vox 5469).

Fantasy, violin and orchestra, op. 131, C major. cf BRAHMS: Con-
    certo, violin and violoncello, op. 102, A minor.

Faschingsschwank aus Wien, op. 26. cf Piano works (Telefunken
    SKA 25085 T/1-4).

Faschingsschwank aus Wien, op. 26. cf Piano works (Vox SVBX 5468).

2071 Faust. Elizabeth Harwood, Felicity Palmer, Jennifer Vyvyan, s;
        Pauline Stevens, con; Peter Pears, t; Dietrich Fischer-Dieskau,
        bar; Robert Lloyd, John Shirley-Quirk, bs; Aldeburgh Festival
        Chorus, Wandsworth School Boys' Choir; ECO; Benjamin Britten.
        London OSA 12100 (2). Tape (c) J 512 100 (r) K 412 100. (also
        Decca SET 567/8)
            +Gr 12-73 p1237   +NR 10-74 p9
            ++HF 8-74 p73   ++NYT 7-21-74 pD16
            +HFN 12-73 p2616   +ON 1-11-75 p35
            +MJ 10-74 p42   +-Op 3-74 p224
            +-MQ 10-74 p681   ++St 7-74 p73

2072 Frauenliebe und Leben, op. 42. Liederkreis, op. 39. Elisabeth
        Schwarzkopf, ms; Geoffrey Parsons, pno. Angel S 37043. (also
        HMV ASD 3037)
            -Gr 1-75 p1379   +ON 5-75 p36
            -HF 5-75 p90   -RR 12-74 p73
            +NR 3-75 p9   +-St 5-75 p103

Frauenliebe und Leben, op. 42. cf SCHUBERT: Songs (Philips 6500
    706).

Frauenliebe und Leben, op. 42. cf Decca 6BB 197/8.

Fünf Stücke im Volkston, op. 102. cf CBS 61579.

Genoveva overture, op. 81. cf Symphonies, nos. 1-4.

Genoveva overture, op. 81. cf Symphony, no. 2, op. 61, C major.

Gesange der Frühe, op. 133. cf Piano works (Telefunken SKA 25085
    T/1-4).

2073 Humoreske, op. 20, B flat major. Kreisleriana, op. 16. Vladimir
        Ashkenazy, pno. London CS 6859. (also Decca SXL 6642)
            +Gr 10-74 p728   +RR 11-74 p83
            +-HF 11-75 p122   +SFC 8-10-75 p26
            ++NR 9-75 p13   ++St 10-75 p118

Humoreske, op. 20, B flat major. cf Piano works (DG 2740 133).

2074 Impromptus on a theme by Clara Wieck, op. 5. Sonata, piano, no.
      3, op. 14, F minor. Jean-Philippe Collard, pno. Connoisseur
      Society CS 2081.
              +HF 12-75 p106   +NR 11-75 p14
      Introduction and allegro, op. 92, G major. cf Concerto, piano,
      op. 54, A minor (HMV 3053).
      Introduction and allegro, op. 92, G major. cf Concerto, piano,
      op. 54, A minor (DG 2530 484).
2075 Introduction and allegro, op. 134, D minor. Concerto, piano, op.
      54, A minor. Konzertstück, op. 92, G major. Peter Frankl,
      pno; Bamberg Symphony Orchestra; Janos Fürst. Turnabout TV
      34559. (Q) QTV 34559.
              /NR 12-74 p5   ++SFC 6-29-75 p26
              /RR 12-75 p61
      Julius Caesar overture, op. 128. cf Symphony, no. 2, op. 61, C
      major.
      Kinderscenen, op. 15. cf Piano works (Da Capo 1C 147 01544/5).
      Kinderscenen, op. 15. cf Piano works (DG 2740 133).
      Kinderscenen, op. 15. cf Piano works (Telefunken SKA 25085 T/1-4).
      Kinderscenen, op. 15. cf Piano works (Vox 5469).
      Kinderscenen, op. 15. cf BEETHOVEN: Sonata, piano, no. 29, op.
      106, B flat major.
      Kinderscenen, op. 15. cf NILSSON: Quantitaten.
      Kinderscenen, op. 15. cf SCHUBERT: Sonata, piano, no. 21, op.
      posth., D 960, B flat major.
      Kinderscenen, op. 17, no. 7: Träumerei. cf Discopaedia MB 1002.
      Kinderscenen, op. 15, no. 7: Träumerei. cf Discopaedia MB 1009.
      Kinderscenen, op. 15, no. 7: Träumerei. cf Rococo 2072.
      Kinderscenen, op. 15, no. 7: Träumerei. cf Turnabout TV 37033.
      Klavierstücke in Fughettenform, op. 126 (7). cf Piano works (Tele-
      funken SKA 25085 T/1-4).
      Konzertstück, op. 92, G major. cf Introduction and allegro, op.
      134, D minor.
2076 Kreisleriana, op. 16. Presto, op. posth., G major. Sonata, piano,
      no. 2, op. 22, G minor. Claudio Arrau, pno. Philips 6500 394.
              +-Gr 6-75 p70   +RR 5-75 p56
              +-HF 2-74 p102   ++SFC 1-27-74 p30
              ++HFN 6-75 p93   ++St 3-74 p119
              +-NR 1-74 p14
      Kreisleriana, op. 16. cf Fantasia, op. 17, C major.
      Kreisleriana, op. 16. cf Humoreske, op. 20, B flat major.
      Kreisleriana, op. 16. cf Piano works (DaCapo 1C 147 01544/5).
      Kreisleriana, op. 16. cf Piano works (DG 2740 133).
      Kreisleriana, op. 16. cf Piano works (Telefunken SKA 25085 T/1-4).
      Kreisleriana, op. 16. cf LISZT: Concerto, piano, no. 1, G 124,
      E flat major.
      Kreisleriana, op. 16. cf International Piano Archives IPA 5007/8.
      Liederkreis, op. 39. cf Dichterliebe, op. 48.
      Liederkreis, op. 39. cf Frauenliebe und Leben, op. 42.
      Liederkreis, op. 39. cf Columbia M 32231.
      Manfred overture, op. 115. cf Symphonies, nos. 1-4.
      Manfred overture, op. 115. cf Symphony, no. 2, op. 61, C major.
      Manfred overture, op. 115. cf Symphony, no. 3, op. 97, E flat major.
      Marches, op. 76 (4). cf Piano works (Telefunken SKA 25085 T/1-4).
      Myrthen, op. 25, no. 1: Widmung. cf RCA ARM 4-0942/7.
      Nachtstücke, op. 23. cf Piano works (DG 2740 133).

2077 Novelletten, op. 21. Claudio Arrau, pno. Philips 6500 396.
        +Gr 2-75 p1519   +RR 3-75 p52
        +HF 12-74 p115   +SFC 9-22-74 p22
        +MJ 11-74 p48   -St 12-74 p145
        +-NR 11-74 p10
    Overture, scherzo and finale, op. 52, E major. cf Symphony, no. 1,
        op. 38, B flat major.
    Overture, scherzo and finale, op. 52, E major. cf Symphony, no. 2,
        op. 61, C major.
    Papillons, op. 2. cf Carnaval, op. 9.
    Papillons, op. 2. cf Piano works (DaCapo 1C 147 01544/5).
    Papillons, op. 2. cf Piano works (DG 2740 133).
    Papillons, op. 2. cf DG 2548 137.
2078 Piano works: Carnaval, op. 9. Davidsbündlertänze, op. 6. Kinder-
        scenen, op. 15. Kreisleriana, op. 16. Papillons, op. 2.
        Alfred Cortot, pno. DaCapo 1C 147 015445/5 (2). (Recorded,
        1928-47)
        +-HF 11-75 p122
2079 Piano works : Arabeske, op. 18, C major. Carnaval, op. 9. Davids-
        bündlertänza, op. 6. Fantasie, op. 17, C major. Humoreske, op.
        20, B flat major. Kinderscenen, op. 15. Kreisleriana, op. 16.
        Nachtstücke, op. 23. Papillons, op. 2. Romances, op. 28 (3).
        Sonata, piano, no. 2, op. 22, G minor. Symphonic etudes, op.
        13. Waldscenen, op. 82. Wilhelm Kempff, pno. DG 2740 133 (6).
        (Reissues from 139316, 2530185, 2530 317, 2530 185, 2530 410,
        2530 348, 2530 321)
        +Gr 11-75 p909   +RR 11-75 p83
        +HFN 11-75 p173
2080 Piano works: Arabeske, op. 18, C major. Blumenstück, op. 19, D
        major. Fantasia, op. 17, C major. Fantasiestücke, opp. 12, 111.
        Faschingsschwank aus Wien, op. 26. Gesange der Frühe, op. 133.
        Kinderscenen, op. 15. Klavierstücke in Fughettenform, op. 126
        (7). Kreisleriana, op. 16. Marches, op. 76 (4). Romances, op.
        28 (3). Karl Engel, pno. Telefunken SKA 25085 T/1-4 (4).
        +-Gr 6-74 p80   +-NR 9-74 p9
        +-HF 11-74 p122   /RR 5-74 p58
        +-HFN 3-74 p113   +-St 1-75 p119
2081 Piano works: Album für die Jugend, op. 68. Presto passionato, op.
        posth., G minor. Scherzi, op. posth., F minor, B flat major.
        Sonatas, nos. 1-3. Sonaten für die Jugend, op. 118 (3). Karl
        Engel, pno. Telefunken SKA 25112 T/1-4 (4).
        +Gr 10-74 p728   +-NR 12-74 p15
        +HF 11-75 p122   +RR 12-74 p64
2082 Piano works: Album für die Jugend, op. 68. Arabeske, op. 18, C
        major. Blumenstück, op. 19, D major. Carnaval, op. 9. Fasch-
        ingsschwank aus Wien, op. 26. Symphonic etudes, op. 13. Toc-
        cata, op. 7, C major. Peter Frankl, pno. Vox SVBX 5468 (2).
        +HF 12-74 p115   ++SFC 1-12-75 p26
        +NR 9-74 p10   ++St 1-75 p119
2083 Piano works: Albumblätter, op. 124. Fantasiestücke, op. 12 and
        op. 111. Kinderscenen, op. 15. Romances, op. 28 (3).
        Sonata, piano, no. 2, op. 22, G minor (with Presto passionato,
        op. posth.). Sonatas for the young, op. 118 (3). Variations
        on A-B-E-G-G, op. 1. Peter Frankl, pno. Vox SVBX 5469 (3).
        +-HF 11-75 p122   +NR 7-75 p14
    Presto, op. posth., G major. cf Kreisleriana, op. 16.

Presto passionato, op. posth., G minor. cf Piano works (Telefunken
SKA 25112 T/1-4).

2084 Quartet, strings, no. 3, op. 41, A major. SIBELIUS: Quartet,
strings, op. 56, D minor. Voces Intimae Quartet. BIS LP 10.
    -RR 11-75 p66   +-St 12-75 p132

Quartet, strings, no. 3, op. 41, A major. cf MENDELSSOHN: Quartet,
strings, no. 2, op. 13, A minor.

Romances, op. 28. cf Piano works (DG 2740 133).

Romances, op. 28. cf Piano works (Vox 5469).

Romances, op. 28 (3). cf Piano works (Telefunken SKA 25085 T/1-4).

Romances, op. 28, nos. 1-3. cf BRAHMS: Ballades, op. 10, nos. 1-4.

Romance, op. 28, no. 2, F sharp major. cf Decca SPA 372.

Romance, op. 28, no. 2, F sharp major. cf HMV HQS 1353.

Scherzi, op. posth., F minor, B flat major. cf Piano works
(Telefunken SKA 25112 T/1-4).

Scherzo and presto passionata. cf BRAHMS: Capriccio, op. 116.

Sonatas, piano, nos. 1-3. cf Piano works (Telefunken SKA 25112
T/1-4).

Sonata, piano, no. 1, op. 11, F sharp minor. cf BRAHMS: Sonata,
piano, no. 2, op. 2, F sharp minor.

Sonata, piano, no. 2, op. 22, G minor. cf Kreisleriana, op. 16.

Sonata, piano, no. 2, op. 22, G minor. cf Piano works (DG 2740 133).

Sonata, piano, no. 2, op. 22, G minor (with Presto passionato,
op. posth.). cf Piano works (Vox 5469).

Sonata, piano, no. 2, op. 22, G minor. cf SCHUBERT: Fantasia,
op. 15, D 760, C major.

Sonata, piano, no. 3, op. 14, F minor. cf Impromptus on a theme
by Clara Wieck, op. 5.

Sonaten für die Jugend, op. 118 (3). cf Piano works (Telefunken
SKA 25112 T/1-4).

Sonatas for the young, op. 118. cf Piano works (Vox 5469).

2085 Songs: Aufträge, op. 77, no. 5; Er ist's, op. 79, no. 23; Erstes
Grün, op. 35, no. 4; Frage, op. 35, no. 9; Jasminenstrauch, op.
27, no. 4; Die Kartenlegerin, op. 31, no. 2; Das Käuzlein, op.
79, no. 10; Die letzten Blumen starben, op. 104, no. 6; Loreley,
op. 53, no. 2; Marienwürchen, op. 79, no. 13; Mein schöner
Stern; Die Meerfee, op. 125, no. 3; Der Nussbaum, op. 25, no. 3;
Der Sandmann, op. 79, no. 12; Schmetterling, op. 79, no. 2;
Schneeglöckchen, op. 79, no. 26; Sehnsucht, op. 11, no. 1;
Sehnsucht nach der Waldgegend, op. 35, no. 5; Die Sennerin, op.
90, no. 4; Waldesgespräch, op. 39, no. 3. Elly Ameling, s;
Jörg Demus, pno. BASF HB 29369.
    +HF 1-74 p80   -ON 5-75 p36
    ++NR 1-74 p12   ++SFC 9-23-73 p37
    +NYT 10-14-73 pD33   +St 2-74 p127

2086 Songs: Liederkreis, op. 24. Myrthen, op. 25. Dietrich Fischer-
Dieskau, bar; Christoph Eschenbach, pno. DG 2530 543.
    +Gr 8-75 p355   +NR 12-75 p12
    +HF 12-75 p106   ++SFC 11-16-75 p32

Songs: Herbstlied, op. 43, no. 2; Schön Blümelein, op. 43, no. 3;
So wahr die Sonne scheinet, op. 37. cf BRAHMS: Songs (Musical
Heritage Society 1896).

Songs: Zwölf Gedichte von Justinus Kerner, op. 35. cf KILPINEN:
Lieder and Tunturilauluja.

Songs: Abendlied, op. 107, no. 6; Aus den Hebräischen Gesangen, op.
25, no. 15; Der Kartenlegerin, op. 31, no. 2; Die Lotusblume,

op. 25, no. 7. cf SCHUBERT: Songs (Decca SXL 6578).

Songs: Der Hidalgo, op. 30, no. 3; Provencalisches Lied. cf WAGNER: Lohengrin.

Songs: Botschaft, op. 74, no. 8; Das Glück, op. 79, no. 15. cf Columbia M 33307.

Songs: The snowdrop; Ladybird. cf HMV RLS 714.

Songs: Lust der Sturmnacht, op. 35, no. 1; Mein schöner Stern; Stille Liebe, op. 35, no. 8; Stille Tränen, op. 35, no. 10; Widmung, op. 25, no. 1. cf Seraphim S 60251.

Songs: Die beiden Granadiere, op. 49, no. 1; Frühlingsfahrt, op. 45, no. 2; Intermezzo, op. 39, no. 2; Schöne Fremde, op. 39, no. 6; Zum Schluss, op. 25, no. 26. cf Telefunken DLP 6-48064.

Spanisches Liederspiel, op. 74. cf SCHUBERT: Songs (Hungaroton SHLX 90050).

2087 Symphonic etudes, op. 13. Fantasia, op. 17, C major. Géza Anda, pno. DG 2538 317. (Reissue from SLPM 138868)
    +-Gr 6-75 p70   -RR 11-75 p83
    +-HFN 11-75 p173

Symphonic etudes, op. 13. cf Concerto, piano, op. 54, A minor.

Symphonic etudes, op. 13. cf Piano works (DG 2740 133).

Symphonic etudes, op. 13. cf Piano works (Vox SVBX 5468).

2088 Symphonies, nos. 1-4. Genoveva overture, op. 81. Manfred overture, op. 115. BPhO; Rafael Kubelik. DG 2535 116/8 (3). (Reissues from 138860, 138955, 138908)
    +Gr 9-75 p471   +HFN 8-75 p89

2089 Symphonies, nos. 1-4. Leipzig Gewandhaus Orchestra; Franz Konwitschny. Philips 6780 015 (2). (Reissue from GL 5794/7)
    -Gr 6-75 p49   +-RR 6-75 p47
    +-HFN 9-75 p104

2090 Symphonies, nos. 1-4. LPO; Adrian Boult. Pye GGCD 302 (2). (Reissues from GSGC 14066/7)
    +-Gr 8-75 p335   +-RR 7-75 p34

2091 Symphony, no. 1, op. 38, B flat major. Symphony, no. 3, op. 97, E flat major. CO; Georg Szell. CBS 61595. (Reissues from Columbia SAX 2475, 2506)
    +Gr 3-75 p1663   ++RR 3-75 p34

2092 Symphony, no. 1, op. 38, B flat major. Symphony, no. 4, op. 120, D minor. LSO; Josef Krips. Decca ECS 758. (Reissue from SXL 2223) (also London STS 15019)
    +Gr 3-75 p1663   +RR 3-75 p34
    -HFN 7-75 p88

2093 Symphony, no. 1, op. 38, B flat major. Symphony, no. 4, op. 120, D minor. BPhO; Rafael Kubelik. DG 2535 116. (also DG 2530 169. Tape (c) 3300 419 (ct) 89479)
    +RR 6-75 p47

2094 Symphony, no. 1, op. 38, B flat major. Overture, scherzo and finale, op. 52, E major. VPO; Georg Solti. London CS 6696. (also Decca SXL 6486. Tape (c) KSXC 6486)
    +-Gr 8-75 p376 tape   +RR 6-75 p93 tape
    +HFN 7-75 p91 tape   ++St 2-74 p120
    +NR 11-73 p3

2095 Symphony, no. 2, op. 61, C major. Manfred overture, op. 115. OSR; Ernest Ansermet. Decca ECS 759. (Reissue from SXL 6205)
    +Gr 3-75 p1663   +RR 3-75 p34
    +HFN 7-75 p88

2096 Symphony, no. 2, op. 61, C major. Julius Caesar overture, op. 128.

VPO; Georg Solti. Decca SXL 6487. Tape (c) KSXC 6487.
+Gr 8-75 p376 tape   +-HFN 7-75 p91 tape
2097 Symphony, no. 2, op. 61, C major. Overture, scherzo and finale,
op. 52, E major. BPhO; Herbert von Karajan. DG 2530 170. Tape
(c) 3300 482. (Reissue from 2720 046)
+Gr 10-75 p626   +RR 8-75 p42
+HFN 8-75 p89   +RR 10-75 p97 tape
2098 Symphony, no. 2, op. 61, C major. Genoveva overture, op. 81.
BPhO; Rafael Kubelik. DG 2535 117. (also DG 138955)
+RR 6-75 p48
2099 Symphony, no. 3, op. 97, E flat major. MENDELSSOHN: A midsummer
night's dream, op. 21: Overture. LSO; Rafael Frühbeck de Burgos.
Decca ECS 760. (Reissue from SXL 6213) (also London STS 15246)
++Gr 3-75 p1663   +-RR 3-75 p34
+HFN 5-75 p134
2100 Symphony, no. 3, op. 97, E flat major. BPhO; Herbert von Karajan.
DG 2530 447. Tape (c) 3300 404. (Reissue from 2720 046).
+Gr 9-74 p517   +RR 8-74 p40
++Gr 9-74 p592 tape   +RR 8-74 p86 tape
++NR 1-75 p3
2101 Symphony, no. 3, op. 97, E flat major. Manfred overture, op. 115.
BPhO; Rafael Kubelik. DG 2535 118. (also DG 138908)
+RR 6-75 p48
Symphony, no. 3, op. 97, E flat major. cf Symphony, no. 1, op.
38, B flat major.
Symphony, no. 4, op. 120, D minor. cf Symphony, no. 1, op. 38,
B flat major (Decca 758).
Symphony, no. 4, op. 120, D minor. cf Symphony, no. 1, op. 38,
B flat major (DG 2535 116).
Symphony, no. 4, op. 120, D minor. cf MENDELSSOHN: Symphony, no.
4, op. 90, A major.
Symphony, no. 4, op. 120, D minor. cf Toscanini Society ATS GC
1201/6.
Toccata, op. 7, C major. cf Piano works (Vox SVBX 5468).
Toccata, op. 7, C major. cf Westminster WGM 8309.
2102 Trio, piano, no. 1, op. 63, D minor. Henryk Szeryng, vln; Pierre
Fournier, vlc; Artur Rubinstein, pno. RCA LRL 1-7529.
++Gr 7-75 p205   +-RR 8-75 p53
+HFN 8-75 p83
Trio, piano, no. 1, op. 63, D minor. cf BRAHMS: Trios, piano, nos.
1-3.
Trio, piano, no. 1, op. 63, D minor. cf BRAHMS: Trio, piano, no.
2, op. 87, C major.
Variations on A-B-E-G-G, op. 1. cf Piano works (Vox 5469).
Variations on A-B-E-G-G, op. 1. cf BEETHOVEN: Sonata, piano, no.
29, op. 106, B flat major.
Waldscenen, op. 82. cf Carnaval, op. 9.
Waldscenen, op. 82. cf Piano works (DG 2740 133).
Waldscenen, op. 82, no. 6. cf Fantasiestücke, op. 12.
Waldscenen, op. 82, no. 6. cf BRAHMS: Ballades, op. 10, nos. 1-4.
Whims. cf Canon CNN 4983.
SCHUTZ, Heinrich
Was hast du verwirket. cf Telefunken AJ 6-41867.
The words of the Institution of the Lord's supper. cf BASF BAC
3087.
SCHWANTER, Joseph
Modus caelestis. cf PENN: Ultra mensuram.

SCHWARZKOPF, ELISABETH, SINGS OPERETTA. cf Angel S 35696.
SCOTT
    Ame nasceri; En prière. cf Rubini RS 300.
    The gentle maiden. cf RCA ARM 4-0942/7.
    Tallahassee suite. cf RCA ARM 4-0942/7.
SCOTT, Cyril
    Blackbird's song. cf HMV RLS 714.
2103 Concerto, piano, no. 1, C major. John Ogdon, pno; LPO; Bernard
        Herrmann. Lyrita SRCS 81.
            ++Gr 9-75 p471   ++RR 9-75 p46
            +HFN 10-75 p149
2104 Danse nègre. Lotus land. Poëms (5). Sonata, piano, no. 1, op.
        66. Martha Anne Verbit, pno. Genesis GS 1049.
            +NR 5-75 p13   +St 8-75 p104
    Lotus land. cf Danse nègre.
    Poëms. cf Danse nègre.
    Sonata, piano, no. 1, op. 66. cf Danse nègre.
SCRIABIN, Alexander
    Caress dansée, op. 57, no. 2. cf Piano works (Connoisseur Society
        CS 2035).
2105 Concerto, piano, op. 20, F sharp minor. Prometheus, Poem of fire,
        op. 60. Vladimir Ashkenazy, pno; LPO; Loren Maazel. London CS
        6732. Tape (c) M 10251 (r) L 80251. (also Decca SXL 6527. Tape
        (c) KSXC 6527)
            +Gr 1-72 p1220   +NR 4-72 p3
            ++Ha 11-72 p126   ++SFC 2-20-72 p32
            ++HF 6-72 p92   /SR 4-22-72 p52
            +HF 11-72 p132 tape   +St 6-72 p92
            ++HFN 1-72 p113   +STL 2-13-72 p29
            ++HFN 8-75 p110 tape
    Désir, op. 57, no. 1. cf Piano works (Connoisseur Society CS 2035).
    Etude, op. 2, no. 1, C sharp minor. cf Piano works (Connoisseur
        Society CS 2035).
    Etude, op. 2, no. 1, C sharp minor. cf Draco DR 1333.
    Etudes, op. 8. cf RACHMANINOFF: Lilacs, op. 21, no. 5.
    Etudes, op. 8, nos. 2, 8, 12. cf BACH: Prelude and fugue, B flat
        minor.
    Etude, op. 8, no. 12, D sharp minor. cf Columbia Special Products
        AP 12411.
2106 Etudes, op. 42, nos. 1-8. Sonatas, nos. 5, 7, 9. Ruth Laredo,
        pno. Connoisseur Society CS 2032.
            +-Gr 2-75 p1519
    Nocturne for the left hand alone, op. 9, no. 2. cf Columbia Special
        Products AP 12411.
2107 Piano works: Caress dansée, op. 57, no. 2. Etude, op. 2, no. 1,
        C sharp minor. Désir, op. 57, no. 1. Sonatas, piano, nos. 1,
        2, 8. Ruth Laredo, pno. Connoisseur Society CS 2035.
            +RR 3-75 p55
    Poem, op. 32, no. 1. cf Preludes, op. 11 and op. 74.
    Poème de l'extase, op. 54. cf DVORAK: Slavonic dance, op. 72,
        no. 2, E minor.
2108 Preludes, op. 11 (24); op. 74 (5). Poem, op. 32, no. 1. Ruth
        Laredo, pno. Desto DC 7145. (also Peerless PRCM 201)
            +Gr 7-75 p220   +SFC 3-4-73 p33
            +HF 3-73 p92   +SR 4-73 p72
            +NR 3-75 p11   ++St 7-73 p108

Prelude, op. 11, no. 14. cf Draco DR 1333.
Prelude, op. 45, no. 3, E flat major. cf Ember GVC 40.
Prometheus, Poem of fire, op. 60. cf Concerto, piano, op. 20,
    F sharp minor.
Sonatas, piano, nos. 1, 2, 8. cf Piano works (Connoisseur Society
    CS 2035).
2109 Sonatas, piano, nos. 3-5, 9. Vladimir Ashkenazy, pno. Decca SXL
    6705.
            +Gr 11-75 p864   ++RR 11-75 p84
            +HFN 12-75 p164
2110 Sonatas, piano, nos. 3, 4, 6, 10. Vers la flamme, op. 72. Ruth
    Laredo, pno. Connoisseur Society CS 2034.
            ++RR 3-75 p55
Sonatas, piano, nos. 5, 7, 9. cf Etudes, op. 42, nos. 1-8.
Vers la flamme, op. 72. cf Sonatas, piano, nos. 3, 4, 6, 10.
Winged poem (Poeme aile), op. 51, no. 3. cf Ember GVC 40.
SEARLE, Humphrey
Aubade, op. 28. cf BANKS: Concerto, horn.
The barber of Darmstadt: Duet. cf HMV SLA 870.
Lochinvar, speakers and percussion. cf HMV SLA 870.
Punkt Contrapunkt. cf HMV SLA 870.
2111 Symphony, no. 1, op. 23. Symphony, no. 2, op. 33. LPO; Adrian
    Boult, Josef Krips. Lyrita SRCS 72. (Reissue from Decca SXL
    2232)
            +Gr 6-75 p49   +RR 5-75 p39
            ++HFN 5-75 p139
Symphony, no. 2, op. 33. cf Symphony, no. 1, op. 23.
Toccata alla passacaglia. cf Gemini GM 2022.
SECKENDORFF, Siegmund von
Romance. cf DG Archive 2533 149.
SEGERSTAM, Leif
Divertimento. cf RAUTAVAARA: Pelimannit, op. 1.
Poem. cf BIS LP 18.
SEIBER, Matyas
The famous Tay whale. cf HMV SLA 870.
SEIFERT
Kärtner Liedermarsch. cf BASF 292 2116-4.
Kärtner Liedermarsch. cf DG 2721 077.
SENFL, Ludwig
Lieder (5). cf BIS LP 3.
THE SERAPHIM GUIDE TO THE INSTRUMENTS OF THE ORCHESTRA. cf
    Seraphim S 60234.
SERMISY, Claude de
Allez souspirs. cf HMV SLS 5022.
Amour me voyant. cf HMV SLS 5022.
Content désir; La, je m'y plains. cf L'Oiseau-Lyre 12BB 203/6.
Tant que vivray. cf Harmonia Mundi 204.
SEROCKI, Kazimierz
Suita preludiów. cf BRAHMS: Variations on a theme by Paganini,
    op. 35, A minor.
SESSIONS, Roger
Rhapsody, orchestra. cf MUSGRAVE: Night music, chamber orchestra.
2112 Sonata, piano, no. 3. WUORINEN: Duo, violin and piano. Robert
    Helps, Charles Wuorinen, pno; Paul Zukofsky, vln. DG/Acoustic
    Research 064 086.
            +HF 2-71 p97   +SFC 4-18-71 p31

```
        +MJ  2-75 p29                    +SR 12-26-70 p47
        +MQ  7-72 p501                   +St  5-71 p95
        *NYT 1-3-71 pD22
```
    Symphony, no. 2. cf MENNIN: Symphony, no. 3.
    Symphony, no. 8. cf MUSGRAVE: Night music, chamber orchestra.
SEYFERT, Johann Caspar
    Songs and quodlibets: Sie frohe Compagnie; Die Bescherlichkeiter
        des Ehestandes; Die lustige Tyrolerin; Wir haben drey Katzen;
        Der verachtete Liebhaber. cf RATHGEBER: Songs and quodlibets
        (Angel 7107).
SHAW, Christopher
    A lesson from Ecclesiastes. cf DALLAPICCOLA: Tempus destruendi,
        Tempus aedificandio.
    Music when soft voices die. cf DALLAPICCOLA: Tempus destruendi,
        Tempus aedificandi.
    Peter and the lame man. cf DALLAPICCOLA: Tempus destruendi, Tempus
        aedificandi.
    To the Bandusian spring. cf DALLAPICCOLA: Tempus destruendi, Tem-
        pus aedificandi.
SHAW, Geoffrey
    Anglican folk mass: The Creed. cf HMV HQS 1350.
SHELLEY
    The King of love my Shepherd is. cf RCA ARL 1-0562.
SHERLAW JOHNSON, Robert
    Sonata, piano, no. 2. cf HMV HQS 1337.
SHERWIN
    Sign tonight. cf New England Conservatory NEC 111.
SHOSTAKOVICH, Dmitri
    The age of gold, op. 22. cf Symphonies, nos. 1-15.
    The bolt, op. 27a. cf Symphonies, nos. 1-15.
    Concerto, piano, no. 2, op. 102, F major. cf Concerto, piano and
        trumpet, no. 1, op. 35, C minor (CBS 73400).
    Concerto, piano, no. 2, op. 102, F major. cf Concerto, piano and
        trumpet, no. 1, op. 35, C minor (HMV ASD 3081).
    Concerto, piano, no. 2, op. 102, F major. cf Works selections
        (CBS 77394).
2113 Concerto, piano and trumpet, no. 1, op. 35, C minor. Concerto,
        piano, no. 2, op. 102, F major. André Previn, Leonard Bernstein,
        pno; William Vacchiano, tpt; NYP; Leonard Bernstein. CBS 73400.
        (Reissues from CBS SBRG 72349/50, Philips SABL 134)
            +-Gr 5-75 p1974   +-RR 5-75 p39
            +HFN 6-75 p93
2114 Concerto, piano and trumpet, no. 1, op. 35, C minor. Concerto,
        piano, no. 2, op. 102, F major. Fantastic dances, op. 5 (3).
        Cristina Ortiz, pno; Rodney Senior, tpt; Bournemouth Symphony
        Orchestra; Paavo Berglund. HMV ASD 3081. (also Angel S 37109)
            +Gr 6-75 p55   +NR 7-75 p6
            +-HF 9-75 p90   +-RR 6-75 p53
            ++HFN 6-75 p93   +St 11-75 p136
    Concerto, piano and trumpet, no. 1, op. 35, C minor. cf Works,
        selections (CBS 77394).
    Concerto, violin, no. 1, op. 99, A minor. cf Works, selections
        (CBS 77394).
    Concerto, violoncello, op. 107, E flat major. cf Works, selections
        (CBS 77394).
    Fantastic dances, op. 5. cf Concerto, piano and trumpet, no. 1,
        op. 35, C minor.

The gadfly, op. 97a. cf Works, selections (CBS 77394).

2115 Hamlet: Suite. ROZSA: Julius Caesar: Suite. WALTON: Richard III:
     Prelude. National Philharmonic Orchestra; Bernard Herrmann.
     Decca PFS 4315. (also London 21132. Tape (c) 05 21132 (ct)
     08 21132)
              +HFN 8-75 p80   +NR 10-75 p2

     Hamlet: Suite. cf ROZSA: Film music to Julius Caesar: Suite.
     Octet, strings, op. 11. cf Quintet, piano, op. 57, G minor.

2116 Preludes and fugues, op. 87 (24). Roger Woodward, pno. RCA LRL
     2-5100 (2).
              +Gr 12-75 p1082

2117 Quartet, strings, no. 1, op. 49, C major. Quartet, strings, no. 3,
     op. 73, F major. Gabrieli Quartet. Decca SDD 453.
              ++Gr 4-75 p1829   ++RR 4-75 p41
              ++HFN 6-75 p93

     Quartet, strings, no. 1, op. 49, C major: Scherzo. cf Amon Ra
     SARB 01.
     Quartet, strings, no. 2, op. 68. cf ARISTAKESIAN: Sinfonietta,
     piano, xylophone and strings.
     Quartet, strings, no. 3, op. 73, F major. cf Quartet, strings,
     no. 1, op. 49, C major.
     Quartet, strings, no. 3, op. 73, F major: Moderato con moto. cf
     HMV SEOM 20.

2118 Quartet, strings, no. 4, op. 83. WELIN: Quartet, strings, no. 2.
     WIREN: Quartet, strings, no. 5, op. 41. Saulesco Quartet.
     Caprice RIKSLP 24.
              +Gr 12-74 p1167   +-RR 1-75 p44

2119 Quartet, strings, no. 7, op. 108, F sharp minor. Quartet, strings,
     no. 13, op. 138, B flat minor. Quartet, strings, no. 14, op.
     142, F sharp major. Fitzwilliam Quartet. L'Oiseau-Lyre DSLO 9.
              +Gr 12-75 p1065   +RR 11-75 p65
              +HFN 12-75 p164

     Quartet, strings, no. 13, op. 138, B flat minor. cf Quartet,
     strings, no. 7, op. 108, F sharp minor.
     Quartet, strings, no. 14, op. 142, F sharp major. cf Quartet,
     strings, no. 7, op. 108, F sharp minor.

2120 Quintet, piano, op. 57, G minor. Octet, strings, op. 11 (2).
     STRAVINSKY: Pieces, string quartet (3). Borodin Quartet,
     Prokofiev Quartet; Lyubov Yedlina, pno. HMV ASD 3072.
              +-Gr 5-75 p1991   +MT 10-75 p888
              +HFN 5-75 p139   +-RR 5-75 p51

     Romances on words of Alexander Block, op. 127 (7). cf ARISTAKESIAN:
     Sinfonietta, piano, xylophone and strings.
     The sun shines over our Motherland, op. 90. cf Symphonies, nos.
     1-15.

2121 Symphonies, nos. 1-15. The bolt, op. 27a. The age of gold, op.
     22. The sun shines over our Motherland, op. 90. MPO, USSR,
     MRSO, Leningrad Philharmonic Orchestra, Moscow Chamber Orches-
     tra; Kiril Kondrashin, Maxim Shostakovich, Yevgeny Svetlanov,
     Yevgeny Mravinsky, Rudolf Barshai. HMV/Melodiya SLS 5025 (3).
     (Reissues)
              +Gr 12-75 p1046   +RR 10-75 p50
              +HFN 10-75 p149   +STL 9-7-75 p37

2122 Symphony, no. 1, op. 10, F major. Symphony, no. 3, op. 20, E flat
     major. MPO; RSFR Russian Chorus; Kyril Kondrashin. HMV Melo-
     diya ASD 3045.
              +Gr 2-75 p1499   +RR 2-75 p38

Symphony, no. 2, op. 14, C major. cf PROKOFIEV: Scythian suite,
   op. 20.
Symphony, no. 3, op. 20, E flat major. cf Symphony, no. 1, op. 10,
   F major.
2123 Symphony, no. 5, op. 47, D minor. PO; Eugene Ormandy. CBS 61643.
   (Reissue from CBS 72811) (also Columbia MS 7279)
       +-Gr 7-75 p195   +RR 6-75 p53
       +HFN 8-75 p89
2124 Symphony, no. 5, op. 47, D minor. OSR; István Kertész. London
   6327. (also Decca ECS 767. Reissue from SXL 6018)
       /Gr 9-75 p471   ++RR 8-75 p42
      +-HFN 8-75 p89
2125 Symphony, no. 5, op. 47. PO; Eugene Ormandy. RCA ARL 1-1149.
   Tape (c) ARK 1-1149 (ct) ARS 1-1149.
       +MJ 11-75 p20   ++SFC 9-14-75 p28
      ++NR 11-75 p2
2126 Symphony, no. 5, op. 47. CPhO; Karel Ančerl. Vanguard/Supraphon
   SU 1. (Reissue from Parliament PLPS 168)
       +Audio 8-75 p83   +SFC 6-15-75 p24
      +-HF 5-75 p74   +St 5-75 p102
      +NR 3-75 p3
Symphony, no. 5, op. 47, D minor: Allegretto. cf HMV SEOM 20.
2127 Symphony, no. 6, op. 54, B minor. PROKOFIEV: Lieutenant Kije
   suite, op. 60. LSO; André Previn. HMV ASD 3029.   Tape (c)
   TC ASD 3029. (also Angel S 37026)
       ++Gr 12-74 p1143   +RR 12-74 p39
       +-Gr 10-75 p721 tape   +RR 10-75 p98 tape
       -HF 8-75 p100   ++SFC 6-15-75 p24
       +HFN 10-75 p155 tape   +Te 3-75 p44
       +NR 5-75 p5
2128 Symphony, no. 8, op. 65, C minor. LSO; André Previn. HMV ASD
   2917. (also Angel 36980)
       ++Gr 10-73 p689   ++RR 11-73 p50
       ++HF 10-74 p84   ++St 1-75 p75
       +NR 11-74 p2
2129 Symphony, no. 10, op. 93, E minor. LPO; Andrew Davis. Classics
   for Pleasure CFP 40216.
       +Gr 5-75 p1973   +RR 6-75 p53
       +-HFN 6-75 p93
2130 Symphony, no. 11, op. 103, G minor. Houston Symphony Orchestra;
   Leopold Stokowski. Seraphim S 20228. (Reissue from Capitol
   SPBR 8448, SPBO 8700)
       +-HF 8-75 p100
2131 Symphony, no. 14, op. 135. Galina Vishnevskaya, s; Mark Reshetin,
   bs; MPO; Mstislav Rostropovich. HMV Melodiya ASD 3090.
       +Gr 12-75 p1046
2132 Symphony, no. 15, op. 141, A major. PO; Eugene Ormandy. RCA (Q)
   ARD 1-0014. Tape (c) ARK 1-0014 (ct) ARS 1-0014.
       ++Gr 2-73 p1515   +NR 2-73 p4
       +Gr 5-75 p1974   +-NYT 1-21-73 pD26
       +-HF 3-73 p74   ++RR 2-73 p63
       ++HFN 3-73 p570   +RR 5-75 p40
       ++HFN 5-75 p139   +SR 2-3-73 p53
       -LJ 5-73 p33   +St 2-73 p83
2133 Works, selections: Concerto, piano, no. 2, op. 102, F major. Concerto, piano and trumpet, no. 1, op. 35, C minor. Concerto,

violin, no. 1, op. 99, A minor. Concerto, violoncello, op. 107, E flat major. The gadfly, op. 97a. André Previn, Leonard Bernstein, pno; David Oistrakh, vln; William Vacchiano, tpt; NYP, PO; Leonard Bernstein, Dmitri Mitropoulos, Eugene Ormandy, André Kostelanetz. CBS 77394 (3). (Reissues from SBRG 72349/50, 72504, Philips SABL 134, 165, ABL 3101)

+Gr 10-75 p626   +-RR 10-75 p49
+-HFN 10-75 p152

SIBELIUS, Jean

Belshazzar's feast, op. 51. cf Symphony, no. 1, op. 39, E minor.

Canzonetta, op. 62, no. 1. cf RAUTAVAARA: Pelimannit, op. 1.

Concerto, violin, op. 47, D minor. cf BEETHOVEN: Romances, nos. 1 and 2, opp. 40, 50.

Concerto, violin, op. 47, D minor. cf BRUCH: Concerto, violin, no. 1, op. 26, G minor.

Concerto, violin, op. 47, D minor. cf DVORAK: Concerto, violin and piano, op. 53, A minor.

Concerto, violin, op. 47, D minor. cf HMV SLS 5004.

Concerto, violin, op. 47, D minor. cf RCA CRL 6-0720.

Concerto, violin, op. 47, D minor: 2nd movement. cf BEETHOVEN: Concerto, violin, op. 61, D major.

Finlandia, op. 26. cf ALFVEN: Midsommarvaka, no. 1, op. 19.

Finlandia, op. 26. cf HMV SLS 5019.

Finlandia, op. 26. cf Philips 6747 204.

In memoriam, op. 59. cf The tempest, op. 109.

Karelia suite, op. 11. cf Symphony, no. 5, op. 82, E flat major.

Karelia suite, op. 11, no. 3: March. cf BERLIOZ: Les Troyens: Overture, March Troyenne.

King Christian suite, op. 27: Musette. cf Discopaedia MB 1005.

Kuolema, op. 44: Valse triste. cf ALFVEN: Midsommarvaka, no. 1, op. 19.

Kuolema, op. 44: Valse triste. cf HMV SLS 5019.

2134 Legends, op. 22. Royal Liverpool Philharmonic Orchestra; Charles Groves. Angel S 37106. (also HMV ASD 3092)

+Gr 9-75 p472   ++NR 8-75 p6
+HF 11-75 p126   +RR 8-75 p43
+HFN 9-75 p104   +SFC 8-31-75 p20

Legends, op. 22: Swan of Tuonela. cf ALFVEN: Midsommarvaka, no. 1, op. 19.

Quartet, strings, op. 56, D minor. cf SCHUMANN: Quartet, strings, no. 3, op. 41, A major.

Rakastava, op. 14. cf RAUTAVAARA: Pelimannit, op. 1.

En saga, op. 9. cf ALFVEN: Midsommarvaka, no. 1, op. 19.

En saga, op. 9. cf Symphony, no. 3, op. 52, C major.

En saga, op. 9. cf Symphony, no. 5, op. 82, E flat major (Classics for Pleasure 40218).

En saga, op. 9. cf Symphony, no. 5, op. 83, E flat major (HMV 3038).

Scaramouche, op. 71. cf Tempest, op. 109: Suites, nos. 1 and 2.

2135 Songs: Flickan kom ifrän, op. 37, no. 5; Den första kyssen, op. 37, no. 1; Illalle, op. 17, no. 6; Im Feld ein Mädchen singt, op. 50, no. 3; Pa veradan, op. 38, no. 2; Säf, säf susa, op. 36, no. 4; Se'n har jag, op. 17, no. 1; Svarta Rosor, op. 36, no. 1; Var det en Drom, op. 37, no. 4; Varen flyktar hastigt, op. 13 no. 4. STRAUSS, R.: Songs: Allerseelen, op. 10, no. 8; Befreit, op. 39, no. 4; Cäcilie, op. 27, no. 2; Ruhe, meine Seele, op. 27, no. 1; Ständchen, op. 17, no. 2; Wiegenlied, op. 41, no. 1;

Zueignung, op. 10, no. 1. Birgit Nilsson, s; János Solyom, pno.
   BIS LP 15.
         +-Gr 11-75 p882
Songs: Blackroses, op. 36, no. 1; The kiss, op. 37, no. 1; The
   tryst, op. 37, no. 5; Varen flyktar hastigt, op. 13, no. 4.
   cf GRIEG: Af Haugtussa, op. 67.
Songs: Diamonds on the March snow; Sigh, sigh, sedges. cf RCA SER
   5719.
Suite mignonne, op. 98. cf RAUTAVAARA: Pelimannit, op. 1.
Svarta Rosor, op. 36, no. 1. cf RCA SER 5704/6.
2136 Symphony, no. 1, op. 39, E minor. Belshazzar's feast, op. 51.
   Orchestra; Robert Kajanus. Turnabout THS 65045.
         +NR 8-75 p6
2137 Symphony, no. 2, op. 43, D major. Stockholm Philharmonic Orchestra;
   Antal Dorati. Camden CCV 5029.
         +RR 11-75 p55
2138 Symphony, no. 2, op. 43, D major. LPO; John Pritchard. Pye GSGC
   15003. (Reissue from Virtuoso TPLS 13032/3)
         -Gr 8-75 p335   +-RR 8-75 p43
2139 Symphony, no. 3, op. 52, C major. En saga, op. 9. Helsinki Radio
   Symphony Orchestra; Okko Kamu. DG 2530 426. (Reissue from
   2720 067)
         +Gr 12-74 p1144   ++RR 11-74 p50
         +NR 1-75 p3   +SFC 3-16-75 p25
2140 Symphony, no. 5, op. 82, E flat major. En saga, op. 9. Scottish
   National Orchestra; Alexander Gibson. Classics for Pleasure
   CFP 40218.
         +Gr 9-75 p466   +RR 9-75 p47
         /HFN 8-75 p83
2141 Symphony, no. 5, op. 82, E flat major. En saga, op. 9. Bournemouth Symphony Orchestra; Paavo Berglund. HMV ASD 3038. (also
   Angel S 37104)
         +Gr 1-75 p1354   +-RR 1-75 p33
         /HF 9-75 p91   +St 8-75 p105
         /NR 8-75 p2
2142 Symphony, no. 5, op. 82, E flat major. Karelia suite, op. 11.
   LSO; Alexander Gibson. London STS 15189. (Reissues from RCA
   LSC 2045, VICS 1016)
         /HF 9-75 p91   ++St 8-75 p105
         ++NR 7-75 p4
2143 Symphony, no. 5, op. 82, E flat major. Symphony, no. 7, op. 105,
   C major. BSO; Colin Davis. Philips 6500 959. Tape (c) 7300
   415.
         +-Gr 11-75 p833   +-NR 12-75 p3
         +HF 12-75 p107   +RR 11-75 p55
         +-HFN 11-75 p165   ++SFC 9-28-75 p30
   Symphony, no. 7, op. 105, C major. cf Symphony, no. 5, op. 82,
   E flat major.
2144 The tempest, op. 109. In memoriam, op. 59. Royal Liverpool
   Philharmonic Orchestra; Charles Groves. HMV ASD 2916.
         ++Gr 2-74 p1566   +RR 2-74 p35
         +HF 2-75 p101
   The tempest, op. 109: Miranda, The Naiads, The storm. cf BERLIOZ:
   Les Troyens: Overture, March Troyenne.
2145 The tempest, op. 109: Suites, nos. 1 and 2. Scaramouche, op. 71.
   HSO; Jussi Jales. London 6824. Tape (r) L 46824. (also Decca

SDD 467)
+-Gr 4-75 p1817   ++SFC 9-29-74 p26
+HF 2-75 p101   ++SFC 8-31-75 p20 tape
+NR 11-74 p2   +St 2-75 p111
+RR 2-75 p39

SIECZYNSKY, Richard
Vienna, city of my dreams. cf Columbia D3M 33448.
Wien, du Stadt meiner Traume. cf Angel S 35696.

SILVERMAN, Stanley
2146 Elephant steps. Pop singers, opera singers, orchestra, rock band,
electronic tape, raga group, tape recorder, gypsy ensemble and
elephants; Michael Tilson Thomas. Columbia M2X 33044 (2).
/HF 5-75 p90

SIMMES, William
Fantasie. cf Turnabout TV 37071.

SIMON, Anton
Quatuor en forme de sonatine, op. 23, no. 1. cf Desto DC 6474/7.

SIMON, Frank
Willow echoes. cf Nonesuch H 71298.

SIMPSON, Robert
2147 Symphony, no. 3. LSO; Jascha Horenstein. Unicorn UNS 225.
++NR 9-75 p3   ++St 4-75 p106

SIMPSON, Thomas
Ballet. cf Oryx BRL 1.
Mascarada. cf Oryx BRL 1.
Volta. cf Oryx BRL 1.

SINDING, Christian
Elegiac pieces: Andante religioso. cf Discopaedia MB 1003.

SINIGAGLIA, Leone
Capriccio all'antica. cf Discopaedia MB 1002.
Capriccio all'antica. cf Rococo 2072.

SJOBERG, Emil
Erotikon, op. 10. cf KJERULF: Albumblatt, op. 24, no. 1.
The music. cf RCA SER 5719.

SKALKOTTAS, Nikos
2148 Octet. Quartet, strings, no. 3. Variations on a Greek folk tune
(8). Robert Masters, vln; Derek Simpson, vlc; Marcel Gazelle,
pno; Dartington String Quartet. Argo ZRG 753. (Reissue from
HMV ASD 2289)
+Gr 11-74 p921   ++RR 11-74 p73
++NR 2-75 p6   ++St 6-75 p107
Quartet, strings, no. 3. cf Octet.
Suite, piano, no. 3. cf HADZIDAKIS: For a little white seashell,
op. 1.
Variations on a Greek folk tune. cf Octet.

SLAVONIC ORTHODOX LITURGY. cf Monitor MFS 757.

SMETANA, Bedrich
The bartered bride: Overture, Polka, Furiant. cf DVORAK: Slavonic
dances, op. 72, nos. 9, 10, 15.
The bartered bride: Polka, Furiant. cf BRAHMS: Hungarian dances,
nos. 5, 6.
2149 Bettina polka. Dahlia polka. Furiant. The lancer. Little hen.
Louisa's polka. Obrocák. Skocná. The student's life: Polka,
G major. To our girls polka. Brno State Philharmonic Orchestra;
Frantisek Jílek. Supraphon 110 1225.
+Gr 12-74 p1149   ++SFC 8-31-75 p20

/NR 11-74 p3   +St 2-75 p112
+RR 9-74 p62
Dahlia polka. cf Bettina polka.
Furiant. cf Bettina polka.
Hubička: Trio. cf Rubini RS 300.
The lancer. cf Bettina polka.
Little hen. cf Bettina polka.
Louisa's polka. cf Bettina polka.
2150 Má Vlast. CSO; Rafael Kubelik. Mercury SRI 2-77006 (2).
+SFC 12-7-75 p31
2151 Má Vlast. CPhO; Karel Ancerl. Supraphon 10521/2.
+-Gr 7-73 p202   ++RR 1-75 p34
+NR 10-74 p2
2152 Má Vlast. Leipzig Gewandhaus Orchestra; Václav Neumann. Telefun-
ken KT 11043/1-2 (2).
++Gr 1-75 p1357   +RR 12-74 p40
2153 Má Vlast. CPhO; Karel Ancerl. Vanguard SU 9/10 (2).
+NR 11-75 p3
Má Vlast: Vlatava. cf DVORAK: Symphony, no. 9, op. 95, E minor.
Má Vlast: Vltava. cf DVORAK: Slavonic dances, op. 72, nos. 9,
10, 15.
Má Vlast: Vltava; From Bohemia's woods and fields. cf DVORAK: In
nature's realm overture, op. 91.
Má Vlast: Vltava. cf Decca DPS 519/20.
March for the Shakespeare festivities. cf Panton 110 361.
March of the students' legion. cf Supraphon 110 1429.
Obrocák. cf Bettina polka.
Skocná. cf Bettina polka.
2154 Songs: Choruses for female voices (3); The dedication; Festive
chorus; Our song; The peasant; The prayer; The renegade (2
versions): Slogans (2); Song of the sea; The three riders.
Miroslav Svejda, t; Jindrich Jindrák, bar; Jaroslav Horácek,
bs; Czech Philharmonic Chorus; Josef Veselka. Supraphon 112 1143.
+Gr 2-74 p1595   +NR 3-74 p8
+HF 6-74 p101   +RR 2-74 p58
+HFN 2-74 p345   +St 8-74 p122
+-LJ 5-75 p44
Songs: Aus der Heimat. cf Ember GVC 46.
The students' life: Polka, G major. cf Bettina polka.
To our girls polka. cf Bettina polka.
To our girls polka. cf Supraphon 110 1429.
SMITH, Gregg
2155 Beware of the soldier. Rosalind Rees, s; Chuck Garretson, boy
soprano; Douglas Perry, t; Charles Greenwell, bs; Texas Boys'
Choir, Columbia Men's Glee Club; Orchestra Ensemble; Gregg
Smith. CRI SD 341.
+NR 11-75 p9
SMITH, Hale
Contours. cf Century/Advent OU 97 734.
SODERMAN
King Heimer and Aslög. cf RCA SER 5719.
SOLAGE (14th century France)
Femeux fume. cf Telefunken ER 6-35257.
SOLER, Antonio
The emperor's fanfare. cf Columbia M 33514.
Sonata, harp, E minor. cf RCA LRL 1-5087.

2156 Sonatas, harpsichord (10).   Fernando Valenti, hpd.   Desmar 1001.
        +-MJ 12-75 p38
SOLLBERGER, Harvey
    Grand quartet, flutes. cf BERGER: Septet.
SOLOVYEV, Vasilii
    Ballad about a soldier; Evening out in the bay; Nightingales.   cf
        HMV ASD 3116.
    SONGS AND ARIAS, Peter Schreier. cf Telefunken AJ 6-41867.
    SONGS AND INSTRUMENTAL WORKS FROM THE MANNHEIM COURT. cf Oryx
        ORYX 1720.
    SONGS AND OPERATIC ARIAS, Renata Tebaldi. cf Decca SXL 6629.
    SONNETS OF PETRARCH. Cf DG 2530 332.
SOPRONI, József
    Invenzioni sul B-A-C-H. cf Hungaroton SLPX 11692.
SOR, Fernando
    Andante, C minor. cf RCA ARL 1-0864.
    Andante pastorale, op. 32, no. 4, D major. cf Amberlee ACL 501X.
    Cantabile, no. 22, op. 35, B minor. cf Amberlee ACL 501X.
    Estudio, no. 9 en si menor. cf RODRIGO: Concierto de Aranjuez,
        guitar and orchestra.
    Study, no. 13, op. 35, C major. cf Amberlee ACL 501X.
    Study, no. 15, op. 31, D major. cf Amberlee ACL 501X.
    Fantasia elegiaca, op. 59: Marcha fúnebra. cf GIULIANI: Sonata
        eroica, op. 150, A major.
    Folias de España. cf RCA ARL 1-0485.
    Minuets, C major, A major, C major. cf RCA ARL 1-0864.
    Minuetto, op. 22. cf RODRIGO: Concierto de Aranjuez, guitar and
        orchestra.
    Minuetto en re mayor y variaciones sobre un tema de la "Flauta
        encantada" de Mozart. cf Everest SDBR 3380.
    Sonata, op. 22, C major. cf GIULIANI: Sonata eroica, op. 150,
        A major.
    Sonata, guitar, op. 25, C major. cf GIULIANI: Le Rossiniane, op.
        121 and 119.
    Variations on a theme by Mozart, op. 9. cf CBS 72950.
    Variations on a theme by Mozart, op. 9. cf London STS 15224.
SOTHCOTT, John
    Fanfare. cf CRD CRD 1019.
    THE SOUND OF MUSICAL INSTRUMENTS, a demonstraton record. cf
        Acoustic Research, Inc., unnumbered.
    THE SOUND OF THE ITALIAN HARPSICHORD. cf Pandora PAN 101.
SOUSA, John Philip
    El Capitán. cf Michigan University SM 0002.
    El Capitán; The gallant Seventh; The liberty bell; Semper Fidelis;
        Stars and stripes forever; The thunderer. cf Columbia 33513.
2157 Marches: The belle of Chicago; El Capitán; George Washington bi-
        centennial; The liberty bell; The New York hippodrome; Our
        flirtations; The pathfinder of Panama; The stars and stripes
        forever; The thunderer; U. S. field artillery; The Washington
        Post; We are coming. Detroit Concert Band; Leonard B. Smith.
        Sousa American Bicentennial Collection 1.
            -HF 9-75 p92
    Preussen's Gloria. cf BASF 292 2116-4.
    Stars and stripes forever. cf BASF 292 2116-4.
    Stars and stripes forever. cf Columbia M 31726.
    Stars and stripes forever. cf RCA VH 020.

SOWANDE, Fela
2158 African suite: Akinla, Joyful day, Nostalgia. STILL: Sahdji.
      WALKER: Lyric for strings. Morgan State College Choir; LSO;
      Paul Freeman. Columbia M 33433.
              ++Audio 12-75 p106   ++NR 9-75 p5
              +HF 10-75 p65   ++St 12-75 p132
    Go down, Moses. cf Wealden WS 110.
SOWERBY, Leo
    Pageant. cf Pye GSGC 14153.
    Toccata. cf Orion ORS 74161.
SPALDING, Albert
    Dragonfly. cf Discopaedia MB 1009.
    SPANISH SONGS OF THE MIDDLE AGES AND RENAISSANCE. cf DG 2530 504.
SPEAKS, Oley
    In the end of the Sabbath. cf RCA ARL 1-0562.
    SPECTRUM: New American music. cf Nonesuch H 71302/3.
SPENDIAROV, Alexander
    Almast. cf Crimean sketches, op. 9: Elegiac song, Drinking song.
2159 Crimean sketches, op. 9: Elegiac song, Drinking song. Almast.
      TANEIEV: Symphony, op. 12, C minor. MRO, Bolshoi Theatre Orch-
      estra; Gennady Rozhdestvensky. HMV Melodiya ASD 3106.
              +Gr 9-75 p472   +RR 9-75 p50
              +HFN 9-75 p104
SPETH, Johann
    Ars magna consoni et disone: Toccata septima oder Sibendtes musik-
      alisches Blumen-Feld. cf Pelca PSRK 41013/6.
    Toccata quarta, E minor. cf Telefunken TK 11567/1-2.
    Toccata quinta. cf Telefunken TK 11567/1-2.
SPIEGELMAN, Joel
    Kousochki. cf Desto DC 7131.
SPINACINO, Francesco
    Ricercare. cf DG Archive 2533 173.
SPISAK, Michal
    Sonatina, oboe, clarinet and bassoon. cf ARNOLD: Divertimento,
      flute, oboe and clarinet, op. 37.
SPOHR, Ludwig
    Concerto, violin, no. 8, op. 47, A major. cf Discopaedia MB 1009.
    Fantasie, C minor. cf RCA LRL 1-5087.
    Fantasie, op. 35, A flat major. cf GLINKA: Nocturne.
    Gruss an Kiel. cf BASF 292 2116-4.
    Zemira et Azor: Rose softly blooming. cf DONIZETTI: Lucia di
      Lammermoor: Sulla tomba to end of Act 1.
SPOLIANSKY
    Heute Nachte oder nie. cf Polydor 2489 519.
SPONTINI, Gasparo
    La Vestale: Overture. cf Erato STU 70880.
STAINER, John
2160 On a bass. Prelude and fughetta. Songs: Drop down ye heavens;
      Lead kindly light; I saw the Lord; Magnificat and nunc dimittis,
      B flat major; Let Christ the King. Magdalen College Choir,
      Oxford; Bernard Rose. Argo ZRG 811.
              +Gr 7-75 p233   +RR 7-75 p60
              +HFN 8-75 p84
    Prelude and fughetta. cf On a bass.
    Songs: Drop down ye heavens; Lead kindly light; I saw the Lord;
      Magnificat and nunc dimittis, B flat major; Let Christ the King.
      cf On a bass.

STANFORD, Charles
    Beati quorum via, op. 51, no. 3. cf HMV HQS 1350.
    Irish dances. cf HARRISON: Grand defile, organ and percussion.
    Psalm 23: The Lord is my Shepherd. cf Polydor 2460 250.
    Te Deum, op. 66, B flat major. cf Argo ZRG 785.
STANLEY, John
    Trumpet tune, D major. cf London SPC 21100.
    Trumpet tunes, nos. 1 and 2. cf Argo ZDA 203.
    Voluntary, D major. cf CRD CRD 1008.
    Voluntary, D minor: Adagio. cf Argo ZDA 203.
    Voluntary, C major. cf Argo ZRG 807.
    Voluntary, no. 7. cf Vista VPS 1021.
    Voluntary, organ, A major. cf Orion ORS 74161.
    Voluntary, organ, op. 5, no. 6. cf Wealden WS 160.
STARER, Robert
    Miniatures (5). cf Desto DC 6474/7.
STARK
    Pastorale. cf Hungaroton SLPX 11601/2.
STATHAM, Heathcote
    Allegro. cf Decca SDD 463.
STEBBINS
    Songs: If the apples bloom; The longest day in June; Pretty rally
        Oliver. cf Discophilia KGS 3.
STEENWICK, Gisbert
    Heyligh saligh Bethlehem, variations. cf Telefunken DX 6-35265.
    More Palatino variations. cf Telefunken DX 6-35265.
STEIBELT, Daniel
    Bacchanales, op. 53. cf Angel S 36080.
STENHAMMER, Wilhelm
2161 Florez och Blanzeflor, baritone and orchestra, op. 3. Serenade,
        orchestra, F major. Swedish Radio Symphony Orchestra; Ingvar
        Wixell, bar; Stig Westerberg. HMV 4E 061 35148.
            ++Gr 12-75 p1061
2162 Quartet, strings, no. 6, op. 35, D minor. Erik Saedén, bar; Frydén
        Quartet. HMV 4E 053 35116.
            +Gr 12-75 p1061
    Serenade, orchestra, F major. cf Florez och Blanzeflor, baritone
        and orchestra, op. 3.
    Sweden. cf RCA SER 5719.
2163 Symphony, no. 2, op. 17, G minor. Stockholm Philharmonic Orches-
        tra; Tor Mann. Swedish Discofil SLT 33198.
            +RR 1-75 p29   /St 6-75 p52
STERN
    Coquette. cf Rubini RS 300.
STEVENS, Halsey
    Symphony, no. 1. cf COPLAND: Dance symphony.
STILL, William Grant
2164 Afro-American symphony. Highway 1, USA: Arias (2). COLERIDGE-
        TAYLOR: Danse negre. Hiawatha's wedding feast: Awake, beloved.
        Onaway. William Brown, t; LSO; Paul Freeman. Columbia M 32782.
            ++HF 6-74 p71   +NYT 5-12-74 pD26
            ++MQ 10-75 p645   *St 7-74 p105
            +NR 7-74 p13
    Darker America. cf KAY: Dances for string orchestra.
    From the black belt. cf KAY: Dances for string orchestra.
    Highway 1, USA: Arias (2). cf Afro-American symphony.

Sahdji. cf SOWANDE: African suite.
STOCK, Dave
    Quintet, clarinet and strings. cf BLANK: Rotation.
STOCKHAUSEN, Karlheinz
2165 Adieu, wind quintet. Kontro-punkte, ten instruments. Kreuzspiel,
        oboe, bass clarinet, piano and 3 percussionists. Zeitmasse, 5
        woodwinds. London Sinfonietta; Karlheinz Stockhausen. DG 2530
        443.
                +Gr 6-75 p50   +RR 4-75 p32
                ++HFN 5-75 p140
2166 Aus den sieben Tagen (From the seven days): It; Upwards; Communion;
        Intensity; Right durations; Connection; Unlimited; Meeting point;
        Night music; Downwards; Set sail for the sun; Gold dust. Aloys
        Kontarsky, Johannes Fritsch, Alfred Alings, Rolf Dehlhaar,
        Harald Bojé, Carlos Alsina, Jean-François Jenny-Clark, Michel
        Portal, Jean-Pierre Drouet, Peter Eötvös, Herbert Henck, Michael
        Vetter, Karlheinz Stockhausen. DG 2720 073 (7).
                *Gr 1-74 p1409   +Te 3-75 p48
                ++RR 1-74 p65
    Kontra-punkte, ten instruments. cf Adieu, wind quintet.
    Kreuzspiel, oboe, bass clarinet, piano and 3 percussionists. cf
        Adieu, wind quintet.
2167 Stop. Ylem. London Sinfonietta; Karlheinz Stockhausen. DG 2530
        422.
                +Gr 6-75 p50   +-MT 12-75 p1072
                +HF 6-75 p103   +-NR 6-75 p2
                ++HFN 5-75 p139   +-RR 4-75 p32
    Ylem. cf Stop.
    Zeitmasse, 5 woodwinds. cf Adieu, wind quintet.
    Zyklus. cf Classics for Pleasure CFP 40205.
STOKOWSKI, Leopold
    Caprice oriental. cf International Piano Archives IPA 5007/8.
STOLZ, Robert
    Songs: Mein herz ruft immer nur nach dir, Oh Marta; Ob blond, ob
        braun ich liebe alle Frau'n. cf Polydor 2489 519.
    Wiener Cafe. cf Olympic 8136.
STORACE, Bernardo
    Ballo della battaglia. cf Philips 6775 006.
STRADELLA, Alessandro
    Per Pietá. cf Desto DC 7118/9.
    Sonata, trumpet and strings. cf MARCELLO, A.: Concerto, oboe and
        strings.
STRAIGHT, Willard
    Development. cf HOVHANESS: Triptych.
STRANDBERG, Newton
    Sea of tranquility. cf FUSSELL: Processionals, orchestra.
STRAUSS, Eduard
    Ohne bremse, op. 238. cf HMV SLS 5017.
    Unter den Enns. cf Olympic 8136.
STRAUSS, Franz
2168 Concerto, horn, op. 8. STRAUSS, R.: Concerti, horn, nos. 1 and 2.
        Barry Tuckwell, hn; LSO; István Kertész. Decca SXL 6285. Tape
        (c) KSXC 6285.
                ++RR 10-75 p98 tape
STRAUSS, J.
    Man lebt nur einmal. cf RCA ARL 2-0512.

Nachtfalter. cf RCA ARL 2-0512.
STRAUSS, Johann I
2169 Loreley Rheinklänge, op. 154. STRAUSS, J. II: Die Fledermaus, op.
    363: Quadrille. Emperor waltz, op. 437. Napoleon march, op.
    156. Wo di Zitronen blühn, op. 364. Thousand and one nights,
    op. 346. STRAUSS, Josef: Feuerfest polka, op. 269. Jockey
    polka, op. 278. ZIEHRER: Fächerpolonaise. VPO; Willi Boskovsky.
    Decca SDD 473.
        +RR 9-75 p48
    Radetzky march, op. 228. cf STRAUSS, J. II: Works, selections
    (Decca SXL 6740).
    Radetzky march, op. 228. cf STRAUSS, J. II: Works, selections
    (Pye GSGC 15024).
    Radetzky march, op. 228. cf STRAUSS, J. II: Works, selections
    (Saga 5408).
    Radetzky Marsch, op. 228. cf BASF 292 2116-4.
    Radetzky march, op. 228. cf DG 2548 148.
    Radetzky march, op. 228. cf Olympic 8136.
    Sperl, Piefke und Pufke. cf Decca SXL 6572.
    Vienna carnival. cf Saga 5411.
    Wettrennen Galopp, op. 29. cf STRAUSS, J. II: Ker Karneval in
    Rom: Overture.
STRAUSS, Johann II
    An der schönen blauen Donau, op. 314. cf Der Karneval in Rom:
    Overture.
    Le beau Danube, op. 314. cf Die Fledermaus, op. 363: Overture.
    The beautiful blue Danube, op. 314. cf Works, selections (Angel
    S 37070).
    The beautiful blue Danube, op. 314. cf Works, selections (CBS
    61510).
    An der schönen blauen Donau, op. 314. cf Works, selections (Decca
    DKPA 513/4).
    An der schönen Blauen Donau, op. 314. cf Works, selections
    (Olympic 8132).
    The beautiful blue Danube, op. 314. cf Works, selections (Pye
    GSGC 15024).
    An der schönen blauen Donau, op. 314. cf Works, selections (RCA
    SMA 7012).
    The beautiful blue Danube, op. 314. cf Works, selections (Saga
    5408).
    An der schönen blauen Donau, op. 314. cf CBS 77513.
    An der schönen blauen Donau (Blue Danube waltzes), op. 314. cf
    Columbia 31726.
    An der schonen blauen Donau, op. 314. cf Rediffusion 15-16.
    The beautiful blue Danube, op. 314. cf Supraphon 114 1458.
    Annen polka, op. 117. cf Works, selections (Decca SXL 6740).
    Annen polka, op. 117. cf Works, selections (Fontana 6747 176).
    Annen polka, op. 117. cf Works, selections (Music for Pleasure
    MFP 57016).
    Annen polka, op. 117. cf Works, selections (Olympic 8138).
    Annen polka, op. 117. cf Works, selections (Saga 5408).
    Annen polka, op. 117. cf Works, selections (Pye GSGC 15024).
    Arabesques on "The beautiful blue Danube". cf RCA ARL 2-0512.
    Auf der Jagd, op. 373. cf Works, selections (Decca SXL 6740).
    Aufs Korn, op. 478. cf Der Karneval in Rom: Overture.
    Aus der Heimat, op. 347. cf Works, selections (BASF BAC 3069/70).

Austrian march, op. 20. cf Works, selections (BASF BAC 3069/70).
Bei uns zu Haus, op. 361. cf Works, selections (BASF BAC 3069/70).
Bei uns zu Haus, op. 361. cf Works, selections (Decca SXL 6740).
Bijouterie Quadrille, op. 169. cf Works, selections (RCA SMA 7012).
2170 Blindekuh. A night in Venice. SUPPE: Overtures: Boccaccio. Light
     cavalry. A morning, noon and night in Vienna. Poet and peasant.
     Johann Strauss Orchestra, Vienna; Willi Boskovsky. Angel S
     37099. Tape (c) 4XS 37099 (ct) 8XS 37099. (Some reissues from
     S 36826, 36887, 36956)
          +HF 7-75 p85   +St 11-75 p140
          -NR 8-75 p3
Blindekuh overture. cf HMV SLS 5017.
Blumenfest, op. 111. cf Works, selections (BASF BAC 3069/70).
Bouquet Quadrille, op. 135. cf Works, selections (RCA SMA 7012).
Brünn National Gaurd march, op. 58. cf Works, selections (BASF
     BAC 3069/70).
Casanova: Nun's chorus. cf Works, selections (Fontana 6747 176).
Casanova: Nun's chorus. cf Angel S 35696.
Champagne polka, op. 211. cf Works, selections (Fontana 6747 176).
Champagne polka, op. 211. cf Works, selections (Olympic 8138).
Deutschmeister Jubilee march, op. 470. cf Works, selections (BASF
     BAC 3069/70).
Egyptischer Marsch, op. 335. cf Works, selections (Fotana 6747 176).
Egyptischer Marsch, op. 335. cf Works, selections (Olympic 8138)
Elektro-Magnetische, op. 110. cf Works, selections (Olympic 8138).
Eljen a Magyar, op. 332. cf Works, selections (Fontana 6747 176).
Eljen a Magyar, op. 332. cf HMV SLS 5017.
Emperor Franz Joseph march, op. 67. cf Works, selections (BASF
     BAC 3069/70).
Erinnerung an Covent Garden, op. 329. cf Der Karneval in Rom:
     Overture.
Es gibt nur a Kaiserstadt, op. 291. cf Works selections (BASF BAC
     3069/70).
Es gibt nur a Kaiserstadt, es gibt nur ein Wien, op. 291. cf
     Works, selections (RCA SMA 7012).
Explosions, op. 43. cf Works, selections (Decca SXL 6740).
Explosions, op. 43. cf Decca SXL 6572.
2171 Die Fledermaus, op. 363. Gundula Janowitz, Renate Holm, s; Wolf-
     gang Windgassen, Waldemar Kmentt, t; Erich Kuchar, Eberhard
     Wächter, Heinz Holecek, Erich Kunz, bar; Vienna State Opera
     Chorus; VPO; Karl Böhm. London OSA 1296 (2). (also Decca SET
     540/1)
          +Gr 4-75 p1864   -ON 12-20-75 p38
          +-HF 7-75 p84   +RR 4-75 p19
          +NR 7-75 p11   -St 8-75 p105
     Die Fledermaus, op. 363: Du und du. cf Works, selections (Decca
     DKPA 513/4).
2172 Die Feldermaus, op. 363: Overture. Le beau Danube, op. 314 (arr.
     Desormier). STRAUS, J. I and J. II: Bal de Vienne (arr. Gamley).
     National Philharmonic Orchestra; Richard Bonynge. Decca SXL
     6701.
          +Gr 12-75 p1098   +RR 12-75 p62
          +HFN 12-75 p165
     Die Fledermaus, op. 363: Overture. cf Works, selections (Pye GSGC
     15024).
     Die Fledermaus, op. 363: Overture. cf Works, selections (CBS 61510).

Die Fledermaus, op. 363: Overture. cf Works, selections (Fontana
6747 176).
Die Fledermaus, op. 363: Overture. cf Works, selections (Saga 5408).
Die Fledermaus, op. 363: Overture. cf LEHAR: Gold and silver waltz,
op. 79.
Die Fledermaus, op. 363: Overture. cf Decca SXL 6643.
Die Fledermaus, op. 363: Overture. cf London CS 6856.
Die Fledermaus, op. 363: Overture. cf Olympic 8136.
Die Fledermaus, op. 363: Overture. cf Telefunken DX 6-35262.
Die Fledermaus, op. 636: Overture; Mein Herr Marquis. cf LEHAR:
Eva: Wär es auch nichts al ein Traum vom Gluck.
Die Fledermaus, op. 363: Quadrille. cf STRAUSS, J. I: Loreley
Rheinklänge, op. 154.
Frauenherz, op. 166. cf Works, selections (Olympic 8138).
Freikugeln, op. 326. cf Works, selections (RCA SMA 7012).
Frühlingstimmen, op. 410. cf Works, selections (Olympic 8132).
Frühlingstimmen, op. 410. cf CBS 77513.
Frühlingstimmen, op. 410. cf Discophilia KGS 3.
Galopin, op. 237. cf Works, selections (Olympic 8138).
Geschichten aus dem Wienerwald, op. 325. cf Works, selections
(Fontana 6747 176).
Tales from the Vienna Woods, op. 325. cf Works, selections (Angel
S 37070).
Tales from the Vienna Woods, op. 325. cf Works, selections (BASF
BAC 3069/70).
Tales from the Vienna Woods, op. 325. cf Works, selections (CBS
61510).
Tales from the Vienna Woods, op. 325. cf Works, selections (Decca
DKPA 513/4).
Geschichten aus dem Wienerwald, op. 325. cf Works, selections
(Olympic 8132).
Tales from the Vienna Woods, op. 325. cf Works, selections (Pye
GSGC 15024).
Geschichten aus dem Wienerwald, op. 325. cf Works, selections (RCA
SMA 7012).
Tales from the Vienna woods, op. 325. cf LEHAR: Gold and silver
waltz, op. 79.
Geschichten aus dem Wienerwald, op. 325. cf CBS 77513.
Geschichten aus dem Wienerwald, op. 325. cf Discophilia KGS 3.
Graduation ball, op. 97. cf Works, selections (Fontana 6747 176).
Heiligenstad rendezvous, op. 78. cf Works, selections (BASF BAC
3069/70).
Hofballtänze, op. 298. cf Works, selections (Philips 6833 162).
Im Krapfenwald, op. 336. cf Works, selections (BASF BAC 3069/70).
Im Krapfenwald, op. 336. cf HMV SLS 5017.
Immer Heiterer, op. 235. cf HMV SLS 5017.
Jubilee waltz. cf Works, selections (RCA SMA 7012).
Kaiserwälzer (Emperor waltz), op. 437. cf Works, selections (Angel
S 37070).
Emperor waltz, op. 437. cf Works, selections (CBS 61510).
Kaiserwälzer, op. 437. cf Works, selections (Music for Pleasure
MFP 57016).
Kaiserwälzer, op. 437. cf Works, selections (Olympic 8132).
Emperor waltz, op. 437. cf Works, selections (Decca DKPA 513/4).
Emperor waltz, op. 437. cf Works, selections (Saga 5408).
Kaiserwälzer, op. 437. cf CBS 77513.

Kaiserwälzer, op. 437. cf HMV SLS 5017.
Emperor waltz, op. 437. cf STRAUSS, J. I: Loreley Rheinklänge,
    op. 154.
2173 Der Karneval in Rom: Overture. Erinnerung an Covent Garden, op.
    329. Sängerslust, op. 328. An der schönen blauen Donau, op.
    314. Aufs Korn, op. 478. Tik-Tak, op. 365. Perpetuum Mobile,
    op. 257. STRAUSS, J. I: Wettrennen Galop, op. 29. STRAUSS,
    Josef: Rudolfsheimer, op. 152. Sphärenklänge, op. 235. VPO;
    Vienna State Opera Chorus; Willi Boskovsky. Decca SXL 6692.
    Tape (c) KSXC 6692.
       +Gr 12-74 p1158   ++RR 12-74 p40
       ++Gr 3-75 p1712 tape   +RR 5-75 p77 tape
Kreuzfidel, op. 301. cf Works, selections (RCA SMA 7012).
Künstlerleben (Artist's life), op. 316. cf Works, selections
    (Olympic 8132).
Artist's life, op. 316. cf Works, selections (Decca DKPA 513/4).
Künstlerleben, op. 316. cf HMV SLS 5017.
Leichtes Blut, op. 319. cf Works, selections (Decca SXL 6740).
Leichtes Blut, op. 319. cf Works, selections (Olympic 8138).
Leichtes Blut, op. 319. cf LEHAR: Gold and silver waltz, op. 79.
Liebeslieder, op. 114. cf Works, selections (Decca SXL 6740).
Lob der Frauen, op. 315. cf Decca SXL 6572.
Morgenblätter (Morning papers), op. 279. cf Works, selections
    (Decca DKPA 513/4).
Morgenblätter, op. 279. cf Works, selections (Olympic 8132).
Morning papers, op. 279. cf Works, selections (Philips 6833 162).
Morgenblätter, op. 279. cf Works, selections (RCA SMA 7012).
Morgenblätter, op. 279. cf Decca SXL 6572.
Eine Nacht in Venedig: Overture. cf Works, selections (Fontana
    6747 176).
A night in Venice: Overture. cf Blindekuh.
A night in Venice: Overture. cf HMV SLS 5017.
Napoleon march, op. 156. cf STRAUSS, J. I: Loreley Rheinklänge,
    op. 154.
Neuhauser, op. 137. cf Works, selections (BASF BAC 3069/70).
Neue Pizzicato Polka, op. 449. cf HMV SLS 5017.
Neu Wien, op. 342. cf HMV SLS 5017.
Perpetuum Mobile, op. 257. cf Der Karneval in Rom: Overture.
Perpetuum Mobile, op. 257. cf Works, selections (Decca SXL 6740).
Perpetuum Mobile, op. 257. cf Works, selections (Pye GSGC 15024).
Persische Marsch (Persian march), op. 289. cf Decca SXL 6572.
Ritter Pasman, op. 441: Csardas. cf Works, selections (Decca SXL
    6740).
Rosen aus dem Süden (Roses from the South), op. 388. cf Works,
    selections (Angel S 37070).
Roses from the South, op. 388. cf Works, selections (Decca DKPA
    513/4).
Rosen aus dem Süden, op. 388. cf Works, selections (Music for
    Pleasure MFP 57016).
Rosen aus dem Süden, op. 388. cf Works, selections (Olympic 8132).
Roses from the South, op. 388. cf Works, selections (Philips
    6833 162).
Rosen aus dem Süden, op. 388. cf HMV SLS 5017.
Sängerslust, op. 328. cf Der Karneval in Rom: Overture.
Stadt und Land, op. 322. cf Works, selections (BASF BAC 3069/70).
Stadt und Land, op. 322. cf Works, selections (Decca SXL 6740).

A thousand and one nights, op. 346. cf Works, selections (Decca
  DKPA 513/4).
A thousand and one nights, op. 346. cf STRAUSS, J. I: Loreley
  Rheinklänge, op. 154.
A thousand and one nights, op. 346: Intermezzo. cf Works, selec-
  tions (Saga 5408).
Tik-Tak, op. 365. cf Der Karneval in Rom: Overture.
Tik-Tak, op. 365. cf Works, selections (BASF BAC 3069/70).
Tik-Tak, op. 365. cf Works, selections (Olympic 8138).
Tritsch-Tratsch, op. 214. cf Works, selections (Music for Pleasure
  MFP 57016).
Tritsch-Tratsch, op. 214. cf Works, selections (Olympic 8138).
Tritsch-Tratsch, op. 214. cf Works, selections (Saga 5408).
Tritsch-Tratsch, op. 214. cf CBS 77513.
Unter Donner und Blitz (Thunder and lightning), op. 324. cf
  Works, selections (Olympic 8138).
Thunder and lightning, op. 324. cf Works, selections (Saga 5408).
Thunder and lightning, op. 324. cf Columbia M 31726.
Unter Donner und Blitz, op. 324. cf Decca SXL 6572.
Unter Donner und Blitz, op. 324. cf Sound Superb SPR 90049.
Verbrüderungs-Marsch, op. 287. cf Works, selections (BASF BAC
  3069/70).
Vergnugungzug, op. 281. cf Works, selections (Decca SXL 6740).
Vibrations. cf Works, selections (Philips 6833 162).
Vienna bonbons, op. 307. cf Works, selections (BASF BAC 3069/70).
Vienna bonbons, op. 307. cf Works, selections (Decca DKPA 513/4).
Vienna bonbons, op. 307. cf Works, selections (Philips 6833 162).
Voices of spring, op. 410. cf Works, selections (BASF BAC 3069/70).
Voices of spring, op. 410. cf Works, selections (Decca DKPA 513/4).
Voices of spring, op. 410. cf HMV RLS 717.
Vöslau polka, op. 100. cf Works, selections (BASF BAC 3069/70).
Waldmeister overture. cf Works, selections (Decca SXL 6740).
Waldmeister overture. cf Telefunken DX 6-35262.
Wein, Weib und Gesang (Wine, women and song), op. 333. cf Works,
  selections (Angel S 37070).
Wein, Weib und Gesang, op. 333. cf Works, selections (Decca DKPA
  513/4).
Wein, Weib und Gesang, op. 333. cf Works, selections (Music for
  Pleasure MFP 57016).
Wein, Weib und Gesang, op. 333. cf Works, selections (Olympic
  8132).
Wine, women and song, op. 333. cf Works, selections (Philips
  6833 162).
Wein, Weib und Gesange, op. 333. cf HMV RLS 717.
Wine, women and song, op. 333. cf L'Oiseau-Lyre DSLO 7.
Wiener Blut (Vienna blood), op. 354. cf Works, selections (Angel
  S 37070).
Vienna blood, op. 354. cf Works, selections (CBS 61510).
Vienna blood, op. 354. cf Works, selections (Decca DKPA 513/4).
Wiener Blut, op. 354. cf Works, selections (Fontana 6747 176).
Wiener Blut, op. 354. cf Works, selections (Music for Pleasure
  MFP 57016).
Wiener Blut, op. 354. cf Works, selections (Olympic 8132).
Wiener Blut, op. 354. cf HMV SLS 5017.
Wiener Blut, op. 354 (arr. Urbanec). cf Rediffusion 15-16.
Wiener Blut, op. 354. cf Supraphon 114 1458.

Wiener Frauen, op. 423. cf Works, selections (BASF BAC 3069/70).
Wo die Zitronen blühn, op. 364. cf STRAUSS, J. I: Loreley Rhein-
    klänge, op. 154.
Wo die Zitronen blühn, op. 364. cf HMV SLS 5017.
2174 Works, selections: Emperor waltz, op. 437. The beautiful blue
    Danube, op. 314. Roses from the South, op. 388. Tales from
    the Vienna Woods, op. 325. Vienna blood, op. 354. Wine, women
    and song, op. 333. Johann Strauss Orchestra; Willi Boskovsky.
    Angel S 37070.
        +NR 6-75 p3
2175 Works, selections: Aus der Heimat, op. 347. Austria march, op. 20.
    Bei uns zu Haus, op. 361. Blumenfest, op. 111. Brünn National
    Guard march, op. 58. Deutschmeister Jubilee march, op. 470.
    Emperor Franz Joseph march, op. 67. Heiligenstad rendezvous,
    op. 78. Im Krapfenwald, op. 336. Neuhauser, op. 137. Es
    gibt nur a Kaiserstadt, op. 291. Stadt und Land, op. 322.
    Tales from the Vienna Woods, op. 325. Tik-Tak, op. 365. Ver-
    brüderungs-Marsch, op. 287. Vienna bonbons, op. 307. Voices
    of spring, op. 410. Vöslau polka, op. 100. Wiener Fraunen,
    op. 423. STRAUSS, J. II/Josef: Pizzicato polka. BeSO; Robert
    Stolz. BASF BAC 3069/70 (2).
        +-Gr 1-75 p1357   +RR 1-75 p34
2176 Works, selections: The beautiful blue Danube, op. 314. Emperor
    waltz, op. 437. Die Fledermaus, op. 363: Overture. The gypsy
    baron, op. 420: Overture. Tales from the Vienna Woods, op. 325.
    Vienna blood, op. 354. Columbia Symphony Orchestra; Bruno Wal-
    ter. CBS 61510. (Reissue, 1956)
        +RR 9-75 p48
2177 Works, selections: An der schönen blauen Donau, op. 314. Artist's
    life, op. 316. Emperor waltz, op. 437. Die Fledermaus, op.
    363: Du und du. Morning papers, op. 279. Tales from the Vienna
    Woods, op. 325. Thousand and one nights, op. 346. Roses from
    the South, op. 388. Vienna bonbons, op. 307. Vienna blood,
    op. 354. Voices of spring, op. 410. Wein, Weib und Gesang,
    op. 333. VPO; Willi Boskovsky. Decca DKPS 513/4.
        +RR 10-75 p57
2178 Works, selections: Annen polka, op. 117. Auf der Jagd, op. 373.
    Bein uns zu Haus, op. 361. Leichtes Blut, op. 319. Liebes-
    lieder, op. 114. Explosions, op. 43. Ritter Pasman, op. 441:
    Csardas. Perpetuum Mobile, op. 257. Stadt und Land, op. 322.
    Vergnugungzug, op. 281. Waldmeister overture. STRAUSS, J. I:
    Radetzky march, op. 228. VPO; Willi Boskovsky. Decca SXL 6740.
        ++RR 12-75 p62
2179 Works, selections: Annen polka, op. 117. Casanova: The nun's
    chorus. Champagne polka, op. 211. Egyptischer Marsch, op. 335.
    Eljen a Magyar, op. 332. Die Fledermaus, op. 363: Overture.
    Geschichten aus dem Wienerwald, op. 325. Graduation ball, op.
    97 (arr. Dorati). Eine Nacht in Venedig: Overture. Wiener
    Blut, op. 354. Der Zigeunerbaron, op. 420: Overture. German
    Radio Orchestra, VSO, Dresden Philharmonic Orchestra, LSO,
    Minneapolis Symphony Orchestra; Berlin Radio Choir; Paul Walter,
    Herbert Kegel, Charles Mackerras, Wolfgang Sawallisch, Antal
    Dorati. Fontana 6747 176 (2).
        +RR 7-75 p35
2180 Works, selections: Annen polka, op. 117. Kaiserwälzer, op. 437.
    Rosen aus dem Süden, op. 388. Tritsch-Tratsch, op. 214. Wein,
    Weib und Gesang, op. 333. Wiener Blut, op. 354. Scottish

National Orchestra; Alexander Gibson. Music for Pleasure MFP
57016.
    +RR 2-75 p40
2181 Works, selections: An der schönen Blauen Donau, op. 314. Frühlingstimmen, op. 410. Geschichten aus dem Wienerwald, op. 325.
Kaiserwälzer, op. 437. Künstlerleben, op. 316. Morgenblätter,
op. 279. Rosen aus dem Suden, op. 388. Wein, Weib und Gesang,
op. 333. Wiener Blut, op. 354. VSO; Robert Stolz. Olympic
8132.
    +NR 6-75 p3
2182 Works, selections: Annen polka, op. 117. Egyptischer Marsch, op.
335. Elektro-Magnetische, op.110. Champagne polka, op. 211.
Frauenherz, op. 116. Galopin, op. 237. Leichtes Blut, op.
319. Tik-Tak, op. 365. Tritsch-Tratsch, op. 214. Unter Donner
und Blitz, op. 324. STRAUSS, J. II/Josef: Pizzicato polka.
VSO; Robert Stolz. Olympic 8138.
    +NR 11-75 p4
2183 Works, selections: Hofballtänze, op. 298. Morning papers, op. 279.
Roses from the South, op. 388. Vibrations. Vienna bonbons,
op. 307. Wine, women and song, op. 333. VSO; Eduard Strauss.
Philips 6833 162.
    +-RR 10-75 p57
2184 Works, selections: Annen polka, op. 117. The beautiful blue
Danube, op. 314. Die Fledermaus, op. 363: Overture. Perpetuum
Mobile, op. 257. Tales from the Vienna Woods, op. 325. Der
Zigeunerbaron, op. 420: Overture. STRAUSS, J. I: Radetzky
march, op. 228. STRAUSS, J. II/Josef: Pizzicato polka. Hallé
Orchestra; John Barbirolli. Pye GSGC 15024.
    +RR 12-75 p67
2185 Works, selections: An der schönen blauen Donau, op. 314. Bouquet
Quadrille, op. 135. Bijouterie Quadrille, op. 169. Es gibt nur
a Kaiserstadt, es gibt nur ein Wien, op. 291. Freikugeln, op.
326. Geschichten aus dem Wienerwald, op. 325. Kreuzfidel, op.
301. Jubilee waltz. Morgenblätter, op. 279. STRAUSS, J. II/
Josef: Pizzicato polka. BPO; Arthur Fiedler. RCA SMA 7012.
    +HFN 12-75 p165   +RR 12-75 p62
2186 Works, selections: Anen polka, op. 117. The beautiful blue Danube,
op. 314. Emperor waltz, op. 437. Die Fledermaus, op. 363: Overture. A thousand and one nights, op. 346: Intermezzo. Thunder
and lightning, op. 324. Tritsch-Tratsch, op. 214. Der Zigeun-
erbaron, op. 420: Overture. STRAUSS, J. II/Josef: Pizzicato
polka. STRAUSS, J. I: Radetzky march, op. 228. VSO; Robert
Stolz. Saga 5408.
    +HFN 12-75 p165   +RR 12-75 p52
Der Zigeunerbaron (The gypsy baron), op. 420: Einzugsmarsch. cf
DG 2721 077.
The gypsy baron, op. 420: Overture. cf Works, selections (CBS
61510).
Der Zigeunerbaron, op. 420: Overture. cf Works, selections
(Fontana 6747 176).
Der Zigeunerbaron, op. 420: Overture. cf Works, selections (Pye
GSGC 15024).
Der Zigeunerbaron, op. 420: Overture. cf Works, selections (Saga
5408).
The gypsy baron, op. 420: The pig breeder's song. cf Pye NSPH 6.

STRAUSS, Johann II/Johann I
    Bal Vienne. cf STRAUSS, J. II: Die Fledermaus, op. 363: Overture.
STRAUSS, Johann II/Josef
    Pizzicato polka. cf STRAUSS, J. II: Works, selections (BASF BAC
        3069/70).
    Pizzicato polka. cf STRAUSS, J. II: Works, selections (Olympic 8138).
    Pizzicato polka. cf STRAUSS, J. II: Works, selections (Pye GSGC
        15024).
    Pizzicato polka. cf STRAUSS, J. II: Works, selections (RCA SMA
        7012).
    Pizzicato polka. cf STRAUSS, J. II: Works, selections (Saga 5408).
    Pizzicato polka. cf CBS 77513.
    Pizzicato polka. cf Olympic 8136.
STRAUSS, Josef
    Allerlei, op. 219. cf HMV SLS 5017.
    Dorfschwalben aus Osterreich, op. 164. cf Olympic 8136.
    Eingesendent, op. 240. cf Decca SXL 6572.
    Feuerfest polka, op. 269. cf STRAUSS, J. I: Loreley Rheinklänge,
        op. 154.
    Jockey polka, op. 278. cf STRAUSS, J. I: Loreley Rheinklänge,
        op. 154.
    Künstlergruss, op. 274. cf HMV SLS 5017.
    Perlen der Liebe, op. 39. cf HMV SLS 5017.
    Rudolfsheimer, op. 152. cf STRAUSS, J. II: Der Karneval in Rom:
        Overture.
    Sphärenklänge, op. 235. cf LEHAR: Gold and silver waltz, op. 79.
    Sphärenklänge, op. 235. cf STRAUSS, J. II: Der Karneval in Rom:
        Overture.
STRAUSS, Richard
2187 Der Abend, op. 34, no. 1. Deutsche Motette, op. 62. Hymne, op.
        34, no. 2. Jessica Cash, s; Jean Temperley, ms; Wynford Evans,
        t; Stephen Varcoe, bar; Schütz Choir; Roger Norrington. Argo
        ZRG 803.
                +Gr 4-75 p1856   ++RR 5-75 p67
                +-MT 12-75 p1072   ++SFC 8-24-75 p28
                +NR 10-75 p6   ++STL 5-4-75 p37
                ++NYT 9-21-75 pD18
    Die Agyptische Helena: Bei jener Nacht, Zweite Nacht; Zaubernacht.
        cf Discophilia KGP 4.
2188 Also sprach Zarathustra, op. 30. VPO; Herbert von Karajan. London
        15083. Tape (c) A 30683. (also Decca SDD 175. Tape (c) DSDC
        175)
                -Gr 10-74 p766 tape   -RR 1-75 p60 tape
2189 Also sprach Zarathustra, op. 30. Till Eulenspiegels lustige
        Streiche, op. 28. RPO; Henry Lewis; Neville Taweel, vln.
        Decca SPA 397. (Reissue from PFS 4202, 4215)
                +Gr 7-75 p195   -RR 8-75 p43
                /HFN 8-75 p89
2190 Also sprach Zarathustra, op. 30. Don Juan, op. 20. Till Eulen-
        spiegels lustige Streiche, op. 28. Samuel Magad, vln; CSO;
        Georg Solti. Decca SXL 6749.
                +Gr 12-75 p1053   +RR 12-75 p62
2191 Also sprach Zarathustra, op. 30. CPhO; Herbert von Karajan. DG
        Tape (c) 3300 375.
                +RR 10-75 p98 tape
2192 Also sprach Zarathustra, op. 30. Paul Crépel, vln; Strasbourg

Philharmonic Orchestra; Alain Lombard. Erato STU 70873.
    +–Gr 2-75 p1500   /St 9-75 p114
    /RR 2-75 p39

2193 Also sprach Zarathustra, op. 30. Schlagobers, op. 70: Waltzes.
    VPO; Richard Strauss. Everest-Olympic 8111.
        +NR 6-75 p5   +St 1-75 p119
        +–SFC 9-1-74 p19

2194 Also sprach Zarathustra, op. 30. COA; Bernard Haitink. Philips
    6500 624. Tape (c) 7300 280 (r) PHIE 45264.
        /Gr 12-74 p1149   +RR 10-75 p98 tape
        ++HF 12-74 p118   ++SFC 9-1-74 p19
        ++NR 10-74 p2   ++SFC 5-11-75 p23 tape
        +RR 11-74 p59   ++St 1-75 p119
        ++RR 5-75 p77 tape

2195 Also sprach Zarathustra, op. 30. PO; Eugene Ormandy. RCA ARL
    1-1120.
        ++SFC 12-7-75 p31

2196 Also sprach Zarathustra, op. 30. St. Louis Symphony Orchestra;
    Walter Susskind. Turnabout (Q) QTVS 34584.
        +HF 9-75 p92   +–SFC 8-3-75 p30
        +NR 8-75 p2   +St 9-75 p114

2197 Also sprach Zarathustra, op. 30. The Schlagobers, op. 70: Whipped
    cream waltz. VPO; Richard Strauss. Turanbout THS 65021.
        +–NR 1-75 p3   +St 9-75 p114

    Also sprach Zarathustra, op. 30. cf Works, selections (DG 2726 028)
    Also sprach Zarathustra, op. 30. cf Works, selections (DG 2740 111)

2198 Ariadne auf Naxos, op. 60. Erna Berger, Viorica Ursuleac, Meliza
    Korjus, s; Helge Roswänge, t; Stuttgart Radio Orchestra; Clemens
    Krauss. BASF KBF 21806 (2).
        +–HF 6-75 p104   +SFC 3-30-75 p16
        +NR 6-75 p11   +–St 8-75 p105

2199 Aus Italien, op. 16, G major. Josephslegende, op. 63. Salome,
    op. 54: Dance of the seven veils. Symphonia domestica, op. 53.
    Till Eulenspiegels lustige Streiche, op. 28. Dresden Staats-
    kapelle Orchestra; Rudolf Kempe. HMV SLS 894 (3).
        +Gr 3-75 p1663   ++RR 4-75 p35
        +MT 10-75 p888

2200 Le bourgeois gentilhomme, op. 60. Concerto, horn, no. 1, op. 11,
    E flat major. Mason Jones, hn; PO; Eugene Ormandy. Columbia
    M 32233.
        ++HF 8-75 p100   ++SFC 7-20-75 p27
        ++NR 6-75 p3

2201 Capriccio, op. 85: Bezaubernd ist sie heute wieder; Ein schönes
    Gedicht...Kein and'res, das mir so im Herzen loht...Des Dichters
    Wort; Verraten hab ich mien Gefühle; Hola, Ihr Streiter in Apoll;
    Wo ist mein Bruder...Morgen mittag um elf. Viorica Ursuleac, s;
    Franz Klarwein, Hans Hotter, Georg Wieter, bar; George Hann,
    bs-bar; Bavarian State Opera Orchestra; Clemens Krauss. BASF
    102 1363-3.
        +–Gr 10-75 p680   +–RR 10-75 p30
        +–HFN 10-75 p149

2202 Capriccio, op. 85: Closing scene. Four last songs. Elisabeth
    Schwarzkopf, s; PhO; Otto Ackermann. Angel 35084.
        +St 4-75 p70

    Capriccio, op. 85: Sextet. cf JANACEK: Suite, strings.
    Concerti, horn, nos. 1, 2. cf STRAUSS, F.: Concerto, horn, op. 8.

Concerto, horn, no. 1, op. 11, E flat major. cf Le bourgeois
    gentilhomme, op. 60.
2203 Concerto, horn, no. 2, E flat major. Concerto, oboe, D major.
    Lothar Koch, ob; Norbert Hauptmann, hn; BPhO; Herbert von Kara-
    jan. DG 2530 439.
                /Gr 6-74 p60   +-RR 6-74 p53
                +HF 9-74 p103   +SFC 6-23-74 p26
                ++MJ 7-75 p34   +SR 10-5-74 p11
                +NR 9-74 p5   ++St 10-74 p138
    Concerto, horn, no. 2, E flat major. cf HAYDN: Concerto, horn,
    no. 1, D major.
    Concerto, oboe, D major. cf Concerto, horn, no. 2, E flat major.
    Deutsche Motette, op. 62. cf Der Abend, op. 34, no. 1.
    Don Juan, op. 20. cf Also sprach Zarathustra.
    Don Juan, op. 20. cf Works, selections (DG 2726 028).
    Don Juan, op. 20. cf Works, selections (DG 2740 111).
    Don Juan, op. 20. cf ELGAR: Enigma variations, op. 36.
    Don Juan, op. 20. cf RESPIGHI: Feste Romane.
2204 Don Quixote, op. 35. Der Rosenkavalier, op. 59: Waltzes. Paul
    Tortelier, vlc; Max Rostal, vla; Dresden State Orchestra;
    Rudolf Kempe. Angel S 37046. (also HMV ASD 3074. Reissue
    from SLS 880)
                +Gr 7-75 p195   +-SFC 5-11-75 p23
                ++HF 3-75 p86   ++St 3-75 p106
                ++NR 2-75 p3
2205 Don Quixote, op. 35. Kurt Reher, vlc; Jan Hlinka, vla; LAPO;
    Zubin Mehta. Decca SXL 6634. (also London CS 6849)
                -Gr 4-74 p1861   ++RR 5-74 p40
                +-HF 9-75 p92   ++SFC 4-27-74 p23
                +MJ 9-75 p51   -St 7-75 p106
                +NR 6-75 p5
    Don Quixote, op. 35. cf Works, selections (DG 2740 111).
    Don Quixote, op. 35: Finale. cf HMV SEOM 19.
    Festival prelude, op. 61. cf Works, selections (DG 2726 028).
    Die Frau ohne Schatten, op. 65: Empress' awakening scene. cf
    Songs (RCA ARL 1-0333).
    Guntram, op. 25: Freihild's aria. cf Songs (RCA ARL 1-0333).
2206 Ein Heldenleben, op. 40. BPhO; Herbert von Karajan. HMV (Q) ASD
    3126. (also Angel S 37060)
                ++Gr 10-75 p633   -NR 12-75 p4
                +HFN 9-75 p105   +-RR 9-75 p48
2207 Ein Heldenleben, op. 40. NYP; Willem Mengelberg. RCA SMA 7001.
    (Reissue from HMV D 1711/5)
                +-Gr 10-75 p634   +RR 8-75 p43
    Ein Heldenleben, op. 40. cf Works, selections (DG 2740 111).
    Hymne, op. 34, no. 2. cf Der Abend, op. 34, no. 1.
    Josephslegende, op. 63. cf Aus Italien, op. 16, G major.
    Metamorphosen, 23 solo strings. cf Works, selections (DG 2740 111).
    Morgen, op. 27, no. 4. cf Bel Canto Club No. 3.
2208 Quartet, piano, op. 13, C minor. Los Angeles String Trio; Irma
    Vallecillo, pno. Desmar DSM 1002.
                ++MJ 12-75 p38   ++NR 12-75 p9
    Quartet, piano, op. 13, C minor. cf REGER: Quartet, piano, op.
    133, A minor.
2209 Der Rosenkavalier, op. 59. Elisabeth Schwarzkopf, Teresa Stich-
    Randall, s; Christa Ludwig, ms; Nicolai Gedda, t; Eberhard

Wächter, bar; Otto Edelmann, bs-bar; PhO and Chorus; Herbert von
Karajan. Angel S 3563.
+St 4-75 p70

2210 Der Rosenkavalier, op. 59: Wie Du Warst, wie Du bist, da; Weiss
keiner; Da gent er hin, der aufgeblas 'ns schlechte Kerl; Mir
ist die Ehre widerfahren das lieg ich; Hab mir's gelobt, ihn
liebzuhaben; Ist ein Traum, kann nicht wirklich sein. Viorica
Ursuleac, Adele Kern, s; Georgine von Milinkovic, ms; Luise Willer con, Ludwig Weber, Georg Hann, bs; Bavarian State Opera Orchestra and Chorus; Clemens Krauss. BASF 10 22322/1.
-Gr 12-75 p1094

2211 Der Rosenkavalier, op. 59: Wie du Warst; Di rigori armato; Da geht
er hin; Mir ist die Ehre Widerfahren; Ohne mich, ohne mich;
Euer Gnaden sind die Güte selbst. Helga Dernesch, Anne Howells,
Teresa Cahill, s; Derek Blackwell, t; Michael Langdon, bs;
Clair Livingstone; Gordon Sandison; Scottish National Orchestra;
Alexander Gibson. Classics for Pleasure CFP 40217.
+Gr 7-75 p238   +RR 6-75 p25
+HFN 6-75 p93

Der Rosenkavalier, op. 59: Arias. cf London OS 26373.
Der Rosenkavalier, op. 59: Marschallin's monolog. cf Songs (RCA
ARL 1-0333).
Der Rosenkavalier, op. 59: Waltzes. cf Don Quixote, op. 35.
Der Rosenkavalier, op. 59: Waltzes. cf Works, selections (DG 2726
028).
Salome, op. 54: Dance of the seven veils. cf Aus Italien, op. 16,
G major.
Salome, op. 54: Dance of the seven veils. cf Works, selections
(DG 2726 028).
Salome, op. 54: Dance of the seven veils. cf Works, selections
(DG 2740 111).
Salome, op. 54: Final scene. cf BERG: Lulu: Suite.
Schlagobers, op. 70: Waltzes. cf Also sprach Zarathustra, op. 30.
Schlagobers, op. 70: Whipped cream waltz. cf Also sprach Zara-
thustra, op. 30.
Sonata, violin and piano, op. 18, E flat major. cf RCA ARM 4-0942/7
Sonata, violoncello, op. 6, F major. cf BEETHOVEN: Variations on
Handel's "See the conquering hero comes".

2212 Songs: Four last songs, op. posth; Befreit, op. 39, no. 4; Freund-
liche Vision, op. 48, no. 1; Meinem Kinde, op. 37, no. 3; Morgen,
op. 27, no. 4; Zueignung, op. 10, no. 1. Anneliese Rothenberger,
s; LSO; André Previn. HMV ASD 3082.
-Gr 10-75 p675   +-RR 9-75 p72
+-HFN 9-75 p105

2213 Songs: Four last songs: Frühling, September, Beim Schlafengehen,
Im Abendrot. Die Frau ohne Schatten, op. 65: Empress' awakening
scene. Der Rosenkavalier, op. 59: Marschallin's monolog. Gun-
tram, op. 25: Freihild's aria. Leontyne Price, s; NPhO; Erich
Leinsdorf. RCA ARL 1-0333.
+Gr 3-75 p1701   +-ON 1-11-75 p35
-HF 3-75 p87   +-RR 3-75 p66
++HFN 5-75 p140   +-SR 1-25-75 p51
+-MJ 2-75 p30   +St 3-75 p70
-NR 12-74 p11

2214 Songs: Heimliche Aufforderung, op. 27, no. 3; Ich trage meine
Minne, op. 32, no. 1; Ständchen, op. 17, no. 2; Traum durch die

Dämmerung, op. 29, no. 1. Peter Anders, t; Hubert Giesen, pno;
   BPhO; Walter Lutze. Telefunken DP 6-48064 (2).
      -RR 10-75 p89
Songs: Befreit, op. 39, no. 4; Für funfzehn Pfennige, op. 36, no.
   2; Schon sind, doch Kalt, op. 19, no. 3. cf SCHUBERT: Songs
   (Decca SXL 6578).
Songs: Allerseelen, op. 10, no. 8; Befreit, op. 39, no. 4; Cäcilie,
   op. 27, no. 2; Ruhe, meine Seele, op. 27, no. 1; Ständchen, op.
   17, no. 2; Wiegenlied, op. 41, no. 1; Zueignung, op. 10, no. 1.
   cf SIBELIUS: Songs (BIS LP 15).
Songs: Vier letze Lieder (Four last songs): Frühling, September,
   Beim Schlafengehen, Im Abendrot. cf Tod und Verklärung, op. 24.
Songs: Four last songs. cf Capriccio, op. 85: Closing scene.
Songs: Breit uber mei Haupt, op. 19, no. 2; Die Georgine, op. 10,
   no. 4; Heimliche Aufforderung, op. 27, no. 3; Ich trage meine
   Minne, op. 32, no. 1; Nachtgang, op. 29, no. 3; Ständchen, op.
   17, no. 2; Traum durch die Dämmerung, op. 29, no. 1. cf Tele-
   funken DLP 6-48064.
Stimmungsbilder: An einsamer Quelle. cf RCA ARM 4-0942/7.
Symphonia domestica, op. 53. cf Aus Italien, op. 16, G major.
Till Eulenspiegels lustige Streiche, op. 28. cf Also sprach
   Zarathustra, op. 30 (Decca 397).
Till Eulenspiegels lustige Streiche, op. 28. cf Also sprach
   Zarathustra (Decca SXL 6749).
Till Eulenspiegels lustige Streiche, op. 28. cf Aus Italien, op.
   16, G major.
Till Eulenspiegels lustige Streiche, op. 28. cf Works, selections
   (DG 2726 028).
Till Eulenspiegels lustige Streiche, op. 28. cf Works, selections
   (DG 2740 111).
2215 Tod und Verklärung (Death and transfiguration), op. 24.* Vier letze
   Lieder (Four last songs): Frühling, September, Beim Schlafengehen,
   Im Abendrot. Gundula Janowitz, s; BPhO; Herbert von Karajan.
   DG 2530 368. Tape (c) 3300 421. (*Reissue from 2740 111)
      ++Gr 12-74 p1149   ++NYT 11-10-74 pD1
      ++HF 3-75 p87   +ON 1-11-75 p35
      +HF 6-75 p122 tape   +-RR 11-74 p60
      ++MJ 1-75 p49   +SFC 11-24-74 p31
      ++NR 1-75 p3   +St 3-75 p70
Tod und Verklärung, op. 24. cf Works, selections (DG 2740 111).
Death and transfiguration, op. 24. cf HINDEMITH: Mathis der
   Maler symphony.
2216 Works, selections: Also sprach Zarathustra, op. 30. Don Juan, op.
   20. Festival prelude, op. 61. Der Rosenkavalier, op. 59:
   Waltzes. Salome, op. 54: Dance of the seven veils. Till Eulenspiegels lustige Streiche, op. 28. BPhO; Karl Böhm. DG 2726
   028 (2). (Reissues from SLPEM 136001, 138866, 138040-3)
      +-Gr 2-75 p1500   +RR 2-75 p39
      +HFN 5-75 p140
2217 Works, selections: Also sprach Zarathustra, op. 30. Don Juan, op.
   20. Don Quixote, op. 35. Ein Heldenleben, op. 40. Metamor-
   phosen, 23 solo strings. Salome, op. 54: Dance of the seven veils.
   Till Eulenspiegels lustige Streiche, op. 28. Tod und Verklärung,
   op. 24. Pierre Fournier, vlc; Giusto Cappone, vla; BPhO; Herbert von Karajan. DG 2740 111 (5). (Reissues from 2530 349,
   2530 402, 139 009, 138 025, 2530 066)

```
        +Gr 10-74 p707                    +-RR 10-74 p57
        +-NR 6-75 p3                      +-SFC 5-25-75 p29
    Zueignung, op. 10, no. 1.  cf RCA SER 5704/6.
```
STRAVINSKY, Igor
    Canticum sacrum. cf Symphony of psalms.
    Canticum sacrum. cf BOULEZ: e. e. cummings.
    Le chant du rossignol (Song of the nightingale). cf The firebird:
        Suite.
    Song of the nightingale. cf Symphony in three movements.
    Song of the nightingale. cf Les noces.
    Circus polka. cf Mavra.
    Circus polka. cf Works, selections (Columbia M 31729).
    Concertino. cf HAYDN: Quartet, strings, op. 74, no. 3, G minor.
    Concertino, 12 instruments. cf Works, selections (DG 2530 551).
    Concerto, E flat major. cf Works, selections (Columbia M 31729).
2218 Concerto, piano and wind instruments. Ebony concerto. Octet,
        wind instruments. Symphonies, wind instruments. Theo Bruins,
        pno; George Pieterson, clt; NWE; Edo de Waart. Philips 6500 841.
            ++NR 12-75 p7   ++SFC 12-14-75 p40
2219 Concerto, 16 wind instruments, E flat major. Octet, wind instru-
        ments. The soldier's tale: Suite. Nash Ensemble; Elgar Howarth.
        Classics for Pleasure CFP 40098.
            +Gr 2-75 p1510   ++RR 3-75 p35
            ++HFN 5-75 p140
    Concerto, 16 wind instruments, E flat major. cf Concerto, string
        orchestra, D major.
2220 Concerto, string orchestra, D major. Concerto, 16 wind instruments,
        E flat major. Danses concertantes. Los Angeles Chamber Orches-
        tra; Neville Marriner. Angel S 37081. (also HMV ASD 3077)
            +Gr 7-75 p195   ++NR 5-75 p3
            +HF 6-75 p104   ++RR 7-75 p36
            +HFN 7-75 p88   ++St 11-75 p139
2221 Concerto, violin, D major. WALTON: Concerto, violin, B minor.
        Kyung-Wha Chung, vln; LSO; André Previn. Decca SXL 6601. Tape
        (c) KSXC 6601. (also London CS 6819. Tape (c) M5 6819 (ct)
        M8 6819)
            +Gr 5-73 p2052   ++RR 5-73 p71
            ++Gr 9-73 p563 tape   +SFC 4-28-74 p29
            +HF 8-74 p104   ++SR 1-11-75 p51
          . +HFN 6-73 p1184   ++St 11-74 p139
            +NR 5-74 p7
    Danses concertantes. cf Concerto, string orchestra, D major.
    Ebony concerto. cf Concerto, piano and wind instruments.
    Etudes, orchestra (4). cf Works, selections (Columbia M 31729).
2222 The firebird. NYP; Pierre Boulez. Columbia M 33508. (Q) MQ 33508.
        (also CBS 76418. Tape (c) 40-76418)
            +-Gr 12-75 p1053   ++NR 11-75 p2
            +-HF 11-75 p126   +RR 12-75 p67
            +HFN 12-75 p165   -SFC 9-28-75 p30
            -MJ 12-75 p38
2223 The firebird. Igor Stravinsky, pno. Klavier KS 126. (Reissue
        from Duo-Art piano rolls)
            /-HF 12-75 p109   *NR 10-75 p8
2224 The firebird. LSO; Antal Dorati. Mercury SRI 75058. (Reissue
        from SR 90226)
            +Audio 10-75 p120   +-HF 11-75 p126
```
```

2225 The firebird. LPO; Bernard Haitink. Philips 6500 483. (Reissue
      from 6747 094)
              ++Gr 11-75 p834   ++RR 11-75 p55
2226 The firebird. Petrouchka. The rite of spring. LPO; Bernard
      Haitink. Philips 6599 619/21.
              ++HFN 11-75 p173   +Te 6-75 p50
        L'Oiseau de fue: Rondo, Infernal dance, Lullaby, Finale. cf Decca
        SDDJ 393/95.
2227 The firebird: Suite. Jeu de cartes. LSO; Claudio Abbado. DG
      2530 537. Tape (c) 3300 483.
                +Gr 8-75 p331   +RR 8-75 p44
                +HFN 8-75 p84   ++SFC 12-14-75 p40
                +HFN 10-75 p155 tape
2228 L'Oiseau de feu: Suite. Petrouchka. OSCCP; Pierre Monteux. Cam-
      den CCV 5034.
                +RR 10-75 p57
2229 The firebird: Suite. Le chant du rossignol. Berlin Radio Sym-
      phony Orchestra; Lorin Maazel. DG 2548 145. (Reissue from
      138006)
              +Gr 4-75 p1867   ++RR 4-75 p35
        The firebird: Suite. cf BERLIOZ: Les Troyens: Royal hunt and storm.
        Fireworks, op. 5. cf SABATA: Juventus.
        Greeting prelude. cf Works, selections (Columbia M 31729).
        L'Histoire du soldat. cf IVES: Largo.
        The soldier's tale: Suite. cf Concerto, 16 wind instruments, E
          flat major.
        Instrumental miniatures. cf Works, selections (Columbia M 31729).
        Jeu de cartes. cf The firebird: Suite.
2230 Mavra. Circus polka. Norwegian moods (4). Scherzo à la Russe.
      Ludmilla Belobragina, s; Anna Matyushina, Nina Postavnicheva,
      ms; Nicolai Gutorovich, t; MRSO; Gennady Rozhdestvensky, Neimi
      Järvi. HMV Melodiya ASD 3104.
              +-Gr 9-75 p505   +RR 8-75 p27
              +HFN 8-75 p84
        Miniatures, 15 players. cf Le sacre du printemps.
2231 Les noces (The wedding) (sung in Russian). Song of the nightingale.
      Symphonies, wind instruments. Rosalind Rees, s; Rose Taylor,
      ms; Richard Nelson, t; Bruce Fifer, bar; Gregg Smith Singers;
      Orpheus Chamber Ensemble, Columbia Symphony Orchestra; Robert
      Craft. Columbia M 33201.
              +-Audio 8-75 p80   ++SR 6-14-75 p46
              +-HF 5-75 p61   *St 8-75 p106
              +-NR 4-75 p3   +-Te 6-75 p50
              ++SFC 2-16-75 p23
        Norwegian moods. cf Mavra.
        Octet, wind instruments. cf Concerto, 16 wind instruments, E
          flat major.
        Octet, wind instruments. cf Concerto, piano and wind instruments.
        Octet, wind instruments. cf Works, selections (DG 2530 551).
        Oedipus Rex. cf Columbia M2X 33014.
        Pastorale. cf Works, selections (DG 2530 551).
2232 Petrouchka. LSO; Eugene Goossens. Ember ECL 9005. (Reissue from
      Top Rank BUY 007; Vox GBYE 15010)
              -Gr 7-75 p195   -RR 6-75 p54
              -HFN 6-75 p95
2233 Petrouchka (1911 version). LPO; Bernard Haitink. Philips 6500 458.

```
              Tape (c) 7300 354.  (Reissue from 6747 094)
                    +Audio 10-75 p120              +NR 9-75 p3
                    +Gr 12-75 p1053               +-RR 10-75 p57
                    +HF 10-75 p84                  +SFC 7-27-75 p22
                    ++HFN 10-75 p152
2234 Petrouchka.  BSO; Pierre Monteux.  RCA AGL 1-1272.  (Reissue)
                    -SFC 12-14-75 p40
2235 Petrouchka (1911 version).  LSO; Charles Mackerras.  Vanguard VSD
        71177.  Tape (c) ZCVSM 71177  (Q) VSQ 30021.
                    +-Gr 2-74 p1566              ++RR 11-73 p53
                    +-HF 8-73 p104               +-RR 2-75 p76
                    +-HFN 12-73 p2617            ++St 6-73 p123
     Petrouchka.  cf The firebird.
     Petrouchka.  cf BARTOK: Out of doors.
     Petrouchka.  cf L'Oiseau de feu: Suite.
     Piano rag music.  cf Golden Crest CRS 4132.
     Pieces, string quartet (3).  cf SHOSTAKOVICH: Octet, op. 11.
     Pulcinella: Suite.  cf RAVEL: Le tombeau de Couperin.
     Ragtime.  cf Golden Crest CRS 4132.
     Ragtime, 11 instruments.  cf Works, selections (DG 2530 551).
2236 The rite of spring.  CO; Pierre Boulez.  CBS 72807.  Tape (c)
        40-72807.
                    -RR 11-75 p94 tape
2237 The rite of spring.  LPO; Erich Leinsdorf.  Decca PFS 4307.
                    +Gr 10-75 p634               +RR 9-75 p48
                    +HFN 9-75 p105
2238 Le sacre du printemps.  Miniatures, 15 players (8).  LAPO; Zubin
        Mehta.  Decca SXL 6444.  Tape (c) KSXC 6444.
                    ++Gr 6-75 p106 tape          +-RR 7-75 p72 tape
                    ++HFN 7-75 p91 tape
2239 The rite of spring.  CSO; Georg Solti.  London CS 6885.  (also
        Decca SXL 6691.  Tape (c) KSXC 6691)
                    ++Audio 9-75 p70             +-NYT 12-15-74 pD21
                    +Gr 11-74 p909               +RR 4-75 p78 tape
                    +Gr 2-75 p1562 tape          ++RR 11-74 p60
                    +HF 2-75 p102                ++SFC 12-8-74 p36
                    +NR 2-75 p4                  ++St 3-75 p106
2240 The rite of spring.  LPO; Bernard Haitink.  Philips 6500 482.
        (Reissue from 6747 094)
                    +Gr 11-75 p834               +RR 11-75 p55
     The rite of spring.  cf The firebird.
     Scherzo à la Russe.  cf Mavra.
     Septet.  cf Works, selections (DG 2530 551).
     Serenade, A major.  cf NILSSON: Quantitaten.
2241 Sonata, piano.  TCHAIKOVSKY: Sonata, piano, op. 37, G major.
        Paul Crossley, pno.  Philips 6500 884.
                    +NR 12-75 p13                +SR 11-29-75 p50
                    +SFC 12-21-75 p42
     Suites, small orchestra, nos. 1 and 2.  cf Works, selections (Col-
        umbia M 31729).
     Symphonies, wind instruments.  cf Concerto, piano and wind instru-
        ments.
     Symphonies, wind instruments.  cf Les noces.
2242 Symphony in three movements.  Song of the nightingale.  PhO; Con-
        stantin Silvestri.  Classics for Pleasure CFP 40094.  (Reissue
        from HMV ASD 401)
                    +-Gr 6-75 p50                +-RR 6-75 p54
```

2243 Symphony of psalms. Canticum sacrum. Richard Morton, t; Marcus
      Creed, bar; Christ Church Cathedral Choir; Philip Jones Ensemble;
      Simon Preston. Argo ZRG 799.
              +Gr 10-75 p675   +RR 10-75 p90
              +-HFN 10-75 p150
    Tango. cf Golden Crest CRS 4132.
2244 Works, selections: Concerto, E flat major. Etudes, orchestra (4).
      Circus polka. Instrumental miniatures. Greeting prelude.
      Suites, small orchestra, nos. 1 and 2. Columbia Symphony Orch-
      estra, C. B. S. Symphony Orchestra; Igor Stravinsky. Columbia
      M 31729. Tape (ct) MT 31729. (Reissues)
              ++LJ 9-75 p37   ++SFC 4-29-73 p30
2245 Works, selections: Concertino, 12 instruments. Octet, wind
      instruments. Ragtime, 11 instruments. Septet. Boston Symphony
      Chamber Players. DG 2530 551.
              ++HFN 12-75 p165   ++SFC 12-14-75 p40
              ++RR 11-75 p66
STREET, Tison
    Quartet, strings. cf DAVIDOVSKY: Chacona (1972), violin, violon-
      cello and piano.
STROKIN
    Nini otpuchtalechi. cf Harmonia Mundi HMU 133.
SUDERBURG, Robert
    Chamber music II. cf ROCHBERG: Quartet, strings, no. 2.
SUK, Josef
    Elegie, op. 23. cf DVORAK: Trio, piano, op. 90, E minor.
2246 Meditation on St. Wenceslas chorale, op. 35. Quartet, strings,
      no. 2, op. 31. Suk Quartet. Panton 110 395.
              +Gr 8-75 p342   +RR 8-75 p53
              +HFN 9-75 p105
    Pieces, piano, op. 7, no. 1: Love song. cf Supraphon 110 1429.
    Pieces, violin and piano, op. 17 (4). cf Orion ORS 74160.
    Pieces, violin and piano, op. 17: Burleska. cf Discopaedia MB 1009.
    Quartet, strings, no. 2, op. 31. cf Meditation on St. Wenceslas
      chorale, op. 35.
    Serenade, strings, op. 6, E flat major. cf JANACEK: Suite, strings.
SULLIVAN, Arthur
2247 Cox and box. Trial by jury. Soloists; Gilbert and Sullivan Festi-
      val Orchestra and Chorus; Peter Murray. Pye NSPH 15.
              +-HFN 12-75 p171   +-RR 12-75 p31
2248 The gondoliers: Overture; For the merriest fellows are we; We're
      called gondolieri; Thank you, gallant gondolieri...Duke of
      Plaza Toro's song; I stole the prince; Then one of us will be
      a queen; Come let's away and Act 1 finale; Rising early in the
      morning...Take a pair of sparkling eyes; Dance a cachucha;
      Duchess's song; Gavotte song...Here is a case unprecedented
      and Act 2 finale. Soloists; Gilbert and Sullivan Festival Orch-
      estra and Chorus; Peter Murray. Pye NSPH 8.
              +-HFN 12-75 p171   +RR 12-75 p31
    The gondoliers, excerpts. cf Works, selections (Pye NSPH 16).
    The gondoliers: But, bless my heart...Try we life-long; Kind sir,
      you cannot have the heart; Rising early in the morning...Take
      a pair of sparkling eyes; Dance a cachucha. cf Works, selections
      (Decca STBB 4-6).
    The gondoliers: Overture. cf Overtures (Pye NSPH 7).
    Henry VIII: March; Graceful dance. cf Trail by jury.

Henry VIII: Orpheus with his lute. cf BBC REB 191.

2249 HMS Pinafore, highlights. Soloists; Gilbert and Sullivan Festival
      Orchestra and Chorus; Peter Murray. Pye NSPH 9.
        +-HFN 12-75 p171   +RR 12-75 p31
    HMS Pinafore, excerpts. cf Works, selections (Pye NSPH 16).
    HMS Pinafore: Hail...Buttercup's song; Captain's greeting and song;
      Sir Joseph Porter's entrance and song; Never mind the why and
      wherefore. cf Works, selections (Decca STBB 4-6).
    HMS Pinafore: Overture. cf Overtures (Pye NSPH 7).

2250 Iolanthe, highlights. Soloists; Gilbert and Sullivan Festival
      Orchestra and Chorus; Peter Murray. Pye NPHS 11.
        +-HFN 12-75 p171   +-RR 12-75 p31
    Iolanthe, excerpts. cf Works, selections (Pye NSPH 16).
    Iolanthe: Overture. cf Overtures (Pye NSPH 7).
    Iolanthe: Overture; Tripping hither; Loudly let the trumpet bray;
      Sentry's song; When Britain really ruled the waves; Oh, foolish
      fay; Nightmare song; My Lord, a suppliant at your feet I kneel.
      cf Works, selections (Decca STBB 4-6).
    Macbeth: Overture. cf Trial by jury.

2251 The Mikado. Valerie Masterson, Pauline Wales, s; Peggy Ann Jones,
      ms; Lyndsie Holland, con; Colin Wright, t; Michael Rayner, John
      Reed, bar; John Ayldon, Kenneth Sandford, John Broad, bs;
      D'Oyly Carte Opera Chorus; RPO; Royston Nash. Decca SKL 5158/9
      (2). Tape (c) KSKC 5158/9 (ct) ESKC 5158/9. (also London OSA
      12103)
          +-Gr 1-74 p1409   +NR 5-75 p12
          +Gr 4-74 p1919 tape   ++RR 4-74 p93 tape
          +HF 3-75 p74   ++SFC 2-23-75 p23
          +HFN 2-74 p345   +St 7-75 p99
          +MJ 10-75 p45

2252 The Mikado, highlights. Soloists; Gilbert and Sullivan Festival
      Orchestra and Chorus; Peter Murray. Pye NSPH 13.
        +-HFN 12-75 p171   +RR 12-75 p31
    The Mikado, excerpts. cf Works, selections (Pye NSPH 16).
    The Mikado: Opening chorus...A wand'ring minstrel I; Three little
      maids; I am so proud; The sun, whose rays...Brightly dawns our
      wedding day...Here's a how-de-do; Mikado's song; The flowers
      that bloom; Tit-willow; For he's gone and married Yum-Yum. cf
      Works, selections (Decca STBB 4-6).
    The Mikado: Overture. cf Overtures (Pye NSPH 7).
    The Mikado: Tit willow. cf RCA ARL 1-5007.

2253 Overtures; The gondoliers. HMS Pinafore. Iolanthe. The Mikado.
      The pirates of Penzance. Ruddigore. The yeomen of the guard.
      Soloists; Gilbert and Sullivan Festival Orchestra; Peter Murray.
        +-HFN 12-75 p171   +RR 12-75 p31
    Patience: Am I alone...Bunthorne's song; A magnet hung in a hard-
      ware shop. cf Works, selections (Decca STBB 4-6).

2254 Pineapple Poll (arr. Mackerras). Pro Arte Orchestra; John Hollings-
      worth. Pye GSGC 15023. (Reissue)
        +RR 12-75 p67

2255 The pirates of Penzance, highlights. Soloists; Gilbert and Sulli-
      van Festival Orchestra and Chorus; Peter Murray. Pye NSPH 14.
        +-HFN 12-75 p171   +RR 12-75 p31
    The pirates of Penzance, excerpts. cf Works, selections (Pye NSPH
      16).
    The pirates of Penzance: Pirate King's song; Poor wand'ring one;

Major General's song; Ah, leave me not to pine; Police
Sargeant's song. cf Works, selections (Decca STBB 4-6).
The pirates of Penzance: Overture. cf Overtures (Pye NSPH 7).
2256 Princess Ida. Winifred Lawson, Kathleen Anderson, s; Eileen Sharp,
ms; Bertha Lewis, con; Derek Oldham, Leo Darnton, t; Sydney
Granville, Henry Lytton, bar; Leo Sheffield, Darrell Fancourt,
Leonard Hubbard, Edward Halland, bs; Orchestra; Harry Norris,
George W. Byng. Pearl GEM 129/30 (2). (Reissue from HMV D
977/86)
        *Gr 6-75 p84   +RR 5-75 p24
Princess Ida: Gama's song; Minerva; O hear me; The world is but a
broken toy. cf Works, selections (Decca STBB 4-6).
2257 The rose of Persia. Soloists; Orphean Singers and Orchestra; James
Walters. Rare Recorded Editions SRRE 152/3.
        +-HFN 9-75 p105
2258 Ruddigore: Overture; Sir Rupert Murgatroyd his leisure; I know a
youth...(entire score to, and including) You understand; I
think you do; When the buds are blossoming; Within this breast
there beats a heart; Happily coupled are we; Painted emblems...
(entire score to, and including) My eyes are fully open; Oh,
happy the lily. Soloists; Gilbert and Sullivan Festival Orches-
tra and Chorus; Peter Murray. Pye NSPH 12.
        +-HFN 12-75 p171   +RR 12-75 p31
Ruddigore, excerpts. cf Works, selections (Pye NSPH 16).
Ruddigore: When the buds are blossoming; When the night wind howls;
I once was a very abandon'd person. cf Works, selections
(Decca STBB 4-6).
Ruddigore: Overture. cf Overtures (Pye NSPH 7).
Songs: The lost chord. cf RCA ARL 1-0562.
Songs: Onward Christian soldiers. cf CBS 71599.
The sorcerer: The air is charged...Time was, when love and I; My
kindly friends...Oh, happy young heart; My name is John Welling-
ton Wells. cf Works, selections (Decca STBB 4-6)
2259 Trial by jury. Macbeth: Overture. Henry VIII: March; Graceful
dance. Julia Goss, s; Colin Wright, t; John Reed, Michael
Rayner, John Ayldon, Kenneth Sandford, bar; D'Oyly Carte Opera
Chorus, Members; RPO; Royston Nash. Decca TXS 113. Tape (c)
KTXC 113 (ct) EDXC 113.
        +-Gr 5-75 p2014   +RR 5-75 p24
        /HFN 7-75 p91 tape
Trial by jury. cf Cox and box.
Trial by jury: All hail, great Judge...Judge's song; Comes the
broken flower; Oh gentlemen, listen, I pray. cf Works, selections (Decca STBB 4-6).
2260 Works, selections: The gondoliers: But, bless my heart...Try we
life-long; Kind sir, you cannot have the heart; Rising early in
the morning...Take a pair of sparkling eyes; Dance a cachucha.
HMS Pinafore: Hail...Buttercup's song; Captain's greeting and
song; Sir Joseph Porter's entrance and song; Never mind the why
and wherefore. Iolanthe: Overture; Tripping hither; Loudly let
the trumpet bray; Sentry's song; When Britain really ruled the
waves; Oh, foolish fay; Nightmare song; My lord, a suppliant
at your feet I kneel. The Mikado: Opening chorus...A wand'ring
minstrel I; Three little maids; I am so proud; The sun, whose
rays...Brightly dawns our wedding day...Here's a how-de-do;
Mikado's song; The flowers that bloom; Tit-willow; For he's gone

and married Yum-Yum. Patience: Am I alone...Bumthorne's song;
A magnet hung in a hardware shop. The pirates of Penzance:
Pirate King's song; Poor wand'ring one; Major General's song;
Ah, leave me not to pine; Police Sergeant's song. Princess Ida:
Gama's song; Minerva; Oh hear me; The world is but a broken toy.
Ruddigore: When the buds are blossoming; When the night wind
howls; I once was a very abandon'd person. The sorcerer: The
air is charged...Time was, when love and I; My kindly friends...
Oh, happy young heart; My name is John Wellington Wells. Trial
by jury: All hail, great Judge...Judge's song; Comes the broken
flower; Oh gentlemen, listen, I pray. The yeomen of the guard:
When our gallant Norman foes; I have a song to sing; Oh, were
I thy bride; Free from his fetters grim...Strange adventure.
D'Orly Carte Opera Company, RPO, NSL, ROHO; Royston Nash, Malcolm Sargent, Isidore Godfrey. Decca STBB 4-6. (Tape (c) K3G
13.
          ++Gr 12-75 p1098   +RR 12-75 p32
2261 Works, selections: The gondoliers, excerpts. HMS Pinafore, ex-
     cerpts. The Mikado, excerpts. The pirates of Penzance, ex-
     cerpts. Ruddigore, excerpts. Iolanthe, excerpts. The yeomen
     of the guard, excerpts. Soloists; Gilbert and Sullivan Festival
     Orchestra; Peter Murray. Pye NSPH 16.
          +-HFN 12-75 p171   +-RR 12-75 p31
     The yeomen of the guard, excerpts. cf Works, selections (Pye NSPH
     16).
     The yeomen of the guard: When our gallant Norman foes; I have a
     song to sing; Oh were I thy bride; Free from his fetters grim...
     Strange adventure. cf Works, selections (Decca STBB 4-6).
     The yeomen of the guard: Overture. cf Overtures (Pye NSPH 7).
SULZER
     Sarabande. cf Pearl GEM 132.
SUPPE, Franz von
     The beautiful Galathea: Overture. cf Overtures (DG 2548 030).
     Die schöne Galatea: Overture. cf Decca SXL 6572.
     The beautiful Galathea: Overture. cf Telefunken DX 6-35262.
     Boccaccio: Hab' ich nur deine Liebe. cf Angel S 35696.
     Boccaccio: Overture. cf Overtures (DG 2548 030).
     Boccaccio: Overture. cf STRAUSS, J. II: Blindekuh.
     Light cavalry: Overture. cf Overtures (Decca 374).
     Light cavalry: Overture. cf Overtures (DG 2548 030).
     Light cavalry: Overture. cf HEROLD: Zampa: Overture.
     Light cavalry: Overture. cf STRAUSS, J. II: Blindekuh.
     Morning, noon and night in vienna: Overture. cf Overtures (Decca
     374).
     Morning, noon and night in Vienna: Overture. cf Overtures (DG 2548
     030).
     Morning, noon and night in Vienna: Overture. cf STRAUSS, J. II:
     Blindekuh.
     Morning, noon and night in Vienna: Overture. cf LEHAR: Gold and
     silver waltz, op. 79.
     Morning, noon and night in Vienna: Overture. cf HMV SLS 5017.
2262 Overtures: Poet and peasant. Light cavalry. Pique Dame. Morning,
     noon and night in Vienna. VPO; Georg Solti. Decca SPA 374.
     Tape (c) KSDC 194. (Reissue from SXL 2174; SDD 194)
          +HFN 9-75 p108   +-RR 10-75 p57
2263 Overtures: The beautiful Galathea. Boccaccio. Light cavalry.

Morning noon and night in Vienna. Pique Dame. Poet and peas-
ant. Polish Radio Symphony Orchestra; Stefan Rachon. DG 2548
030.
    +-HFN 11-75 p173   ++RR 10-75 p57
Pique Dame: Overture. cf Overtures (Decca 374).
Pique Dame: Overture. cf Overtures (DG 2548 030).
Poet and peasant: Overture. cf Overtures (Decca 374).
Poet and peasant: Overture. cf Overtures (DG 2548 030).
Poet and peasant: Overture. cf HEROLD: Zampa: Overture.
Poet and peasant: Overture. cf STRAUSS, J. II: Blindekuh.
Poet and peasant: Overture. cf HMV SLS 5017.

SUSATO, Tielman
Basse danse. cf L'Oiseau-Lyre SOL 329.
Le cueur est bon. cf L'Oiseau-Lyre SOL 329.
Dances. cf BIS LP 3.
Danserye: Dances (12). cf MORLEY: Dances for broken consort.
Entrée du fol. cf L'Oiseau-Lyre SOL 329.
Flemish dances (5). cf Desto DC 6474/7.
La Mourisque. cf L'Oiseau-Lyre SOL 329.
Pour quoy. cf L'Oiseau-Lyre SOL 329.
Ronde. cf DG Archive 2533 111.
Ronde. cf DG Archive 2533 184.
Ronde-Allemande-Gaillarde. cf Oryx BRL 1.
Salterelle. cf L'Oiseau-Lyre SOL 329.

SVETLANOV, Yevgeny
Festive poem, op. 9. cf MIASKOVSKY: Symphony, no. 22, op. 54,
    B minor.

SWAYNE, Giles
Canto 1, Mr. Timothy's troubles. cf L'Oiseau-Lyre DSLO 3.
SWEDISH SONGS. cf RCA SER 5719.

SWELLINCK, Jan
Hodie Christus natus est. cf HMV Tape (c) TC MCS 13.
More Palatino, variations. cf Argo ZRG 783.
Toccata. cf BASF BAB 9005.
Variations on "Est-ce mars". cf Wealden WS 139.
Variations on "Mein Junges Leben Hatt Ein End". cf BACH: Chorale
    preludes, S 669-671.

SWIFT, Richard
Sonata, solo violin, op. 15. cf DALLAPICCOLA: Studi, violin and
    piano.

SYDEMAN, William
Concerto, piano, 4 hands and chamber orchestra. cf Desto DC 7131.
SZEKELY, Endre
2264 Concerto, trumpet and orchestra. Fantasma, orchestra. Quartet,
    strings, no. 4. Trio, percussion, piano and violoncello. György
    Geiger, tpt; Ferenc Petz, perc; Adám Fellegri, pno; László Mezö,
    vlc; Kodály Quartet, HRT Orchestra; György Lehel. Hungaroton
    SLPX 11666.
        +Gr 7-75 p182   ++NR 12-75 p5
        +-HFN 12-75 p165   ++RR 5-75 p40
Fantasma, orchestra. cf Concerto, trumpet and orchestra.
Quartet, strings, no. 4. cf Concerto, trumpet and orchestra.
Sonata, piano, no. 3. cf Hungaroton SLPX 11692.
Trio, percussion, piano and violoncello. cf Concerto, trumpet and
    orchestra.

SZULC, Jozsef
  Clair de lune. cf DEBUSSY: Songs (Cambridge CRS 2774).
SZYMANOWSKA, Maria
  Etudes, F major, C major, E major. cf Avant AV 1012.
  Nocturne, B flat major. cf Avant AV 1012.
SZYMANOWSKI, Karol
2265 Etude, op. 4, no. 3, B flat minor. King Roger, op. 46: Roxane's
      song. Symphony, no. 2, op. 19, B flat major. Łódź Philharmonic
      Symphony Orchestra; Henryk Czyż. Muza SXL 0981.
            +Gr 7-75 p196
2266 Mythes, op. 30 (3). FRANCK: Sonata, violin and piano, A major.
      Wanda Wilkomirska, vln; Antonio Barbosa, pno. Connoisseur
      Society CS 2050. (Q) CSQ 2050.
            +Gr 2-75 p1509   +RR 2-75 p55
            +HF 5-74 p83   ++SFC 11-25-73 p32
            +NR 2-74 p6   ++St 1-74 p108
  Le Roi Roger, op. 46: Chant do Roxane. cf RCA ARM 4-0942/7.
  King Roger, op. 46: Roxane's song. cf Etude, op. 4, no. 3, B
    flat minor.
  King Roger, op. 46: Roxane's song. cf Coinnoisseur Society CS
    2070.
  Symphony, no. 2, op. 19, B flat major. cf Etude, op. 4, no. 3,
    B flat minor.
TAGLIAFERRI
  Passione; Piscatore e pusilleco. cf Musical Heritage Society MHS 195
TAKEMITSU, Toru
2267 Corona. Far away. Piano distance. Undisturbed rest. Roger Wood-
      ward, pno, hpd, org. Decca Headline HEAD 4. (also London HEAD
      4)
            +HF 4-75 p87   ++SFC 12-8-74 p36
            +NR 10-75 p5   ++SR 2-8-75 p37
            +RR 5-74 p18   +St 11-75 p140
  Far away. cf Corona.
  Piano distance. cf Corona.
  Undisturbed rest. cf Corona.
TALLIS, Thomas
  Jesu salvator saeculi. cf Missa salve intemerata.
2268 Missa salve intemerata. Jesu salvator saeculi. Nunc dimittis.
      Sancte Deus. PLAINSONG: Kyrie trope: Orbis factor. Salve nos
      Domine. Salve regina (with tropes). Peter Hall, Edgar Fleet,
      t; Brian Etheridge, bar; Cantores in Ecclesia; Michael Howard.
      L'Oisea-Lyre SOL 337.
            +Gr 7-74 p254   +RR 6-74 p73
            +HF 1-75 p90   ++St 3-75 p106
            ++NR 12-74 p9
  Nunc dimittis. cf Missa salve intemerata.
  A poynct. cf Angel SB 3816.
  Sancte Deus. cf Missa salve intemerata.
TALMA, Louise
  Alleluia in form of toccata. cf Avant AV 1012.
TANEIEV, Sergei
  Suite, op. 28. cf HMV SLS 5004.
  Symphony, op. 12, C minor. cf SPENDIAROV: Crimean sketches, op. 9:
 -    Elegiac song; Drinking song.
TANENBAUM
  Improvisations and patterns, brass quintet and tape. cf Desto DC
    6476/7.

TANSMAN, Alexandre
    Mazurka. cf CBS 61654.
    Mouvement perpetuel. cf RCA ARM 4-0942/7.
TARREGA, Francisco
    Adelita. cf Amberlee ACL 501X.
    Capricho arabe. cf RCA ARL 1-0485.
    Mazurka. cf RODRIGO: Concierto de Aranjuez, guitar and orchestra.
    Preludes (6). cf CBS 61654.
    Preludes, nos. 10 and 15, D major. cf Amberlee ACL 501X.
    Recuerdos de la Alhambra. cf RODRIGO: Concierto de Aranjuez,
        guitar and orchestra.
    Recuerdos de la Alhambra. cf CBS 72950.
    Recuerdos de la Alhambra. cf London STS 15224.
    Recuerdos de la Alhambra. cf Philips 6833 159.
    Recuerdos de la Alhambra. cf Pye GSGC 14154.
    Sueño. cf Everest SDBR 3380.
TARTINI, Giuseppe
    Concerto, flute and orchestra, F major. cf RCA CRL 2-7003.
    Concerto, flute, a 5, G major. cf BOCCHERINI: Concerto, flute,
        op. 27, D major.
    Concerto, trumpet, D major. cf Erato STU 70871.
2269 Concerti, violin and strings, A major, B flat major, G major.
        Salvatore Accardo, vln; I Musici. Philips 6500 784. Tape
        (c) 7300 333.

+-Gr 5-75 p1974	++NR 2-75 p5
-HF 6-75 p105	+RR 5-75 p45
++HFN 6-75 p95	+RR 10-75 p98 tape
++HFN 7-75 p91	+SR 2-8-75 p37
+MT 12-75 p1072	++St 8-75 p106

    Concerto, violoncello, D minor: Adagio. cf Ember GVC 42.
    Devil's trill. cf Camden CCV 5000-12, 5014-24.
    Sonata, violin, G minor. cf Discopaedia MB 1002.
    Sonata, violin, G minor. cf Discopaedia MB 1004.
    Sonata, violin, no. 4: Finale. cf Rococo 2072.
    Variations on a theme by Corelli. cf Argo ZRG 805.
TAUBERT
    Der Vogel im Walde. cf Pearl GEM 121/2.
TAUSINGER, Jan
    Canto di speranza, piano quartet. cf Panton 110 253.
TAYLOR
    Glorious things of Thee are spoken. cf Abbey LPB 734.
    Toy symphony: Adagio and finale. cf Angel S 36080.
TAYLOR, Silas
    Pavan and galliard. cf Philips 6775 006.
TCHAIKOVSKY, Peter
2270 Capriccio italien, op. 45. Francesca da Rimini, op. 32. March
        slav, op. 31. Overture, the year 1812, op. 49. Romeo and
        Juliet. Symphony, no. 5, op. 64, E minor. NYP; Leonard Bern-
        stein. CBS 77391 (3). (Reissues)

+-HFN 10-75 p152	+-RR 10-75 p60

2271 Capriccio italien, op. 45. The nutcracker, op. 71a: Suite. Eugene
        Onegin, op. 24: Waltz and polonaise. LPO; Leopold Stokowski.
        Philips 6500 766. Tape (c) 7300 332.

+Gr 7-75 p245	+NR 12-75 p2
+Gr 5-75 p2037 tape	+RR 7-75 p36
+HFN 6-75 p109	-SFC 11-9-75 p22
++HFN 7-75 p89	

Capriccio italien, op. 45. cf Works, selections (DG 2740 126).
Capriccio italien, op. 45. cf Works, selections (Philips 7517 012).
Capriccio italien, op. 45. cf RIMSKY-KORSAKOV: Capriccio espagnol,
  op. 34.
Capriccioso, op. 62. cf Command COMS 9006.
Chant sans paroles. cf Pearl GEM 132.
Chant sans paroles, op. 2, no. 3, F major. cf London STS 15160.
Chant sans paroles, op. 2, no. 3, F major. cf Turnabout TV 37033.
Concert fantasia, op. 56, G major. cf Concerto, piano, no. 1,
  op. 23, B flat minor.
2272 Concerto, piano, no. 1, op. 23, B flat minor. Andre Watts, pno;
  NYP; Leonard Bernstein. Columbia M 33071. Tape (c) MT 33071
  (ct) MA 33071 (Q) MQ 33071 Tape (ct) MAQ 33071.
           +HF 6-75 p122 tape   -SFC 11-17-74 p32
           -HF 4-75 p84   /St 3-75 p108
          ++NR 12-74 p6
2273 Concerto, piano, no. 1, op. 23, B flat minor. Concert fantasia,
  op. 56, G major. Peter Katin, pno; LSO; Edric Kundell; LPO;
  Adrian Boult. Decca SPA 168. Tape (c) KCSP 168. (also London
  STS 15227)
           +Gr 12-72 p1214 tape   +NR 7-75 p5
           -HFN 2-72 p317   +RR 1-74 p78 tape
           -HFN 12-72 p2456 tape   +SFC 8-3-75 p30
2274 Concerto, piano, no. 1, op. 23, B flat minor. Concerto, piano,
  no. 3, op. 75, E flat major. Emil Gilels, pno; NPhO; Lorin
  Maazel. HMV ASD 3067. (Reissue from SLS 865)
          ++Gr 4-75 p1817   ++RR 4-75 p36
2275 Concerto, piano, no. 1, op. 23, B flat minor. Concerto, violin.
  op. 35, D major. Byron Janis, pno; Henryk Szeryng, vln; LSO;
  Herbert Menges, Antal Dorati. Philips 6582 009. (Reissues
  from Mercury AMS 16086)
           +-Gr 12-75 p1054   +-RR 10-75 p58
           +-HFN 10-75 p152
2276 Concerto, piano, no. 1, op. 23, B flat minor. Vladimir Horowitz,
  pno; NBC Symphony Orchestra; Arturo Toscanini. RCA VH 015.
  (Reissue)
           +RR 12-75 p68
Concerto, piano, no. 1, op. 23, B flat minor. cf Works, selections
  (DG 2740 126).
Concerto, piano, no. 1, op. 23, B flat minor. cf BEETHOVEN:
  Concerto, piano, no. 5, op. 73, E flat major.
Concerto, piano, no. 1, op. 23, B flat minor. cf PROKOFIEV:
  Concerto, piano, no. 2, op. 16, G minor.
Concerto, piano, no. 1, op. 23, B flat minor. cf RACHMANINOFF:
  Concerto, piano, no. 2, op. 18, C minor (Mercury SRI 75032).
Concerto, piano, no. 1, op. 23, B flat minor. cf RACHMANINOFF:
  Concerto, piano, no. 2, op. 18, C minor (Philips 6833 156).
Concerto, piano, no. 1, op. 23, B flat minor. cf Camden CCV 5000-12
  5014-24.
Concerto, piano, no. 1, op. 23, B flat minor. cf HMV SLS 5033.
Concerto, piano, no. 1, op. 23, B flat minor: Opening. cf Works,
  selections (HMV Tape (c) TC EXE 5004).
2277 Concerto, piano, no. 2, op. 44, G major. Sylvia Kersenbaum, pno;
  ORTF; Jean Martinon. Connoisseur Society CS 2076. (Q) CSQ 2076.
          ++NR 12-75 p2   +St 10-75 p118
          ++SFC 11-9-75 p22

2278 Concerto, piano, no. 2, op. 44, G minor. Paraphrases on Waltz of
      the flowers, by Grainger and Eugene Onegin, polonaise by Liszt.
      Michael Ponti, pno. Prague Symphony Orchestra; Richard Kapp.
      Turnabout TVS 34560.
            -SFC 10-5-75 p38
   Concerto, piano, no. 2, op. 44, G major. cf GRAINGER: Paraphrase
      on TCHAIKOVSKY: Nutcracker: Waltz of the flowers.
   Concerto, piano, no. 3, op. 75, E flat major. cf Concerto, piano,
      no. 1, op. 23, B flat minor.
2279 Concerto, violin, op. 35, D major. BRUCH: Concerto, violin, no. 1,
      op. 26, G minor. Mayumi Fujikawa, vln; Rotterdam Philharmonic
      Orchestra; Edo de Waart. Philips 6500 708. Tape (c) 7300 292.
            +-Gr 1-75 p1341   +RR 1-75 p35
            +HF 3-75 p88   ++SFC 2-2-75 p25
            +MJ 3-75 p26   +-SR 1-11-75 p51
            +NR 2-75 p5   +St 3-75 p109
2280 Concerto, violin, op. 35, D major. SAINT-SAENS: Introduction and
      rondo capriccioso, op. 28. Eugene Fodor, vln; NPhO; Eric Leins-
      dorf. RCA ARL 1-0781. Tape (c) ARK 1-0781 (ct) ARS 1-0781.
            +HF 6-75 p122 tape   +-RR 5-75 p45
            +-HFN 6-75 p95   +SR 1-11-75 p51
   Concerto, violin, op. 35, D major. cf Concerto, piano, no. 1,
      op. 23, B flat minor.
   Concerto, violin, op. 35, D major. cf Works, selections (DG 2740
      126).
   Concerto, violin, op. 35, D major. cf MENDELSSOHN: Concerto,
      violin, op. 64, E minor.
   Concerto, violin, op. 35, D major. cf MOZART: Concerto, violin,
      no. 5, K 219, A major.
   Concerto, violin, op. 35, D major. cf SAINT-SAENS: Introduction
      and rondo capriccioso, op. 28.
   Concerto, violin, op. 35, D major. cf Camden CCV 5000-12, 5014-24.
   Concerto, violin, op. 35, D major. cf RCA ARM 4-0942/7.
   Concerto, violin, op. 35, D major: Canzonetta. cf RCA ARM 4-0942/7.
   Dmitri the imposter and Vassili Shuisky: Introduction and mazurka.
      cf ARENSKY: The fountain of Bakhchisarai, op. 46.
2281 Eugene Onegin, op. 24. Teresa Kubiak, s; Anna Reynolds, Julia
      Hamari, Enid Hartle, ms; Stuart Burrows, Michel Sénéchal, t;
      Bernd Weikl, bar; William Mason, Nicolai Ghiaurov, Richard Van
      Allan, bs; John Alldis Choir; ROHO; Georg Solti. Decca SET
      596/8 (3). (also London OSA 13112)
            +-Gr 6-75 p84   +RR 5-75 p18
            ++HFN 6-75 p96   ++SFC 11-2-75 p28
            +ON 12-6-75 p52
2282 Eugene Onegin, op. 24: Polonaise. Sleeping beauty, op. 66a: Suite.
      Swan Lake, op. 20: Suite. Warsaw National Philharmonic Orches-
      tra; Witold Rowicki. DG 2548 125. (Reissue from 136036)
            +Gr 4-75 p1867   -RR 4-75 p36
   Eugene Onegin, op. 24: Beautiful maidens. cf HMV Melodiya SXL 30190.
   Eugene Onegin, op. 24: Final scene. cf PROKOFIEV: War and peace,
      op. 91: Duets, Scenes 1 and 3.
   Eugene Onegin, op. 24: Lenski's aria. cf RCA SER 5704/6.
   Eugene Onegin, op. 24: Waltz and polonaise. cf Capriccio italien,
      op. 45.
2283 Fate, op. 77. Romeo and Juliet, fantasy overture. The tempest,
      op. 18. Washington National Symphony Orchestra; Antal Dorati.

Decca SXL 6694. (also London CS 6891)
        +Gr 2-75 p1500   +RR 2-75 p40
        +NR 9-75 p6

2284 Francesca da Rimini, op. 32. Hamlet overture, op. 67a. LPO; Vernon
        Handley. Classics for Pleasure CFP 40223.
        +-HFN 9-75 p106   +RR 9-75 p50

2285 Francesca da Rimini, op. 32. Hamlet overture, op. 67a. The
        Voyevode (The landowner), op. 78. Washington National Symphony
        Orchestra; Antal Dorati. Decca SXL 6627. (also London CS 6841.
        Tape (c) 056841 (ct) 086841 (r) E 46841)
        ++Gr 2-74 p1575   +MJ 1-75 p49
        +HF 6-75 p122 tape   +-NR 11-74 p3
        /-HF 7-75 p63   ++RR 2-74 p36
        +HFN 3-74 p115   ++St 12-74 p145

2286 Francesca da Rimini, op. 32. Romeo and Juliet: Fantasy overture.
        USSR Symphony Orchestra; Yevgeny Svetlanov. HMV SXLP 30183.
        +Gr 3-75 p1664   +-RR 3-75 p36

2287 Francesca da Rimini, op. 32. Overture, the year 1812, op. 49.
        Marche slav, op. 31. COA, Royal Military Band; Bernard Haitink,
        Rocus van Yperen. Philips 6500 643. Tape (c) 7300 253.
        ++Gr 2-74 p1575   +-NR 5-74 p3
        +Gr 9-74 p592 tape   +RR 2-74 p36
        ++HF 8-74 p111   +RR 2-75 p76 tape
        ++HFN 3-74 p115   +-SFC 10-6-74 p26

     Francesca da Rimini, op. 32. cf Capriccio italien, op. 45.
     Francesca da Rimini, op. 32. cf Works, selections (Vox QSVBX
        5129/31).
     Francesca da Rimini, op. 32. cf Camden CCV 5000-12, 5014-24.
     Hamlet overture, op. 67a. cf Works, selections (Vox QSVBX 5129/31).
     Hamlet overture, op. 67a. cf Francesca da Rimini, op. 32 (Classics
        for Pleasure 40223).
     Hamlet overture, op. 67a. cf Francesca da Rimini, op. 32 (Decca
        SXL 6627).
     Humoreske, op. 10, no. 2, G major. cf Canon CNN 4983.
     Humoresque, op. 10, no. 2, G major. cf Decca SPA 372.

2288 The maid of Orleans. Claudia Radchenko, s; Irina Arkhipova, ms;
        Vladimir Makhov, Andrei Sokolov, t; Vladimir Valaitis, Sergei
        Yakovenko, Alexander Sibirtsev, bar; Lev Vernigora, Yevgeny
        Vladimirov, Victor Selivanov, Vartan Makelian, bs; MRSO and
        Chorus; Bolshoi Theatre Wind Ensemble; Gennady Rozhdestvensky,
        Roman Altaev. HMV Melodiya SLS 852 (4). (also Columbia/Melo-
        diya M4 33210)
             /Gr 9-73 p528   +-Op 11-73 p1005
             +NR 3-75 p8   ++RR 9-75 p35
             +-ON 3-22-75 p33

     The maid of Orleans: Hymn. cf HMV SEOM 20.

2289 Manfred symphony, op. 58. MRSO: Gennady Rozhdestvensky. Melodiya/
        Angel S 40267.
             /HF 7-75 p66   -SFC 4-13-75 p23
             +NR 3-75 p2

2290 Manfred Symphony, op. 58. LSO; André Previn. HMV ASD 3018. (also
        Angel S 37018)
             +Gr 12-74 p1150   +NR 8-75 p2
             ++HF 11-75 p126   +-RR 12-74 p40

     Manfred symphony, op. 58. cf Symphonies, nos. 1-6.
     Manfred symphony, op. 58. cf Works, selections (Vox QSVBX 5129/31).

March, B flat major.   cf PROKOFIEV: Marches, nos. 1-3, op. 69.
2291 Marche slav, op. 31.   Overture, the year 1812, op. 49.   Romeo and
    Juliet, fantasy overture.   NPhO; Norman Del Mar.   Contour 2870
    419.
        +Gr 1-75 p1357   +-RR 1-75 p35
    Marche slav, op. 31.   cf Capriccio italien, op. 45.
    Marche slav, op. 31.   cf Francesca da Rimini, op. 32.
    Marche slav, op. 31.   cf Works, selections (DG 2740 126).
    Marche slav, op. 31.   cf Works, selections (London SPX 21108).
    Marche slav, op. 31,   cf Works, selections (Philips 7517 012).
    Marche slav, op. 31.   cf RIMSKY-KORSAKOV: Capriccio espagnol, op.
    34.
    Mazeppa: Ariosa.   cf Columbia/Melodiya M 33120.
    Mazeppa: Gopak.   cf Works, selections (Philips 7517 012).
    Mazeppa: Gopak.   cf RIMSKY-KORSAKOV: Capriccio espagnol, op. 34.
2292 Mazeppa: Introduction; Cossack dance.   The Oprichnik: Overture; Dance.
    The Voyevoda, op. 78: Overture; Entr'acte; Dances of the maids.
    Bamberg Symphony Orchestra; Janos Fürst.   Turnabout TV 34548.
    (Q) QTV 34548.
        +-HFN 12-75 p167   +NR 12-74 p4
    Mazeppa: No, oh no, there is no bridge.   cf HMV Melodiya SXL 30190.
    Nocturne, op. 19, no. 4.   cf Command COMS 9006.
    Nocturne, op. 19, no. 4.   cf L'Oiseau-Lyre DSLO 7.
2293 The nutcracker, op. 71.   Bolshoi Theatre Orchestra; Gennady Rozh-
    destvensky.   Columbia/Melodiya M2 33116 (2).
        +-HF 3-75 p72   +-St 3-75 p109
        +-NR 2-75 p2
2294 The nutcracker, op. 71.   National Philharmonic Orchestra; Richard
    Bonynge.   London CSA 2239 (2).   (also Decca SXL 6688/9)
        +Gr 11-75 p833   ++NYT 3-9-75 pD23
        +HF 3-75 p71   +RR 11-75 p55
        +HFN 11-75 p169   +St 3-75 p109
        ++NR 2-75 p2
2295 The nutcracker, op. 71.   LSO and Chorus; Antal Dorati.   Philips
    6780 250 (2).   (Reissue from Mercury AMS 16143/4)
        +Gr 11-75 p833   +RR 11-75 p55
        +HFN 11-75 p173
2296 The nutcracker, op. 71, excerpts.   The sleeping beauty, op. 66,
    excerpts.   OSR; Ernest Ansermet.   Decca SPA 357/8.   Tape (c)
    KCSP 357/8 (ct) ECSP 357.
        +Gr 2-75 p1561 tape   +RR 4-75 p78 tape
        +RR 11-74 p61
2297 The nutcracker, op. 71, excerpts.   Ambrosian Singers; LSO; André
    Previn.   HMV ASD 3051.
        +Gr 2-75 p1556   ++RR 3-75 p36
    The nutcracker, op. 71, excerpts.   cf Works, selections (HMV Tape
    (c) TC EXE 5004).
    The nutcracker, op. 71: Chinese dance; Trepak; Dance of the sugar-
    plum fairy; Waltz of the flowers; Pas de deux.   cf Decca SDDJ
    393/5.
    The nutcracker, op. 71: Chinese dance; Trepak; Dance of the sugar-
    plum fairy; Waltz of the flowers; Pas de deux, Act 2.   cf Decca
    DPA 515/6.
    Casse-Noisette (The nutcracker), op. 71: March.   cf DG 2548 148.
    The nutcracker, op. 71: Russian dance; Dance of the sugarplum fairy;
    Dance of the toy flutes.   cf Columbia CXM 32088.

The nutcracker, op. 71: Serenade; Waltz. cf Works, selections
    (Philips 7517 012)
2298 The nutcracker, op. 71a: Suite. The sleeping beauty, op. 66:
    Suite. Orchestre de Paris; Seiji Ozawa. Philips 6500 851
             +Gr 5-75 p2023   -RR 5-75 p45
             +-HF 3-75 p72   ++St 3-75 p109
             +HFN 5-75 p140
2299 The nutcracker, op. 71a: Suite. Romeo and Juliet. NBC Symphony
    Orchestra; Arturo Toscanini. RCA AT 119. (Reissue from HMV
    ALP 1441, ALP 1469)
             +Gr 8-75 p336   +RR 6-75 p54
    The nutcracker, op. 71a: Suite. cf Capriccio italien, op. 45.
    The nutcracker, op. 71a: Suite. cf Works, selections (DG 2740 126).
    The nutcracker, op. 71a: Suite. cf MENDELSSOHN: A midsummer night's
    dream: Overture, nocturne and scherzo.
    The nutcracker, op. 71a: Suite, excerpts. cf Camden CCV 5000-12,
    5014-24.
    The nutcracker, op. 71: Waltz of the flowers; Pas de deux. cf
    HMV SEOM 14.
    The Oprichnik: Overture; Dance. cf Mazeppa: Introduction; Cossack
    dance.
2300 Overture, the year 1812, op. 49. BEETHOVEN: Wellington's victory,
    op. 91. BPhO; Don Cossack Choir; Herbert von Karajan. DG
    2538 142.
             +NR 5-75 p6   ++SFC 3-9-75 p27
    Overture, the year 1812, op. 49. cf Capriccio italien, op. 45.
    Overture, the year 1812, op. 49. cf Francesca da Rimini, op. 32.
    Overture, the year 1812, op. 49. cf Marche slav, op. 31.
    Overture, the year 1812, op. 49. cf Works, selections (HMV Tape
    (c) TC EXE 5004).
    Overture, the year 1812, op. 49. cf Works, selections (London
    SPC 21108).
    Overture, the year, 1812, op. 49. cf Works, selections (Philips
    7517 012).
    Overture, the year 1812, op. 49. cf Works, selections (Vox QSVBX
    5129/31).
    Paraphrases on Waltz of the flowers by Grainger and Eugene Onegin
    polonaise by Liszt. cf Concerto, piano, no. 2, op. 44, G minor.
2301 Pique Dame (Queen of spades), op. 68. Tamara Milashkina, s; Valen-
    tina Levko, ms; Galina Borisova, con; Vladimir Atlantov, t;
    Vladimir Valaitis, bar; Andrei Fedoseyev, bs; Bolshoi Opera
    Orchestra; Mark Ermler. Columbia/Melodiya M 33328 (3).
             +ON 12-6-75 p52
2302 The Queen of Spades, op. 68. Tamara Milashkina, Margarita Miglayu,
    Vera Firsova, s; Valentina Levko, Marya Matukova, ms; Zurab
    Andzhaparidzhe, Andrei Sokolov, Anatoly Mishutin, Vitaly Vlas-
    sov, t; Irina Arkhipova, con; Mikhail Kiselev, Yuri Mazurok,
    Viktor Nechinailo, bar; Valery Yaroslavtsev, Yuri Dementiev,
    bs; Bolshoi Theatre Orchestra and Chorus; Boris Khaikin. HMV
    Melodiya SLS 5005 (3). (also Melodiya Angel S 4104)
             +Gr 5-75 p2014   +STL 6-8-75 p36
             +HFN 5-75 p141
    Quartet, strings, no. 1, op. 11, D major: Andante cantabile. cf
    Works, selections (HMV Tape (c) TC EXE 5004).
    Quartet, strings, no. 1, op. 11, D major: Andante cantabile. cf
    Amon Ra SARB 01.

Quartet, strings, no. 1, op. 11, D major: Andante cantabile. cf
    HMV SEOM 20.
Quartet, strings, no. 1, op. 11, D major: Andante cantabile. cf
    London STS 15160.
Quartet, strings, no. 1, op. 11, D major: Andante cantabile. cf
    BORODIN: Quartet: Quartet, strings, no. 2, D major: Nocturne.
Quartet, strings, no. 1, op. 11, D major: Andante cantabile. cf
    Philips 6850 066.
Romeo and Juliet. cf Capriccio italien, op. 45.
Romeo and Juliet. cf Works, selections (London SPC 21108).
Romeo and Juliet. cf The nutcracker, op. 71a: Suite.
Romeo and Juliet. cf Works, selections (Vox QSVBX 5129/31).
Romeo and Juliet. cf Camden CCV 5000-12, 5014-24.
Romeo and Juliet: Fantasy overture. cf Fate, op. 77.
Romeo and Juliet: Fantasy overture. cf Francesca da Rimini, op. 32.
Romeo and Juliet: Fantasy overture. cf Marche slav, op. 31.
Romeo and Juliet: Overture. cf Works, selections (DG 2740 126).
Romeo and Juliet: Overture. cf MUSSORGSKY: Pictures at an exhib-
    ition.
Serenade, strings, op. 48, C major. cf DVORAK: Serenade, strings,
    op. 22, E major (Decca 375).
Serenade, strings, op. 48, C major. cf DVORAK: Serenade, strings,
    E major (DG 2548 121).
Serenade, strings, op. 48, C major. cf GRIEG: Holberg suite,
    op. 40.
Serenade, strings, op. 48, C major. cf DVORAK: Serenade, strings,
    op. 22, E major (HMV ASD 3036).
Serenade, strings, op. 48, C major: Waltz. cf Angel S 36076.
Serenade, strings, op. 48, C major: Waltz. cf Works, selections
    (HMV Tape (c) TC EXE 5004).
Serenade, strings, op. 48, C major: Valse. cf Discopaedia MB 1010.
Serenade, strings, op. 48, C major: Valse. cf RCA ARM 4-0942/7.
Sérénade mélancolique, op. 26. cf LALO: Symphonie espagnole, op.
    21.
Sérénade mélancolique, op. 26. cf RCA ARL 1-0735.
Sérénade mélancolique, op. 26. cf RCA ARM 4-0942/7.
Slavonic march, op. 31. cf DG 2548 148.
2303 The sleeping beauty, op. 66. LSO; André Previn. Angel SCLX 3812
    (3). (also HMV SLS 5001)
            ++HF 3-75 p71   ++RR 12-74 p41
            +-NR 2-75 p3   ++St 5-75 p103
            ++NYT 3-9-75 pD23
Sleeping beauty, op. 66. cf Works, selections (Philips 7517 012).
2304 Sleeping beauty, op. 66, excerpts. Swan Lake, op. 20, excerpts.
    LSO; Pierre Monteux, Anatole Fistoulari. Philips Tape (c) 7505
    040 (2).
            +Gr 4-75 p1872 tape   +-RR 5-75 p77 tape
Sleeping beauty, op. 66, excerpts. cf The nutcracker, op. 71,
    excerpts.
Sleeping beauty, op. 66: Introduction; Garland waltz; Rose adagio;
    Panorama. cf Decca SDDJ 393/5.
Sleeping beauty, op. 66: Introduction; Garland waltz; Rose adagio;
    Panorama. cf Decca DPA 515/6.
Sleeping beauty, op. 66: Suite. cf Eugene Onegin, op. 24: Polonaise.
Sleeping beauty, op. 66: Suite. cf The nutcracker, op. 71a: Suite.
Sleeping beauty, op. 66: Waltz. cf Works, selections (HMV Tape
    (c) TC EXE 5004)

Sleeping beauty, op. 66: Waltz. cf Works, selections (London SPC 21108).

Sonata, piano, op. 37, G major. cf STRAVINSKY: Sonata, piano.

2305 Songs (15). Irina Arkhipova, s; Stuchevsky, pno. Westminster/ Melodiya WGS 8257.
        +ON 1-25-75 p32

Songs: None but the lonely heart, op. 6, no. 6. cf Works, selections (HMV Tape (c) TC EXE 5004).

Songs: Au jardin, prés du ruisseau, op. 46, no. 4; L'aube, op. 46, no. 6; Larmes humaines, op. 46, no. 3. cf BIS LP 17.

Songs: Wait; Night. cf Columbia M 32231.

Songs: None but the lonely heart, op. 6, no. 6. cf Desto DC 7118/9.

Songs: Don Juan's serenade, op. 38, no. 1; I bless you woods, op. 47, no. 5; It was in the early spring, op. 38, no. 2; Mid the noisy stir of a ball, op. 38, no. 3; None but the lonely heart, op. 6, no. 6; Not a word, oh my love, op. 6, no. 2. cf London OS 26249.

Songs: Einst zum Narr'n der Weise sprach, op. 25, no. 6; Inmitten des Balles, op. 38, no. 3; Nur wer die Sehnsucht kennt, op. 6, no. 6; O singe mir, Mutter, diese Weise, op. 16, no. 4; Töte mich, aber liebe mich; Unendliche Liebe; Warum sind denn die Rosen so blass. cf Telefunken DLP 6-48064.

Souvenir d'un lieu cher, op. 42: Meditation. cf LALO: Symphonie espagnole, op. 21.

Souvenir d'un lieu cher, op. 42: Melodie. cf Discopaedia MB 1003.

Souvenir d'un lieu cher, op. 42: Scherzo. cf RCA ARM 4-0942/7.

2306 Suite, no. 3, op. 55, G major. OSCCP: Adrian Boult. Decca ECS 766.
        +HFN 8-75 p89

2307 Suite, no. 3, op. 55, G major. LPO; Adrian Boult. HMV (Q) ASD 3135.
        ++Gr 11-75 p834   ++RR 10-75 p58
        ++HFN 10-75 p150

2308 Suite, no. 4, op. 61. Variations on a rococo theme, op. 33. Stanislav Apolin, vlc; Jiři Tomášek, vln; PSO; Václav Smetáček. Supraphon 110 1454.
        +NR 1-75 p2   +-RR 10-74 p63

Swan Lake, op. 20. cf Works, selections (Philips 7517 012).

Swan Lake, op. 20, excerpts. cf Sleeping beauty, op. 66, excerpts.

Swan Lake, op. 20, excerpts. cf Works, selections (HMV Tape (c) TC EXE 5004).

Swan Lake, op. 20: Suite. cf Eugene Onegin, op. 24: Polonaise.

Swan Lake, op. 20: Waltz, Act 1. cf Works, selections (London SPC 21108).

Swan Lake, op. 20: Waltz; Scene; Dance of the little swans; Czardas. cf Decca SDDJ 393/5.

Swan Lake, op. 20: Waltz, Act 1; Scene; Dance of the little swans, Act 2; Czardas, Act 3. cf Decca DPA 515/6.

2309 Symphonies, nos. 1-6. Manfred symphony, op. 58. MRSO; Gennady Rozhdestvensky. Melodiya/Angel SR 40261/7 (7).
        +-HF 7-75 p63

2310 Symphonies, nos. 1-6. VPO: Lorin Maazel. London Tape (r) LC 1-80200 (3).
        ++LJ 9-75 p37

Symphonies, nos. 1-6. cf Works, selections (Vox QSVBX 5129/31).

2311 Symphony, no. 1, op. 13, G minor. MRSO: Gennady Rozhdestvensky.

Melodiya/Angel SR 40261.
    +–HF 7-75 p63   ++St 5-75 p104
    +NR 1-75 p2
2312 Symphony, no. 2, op. 17, C minor. MRSO; Gennady Rozhdestvensky.
    Melodiya/Angel SR 40262.
        +–HF 7-75 p63   ++SFC 4-13-75 p23
        +NR 1-75 p1   +St 5-75 p104
2313 Symphony, no. 3, op. 29, D major. VSO; Moshe Atzmon.   DG 2530
    401. (Reissue from 2720 065)
        +–Gr 8-74 p364   +SFC 2-17-74 p28
        +–MJ 7-74 p48   +–St 8-74 p123
        +NR 5-74 p3   +–St 7-75 p63
        -RR 8-74 p42
2314 Symphony, no. 3, op. 29, D major. MRSO; Gennady Rozhdestvensky.
    Melodiya/Angel SR 40263.
        +–HF 7-75 p63   ++St 5-75 p104
        +NR 1-75 p2
2315 Symphony, no. 3, op. 29, D major. VSO; Hans Swarowsky. Olympic
    8114.
        /NR 5-75 p4
2316 Symphonies, nos. 4-6. BSO; Pierre Monteux. RCA LVL 3-7530 (3).
    (Reissues from SB 2045, 2093, HMV ALP 1356)
        +–Gr 12-75 p1054   /-RR 11-75 p56
        +–HFN 11-75 p173
    Symphonies, nos. 4-6. cf Works, selections (DG 2740 126).
2317 Symphony, no. 4, op. 36, F minor. MRSO; Gennady Rozhdestvensky.
    Angel SR 40264.
        +NR 5-75 p4
2318 Symphony, no. 4, op. 36, F minor. BSO; Charles Munch. Camden
    CCV 5036.
        +RR 10-75 p58
2319 Symphony, no. 4, op. 36, F minor. Scottish National Orchestra;
    Alexander Gibson. Classics for Pleasure CFP 40228.
        +–HFN 11-75 p169   -RR 11-75 p56
2320 Symphony, no. 4, op. 36, F minor. BPhO; Lorin Maazel. DG 2548
    176. (Reissue from 138789)
        +–Gr 11-75 p840   +RR 10-75 p58
        +HFN 10-75 p152
2321 Symphony, no. 4, op. 36, F minor. Symphony, no. 6, op. 74, B
    minor. BPhO; Herbert von Karajan. HMV ASD 2814, 2816 (4).
    Tape (c) TC ASD 2814, TC 2816. (Reissues from SLS 833)
        +Gr 6-75 p50   ++RR 4-75 p36
        +Gr 5-75 p2032 tape
2322 Symphony, no. 4, op. 36, F minor. LSO; Antal Dorati. Mercury
    SRI 75044.
        +NR 1-75 p2
2323 Symphony, no. 4, op. 36, F minor. PO; Eugene Ormandy. RCA ARL
    1-0665. Tape (c) ARK 1-0665 (ct) ARS 1-0665 (Q) ARD 1-0665
    Tape (ct) ART 1-0665.
        +Gr 5-75 p1974   +NR 11-74 p3
        -HF 7-75 p63   +–RR 6-75 p55
        +HF 1-75 p110 tape   +St 1-75 p114
        +HFN 8-75 p84
    Symphony, no. 4, op. 36, F minor. cf Toscanini Society ATS GC
    1201/6.
2324 Symphony, no. 5, op. 64, E minor. MRSO; Gennady Rozhdestvensky.

Angel SR 40265.
　　++NR 6-75 p5
2325 Symphony, no. 5, op. 64, E minor. COA; Willem Mengelberg. Bruno
　　Walter Society RR 421.
　　　　+-NR 11-75 p4
2326 Symphony, no. 5, op. 64, E minor. Leningrad Philharmonic Orches-
　　tra; Yevgeny Mravinsky. DG 2538 179. (Reissue from SLPX 138
　　658)
　　　　+Gr 3-75 p1664　　　　　　　　　　+RR 1-75 p35
2327 Symphony, no. 5, op. 64, E minor. BPhO; Herbert von Karajan. HMV
　　Q4ASD 2815. Tape (c) TC ASD 2815. (Reissue from SLS 833) (also
　　Angel S 36855. Tape (c) 4XS 36885 (ct) 8XS 36885)
　　　　+Gr 1-75 p1357　　　　　　　　　　+Gr 5-75 p2032 tape
　　　　++Gr 6-74 p112 Quad　　　　　　　++RR 2-74 p36 Quad
2328 Symphony, no. 5, op. 64, E minor. RAI Rome Orchestra; Wilhelm
　　Furtwängler. Olympic 8137.
　　　　+NR 11-75 p4
2329 Symphony, no. 5, op. 64, E minor. PO; Eugene Ormandy. RCA ARL
　　1-0664. Tape (c) ARK 1-0664 (ct) ARS 1-0774 (Q) ARD 1-0664
　　Tape (ct) ART 1-0664.
　　　　+-Gr 6-75 p55　　　　　　　　　　+NR 11-74 p3
　　　　-HF 7-75 p63　　　　　　　　　　+-RR 6-75 p55
　　　　+HFN 6-75 p95　　　　　　　　　　+St 1-75 p114
　　Symphony, no. 5, op. 64, E minor. cf Capriccio italien, op. 45.
　　Symphony, no. 5, op. 64, E minor: 3rd movement. cf Sound Superb
　　SPR 90049.
2330 Symphony, no. 6, op. 74, B minor. VPO; Claudio Abbado. DG 2530
　　350. Tape (c) 3300 405.
　　　　+-Gr 7-74 p215　　　　　　　　　　+-NR 12-74 p4
　　　　++HF 7-75 p66　　　　　　　　　　/RR 8-74 p47
　　　　++HF 6-75 p122 tape　　　　　　　+St 1-75 p114
2331 Symphony, no. 6, op. 74, B minor. BPhO; Herbert von Karajan. HMV
　　Tape (c) TC ASD 2816. (also Angel S 36886. Tape (c) 4XS 36886
　　(ct) 8XS 36886)
　　　　+Gr 2-75 p1562 tape
2332 Symphony, no. 6, op. 74, B minor. MRSO; Gennady Rozhdestvensky.
　　Melodiya/Angel SR 40266.
　　　　-NR 8-75 p4　　　　　　　　　　　+St 9-75 p115
2333 Symphony, no. 6, op. 74, B minor. LSO; Jascha Horenstein. Music
　　for Pleasure MFP 57017. (Reissue from HMV ASD 2332)
　　　　+Gr 5-75 p1979　　　　　　　　　　+RR 2-75 p40
　　　　+HFN 8-75 p89
2334 Symphony, no. 6, op. 74, B minor. Orchestra de Paris; Seiji Ozawa.
　　Philips 6500 850.
　　　　+-Gr 4-75 p1817　　　　　　　　　-RR 4-75 p37
　　　　+HF 7-75 p63　　　　　　　　　　++SFC 3-9-75 p26
　　　　+-NR 6-75 p5　　　　　　　　　　+St 9-75 p114
2335 Symphony, no. 6, op. 74, B minor. LSO; Leopold Stokowski. RCA
　　ARL 1-0426. Tape (c) ARK 1-0426 (r) ARS 1-0426 (Q) ARD 1-0426
　　Tape (ct) ART 1-0426.
　　　　+-Gr 4-75 p1817　　　　　　　　　+RR 4-75 p37
　　　　++HF 7-75 p63　　　　　　　　　　+-SFC 9-29-74 p28
　　　　+-HF 6-75 p122 tape　　　　　　　-St 1-75 p114
2336 Symphony, no. 6, op. 74, B minor. BPhO; Wilhelm Furtwängler.
　　Seraphim 60231. (Reissue from Angel COLH 21)
　　　　+HF 7-75 p63　　　　　　　　　　+-St 1-75 p114

Symphony, no. 6, op. 74, B minor. cf Symphony, no. 4, op. 36,
   F minor.
Symphony, no. 6, op. 74, B minor. cf Camden CCV 5000-12, 5014-24.
Symphony, no. 6, op. 74, B minor. cf HMV SLS 5003.
Symphony, no. 6, op. 74, B minor: 3rd movement. cf Works, selec-
   tions (HMV Tape (c) TC EXE 5004).
The tempest, op. 18. cf Fate, op. 77.
Undine, excerpts. cf ARENSKY: The fountain of Bakhchisarai, op.
   46.
Valse scherzo, op. 34. cf Hungaroton SLPX 11677.
Valse scherzo. cf RCA ARL 1-0735.
Variations on a rococo theme, op. 33. cf Suite, no. 4, op. 61.
Variations on a rococo theme, op. 33. cf Works, selections
   (DG 2740 126).
Variations on a rococo theme, op. 33. cf DVORAK: Concerto, violon-
   cello, op. 104, B minor.
The Voyevode (The landowner), op. 78. cf Francesca da Rimini,
   op. 32.
The Voyevoda, op. 78: Overture; Entr'acte; Dance of the maids. cf
   Mazeppa: Introduction; Cossack dance.
2337 Works, selections: Capriccio italien, op. 45. Concerto, piano,
   no. 1, op. 23, B flat minor. Concerto, violin, op. 35, D major.
   Marche slav, op. 31. The nutcracker, op. 71a: Suite. Romeo
   and Juliet: Overture. Variations on a rococo theme, op. 33.
   Symphonies, nos. 4-6. Sviatoslav Richter, pno; Christian Ferras,
   vln; Mstislav Rostropovich, vlc; BPhO, VSO; Herbert von Karajan.
   DG 2740 126. (Reissue from SKL 922/8)
        ++Gr 11-75 p834   +RR 10-75 p58
        ++HFN 10-75 p152
2338 Works, selections: Quartet, strings, no. 1, op. 11, D major: Andante cantabile. Concerto, piano, no. 1, op. 23, B flat minor:
   Opening. The nutcracker, op. 71, excerpts. None but the lonely heart, op. 6, no. 6. Serenade, strings, op. 48, C major:
   Waltz. The sleeping beauty, op. 66: Waltz. Symphony, no. 6,
   op. 74, B minor: 3rd movement. Swan Lake, op. 20, excerpts.
   Overture, the year 1812, op. 49. Various artists. HMV Tape
   (c) TC EXE 5004.
        +RR 4-75 p78 tape
2339 Works, selections: Marche slav, op. 31. Overture, the year 1812,
   op. 49. Romeo and Juliet. Sleeping beauty, op. 66: Waltz,
   Act 1. Swan Lake, op. 20: Waltz, Act 1. OSR, RPO, Band and
   Choruses; Leopold Stokowski. London SPC 21108. Tape (c) M5
   21108 (ct) M8 21108 (r) L 475108. (also Decca SDD 454)
        +-Gr 5-75 p2023   +RR 5-75 p46
        +HF 11-74 p94   +-SFC 4-28-74 p29
        +-HFN 6-75 p95
2340 Works, selections: Capriccio italien, op. 45. Overture, the year
   1812, op. 49. Marche slav, op. 31. Swan Lake, op. 20. Maz-
   zeppa: Gopak. Sleeping beauty, op. 66. The nutcracker, op.
   71: Serenade, Waltz. COA, VSO, LSO; Igor Markevitch, Karel
   Ančerl, Pierre Monteux, Charles Mackerras, Anatole Fistoulari,
   Bernard Haitink. Philips Tape (c) 7517 012 (2).
        +-Gr 4-75 p1872   /HFN 6-75 p109
2341 Works, selections: Symphonies, nos. 1-6. Hamlet, op. 67a. Overture, the year 1812, op. 49. Manfred symphony, op. 58. Francesca da Rimini, op. 32. Romeo and Juliet. Utah Symphony

Orchestra; Maurice Abravanel. Vox QSVBX 5129/31 (3).
        +-HF 7-75 p63   +NR 1-75 p2

TEIKE
    Alte Kameraden. cf BASF 292 2116-4.
    Alte Kameraden. cf DG 2721 077.

TELEMANN, Georg Philipp
2342 Concerto, oboe d'amore and strings, G major. Concerto, viola and
        strings, G major. Kurt Hausmann, oboe d'amore; Karl Bender,
        vla; Würzburg Camerata Accademica; Hans Reinartz. Pye GSGC
        14149.
            +Gr 3-75 p1665   +RR 1-75 p35
    Concerto, trumpet. cf ARUTANIAN: Concerto, trumpet.
    Concerto, trumpet, D major. cf ALBINONI: Concerto, trumpet, C
        major.
    Concerto, trumpet, D major. cf HUMMEL: Concerto, trumpet, E flat
        major (DG 2530 289).
    Concerto, trumpet, D major. cf HUMMEL: Concerto, trumpet, E flat
        major (HMV 3044).
    Concerto, trumpet, D major. cf Erato STU 70871.
    Concerto, trumpet, D major. cf HMV ASD 2938.
    Concerto, trumpet, D major. cf RCA CRL 2-7002.
    Concerto, viola and strings, G major. cf Concerto, oboe d'amore
        and strings, G major.
    Concerto, 3 violins, strings and continuo, F major. cf BACH: Con-
        certo, 3 violins, strings and continuo, D minor.
    Fantasies, solo recorder, A minor and C major. cf Telefunken
        Tape (c) 4-41203.
    Der Gedultige Socrates: Rodisetta's air. cf Strobe SRCM 123.
    Sonata, flute, F major. cf London STS 15198.
    Sonata, flute, G major. cf Orion ORS 75199.
    Sonata, flute, no. 3, E minor. cf Pandora PAN 103.
    Sonata, flute and continuo, op. 2, E minor. cf Orion ORS 73114.
    Sonata, 2 flutes, E major. cf Ember ECL 9040.
    Sonata, organ, G minor. cf RCA CRL 2-7001.
    Sonata, viola da gamba, E minor. cf Musical Heritage Society MHS
        1853.
    Sonata, viola da gamba and harpsichord, A minor. cf BACH: Sonatas,
        viola da gamba and harpsichord, nos. 1-3, S 1027-29.
    Sonate de concert, trumpet, D major. cf RCA CRL 2-7002.
    Suite, 4 horns and strings, F major. cf ROSETTI: Concerto, horn,
        D minor.
    Suite, recorder and strings, A minor. cf HANDEL: Concerto, recorder
        and strings, B flat major.
    10 PLUS 2 EQUALS 12 AMERICAN TEXT SOUND PIECES. cf 1750 Arch 1752.

TERZI, Giovanni
    Ballo tedesco e françese. cf DG Archive 2533 173.
    Canzona françese. cf L'Oiseau-Lyre 12BB 203/6.
    Tre parti di gagliarde. cf DG Archive 2533 173.

THALBERG, Sigismond
    Les capricieuses, op. 64. cf Piano works (Candide 31084).
    Concerto, piano, op. 5, F minor. cf Piano works (Candide 31084).
    Fantasy on Meyerbeer's "Les Huguenots", op. 20. cf Piano works
        (Candide 31084).
    Fantasy on Rossini's "Barber of Seville", op. 63. cf LISZT: Ballade
        no. 2, G 171, B minor.
    Fantasy on Rossini's "Moise", op. 33. cf LISZT: Ballade, no. 2,
        G 171, B minor.

2343 Piano works: Concerto, piano, op. 5, F minor. Les capricieuses,
        op. 64. Fantasy on Meyerbeer's "Les Huguenots", op. 20.
        Variations on "Home, sweet home", op. 72. Variations on "The
        last rose of summer", op. 73. Michael Ponti, pno; Westphalian
        Symphony Orchestra; Richard Kapp. Candide CE 31084.
            +HF 9-75 p93   +-NR 4-75 p6
    Variations on "Home, sweet home", op. 72. cf Piano works (Candide
        31084).
    Variations on "The last rose of summer", op. 73. cf Piano works
        (Candide 31084).
    THIRTEENTH CENTURY POLYPHONY. cf Pleiades P 249.
THOMAS
    Close of day. cf Wayne State University, unnumbered.
THOMAS, Ambroise
    Le Caid: Je comprends que la belle aime le militaire. cf Decca
        SXL 6637.
    Hamlet: A vos jeux, mes amis. cf ABC ATS 20019.
    Hamlet: A vox veux mes amis...Partagez vous mes fleurs. cf Saga
        7029.
    Hamlet: Mad scene. cf RCA ARL 1-0844.
    Mignon: Entr'acte. cf Discopaedia MB 1005.
    Mignon: Kennst du das Land. cf Discophilia KGP 4.
    Mignon: Styrienne. cf Pearl GEM 121/2.
    Mignon: Styrienne, Schwalbenduett. cf Discophilia KGM 2.
THOMAS, Mansel
    Prayers from the Gaelic (4). cf Argo ZRG 769.
THOME, Francis
    Simple aveu, op. 25. cf Discopaedia MB 1006.
THOMSON, Virgil
    At the beach: Concert waltz. cf Nonesuch H 71298.
    Parallel chords (tango). cf Golden Crest CRS 4132.
    Ragtime bass. cf Golden Crest CRS 4132.
TIBURTINO, Giuliano
    Ricercare a 3. cf L'Oiseau-Lyre 12BB 203/6.
TIERSOT, Julien (Jean Baptiste Elesee)
    Chants de la vieille France (4). cf FAURE: Songs (1750 Arch S
        1754).
TIPPETT, Michael
2344 A child of our time. Jessye Norman, s; Janet Baker, ms; Richard
        Cassilly, t; John Shirley-Quirk, bs; BBC Singers, BBC Choral
        Society; BBC Symphony Orchestra; Colin Davis. Philips 6500 985.
            +Gr 11-75 p882   ++NYT 12-21-75 pD18
            +-HFN 12-75 p167   +RR 11-75 p88
    Concerto, orchestra. cf The midsummer marriage: Four ritual dances.
    Fantasia concertante on a theme by Corelli. cf ELGAR: Introduction
        and allegro, op. 47.
2345 The knot garden. Josephine Barstow, Jill Gomez, s; Yvonne Minton,
        ms; Robert Tear, t; Thomas Hemsley, Raimund Herincx, bar;
        Thomas Carey, bs-bar; ROHO; Colin Davis. Philips 6700 063 (2).
            ++Gr 4-74 p1888   ++NYT 2-17-74 pD33
            ++HF 5-74 p76   ++ON 5-74 p42
            ++LJ 11-75 p44   ++RR 4-74 p24
            ++MJ 5-74 p42   +St 3-74 p80
            +-NR 6-74 p8
    Little music for strings. cf ELGAR: Introduction and allegro,
        op. 47.

2346 The midsummer marriage: Four ritual dances. Concerto, orchestra.
　　　Joan Carlyle, Elizabeth Bainbridge, s; Alberto Remedios, t;
　　　Stafford Dean, bs; ROHO and Chorus, LSO; Colin Davis. Philips
　　　6580 093. (Reissues from 6703 027, SAL 3497)
　　　　　　+Audio 9-75 p69　　　　　　+NR 4-75 p2
　　　　　　+-Gr 1-75 p1358　　　　　　+RR 1-75 p36
　　　　　　++HF 4-75 p91　　　　　　　+SR 3-22-75 p36
2347 Quartets, strings, nos. 1-3. Lindsay Quartet. L'Oiseau-Lyre DSLO
　　　10.
　　　　　　++Gr 12-75 p1065　　　　　　++RR 12-75 p74
　　　　　　+HFN 12-75 p167
2348 Sonatas, piano, nos. 1-3. Paul Crossley, pno. Philips 6500 534.
　　　　　　++Audio 9-75 p69　　　　　　+SFC 2-16-75 p24
　　　　　　+Gr 11-74 p925　　　　　　　+SR 3-22-75 p36
　　　　　　++HF 4-75 p91　　　　　　　+St 6-75 p143
　　　　　　+NR 3-75 p10　　　　　　　++Te 9-75 p43
　　　　　　++RR 11-75 p84
　　　Sonata, piano, no. 1. cf HAMILTON: Sonata, piano, op. 13.
2349 Symphony, no. 3. Heather Harper, s; LSO; Colin Davis. Philips
　　　6500 662.
　　　　　　++Gr 1-75 p1358　　　　　　+RR 1-75 p36
　　　　　　+HF 5-74 p76　　　　　　　++St 6-74 p84
　　　　　　+NR 5-74 p2　　　　　　　　+St 6-75 p143
　　　　　　+NYT 2-17-74 pD33　　　　　+STL 2-9-75 p37
　　　　　　+ON 5-74 p42

TJEKNAVORIAN, Loris
2350 Armenian bagatelles. Requiem for the massacred. Howard Snell,
　　　tpt; London Percussion Virtuosi, LSO, Members; Loris Tjekna-
　　　vorian. Unicorn RHS 334.
　　　　　　++RR 12-75 p92
　　　Requiem for the massacred. cf Armenian bagatelles.
2351 Simorgh. Roudaki Hall Soloists; Loris Tjeknavorian. Unicorn RHS
　　　333.
　　　　　　+HFN 12-75 p167

TOCH, Ernest
2352 Big Ben variations, op. 62. Pinocchio overture. Symphony, no. 1,
　　　op. 72. VSO, Berlin Radio Symphony Orchestra, RAI Orchestra;
　　　Herbert Haefner, Ljubomir Romanski, Rudolf Kempe. Bruno Walter
　　　Society EMA 101.
　　　　　　+NR 8-75 p5
　　　Geographische Fugue für Sprechchor. cf Caprice RIKS LP 46.
　　　Impromptu, op. 90c. cf Command COMS 9006.
　　　Pinocchio overture. cf Big Ben variations, op. 62.
　　　Symphony, no. 1, op. 72. cf Big Ben variations, op. 62.

TOLDRA
　　　Maig; Anacreontica. cf HMV SLS 5012.

TOMASEK, Jan Vaclav
　　　Concerto, piano, op. 18, C major. cf KALLIWODA: Symphony, no. 1,
　　　op. 7, F minor.
　　　Eglogue, no. 2, op. 35. F major. cf BENDA: Sonata, piano, no. 9,
　　　A minor.

TOMASI, Henri
　　　Concerto, trumpet. cf HUMMEL: Concerto, trumpet, E flat major.

TOMKINS, Thomas
　　　Alman à four viols (Musica britannica, no. 33). cf Turnabout TV
　　　37079.

Barafostus' dream. cf BASF BAC 3075.
A hunting galliard. cf Angel SB 3816.
Pavan, A minor. cf Angel SB 3816.
A sad pavan for these distracted times. cf Angel SB 3816.
Songs: See, see the shepherd's queen; Weep no more, thou sorry
    boy. cf Harmonia Mundi 593.
Songs: Music divine; See, see the shepherd's queen; When I observe;
    Adieu, ye city-prisoning towers; When David heard; Fusca, in thy
    starry eyes; Weep no more, thou sorry boy. cf GIBBONS: Songs
    (Decca SXL 6639).
To the shady woods. cf Advent 5012.
TORELLI, Giuseppe
    Concerto, trumpet, D major. cf RCA CRL 2-7002.
    Suite, no. 7, a cinque, D major. cf HMV ASD 2938.
    Tu lo sai. cf Columbia D3M 33448.
TORRA
    Sardana (Viladrau). cf International Piano Library IPL 5005/6.
TORRE, Francisco de la
    Adormas te, Señor; La Spagna. cf L'Oiseau-Lyre 12BB 203/6.
    Alta danza a 3. cf DG Archive 2533 111.
    Dime, triste corazon. cf DG 2530 504.
    Pámpano verde, racimo albar. cf DG 2530 598.
TORROBA, Federico
    Aires de la Mancha. cf CBS 72526.
    Aires de la Mancha. cf CBS 72950.
    Burgalesa. cf RCA ARL 1-0485.
    Madroños. cf London STS 15224.
    Nocturno. cf Philips 6833 159.
    Nocturno. cf RCA ARL 1-0485.
    Piezas características: Albada. cf RCA ARL 1-0485.
    Rumor de Copla. cf RODRIGO: Concierto de Aranjuez, guitar and
        orchestra.
    Sonatina, A major: Allegretto. cf RCA ARL 1-0485.
    Suite castellana: Fandanguillo, Arada. cf RCA ARL 1-0485.
    Zapateado. cf RODRIGO: Concierto de Aranjuez, guitar and orchestra.
TORTELIER, Paul
    Pishnetto. cf HMV SEOM 19.
TOSELLI, Enrico
    Serenade. cf Pearl GEM 121/2.
    Serenade, violin, no. 1, op. 6. cf Discopaedia MB 1004.
TOSTI, Francesco
    Consolazione. cf Pearl SHE 511.
    Ridonami la calma. cf Rubini RS 300.
    Songs: La serenata; Luna d'estate; Malia; Non t'amo più. cf London
        OS 26391.
    Songs: Ideale; L'alba separa dalla luce ombra. cf RCA SER 5704/6.
    TOUREL, JENNIE, AND LEONARD BERNSTEIN AT CARNEGIE HALL. cf Colum-
        bia M 32231.
TOURNIER, Marcel
    Vers la source danse le bois. cf RCA LRL 1-5087.
TRABACI, Giovanni
    Canzona francesa cromatica VII. cf Telefunken AW 6-41890.
    Libero secondo: Capriccio sopra la, fa, sol, la. cf Pelca PSRK
        41013/6.
TRIANA (17th century Spain)
    Dinos, madre del donsel. cf DG 2530 504.

TRIBUTE TO JENNIE TOUREL. cf Odyssey Y2 32880.
TRIMBLE, Lester
    Songs: Love seeketh not itself to please; Telle me where is fancy
      bred. cf Duke University Press DWR 7306.
TROMBLY, Preston
    Kinetics III. cf Nonesuch HB 73028.
TROMBONCINO, Bartolomeo
    Ave Maria; Hor ch'el ciel e la terra; Ostinato vo seguire. cf
      L'Oiseau-Lyre 12BB 203/6.
    TRUMPET AND ORGAN WORKS. cf CRD CRD 1008.
TRYTHALL, Richard
    Coincidences, piano. cf DAVIDOVSKY: Chacona (1972), violin, violon-
      cello and piano.
TULOU, Jean Louis
    Grand solo, no. 13, op. 96, A minor. cf BOHM: Introduction and
      variations on "Nel cor piu", op. 4.
TUMA, František
    Partita, D minor. cf Works, selections (Supraphon 101 1444).
    Sinfonias, A major, B flat major. cf Works, selections (Supraphon
      101 1444).
    Sonata, 2 trombones, E minor. cf Works, selections (Supraphon 101
      1444).
    Suite, A major. cf Works, selections (Supraphon 101 1444).
2353 Works, selections: Partita, D minor. Sinfonias, A major, B flat
      major. Sonata, 2 trombones, E minor. Suite, A major. Zdeněk
      Pulec, Ladislav Odcházel, trom; PCO. Supraphon 101 1444.
            +Gr 7-75 p196   ++NR 9-75 p7
            +-HFN 9-75 p106   +RR 8-75 p44
TUNDER, Franz
    In Dich hab' ich gehoffet, Herr, chorale prelude. cf Saga 5374.
TURGES
    Enforce yourself as Goddës knight. cf BASF 25 22286-1.
TURINA, Joaquin
    Fandanguillo, op. 36. cf ALBENIZ: Suite española, no. 1: Granada;
      no. 3: Sevillanas.
    Fandanguillo, op. 36. cf CBS 72950.
    Farruca jota. cf HMV SLS 5012.
    Garrotin y soleares (Homenaje a Tárrega). cf Everest SDBR 3380.
    Garrotin y soleares. cf London STS 15224.
    La giralda. cf HMV SLS 5012.
    Homenaje a Tárrega: Garrotin, Soleares. cf ALBENIZ: Suite española,
      no. 1: Granada; no. 3: Sevillanas.
    Poema en forma de canciones, op. 19. cf FALLA: Mélodies.
    Ráfaga. cf London STS 15224.
    Ráfaga, op. 53. cf ALBENIZ: Suite española, no. 1: Granada; no. 3:
      Sevillanas.
    Sevillana, op. 29. cf ALBENIZ: Suite española, no. 1: Granada;
      no. 3: Sevillanas.
    Sonata, guitar, op. 61: Andante. cf ALBENIZ: Suite española, no.
      1: Granada; no. 3: Sevillanas.
    Songs: Saeta en forma de Salve a la Virgen de la Esperanza; El
      fantasma; Cantares. cf DG 2530 598.
TUROK, Paul
    Lyric variations, oboe and strings, op. 32. cf BRICCETTI: The
      fountain of youth overture.
    TWO HUNDRED YEARS OF AMERICAN MARCHES. cf Michigan University SM
      0002.

TYE, Christopher
2354 Euge Bone mass. Western Wynde mass. Cambridge, King's College
        Choir; David Willcocks. Argo ZRG 740.
                +Gr 12-73 p1238   +RR 11-73 p85
                ++MQ 1-75 p169   ++St 9-74 p132
                ++NR 7-74 p7
     In Nomine. cf L'Oiseau-Lyre 12BB 203/6.
     Western Wynde mass. cf Euge Bone mass.
UHL, Alfred
     Symphonie concertante, clarinet and orchestra. cf BRUCH: Concerto,
        clarinet, viola and orchestra, op. 88.
URREDA (15th century Spain)
     Songs: De vos i de mi; Muy triste será mi vida; Nunca fue pena
        mayor. cf Telefunken SAWT 9620/1.
VAILLANT, Jean
     Par maintes foys. cf 1750 Arch S 1753.
     Par maintes fois. cf Telefunken TK 11569/1-2.
     Par maintes foys. cf Vanguard VSD 71179.
VALDERRABANO, Enriquez de
     De dónde venís, amore. cf DG 2530 504.
     Sonatas, D major. cf RCA ARL 1-0485.
VALENTINO
     Sonata, trumpet, D minor. cf ALBINONI: Concertos, trumpet, F
        major, D minor.
VAN BIENE
     Broken melody. cf Discopaedia MB 1008.
VAN HEMEL, Oscar
     Concerto, violin, no. 2. cf DRESDEN: Concerto, oboe.
VAN VACTOR, David
     Fantasia, chaconne and allegro. cf BENTZON: Pezzi sinfonici, op.
        109.
VANHAL, Jan
     Fugue, G major. cf Hungaroton SLPX 11601/2.
     Sonata, clarinet, B flat major. cf BERNSTEIN: Sonata, clarinet
        and piano.
VARESE, Edgard
     Density 21.5. cf Nonesuch HB 73028.
VARGA, Ruben
     Prelude and caprices (4). cf PAGANINI: Caprices, op. 1, no.
        17, E major; no. 24, A minor.
     Sonata, violin, G minor. cf PAGANINI: Caprices, op. 1, no. 17,
        E major; no. 24, A minor.
VASQUEZ, Juan
     En la fuente del rosal. cf DG 2530 504.
     Lindos ojos aveys, señora. cf L'Oiseau-Lyre 12BB 203/6.
     Ojos morenos; Não me firays. cf Advent 5012.
     Se me llaman. cf Telefunken SAWT 9620/1.
     Vos me matastes (arr. Fuenllano). cf DG 2530 504.
VAUGHAN THOMAS, David
     Saith O Ganeuon. cf Argo ZRG 769.
VAUGHAN WILLIAMS, Ralph
     Along the field. cf Musical Heritage Society MHS 1976.
     Concerto, oboe. cf Symphony, no. 6, E minor.
     Concerto, tuba, F minor. cf Symphony, no. 3.
2355 Fantasia on a theme by Thomas Tallis. Fantasia on "Greensleeves".
        Folk dance suite (arr. Jacobs). Old King Cole: Ballet suite.

The wasps: Overture. LPO; Adrian Boult. Pye GSGC 15019.
+RR 12-75 p69

Fantasia on "Greensleeves". cf Fantasia on a theme by Thomas
Tallis.

Fantasia on "Greensleeves". cf DELIUS: Aquarelles.

Folk dance suite. cf Fantasia on a theme by Thomas Tallis.

2356 Folksong arrangements. London Madrigal Singers; Christopher Bishop.
Seraphim S 60249.
+HF 11-75 p126

Folk song suite. cf Philips 6747 177.

2357 The house of life. Songs of travel. Anthony Rolfe Johnson, t;
David Willison, pno. Polydor 2460 236.
+-Gr 2-75 p1541   /-RR 2-75 p60

The lark ascending. cf DELIUS: Aquarelles.

Mass, G minor: Kyrie. cf HMV Tape (c) TC MCS 14.

Mystical songs: Antiphon. cf Argo ZRG 789.

Mystical songs: Let all the world. cf HMV Tape (c) TC MCS 13.

O taste and see. cf HMV HQS 1350.

Old King Cole. cf Songs (Pearl 127).

Old King Cole: Ballet suite. cf Fantasia on a theme by Thomas
Tallis.

Preludes on Welsh hymn tunes, no. 2: Rhosymedre, O taste and see.
cf Gemini GM 2022.

Rhosymedre. cf Decca SDD 463.

Rhosymedre. cf Wealden WS 139.

2358 Sir John in love. Felicity Palmer, Wendy Eathorne, s; Elizabeth
Bainbridge, Helen Watts, ms; Robert Tear, Gerald English, t;
Raimund Herincx, John Noble, Rowland Jones, bar; Robert Lloyd,
bs; John Alldis Choir; NPhO; Meredith Davies. Angel SCLX 3822
(3). (also HMV SLS 980)
+Gr 9-75 p513   +NYT 8-10-75 pD14
+-HF 9-75 p76   +ON 7-75 p30
++HFN 9-75 p106   ++RR 9-75 p16
+MT 12-75 p1072   ++St 9-75 p83
++NR 7-75 p9

Six studies in English folk song. cf Crystal S 331.

2359 Songs: On Wenlock edge: On Wenlock edge; From far, from eve and
morning; Is my team ploughing; Oh, when I was in love with you;
Bredon Hill; Clun. Old King Cole. Songs of travel: The roadside
fire. The wasps: Overture. Gervase Elwes, t; Frederick Kiddie,
pno; London String Quartet, Aeolian Orchestra; Ralph Vaughan
Williams. Pearl GEM 127. (Reissues from Vocalion AO 249, 247/8,
Columbia 7146/50, 7365)
+-Gr 8-75 p331   +-RR 8-75 p44

2360 Songs: Along the field. Blake songs (1). Merciless beauty. Lois
Winter, s; John Langstaff, bar; Marvin Morgenstern, Hiroko Yaj-
ima Rhodes, vln; John Goberman, vlc; Ronald Roseman, ob. Peerless PRCM 200.
+Gr 4-75 p1856   +-HFN 5-75 p141

Songs of travel. cf The house of life.

Songs of travel: The roadside fire. cf Songs (Pearl GEM 127).

Studies in English folksong (6). cf BERNSTEIN: Sonata, clarinet
and piano.

Suite de ballet. cf Armstrong 721-4.

2361 Symphony, no. 2. Hallé Orchestra; John Barbirolli. HMV SXLP 30180.
(Reissue from ASD 2360) (also Angel S 36478)

          +Gr 5-75 p1979   ++RR 5-75 p46
          ++HFN 6-75 p96
2362 Symphony, no. 2. LSO; André Previn. RCA LSC 3282. Tape (r) ERPA
     3282. (also RCA SB 6860. Tape (c) RK 6860)
          +ARG 10-72 p694   +LJ 2-73 p43
          ++Gr 8-72 p342   +NR 9-72 p3
          +Gr 6-75 p111 tape   ++RR 8-72 p66
          /HF 11-72 p104   +-RR 6-75 p93 tape
          +HF 3-75 p108 tape   ++St 11-72 p128
          +HFN 8-72 p1468
2363 Symphony, no. 3. Concerto, tuba, F minor. Heather Harper, s;
     John Fletcher, tuba; LSO; André Previn. RCA LSC 3281. Tape (r)
     ERPA 3281C. (also RCA SB 6861. Tape (c) RK 6861)
          +ARG 10-72 p690   +HFN 10-72 p1918
          +Gr 9-72 p502   ++NR 8-72 p2
          +-Gr 6-75 p111 tape   ++RR 9-72 p72
          /HF 11-72 p104   +-RR 6-75 p93
          ++HF 3-75 p108 tape   ++St 9-72 p103
2364 Symphony, no. 6, E minor. Concerto, oboe. John Williams, ob;
     Bournemouth Symphony Orchestra; Paavo Berglund. HMV (Q) ASD
     3127.
          +-Gr 10-75 p634   +RR 10-75 p61
          +-HFN 10-75 p150
2365 Symphony, no. 7. LPO; Adrian Boult. HMV ASD 2631. Tape (c)
     TC ASD 2631.
          ++Gr 8-75 p376 tape   +RR 9-75 p79 tape
          ++HFN 7-75 p91 tape
     Symphony, no. 8, D minor. cf BAX: The garden of Fand.
     Toccata marziale. cf Philips 6747 177.
     The wasps: Overture. cf Fantasia on a theme by Thomas Tallis.
     The wasps: Overture. cf Songs (Pearl 127).
VAUTOR, Thomas
     Mother, I will have a husband. cf Harmonia Mundi 204.
     Shepherds and nymphs. cf Harmonia Mundi 593.
     Weep, weep mine eyes. cf Turnabout TV 37079.
VECCHI, Orazio
2366 L'Amfiparnaso. Western Vocal Wind. Nonesuch H 71286. (Q) HQ 1286.
          +HF 4-74 p108   +ON 6-74 p42 Quad
          +HF 2-75 p110 Quad   ++St 2-74 p84
          ++NR 4-74 p10
     Gioite tutti. cf University of Illinois, unnumbered.
     L'Humor svegghiato; Applauso. cf Advent 5012.
     Saltarello. cf Kendor KE 22174.
VECSEY, Ferenc von
     Foglio d'Album. cf Discopaedia MB 1002.
     Foglio d'Album. cf Rococo 2072.
VEDEL
     Na rekah Vavilonskih. cf Harmonia Mundi HMU 133.
VEDEL, Artemii
     How long, O Lord, how long. cf HMV Melodiya ASD 3102.
VENTADORN, Bernart de
     Be M'an perdut. cf Argo ZRG 765.
VERACINI, Francesco
     Sonata, recorder and continuo, A minor. cf Telefunken SAWT 9589.
     Sonata, recorder and continuo, A major, G major. cf Telefunken
     SMA 25121 T/1-3.

    Sonata prima, F major. cf BIGAGLIA: Sonata, recorder, bassoon
        and harpsichord, A minor.
VERDELOT, Philippe
    Madonna il tuo bel viso. cf Telefunken TK 11569/1-2.
    Madonna, qual certezza. cf L'Oiseau-Lyre 12BB 203/6.
VERDI, Giuseppe
2367 Aida. Monteserrat Caballé, Esther Casas, s; Fiorenza Cossotto, ms;
        Placido Domingo, Nicola Martinucci, t; Piero Cappuccilli, bar;
        Nicolai Ghiaurov, Luigi Roni, bs; NPhO; ROHO Chorus; Riccardo
        Muti. Angel SCLX 3815 (3). (also HMV SLS 977. Tape (c) TC
        SLS 977)
                +Gr 2-75 p1548   +-ON 1-18-75 p32
                +Gr 10-75 p721 tape   +-RR 2-75 p23
                +-HF 2-75 p102   +RR 9-75 p79 tape
                +HFN 9-75 p110 tape   ++St 1-75 p118
                +NR 1-75 p11   +-STL 2-9-75 p37
2368 Aida. Dusolina Giannini, s; Irene Minghini-Cattaneo, ms; Aureliano
        Pertile, Giuseppe Nessi, t; Giovanni Inghilleri, bar; Luigi
        Manfrini, Guglielmo Masini, bs; La Scala Orchestra and Chorus;
        Carlo Sabajno. Discophilia KS 7/9 (3). (Reissue from HMV/
        Victor 78 rpm originals)
                +-HF 2-75 p103   +ON 3-22-75 p33
    Aida: L'abboritta rivale...Gia i sacerdoti adunanso. cf London OS
        26315.
    Aida: Celeste Aida. cf BASF BC 21723.
    Aida: Celeste Aida. cf Columbia D3M 33448.
    Aida: Celeste Aida. cf Decca SXL 6649.
    Aida: Gloria all'egitto. cf Works, selections (DG 2530 549).
    Aida: Grand chorus, Act 3. cf Decca ECS 2159.
    Aida: Grand march, Act 2. cf Decca PFS 4323.
    Aida: O patria mia...Ciel, mio padre...Pur ti reveggo...Ma dimmi,
        per qual via. cf Arias (Discophilia KGR 5).
    Aida: O terra, addio. cf Ember GVC 41.
    Aida: Ritorna vincitor; O patria mia. cf PROKOFIEV: War and peace,
        op. 91: Duets, Scenes 1 and 3.
    Aida: Se quel guerrier io fossi...Celeste Aida. cf Saga 7206.
    Aida: Se quel guerrier io fossi...Celeste Aida: Tu, Amonasro, Io
        son disonorato. cf RCA SER 5704/6.
2369 Arias: La battaglia di Legnano: Voi lo deceste...Quante volte come
        un dono. I Lombardi: Se vano è il pregare. Nabucco: Ben io
        t'invenni...Anch'io dischiuso. Otello: Willow song; Ave Maria.
        La traviata: Addio del passato. I vespri siciliani: Arrigo;
        Ah, parli, a un core; Mercé dilette amiche. Renata Scotto, s;
        Elizabeth Bainbridge, ms; William Elvin, bar; LPO; Gianandrea
        Gavazzeni. Columbia M 33516.
                +-HF 10-75 p87   +-ON 12-6-75 p52
                +-NR 12-75 p11   +St 9-75 p116
2370 Arias: Aida: O patria mia...Ciel, mio padre...Pur ti reveggo...Ma
        dimmi, per qual via. Attila: Te sol quest'anima. Un ballo in
        maschera: Ma dall'arido stelo; Morrò, ma prima in grazia. I
        Lombardi: Qual volutta trascorrere. Otello: Willow song; Ave
        Maria. Elisabeth Rethberg, s; Giacomo Lauri-Volpi, Beniamino
        Gigli, t; Giuseppe de Luca, bar; Ezio Pinza, bs; Orchestral
        accompaniment. Discophilia KGR 5. (Recorded 1927-30)
                +-HF 2-75 p107   +NR 7-75 p9
2371 Arias: Aida: Se quel guerrier fossi...Celeste Aida. Alzira: Miser-
        andi avanzi...Irne lungi. Aroldo: Sotto il sol di Siria ardent.

Attila: Qual notte...Qui, qui sostiamo...Ella in poter del
barbaro...Cara patria; Qui del convegno è il loco... Che non
avrebbe il misero. Un ballo in maschera: Di tu se fedele;
Forse la soglia attinse...Ma se m'e forza perderti. La battaglia
di Legnano: O magnanima...La pia materna mano. Il Corsaro:
Eccomi prigioniero...Al mio stanco cadavere. Don Carlo: Fon-
tainebleau...Io la vidi e al suo sorriso. Ernani: Mercè,
diletti amici...Come rugiada al cespite. La forza del destino:
La vita è inferno...O tu che in seno agli angeli. Falstaff:
Dal labbro il canto estasiato vola. Un giorno di regno: Piet-
oso al lungo pianto. Giovanna d'Arco: Nel suo bel volta...
Sotto una guercia. I Lombardi: La mia Letizia infondere...Sien
miei sensi...Come poteva un angelo. Luisa Miller: Oh, Fede
negar potessi...Quando le sere al placido. Macbeth: O figli...
Ah, la paterna mano. I Masnadieri: Quando io leggo in Plutarcho.
...O mio castel paterno...Ecco un foglio a te diretto...Fiere
umane, umane fieri...Senti O moor...Nell'argilla maledetta; Come
splendido...Di ladroni attorniato. Oberto, Conte di San Bonifacio: Ciel che feci...Ciel pietosa. Otello: Dio mi potevi
scagliar; Niun mi tema. Rigoletto: Questa o quella; Ella mi
fu rapita...Parmi veder le lagrime; La donna è mobile. Simone
Boccanegra: O inferno...Sento avvampar nel l'amina. La traviata:
Lunge a lei...De miei bollenti spiriti. Il trovatore: Il presagio funesto...Ah si, ben mio...Di quella pira. I vespri
siciliani: E' di monforte il cenno...Giorno di pianto. I due
Foscari: Notte, perputua notte...Non maledirmi, O prode. Carlo
Bergonzi, t; NPhO; Nello Santi, Lamberto Gardelli. Philips
6747 193 (3).
    +Gr 9-75 p520   +RR 10-75 p35
    +HFN 9-75 p107

Attila: Arias and scenes. cf London OS 26306.

Attila: Te sol quest'anima. cf Arias (Discophilia KGR 5).

2372 Un ballo in maschera. Martina Arroyo, Reri Grist, s; Fiorenza
Cossotto, ms; Placido Domingo, Kenneth Collins, t; Giorgio
Giorgetti, Piero Cappuccilli, bar; Gwynne Howell, bs-bar;
Richard Van Allan, bs; ROHO Chorus; NPhO; Riccardo Muti. HMV
SLS 984 (3). (also Angel SX 3762)
    ++Gr 12-75 p1094   ++RR 12-75 p32

2373 Un ballo in maschera. Leontyne Price, Reri Grist, s; Shirley
Verrett, ms; Carlo Bergonzi, Fernando Iacopucci, t; Robert
Merrill, Mario Basiola, bar; Ezio Flagello, Ferruccio Mazzoli,
bs; Renata Cortiglioni Children's Chorus; RCA Italiana Opera
Chorus and Orchestra; Erich Leinsdorf. RCA SER 5710/2 (3).
(Reissue from SER 5556/8) (also RCA LSC 6179)
    +-Gr 2-75 p1551   +RR 2-75 p19

2374 Un ballo in maschera, excerpts. Paula Takács, Margit László, s;
Magda Tiszay, ms; Robert Ilosfalvy, t; György Radnai, bar;
András Faragó, Sándor Mészáros, bs; Hungarian Radio and Television Chorus; Hungarian State Opera Orchestra; Lamberto Gardelli. Hungaroton SLPX 11665.
    +NR 6-75 p11

Un ballo in maschera: Alzati...Eri tu. cf RCA ARL 1-0851.

Un ballo in maschera: Arias. cf BELLINI: La sonnambula: Arias.

Un ballo in maschera: Eri tu. cf Columbia/Melodiya M 33120.

Un ballo in maschera: Ma dall'arido stelo; Morrò, ma prima in
grazia. cf Arias (Discophilia KGR 5)

Un ballo in maschera: Morrò, ma prima in grazia. cf CILEA:
   Adriana Lecouvreur: Io sono l'umile ancella.
La battaglia di Legnano: Voi lo deceste...Quante volte come un
   dono. cf Arias (Columbia M 33516).
2375 Don Carlo: Ballo della regina. Otello: Ballet music, Act 3. I
   vespri siciliani: Le quattro stagioni. CO; Lorin Maazel. Decca
   SXL 6727.
         ++Gr 8-75 p331   ++RR 8-75 p45
Don Carlo: Ascolta; Le porte dell'asil s'apron. cf Decca DPA 517/8.
Don Carlo: Il ballo della regina. cf Works, selections (Philips
   6747 093).
Don Carlo: Canzone de veli. cf Rococo 5368.
Don Carlo: Dio che nell'alma. cf Discophilia KGC 2.
Don Carlo: Doman dal ciel...Dio che nell'alma infondere. cf Saga
   7206.
Don Carlo: Elle ne m'aime pas. cf Decca SXL 6637.
Don Carlo: Io l'ho perduta...Qual pallor. cf RCA SER 5704/6.
Don Carlo: O don fatale. cf Decca GOSC 666/8.
Don Carlo: Spuntato ecco il di. cf Works, selections (DG 2530 549).
Ernani: Ernani involami. cf Discophilia KGS 3.
Ernani: Ernani involami. cf Pearl GEM 121/2.
Ernani: Si ridesti il Leon di Castiglia. cf Works, selections
   (DG 2530 549).
Ernani: Surta è la notte...Ernani, Ernani involami. cf DONIZETTI:
   Linda di Chamonix: Ah, tardai troppo...O luce di quest'anima.
2376 Falstaff. Elisabeth Schwarzkopf, Anna Moffo, s; Nan Merriman,
   Fedora Barbieri, ms; Luigi Alva, t; Tito Gobbi, Rolando Panerai,
   bar; PhO and Chorus; Herbert von Karajan. Angel CL 3552.
         +St 4-75 p69
Falstaff: Ehi, Paggio...L'onore. cf Decca GOSC 666/8.
Falstaff: L'onore. cf Ember GVC 37.
2377 La forza del destino. Zinka Milanov, Luisa Gioia, s; Rosalind
   Elias, ms; Giuseppe di Stefano, Angelo Mercuriali, t; Leonard
   Warren, Virgilio Carbonari, Sergio Liviabella, bar; Dino Manto-
   vani, Paolo Washington, bs; Rome, Santa Cecilia Orchestra and
   Chorus; Fernando Previtali. Decca GOS 660/2 (3). (Reissue
   from RCA SER 4516/9)
         ++Gr 5-75 p2019   +RR 5-75 p25
         +HFN 5-75 p141
La forza del destino: O tu che in seno. cf Supraphon 012 0789.
La forza del destino: Pace, pace, mio Dio. cf Decca GOSC 666/8.
La forza del destino: Solenne in quest'ora. cf Decca DPA 517/8.
La forza del destino: Solenne in quest'ora. cf RCA SER 5704/6.
2378 Un Giorno di Regno (King for a day). Fiorenza Cossotto, ms; Jessye
   Norman, s; José Carreras, Ricardo Cassinelli, t; Ingvar Wixell,
   Vincenzo Sardinero, Wladimiro Ganzarolli, William Elvin, bar;
   Ambrosian Singers; RPO; Lamberto Gardelli. Philips 6703 055 (3).
         +-Gr 9-74 p587   +Op 11-74 p988
         +-HF 1-75 p73   +RR 9-74 p21
         +MJ 1-75 p48   +-SFC 11-10-74 p27
         +NR 12-74 p10   +SR 11-30-74 p41
         +ON 11-74 p64   +St 12-74 p95
Giovanna d'Arco: Overture. cf Erato STU 70880.
I Lombardi: Jerusalem. cf Works, selections (Philips 6747 093).
I Lombardi: Jerusalem, Jerusalem; O Signore. cf Works, selections
   (DG 2530 549).

I Lombardi: La mia Letizia infondere...Sien miei sensi...Come
    poteva un angelo. cf Arias (Philips 6747 193).
I Lombardi: Qual prodigo, son fu sogno. cf RCA ARL 1-0702.
I Lombardi: Qual volutta trascorrere. cf Arias (Discophilia KGR 5).
I Lombardi: Qual volutta trascorrere. cf Discophilia KGC 2.
I Lombardi: Qui posa il fianco. cf Decca SXL 6719.
I Lombardi: Se vano è il pregare. cf Arias (Columbia M 33516).
2379 Luisa Miller. Anna Moffo, s; Shirley Verrett, Gabriella Carturan,
    ms; Carlo Bergonzi, t; Cornell MacNeil, bar; Giorgio Tozzi,
    Ezio Flagello, bs; RCA Italiana Opera Orchestra and Chorus;
    Fausto Cleva. RCA SER 5713/5 (3). (Reissue from SER 5534/6)
    (also RCA LSC 6168)
        +-Gr 3-75 p1701   +RR 3-75 p19
Luisa Miller: Ferma ed ascolta...Sacra la scalta e d'un consorte.
    cf RCA ARL 1-0851.
Luisa Miller: Oh, Fede negar potessi...Quando le sere al placido.
    cf Arias (Philips 6747 193).
Luisa Miller: Quando le sere al placido. cf BASF BC 21723.
Luisa Miller: Quando le sere al placido. cf Columbia D3M 33448.
Macbeth: Ah la paterna mano. cf Discophilia KGC 2.
Macbeth: Ballet music. cf Works, selections (Philips 6747 093).
Macbeth: O figli...Ah, la paterna mano. cf Arias (Philips 6747 193).
Macbeth: Patria oppressa. cf Works, selections (DG 2530 549).
Macbeth: Perfidi...Pietà, rispetto, amore...Mal per me. cf RCA
    ARL 1-0851.
2380 I Masnadieri (The brigands). Montserrat Caballé, s; Carlo Bergonzi,
    John Sandor, t; Piero Cappuccilli, William Elvin, bar; Ruggero
    Raimondi, Maurizio Mazzieri, bs; Ambrosian Singers; NPhO;
    Lamberto Gardelli. Philips 6703 064 (3).
        +Gr 9-75 p514   +RR 9-75 p24
        +HF 12-75 p80   +SFC 8-31-75 p20
        +HFN 9-75 p106   +SR 10-4-75 p50
        +NR 10-75 p7   ++St 11-75 p87
        ++OC 12-75 p49   +STL 11-2-75 p38
        +ON 10-75 p56
I Masnadieri: Arias and scenes. cf London OS 26306.
I Masnadieri: Come un bacio. cf Rubini RS 300.
I Masnadieri: Quando io leggo in Plutarcho...O mio castel paterno...
    Ecco un foglio a te diretto...Fiere umane, umane fieri...Senti
    O moor...Nell'argilla maledetta; Come splendido...Di ladroni
    attorniato. cf Arias (Philips 6747 193).
Motets: Ave Maria; Laudi alla Vergine Maria; Pater Noster. cf
    BRAHMS: Motets (CRD CRD 1009).
Nabucco: Ben io t'invenni...Anch'io dischiuso. cf Arias (Columbia
    M 33516).
Nabucco: Gli arredi festivi; Va pensiero. cf Works, selections
    (DG 2530 549).
Nabucco: O di qual onta. cf Rubini RS 300.
Nabucco: Speed your journey. cf Decca ECS 2159.
Nabucco: Va pensiero, Act 3. cf Decca PFS 4323.
Oberto, Conte di San Bonifacio: Ciel che feci...Ciel pietosa. cf
    Arias (Philips 6747 193).
2381 Otello. Mirella Freni, s; Stefania Malagu, ms; Jon Vickers, Aldo
    Bottion, Michel Senechal, t; Peter Glossop, Hans Helm, bar;
    José van Dam, Mario Machi, bs; BPhO; Berlin Deutsche Oper Chorus;
    Herbert von Karajan. Angel SCLX 3809 (3). (also HMV SLS 975.

Tape (c) TC SLS 975)
  +Gr 10-74 p755   +-ON 12-28-74 p48
  +-Gr 1-75 p1402 tape   +-RR 10-74 p28
  +-HF 1-75 p90   +RR 3-75 p73 tape
  +NR 12-74 p11   +-St 12-74 p132
  -NYT 11-10-74 pD1

2382 Otello. Dragica Martinis, s; Sieglinde Wagner, ms; Ramon Vinay,
 Anton Dermota, August Jaresch, t; Paul Schöffler, bar; VSOO
 Chorus; VPO; Wilhelm Furtwängler. Bruno Walter Society IGI
 342 (3).
  +NR 7-75 p11

2383 Otello. Renata Tebaldi, s; Luisa Ribacchi, ms; Mario del Monaco,
 Piero de Palma, Angelo Mercuriali, t; Aldo Protti, bar; Fernando
 Corena, Pier Luigi Latinucci, Dario Caselli, bs; Rome, Santa
 Cecilia Orchestra and Chorus; Alberto Erede. Decca ECS 732/4
 (4). (Reissue from LXT 5009/11)
  +-Gr 2-75 p1548   +-RR 5-75 p25
  +HFN 6-75 p97

Otello: Ave Maria. cf Discophilia KGP 4.
Otello: Ballet music. cf Works, selections (Philips 6747 093).
Otello: Ballet music, Act 3. cf Don Carlo: Ballo della regina.
Otello: Credo in un Dio crudel. cf Ember GVC 37.
Otello: Dio, mi potevi scagliar; Niun mi tema. cf Arias (Philips
 6747 193).
Otello: Era più calmo...Piangea cantando...Ave Maria piena. cf
 HMV SLS 5012.
Otello: Fuoco di gioia. cf Works, selections (DG 2530 549).
Otello: Già nella notte densa. cf PUCCINI: Arias (BASF KBC 22007).
Otello: Non ti crucciar...Credo. cf Decca GOSC 666/8.
Otello: O mostruosa colpa...Si pel ciel marmoreo. cf Saga 7206.
Otello: Si pel ciel marmoreo giuro. cf RCA SER 5704/6.
Otello: Si pel ciel marmoreo giuro. cf Supraphon 012 0789.
Otello: Willow song; Ave Maria. cf Arias (Columbia M 33516).
Otello: Willow song; Ave Maria. cf Arias (Discophilia KGR 5).
Pater Noster. cf BRAHMS: Warum ist das Licht gegeben dem Müseligen,
 op. 74, no. 1.
Pezzi sacri: Ave Maria; Laudi alla Vergine Maria. cf BRAHMS: Warum
 ist das Licht gegeben dem Müseligen, op. 74, no. 1.
Pezzi sacri, no. 4: Te Deum. cf BOITO: Mefistofele: Prologue.
Sacred pieces: Te Deum. cf Requiem.

2384 Quartet, strings, E minor. WIKMANSON: Quartet, strings, op. 1,
 no. 2, E minor. Saulesco Quartet. Caprice RIKSLP 65.
  +Gr 4-75 p1829   +RR 12-74 p51
Quartet, strings, E minor. cf MENDELSSOHN: Quintet, strings, no.
 2, op. 87, B flat major.
Quartet, strings, E minor. cf ROSSINI: Sonata, strings, no. 1,
 G major.

2385 Requiem. Sacred pieces: Te Deum. Herva Nelli, s; Claramae Turner,
 con; Eugene Conley, t; Nicola Moscona, bs; New England Conservatory Chorus, Westminster Cathedral Choir; BSO, NYP; Guido
 Cantelli. Bruno Walter Society IGI 340.
  -NR 9-75 p10

2386 Requiem. Heather Harper, s; Josephine Veasey, ms; Carlo Bini, t;
 Hans Sotin, bs; London Philharmonic Chorus; RPO; John Alldis,
 Carlos Paita. Decca OPFS 5-6.
  +-Gr 12-75 p1090   /RR 12-75 p94
  +-HFN 12-75 p167

Requiem: Ingemisco tamquam reus. cf Columbia D3M 33448.
Requiem: Sanctus. cf HMV Tape (c) TC MCS 14.
2387 Requiem mass, in memory of Manzoni. Elisabeth Schwarzkopf, s;
    Christa Ludwig, ms; Nicolai Gedda, t; Nicolai Ghiaurov, bs;
    PhO and Chorus; Carlo Maria Giulini. Angel SBL 3649.
        +St 4-75 p69
2388 Rigoletto. Joan Sutherland, Luisa Valle, Maria Fiori, s; Stefania
    Malagù, Anna di Stasio, ms; Angelo Mercuriali, Renato Cioni, t;
    Cornell MacNeil, Giuseppe Morresi, bar; Cesare Siepi, Fernando
    Corena, Giulio Corti, bs; Rome, Santa Cecilia Orchestra and
    Chorus; Nino Sanzogno. London 1332. (also Decca GOS 655/7.
    Reissue from SET 244/6)
        +Gr 9-75 p519   -RR 8-75 p27
        +-HFN 9-75 p106
2389 Rigoletto. Joan Sutherland, s; Huguette Tourangeau, ms; Luciano
    Pavarotti, t; Sherrill Milnes, bar; Martti Talvela, Clifford
    Grant, bs; The Ambrosian Opera Chorus; LSO; Richard Bonynge.
    London OSA 13105 (3).   Tape (r) R 490225. (also Decca SET 542/4.
    Tape (c) K2A3)
        ++Gr 5-73 p2090   +-ON 6-73 p34
        +Gr 5-75 p2031 tape   +-Op 7-73 p620
        +-HF 6-73 p108   +-RR 5-73 p36
        +HFN 5-73 p990   +-RR 5-75 p76 tape
        ++HFN 6-75 p109 tape   ++SFC 4-1-73 p30
        +-LJ 2-75 p38 tape   ++SR 4-73 p72
        +NR 5-73 p9   ++St 6-73 p81
        /NYT 3-4-73 pD28
2390 Rigoletto. Erna Berger, Joyce White, s; Nan Merriman, Mary Kreste,
    ms; Jan Peerce, Nathaniel Sprinzena, t; Leonard Warren, Arthur
    Newman, bar; Italo Tajo, Richard Wentworth, Paul Ukena, bs;
    Robert Shaw Chorale; RCA Symphony Orchestra; Renato Cellini.
    RCA AVM 2-0698 (2). (Reissue from LM 6101, 6021)
        +-HF 4-75 p92   +NR 5-75 p10
  Rigoletto: Arias. cf BELLINI: La sonnambula: Arias.
  Rigoletto: Caro nome. cf BELLINI: Capuletti ed i Montecchi: O,
    qaunte volte, O quante.
  Rigoletto: Caro nome. cf CILEA: Adriana Lecouvreur: Io sono l'umile
    ancella.
  Rigoletto: Caro nome. cf Pearl GEM 121/2.
  Rigoletto: Caro nome; Bella figlia dell'amore. cf Saga 7029.
  Rigoletto: La donna è mobile. cf Decca SXL 6649.
  Rigoletto: La donna è mobile. cf Supraphon 10924.
  Rigoletto: Pari siamo. cf Ember GVC 37.
  Rigoletto: Parmi veder le lagrime. cf Columbia D3M 33448.
  Rigoletto: Questa o quella. cf Saga 7206.
  Rigoletto: Questa o quella; Ella mi fu rapita...Parmi veder le
    lagrime; La donna è mobile. cf Arias (Philips 6747 193).
  Rigoletto: Questa o quella; Ella mi fu rapita...Parmi veder la
    lagrime; La donna è mobile; Un dì, se ben rammentomi; Bella
    figlia dell'amore. cf RCA SER 5704/6.
  Rigoletto: Rede wir sind allein...Wein weine. cf Discophilia KGB 2.
  Rigoletto: Un dì sen ben rammentomi. cf Decca SXL 6719.
2391 Simon Boccanegra. Katia Ricciarelli, Ornella Jachetti, s; Placido
    Domingo, Piero de Palma, t; Piero Cappuccilli, Gian Piero Mas-
    tromei, bar; Ruggero Raimondi, Maurizio Mazzieri, bs; RCA Orches-
    tra and Chorus; Gianandrea Gavazzeni. RCA SER 5696/8 (3). (also

RCA ARL 3-0564)
   ++Gr 2-74 p1599   -NYT 10-6-74 pD26
   +-HF 12-74 p120   +-ON 1-18-75 p32
   +HFN 3-74 p116   +-Op 4-74 p323
   +-MJ 12-74 p44   +RR 2-74 p12
   +-NR 11-74 p9   +-St 11-74 p138

Simon Boccanegra: Come in quest'ora bruna. cf Decca GOSC 666/8.
Simon Boccanegra: O inferno...Sento avvampar nel l'anima. cf
   Arias (Philips 6747 193).
2392 La traviata. Mirella Freni, s; Hania Kovicz, ms; Franco Bonisolli,
   Peter Bindszus, t; Sesto Bruscantini, Rudolf Jedlicka, bar;
   Berlin State Opera Orchestra and Chorus; Lamberto Gardelli.
   BASF KBL 21644 (3).
      +HF 6-75 p105   -ON 1-18-75 p32
      +-NR 3-75 p7   +-St 6-75 p107
2393 La traviata: Libiamo ne' lieti calici; Un dì felice...Ah, fors'e
   lui; Lunge da lei; Pura siccome un angelo; Dammi tu forza; In-
   vitato a que seguirmi; Teneste la promessa; Parigi, O cara...
   Ah, Violetta. Renata Tebaldi, s; Rina Cavalleri, Angela Ver-
   celli, ms; Gianni Poggi, Piero di Palma, t; Aldo Protti, bar;
   Antonio Sacchetti, Dario Caselli, Ivan Sardi, bs; Rome, Santa
   Cecilia Orchestra and Chorus; Francesco Molinari-Pradelli.
   Decca ECS 777. (Reissue from LXT 2992/4)
      -Gr 9-75 p519   -RR 8-75 p27
      /HFN 8-75 p99
La traviata: Arias. cf London OS 26381.
La traviata: Addio del passato. cf Arias (Columbia M 33516).
La traviata: Ah fors'e liu (2 versions); Sempre libera. cf Disco-
   philia KGS 3.
La traviata: Ah fors'e liu...Sempre libera. cf Saga 7029.
La traviata: Addio del passato. cf Pearl GEM 121/2.
La traviata: Di provenza il mar. cf Columbia /Melodiya M 33120.
La traviata: Di provenza. cf Decca GOSC 666/8.
La traviata: Fraulein Valery...Ach ich verstehe...Saget der Jung-
   frau...Ich sterbe. cf Discophilia KGB 2.
La traviata: Lunge a lei...De miei bollenti spiriti. cf Arias
   (Philips 6747 193).
La traviata: Parigi o cara. cf PUCCINI: Arias (BASF KBC 22007).
La traviata: Prelude, Act 1. cf Philips 6747 204.
La traviata: Prelude, Act 3. cf HMV SLS 5019.
La traviata: Un dì felice. cf Decca DPA 517/8.
2394 Il trovatore. Zinka Milanov, s; Fedora Barbieri, Margaret Roggero,
   ms; Jussi Björling, Nathaniel Sprinzena, t; Leonard Warren,
   George Cehanovsky, bar; Nicola Moscona, bs; Robert Shaw Chorale;
   RCA Orchestra; Renato Cellini. RCA AVM 2-0699 (2). (Reissue
   from LM 6008)
      +-HF 4-75 p92   +NR 5-75 p10
2395 Il trovatore. Caterina Mancini, Graziella Sciutti, s; Miriam Pira-
   zzini, ms; Giacomo Lauri-Volpi, t; Carlo Tagliabue, bar; Alfredo
   Colella, bs; Italian Radio and Television Orchestra and Chorus;
   Fernando Previtali. Turnabout THS 65037/9 (3).
      +-NR 10-75 p7
Il trovatore: Arias. cf BELLINI: La sonnambula: Arias.
Il trovatore: Arias. cf London OS 26373.
Il trovatore: Ai nostri monti. cf Decca DPA 517/8.
Il trovatore: Il balen del suo sorriso. cf Columbia/Melodiya M 3312

Il trovatore: Ballet music. cf Works, selections (Philips 6747 093).
Il trovatore: D'amor sull'ali rosee. cf CILEA: Adriana Lecouvreur:
    Io sono l'umile ancella.
Il trovatore: Deserto sulla terra. cf Ember GVC 41.
Il trovatore: Deserto sulla terra; De qual tetra luce. Ah si,
    ben mio; Di quella pira. cf RCA SER 5704/6.
Il trovatore: Di quella pira. cf BASF BC 21723.
Il trovatore: Di quella pira. cf Decca SXL 6649.
Il trovatore: Di quella pira. cf Rubini RS 300.
Il trovatore: Duets and solos. cf MASCAGNI: Cavalleria rusticana:
    Duets and solos.
Il trovatore: Gern will ich schlieben; Nicht darfst du von meiner
    Seite; Lodernde Flammen. cf Discophilia KGM 2.
Il trovatore: Il presagio funesto...Ah si, ben mio...Di quella pira.
    cf Arias (Philips 6747 193).
Il trovatore: Tacea la notte. cf Decca GOSC 666/8.
Il trovatore: Vedi le fosche. cf Works, selections (DG 2530 549).
Il trovatore: Vivra contende. cf Pearl GEM 121/2.
I vespri siciliani: Arias. cf BELLINI: La sonnambula: Arias.
I vespri siciliani: Arrigo, Ah, parli a un core; Mercé dilette
    amiche. cf Arias (Columbia M 33516).
I vespri siciliani: Arrigo...Ah, parli a un core; Mercé, dilette
    amiche. cf CILEA: Adriana Lecouvreur: Io sono l'umile ancella.
I vespri siciliani: Ballet music. cf Works, selections (Philips
    6747 093).
I vespri siciliani: E' di monforte il cenno...Giorno di pianto.
    cf Arias (Philips 6747 193).
I vespri siciliani: Mercé, diletti amiche. cf DONIZETTI: Linda
    di Chamonix: Ah, tardai troppo...O luce di quest'anima.
Les vespres siciliennes: Palerme, O mon pays...O toi Palerme. cf
    Decca SXL 6637.
I vespri siciliani: Le quattro stagioni. cf Don Carlo: Ballo
    della regina.
2396 Works, selections: Aida: Gloria all'egitto. Don Carlo: Spuntato
    ecco il di. Ernani: Si ridesti il Leon de Castiglia. I
    Lombardi: Jerusalem, Jerusalem; O Signore. Macbeth: Patria
    oppressa. Nabucco: Gli arredi festivi; Va pensiero. Otello:
    Fuoco di gioia. Il trovatore: Vedi, le fosche. La Scala
    Orchestra and Chorus; Claudio Abbado. DG 2530 549.
        +-Gr 11-75 p903   +RR 11-75 p35
        +HFN 11-75 p171
2397 Work, selections: Don Carlo: Il ballo della regina. Macbeth:
    Ballet music. I Lombardi: Jerusalem. Otello: Ballet music.
    Il trovatore: Ballet music. I vespri siciliani: Ballet music.
    Monte Carlo Opera Orchestra, LSO; Antonio de Almeida. Philips
    6747 093 (2).
        +Gr 9-74 p521   +-NYT 10-6-74 pD26
        /HF 2-75 p105   /ON 11-74 p64
        +MJ 2-75 p30   +RR 9-74 p63
        +NR 12-74 p4   ++SFC 10-20-74 p26
VIADANA, Ludovico
    Missa "L'Hora passa": Kyrie, Gloria, Credo, Sanctus, Benedictus,
    Agnus Dei. cf University of Illinois, unnumbered.
VIANNA, Fructuoso
    Dansa de negros. cf Angel S 37110.
    Jogos pueris. cf Angel S 37110.

VICTORIA, Tomas Luis de
    Estote fortes in bello. cf Gemini GM 2022.
    Motets: Ascendens Christus; Ave Maria; O magnum mysterium. cf
      Officium defunctorum.
2398 Officium defunctorum. Motets: Ascendens Christus; Ave Maria;
     O magnum mysterium. Prague Madrigal Singers; Miroslav Venhoda.
     Telefunken AW 6-41273.
        +-Gr 10-75 p675   +-RR 9-75 p72
        +-HFN 9-75 p107
    O magnum mysterium. cf Advent 5012.
    O magnum mysterium. cf Turnabout TV 34569.
2399 Tenebrae responsories. Westminster Cathedral Choir; George Malcolm.
     Decca ECS 747. (Reissue from Argo ZRG 5149)
        +Gr 3-75 p1681   +RR 4-75 p68
        +-HFN 5-75 p141
    VIENNESE DANCE MUSIC FROM THE CLASSICAL PERIOD. cf DG Archive 2533
     182.
VIERNE, Luis
    Berceuse. cf Wealden WS 160.
    Carillon. cf Wealdon WS 160.
2400 Symphony, no. 1, op. 14, D minor. Rollin Smith, org. Repertoire
     RRS 15.
        +CL 12-75 p5   +-MT 9-75 p797
    Symphony, no. 1, op. 14, D minor: Andante. cf Stentorian SC 1685.
    Symphony, no. 1, op. 14, D minor: Final. cf Decca 5BBA 1013-5.
VIEUXTEMPS, Henri
    Ballade et polonaise, op. 38. cf LEKEU: Sonata, violin and piano,
     G major.
    Concerto, violin, no. 4, op. 31, D minor. cf RCA ARM 4-0942/7.
    Concerto, violin, no. 5, op. 37, A minor. cf RCA ARM 4-0942/7.
    Rondino, op. 32. cf Rococo 2035.
VILLA-LOBOS, Heitor
    Bachianas brasileiras, no. 2; no. 4: Prelude and introduction;
     no. 5: Aria; Modhina preludio. cf Magdalena: Suite.
2401 Bachianas brasileiras, no. 4: Prelude. Prôle do bêbê, Bk 1.
     Rudepoêma. The three Marias. Nelson Freire, pno. Telefunken
     SAT 22547.
        +Gr 9-74 p553   ++RR 8-74 p55
        ++NR 6-75 p13   ++St 9-75 p115
    Bachianas brasileiras, no. 5: Aria. cf HMV SLS 5012.
    Bachianas brasileiras, no. 5. cf CANTELOUBE: Songs of the Auvergne.
    Cancão do carreiro. cf Odyssey Y2 32880.
    Chôro typico. cf LAURO: Danza negra.
    Chôros, no. 1, E minor. cf CBS 61654.
    Chôros, no. 1, E minor. cf London STS 15224.
    Chôros, no. 5 (Alma brasileira). cf Angel S 37110.
    Ciclo brasileiro. cf Angel S 37110.
    Concerto, guitar and small orchestra. cf RODRIGO: Concierto de
     Aranjuez, guitar and orchestra.
    Concerto, harp. cf CASTELNUOVO-TEDESCO: Concertino, harp and
     chamber orchestra, op. 93.
    Etude, guitar, no. 11. cf Philips 6833 159.
    Festo no sertão. cf Angel S 37110.
    Impressões seresteiras. cf Angel S 37110.
2402 Magdalena: Suite. Bachianas brasileiras, no. 2; no. 4: Prelude
     and introduction; no. 5: Aria; Modhina preludio. André Kostel-

anetz. CBS 61564. (also Columbia (Q) Tape (c) MAQ 32821)
    /Gr 9-74 p522   +RR 10-74 p64
    +—HF 9-75 p116 Quad tape
Preludes, nos. 1-5. cf LAURO: Danza negra.
Preludes, nos. 1, 2. cf Classics for Pleasure CFP 40205.
Prelude, no. 1, E minor. cf Amberlee ACL 501X.
Prelude, no. 1, E minor. cf London STS 15224.
Preludes, no. 2, E major; no. 4, E minor. cf CBS 72526.
Preludes, no. 2, E major; no. 4, E minor. cf CBS 72950.
Preludio estudio chôros. cf Everest SDBR 3380.
A prôle do bêbê, Bk 1. cf Bachianas brasileiras, no. 4: Prelude.
A prôle do bêbê, no. 1. cf Angel S 37110.
Rudepoêma. cf Bachianas brasileiras, no. 4: Prelude.
Suite for voice and violin. cf Musical Heritage Society MHS 1976.
The three Marias. cf Bachianas brasileiras, no. 4: Prelude.
VINCI, Leonardo
Sonata, flute, D major. cf London STS 15198.
VISEE, Robert de
Suite, guitar. cf GIULIANI: Sonata, flute and guitar, op. 85.
VIVALDI, Antonio
2403 Il cimento dell'armonia e dell'invenzione, op. 8, nos. 5-12. ECO;
    Pinchas Zukerman, vln and cond. Columbia M 32693, 32840. Tape
    (c) 32693, MT 32840 (ct) MA 32693, MA 32840 (Q) MQ 32693, MT
    32840. Tape (ct) MAQ 32693, 32840.
        +—HF 1-75 p94.
Concerto, P 77, A minor. cf HMV SLS 5022.
Concerto, bassoon, oboe and strings, C major. cf Concerto, bassoon
    and strings, P 70, A minor.
Concerto, op. 3, no. 11, D minor. cf CASALS: Sardana, celli.
2404 Concerti, op. 4 (La stravaganza). Carmel Kaine, Alan Loveday,
    vln; AMF; Neville Marriner. Argo ZRG 800/1 (2).
        +Gr 4-75 p1818   +RR 4-75 p38
        ++HF 7-75 p86   ++St 6-75 p147
        +MT 9-75 p799   +St 8-75 p106
        ++NR 8-75 p6
2405 Concerto, bassoon and strings, P 70, A minor. Concerto, bassoon,
    oboe and strings, C major. Concerto, oboe and strings, C major.
    Concerto, 2 oboes, 2 clarinets and strings, C major. Soloists;
    Gli Academy; Nello Santi. Turnabout 34024. Tape (c) KTVC 34025.
    (Reissue from Vox)
        +RR 3-75 p75 tape
2406 Concerti, bassoon and strings, P 137, E minor; P 70, A minor; P
    305, F major; P 382, B flat major. Klaus Thunemann, bsn; I
    Musici. Philips 6500 919.
        ++Gr 11-75 p834   +RR 11-75 p56
        +HFN 12-75 p169
2407 Concerto, flute, op. 10, no. 1, F major. Concerti, flute and
    strings, P 77, P 118, P 440. Concerto, 2 flutes and strings,
    P 76, C major. Severino Gazzelloni, Marja Steinberg, flt;
    Bernard Schenkel, ob; Jiri Staviček, bsn; I Musici. Philips
    6500 820.
        +Gr 8-75 p332   +RR 7-75 p36
        +HFN 8-75 p85
    Concerto, flute and bassoon, op. 10, no. 2, G minor. cf Concerti,
    flute, P 140, G major; P 203, D major; P 80, A minor; P 205,
    D major.

Concerto, flute and bassoon, op. 10, no. 2, G minor. cf BIBER:
   Serenade, C major.
2408 Concerti, flute, P 140, G major; P 203, D major; P 80, A minor;
   P 205, D major. Concerto, flute and bassoon, op. 10, no. 2, G
   minor. Severino Gazzelloni, flt; Jiri Staviček, bsn; I Musici.
   Philips 6500 707.
     +Gr 4-75 p1817   +fNR 8-75 p6
     ++HF 10-75 p86   ++RR 4-75 p38
     +HFN 8-75 p85
Concerti, flute and strings, P 177, P 118, P 440. cf Concerto,
   flute, op. 10, no. 1, F major.
Concerto, flute and strings, op. 10, no. 2, G minor. cf HANDEL:
   Concerto, oboe, no. 3, G minor.
Concerto, flute and strings, op. 10, no. 3, G major. cf DALL'ABACO:
   Concerto da chiesa, a 4, op. 2.
Concerto, 2 flutes and strings, P 76, C major. cf Concerto, flute,
   op. 10, no. 1, F major.
2409 Concerto, guitar, C major. Concerto, 2 guitars, G major. Concerto,
   4 guitars, B minor. Concerto, guitar, violin, viola and violon-
   cello, A major. The Romeros, gtr; John Corigliano, vln; Domen-
   ick Saltarelli, vla; Margaret Bella, vlc; San Antonio Symphony
   Orchestra; Victor Alessandro. Mercury SRI 75054. (Reissue from
   SR 90487)
     ++NR 7-75 p7
Concerto, guitar, violin, viola and violoncello, A major. cf Con-
   certo, guitar, C major.
Concerto, 2 guitars, G major. cf Concerto, guitar, C major.
Concerto, 2 guitars, P 134, C major. cf ALBINONI: Adagio, organ
   and strings, G minor.
Concerto, 4 guitars, B minor. cf Concerto, guitar, C major.
Concerto, 2 horns, P 320, F major. cf ROSETTI: Concerto, horn,
   D minor.
Concerto, lute, D major. cf HANDEL: Concerto, 2 lutes, strings
   and recorders, op. 4, no. 6, B flat major.
Concerto, 2 mandolins, P 133, G major. cf BACH: Concerto, violin
   and oboe, S 1060, D minor.
Concerto, oboe and strings, C major. cf Concerto, bassoon and
   strings, P 70, A minor.
Concerto, 2 oboes, 2 clarinets and strings, C major. cf Concerto,
   bassoon and strings, P 70, A minor.
2410 Concerto, orchestra, P 84, C major. Concerto, orchestra, P 383,
   G minor. Concerto, orchestra, P 267, F major. Concerto, orch-
   estra, P 359, G minor. I Solisti Veneti; Claudio Scimone.
   Erato STU 70818.
     /Gr 12-75 p1054   +RR 12-75 p69
     +HFN 12-75 p169
Concerto, orchestra, P 267, F major. cf Concerto, orchestra, P 84,
   C major.
Concerto, orchestra, P 359, G minor. cf Concerto, orchestra, P 84,
   C major.
Concerto, orchestra, P 383, G minor. cf Concerto, orchestra, P 84,
   C major.
2411 Concerti, 2 orchestras, D major; A major; C major; B major. Franco
   Fantini, Lola Bobesco, vln; Brussels Soloists; I Solisti di
   Milano; Angelo Ephrikian. Telefunken SAWT 9600. (Reissue from
   HMV HQS 1060).

```
     ++AR 2-75 p22                 +-HFN 12-73 p2617
     +-Gr 1-74 p1419               +-NR 3-74 p3
     +-HF 10-74 p116               +RR 12-73 p76
```

Concerto, organ, A minor. cf ALBINONI: Adagio, organ and strings,
   G minor.

Concerto, organ, D minor. cf ALBINONI: Adagio, organ and strings,
   G minor.

Concerto, recorder and strings, A minor. cf Telefunken SMA 25121
   T 1-3 (3).

Concerto, recorder, op. 44, no. 11, C major. cf Transatlantic TRA
   292.

Concerti, alto recorder and strings, A minor (2), F major. cf
   Concerto, soprano recorder and strings, D major.

Concerto, alto recorder, violin, oboe, bassoon and continuo, G
   minor. cf Strobe SRCM 123.

Concerto, sopranino recorder and strings, C major. cf Concerto,
   soprano recorder and strings, D major.

2412 Concerto, soprano recorder and strings, D major. Concerti, alto
   recorder and strings, A minor (2), F major. Concerto, sopranino recorder and strings, C major. Laszlo Czidra, rec; Franz
   Liszt Academy Chamber Orchestra; Frigyes Sándor. Hungaroton
   SLPX 11671.
        +RR 3-75 p36

Concerto, trumpet, A flat major. cf HUMMEL: Concerto, trumpet,
   E flat major.

Concerto, trumpet, A flat major. cf RCA CRL 2-7002.

Concerto, trumpet, B flat major. cf RCA CRL 2-7002.

Concerto, trumpet and organ, P 75, C major. cf RCA CRL 2-7001.

2413 Concerti, viola d'amore, P 37, D minor; P 166, D major; P 233,
   A major; P 287, D minor. Michel Pons, vla d'amore; Toulouse
   Chamber Orchestra; Georges Armand. Seraphim S 60244.
        +HF 11-75 p126   ++NR 8-75 p6

2414 Concerto, viola d'amore and strings, P 166, D major. Concerto,
   2 violins and strings, P 7, E flat major. Concerto, violoncello
   and strings, P 120, G major. Concerto, 2 violoncellos and
   strings, P 411, G minor. Concerto, 2 violoncellos and orchestra,
   P 402, G minor. Sonata, 2 violins and strings, C major. Sonata,
   violoncello and strings, op. 14, G minor. Collegium Aureum.
   BASF Tape (c) KBAC 2-7025.
        ++RR 3-75 p75 tape   +RR 4-75 p78 tape

2415 Concerti, viola d'amore, P 166, D major; P 37, A minor; P 287,
   D minor; P 233, A major; P 288, D minor; P 289, D minor; P
   286, F major; P 266, D minor. Orlando Cristoforetti, lt; Nane
   Calabrese, vla d'amore; I Solisti Veneti; Claudio Scimone.
   Erato STU 70826/7 (2).
        +-Gr 2-75 p1503   ++STL 2-9-75 p37
        ++RR 2-75 p45

Concerto, viola d'amore, A major. cf PORPORA: Concerto, violoncello, G major.

Concerto, violin, P 10, A minor. cf Concerto, violin, P 106, E
   minor.

2416 Concerto, violin, P 106, E minor. Concerto, violin, P 10, A minor.
   Concerto, violin, P 343, G minor. Concerto, violin, P 246, E
   major. Arthur Grumiaux, vln; Dresden State Orchestra; Vittorio
   Negri. Philips 6500 690.
        +Gr 2-75 p1503   ++RR 2-75 p45

+HF 3-75 p88   ++SFC 1-19-75 p28
++NR 10-74 p6
Concerto, violin, P 246, E major. cf Concerto, violin, P 106, E
    minor.
Concerto, violin, P 343, G minor. cf Concerto, violin, P 106, E
    minor.
Concerto, violin and strings, op. 3, no. 6, A minor. cf Concerto,
    violin and strings, op. 12, no. 1, G minor.
2417 Concerto, violin and strings, op. 12, no. 1, G minor. Concerto,
    violin and strings, op. 3, no. 6, A minor. Concerto, 2 violins
    and strings, op. 3, no. 8. Concerto, 4 violins and strings, op.
    3, no. 10, B minor. Edith Volckaert, vln; Belgian Chamber
    Orchestra; Georges Maes. Decca SDDR 466.
        +Gr 9-75 p479   +RR 8-75 p45
        +HFN 9-75 p109
2418 Concerti, violin, flute, oboe, bassoon and continuo, P 82, P 204,
    P 207, P 323, P 342, P 360, P 403. Secolo Barocco. Arion
    ARN 38264.
        ++RR 10-75 p61
Concerto, violins, strings and harpsichord, F major. cf BACH:
    Concerto, 3 violins, strings and continuo, D minor.
Concerto, 2 violins and strings, P 7, E flat major. cf Concerto,
    viola d'amore and strings, P 166, D major.
2419 Concerti, 2 violins, P 281, D minor; P 436, C minor; P 366, G
    minor; P 189, D major. Isaac Stern, David Oistrakh, vln; PO
    Members; William R Smith, hpd; Eugene Ormandy. CBS 61629.
    (Reissue from Fontana SCFL 136)
        +-Gr 10-75 p639   ++RR 9-75 p51
        +-HFN 9-75 p109   ++STL 10-5-75 p36
Concerto, 2 violins and strings, op. 3, no. 8. cf Concerto, violin
    and strings, op. 12, no. 1, G minor.
Concerto, 4 violins and strings, op. 3, no. 10, B minor. cf Con-
    certo, violin and strings, op. 12, no. 1, G minor.
Concerto, 4 violins and strings, no. 10, P 148, B minor. cf BACH:
    Concerto, violin and oboe, S 1060, D minor.
Concerto, violoncello, D major. cf DALL'ABACO: Concerto da chiesa,
    a 4, op. 2.
Concerto, violoncello and strings, P 120, G major. cf Concerto,
    viola d'amore and strings, P 166, D major.
Concerto, 2 violoncellos and strings, P 411, G minor. cf Concerto,
    viola d'amore and strings, P 166, D major.
Concerto, 2 violoncellos and orchestra, P 402, G minor. cf Con-
    certo, voila d'amore and strings, P 166, D major.
2420 L'Estro armonico, op. 3. Jan Tomasow, Willi Boskovsky, Philipp
    Matheis, Walter Hintermeyer, vln; Vienna State Opera Chamber
    Orchestra; Mario Rossi. Vanguard HM 37/39SD.
        +MJ 11-74 p48   +SFC 10-27-74 p6
        +-RR 1-75 p37
L'Estro armonico: Concerto, violoncello, op. 3, no. 9, D major.
    cf BOCCHERINI: Concerto, violoncello, B flat major.
2421 The four seasons, op. 8, nos. 1-4. Jean-Jacques Kantorow, vln;
    Bernard Thomas Chamber Orchestra; Bernard Thomas. Arion ARN
    38249.
        /RR 2-75 p45
2422 The four seasons, op. 8, nos. 1-4. Igor Kipnis, hpd; Suttgart
    Chamber Orchestra; Milan Munclinger. Decca SXL 6557. Tape (c)

KSXC 6557.
    +-Gr 8-75 p376 tape   ++RR 9-75 p79 tape
    +-HFN 7-75 p91 tape
2423 The four seasons, op. 8, nos. 1-4. Monique Frasca-Colombier, vln;
    Paul Kuentz Chamber Orchestra; Paul Kuentz. DG 2548 005.
        +Gr 11-75 p839   -RR 10-75 p61
        ++HFN 10-75 p153
2424 The four seasons, op. 8, nos. 1-4. Yoshio Unno, vln; NHK String
    Ensemble; Yoshio Unno. Odyssey Y 32884.
        +HF 1-75 p94   +-St 10-74 p139
2425 The four seasons, op. 8, nos. 1-4. Felix Ayo, vln; I Musici.
    Philips Tape (c) 7300 176.
        +RR 7-75 p72
    The four seasons, op. 8: Spring. cf Sound Superb SPR 90049.
    Piango gemo sospiro (Songs arr. by Gamley). cf Decca SXL 6629.
2426 Juditha triumphans. Elly Ameling, s; Ingeborg Springer, Julia
    Hamari, ms; Birgit Finnilä, Annelies Burmeister, con; Berlin
    Radio Soloists Ensemble; Berlin Chamber Orchstra; Vittorio Negri.
    Philips 6747 173 (3).
        +-Gr 10-75 p676   +RR 9-75 p72
        +-HFN 11-75 p171
2427 Il Pastor Fido, op. 13. Eduard Melkus, vln; Alfred Sous, ob; René
    Zosso, hurdy-gurdy; Hans-Martin Linde, flt and rec; Walter
    Stiftner, bsn; Garo Altmacayan, vlc; Huguette Dreyfus, hpd.
        +AR 5-75 p52   +NYT 4-12-73 pC58
        +-Gr 3-73 p1706   ++RR 3-73 p71
        +HFN 3-73 p573   +St 2-73 p129
    Sinfonia, G major. cf Grange SGR 1124.
    Sonata, no. 6, G minor. cf BIGAGLIA: Sonata, recorder, bassoon
        and harpsichord, A minor.
    Sonata, oboe, bassoon and harpsichord, C minor. cf BIGAGLIA:
        Sonata, recorder, bassoon and harpsichord, A minor.
    Sonata, recorder, G minor. cf Telefunken SMA 25121 T/1-3.
    Sonata, violin, op. 2, no. 2, A major. cf RCA ARM 4-0942/7.
    Sonata, 2 violins and strings, C major. cf Concerto, viola d'amore
        and strings, P 166, D major.
    Sonata, violoncello and strings, op. 14, G minor. cf Concerto,
        viola d'amore and strings, P 166, D major.
VIVES
    El retrato de Isabela. cf HMV SLS 5012.
VIVIANI, Giovanni
    Sonatas, trumpet, nos. 1 and 2. cf RCA CRL 2-7001.
    Sonata, trumpet, no. 1, C major. cf CRD CRD 1008.
    Sonata, trumpet, no. 1, C major. cf HMV ASD 2938.
    Sonata, trumpet, no. 2, D major. cf HMV ASD 2938.
VLIJMEN, Jan van
    Omaggio a Gesualdo, violin and 6 instrumental groups. cf KUNST:
        Trajectoire, 16 singers and 11 instrumentalists.
VOGLER, Georg
    Comic ballets: 4 movements. cf CANNABICH: Sinfonia concertante,
        F major.
VOGRICH
    Dans le bois. cf Discopaedia MB 1006.
VOLKMANN, Robert
2428 Concerto, violoncello, op. 33. Konzertstück, piano and orchestra,
        op. 42. Thomas Blees, vlc; Jerome Rose, pno; Hamburg Symphony

Orchestra, Luxembourg Radio Orchestra; Alois Springer, Pierre
Cao. Turnabout TVS 34576.
+HF 9-75 p96   +NR 7-75 p5
Konzertstück, piano and orchestra, op. 42. cf Concerto, violon-
cello, op. 33.

VON RUGEN
Loybere risen. cf BIS LP 3.

VORISEK, Jan Hugo
Impromptu, no. 4, op. 7, A major. cf BENDA: Sonata, piano, no. 9,
A minor.

VOSTRAK, Zbynek
Scale of light. cf ISTVAN: Isle of toys.

WAELRANT, Hubert
Als ic u vinde. cf Telefunken TK 11569/1-2.

WAGNER, J. F.
Unter dem Doppeladler. cf BASF 292 2116-4.
Unter dem Doppeladler. cf DG 2721 077.
Tiroler Holzhackerbaub'n. cf DG 2721 077.

WAGNER, Richard
Adagio. cf BERNSTEIN: Sonata, clarinet and piano.
Albumblatt, C major. cf Discopaedia MB 1007.
Albumblatt, C major. cf Rococo 2035.

2429 A Faust overture. Der fliegende Holländer: Overture. Das Rhein-
gold: Entrance of the Gods. Rienzi: Overture. Tannhäuser:
Grand march. LPO; Adrian Boult. HMV ASD 3071.
++Gr 4-75 p1818   -RR 4-75 p38
A Faust overture. cf Symphony, C major.
Die Feen: Overture. cf Symphony, C major.

2430 Der fliegende Holländer (The flying Dutchman). Gwyneth Jones,
s; Sieglinde Wagner, ms; Hermin Esser, Harald Ek, t; Thomas
Stewart, bar; Karl Ridderbusch, bs; Bayreuth Festival Orchestra
and Chorus; Karl Böhm. DG 2709 040 (3). Tape (r) 47040.
(also DG 2520 052)
+-Gr 10-72 p747   +-NR 11-72 p12
+-HF 11-72 p106   +-ON 3-24-73 p36
+-HFN 10-72 p1919   +-Op 11-72 p1000
+LJ 4-75 p70 tape   +-RR 10-72 p52
+-MJ 2-73 p36   +-St 12-72 p141

2431 Der fliegende Holländer, excerpts. Viorica Ursuleac, s; Luise
Willer, alto; Karl Ostertag, Franz Klarwein, t; Hans Hotter,
bar; Georg Hanns, bs; Bavarian State Opera Orchestra and Chorus;
Clemens Krauss. BASF KBF 21538. (Recorded 1944)
+HF 7-75 p86   +-ON 4-12-75 p36
+NR 6-75 p11   +St 8-75 p107

2432 Der fliegende Holländer, highlights. Kirsten Flagstad, s; Mary
Jarred, ms; Max Lorenz, Ben Williams, t; Herbert Janssen, bar;
Ludwig Weber, bs; Orchestra and Chorus; Fritz Reiner. Rococo
1008 (2).
+NR 2-75 p9

2433 Der fliegend Holländer: Die Frist ist um; Summ und Brumm; Jo ho
hoe...Traft ihr das Schiff; Mögst du, mein Kind; Wie aus der
Ferne; Steuermann, lass die Wacht; Willst jenes Tags; Verloren,
Ach, verloren; Erfahre das Geschick. Leonie Rysanek, s; Rosalind Elias, ms; Karl Liebl, t; George London, bs-bar; Giorgio
Tozzi, bs; ROHO and Chorus; Antal Dorati. Decca SDD 439. (Reissue from RCA SER 4535)

-Gr 5-75 p2019   +RR 6-75 p26
+HFN 8-75 p91
2434 Der fliegende Holländer: Die Frist ist um. Die Meistersinger von
     Nürnberg; Was dufet doch der Flieder, Whan, Wahn; Uberall Wahn.
     Die Walküre: Leb wohl, du kühnes, herrliches Kind. George
     London, bs; VPO; Hans Knappertsbusch. Richmond SR 33198. (Re-
     issue)
           +NR 6-75 p11
2435 The flying Dutchman: Overture. Lohengrin: Preludes, Acts 1, 3.
     Parsifal: Preludes, Acts 1, 3. Die Meistersinger von Nürnberg:
     Prelude. BPhO; Herbert von Karajan. Angel S 37098.
           +SFC 12-21-75 p39
2436 Der fliegende Holländer, Overture. Die Götterdämmerung: Sunrise,
     Siegfried's Rhine journey, Funeral music. Tannhäuser: Over-
     ture, Entry march and chorus. Baden State Opera Orchestra and
     Chorus; Carl Bamberger. Audio Fidelity FCS 50059.
           -NR 9-75 p2
2437 Der fliegende Holländer: Overture. Lohengrin: Preludes, Acts 1,
     3. Die Meistersinger von Nürnberg: Prelude. Tristan und Isolde:
     Prelude and Liebestod. VPO: Horst Stein. Decca SXL 6656.
     (also London CS 6860)
           +Gr 7-74 p216   /NR 9-75 p2
           +HF 11-75 p127   /RR 7-74 p48
     Der fliegende Holländer: Overture. cf A Faust overture.
     Der fliegende Holländer: Overture. cf Works, selections (CBS 78252).
     Der fliegende Holländer: Overture. cf Decca 396.
2438 Götterdämmerung: Dawn and Siegfried's Rhine journey; Siegfried's
     funeral music. Tristan und Isolde: Prelude, Act 1. Die
     Walküre: Wotan's farewell and Magic fire music. George London,
     bs-bar; VPO; Hans Knappertsbusch. Decca SDD 426. (Reissues
     from LXT 5255/6, SXL 2068, 2184)
           +-Gr 4-75 p1818   -RR 3-75 p19
     Götterdämmerung: Brunhilde's immolation. cf BEETHOVEN: Fidelio,
     op. 72.
     Götterdämmerung: Erzahlung der Waltraute. cf Discophilia KGM 2.
     Götterdämmerung: Immolation scene. cf Decca GOSC 666/8.
     Götterdämmerung: Sunrise, Siegfried's Rhine journey, Funeral music.
     cf Der fliegende Holländer: Overture.
     Das Liebesverbot: Overture. cf Symphony, C major.
2439 Lohengrin. Songs: PFITZNER: Mailied, Trauerstille. SCHUBERT:
     Schwanengesang: Die Taubenpost. SCHUMANN: Der Hidalgo, op. 30,
     no. 3; Provencalisches Lied. WOLF: Liebesfrühling, Nachtgruss,
     Der Musikant, Wohin mit der Freud. Trude Eipperle, s; Helena
     Braun, con Peter Anders, t; Carl Kronenberg, bar; Josef Greindl,
     bs; Cologne Radio Symphony Orchestra and Chorus; Richard Kraus.
     Rococo 1015 (4).
           +NR 10-75 p7
2440 Lohengrin. Leonore Kirschstein, s; Ruth Hesse, ms; Herbert Schacht-
     schneider, t; Heinz Imdahl, bar; Hans Helm, Walter Kreppel, bs;
     Vienna State Opera Chorus; South German Philharmonic Orchestra;
     Hans Swarowsky. Westminster WGSO 8285 (4).
           -HF 7-75 p86   -ON 4-12-75 p36
     Lohengrin: Bridal march. cf HMV HLM 7065.
     Lohengrin: Einsam in trüben Tagen. cf HMV SLS 5012.
     Lohengrin: Einsam in trüben Tagen. cf Saga 7029.
     Lohengrin: Entweihte Götter. cf Discophilia KGM 2.

Lohengrin: In fernem Land. cf RCA SER 5704/6.
2441 Lohengrin: Prelude, Act 1. Tannhäuser: Overture and bacchanale.
     Tristan und Isolde: Prelude and Liebestod. BPhO; Herbert von
     Karajan. Angel S 37097. Tape (c) 4XS 37097 (ct) 8XS 37097.
     (also HMV (Q) ASD 3130)
              +Gr 12-75 p1054   ++NR 9-75 p2
              +HF 12-75 p109   +-RR 12-75 p70
              +-HFN 12-75 p169
     Lohengrin: Prelude, Act 1. cf Works, selections (CBS 78252).
2442 Lohengrin: Preludes, Acts 1, 3. Die Meistersinger von Nürnberg:
     Prelude, Act 1. Tristan und Isolde: Prelude and Liebstod.
     Parsifal: Prelude. COA; Bernard Haitink. Philips 6500 932.
     Tape (c) 7300 391. (also 6500 032)
              +-Gr 11-75 p839   ++RR 11-75 p58
              +HFN 12-75 p169   ++SFC 12-21-75 p39
     Lohengrin: Preludes, Acts 1, 3. cf The flying Dutchman: Overture.
     Lohengrin: Preludes, Acts 1, 3. cf Der fliegende Holländer: Over-
     ture.
     Lohengrin: Preludes, Acts 1, 3. cf Works, selections (London SPC
     21051).
2443 Lohengrin: Prelude, Act 3. Tannhäuser: Overture, Grand march. Die
     Walküre: Magic fire music; Ride of the Valkyries. Tristan und
     Isolde: Liebestod. PO; Eugene Ormandy. CBS 30065. (Reissue)
              +-RR 12-75 p70
     Lohengrin: Wedding march. cf Columbia CXM 32088.
     Lohengrin: Winterstürme. cr Rubini RS 300.
2444 Die Meistersinger von Nürnberg. Hannelore Bodes, s; Anna Reynolds,
     ms; Heribert Steinbach, Robert Licha, Wolf Appel, Norbert Orth,
     Jean Cox, Frieder Stricker, t; Karl Ridderbusch, bar; Hans
     Sotin, Jozsef Dene, Klaus Hirte, Gerd Nienstedt, Heinz Feldhoff,
     Hartmut Bauer, Nikolaus Hillebrand, Bernd Weikl, bs; Bayreuth
     Festival Orchestra and Chorus; Silvio Varviso. Philips 6747
     167 (5).
              +Gr 9-75 p519   +-RR 9-75 p25
              /HFN 9-75 p107
     Die Meistersinger von Nürnberg: Overture. cf Das Rheingold: Ent-
     rance of the Gods into Valhalla.
     Die Meistersinger von Nürnberg: Preizlied. cf Discopaedia MB 1009.
     Die Meistersinger von Nürnberg: Preizlied. cf Rococo 2035.
     Die Meistersinger von Nürnberg: Prelude. cf The flying Dutchman:
     Overture.
     Die Meistersinger von Nürnberg: Prelude. cf Der fliegende Holländer:
     Overture.
     Die Meistersinger von Nürnberg: Prelude, Act 1. cf Lohengrin:
     Preludes, Acts 1, 3.
     Die Meistersinger von Nürnberg: Prelude, Act 1. cf Works, selec-
     tions (CBS 78252).
     Die Meistersinger von Nürnberg: Prelude, Act 1. cf BEETHOVEN:
     Egmont overture, op. 84.
     Die Meistersinger von Nürnberg: Was dufet doch der Flieder, Wahn,
     Wahn; Uberall Wahn. cf Der fliegende Holländer: Die Frist ist
     um.
     Nibelungen-Marsch (arr. Sonntag). cf DG 2721 077.
2445 Parsifal, excerpts. Gwyneth Jones, s; James King, t; Thomas Stew-
     art, Donald McIntyre, bar; Franz Crass, bs; Bayreuth Festival
     Orchestra and Chorus; Pierre Boulez. DG 2536 023.
              +-St 8-75 p107

2446 Parsifal: Duets. Die Walküre: Duets. Birgit Nilsson, s; Helge
      Brillioth, t; Norman Bailey, bs-bar; ROHO: Leif Segerstam.
      Philips 6500 661.
            +MJ 1-75 p49
2447 Parsifal: Dies alles...hab ich nun geträumt; Erlösung Frevlerin
      biet ich auch dir. Die Walküre: Schläftst du, Gast. Birgit
      Nilsson, s; Helge Brillioth, t; Norman Bailey, bar; ROHO; Leif
      Segerstam. Philips 6500 661.
            +Gr 1-75 p1391   +-ON 12-21-74 p28
            -HF 1-75 p96   +-RR 1-75 p24
            +MJ 1-75 p49   ++SFC 9-22-74 p22
            +-NR 11-74 p9   +St 2-75 p112
            +-NYT 9-8-74 pD36
2448 Parsifal: Good Friday music; Preludes, Acts 1, 3; Transformation
      music, Acts 1, 3. Siegfried Idyll. LSO; Adrian Boult. HMV
      ASD 3000. (also Angel S 37090)
            +-HF 8-75 p101   +-RR 9-74 p64
            ++NR 8-75 p3
     Parsifal: Prelude. cf Lohengrin: Preludes, Acts 1, 3.
     Parsifal: Prelude. cf Das Rheingold: Entrance of the Gods into
      Valhalla.
     Parsifal: Preludes, Acts 1, 3. cf The flying Dutchman: Overture.
     Parsifal: Prelude, Good Friday music. cf Works, selections (CBS
      78252).
     Parsifal: Prelude, Good Friday music. cf BRUCKNER: Symphony, no.
      7, E major.
2449 The Rheingold (sung in English). Valerie Masterson, Shelagh Squires,
      Lois McDonall, s; Helen Attfield, Anne Collins, ms; Katherine
      Pring, con; Robert Ferguson, Emile Belcourt, Gregory Dempsey, t;
      Derek Hammond-Stroud, Norman Bailey, Norman Welsby, bar; Robert
      Lloyd, Clifford Grant, bs; English Opera Group Orchstra; Regi-
      nald Goodall. HMV (Q) SLS 5032 (4).
            +Gr 11-75 p903   ++RR 12-75 p33
            +-HFN 12-75 p161
     Das Rheingold: Alberich's Fluch. cf Rubini RS 300.
2450 Das Rheingold: Entrance of the Gods into Valhalla. Die Meister-
      singer von Nürnberg: Overture. Parsifal: Prelude. Die Walküre:
      The ride of the Valkyries; Wotan's farewell; Magic fire music.
      Anthony Newman, org. Columbia M 33268.
            -NR 8-75 p2   +St 12-75 p132
     Das Rheingold: Entrance of the Gods. cf A Faust overture.
     Das Rheingold: Weiche Wotan. cf Discophilia KGM 2.
     Rienzi: Overture. cf A Faust overture.
     Rienzi: Overture. cf Symphony, C major.
     Rienzi: Overture. cf Decca SXL 6643.
     Rienzi: Overture. cf HMV RLS 717.
     Rienzi: Overture. cf London CS 6856.
     Rienzi: Overture. cf Toscanini Society ATS GC 1201/6.
     Siegfried: Forest murmurs. cf Works, selections (London SPC 21051).
     Siegfried Idyll. cf Parsifal: Good Friday music.
     Siegfried Idyll. cf Works, selections (CBS 78252).
     Siegfried Idyll. cf MENDELSSOHN: Symphony, no. 3, op. 56, A minor.
     Songs: Lied des Mephistopheles; Branders Lied. cf DG Archive 2533
      149.
2451 Symphony, C major. A Faust overture. Reinzi: Overture. Bamberg
      Symphony Orchestra; Otto Gerdes. DG 2530 194.

              +Gr 9-72 p502   -RR 9-72 p75
              /HF 9-75 p97   +SFC 12-21-75 p39
              +HFN 9-72 p1673   +SR 6-14-75 p46
              +-NR 7-75 p2
2452 Symphony, C major. Die Feen: Overture. Das Liebesverbot: Overture.
     Hamburg Symphony Orchestra, Luxembourg Radio Orchestra; Heribert
     Beissel, Alois Springer. Turnabout TVS 34497.
              +HF 9-75 p97
     Tannhäuser: Dich teure Halle. cf HMV SLS 5012.
     Tannhäuser: Dich teure Halle. cf Rubini RS 300.
2453 Tannhäuser: Dir töne Lob; Dich, teure Halle; Ihr habts gehört; Wohl
     wüsst ich hier; O du mein holder Abendstern...Hör an, Wolfram.
     Helga Dernesch, s; René Kollo, t; Christa Ludwig, ms; Victor
     Braun, bar; Hans Sotin, bs; Vienna State Opera Chorus, Vienna
     Boys' Choir; VPO; Georg Solti. London OS 26299.
              +MJ 9-75 p51   /NR 11-74 p9
     Tannhäuser: Dir töne Lob. cf Rubini RS 300.
     Tannhäuser: Entry of the guests. cf Philips 6580 107.
     Tannhäuser: Grand march. cf A Faust overture.
     Tannhäuser: Grand march. cf HMV HLM 7065.
     Tannhäuser: O star of eve. cf Ember GVC 42.
     Tannhäuser: Overture. cf Works, selections (London SPC 21051).
     Tannhäuser: Overture. cf RCA ARL 2-0512.
     Tannhäuser: Overture and bacchanale. cf Lohengrin: Prelude, Act 1.
     Tannhäuser: Overture, Entry march and chorus. cf Der fliegende
       Holländer: Overture.
     Tannhäuser: Overture and Venusberg music. cf Works, selections
       (CBS 78252).
     Tannhäuser: Overture and Venusberg music. cf Camden CCV 5000-12,
       5014-24.
     Tannhäuser: Overture, Grand march. cf Lohengrin: Prelude, Act 3.
     Tannhäuser: Pilgrims chorus, Act 3. cf Decca PFS 4323.
     Tannhäuser: Prelude, Act 3. cf HMV RLS 717.
     Tristan und Isolde: Liebestod. cf Lohengrin: Prelude, Act 3.
     Tristan und Isolde: Liebestod. cf Works, selections (London SPC
       21051).
     Tristan und Isolde: Love music, Acts 2 and 3 (arr. Stokowski). cf
       FALLA: El amor brujo.
     Tristan und Isolde: Prelude, Act 1. cf Götterdämmerung: Dawn and
       Siegfried's Rhine journey; Siegfried's funeral music.
     Tristan und Isolde: Prelude, Act 3. cf Camden CCV 5000-12, 5014-24.
     Tristan und Isolde: Prelude, Act 3. cf HMV RLS 717.
     Tristan und Isolde: Prelude and Liebestod. cf Der fliegende
       Holländer: Overture.
     Tristan und Isolde: Prelude and Liebestod. cf Lohengrin: Prelude,
       Act 1 (Angel 37097).
     Tristan und Isolde: Prelude and Liebestod. cf Lohengrin: Prelude,
       Act 1 (HMV 3130).
     Tristan und Isolde: Prelude and Liebestod. cf Lohengrin: Preludes,
       Acts 1, 3.
     Tristan und Isolde: Prelude and Liebestod. cf Columbia M2X 33014.
     Die Walküre: Du bist der Lenz. cf Rubini RS 300.
     Die Walküre: Duets. cf Parsifal: Duets.
     Die Walküre: Leb wohl, du kühnes, herrliches Kind. cf Der flieg-
       enge Holländer: Die Frist ist um.
     Die Walküre: Magic fire music; Ride of the Valkyries. cf Lohengrin
       Prelude, Act 3.

Die Walküre: Der Männer Sippe. cf Decca GOSC 666/8.
Die Walküre: Ride of the Valkyries. cf Works, selections (London
    SPC 21051).
Die Walküre: Ride of the Valkyries. cf Camden CCV 5000-12,
    5014-24.
Die Walküre: Ride of the Valkyries. cf Decca DPA 519/20.
Die Walküre: Ride of the Valkyries. cf HMV HLM 7065.
Die Walküre: Ride of the Valkyries; Wotan's farewell; Magic fire
    music. cf Das Rheingold: Entrance of the Gods into Valhalla.
Die Walküre: Schläfst du, Gast. cf Parsifal: Dies alles...hab ich
    nun geträumt; Erlösung Frevlerin biet ich auch dir.
Die Walküre: Walkurenruf; Nich weise bin ich. cf Discophilia KGM 2.
Die Walküre: Winterstürme. cf Ember GVC 41.
Die Walküre: Wotan's farewell and Magic fire music. cf Götterdäm-
    merung: Dawn and Siegfried's Rhine journey; Siegfried's funeral
    music.
Wesendonck Lieder (5). cf BEETHOVEN: Fidelio. op. 72.
2454 Works, selections: Der fliegende Holländer: Overture. Lohengrin:
    Prelude, Act 1. Die Meistersinger von Nürnberg: Prelude, Act 1.
    Parsifal: Prelude, Good Friday music. Siegfried Idyll. Tann-
    häuser: Overture and Venusberg music. Columbia Symphony Orches-
    tra; Bruno Walter. CBS 78252 (2). (Reissue from Philips
    SABL 114, CBS SBRG 72143)
            +-Gr 8-75 p335   +-RR 8-75 p46
            +-HFN 8-75 p89
2455 Works, selections: Lohengrin: Preludes, Acts 1, 3. Siegfried:
    Forest murmurs. Tannhäuser: Overture. Tristan und Isolde:
    Liebestod. Die Walküre: Ride of the Valkyries. LSO, NPhO;
    Leopold Stokowski, George Hurst, Erich Leinsdorf, Stanley Black,
    Carlos Paita. London SPC 21051.
            +-NR 9-75 p2
WAGNES
    Die Bosniaken kommen. cf DG 2721 077.
WALDTEUFEL, Emile
2456 Dolores. España. The Grenadiers. Mon rêve. Pomone. Les pati-
    neurs. Toujours ou jamais. National Philharmonic Orchestra;
    Douglas Gamley. Decca SXL 6704.
            +HFN 12-75 p169   +RR 12-75 p71
    España. cf Dolores.
    The Grenadiers. cf Dolores.
    Mon rêve. cf Dolores.
    Les patineurs. cf Dolores.
    Pomone. cf Dolores.
    Toujours ou jamais. cf Dolores.
WALKER
    But for the grace of God. cf RCA ARL 1-0561.
WALKER, Ernest
    I will lift up mine eyes. cf Argo ZRG 789.
WALKER, George
    Lyric for strings. cf SOWANDE: African suite.
    Concerto, trombone and orchestra. cf KAY: Markings.
WALKER, J. F.
    Clear vault of Heav'n. cf Argo ZRG 785.
WALKER, Timothy
    Etude. cf L'Oiseau-Lyre DSLO 3.
    Lorelei. cf L'Oiseau-Lyre DSLO 3.

WALOND, William
    Voluntary, no. 5, G major. cf Pye TPLS 13066.
WALTHER, Johann
    Jesu, meine Freude, chorale partita. cf Telefunken DX 6-35265.
    Joseph, lieber Joseph mein. cf Turnabout TV 34569.
    Warum sollt ich mich denn gramen. cf Pelca PSR 40598.
WALTON, William
    Belshazzar's feast, excerpt. cf HMV SLA 870.
    Concerto, violin, B minor. cf STRAVINSKY: Concerto, violin, D
       major.
    Concerto, violin, B minor. cf RCA ARM 4-0942/7.
    Concerto, violoncello: 2d movement. cf HMV SEOM 19.
    Crown imperial. cf BRITTEN: Simple symphony, strings, op. 4.
    Crown imperial. cf Decca 419.
    Crown imperial. cf Philips 6747 177.
    Death of Falstaff: Passacaglia. cf DELIUS: Aquarelles.
    Music for children. cf BERKELEY: Mont Juic, op. 9.
    Orb and sceptre. cf HMV (Q) ASD 3131.
    Richard III: Prelude. cf ROZSA: Film music to Julius Caesar: Suite.
    Richard III: Prelude. cf SHOSTAKOVICH: Hamlet: Suite.
    Set me as a seal upon thine heart. cf HMV HQS 1350.
    Touch her soft lips and part. cf DELIUS: Aquarelles.
    WALTZES OF JOHANN STRAUSS II, paraphrases. cf Musical Heritage
       Society MHS 1959.
    WANDA WILKOMIRSKA, recital. cf Connoisseur Society CA 2070.
WARLOCK, Peter
    Benedicamus Domino. cf Abbey LPB 734.
WARREN
    God of our Fathers. cf RCA ARL 1-0561.
WATKINS, David
    Petite suite. cf RCA LRL 1-5087.
WEBER
    Sonata, violoncello, A major. cf Command COMS 9006.
WEBER, Bedrich A.
    An den Monde. cf Oryx ORYX 1720.
WEBER, Carl Maria von
    Concerto, clarinet, no. 2, op. 74, E flat major. cf SCHUBERT:
       Symphony, no. 3, D 200, D major.
2457 Euryanthe. Jessye Norman, Rita Hunter, Renate Krahmer, s; Nicolai
      Gedda, Harald Neukirch, t; Tom Krause, bar; Siegfried Vogel,
      bs; Leipzig Radio Chorus; Dresden Staatskapelle; Marek Janowski.
      HMV (Q) SLS 983 (4). (also Angel SDL 3746)
         +Gr 10-75 p687   +NYT 11-16-75 pD1
        +—HFN 11-75 p171   +RR 11-75 p22
         +NR 12-75 p11   +SFC 12-7-75 p31
    Euryanthe: Overture. cf Toscanini Society ATS GC 1201/6.
    Euryanthe: Wo berg ich mich. cf RCA ARL 1-0851.
2458 Der Freischütz. Elisabeth Grümmer, Rita Streich, s; Hans Hopf, t;
      Alfred Poell, Oskar Czerwenka, Kurt Böhme, bar; VSOO and Chorus;
      Wilhelm Furtwängler. Bruno Walter Society IGI 338 (3).
        +NR 7-75 p11
2459 Der Freischütz: Overture; Nein, länger trag ich nicht die Qualen...
      Durch die Wälder; Schweig, damit dich niemand warnt; Schelm,
      Halt fest; Kommt ein schlanker Bursch gegangen; Wie? Was? Ent-
      setzen; Und ob die Wolke sie verhülle; Wir winden dir den Jung-
      fernkranz, Was Gleicht wohl auf Erden. Edith Mathis, s; Peter

Schreier, t; Bernd Weikl, bar; Theo Adam, bs-bar; Franz Crass,
Siegfried Vogel, bs; Leipzig Radio Chorus; Dresden State Opera
Orchestra; Carlos Kleiber. DG 2530 661. (Reissue from DG
2720 071)
      ++RR 12-75 p34
Der Freischütz: Arias. cf London OS 26381.
Der Freischütz: Overture. cf Decca SXL 6643.
Der Freischütz: Overture. cf London CS 6856.
Invitation to the dance, op. 63. cf HMV SLS 5019.
Konzertstück, F major. cf BERLIOZ: Le carnaval romain, op. 9.
Silvana: So soll denn dieses Herz. cf Bel Canto Club No. 3.

WEBERN, Anton
2460 Bagatelles, string quartet, op. 9 (6). Movements, string quar-
    tet, op. 5 (5). Quartet, strings (1905). Quartet, strings,
    op. 28. LaSalle Quartet. DG 2530 284.
      ++MJ 9-75 p51   ++SFC 9-1-74 p20
      ++NR 8-74 p6   +St 10-74 p140
Kleine Stücke, violoncello, op. 11. cf DEBUSSY: Sonata, violon-
cello and piano, no. 1, D minor.
Movements, string quartet, op. 5 (5). cf Bagatelles, string quar-
tet, op. 9.
Movements, string quartet, op. 5 (5). cf BERG: Lyric suite:
Pieces.
Movements, string quartet, op. 5 (5). cf BERG: Sonata, piano, op. 1.
Movements, string quartet, op. 5 (5). cf HINDEMITH: Pieces,
string orchestra, op. 44, no. 4.
Passacaglia, op. 1. cf BERG: Lyric suite: Pieces.
Passacaglia, op. 1. cf SCHONBERG: Pelleas und Melisande, op. 5.
Pieces, op. 6 (6). cf BERG: Lyric suite: Pieces.
Pieces, op. 7 (4). cf DALLAPICCOLA: Studi, violin and piano (2).
Quartet, strings (1905). cf Bagatelles, string quartet, op. 9.
Quartet, strings, op. 28. cf Bagatelles, string quartet, op. 9.
Sonata, violoncello. cf DG 2530 562.
Stücke, op. 11 (3). cf DG 2530 562.
Symphony, op. 21. cf BERG: Lyric suite: Pieces.

WECKERLIN, Jean Baptiste
Maman, dites-moi. cf RCA ARL 1-5007.

WEELKES, Thomas
All people clap your hands. cf Gemini GM 2022.
As wanton birds. cf Turnabout TV 37071.
Cease sorrows now. cf Harmonia Mundi 204.
The cryes of London. cf Turnabout TV 37079.
Death hath deprived me. cf Turnabout TV 37071.
Hark, all ye lovely saints above; Say dear, when will your frown-
ing leave. cf Harmonia Mundi 593.
In pride of May. cf Turnabout TV 37071.
O care, thou wilt despatch me. cf Turnabout TV 37071.
Since Robin Hood. cf Turnabout TV 37079.
Sit down and sing. cf Turnabout TV 37071.
Thule, the period of cosmography. cf Turnabout TV 37079.

WEIGL, Vally
Nature moods. cf HOVHANESS: Tumburu, op. 264, no. 1.
New England suite. cf HOVHANESS: Tumburu, op. 264, no. 1.

WEILL, Kurt
2461 Die Dreigroschenoper (The threepenny opera). Mahagonny: Havana-
    Lied; Alabama-song; Wie mann sich bettet. Der Silbersee: Lied

der Fennimore; Cäsars Tod. Songs: Das Berliner Requiem: Ballade
vom ertrunken Mädchen; Happy end: Bilbao-song; Surabaya-Johnny;
Matrosen-tango. Lotte Lenya, s; Berlin Radio Free Orchestra;
Orchestra; Roger Bean, Wilhelm Bruckner-Ruggeberg. CBS 78279
(2). (Reissue)
+Gr 4-75 p1864   +STL 5-4-75 p37
+-HFN 5-75 p141   +-Te 6-75 p51
+-RR 4-75 p19

2462 The threepenny opera. Lotte Lenya, s; other soloists; Gunther-
Arndt Choir; Berlin Radio Free Orchestra; Wilhelm Bruckner-
Ruggeberg. CBS 77268 (2). (Reissue from 62264/5, ABL 3363)
(also Odyssey Y2 32977)
+HFN 1-73 p125   +-RR 12-72 p46
+NR 2-75 p10   /SFC 11-10-74 p27
The threepenny opera: Suite. cf KURKA: The good soldier Schweik:
Suite.
Kleine Dreigroschenmusik. cf MILHAUD: La création du monde.
Mahagonny: Havana-Lied; Alabama-song; Wie mann sich bettet. cf
Die Dreigroschenoper.
Der Silbersee: Lied der Fennimore; Cäsars Tod. cf Die Dreigroschen-
oper.
Songs: Das Berliner Requiem: Ballade vom ertrunken Mädchen;
Happy end: Bilbao-song; Surabaya-Johnny; Matrosen-tango. cf
Die Dreigroschenoper.
2463 Symphony, no. 1. Symphony, no. 2. BBC Symphony Orchestra; Gary
Bertini. Argo ZRG 755. (Reissue from HMV ASD 2390; Angel S
36506)
+-Gr 10-74 p713   +RR 9-74 p64
+HF 9-75 p75   +SFC 2-16-75 p24
++NR 1-75 p4   +Te 9-75 p41
2464 Symphony, no. 1. Symphony, no. 2. Leipzig Gewandhaus Orchestra;
Edo de Waart. Philips 6500 642.
+-Gr 7-75 p196   +RR 6-75 p56
+HF 9-75 p75   +SFC 5-18-75 p23
+HFN 6-75 p96   +SR 6-14-75 p46
++NR 6-75 p5   +-Te 9-75 p41
Symphony, no. 2. cf Symphony, no. 1 (Argo ZRG 755).
Symphony, no. 2. cf Symphony, no. 1 (Philips 6500 642).
WEINBERGER, Jaromir
Schwanda the bagpiper: Polka. cf HMV SLS 5019.
WEISGALL, Hugo
End of summer. cf GIDEON: The condemned playground.
WEISS, Sylvius
Bourrée. cf RCA ARL 1-0864.
Fantasie. cf Swedish Society SLT 33189.
Passacaglia. cf Philips 6833 159.
Suite, E major. cf BACH: Bourrée, E minor.
Tombeau sur la mort de M. Comte de Logy. cf BACH: Suite, lute,
S 995, G minor.
WELCHER, Dan
Concerto, flute and orchestra. cf DELLO JOIO: Homage to Haydn.
WELDON
Gate city. cf Michigan University SM 0002.
WELIN, Karl-Erik
Quartet, strings, no. 2. cf SHOSTAKOVICH: Quartet, strings, no. 4,
op. 83.

WELLESZ, Egon
    Octet, winds and strings, op. 67.   cf BADINGS: Octet, winds and
        strings.
    Sonata, solo violoncello, op. 30.   cf CRUMB: Sonata, solo violon-
        cello.
WENNERBERG
    Songs: Flickorna; Marketentersksorna.   cf BIS LP 17.
WERNICK, Richard
    Kaddish-requiem.   cf Nonesuch H 71302/3.
WESLEY
    The church's one foundation.   cf RCA ARL 1-0562.
WESLEY, Charles
    Voluntary III.   cf Wealden WS 160.
WESLEY, Samuel
    Air and gavotte.   cf Wealden WS 160.
WESLEY, Samuel Sebastian
    Blessed be the God and Father.   cf Pye GH 589.
    Choral song.   cf Wealden WS 160.
    Holsworthy church bells.   cf HMV HLM 7065.
    Magnificat and nunc dimittis, E major: Thou wilt keep him.   cf
        Polydor 2460 250.
    O give thanks; Who can express the noble acts of the Lord.   cf
        Abbey LPB 734.
WESTERGAARD, Peter
    Divertimento on Discobbolic fragments. cf Nonesuch HB 73028.
    Variations for six players.   cf BERGER: Septet.
WHITE
    Nobody knows the trouble I've seen.   cf Discopaedia MB 1009.
    Power.   cf Nonesuch H 71276.
WHITE, José Silvestre de los Dolores
    Concerto, violin, F sharp minor.   cf BAKER: Sonata, violoncello
        and piano.
WHITLOCK, Percy
    Allegretto.   cf Wealden WS 110.
    Folk tune.   cf Wealden WS 110.
    Scherzo.   cf Wealden WS 110.
WHITTENBERG, Charles
    Triptych, brass quintet.   cf Desto DC 6474/7.
    Variations for nine players.   cf WOLPE: Piece in two parts for
        solo violin.
WIDMANN, Erasmus
    Agatha.   cf DG Archive 2533 184.
    Canzona, galliard, intrada.   cf Grange SGR 1124.
    Magdalena.   cf DG Archive 2533 184.
    Regina.   cf DG Archive 2533 184.
WIDOR, Charles Marie
    Symphony, organ, no. 1, op. 13, C major: Marche pontificale.   cf
        Polydor 2460 252.
    Symphony, organ, no. 5, op. 42, no. 1, F minor: Toccata.   cf
        Decca SDD 463.
    Symphony, organ, no. 5, op. 42, no. 1, F minor: Toccata.   cf Decca
        5BBA 1013-5.
    Symphony, organ, no. 5, op. 42, no. 1, F minor: Toccata. cf Pye GH
        589.
    Symphony, organ, no. 5, op. 42, no. 1, F minor: Variations.   cf
        HMV HLM 7065.

Symphony, organ, no. 6, op. 42, no. 2, G minor: Allegro.   cf
    HMV HLM 7065.

WIENIAWSKI, Henryk
    Concerto, violin, no. 2, op. 22, D minor.   cf PAGANINI: Concerto,
        violin, no. 1, op. 6, D major.
    Concerto, violin, no. 2, op. 22, D minor.   cf RCA ARM 4-0942/7.
        Concerto, violin, no. 2, op. 22, D minor: Romance. cf RCA ARM
            4-0942/7.
    Fantaisie brillante on themes from "Faust", op. 20.   cf Discopaedia
        MB 1002.
    Mazurkas, op. 19: Obertass, Menetrier.   cf Rococo 2035.
    Mazurka, op. 19, no. 1, G major.   cf Connoisseur Society CS 2070.
    Polonaise.   cf RCA ARL 1-0735.
    Polonaise, op. 4, D major.   cf RCA ARM 4-0942/7.
    Polonaise brillante, no. 2, op. 21, A major.   cf Connoisseur
        Society CS 2070.
    Scherzo tarantelle.   cf RCA ARL 1-0735.
    Scherzo tarantelle (2).   cf RCA ARM 4-0942/7.
    Souvenir de Moscow, op. 6.   cf Discopaedia MB 1002.
    Souvenir de Moscou, op. 6.   cf Rococo 2072.

WIKMANSON, Johann
    Quartet, strings, op. 1, no. 2, E minor.   cf VERDI: Quartet, strings,
        E minor.

WILBYE, John
    Down in a valley.   cf Klavier KS 536.
    Lady, when I behold the roses.   cf Harmonia Mundi 204.
    Songs: Flora gave me fairest flowers; Adieu sweet Amaryllis; Away,
        thou shalt not love me; When shall my wretched life; Lady, when
        I behold; Unkind, O stay thy flying; Lady, your words do spite
        me; Thus saith my Cloris bright.   cf GIBBONS: Songs (Decca SXL
        6639).
    Thus saith my Cloris bright.   cf Advent 5012.

WILDER
    Brassininity.   cf Kendor KE 22174.

WILLAERT, Adrian
    E qui, la dira; Madonna qual certezza.   cf L'Oiseau-Lyre 12BB
        203/6.

WILLIAMS, Mansel
    Cwn pennant; Y Blodau Ger y Drws.   cf Argo ZRG 769.

WILLIAMS, William
    Sonata in imitation of birds.   cf Pelca PSR 40589.

WILLIAMSON, Malcolm
2465 Concerto, organ. Concerto, piano, no. 3, E flat major. Malcolm
        Williamson, org, pno; LPO; Adrian Boult, Leonard Dommett.
        Lyrita SRCS 79.
            +-Gr 5-75 p1979   +RR 5-75 p46
            +HFN 6-75 p96
2466 Concerto, piano and strings. Concerto, 2 pianos and strings.
        Epitaphs for Edith Sitwell. Gwyneth Pryor, Malcolm Williamson,
        pno; ECO Strings; Yuval Zaliouk. HMV EMD 5520.
            +Gr 3-75 p1667   +RR 4-75 p38
    Concerto, piano, no. 3, E flat major.   cf Concerto, organ.
    Concerto, 2 pianos and strings.   cf Concerto, piano and strings.
    Dignus est Agnus.   cf Abbey LPB 734.
    Epitaphs for Edith Sitwell.   cf Concerto, piano and strings.
    The musicians of Bremen.   cf HMV EMD 5521.

Symphony, voices. cf BENNETT: Calendar.
WILLS, Arthur
    Mass of St. Mary and St. Anne. cf Argo ZRG 785.
    Prelude and fugue (Alkmaar). cf BACH: Chorale preludes, S 669-671.
WILSON, Olly Woodrow
    Akwan. cf ANDERSON: Squares.
WILSON, Thomas
    Sinfonietta. cf RCA LFL 1-5072.
WIREN, Dag
    Quartet, strings, no. 5, op. 41. cf SHOSTAKOVICH: Quartet, strings,
        no. 4, op. 83.
    Serenade, strings, op. 11. cf BORODIN: Quartet, strings, no. 2,
        D major: Nocturne.
2467 Serenade, strings, op. 11. GRIEG: Holberg suite, op. 40. Elegaic
        melodies, op. 34, no. 2: The last spring. ECO; Johannes Somary.
        Vanguard Cardinal VCS 10067.
            ++ St 6-75 p52
2468 Symphony, no. 4, op. 27. ROSENBERG: Symphony, no. 2. Swedish
        Radio Symphony Orchestra, Stockholm Philharmonic Orchestra;
        Sixten Ehrling, Herbert Blomstedt. Turnabout TV 34436.
            /NR 12-71 p4   *St 6-75 p52
            /St 2-72 p96
WOLF, Hugo
2469 Intermezzo, string quartet. Italian serenade. Pieces, piano (3).
        Songs from the Schenkenbuch (5). Songs from the Book of
        Suleika (5). Raimund Gilvan, t; Elisabeth Schwarz, Frédéric
        Capon, pno; Keller String Quartet. Musical Heritage Society MHS
        1868.
            +St 7-75 p106
    Intermezzo, E flat major. cf CBS 76267.
    Italian serenade. cf Intermezzo, string quartet.
2470 Italienisches Liederbuch. Elisabeth Schwarzkopf, s; Dietrich
        Fischer-Dieskau, bar; Gerald Moore, pno. Angel SBL 3703 (2).
            +St 4-75 p68
2471 Mörike Lieder: Abschied; An den Schlaf; An die Geliebte; An eine
        Aeolsharfe; Auf ein altes Bild; Auf eine Christblume, I and II;
        Auf einer Wanderung; Auftrag; Begegnung; Bei einer Trauung;
        Denk es o Seele; Er ist's; Der Feuerreiter; Frage und Antwort;
        Fussreise; Der Gärtner; Gebet; Die Geister am Mummelsee; Der
        Genesene and die Hoffnung; Gesang Weylas; Heimweh; Im Fruhling;
        In der Frühe; Der Jäger; Jägerlied; Karwoche; Der Knabe und das
        Immlein; Der König bei der Krönung; Lebewohl; Lied eines Ver-
        liebten; Lied vom Winde; Neue Liebe; Nimmersatte Liebe; Pere-
        grina I and II; Schlafendes Jesuskind; Selbstgeständnis; Seufzer;
        Storchenbotschaft; Der Tambour; Um Mitternacht; Verborgenheit;
        Wo find'ich Trost; Zitronenfalter im April; Zum neuen Jahr; Zur
        Warnung. Daniel Barenboim, pno; Dietrich Fischer-Dieskau, bar.
        DG 2740 113 (3). (also DG 2709 053)
            +Gr 12-74 p1402   ++ON 5-75 p36
            +HF 5-75 p91   ++RR 12-74 p74
            ++NR 1-75 p15   +SR 2-22-75 p47
            ++NYT 1-26-75 pD26   ++St 4-75 p106
    Pieces, piano (3). cf Intermezzo, string quartet.
    Songs: Italienischer Liederbuch: Der Mond hat eine schwere Klag
        erhoben; Und willst du deinen Liebsten. cf BRAHMS: Songs
        (Discophilia KGG 4).

Songs: Auf einer Wanderung; Der Genesene an die Hoffnung; Mein
Liebster; Kennst du das Land. cf SCHUBERT: Songs (Decca SXL
6578).
Songs: Liebesfrühling; Nachtgruss; Der Musikant; Wohin mit der
Freud. cf WAGNER: Lohengrin.
Songs: Auch kleine Dinge; Ich hab in Penna; Wie lange schon war
immer mein Verlangen; Nein, junger Herr; Elfenlied; Das ver-
lassene Mägdlein; Mausfallensprüchlein. cf Philips 6833 105.
Songs from the Schenkenbuch (5). cf Intermezzo, string quartet.
Songs from the Book of Suleika (5). cf Intermezzo, string quartet.
2472 Spanisches Liederbuch. Elisabeth Schwarzkopf, s; Dietrich Fischer-
Dieskau, bar; Gerald Moore, pno. DG 2707 035 (2).
         +St 4-75 p68
2473 Spanish songbook: Tief im Herzen trag ich Pein; Die ihr schwebet;
Führ mich, Kind, nach Bethlehem; Köpfchen, Köpfchen, nich
gewimmert; Wunden trägst du, mein Geliebter; Sie blasen zum
Abmarsch; In dem Schatten meiner Locken; Herr, was trägt der
Boden hier; Sagt, Seid ihr es, feiner Herr; Alle gingen, Herz,
zur Ruh; Geh Geliebter geh jetzt; Ach, des Knaben Augen; Muh-
voll komm ich und beladen; Trau nicht der Liebe; Mögen alle
bösen Zungen; Bedeckt mich mit Blumen. Jan DeGaetani, ms;
Gilbert Kalish, pno. Nonesuch H 71296.
         +HF 9-74 p81   ++SFC 9-22-74 p22
         +-NR 8-74 p9   +-St 10-74 p141
         +-ON 5-75 p36
WOLFENDEN, Guy
Romeo and Juliet: Dance from Capulet's ball. cf BBC REB 191.
WOLF-FERRARI, Ermanno
Quattro rispetti, op. 11, no. 4. cf DEBUSSY: Le promenoir des
deux amants.
WOLKENSTEIN, Oswald von
Car Wunniklaich. cf Telefunken TK 11569/1-2.
In suria. cf Telefunken TK 11569/1-2.
WOLPE, Stefan
Form. cf Piece in two parts for solo violin.
Form. cf PERLE: Toccata.
Form IV. cf PERLE: Toccata.
Piece in two parts, 6 players. cf BUSONI: Berceuse elégiaque,
op. 42.
2474 Piece in two parts for solo violin. Form. RHODES: Duo, violin
and violoncello. WHITTENBERG: Variations for nine players.
Rosemary Harbison, Paul Zukofsky, vln; Russell Sherman, pno;
Robert Sylvester, vlc; Contemporary Chamber Ensemble; Arthur
Weisberg. DG/Acoustic Research 0654 087.
         +HF 2-71 p97   +SFC 4-18-71 p31
         +MJ 2-75 p29   /SR 12-26-70 p47
         +MQ 7-72 p501   +St 5-71 p95
         *NYT 1-3-71 pD22
Quartet, trumpet, tenor saxophone, percussion and piano. cf JONES:
Ambiance.
Quartet, trumpet, tenor saxophone, percussion and piano. cf
Nonesuch H 71302/3.
Trio, flute, violoncello and piano. cf CRUMB: Eleven echoes of
autumn, 1965.
WOOD, Charles
Hail, gladdening light. cf Abbey LPB 734.

Short communion service in the Phrygianmode: Sanctus and benedic-
    tus. cf HMV HQS 1350.
WOOD, Hugh
    Pieces, piano, op. 6 (3). cf BIRTWISTLE: Tragoedia.
WORDSWORTH, William
2475 Ballade. Cheesecombe suite, op. 27. Sonata, piano, op. 13, D
    minor. Margaret Kitchin, pno. Lyrita RCS 13. (Reissue from
    1963)
            +-Gr 9-75 p491   +RR 7-75 p54
            +HFN 12-75 p156
    Cheesecombe suite, op. 27. cf Ballade.
    Sonata, piano, op. 13, D minor. cf Ballade.
    Songs: Clouds, Red skies; The wind. cf Decca 6BB 197/8.
WORK, Henry Clay
    My grandfather's clock. cf RCA ARL 1-5007.
2476 Song of the Civil War era: Agnes by the river; The buckskin bag
    of gold; Come home, father; Crossing the grand Sierras (chorus
    only); Grandfather's clock; Drafted into the army; Kingdom
    coming; The picture on the wall; Now, Moses; Poor Kitty popcorn
    (or The soldier's pet); The silver horn; Take them away, they'll
    drive me crazy; Uncle Joe's "Hail Columbia"; When the evening
    star went down; Who shall rule this American nation. Joan Mor-
    ris, ms; Clifford Jackson, bar; Washington Camerata Chorus;
    William Bolcom, pno. Nonesuch H 71317.
            +NR 11-75 p10
    THE WORLD OF THE VIOLIN, Derek Collier. cf Decca SPA 405.
    THE WORLD'S FAVORITE TENOR ARIAS. cf Decca SXL 6649.
WRANITZKY, Pavel (Paul)
    German dances (10). cf DG Archive 2533 182.
    Quodlibet. cf DG Archive 2533 182.
    Symphony, C major. cf MOZART, L.: Divertimenti, B flat major
        and D major.
WRIGHT, Denis
    Concerto, cornet. cf BRYCE: Promenade.
WUORINEN, Charles
    Duo, violin and piano. cf SESSIONS: Sonata, piano, no. 3.
    Janissary music. cf PARRIS: Concerto, trombone.
    Making ends meet. cf Desto DC 7131.
    Sonata, piano. cf PERLE: Toccata.
2477 Speculum speculi. MARTINO: Notturno. Speculum Musicae; Daniel
    Schulman, Fred Sherry. Nonesuch H 71300.
            ++HF 4-75 p88   *St 6-75 p110
            +NR 1-75 p6
    Variations, flute, I and II. cf Nonesuch HB 73028.
WYNER, Yehudi
    Short fantasies (3). cf PERLE: Toccata.
WYNNE, David
    Evening prayers. cf Argo ZRG 769.
XENAKIS, Iannis
    Nomos alpha. cf DG 2530 562.
    THE YOUNG GILELS. cf Westminster WGM 8309.
YSAYE, Eugene
    Lointain passé. cf Rococo 2035.
    Rêve d'enfant, op. 14. cf LEKEU: Sonata, violin and piano, G major.
    Rêve d'enfant, op. 14. cf Discopaedia MB 1006.
2478 Sonatas, solo violin, op. 27, nos. 1-6. Ruggiero Ricci, vln.

Candide QCE 31085.
   ++Audio 9-75 p69   ++NR 8-75 p15
   +HF 7-75 p89   ++St 9-75 p118
Sonata, solo violin, no. 2. cf Hungaroton SLPX 11677.
Sonata, solo violin, no. 3. cf RCA ARL 1-0735.
Sonata, solo violoncello, op. 28. cf CRUMB: Sonata, solo violon-
   cello.
YTTREHUS, Roly
Sextet. cf BRUN: Gestures for eleven.
YUN, Tsang
Glissées. cf DG 2530 562.
ZACHAU (ZACHOW), Friedrich
Prelude and fugue, C major. cf Pelca PSR 40598.
Prelude and fugue, G major. cf Philips 6775 006.
Vom Himmel hoch da komm ich her. cf Pelca PSR 40598.
ZANDONAI, Riccardo
Francesca da Rimini: No smaragdi no...Inghirlandata di violette.
   cf London OS 26315.
ZANGUIS, Nicolaus
Magnificat secundi toni. cf Panton 010335.
ZANOTTI, Camillo
Tirsi morir volea. cf Panton 010335.
ZAREWUTIUS
Postlude, Benedicamus Dominicale. cf Hungaroton SLPX 11601/2.
ZELEZNY, Lubomir
March. cf Panton 110 361.
ZELLER, Karl
Der Obersteiger: Sei nicht bös. cf Angel S 35696.
Der Vogelhändler: Ich bin die Christel von der Post; Schenkt man
   sich Rosen in Tirol. cf Angel S 35696.
Der Vogelhändler: Rector and director; Roses in Tyrol. cf Pye
   NSPH 6.
ZELTER, Carl
Songs: Rastolose Liebe; Um Mitternacht; Gleich und Gleich; Wo
   geht's Liebchen. cf DG Archive 2533 149.
ZIANI, Marc
Prelude. cf Argo ZRG 746.
ZIEHRER, Karl
Fächerpolonaise. cf STRAUSS, J. I: Loreley Rheinklänge, op. 154.
Fächerpolonaise; Hereinspaziert; Samt und Seide; Singen, Lachen,
   Tanzen; Der Zauber der Montur. cf LANNER: Waltzes: Hofballtänze;
   Pesther Walzer; Die Schönbrunner, op. 200; Steirische Tänze.
Faschingskinder, op. 382. cf HMV SLS 5017.
Fidels Wien. cf HMV SLS 5017.
Hereinspaziert, op. 518. cf Decca SXL 6572.
Hoch und Nieder, op. 372. cf HMV SLS 5017.
Die Landstreicher overture. cf HMV SLS 5017.
Das liegt bei uns im Blut, op. 374. cf HMV SLS 5017.
Loslassen, op. 386. cf HMV SLS 5017.
Schönfeld, op. 422. cf HMV SLS 5017.
Weaner Madeln, op. 388. cf HMV SLS 5017.
Werner March. cf Pye NSPH 6.
Wiener Bürger, op. 419. cf Rediffusion 15-16.
Wiener Bürger, op. 419. cf Supraphon 114 1458.
ZIMBALIST, Efrem
Slavonic dances, no. 2. cf Discopaedia MB 1008.

Suite dans la forme ancienne: Sicilienne, Minuet. cf Discopaedia
    MB 1008.

ZIMMERMAN
    Fugue on Easter Alleluia. cf Hungaroton SLPX 11601/2.
    Prelude, A major. cf Hungaroton SLPX 11601/2.

ZIMMERMAN, Bernd
    Vier kurze Studien. cf DG 2530 562.

ZIPOLI, Domenico
    Largo and gavotte, B minor. cf Pandora PAN 101.
    Pastorale. cf Abbey LPB 738.
    Pastorale. cf Hungaroton SLPX 11548.
    Toccata. cf Telefunken AW 6-41890.

ZUCKERMAN, Mark
    Paraphrases, solo flute. cf CRI SD 342.

ZUNDEL
    Love divine, all loves excelling. cf RCA ARL 1-0561.

Section II

MUSIC IN COLLECTIONS

ABBEY

LPB 734
2479 BULLOCK: Give us the wings of faith. FERGUSON: O Jesus, I have
        promised (Wolvercote). LLOYD: View me, Lord, a work of Thine.
        MATHIAS: O sing unto the Lord a new song. MONTEVERDI: Beatus
        vir. PHILIPS: Ave, Jesu Christe. ROSE: Lord, I have loved
        the habitation of Thy house. TAYLOR: Glorious things of Thee
        are spoken (Abbot's Leigh). WARLOCK: Benedicamus Domino.
        WESLEY, S. S.: O give thanks; Who can express the noble acts
        of the Lord. WILLIAMSON: Dignus est Agnus. WOOD: Hail, glad-
        dening light. TRAD.: Now the green blade riseth (French).
        Michael Smith, org; Salisbury Cathedral Choir; Richard Seal.
            +-RR 4-75 p68

LPB 738
2480 BACH: Prelude and fugue, S 552, E flat major. BOYCE: Voluntary,
        no. 1, D major. FRICKER: Pastorale. GADE: Tre tonestukker,
        op. 22. MATHIAS: Processional. LISZT: Prelude and fugue on
        the name B-A-C-H, G 260. ZIPOLI: Pastorale. Donald Hunt, org.
            +RR 3-75 p40

ABC

ATS 20018
2481 Favorite duets with tenors. DONIZETTI: Anna Bolena: Debole io fu.
        Lucia di Lammermoor: Sulla tomba. MASSENET: Manon: Oui, Je
        fus cruelle et coupable. OFFENBACH: The tales of Hoffmann:
        C'est une chanson d'amour. Beverly Sills, s; Nicolai Gedda,
        Stuart Burrows, Carlo Bergonzi, t; NPhO, LSO; Julius Rudel,
        Thomas Schippers.
            +St 6-75 p109

ATS 20019
2482 The mad scenes. BELLINI: I Puritani: Qui la voce. DONIZETTI:
        Anna Bolena: Pianget voi. Lucia di Lammermoor: Il dolce suono.
        THOMAS: Hamlet: A vous jeux, mes amis. Beverly Sills, Other
        Singers; Ambrosian Opera Chorus; LPO, LSO, RPO; Julius Rudel,
        Thomas Schippers, Charles Mackerras.
            +St 6-75 p109

ACOUSTIC RESEARCH

Unnumbered
2483 Demonstration record, vol. 1: The sound of musical instruments.
            +-Audio 12-75 p105

465

ADVENT

5012
2484 Metropolitan Opera Madrigal Singers: Simple gifts. DOWLAND: What
     if I never speede. MORLEY: Now is the month of maying. TOMKINS
     To the shady woods. VASQUEZ: Ojos morenos. Não me firays.
     VECCHI: L'Humor svegghiato; Applauso. VICTORIA: O magnum mys-
     terium. WILBYE: Thus saith my Cloris bright. ANON.: Simple
     gifts (arr. Luboff). (and 9 others) Metropolitan Opera Madri-
     gal Singers.
          +-Audio 12-75 p104   +St 7-75 p108

AMBERLEE

ACL 501X
2485 ALBERT: Fantasia. BACH: Suite, lute, S 996, E minor: Prelude and
     bourrée. COSTE: Study, no. 22, op. 38, A major. PERNAMBUCO:
     Sons de carrilhoes. RUBIRA: Romance. RUIZ: Galliards. SOR:
     Andante pastorale, op. 32, no. 4, D major. Cantabile, no. 22,
     op. 35, B minor. Study, no. 15, op. 31, D major. Study, no.
     13, op. 35, C major. TARREGA: Adelita. Preludes, nos. 10 and
     15, D major. VILLA-LOBOS: Prelude, no. 1, E minor. TRAD. (arr.
     Llobet): Canco del Ladre. Testament d'Amelia. Edgard Zaldua,
     gtr.
          -Gr 1-75 p1372

AMON RA

SARB 01
2486 BORODIN: Quartet, strings, no. 2, D major: Nocturne. HAYDN: Quar-
     tet, strings, op. 3, no. 5, F major: Serenade. Quartet, strings
     op. 54, no. 1, G major: Allegretto. MENDELSSOHN: Quartet,
     strings, no. 1, op. 12, E flat major: Canzonetta. Quartet,
     strings, no. 2, op. 13, A minor: Intermezzo. MOZART: Quartet,
     strings, no. 15, K 421, D minor: Minuet and trio. SCHUBERT:
     Quartet, strings, no. 10, D 87, E flat major: Scherzo. Quartet,
     strings, no. 13, op. 29, K 804, A minor: Andante. SHOSTAKOVITCH
     Quartet, strings, no. 1, op. 49, C major: Scherzo. TCHAIKOVSKY:
     Quartet, strings, no. 1, op. 11, D major: Andante cantabile.
     Dartington Quartet.
          /HFN 6-75 p103   +RR 5-75 p50

ANGEL

SB 3816 (2)
2487 The English harpsichord. ARNE: Sonata, no. 3, G major. BACH, J. C
     Sonata, op. 5, no. 2, D major. BYRD: Alman. Callina Casturame.
     A fancie. French corantos. Lavolta, Lady Morley. Pavan and
     galliard, no. 4. Rowland or Lord Willoughby's welcome home.
     DUSSEK: Within a mile of Edinburgh. FARNABY: Farnaby's conceit.
     Giles Farnaby's dream. His humour. His rest. The new Sa-hoo.
     The old spagnoletta. Telle me Daphne. A toye. Up tails all.
     HANDEL: Suite, no. 8, F minor. Forest music (arr. Kipnis).
     PEERSON: The fall of the leafe. The primrose. PURCELL: Musick'
     handmaid, Z 648, 655, 656, 653. Suite, no. 8, F major. RED-
     FORD: Eterne rex altissime. TALLIS: A poynct. TOMKINS: The
     hunting galliard. Pavan, A minor. A sad pavan for these dis-
     tracted times. ANON.: My Lady Careys dompe. The short mesure

off My Lady Wynkfylds Rownde.   Igor Kipnis, hpd.
     +HF 4-75 p97   +St 5-75 p95
     ++NR 4-75 p12

S 35696 (also HMV ASD 2807. Tape (c) TC ASD 2807)
2488 Elisabeth Schwarzkopf sings operetta. HEUBERGER: Der Opernball:
     Im chambre separée. LEHAR: Giuditta: Meine Lippen, sie Küssen
     so heiss. Der Graf von Luxembourg: Hoch Evoë, Angèle Didier;
     Heut noch werd ich Ehefrau. Der Zarewitsch: Einer wird kommen.
     MILLOCKER: Die Dubarry: Ich schenk mein Herz; Was ich im Leben
     beginne. SIECZYNSKY: Wien, du Stadt meiner Traüme. STRAUSS, J.:
     Casanova: Nuns chorus. SUPPE: Boccaccio: Hab ich nur deine
     Liebe. ZELLER: Der Obersteiger: Sei nicht bös. Der Vogelhänd-
     ler: Ich bin die Christel von der Post; Schenkt man sich Rosen
     in Tirol. Elisabeth Schwarzkopf, s; PhO and Chorus; Otto
     Ackermann.
     +RR 8-73 p32   +St 4-75 p69
     ++RR 9-75 p78 tape

S 36076
2489 Duets with the Spanish guitar, vol. 3. BACH: Clavierübung: Duetto
     III. Partita, B flat major: Menuets I and II. Partita, B flat
     major: Gigue. FALLA: The three-cornered hat: Farruca. Canción.
     MIGNONE: Passarinho está cantando. OVALLE-BANDEIRO: Modinha.
     RAVEL: Pavane pour une infante défunte. Te tombeau de Couperin:
     Menuet. TCHAIKOVSKY: Serenade, strings, op. 48, C major: Waltz.
     TRAD.: Songs (3). Laurindo Almeida, gtr; Salli Terri, s; Martin
     Ruderman, flt; Vincent de Rosa, Fr hn.
     +HF 6-75 p107   +St 8-75 p108

S 36080
2490 GURLITT: Toy symphony, op. 169, C major. KLING: Kitchen symphony,
     op. 445. MEHUL: Overture burlesque. REINECKE: Toy symphony,
     C major. STEIBELT: Bacchanales, op. 53 (3). TAYLOR: Toy sym-
     phony: Adagio and finale. Instrumental Ensemble: Raymond Lewen-
     thal, pno.
     +NR 10-75 p3

S 37015
2491 BESARD: La bataille de poire. COUPERIN, F.: Les barricades myster-
     ieuses (trans. Diaz). COUPERIN, L.: Pavane, D minor (trans.
     Ghiglia). Le tombeau de M. Blancrocher (trans. Lorimer). FRES-
     COBALDI: Courante (trans. Ghiglia). Courante. Gagliarda (trans.
     Segovia). MILANO: La canzon delli Uccelli (trans. Chilesotti).
     Ricercare. RAMEAU: Menuets (trans. Segovia). RONCALLI: Passa-
     caglia. Gigue (trans. Segovia). SCARLATTI, D.: Sonatas, guitar,
     L 79, G major; L 187, A minor; L 352, E minor; L 483, A major.
     Oscar Ghiglia, gtr.
     ++NR 11-74 p12   ++St 1-75 p120
     +SFC 9-8-74 p30

S 37110 (also HMV HQS 1339)
2492 FERNANDEZ: Brasileira, no. 2: Ponteio, Moda, Cataretè. GUARNIERI:
     Dansa brasileira. Dansa negra. MIGUEZ: Nocturne. VIANNA:
     Dansa de negros. Jogos pueris. VILLA-LOBOS: A próle do bebé,
     no. 1. Ciclo brasileiro. Festa no sertão. Impressões seres-
     teiras. Chôros, no. 5 (Alma brasileira). Cristina Ortiz, pno.
     +Gr 1-75 p1372   ++NR 8-75 p13
     +HF 9-75 p95   +-RR 1-75 p52
     ++HFN 6-75 p97   +St 9-75 p115

S 37143
2493 BIZET: Les pêcheurs de perles: Leila, Leila, Dieu puissant le
        voilà. GLUCK: Orphée et Eurydice: Viens, viens, Eurydice, suis-
        moi. GOUNOD: Mireille: Vincenette à votre âge. Roméo et Juli-
        ette: Madrigal of Juliet and Roméo. LALO: Le Roy d'Ys: Cher
        mylio. MASSENET: Manon: J'ai marqué l'heure du départ. MEYER-
        BEER: Les Huguenots: Duet of Marguerite and Raoul. Mady Mesplé,
        s; Nicolai Gedda, t; Paris Opera Orchestra; Pierre Dervaux.
            /NR 12-75 p11   +SFC 11-2-75 p28
            +ON 12-13-75 p48

                                ARGO

ZDA 203
2494 The baroque sound of the trumpet. CLARKE: Suite, D major. HANDEL:
        Suite, D major. PEARSON: An Elizabethan fantasy: Now is the
        month of Maying; The willow song; The night watch. A medieval
        pageant: Agincourt song; Greensleeves, Summer is icumen in.
        PURCELL: Musick and ayres: Rondeau, Gavotte, Minuet, Trumpet
        tune. STANLEY: Trumpet tunes, nos. 1 and 2. Voluntary, D
        minor: Adagio. John Wilbraham, tpt; Leslie Pearson, org.
            +Gr 8-75 p342   ++HFN 7-75 p73
            ++HF 10-75 p136   +-RR 7-75 p50
ZRG 746
2495 DIEUPART: Sarabande, Gavotte, Menuet en Rondeau. HANDEL: Sonata,
        flute, op. 1, no. 11, F major. MATTEIS: Ground after the Scotch
        humour. PARCHAM: Solo, G major. PEPUSCH: Preludes. PURCELL,
        D.: Sonata, flute, D minor. PURCELL, H.: Prelude. ZIANI: Pre-
        lude. ANON.: Faronells ground. Tunes from The Bird Fancyer's
        Delight (6). Sonata, flute, G major. David Munrow, rec and
        flageolet; Oliver Brookes, bs viol and vlc; Robert Spencer,
        theorbo and gtr; Christopher Hogwood, hpd.
            +-Gr 10-74 p718   +HF 8-75 p107
ZRG 765
2496 BELLMAN (trans. Austin, arr. Best): Song at nightfall. BEST: Four
        songs of love's sorrow. DIBDIN (arr. Best): The Warwickshire
        lad. HOLCOMBE (real. Barlow): Airs for German flutes. JOHNSON,
        J.: Laveche's galliard. MORLEY: The turtle dove. PEASLEE/
        CARROLL (arr. Best): Alice. VENTADORN: Be M'an perdut. TRAD.
        (arr. Best): The cherry tree carol. The streams of lovely
        Nancy. Tomorrow shall be my dancing day. ANON.: A la fontenele.
        Angelus ad virginem. Ductia and danse royale. Le rossignol.
        Martin Best Consort.
            +MJ 2-75 p40   /RR 9-74 p83
ZRG 769
2497 ELWYN-EDWARDS: Coneuom y tri aderun. OWEN: Madonna songs (2).
        THOMAS: Prayers from the Gaelic (4). VAUGHAN THOMAS: Saith O
        Ganeuon. WILLIAMS: Cwn pennat; Y Blodau Ger y Drws. WYNNE:
        Evening prayers. Janet Price, s; Kenneth Bowen, t; Elinor
        Bennett, hp; Anthony Saunders, pno.
            +Gr 5-75 p2009   +RR 6-75 p81
            ++HFN 6-75 p98
ZRG 783
2498 BACH: Concerto, organ, no. 6, S 597, E flat major (attrib.).
        BUXTEHUDE: Toccata and fugue, F major. CABEZON: Diferencias
        sobre el canto Llano 1a alta. CARVALHO: Sonata, D major:

Allegro. DANDRIEU: Noëls: Le Roy des cieux vient de naître;
Adam fut un pauvre homme; A minuit fut fait un reveil; Chrétien
que suivez l'eglise; Joseph est bien Marié. HERON: Cornet voluntary. PESCETTI: Sonata, organ, C minor. RITTER: Sonatina, D
minor. SWEELINCK: More Palatino, variations. Peter Hurford,
org.
          ++Gr 2-75 p1520   +RR 2-75 p55

ZRG 785
2499 Songs: BRAHMS: We love the place, O God. ELGAR: Ave verum, op. 2,
     no. 1. IRELAND: Ex ore innocentium. POSTON: Jesus Christ the
     apple tree. SCHUBERT: Psalm no. 23, The Lord is my Shepherd,
     D 706. STANFORD: Te Deum, op. 66, B flat major. WALKER: Clear
     vault of Heav'n. WILLS: Mass of St. Mary and St. Anne. ANON.:
     Introit. Psalm 122, I was glad. The wind is in the North.
     TRAD.: Gabriel's message. St. Mary and St. Anne School Choir;
     Abbots Bromley, Barry Draycott, org; David Haines, Roy Curran,
     tpt; Llywela Harris.
          +Gr 1-75 p1380   +HFN 6-75 p99

ZRG 789
2500 Five centuries at St. George's. BAINTON: And I saw a new heaven.
     BATTEN: O praise the Lord. BRITTEN: Festival Te Deum. BYRD:
     Exsurge Domine. CAMPBELL: Jubilate Deo. FARRANT: Call to re-
     membrance. GIBBONS: O clap your hands. GREENE: Lord let me
     know mine end. HARRIS: Behold now praise the Lord. MARBECK:
     Credo. MUNDY: Sing joyfully. VAUGHAN WILLIAMS: Mystical songs:
     Antiphon. WALKER: I will lift up mine eyes. Simon Morris,
     Lester Gray, trebles; Timothy Rowe, bar; St. George's Chapel
     Choir; John Porter, org; Sidney Campbell.
          +Gr 4-75 p1856   +RR 3-75 p61
          +HFN 6-75 p100

ZRG 805
2501 ALBENIZ: España, op. 165, no. 2: Tango (arr. Kreisler). FALLA:
     La vida breve: Danse espagnole (arr. Kreisler). GRANADOS: Danse
     espagnole (arr. Kreisler). KREISLER: Old Viennese dances:
     Liebesfreud, Liebeslied, Schon Rosmarin. Syncopation. Tambour-
     in Chinois. Polichinelle sérénade. Rondino on a theme by Beet-
     hoven. POLDINI: Caprice viennois. Poupée valsante. Romance
     (arr. Kreisler). TARTINI: Variations on a theme by Corelli (arr.
     Kreisler). TRAD.: Shepherd's madrigal (arr. Kreisler). Ralph
     Holmes, vln; James Walker, pno.
          +Gr 6-75 p97   +RR 6-75 p66
          +HFN 6-75 p89

ZRG 807
2502 BOSSI: Etude symphonique. FRANCK: Chorale, no. 1. GIGOUT: Scherzo.
     HURFORD: Laudate Dominum. LANGLAIS: Te Deum. MATHIAS: Pro-
     cessional. STANLEY: Voluntary, C major. Peter Hurford, org.
          +-Gr 7-75 p220   +RR 7-75 p46
          ++HFN 6-75 p101

                         ARMSTRONG

721-4
2503 BACH, C. P. E.: Sonata, flute and harpsichord, G major. BOZZA:
     Dialogue. DUBOIS: Preludes faciles (9). MARCELLO: Sonata,
     flute and harpsichord, G major. SAINT-SAENS: Air de ballet.
     VAUGHAN WILLIAMS: Suite de ballet. Mark Thomas, flt; Russell

Woolen, pno, hpd.
    +MJ 7-75 p34

AVANT

AV 1012
2504 BACEWICZ: Sonata, piano, no. 2. BOULANGER: Cortege. D'un vieux
        jardin. JACQUET DE LA GUERRE: Suite, D minor. SZYMANOWSKA:
        Etudes, F major, C major, E major. Nocturne, B flat major.
        TALMA: Alleluia in form of toccata. Nancy Fierro, pno.
            ++NR 3-75 p11

BASF

BAC 3075
2505 English virginalists. BULL: English toy. Fantasia, D minor. The
        King's hunt. BYRD: Pavan and galliard of Mr. Peter. Walsingham
        variations. FARNABY: Maske, G minor. GIBBONS: Fantasia, D
        minor (2). Fancy, D minor. Pavane, G minor. TOMKINS: Bara-
        fostus' dream. Gustav Leonhardt, hpd.
            ++Gr 4-75 p1843   +-RR 4-75 p54
            +HFN 6-75 p99

BAC 3081
2506 Catches and partsongs. LANIERE: Though I am young. LAWES: Gather
        your rosebuds; See how in gathering. PURCELL: As Roger last
        night; He that drinks is immortal; I gave her cakes and ale;
        My Lady's coachman John; Sir Walter; Tom the taylor; Upon
        Christ Church bells in Oxford; When the cock begins to crow.
        RAVENSCROFT: By a bank; The marriage of the frog and the mouse;
        Of all the birds; Remember O thou man; A round of three country
        dances in one; The three ravens; Trudge away quickly; We be
        soldiers three; We be three poor mariners. Pro Cantione Antiqua;
        Mark Brown.
            +Gr 2-75 p1541   /RR 4-75 p64
            +HFN 6-75 p98

BAC 3087
2507 BRUCK: O du armer Judas. BURGK: The Lord with His disciples.
        DEMANTIUS: St. John Passion. LECHNER: Allein zu dir, Herr Jesu
        Christ. PRAETORIUS: O vos omnes. SCHUTZ: The words of the
        Institution of the Lord's supper. Stuttgart Cantata Choir;
        August Langenbeck.
            +Gr 4-75 p1860   +RR 5-75 p58
            +HFN 6-75 p107

BAB 9005
2508 BACH, C. P. E.: Concerto, harpsichord, strings and continuo, W 23,
        D minor: 1st movement. Concerto, harpsichord, W 46, F major:
        1st movement. BACH, J. S.: Concerto, harpsichord, no. 1, S 1052,
        D minor: 2d movement. Prelude, fugue and allegro, S 998, E flat
        major. BULL: The King's hunt. COUPERIN: Prelude, D minor.
        FROBERGER: Toccata, A minor. SWEELINCK: Toccata. Gustav Leon-
        hardt, Alan Curtis, hpd; Collegium Aureum.
            +Gr 5-75 p2024   +RR 4-75 p27
            ++HFN 6-75 p101

BC 21723
2509 Arias from Italian and French operas. DONIZETTI: La favorita: Una
        vergine. GOUNOD: Faust: Salut, demeure chaste et pure. HALEVY:

La Juive: Rachel, quand du Seigneur. MASSENET: Manon: Ah fuyez,
douce image. PUCCINI: La bohème: Che gelida manina. Gianni
Schicchi: Firenze è come un albero. Manon Lescaut: Donna non
vidi ami. Turandot: Nessun dorma. VERDI: Aida: Celeste Aida.
Luisa Miller: Quando le sere al placido. Il trovatore: Di quella
pira. Franco Bonisolli, t; Hamburg State Philharmonic Orchestra;
Leone Magiera.
    +—NR 3-75 p8   +St 5-75 p104

25 22286-1
2510 Love, lust, piety and politics. BROWNE: Woefully array'd. CORNYSSH:
Ah, Robin. Blow thy horn, hunter. Hoyda, jolly Rutterkin.
HENRY VIII, King: Pastime with good company. NEWARK: The farth-
er I go, the more behind. TURGES: Enforce yourself as Goddës
knight. ANON.: Deo gracias Anglia, Agincourt carol. Alas,
departing is ground of woe. And I were a maiden. Synge we to
this mery cumpane. Goday, my Lord, Syr Christenmasse. Tappster,
dryngker. Pro Cantione Antiqua; London Early Music Consort;
Bruno Turner.
    +RR 12-75 p86

292 2116-4 (2)
2511 ARNOLD: Colonel Bogey march. BEETHOVEN: York'scher Marsch. FUCIK:
Einzug der Gladiatoren. HENRION: Dreutzritter Fanfare. Fehr-
belliner Reitermarsch. HERZER: Hoch Heidecksburg. HOLZMANN:
Feuert los. KRETTNER: Tölzer Schütenmarsch. LUBBERT: Helenen
Marsch. MILES: Anchor's away. MOLTKE: Des grossen Kurfürsten
Reitermarsch. PIEFKE: Königgrätzer Marsch. SCHERZER: Bayer-
ischer Defiliermarsch. SEIFERT: Kärtner Liedermarsch. SOUSA:
Preussen's Gloria. Stars and stripes forever. SPOHR: Gruss
an Kiel. STRAUSS, J. I: Radetzky Marsch. TEIKE: Alte Kameraden.
WAGNER, J. F.: Unter dem Doppeladler. ANON.: Parademarsch der
18er Husaren (arr. Gursch). Pariser Einzugsmarsch. Peterburger
Marsch (arr. Gursch). Die grosse Garde-Regimentsmusik; Robert
Stolz.
    +RR 10-75 p56

BBC

REB 191
2512 Music to Shakespeare's plays. ARNE: As you like it: Under the
greenwood tree. CAIN: Much ado about nothing: Suite. HOLBORNE:
Romeo and Juliet: Heartsease. MORLEY: As you like it: It was
a lover and his lass. Twelfth night: O mistress mine. ROTA:
Romeo and Juliet: What is a youth. SULLIVAN: Henry VIII: Orpheus
with his lute. WOLFENDEN: Romeo and Juliet: Dance from Capulet's
ball. ANON.: Hamlet: How should I your true love know. Romeo
and Juliet: Where griping grief. Twelfth night: Peg O'Ramsay.
A winter's tale: Satyr's masque. Salle Le Sage, s; Martyn Hill,
t; Praetorius Consort.
    +Gr 4-75 p1680   +RR 4-75 p57

BEL CANTO CLUB

No. 3 (Reissues)
2513 BIZET: Les pêcheurs de perles: Mi par d'udir ancora. DONIZETTI:
Belisario: A si tremendo annunzio. Don Sebastiano: Deserto in
terra. Il Duca d'Alba: Angelo casto e bel. LEONCAVALLO: La

Bohème: Io non ho che una povera stanzetta. MASCAGNI: Cavalleria
rusticana: Trinklied. MASSENET: Romeo et Juliette: Ach gehe auf.
Werther: O wie süss hier zu weilen. OFFENBACH: Les contes de
Hoffmann: Ha wie in meiner Seele. PUCCINI: La fanciulla del
West: Lasset sie glauben. STRAUSS, R.: Morgen, op. 27, no. 4.
WEBER: Silvana: So soll denn dieses Herz. Alfred Piccaver, t;
Various Accompaniments.
        +Gr 4-75 p1864

                              BIS
LP 2
2514 BURKHART: Tre advetntssanger. CERTON: Psalms and nunc dimittis (3).
     DUFAY: Vergine bella. KUKUCK: Die Brücke. LUNDEN: Lilltåa och
     9 till. MACHAUT: Ballad and plus dure. NOTRE DAME SCHOOL: Flos
     filies and motet. Musica Intima.
        -RR 6-75 p78
LP 3
2515 DUNSTABLE: O rosa bella. ENCINA: Todos los bienes. GLOGAUER LIED-
     ERBUCH: Zwe Lieder. HEINTZ: Da truncken sie. MUSET: Quand je
     voi. PAUMANN: Ellend du hast. PHALESE: Dances. SENFL: Lieder
     (5). SUSATO: Dances. VON RUGEN: Loybere risen. ANON.: Can-
     ciones (2). Estampie and trotto. Sumer is icumen in. Jocula-
     tores Upsaliensis.
        /-Gr 7-75 p234   +RR 6-75 p83
LP 7
2516 ALAIN, J.: Variations on a theme by Jannequin, op. 78. COUPERIN,
     F.: Offertoire sur les grands jeux. DURUFLE: Prélude et fugue
     sur le nom d'Alain, op. 7. EKLUND: Pezzi per organi (3).
     OLSSON: Prelude and fugue, op. 56, D sharp minor. Hans Fagius,
     org.
        +-Gr 10-75 p661   +RR 7-75 p49
LP 17
2517 DVORAK: Die Bescheidene, op. 32, no. 8; Die Gefangene, op. 32, no.
     11; Scheiden ohne Leiden, op. 32, no. 4; Die verlassene, op.
     32, no. 6; Die Zuversicht, op. 32, no. 10. GEIJER: Dansen.
     KODALY: Csillagoknak teremtöje; Kiolvaso. PURCELL: Let us
     wander; Sound the trumpet; Two daughters of this aged stream.
     ROSSINI: Duetto buffo di due gatti; La pesca; La regata Venezi-
     ana. TCHAIKOVSKY: Au jardin, près du ruisseau, op. 46, no. 4;
     L'Aube, op. 46, no. 6; Larmes humaines, op. 46, no. 3. WENNER-
     BERG: Flickorna; Marketenterskorna. Elisabeth Söderström, s;
     Kerstin Meyer, ms; Jan Eyron, pno.
        +-Gr 6-75 p80   +RR 7-75 p58
LP 18
2518 LIGETI: Musica ricercata. MERILAINEN: Opusculum. MERIKANTO: Pre-
     ludio. SALLINEN: Cadenze. SEGERSTAM: Poem. Liisa Pohjola,
     pno; Paavo Pohjola, vln.
        -RR 11-75 p79

                             CAMDEN
CCV 5000-12, 5014-24
2519 Popular classics. BACH: Brandenburg concerti, nos. 1-3, S 1046-48,
     CCV 5007. BEETHOVEN: Overtures (7), CCV 5009. Symphony, no. 5,
     op. 67, C minor. Coriolan overture, op. 26, CCV 5023. Symphony,

no. 9, op. 125, D minor, CCV 5021. BIZET: Carmen: Suite,
CCB 5008. L'Arlesienne: Suites, nos. 1, 2, CCV 5011. BRAHMS:
Symphony, no. 1, op. 68, C minor, CCV 5018. CHOPIN: Preludes,
complete, CCV 5003. DVORAK: Symphony, no. 9, op. 85, E minor.
Carnival overture, op. 92, CCV 5012. GRIEG: Concerto, piano,
op. 16, A minor. Peer Gynt: Suites, CCV 5019. HANDEL: Royal
fireworks music. Water music, CCV 5002. MENDELSSOHN: Concerto,
violin, op. 64, E minor. BRUCH: Concerto, violin, no. 1, op.
26, G minor, CCV 5017. MOZART: Concerto, clarinet, K 622, A
major. Quintet, clarinet, K 581, A major. Serenade, no. 13,
K 525, G major. Symphony, no. 41, K 551, C major, CCV 5000.
RIMSKY-KORSAKOV: Scheherazade, op. 35, CCV 5010. RODRIGO: Concierto de Aranjuez. Fantasia para un gentilhombre, CCV 5004.
ROSSINI: Overture (6), CCV 5020. SCHUBERT: Symphony, no. 5,
D 485, B flat major. Symphony, no. 8, D 759, B minor, CCV 5001.
TCHAIKOVSKY: The nutcracker suite, op. 71a, excerpts, CCV 5022.
Symphony, no. 6, op. 74, B minor, CCV 5024. Concerto, piano,
no. 1, op. 23, B flat minor. Concerto, violin, op. 35, D major.
Romeo and Juliet. Francesca da Rimini, op. 32, CCV 5024, 5016,
5023. TARTINI: Devil's trill, CCV 5016. WAGNER: Tannhäuser:
Overture and Venusberg music. Die Walküre: Ride of the Valky-
ries. Tristan und Isolde: Prelude, Act 3. (also 6 titles,
CCV 5014) Various performers.
    +-HFN 8-75 p86

CANDIDE

CE 31068
2520 Catalan music from Medieval and Renaissance Spain. Ars Musicae
     Ensemble.
        +Audio 9-75 p69   +NR 7-74 p9

CANON

CNN 4983
2521 ALBENIZ: Suite española, no. 3: Sevillanas (arr. Krein). CLERISSE:
     Caravane. Introduction et scherzo. DEBUSSY: Children's corner
     suite, no. 6: Golliwog's cakewalk (arr. Krein). FRANCAIX: Pet-
     it quatuor. GROVLEZ: Les petites litanies de Jésus (arr. Krein).
     HARTLEY: Midnight sun. HAYDN: Quartet, strings, op. 76, no. 3,
     C major: Menuetto (arr. Smith). KREIN: Valse caprice. MENDEL-
     SSOHN: Scherzo a capriccio (arr. Krein). MOSZKOWSKI: Scherzino
     (arr. Krein). MOZART: Quartet, strings, no. 16, K 428, E flat
     major: Menuetto (arr. Krein). SCHUMANN: Whims (arr. Smith).
     TCHAIKOVSKY: Humoreske, op. 10, no. 2, G major (arr. Krein).
     Krein Saxophone Quartet.
        /RR 2-75 p46

CAPRICE

RIKS LP 46
2522 BORTZ: Nightwinds. JANNEQUIN: Le chant des oyseaux. MILHAUD:
     Sonnets composes au secret par Jean Cassou (6). MONTEVERDI:
     Madrigali a cinque voci, Libro V: Ecco Silvio...Ma se con la
     pieta...Dorinda, ah dirò...Ecco piegando...Ferir quel petto.
     TOCH: Geographische Fuge für Sprechchor. Camerata Holmiae;

Lars Edlund.
+-RR 10-75 p86

CBS

61579 (Reissues)
2523 BACH: Concerto, organ, no. 3, S 594, C major: Recitative (trans.
      Rosanoff). Pastorale, S 590, F major: Aria. FALLA: Spanish
      popular songs: Nana. HAYDN: Sonata, piano, no. 9, D major:
      Adagio. SCHUMANN: Fünf Stücke im Volkston, op. 102. TRAD (arr.
      Casals): Cant dels Ocells. Santi Marti del Canigo. Pablo
      Casals, vlc; Eugene Istomin, Leopold Mannes, pno; Prades Fest-
      ival Orchestra, Perpignan Festival Orchestra.
            +Gr 1-75 p1361   ++RR 9-74 p63
61599 (Reissues)
2524 CLAYTON: Come, come ye saints. DRAPER: All creatures of our God
      and King. DYKES: Lead kindly light. SULLIVAN: Onward Christian
      soldiers. TRAD.: A mighty fortress; How firm a foundation; Be-
      hold the Great Redeemer die; More holiness give me; The Lord
      is my Shepherd; O God, our help in ages past; I know that my
      Redeemer lives; Though in the outward church; Father in heaven;
      Abide with me; Tis eventide; Come follow me; God of our Fathers;
      Nearer my God to Thee; Rock of ages; Guide us, O thou great
      Jehovah; Ye simple souls who stray; The morning breaks; The
      shadows flee; Come, thou glorious day of promise. Richard Con-
      die, Alexander Schreiner, Frank Asper, org; Mormon Tabernacle
      Choir.
            -RR 1-75 p58

61654
2525 BARRIOS: Medallón antiguo. CASTELNUOVO-TEDESCO: Tonadilla on name
      of Andrés Segovia. LIBAEK: Musical pictures, guitar, nos. 2
      and 3. PONCE: Preludes, nos. 1-2, 4-6, 12. Thème varié et
      finale. TANSMAN: Mazurka. TARREGA: Preludes (6). VILLA-LOBOS:
      Chôros, no. 1, E minor. José Luis Gonzales, gtr.
            +-Gr 8-75 p351   +RR 7-75 p50
            +HFN 8-75 p89
72526. Tape (c) 40-72526
2526 BACH: Prelude, fugue and allegro, S 998, E flat major. GIULIANI:
      Sonata, guitar, op. 15: 1st movement. MUDARRA: Diferencias
      sobre El Conde claros. Fantasia. PRAETORIUS: Terpsichore:
      Ballet. REUSNER: Suite, no. 2, C minor: Paduana. TORROBA:
      Aires de La Mancha. VILLA-LOBOS: Preludes, no. 2, E major; no.
      4, E minor. John Williams, gtr.
            +RR 7-75 p71 tape
72950. Tape (c) 40-72950
2527 ALBENIZ: Suite española, no. 3: Sevillanas. FALLA: Homenaje.
      GRANADOS: Spanish dance, op. 37, no. 5: Andaluza. MUDARRA:
      Fantasia. SOR: Variations on a theme by Mozart, op. 9.
      TARREGA: Recuerdos de la Alhambra. TORROBA: Aires de La Mancha.
      TURINA: Fandanguillo, op. 36. VILLA-LOBOS: Preludes, no. 2, E
      major; no. 4, E minor. John Williams, gtr.
            +RR 7-75 p71
73396 (Some reissues)
2528 BACH: Concerto, harpsichord, S 971, F major. BEETHOVEN: Fantasia,
      op. 77, G minor. Sonata, piano, no. 27, op. 90, E minor. LISZT:
      Etudes d'exécution transcendente d'après Paganini, no. 3, G 140,

A flat minor. RAVEL: Pavane pour une infante défunte. SCHUMANN:
Fantasiestücke, op. 12, no. 3: Warum. Glenn Gould, Rudolf Serkin, André Watts, Charles Rosen, Murray Perahia, Philippe Entremont, pno.
    +-Gr 4-75 p1844   +-RR 4-75 p42
    +-HFN 6-75 p105

76267
2529 GERSHWIN: Lullaby. HAYDN: Andante and minuet, op. 103. MENDELSSOHN:
Andante and scherzo. PUCCINI: I crisantemi. SCHUBERT: Quartet,
strings, no. 12, D 703, C minor. WOLF: Intermezzo, E flat major.
Juilliard Quartet.
    +Gr 1-75 p1362   ++RR 1-75 p40

76420. Tape (c) 40-76420
2530 BACH: Partita, violin, no. 3, S 1006, E major: Prelude: Loure;
Gigue. BLOCH: Baal Shem suite: Nigun. CASTELNUOVO-TEDESCO:
Etudes d'ondes: Sea murmurs (trans. Heifetz). DEBUSSY: La plus
que lente (trans. Roques). FALLA: Spanish popular songs: Nana
(trans. Kochanski). KREISLER: La chasse (in the style of
Cartier). RACHMANINOFF: Etude-tableaux, op. 33, no. 4, B minor
(trans. Heifetz). RAVEL: Tzigane. Jascha Heifetz, vln; Brooks
Smith, pno.
    +Gr 12-75 p1075   +-RR 12-75 p78
    +HFN 12-75 p159

77513 (3)
2531 BACH: Anna Magdalena notebook, S 508: Little suite. Brandenburg
concerto, no. 3, S 1048, G major: Finale. Cantata, no. 80, A
mighty fortress is our God. Cantata, no. 140, Sleepers awake.
Cantata, no. 147, Jesu, joy of man's desiring. Praeludium, E
major. Suite, orchestra, S 1068, D major: Air on a G string.
Toccata and fugue, D minor. CHOPIN: Etude, op. 10, no. 3, E
major. Fantasie Impromptu, op. 66, C sharp minor. Mazurka,
op. 33, no. 2, D major. Nocturne, op. 9, no. 2, E flat major.
Polonaise, op. 40, no. 1, A major. Polonaise, op. 53, A flat
major. Prelude, op. 28, no. 7, A major. Waltz, op. 18, E flat
major. Waltz, op. 64, no. 2, C sharp minor. Waltz, op. 64,
no. 1, D flat major. Waltz, op. 70, no. 1, G flat major. GRIEG:
Concerto, piano, op. 16, A minor. Lyric pieces, op. 54: March
of the dwarfs. Norwegian dance, op. 35, no. 2. Peer Gynt:
Suite, no. 1, op. 46. Sigurd Jorsalfar, op. 56: Homage march.
Songs: Ich liebe dich, op. 5, no. 3. MOZART: Concerto, piano,
no. 21, K 467, C major: Andante. Divertimento, no. 17, K 334,
D major: Menuetto. Le nozze di Figaro, K 492: Overture. Sere-
nade, no. 13, K 525, G major. Sonata, piano, no. 11, K 311,
A major: Rondo alla Turca. Sonata, piano, no. 15, K 545, C
major: 1st movement. Variations on "Ah, vous dirai-je Maman",
K 265, C major. Don Giovanni, K 527: Menuet. STRAUSS, J. II:
An der schönen blauen Donau, op. 314. Frühlingsstimmen, op.
410. Geschichten aus dem Wienerwald, op. 325. Kaiserwälzer,
op. 437. Tritsch-Tratsch, op. 214. STRAUSS, Johann and Josef:
Pizzicato polka. E. Power Biggs, org; PO, Marlboro Festival
Orchestra, Columbia Chamber Orchestra, NYP, CO, Columbia Symph-
ony Orchestra; Philippe Entremont, Robert Casadesus, Glenn Gould,
André Previn, pno; Eugene Ormandy, Pablo Casals, Andre Kostela-
netz, Georg Szell.
    +HFN 10-75 p153   +-RR 10-75 p56

CENTURY/ADVENT

OU 97 734
2532 All American music concert. AHRENDT: Montage. CHADWICK: Jubilee.
     GERSHWIN: Rhapsody in blue. IVES (Schuman): Variations on
     America. SCHULLER: Dramatic overture. SMITH: Contours. Ohio
     University Symphony Orchestra; Richard Syracuse, pno; Adrian
     Gnam.
          ++MJ 7-75 p34

          CLASSIC RECORD LIBRARY (BOOK OF THE MONTH CLUB)
SQM 80-5731 (4)
2533 Chamber music society of Lincoln Center: BACH: Concerto, 2 harpsi-
     chords, S 1060, C minor. BEETHOVEN: Trio, strings, op. 9, no.
     1, G major. BRAHMS: Songs, piano and viola, op. 91 (2). CARTER:
     Etudes, woodwind quartet (8). FAURE: Sicilienne, op. 78. Fan-
     tasy, op. 79. Dolly, op. 56. HAYDN: Trio, strings, op. 53,
     no. 1, G major. MOSZKOWSKI: Suite, 2 violins and piano, op. 71,
     G minor. MOZART: Quartet, piano, no. 2, K 493, E flat major.
     SAINT-SAENS: Caprice on Danish and Russian airs, op. 79. SCHU-
     MANN: Andante and variations, op. 46, B flat major. Fantasie-
     stücke, op. 73. Charles Treger, Romuald Tecco, Jaime Laredo,
     vln; Walter Trampler, vla; Leslie Parnas, Laurance Lesser, vlc;
     Paula Robison, flt; Leonard Arner, ob; Gervase de Peyer, clt;
     Loren Blickman, bsn; Richard Goode, Charles Wadsworth, pno;
     John Barrows, hn; Alvin Brehm, bs; Anthony Newman, hpd; Maureen
     Forrester, con.
          +-HF 7-75 p90   ++SFC 6-13-75 p21
          +NYT 5-25-75 pD14   ++St 7-75 p100

          CLASSICS FOR PLEASURE
CFP 40205
2534 LISZT: Paraphrase WAGNER: Tannhäuser: Overture, G 442. MOZART:
     Quintet, clarinet, K 581, A major: 4th movement. PAGANINI:
     Caprices, op. 1, nos. 9, 14, 20, 24. PURCELL: The Indian Queen:
     Trumpet overture. STOCKHAUSEN: Zyklus. VILLA-LOBOS: Preludes,
     nos. 1, 2. Michael Laird, tpt; Keith Puddy, clt; Craig Sheppard,
     pno; Desmond Bradley, vln; Julian Byzantine, gtr; Tristan Fry,
     perc; Virtuosi of England, Gabrieli Quartet, Arthur Davison.
          +Gr 1-75 p1398   +RR 1-75 p43
CFP 40230
2535 Music of Wellington's time. BACH, J. C.: Marches for First and
     Second Battalions of Guard Regiment in Hanover. BEETHOVEN:
     Zapfenstreichen, nos. 1-3. EUSTACE: Les grenadiers de la Vieille
     Garde à Waterloo. FURGEOT: Marche des bonnets á poils. La
     favorite, Marche des pupilles de la garde. HAYDN: March for
     the Prince of Wales. Marches for the Derbyshire Cavalry Regi-
     ment (2). MEHUL: Le chant du départ. PAISIELLO: Marche du
     Premier Consul. Marche de la Garde Consulaire à Marengo. TRAD.:
     The British Grenadiers. General salute, bugle call (arr.
     Salzedo). Retreat, bugle call. Sonnerie aux morts. London
     Military Ensemble.
          +Gr 10-75 p640   +HFN 8-75 p80

THOMAS L. CLEAR

TLC 2580 (3)
2536 Augmented history of the violin on records, 1920-1950. Portions
     of violin concertos by Beethoven, Hindemith, Mendelssohn, Paga-
     nini, Tchaikovsky; sonatas by Bach, Beethoven, Grieg; miscellan-
     eous concert pieces. Jacques Thibaud, Jenö Hubay, Carl Flesch,
     Henri Marteau, Cecilia Hansen, Mischa Elman, Duci de Kerekjarto,
     László Szentgyörgi, Harry Soloway, Renee Chemet, Max Strub,
     Louis Zimmerman, Alexander Moguilewsky, Franz von Vecsey; Ibolyka
     Zilzer, Toscha Samaroff, Eddy Brown, Albert Spalding, Alfred
     Dubois, Gerhard Taschner, Alberto Bachmann, Richard Czerwonky,
     Heinz Stanske, Henri Merckel, Juan Manen, Alfredo San Malo,
     Manuel Quiroga, Gregoras Dinicu, Georges Enesco, Miguel Candela,
     René Bendetti, Samuel Gardner, Karl Freund, Joseph Hassid,
     Hugo Kolberg, Tossy Spivakovsky, vln; Various accompaniments.
          ++St 5-75 p105

CMS RECORDS

650/4, 660/4, 670/4 (12). Tape (c) 4650/4, 4660/4, 4670/4
2537 The musical heritage of America: Vol. 1, From Colonial times to the
     beginning of the Civil War: Who is the man; When Jesus wept;
     Tobacco's but an Indian weed; The little Mohee; The Indian
     Christmas carol and 52 others. Vol. 2, The Civil War: I am sold
     and going to Georgia; Run to Jesus; Johnnny, won't you ramble;
     Follow the drinking gourd; No more mourning and 40 others.
     Vol. 3, The winning of the West: Great Grandad; The Sherman
     cyclone; Common Bill; The Lane County batchelor; The housewife's
     lament and 41 others. Tom Glazer, narration, guitar, vocals;
     Kemp Harris, piano and vocals; Patt Moffit, Eileen Gibney,
     Jackie Spector, Pam Goff, Jane Olian, vocals; Dick Weissman,
     guitar, banjo; William Nininger, guitar; Tom Gibney, banjo,
     guitar, autoharp, pennywhistle, vocals.
          +St 8-75 p110

COLUMBIA

Special Products AP 12411
2538 BACH (Siloti): Prelude, organ, G minor. BRAHMS: Capriccio, op.
     76, no. 1, F major. Intermezzo, op. 116, no. 4, E major.
     DEBUSSY: Images: Reflets dans l'eau. Preludes, Bk I, no. 2:
     Voiles. Preludes, Bk II, no. 24: Feux d'artifice. LISZT: Har-
     monies poètiques et réligieuses, G 173: Funerailles. RACHMAN-
     INOFF: Prelude, op. 32, no. 12, G sharp minor. SCRIABIN: Etude,
     op. 8, no. 12, D sharp minor. Nocturne for the left hand alone,
     op. 9, no. 2. Daniel Pollack, pno.
          +NR 10-74 p11   +-SR 6-14-75 p46
M 31726. Tape (c) MT 31726 (ct) MA 31726 (also CBS 73227 (Q) MQ 31726)
2539 Monster concert: 10 pianos, 16 pianists. GOTTSCHALK: La gallina.
     Ojos criollos. JOPLIN: Maple leaf rag. ROSSINI: William Tell:
     Overture (arr. Gottschalk). Semiramide: Overture (arr. Czerny).
     SOUSA: Stars and stripes forever (arr. Gould and Riepe).
     STRAUSS, J. II: Thunder and lightning, op. 324. Blue Danube
     waltzes, op. 314. Eugene List, Frank Glazer, Barry Snyder,
     Maria Luisa Faini, Members of the Eastman School of Music Piano
     Faculty and Eastman School Graduates, pno; Samuel Adler.

                    +Gr 9-73 p507   *NYT 3-11-73 pD27
                    +Gr 7-75 p213   +RR 6-75 p76
                    +HF 5-73 p103   +SFC 3-4-73 p33
                    +HFN 6-75 p103   *St 4-73 p108
                    +-NR 5-73 p14

M 32070 (also CBS 73428)
2540 Philippe Entremont, à la Française. CHABRIER: Scherzo-Valse.
     DEBUSSY: Preludes, Bk I, no. 8: The girl with the flaxen hair.
     Rêverie. FAURE: Impromptu, no. 3, op. 34, A flat major. Noc-
     turne, op. 63, D flat major. POULENC: Pieces: Toccata. RAVEL:
     Miroirs: Alborada del gracioso. Pavane for a dead princess.
     Le tombeau de Couperin: Rigaudon. SATIE: Gymnopédies, nos. 1-3.
     Philippe Entremont, pno.
                    +-Gr 8-75 p351   ++NR 8-73 p15
                    +HF 8-73 p110   +-RR 6-75 p75
                    +-HFN 10-75 p142   +St 7-73 p114

CXM 32088
2541 BACH: Brandenburg concerto, no. 2, S 1047, F major: 1st movement.
     Fugue, S 578, G minor. BACHRACH: What's new pussycat. CARLOS:
     Dialogues, piano and 2 loudspeakers. Episodes, piano and elec-
     tronic sound. Geodesic dance. Pompous circumstances. LENNON:
     Eleanor Rigby. TCHAIKOVSKY: The nutcracker, op. 71: Russian
     dance; Dance of the sugar-plum fairy; Dance of the toy flutes.
     WAGNER: Lohengrin: Wedding march (arr. Carlos). Walter Carlos,
     synthesizer.
                    +NR 12-75 p16

M 32231
2542 Jennie Tourel and Leonard Bernstein at Carnegie Hall. DEBUSSY:
     Fetes galantes, Set I: Fantoches. DUPARC: La vie antérieure.
     LISZT: Oh, quand je dors, G 282. OFFENBACH: La Périchole: O
     mon cher amant; Ah, quel diner. RACHMANINOFF: O cease thy
     singing, maiden fair. SATIE: Le chapelier. SCHUMANN: Lieder-
     kreis, op. 39. TCHAIKOVSKY: Wait; Night. Jennie Tourel, ms;
     Leonard Bernstein, pno.
                    -HF 7-74 p110   +SR 7-13-74 p7
                    -NYT 9-22-74 pD32   +St 8-74 p124
                    +-ON 1-25-75 p32

M2X 33014, 33017, 33020, 33024, 33032 (17). (Reissues from MS 6011, 7029,
     6170, 6549, 6843, M 31011, MS 17141, M3S 776)
2543 Bernstein at Harvard. Vol. 1, Musical phonology, includes MOZART:
     Symphony, no. 40, K 550, G minor. Vol. 2, Musical syntax. Vol.
     3, Musical semantics, includes BEETHOVEN: Symphony, no. 6, op.
     68, F major. Vol. 4, The delights and dangers of ambiguity,
     includes BERLIOZ: Romeo et Juliette, op. 17, Part 1. DEBUSSY:
     Prélude à l'après-midi d'un faune. WAGNER: Tristan und Isolde:
     Prelude and Liebestod. Vol. 5, The twentieth-century crisis,
     includes IVES: The unanswered question. MAHLER: Symphony, no.
     9, D major: Adagio. RAVEL: Rapsodie espagnole: Feria. Vol. 6,
     The poetry of earth, includes STRAVINSKY: Oedipus Rex. Tatiana
     Troyanos, ms; René Kollo, Frank Hoffmeister, t; Tom Krause,
     David Evitts, bar; Ezio Flagello, bs; Harvard Glee Club; BSO,
     NYP; Leonard Bernstein, lecturer and conductor.
                    +-HF 4-75 p71

Melodiya M 33120
2544 CHATTERTON: Tu sola a me. LEONCAVALLO: I Pagliacci: Prologue.
     MOZART: Don Giovanni, K 527: Finch'han dal vino. RACHMANINOFF:

Aleko: The moon is high. RIMSKY-KORSAKOV: The legend of Sadko:
Song of the Venetian guest. The Tsar's bride: Griaznoy's aria.
The snow maiden: Mizghir's arioso. TCHAIKOVSKY: Mazeppa:
Ariosa. VERDI: Un ballo in maschera: Eri tu. La traviata: Di
provenza il mar. Il trovatore: Il balen del suo sorriso. Yuri
Mazurok, bar; Bolshoi Theatre Orchestra; Mark Ermler.
        +NR 1-75 p12   +St 3-75 p112
        +SR 1-25-75 p51

MG 33202 (2)
2545 BEETHOVEN: Sonata, piano, no. 19, op. 49, no. 1, G minor. Sonata,
     piano, no. 20, op. 49, no. 2, G major. CLEMENTI: Sonatinas,
     op. 36, nos. 1-6. DUSSEK: Sonatina, op. 20, no. 1. HAYDN:
     Sonata, piano, no. 35, C major. KUHLAU: Sonatinas, op. 20,
     nos. 1-3. Sonatinas, op. 55, nos. 1-3. MOZART: Sonata, piano,
     no. 15, K 545, C major. Philippe Entremont, pno.
        +HF 10-75 p87   ++SFC 3-2-75 p24
        +-MQ 10-75 p643   +St 7-75 p107
        ++NR 8-75 p13

M 33307
2546 Judith Blegen and Frederica von Stade, recital. BRAHMS: Songs:
     Klänge, op. 66; Klosterfräulein, op. 61, no. 2; Phänomen, op.
     61, no. 3; Walpurgisnacht, op. 75, no. 4; Weg der Liebe, op.
     20, nos. 1 and 2. CHAUSSON: Chanson perpetuelle, op. 37. MOZ-
     ART: Le nozze di Figaro, K 492: Non so più. SAINT-SAENS: Le
     bonheur est chose légère. SCARLATTI, A.: Endimione e Cinta:
     Se geloso e il mio core. SCHUBERT: Die Verschworenen. SCHUMANN:
     Songs: Botschaft, op. 74, no. 8; Das Glück, op. 79, no. 15.
     Judith Blegen, s; Frederica von Stade, ms; Charles Wadsworth,
     pno; Other artists.
        +-Audio 12-75 p104   +-NYT 6-8-75 pD19
        ++HF 7-75 p93   +ON 12-20-75 p38
        /NR 5-75 p12   +St 7-75 p107

D3M 33448 (3)
2547 Richard Tucker: In memoriam. BIZET: Carmen: Air de fleur. The
     pearl fishers: Je crois entendre. GIORDANO: Andrea Chénier: Un
     di all'azzuroo spazio. HALEVY: La Juive: Rachel, quand du
     Seigneur. LEONCAVALLO: I Pagliacci: Vesti la giubba. MASSENET:
     Le Cid: O Souverain, O juge, O père. MEHUL: Joseph: Champs
     paternels. MEYERBEER: L'Africaine: O paradiso. PONCHIELLI: La
     gioconda: Cielo e mar. PUCCINI: La bohème: Che gelida manina.
     La fanciulla del West: Ch'ella mi creda. Manon Lescaut: Donna
     non vidi mai; Guardate passo io son. Tosca: E lucevan le stelle.
     Turandot: Nessun dorma; Non piangere liù. VERDI: Aida: Celeste
     Aida. Luisa Miller: Quando le sere al placido. Requiem: Ingem-
     isco tamquam reus. Rigoletto: Parmi veder le lagrime. GIORDANO:
     Caro mio ben. GLUCK; Paride ed Elena: O del mio dolce ardor.
     GOLDFADEN: Rozshinkes mi Mandlin. HEUBERGER: In our secluded
     rendezvous. LEHAR: Yours is my heart alone. RODGERS: You'll
     never walk alone. ROSSINI: La danza. SIECZYNSKY: Vienna, city
     of my dreams. TORELLI: Tu lo sai. BOCK: Sunrise, sunset.
     FALVO: Dicintencello vuie. GOLD: The Exodus song. ANON.: Kol
     nidre. Kiddush. Tiritomba. Yehi Rotzon. Yir'u eineinu.
     Richard Tucker, t; Orchestras; Fausto Cleva, Nello Santi, Pierre
     Dervaux, Emil Cooper, Alfredo Antonini, Franz Allers, Sholom
     Secunda and others.
        +NR 8-75 p10   ++St 10-75 p120

XM 33513. Tape (c) XMT 33513 (ct) XMA 33513 (Q) XMQ 33153 Tape (ct)
        XQA 33513
2548 Footlifter: A century of American marches in authentic versions.
        ALFORD: Purple carnival. EVANS: Symphonia. FILLMORE: The
        footlifter. HUFFINE: Them basses. IVES: March intercollegiate.
        Omega Lambda Chi. JOPLIN: Combination march (arr. Schuller).
        REEVES: Second Connecticut Regiment. SOUSA: El Capitán; The
        gallant Seventh; The liberty bell; Semper Fidelis; Stars and
        stripes forever; The thunderer. Columbia All-Star Band; Gunther
        Schuller.
            +HF 10-75 p90   +NR 10-75 p13
            +MJ 11-75 p21   +St 12-75 p134

M 33514
2549 The antiphonal organs of the Cathedral of Freiburg. BANCHIERI:
        Dialogo per organo. BUXTEHUDE: Toccata and fugue, F major.
        CAMPRA: Rigaudon, A major. HANDEL: Aylesford pieces: Fugue,
        G major; Saraband; Impertinence. Samson: Awake the trumpet's
        lofty sound. Water music: Pomposo. KREBS: Fugue on B-A-C-H.
        MOZART: Adagio. PURCELL: Ayre, G major. Bonduca. Fanfare,
        C major. Rigaudon. Voluntary, C major. SOLER: The emperor's
        fanfare. E. Power Biggs, org.
            +NR 12-75 p14

                            COMMAND

COMS 9006
2550 DEBUSSY: Song: Ariettes oubliées, no. 2: Il pleure dans mon coeur
        (trans. Hartmann). FAURE: Elégie, op. 24, C minor. Papillon,
        op. 77. Sicilienne, op. 78. RACHMANINOFF: Vocalise, op. 34,
        no. 14. TCHAIKOVSKY: Capriccioso, op. 62. Nocturne, op. 19,
        no. 4. TOCH: Impromptu, op. 90c (arr. Solow). WEBER: Sonata,
        violoncello, A major. Jeffrey Solow, vlc; Doris Stevenson, pno.
            +NR 12-75 p9

                        CONNOISSEUR SOCIETY

CS 2063
2551 Ragas, music for meditation. Ali Akbar Khan, sarod; Accompanied
        by tamboura.
            +MJ 2-75 p40

(Q) CSQ 2065
2552 Great hits you played when you were young, vol. 3. BACH: Prelude,
        C minor. BEETHOVEN: Sonata, piano, no. 20, op. 49, no. 2, G
        major. CHOPIN: Waltz, op. 69, no. 1, A flat major. CHAMINADE:
        Scarf dance. DEBUSSY: Children's corner suite, no. 6: Golliwog's
        cakewalk. GRIEG: Lyric pieces, op. 43, no. 6: To the spring.
        LISZT: Liebestraum, G 541. MASSENET: Thaïs: Méditation. MOZART:
        Allegro, K 3. POLDINI: Dancing doll. SATIE: Gymnopédie, no. 1.
        SCHUBERT: Impromptu, op. 90, no. 3, G flat major. SCHUMANN:
        Album for the young, op. 68: Knight Rupert. Morton Estrin, pno.
            +St 2-75 p115

(Q) CSQ 2066
2553 Great hits you played when you were young, vol. 4. BACH: Musette,
        D major. Minuet, G major. BEETHOVEN: Sonatina, G major.
        CHOPIN: Waltz, op. 34, no. 2, A minor. DEBUSSY: Suite berga-
        masque: Clair de lune. DVORAK: Humoresque, op. 101. FALLA:

El amor brujo: Ritual fire dance. HELLER: L'Avalanche.
MACDOWELL: To a wild rose. MENDELSSOHN: Venetian boat song,
op. 19, no. 6. MOZART: Sonata, piano, no. 15, K 545, C major.
REINHOLD: Impromptu, C sharp minor. SCHUMANN: Fantasiestücke,
op. 12, no. 3: Warum. Morton Estrin, pno.
          +St 2-75 p115

CS 2070. (Q) CSQ 2070
2554 Wanda Wilkomirska, recital. BARTOK: Rumanian folk dances. DEBUSSY:
     Petite suite: En bateau. La plus que lente. KREISLER: London-
     derry air (arr.). MUSSORGSKY: Fair at Sorochinsk: Gopak. SARA-
     SATE: Danzas españolas, op. 22, no. 1: Romanza andaluza. SZYMAN-
     OWSKI: King Roger, op. 46: Roxane's song. WIENIAWSKI: Mazurka,
     op. 19, no. 1, G major. Polonaise brillante, no. 2, op. 21,
     A major. Wanda Wilkomirska, vln; David Garvey, pno.
          +-HF 5-75 p94   ++SFC 12-22-74 p24
          +NR 10-75 p4   +St 4-75 p99

                    CONTINENTAL RECORD DISTRIBUTORS
CRD 1008
2555 BACH: Cantata, no. 147, Jesu joy of man's desiring. BALDASSARE:
     Sonata, trumpet and organ, no. 1, F major. BOYCE: Voluntary,
     no. 1, D major. CLARKE: Trumpet voluntary. FIOCCO: Andante.
     GREENE: Introduction and trumpet tune. CHARPENTIER: Te Deum:
     Prelude. PURCELL: Sonata, trumpet and organ, C major. Trumpet
     tune and air (2). ROMAN: Suite, no. 2. STANLEY: Voluntary,
     op. 5, D major. VIVIANI: Sonata, trumpet, no. 1, C major. Alan
     Stringer, tpt; Noel Rawsthorne, org.
          +Gr 2-75 p1515   +RR 2-75 p50

CRD 1019
2556 ALFONSO X, El Sabio: Rosa das rosas. DOWLAND: Captain Digorie
     piper's galliard. The King of Denmark's galliard. PRAETORIUS:
     Peasant dances. Terpsichore: Fire dance; Stepping dance; Wind-
     mills; Village dance; Sailor's dance; Fishermen's dance; Festive
     march. SOTHCOTT: Fanfare. TRAD. English: Pavan, Good King Wen-
     ceslas. The dressed ship. Staines Morris. Here we come a-
     wassailing. Green garters. Fandango. God rest you merry gen-
     tlemen. I saw three ships. All hail to the days. TRAD. French:
     Branle de l'official. ANON. English: Edi beo thu. Ductia. As
     I lay. ANON. French: Alle psallite cum luya. ANON. Italian:
     La Manfredina. Saltarello. St. George's Canzona; John Sothcott.
          +HFN 12-75 p167   ++RR 12-75 p93

                                  CRI
SD 324
2557 HIBBARD: Trio. JENNI: Cucumber music. LEWIS: Gestes. MARTIRANO:
     Chansons innocentes. PURSWELL: It grew and grew. RILEY: Varia-
     tions II: Trio. Candace Nightbay, s; Jon English, bs trombone;
     Charles West, bs clarinet; Motter Forman, hp; Patrick Purswell,
     flt; Andreas Marchand, pno; Betty Bang Mather, alto flt, pic;
     William Hibbard, vla, toy pno; James Avery, pno, celesta; William
     Parsons, perc; Robert Strava, vln; Carolyn Berdahl, vlc.
          +Gr 12-75 p1066   +NR 2-75 p14

SD 342
2558 BAZELON: Duo, viola and piano. CROSS: Etudes, magnetic tape.

LANSKY: Modal fantasy. PLESKOW: Motet and madrigal. ZUCKERMAN:
Paraphrases, solo flute. Judith Allen, s; Paul J. Sperry, t;
Patricia Spencer, James Winn, flt; Linda Quan, vln; Allen
Blustine, clt; Fred Sherry, vlc; Karen Phillips, vla; Glenn
Jacobsen, Ursula Oppens, Robert Miller, pno; Charles Wuorinen.
Realized at the University of Toronto Electronic Music Studio.
        -NR 10-75 p14

CRYSTAL

S 201
2559 ARNOLD: Quintet, brass. BANCHIERI: Madrigal. BRADE: Allemande.
      GABRIELI: Canzona per sonare, nos. 1-4. GREP: Paduana. HOLBORNE:
      Honey suckle. Night watch. MAURER: Lied. Scherzo. RATHAUS:
      Tower music. SCHEIN: Paduana. ANON.: Die Bankelsangerlieder:
      Sonata. Berlin Brass Quintet.
            +HF 9-75 p98   +NR 1-75 p6
            +MJ 2-75 p39   +RR 1-75 p39

S 202
2560 BACH: Fuga IV. COLEMAN: Pieces, sackbuts and cornets (3) (trans.
      Baines). DAHL: Music, brass instruments. EAST: Desperavi.
      ENGELMANN: Paduana and galliarda (trans. Cran). FINCK: Greiner,
      zanner. HOLBORNE: Pieces (3). JEUNE: Revecy venir du printemps.
      SCHEIN: Banchetto musicale: Suite, no. 3. Annapolis Brass
      Quintet.
            +HF 9-75 p98   +NR 1-75 p6
            +MJ 2-75 p39

S 331
2561 BERG: Pieces, clarinet and piano, op. 5 (4). JEANJEAN: Carnival
      of Venice. Theme and variations. POULENC: Sonata, clarinet and
      piano. SCHUMANN: Fantasiestücke, op. 73. VAUGHAN WILLIAMS:
      Six studies in English folk song. James Campbell, clt; John
      York, pno.
            +NR 12-75 p8

DA CAMERA MAGNA

SM 93399
2562 Pictures from Israel. ACHRON: Hebrew lullaby. Scher. BEN HAIM:
      Sephardic lullaby. BLOCH: Baal shem. CHAJES: Hechassid, op.
      24, no. 1. ENGEL: Chabad melody and Freilachs, op. 20, nos.
      1, 2. KAMINSKI: Recitative and dance. KOUGUELL: Berceuse.
      LAVRY: Jewish dances (3). Theodore Mamlock, vln; Richard Laugs,
      pno.
            ++St 3-75 p114

DECCA

6BB 197/8 (2)
2563 BRAHMS: Songs: Botschaft, op. 47, no. 1; Immer leiser wird mein
      Schlummer, op. 105, no. 2; Der Tod das ist die kühle Nacht, op.
      96, no. 1; Von ewiger Liebe, op. 43, no. 1. FERGUSON: Discovery.
      RUBBRA: Psalm no. 6, O Lord, rebuke me not; Psalm no. 23, The
      Lord is my Shepherd; Psalm no. 150, Praise ye the Lord. SCHUBERT
      Songs: Du bist die Ruh, D 776; Du liebst mich nicht, D 756; Die
      junge Nonne, D 828; Rosamunde, no. 5: Romance, D 797; Suleika,

D 717; Der Tod und das Mädchen, D 531. SCHUMANN: Frauenliebe
und Leben, op. 42. WORDSWORTH, W.: Clouds; Red skies; The wind.
Also an introduction by Kathleen Ferrier on "What the Edinburgh
Festival has meant to me." Kathleen Ferrier, con; Bruno Walter,
Ernest Lush, pno.

  +Gr 11-75 p887   +RR 11-75 p18
  +HFN 12-75-156

SPA 372 Tape (c) KCSP 372 (ct) ECSP 372
2564 ALBENIZ: Cantos de España, op. 232, no. 5: Seguidillas. BEETHOVEN:
  Bagatelle, no. 25, A minor. BRAHMS: Intermezzo, op. 117, no.
  2, B flat minor. CHOPIN: Prelude, op. 28, no. 15, D falt major.
  DEBUSSY: Suite bergamasque: Clair de lune. DVORAK: Humoresque,
  op. 101, no. 7, G flat major. FAURE: Impromptu, no. 2, op. 31,
  F minor. GRANADOS: Goyescas: The maiden and the nightingale.
  GRIEG: Lyric pieces, op. 65, no. 6: Wedding day at Troldhaugen.
  LISZT: Valse oubliée, no. 1, G 215. MENDELSSOHN: Song without
  words, op. 67, no. 4, C major. SCHUBERT: Impromptu, op. 90,
  no. 2, D 899, E flat major. SCHUMANN: Romance, op. 28, no. 2,
  F sharp major. TCHAIKOVSKY: Humoresque, op. 10, no. 2, G major.
  Joseph Cooper, pno.

  +-Gr 2-75 p1523   +RR 2-75 p47
  +Gr 3-75 p1712   +RR 4-75 p76

SDD 390
2565 BIZET: Carmen: La fleur que tu m'avais jetée. Les pêcheurs de
  perles: De mon amie. GIORDANO: Andrea Chénier: Colpito que
  m'avete...Un dì, all'azzurro spacio; Come un bel dì di Maggio.
  GOUNOD: Faust: Quel trouble...Salut, Demeure. MASSENET: Manon:
  En fermant les yeux. Werther: Pourquoi me réveiller. PUCCINI:
  Tosca: Recondita armonia; E lucevan le stelle. Turandot: Non
  piangere liù; Nessun dorma. Giuseppe di Stefano, t; Rome,
  Santa Cecilia Orchestra; Franco Patané.

  +HFN 8-75 p91   +RR 7-75 p49
  /RR 9-73 p46

SDDJ 393/95 (3) (Reissues)
2566 ADAM: Giselle: Peasant pas de deux; Grand pas de deux; Finale.
  CHOPIN (arr. Douglas): Les sylphides: Prélude; Valse; Mazurka;
  Valse; Grand valse brillante. DELIBES: Coppélia: Prelude and
  mazurka; Valse lente; Thème slave varié; Czardas. FALLA: El
  sombrero de tres picos: The neighbours; The miller's dance;
  Final dance. El sombrero de tres picos: Suite, no. 2. HEROLD
  (arr. Lanchbery): La fille mal gardée: Simone; Clog dance; May-
  pole dance; Storm and finale; Spinning; Tambourine dance; Harv-
  esters. PROKOFIEV: Romeo and Juliet: Masks; Balcony scene;
  Death of Tybalt. ROSSINI (arr. Respighi): La boutique fantasque:
  Overture; Tarantella; Mazurka; Can-Can; Galop; Finale. STRAVIN-
  SKY: L'Oiseau de fue: Rondo; Infernal dance; Lullaby; Finale.
  TCHAIKOVSKY: The nutcracker, op. 71: Chinese dance; Trepak;
  Dance of the sugar-plum fairy; Waltz of the flowers; Pas de
  deux. Swan Lake, op. 20: Waltz, Scene; Dance of the little
  swans; Czardas. Sleeping beauty, op. 66: Introduction; Garland
  waltz; Rose adagio; Panorama. ROHO, OSCCP, OSR, LSO, Israel
  Philharmonic Orchestra; John Lanchbery, Jean Morel, Jean Marti-
  non; Peter Maag, Albert Wolff, Ernest Ansermet, Pierre Monteux,
  Georg Solti.

  +HFN 8-75 p91   +RR 10-73 p66

SPA 394
2567 BACH: Suite, orchestra, S 1067, B minor: Rondeau; Minuet; Badinerie.
     BEETHOVEN: Serenade, flute, violin and viola, op. 25, D major:
     Adagio; Allegro. BENEDICT: The gypsy and the moth. CIMAROSA:
     Concerto, 2 flutes and orchestra, G major: Allegro. DEBUSSY:
     Sonata, flute, viola and harp. GLUCK: Orfeo ed Euridice: Dance
     of the blessed spirits. HANDEL: Sonata, flute, op. 1, no. 5,
     G major. MOZART: Concerto, flute, no. 2, K 314, G major: 3rd
     movement. PERGOLESI: Concerto, flute, no. 1, G major. Claude
     Monteux, André Pepin, Alexander Murray, Jean-Pierre Rampal,
     Aurèle Nicolet, Christiane Nicolet, Richard Adeney, flt; Raymond
     Leppard, hpd; Claude Viala, vlc; Emanuel Hurwitz, Pierre Monteux,
     Karl Munchinger, Richard Bonynge.
          +HFN 12-75 p171   ++RR 12-75 p49
SPA 396 (Reissues from PFS 4158, SXL 2268, 6167, 2150/1, 6410, SKL 4663,
          Argo ZRG 5326)
2568 ARNOLD: Shanties, wind quintet (3). BRITTEN: Peter Grimes: Sea
     interludes, nos. 1 and 2. DEBUSSY: La mer: Dialogue between
     the wind and the sea. GRAINGER: Scotch strathspey and reel.
     IRELAND: Sea fever. RIMSKY-KORSAKOV: Scheherazade, op. 35: The s
     and Sinbad's ship. WAGNER: Der fliegende Holländer: Overture.
     Lorand Fenyves, vln; Kenneth McKellar, t; David Woolford, pno;
     NPhO, OSR, ROHO, ECO, London Wind Quintet; Carlos Paita, Ernest
     Ansermet, Benjamin Britten.
          ++Gr 8-75 p363   +-RR 7-75 p36
          +HFN 9-75 p109
SPA 405
2569 The world of the violin, Derek Collier. ALBENIZ: España, op. 165,
     no. 2: Tango (arr. Elman). Suite española, no. 3: Sevillanas
     (arr. Heifetz). BACH: Arioso (arr. Franko). BRANDL: The old
     refrain (arr. Kreisler). BRAHMS: Hungarian dance, no. 17 (arr.
     Kreisler). DVORAK: Slavonic dance, op. 46, no. 1, C major (arr.
     Kreisler). FAURE: Nocturne. KORNGOLD: Much ado about nothing,
     op. 11: Garden scene. KREISLER: Caprice viennois, op. 2. Schön
     Rosmarin. MENDELSSOHN: Song without words (arr. Kreisler).
     PAGANINI: Valse. RAMEAU: Tambourin chinois (arr. Kreisler).
     RIMSKY-KORSAKOV: The legend of Sadko: Chant Hindou. SCHUBERT:
     Cradle song (arr. Elman). Valse sentimentale (arr. Franko).
     TRAD.: Londonderry air (arr. Kreisler). Derek Collier, vln;
     Daphne Ibbot, pno.
          +Gr 8-75 p363   +RR 7-75 p49
          +HFN 8-75 p85
SDD 411 (Reissues from SET 346/8, SXL 2265)
2570 BACH: Christmas oratorio, S 248: Sinfonia. CORELLI: Concerto grosso,
     op. 6, no. 8, C minor. GLUCK: Chaconne. PACHELBEL: Canon (arr.
     Munchinger). RICCIOTTI: Concertino, no. 2, G major. SCO; Karl
     Munchinger.
          +Gr 3-75 p1667   +RR 12-74 p29
SPA 419
2571 BAX: Fanfare for the wedding of Princess Elizabeth, 1948. BLISS:
     Antiphonal fanfare for 3 brass choirs. BRITTEN: Fanfare for
     St. Edmondsbury. COATES: The dambusters march. ELGAR: Imperial
     march, op. 32. Pomp and circumstance marches, op. 39. WALTON:
     Crown imperial. ANON.: God save the Queen (arr. Britten). LSO
     and Chorus, Grenadier Guards Band, Philip Jones Brass Ensemble;
     Arthur Bliss, Rodney Bashford, Benjamin Britten.
          +-HFN 9-75 p109   +RR 10-75 p43

SDD 463
2572 BACH: Suite, orchestra, S 1068, D major. Cantata, no. 147, Jesu,
     joy of man's desiring (arr. Grace). BOELLMANN: Suite gothique,
     op. 25: Toccata. CLARKE: Trumpet voluntary (arr. Ratcliffe).
     DAVIES: Solemn melody. HANDEL: Water music: Air (arr. Blake).
     Serse: Largo (arr. Martin). KARG-ELERT: Nun danket alle Gott,
     op. 65/69. MULET: Carillon-sortie. STATHAM: Allegro. VAUGHAN
     WILLIAMS: Rhosymedre. WIDOR: Symphony, organ, no. 5, op. 42,
     no. 1, F minor: Toccata. Michael Nicholas, org.
          /Gr 1-75 p1375   /RR 2-75 p55
DPA 515/6
2573 ADAM: Giselle: Peasant pas de deux, Act 1; Grand pas de deux and
     finale, Act 2. CHOPIN: Les sylphides: Prélude; Valse; Mazurka;
     Valse; Grande valse brillante (arr. Douglas). DELIBES: Coppélia:
     Prelude and mazurka; Valse lente, Act 1; Theme slav varié;
     Czardas, Act 1. HEROLD: La fille mal gardée: Simone; Clog dance;
     Maypole dance; Storm and finale, Act 1; Spinning; Tambourine
     dance; Harvesters, Act 2 (arr. Lanchbery). ROSSINI: La boutique
     fantasque: Overture; Tarantella; Mazurka; Can-can; Galop; Finale
     (arr. Respighi). TCHAIKOVSKY: The nutcracker, op. 71: Chinese
     dance; Trepak; Dance of the sugar-plum fairy; Waltz of the
     flowers; Pas de deux, Act 2. Sleeping beauty, op. 66: Intro-
     duction; Garland waltz; Rose adagio, Act 1; Panorama, Act 2.
     Swan Lake, op. 20: Waltz, Act 1; Scene; Dance of the little
     swans, Act 2; Czardas, Act 3. ROHO, OSCCP, OSR, LSO; Israel
     Philharmonic Orchestra; John Lanchbery, Jean Morel, Jean Marti-
     non, Peter Maag, Ernest Ansermet, Pierre Monteux, Georg Solti.
          +HFN 10-75 p153   +RR 10-75 p43
DPA 517/8
2574 BERLIOZ: Béatrice et Bénédict: Vous soupirez, madame. BIZET: Car-
     men: C'est toi, C'est moi. Les pêcheurs de perles: C'est toi...
     Au fond du temple saint. PUCCINI: La bohème: O soave fanciulla;
     O Mimì, tu più non torni. Madama Butterfly: Bimba dagli occhi
     pieni di malia; Una nava da guerra...Scuoti quella fronda.
     Tosca: Mario, Mario. ROSSINI: Semiramide: Serbami ognor. VERDI:
     La forza del destino: Solenne in quest'ora. La traviata: Un
     dì felice. Il trovatore: Ai nostri monti. Don Carlos: Ascolta;
     Le porte dell'asil s'apron già. Joan Sutherland, Renata Tebaldi,
     April Cantelo, s; Marilyn Horne, Giuletta Simionato, Fiorenza
     Cossotto, Helen Watts, Regina Resnik, ms; Mario del Monaco,
     Carlo Bergonzi, Giuseppe di Stefano, Libero de Luca, t; Dietrich
     Fischer-Dieskau, Leonard Warren, Ettore Bastianini, bar; Various
     orchestras and conductors.
          +HFN 11-75 p173   +RR 11-75 p35
DPA 519/20 (2)
2575 BERLIOZ: Damnation of Faust, op. 24: Dance of the sylphs. CHABRIER:
     España. DEBUSSY: Prélude à l'après-midi d'un faune. DVORAK:
     Slavonic dance, op. 46, no. 1, C major. DUKAS: The sorcerer's
     apprentice. FAURE: Pavane, op. 50. MUSSORGSKY: A night on the
     bare mountain (orch. Rimsky-Korsakov). RAVEL: Daphnis et Chloe:
     Suite, no. 1. SAINT-SAENS: Danse macabre, op. 40. SMETANA: Ma
     Vlast: Vltava. WAGNER: Die Walküre: Ride of the Valkyries.
     Various orchestras; Leopold Stokowski, Stanley Black, Bernard
     Herrmann.
          +Gr 11-75 p915   +HFN 11-75 p173

GOSC 666/8 (Reissue from SET SXL GOS SDD)
2576 Grand opera gala: BEETHOVEN: Fidelio, op. 72, Ha, Ha, Ha, welch
     ein Augenblick. BELLINI: Norma: Deh, con te...Mira o Norma...
     Sì, fino all'ore streme. I puritani: Qui la voce. BOITO: Mefi-
     stofele: L'altra notte. DONIZETTI: L'Elisir d'amore: Quanto
     è bella, quanto è cara. La fille du regiment: Ah, mes amis...
     Que dire, que faire...Pour mon âme. GLUCK: Paride ed Elena: O
     del mio dolce ardor. MOZART: Don Giovanni, K 527: Crudele...
     Non mir dir; Il quali accessi...Mi tradi. Le nozze di Figaro,
     K 492: Non più andrai. Die Zauberflöte, K 620: Dies Bildnis.
     MUSSORGSKY: Boris Godunov: Death of Boris. PUCCINI: La fanciulla
     del West: Ch'ella mi creda. Tosca: Recondita armonia. Madama
     Butterfly: Con onor muore. ROSSINI: Il barbiere di Siviglia:
     A un dottor della mia sorte. La donna del lago: Tanti affetti.
     VERDI: Don Carlo: O don fatale. Falstaff: Ehi, Paggio...L'onore.
     La forza del destino: Pace, pace, mio Dio. Otello: Non ti
     crucciar...Credo. Simon Boccanegra: Come in quest'ora bruna.
     La traviata: Di provenza. Il trovatore: Tacea la notte. WAGNER:
     Götterdämmerung: Immolation scene. Die Walküre: Der Männer
     Sippe. Régine Crespin, Joan Sutherland, Maria Chiara, Zinka
     Milanov, Renata Tebaldi, Lisa della Casa, Leontyne Price, Kirsten
     Flagstad, Birgit Nilsson, s; Teresa Berganza, Marilyn Horne,
     Regina Resnik, ms; Luciano Pavarotti, Giuseppe di Stefano, Jussi
     Björling, Carlo Bergonzi, Stuart Burrows, t; Geraint Evans,
     Dietrich Fischer-Dieskau, Sherrill Milnes, Tom Krause, bar;
     Fernando Corena, Cesare Siepi, Nicolai Ghiaurov, bs; Various
     choruses, orchestras and conductors.
                +HFN 10-75 p153   +RR 10-75 p28
ECS 774
2577 BRITTEN: Missa brevis, op. 63, D major: Agnus Dei; Gloria in ex-
     celsis. CARTER: Lord of the dance. GOSS: Praise, my soul, the
     King of Heaven. HORSLEY: There is a green hill far away.
     McBAIN: Brother James' air (arr. Jacob). MENDELSSOHN: Song:
     Hear my prayer. TRAD.: Steal away (arr. Trant). Lord of all
     hopefulness. Coventry carol. Robert Weddle, org; Coventry
     Cathedral Boys' Choir; David Lepine.
                +-RR 9-75 p68
5BBA 1013-5 (3) (Reissues from ZRG 5419-20, 5448, 571, 663, 5339, 503,
                5237, 528, ZFA 68, 47)
2578 BACH: Chorale preludes, S 645-50. BRAHMS: Chorale preludes, organ,
     op. 122, nos. 1, 4, 8, 10. BRITTEN: Prelude and fugue on a
     theme by Vittoria. DAVIES: O magnum mysterium. ELGAR: Sonata,
     organ, op. 28, G major. FRANCK: Pièce héroïque. Prelude, fugue
     and variation, op. 18. HINDEMITH: Sonata, organ, no. 3. LISZT:
     Prelude and fugue on the name B-A-C-H, G 260. MESSIAEN: L'asen-
     sion. MOZART: Fantasia, K 594, F minor. PURCELL: Trumpet tune
     (arr. Trevor). REGER: Toccata and fugue, op. 59. VIERNE: Sym-
     phony, no. 1, op. 14, D minor: Final. WIDOR: Symphony, organ,
     no. 5, op. 42, no. 1, F minor: Toccata. Simon Preston, org.
                +Gr 10-75 p661
ECS 2159
2579 BIZET: Agnus Dei. Carmen: Smugglers chorus; March and chorus, Act
     4. ELGAR: The dream of Gerontius, op. 38: Praise to the holiest.
     MENDELSSOHN: Elijah: Thanks be to God; He that shall endure to
     the end; And then shall your light break forth. PUCCINI: Madama
     Butterfly: Humming chorus. VERDI: Aida: Grand chorus, Act 3.

Nabucco: Speed your journey. Charles MacDonald, org; York
Celebrations Choir; John Warburton.
            /RR 3-75 p62

FS 4323
2580 BIZET: Carmen: March of the toreadors, Act 4. GOUNOD: Faust:
     Soldiers' chorus. LEONCAVALLO: I Pagliacci: Bell chorus, Act 1.
     MUSSORGSKY: Boris Godunov: Coronation scene. PUCCINI: Madama
     Butterfly: Humming chorus, Act 2. VERDI: Aida: Grand march,
     Act 2. Nabucco: Va pensiero, Act 3. WAGNER: Tannhäuser: Pil-
     grims' chorus, Act 3. Kingsway Symphony Orchestra and Chorus;
     Salvador Camarata.
            +Gr 6-75 p90   +RR 6-75 p25
            +HFN 6-75 p100

SXL 6572. Tape (c) KSXC 6572. (also London CS 6791)
2581 LEHAR: Gold and silver waltz, op. 79. STRAUSS, J. I: Sperl, Piefke
     und Pufke. STRAUSS, J. II: Explosions, op. 43. Lob der Frauen,
     op. 315. Morgenblätter, op. 279. Persische Marsch, op. 289.
     Unter Donner und Blitz, op. 324. STRAUSS, Josef: Eingesendent,
     op. 240. SUPPE: Die schöne Galatea: Overture. ZIEHRER: Herein-
     spaziert, op. 518. VPO: Willi Boskovsky.
            +HFN 2-73 p348 tape   +RR 11-72 p64
            +NR 10-75 p2

SXL 6629 (also London OS 26376)
2582 Songs and operatic arias, Renata Tebaldi. BONONCINI: Deh più a me
     non vascondete. GLUCK: Alceste: Divinités du Styx. Paride
     ed Elena: O del mio dolce ardor. HANDEL: Alcina: Verdi prati.
     Xerxes: Ombra mai fù. MARTINI il TEDESCO: Plaisir d'amour.
     PAISIELLO: La molinaro: Nel cor più non mi sento. I Zingari in
     in Fiera: Chi vuol la zingarella. PERGOLESI: La serva padrona:
     Stizzoso, mio stizzoso. Songs: Tre giorni son che Nina. SARTI:
     Lungi dal caro bene. SCARLATTI, A.: Le violette. VIVALDI: Pian-
     go gemo sospiro. (Songs arr. by Gamley) Renata Tebaldi, s;
     NPhO; Richard Bonynge.
            -Gr 8-75 p360   -NYT 6-8-75 pD19
            -HF 9-75 p103   +-ON 8-75 p28
            +HFN 9-75 p103   +RR 10-75 p85
            ++NR 8-75 p9   +-St 9-75 p123

SXL 6637 (also London OS 26379)
2583 BIZET: La jolie fille de Perth: La, la, la, la...Quand la flamme
     de l'amour. GOUNOD: Faust: Vous qui faîtes l'endormie. MASSEN-
     ET: Hérodiade: Dors, O cité perverse...Astres étincellants. Le
     jongleur de Notre Dame: La Vierge entend fort bien. MEYERBEER:
     Les Huguenots: Un vieil air Huguenot...Piff, paff. THOMAS: Le
     Caïd: Je comprends que la belle aime le militaire. VERDI: Don
     Carlo: Elle ne m'aime pas. Les vespres siciliennes: Palerme,
     O mon pays...O toi Palerme. Joseph Rouleau, bs; Ambrosian Sing-
     ers; ROHO; John Matheson.
            +Gr 10-74 p756   -ON 4-12-75 p36
            +-HF 1-75 p98   +-Op 12-74 p1087
            +-MJ 2-75 p31   +RR 10-74 p27
            +-NR 1-75 p12

SXL 6643 (also London CS 6858)
2584 Virtuoso overtures. MOZART: Le nozze di Figaro: Overture. ROSSINI:
     La gazza ladra: Overture. STRAUSS, J. II: Die Fledermaus, op.
     363: Overture. WAGNER: Rienzi: Overture. WEBER: Der Freischütz:
     Overture. LAPO: Zubin Mehta.

DECCA (cont.)   488

-Gr 6-74 p66   -RR 6-74 p53
+NR 1-75 p4   +-SFC 1-12-75 p29
SXL 6649 (also London OS 26384)
2585 BIZET: Carmen: Flower song. FLOTOW: Martha: M'appari. GOUNOD:
     Faust: Salut demeure. LEONCAVALLO: I Pagliacci: Vesti la giubba
     PUCCINI: La bohème: Che gelida manina. Tosca: E lucevan le
     stelle. Turandot: Nessun dorma. VERDI: Aida: Celeste Aida.
     Rigoletto: La donna è mobile. Il trovatore: Di quella pira.
     Luciano Pavarotti, t; John Alldis Choir; LPO, NPhO, Vienna
     Volksoper Orchestra and Chorus, BPhO, LSO, RPO; Leone Magiera,
     Richard Bonynge, Herbert von Karajan, Zubin Mehta, Nicola Res-
     cigno.
+-Gr 7-75 p241   +RR 7-75 p16
+-HFN 8-75 p91   ++SFC 10-12-75 p22
++NR 12-75 p11

SXL 6719
2586 Darwin, song for a city. BELLINI: Norma: Dormono entrambi.
     OFFENBACH: Genevieve de Brabant: Gendarmes' duet. VERDI: I
     Lombardi: Qui posa il fianco. Rigoletto: Un dì se ben rammen-
     tomi. Australian national anthem: Advance Australia fair.
     British national anthem: God save the Queen. Joan Sutherland,
     Margreta Elkins, s; Heather Begg, sm; Graham Clark, t; Tom
     McDonnell, Louis Quilico, bar; Clifford Grant, bs; Erich
     Gruenberg, vln; RPO; Richard Bonynge.
+Gr 3-75 p1702   +RR 4-75 p18

                              DESTO
DC 6474/7 (4)
2587 Music for brass, 1500-1970. BERGSMA: Suite, brass quartet. BOZIC:
     Kriki, brass quintet. DAHL: Music, brass instruments. DOWLAND:
     Dances (4). EAST: Desperavi. Triumphavi. EWALD: Quintet,
     op. 5, B flat minor. GLAZUNOV: In modo religioso, op. 38.
     HINDEMITH: Morgenmusik. MANZONI: Quadruplum. PALESTRINA: Ricer-
     car sopra il primo tuono. PEZEL: Seventeenth century dances (6)
     POULENC: Sonata, trumpet, horn and trombone. REICHE: Baroque
     suite. SIMON: Quatuor en forme de sonatine, op. 23, no. 1.
     STARER: Miniatures (5). SUSATO: Flemish dances (5). TANENBAUM:
     Improvisations and patterns, brass quintet and tape. WHITTEN-
     BERG: Triptych, brass quintet. American Brass Quintet.
+-Gr 6-75 p62

DC 7118/9
2588 Jennie Tourel at Alice Tully Hall. BEETHOVEN: Songs: An die Hoff-
     nung, op. 94; Ich liebe dich, G 235. BERLIOZ: Absence. DARGO-
     MIJSKY: Romance. DEBUSSY: Chansons de Bilitis (3). GLINKA:
     Songs: Doubt, Vain temptation. HAHN: Si mes vers avaient des
     ailes. LISZT: Songs: Comment disaient-ils, G 276; O, quand je
     dors, G 282; Uber allen Gipfeln ist Ruh, G 306; Vergiftet sind
     meine Lieder, G 289; Mignon's Lied, G 275. MASSENET: Elégie.
     MONSIGNY: Rose et Colas: La Sagesse est un trésor. OFFENBACH:
     La Périchole: Laughing song. STRADELLA: Per Pietà. TCHAIKOVSKY
     Songs: None but the lonely heart. Jennie Tourel, ms.
+Gr 3-75 p1692   +ON 2-12-72 p34
+HF 12-71 p115   ++St 2-72 p100
+NYT 9-22-74 pD32

DC 7131
2589 Music for piano, four hands. MOSS: Omaggio. SPIEGELMAN: Kousochki.
     SYDEMAN: Concerto, piano, 4 hand and chamber orchestra. WUORI-
     NEN: Making ends meet. Jean and Kenneth Wentworth, pno; Con-
     temporary Music Ensemble; Arthur Weisberg.
          +MQ 4-75 p326   +St 4-73 p130

                    DEUTSCHE GRAMMOPHON

2530 332
2590 Sonnets of Petrarch. LISZT: Songs: Benedetto sia'l giorno; I'vidi
     in terra angelici costumi; Pace non trove. PFITZNER: Voll jener
     süsse. REICHARDT: Songs: Canzon, s'al dolce loco; Di tempo in
     tempo; Erano i capei d'oro; O poggi, O valli, O fiumi; Or ch'il
     ciel; Più volte già dal bel sembiante. SCHUBERT: Songs: Allein,
     nachdenklich, gelähmt, D 629; Apollo, lebet noch dein hold Ver-
     langen, D 628; Nunmehr, da Himmel, Erde schweigt, D 630. Diet-
     rich Fischer-Dieskau, bar; Jörg Demus, Gerald Moore, pno.
          +Gr 8-74 p393   +NYT 4-28-74 pD24
          +-HF 11-74 p121   +-ON 5-75 p36
          +-MJ 7-74 p48   -St 7-74 p115
          +-NR 7-74 p99

2530 504
2591 Spanish songs from the Middle Ages and Renaissance: ALFONSO X, El
     Sabio: Rosa das rosas. Santa Maria. ENCINA: Romerico. FUEN-
     LLANA: Perdida de antequera. MILAN: Toda mi vida hos amé.
     Aquel caballero, madre. MUDARRA: Si me llaman a mi. Claros y
     frescos ríos. Triste estaba el rey David. Isabel, perdiste
     la tu faxa. NARVAEZ: Con qué la lavaré. TRIANA: Dinos, madre
     del donsel. TORRE: Dime, triste corazon. VALDERRABANO: De
     dónde venís, amore. VASQUEZ: Vos me matastes (arr. Fuenllano).
     En la fuente del rosel (arr. Pisador). ANON.: Dindirindin. Los
     hombres con gran plazer. Nuevas te traygo, Carillo. Teresa
     Berganza, ms; Narciso Yepes, gtr.
          +-Gr 3-75 p1692   ++RR 3-75 p66
          +-HF 8-75 p102   +SR 6-14-75 p46
          ++NR 5-75 p11   ++St 7-75 p106
          +NYT 6-8-75 pD19

2530 561
2592 BUSSOTTI: Ultima rara. CAROSO: Laura soave: Balletto, Gagliarda,
     Saltarello (Balleto). CASTELNUOVO-TEDESCO: La guarda cuydadosa.
     Tarantella. GIULIANI: Grande ouverture, op. 61. MURTULA: Tar-
     antella. PAGANINI: Sonata, op. 25, C major. RONCALLI: Suite,
     G major. ANON.: Suite, lute. Pieces, lute (5). Siegfried
     Behrend, gtr; Claudia Brodzinska Behrend, vocalist.
          +Gr 11-75 p869   +RR 11-75 p75
          +HFN 12-75 p155

2530 562
2593 BROWN: Music, violoncello and piano. KAGEL: Unguis incarnatus est.
     PENDERECKI: Capriccio per Siegfried Palm. WEBERN: Sonata, viol-
     oncello. Stücke, op. 11 (3). XENAKIS: Nomos alpha. YUN: Glis-
     sées. ZIMMERMAN: Vier kurze Studien. Siegfried Palm, vlc;
     Aloys Kontarsky, pno.
          ++RR 11-75 p84

2530 598
2594 ANCHIETA: Con amores, la mi madre. ESTEVE Y GRIMAU: Alma, sintamos;

Ojos, llorar. GRANADOS: Tonadillas: La maja dolorosa, nos. 1-3;
El majo discreto; El tra-la-la y el punteado; El majo timido.
GURIDI: Llámle con el pañuelo; No quiero tus avellanas; Cómo
quieres que advine. MONTSALVATGE: Canciones negras. TORRE:
Pámpano verde, racimo albar. TURINA: Saeta en forma de Salve
a la Virgen de la Esperanza; El fantasma; Cantares. Teresa
Berganza, ms; Felix Lavilla, pno.

    +Gr 12-75 p1093   +RR 12-75 p85
    +HFN 12-75 p149

Archive 2533 111
2595 Dance music of the Renaissance. ATTAINGNANT: Basse danse La Brosse:
Tripla, Tourdion; Basse danse La gatta; Basse danse La Magdalena.
DALZA-PETTUCCI: Calata ala Spagnola. GERVAISE: Branle de Bour-
gogne. Branle de Champaigne. GULIELMUS, M.: Bassa danza à 2.
MILAN: Pavana I and II. MUDARRA: Romanesca, Guarda me las vacas.
NEUSIEDLER: Der Judentanz. Welscher Tanz Wascha mesa: Hupfauff.
PAIX: Schirazula Marazula. Ungarescha: Saltarello. PHALESE:
Passamezzo: Saltarello. Passamezzo d'Italie: Reprise, Gaillarde.
LE ROY, A.: Branle de Bourgogne. SCHMID d. "A", B.: Englischer
Tanz. Tanz Du has mich wollen nemmen. TORRE: Alta danza à 3.
SUSATO: Ronde. ANON.: Lamento di Tristano: Rotta, Trotto,
Istampita Ghaetta, Istampita Cominciamento di gioia, Saltarello,
Bassa danza à 2, Bassa danza à 3. Das Ulsamer-Collegium.

    ++AR 5-75 p52   ++HFN 3-73 p575
    +Gr 3-73 p1706   +St 6-73 p129

Archive 2533 149
2596 Goethe Lieder. ARNIM: O schaudre nicht. BEETHOVEN: Mit Mädeln
sich vertragen. HUMMEL: Zur Logenfeier. KREUTZER: Ein Bettler
vor dem Tor. NEEFE: Serenate. REICHARDT: Gott; Feiger Gedanken;
Die schöne Nacht; Einzeiger Augenblick; Einschränkung; Mut;
Rhapsodie; An Lotte; Tiefer liegt die Nacht um mich her. SACHSEI
WEIMAR: Auf dem Land und in der Stadt; Sie scheinen zu spielen.
SECKENDORFF: Romance. WAGNER: Lied des Mephistopheles; Branders
Lied. ZELTER: Rastlose Liebe; Um Mitternacht; Gleich und Gleich
Wo geht's Liebchen. Dietrich Fischer-Dieskau, bar; Jörg Demus,
fortepiano.

    +Gr 3-74 p1733   +NYT 6-8-75 pD19
    +HFN 3-74 p117   +-RR 4-74 p80
    +NR 6-75 p12   ++St 9-75 p122

Archive 2533 157
2597 Lute music of the Renaissance England. BATCHELAR: Mounsiers almaine. BULLMAN: Pavan. CUTTING: Almain. Greensleeves. The
squirrel's toy. Walsingham. DOWLAND: The Earl of Essex galliar
Fantasia. Forlorne hope fancy. The King of Denmark's galliard.
Lachrimae antiquae pavan. Melancholy galliard. Mrs. Winter's
leap. My Lady Hunsdon's puffe. Semper Dowland semper dolens.
HOLBORNE: Galliard. JOHNSON, R.: Alman. MORLEY: Pavan. ANON.:
Sir John Smith his almaine. Konrad Ragossnig, lt.

    +Gr 8-74 p381   -RR 7-74 p67
    +-HF 3-75 p88   ++St 7-75 p108

Archive 2533 172
2598 Dance music of the high baroque. BOUIN: La montauban. CHEDVILLE:
Musette. CORRETTE: Menuets, nos. 1, 2. DESMARETS: Menuet.
Passepied. FISCHER: Bouree. Gigue. HOTTETERRE: Bourrée.
LOEILLET: Corente. Gigue. Sarabande. LULLY: Unce noce de
village: Dernière entrée. PLAYFORD: The English dancing master:

Country dances. POGLIETTI: Balletto. REUSNER: German dances
(arr. Stanley). SANZ: Canarios. Españoleta. Gallard y Villano.
Passacalle de la Cavalleria de Napoles. Konrad Ragossnig, gtr;
Eduard Melkus, Spiros Rantos, baroque vln; Alfred Sous, Helmut
Hucke, baroque ob; René Zosso, hurdy-gurdy; Ulsamer Collegium.

    +Gr 3-75 p1667   +RR 3-75 p37
    +NR 3-75 p6   ++St 11-75 p142

Archive 2533 173

2599 Italian Renaissance lute music. BARBETTA: Moresca detta le Canarie.
CAPIROLA: Ricercare, nos. 1-2, 10, 13. NEGRI: Lo spagnoletto.
Il bianco fiore. MILANO: Fantasia. MOLINARO: Fantasias, nos.
1, 9-10. Ballo detto il Conte Orlando. Saltarello (2). PARMA:
Aria del Gran Duca. Ballo del serenissimo Duca di Parma. La
Cesarina. Corenta. Gagliarda Manfredina. La Mutia. La ne
mente per la gola. SPINACINO: Ricercare. TERZI: Ballo tedesco
e francese. Tre parti di gagliarde. Konrad Ragossnig, lt.

    +Gr 4-75 p1840   +RR 3-75 p45
    +NR 3-75 p15

Archive 2533 182

2600 Viennese dance music from the classical period. BEETHOVEN: Contra-
dances, WoO 14. EYBLER: Polonaise. GLUCK: Orfeo ed Euridice:
Ballet. Don Juan: Allegretto. HAYDN: Minuets (2). MOZART:
Ländler, K 606 (6). Contredances, K 609 (5). SALIERI: Minuet.
WRANITZKY: German dances (10). Quodlibet. Eduard Melkus En-
semble.

    +Gr 8-75 p332   ++NR 11-75 p4
    +HF 11-75 p130   ++RR 7-75 p37
    +HFN 9-75 p107

Archive 2533 184

2601 Golden dance hits of 1600. BEHREND: Tanz im Aicholdinger Schloss.
Eichstatter Hofmühtanz; Riedenburger Tanz. CAROUBEL: Courante
(2). Volte (2). DALZA: Catata ala Spagnola. GERVAISE: Branle
de Bourgogne. Branle de Champaigne. HAUSSMANN: Catkanei.
Galliard. Tantz. MAINERIO: Schiarazula marazula. Ungarescha-
Saltarello. MILAN: Pavan. NEGRI: Balletto. NEUSIEDLER: Wels-
cher Tanz, Wascha mesa: Hupfauff. PHALESE: Reprise, galliard.
PRAETORIUS: La bourrée. Galliarde de la guerre. Galliarde de
Monsieur Wustron. Gavotte. Reprinse. Spagnoletta. SCHEIN:
Allemande-Tripla. SUSATO: Ronde. WIDMANN: Agatha. Magdalena.
Regina. ANON.: Basse danse la Magdalena. Branle de Bourgogne.
Istampita Ghaetta. Italiana. Lamento di Tristano/Rotta.
Saltarello. Siegfried Behrend, gtr; Siegfried Finck, perc;
Das Ulsamer Collegium, Collegium Terpsichore.

    +-NR 8-75 p4   ++St 11-75 p142

Archive 2533 294

2602 BAKFARK: Fantasias (4). CATO: Praeludium, Galliardas I and II.
DLUGORAJ: Carola Polonesa. Finale (2). Fantasia. Kowaly.
Vilanella (2). POLAK: Praeludium. ANON.: Balletto Polacho.
Konrad Ragossnig, lt.

    ++Gr 12-75 p1085   +NR 11-75 p16
    +-HFN 12-75 p159   +RR 11-75 p75

Heliodor 2548 137 (Reissue from 135140)

2603 BEETHOVEN: Bagatelle, no. 25, A minor. Rondo a capriccio, op. 129,
G major. BRAHMS: Intermezzi, op. 117, nos. 1-3. MOZART: Fan-
tasia, K 397, D minor. SCHUBERT: Moments musicaux, op. 94,
D 780. SCHUMANN: Papillons, op. 2. Wilhelm Kempff, pno.

    +-Gr 4-75 p1867   ++RR 4-75 p48

Heliodor 2548 148 (Reissue from 135017)
2604 BERLIOZ: Damnation of Faust, op. 24: Hungarian march. GRIEG: Peer
    Gynt, op. 46, no. 1: In the hall of the mountain king. Sigurd
    Jorsalfar, op. 56: Homage march. MENDELSSOHN: A midsummer
    night's dream, op. 61: Wedding march; Fairies march. MILHAUD:
    Le carnaval d'Aix: Le capitaine Cartuccia. MOZART: March, K 237,
    D major. PROKOFIEV: March, op. 99, B flat major. STRAUSS, J. I:
    Radetzky march, op. 228. TCHAIKOVSKY: Casse-Noisette, op. 71:
    March. Slavonic march, op. 31. Bavarian Radio Symphony Orches-
    tra, Nordmark Symphony Orchestra, Bamberg Symphony Orchestra,
    Berlin Radio Symphony Orchestra, Monte Carlo Opera Orchestra,
    BPhO; Cappella Coloniensis; Rafael Kubelik, Ferdinand Leitner,
    Heinrich Steiner, Richard Kraus, Ferenc Fricsay, Louis Frémaux.
            +Gr 4-75 p1867   +-RR 4-75 p28
Archive 2708 025 (2)
2605 Diabelli variations, excerpts. Jörg Demus, hammerflügel.
            /HF 8-75 p105

2721 077 (2)
2606 Prussian and Austrian marches. BEETHOVEN: Pariser Einzugsmarsch
    (attrib.). York'scher Marsch. ERTL: Hoch-und Deutschmeister-
    Marsch. FUCIK: Florentiner Marsch. Regimentskinder. HAYDN, M.:
    Pappenheimer-Marsch. Coburger-Marsch. HENRION: Kreuzritter-
    Fanfare. Fehrbelliner Reitermarsch. KOMZAK: Vindobona-Marsch.
    Erzherzog-Albrecht-Marsch. LINDEMANN: Unter dem Grillenbanner.
    MOLTKE: Des grossen Kurfürsten Reitermarsch. MUHLBERGER: Mir
    sein die Kaiserjäger. PIEFKE: Königgrätzer Marsch. Preussens
    Gloria. PREIS: O Du mein Oesterreich. RADECK: Fridericus-Rex-
    Grenadiermarsch. SCHOLZ: Torgauer Marsch. SCHRAMMEL: Wien
    bleibt Wien. SEIFERT: Kärtner Liedermarsch. STRAUSS, J. II:
    Der Zigeunerbaron, op. 420: Einzugsmarsch. TEIKE: Alte Kamera-
    den. WAGNER, J. F.: Unter dem Doppeladler. Tiroler Holzhack-
    erbaub'n. WAGNER, R.: Nibelungen-Marsch (arr. Sonntag). WAGNES
    Die Bosniaken kommen. ANON.: Marsch der Finnländischen Reiterei
    Petersburger Marsch. Hohenfriedberger Marsch. BPhO Wind En-
    semble; Herbert von Karajan.
            ++Gr 2-75 p1504   +NR 12-75 p15

2740 105
2607 Free improvisation, New phonic art, 1973. Mike Lewis, org; Mike
    Ranta, perc; Karl-Heinz Böttner, string instruments; Conny Plank
    electronic sounds; New Phonic Art 1973; Iskra 1903.
            -Gr 11-74 p921   +-RR 11-74 p91
            -HFN 1-75 p150

                        DISCOPAEDIA

MB 1002
2608 Masters of the bow, Ferenc von Vecsey. BACH (arr. Wilhelm): Suite,
    violin, no. 3, D major: Air. BEETHOVEN: Sonata, violin and
    piano, no. 3, op. 12, E flat major. HANDEL: Sonata, violin,
    no. 9, B minor. MOSZKOWSKI: Stučke, op. 45, no. 2: Guitarre.
    PAGANINI: Caprice, op. 1, no. 2, B minor. SCHUBERT (arr. Wil-
    helm): Ave Maria, D 839. SCHUMANN: Kinderscenen, op. 15, no. 7:
    Träumerei. SINIGAGLIA: Capriccio all'antica. TARTINI (arr.
    Hubay): Sonata, violin, G minor. VECSEY: Foglio d'Album.
    WIENIAWSKI: Fantaisie brillante on theme from "Faust", op. 20.
    Souvenir de Moscow, op. 6. Ferenc von Vecsey, vln.
            +MT 8-75 p255   +-NR 11-74 p11

MB 1003
2609 Masters of the bow, Leopold Auer, Willy Burmester and Pablo de
     Sarasate. BACH: Partita, violin, no. 3, S 1006, E major: Gav-
     otte, Preludio. Suite, violin, no. 3, D major: Air. BRAHMS:
     Hungarian dance, no. 1, G minor. CHOPIN: Nocturne, op. 9, no.
     2, E flat major. DUSSEK: Minuet. HANDEL: Concerto, violin,
     no. 3, G minor: Arioso, Sarabande. Sonata, violin, no. 5, G
     major: Minuet. RAMEAU: Castor et Pollux: Gavotte. SARASATE:
     Caprice basque, op. 24. Danzas españolas, nos. 2, 6. Intro-
     duction and caprice: Jota. Introduction and tarantelle: Taran-
     telle. Miramar Zortzico, op. 42. Zigeunerweisen, op. 20.
     SINDING: Elegiac pieces: Andante religioso. TCHAIKOVSKY: Souv-
     enir d'un lieu cher, op. 42: Melodie. Leopold Auer, Willy
     Burmester, Pablo de Sarasate, vln.
          +—NR 7-75 p13   +St 8-75 p255
MB 1004
2610 Masters of the bow, Vasa Příhoda. BACH: Suite, violin, no. 3, D
     major: Air. DRDLA: Souvenir. DRIGO: Les millions d'Arlequin:
     Serenade. KREISLER: Liebesfreud. Liebesleid. PAGANINI: Nel
     cor piu non mi sento. PROVAZNIK: Valse joyeuse, op. 137.
     SARASATE: Jota navarra, op. 20, no. 2. Zigeunerweisen, op. 20,
     no. 1. TARTINI: Sonata, violin, G minor. TOSELLI: Serenade,
     violin, no. 1, op. 6. Vasa Příhoda, vln.
          +NR 10-74 p12   +St 9-75 p349
MB 1005
2611 Masters of tbe bow, Maud Powell. BOISDEFFRE: Au Bord du Ruisseau,
     op. 52. BRUCH: Kol Nidrei, op. 47. CADMAN: Little firefly.
     ELGAR: Caprice, op. 51, no. 2, A minor. GRIEG: To the spring.
     HUBAY: Hejre Kati, violin, op. 32. Crespuscule. LEYBACH: Noc-
     turne, no. 5, op. 52, A flat major. MOSZKOWSKI: Serenata.
     NERUDA: Berceuse slav d'apres un chant polonais, op. 11. RAFF:
     Cavatina, D major. SARASATE: Danza española, op. 26, no. 2, C
     major. SAURET: Will-o-the-wisp. SCHUBERT: Rosamunde, op. 26,
     D 797: Entr'acte III. SIBELIUS: King Christian suite, op. 27:
     Musette. THOMAS: Mignon: Entr'acte. Maud Powell, vln.
          +NR 10-74 p12   +St 9-75 p351
MB 1006
2612 Masters of the bow, Mischa Elman. ASCHER: Alice, where art thou.
     BRULL: Scene espagnole. CHOPIN: Nocturne, op. 9, no. 2, E flat
     major. DELIBES: Le Roi s'amuse: Passepied. DVORAK: Humoresque,
     op. 101, no. 7, G flat major. ESPEJO: Airs tziganes, op. 11.
     KREISLER: Rondino on a theme by Beethoven. Sicilienne et Rig-
     audon. Chanson Louis XIII et pavane. MENDELSSOHN: Song with-
     out words, op. 67, no. 6, E major. RUBINSTEIN: Romance, E
     flat major. SAMMARTINI: Sonata, violin, no. 4, A major: Andante.
     SCARLATTI, D.: Sonata, violin, L 413, E minor: Pastorale.
     SCHINDLER: Souvenir poetique. THOME: Simple aveu, op. 25.
     VOGRICH: Dans le bois. YSAYE: Rêve d'enfant, op. 14. Mischa
     Elman, vln.
          +NR 10-74 p12   +St 10-75 p423
MB 1007
2613 Masters of the bow, Toscha Seidel. d'AMBROSIO: Canzonetta, op. 6.
     BAKALEINIKOFF: Brahmsiana. BRAHMS: Hungarian dance, no. 1, G
     minor (arr. Joachim). BURLEIGH: Characteristic pieces: Indian
     snake dance. CUI: Kaleidoscope, op. 50, no. 9. HUBAY: Hejre
     Kati, violin, op. 32. KREISLER: Caprice viennois, op. 2. MARGIS:

Valse bleue. MOZART: Divertimento, no. 17, K 334, D major:
Minuet (arr. Burmester). PADEREWSKI: Minuet, op. 14, no. 1,
G major (arr. Kreisler). PROVOST: Intermezzo. RIMSKY-KORSAKOV:
Scheherazade, op. 35: 3rd movement (arr. Kreisler). SAINT-SAENS:
Le déluge, op. 45: Prelude. SCHUBERT: Ständchen, op. 135, D 920
(arr. Elman). WAGNER: Albumblatt, C major. ANON.: Eli, Eli.
Toscha Seidel, vln.
+NR 9-74 p12   +St 10-75 p423

MB 1008
2614 Masters of the bow, Efrem Zimbalist. AULIN: Aquarelles: Humoresque.
BRAHMS: Hungarian dances, nos. 20, 21 (arr. Joachim). CHOPIN:
Waltz, op. 64, no. 1, D flat major (arr. Zimbalist). Waltz,
op. 70, no. 1, G flat major (arr. Spalding). CUI: Kaleidoscope,
op. 50, no. 9. DRIGO: Les millions d'Arlequin: Serenade. GLIN-
KA: Russlan and Ludmila: Persian song. HUBAY: Blumenleben, op.
30: Der zephir. PIERNE: Serenade, op. 7, A major (arr. Haddock).
REGER: Sonata, violin, no. 2, A major: Andantino. SAINT-SAENS:
Le carnaval des animaux: Le cygne. Le déluge, op. 45: Prelude.
VAN BIENE: Broken melody. ZIMBALIST: Slavonic dances, no. 2.
Suite dans la forme ancienne: Sicilienne, Minuet. Efrem Zim-
balist, vln.
+NR 9-74 p12   +St 10-75 p423

MB 1009
2615 Masters of the bow, Albert Spalding. BOULANGER: Cortege. CASSADO:
Danse due diable vert. CHOPIN: Nocturne, op. 9, no. 2, E flat
major. Waltz, op. 69, no. 2, B minor. MENDELSSOHN: Auf Flügeln
des Gesanges, op. 34, no. 2. MOSZKOWSKI: Stücke, op. 45, no. 2:
Guitarre. SARASATE: Introduction and tarantelle, op. 43.
SCHUBERT: Ave Maria, D 839. SCHUMANN: Abendlied, op. 107, no.
6. Kinderscenen, op. 15, no. 7: Träumerei. SPALDING: Dragonfly.
SPOHR: Concerto, violin, no. 8, op. 47, A major. SUK: Pieces,
op. 17: Burleska. WHITE: Nobody knows the trouble I've seen.
WAGNER: Die Meistersinger von Nürnberg: Preizlied.
+NR 7-75 p14   +St 9-75 p349

MB 1010
2616 Masters of the bow, Jascha Heifetz. ACHRON: Hebrew dance, op. 35,
no. 1. Hebrew melody, op. 33. BACH: Partita, violin, no. 3,
S 1006, E major: 2 minuets. BAZZINI: La ronde des lutins, op.
25. BOULANGER: Cortege. Nocturne. COUPERIN, F.: Livres de
clavecin, Bk III, Ordre no. 17: Les petits moulins à vent.
FALLA: Canciones populares españolas: Jota. GLAZUNOV: Raymonda,
op. 57: Valse (arr. Heifetz). GRIEG: Lyric pieces, op. 71,
no. 3: Puck. JUON: Pieces, violin and piano, op. 28, no. 3.
LALO: Symphonie espagnole, op. 21: 4th movement. MENDELSSOHN:
Auf Flügeln des Gesange, op. 34, no. 2. SAINT-SAENS: Havanaise,
op. 83. SARASATE: Carmen fantasy, op. 25. Danzas españolas,
nos, 2, 6. TCHAIKOVSKY: Serenade, strings, op. 48, C major:
Valse (arr. Auer). ANON.: Gentle maiden. Jascha Heifetz, vln.
-NR 9-74 p12   +St 11-75 p497

DISCOPHILIA

KGC 1
2617 BEMBERG: Ca fait peur aux oiseaux. BIZET: Les pêcheurs de perles:
Au fond du temple saint. BOITO: Mefistofele: Lontano, lontano.
GODARD: Jocelyn: Berceuse. GOUNOD: Roméo et Juliette: Ange

adorable. FAURE: Chanson Lorraine; Les rameaux. LALO: Le Roi
d'Ys: Vainement ma vien aimee. LULLY: Au clair de la lune.
MASSENET: Manon: Le rêve. Werther: Pourquoi me reveiller.
MEYERBEER: Robert de teufel: Du rendezvous; Le bonheur. PESSARD:
L'Adieu du matin; Bergère legère (Weckerlin). Edmond Clement,
t; Various accompaniments.
      +NR 7-75 p10   +ON 9-75 p60

KGB 2
2618 BEETHOVEN: Fidelio, op. 72: O war ich schön mit der vereint. BIZET:
Carmen: Ich sprach dass ich furchtlos. FLOTOW: Martha: Nancy,
Julia verweille. MOZART: Don Giovanni, K 527: Wenn du fein
fromm bist. Le nozze di Figaro, K 492: So lang hab ich geschmachtet. PUCCINI: Madama Butterfly: Eines Tages sehn wir.
VERDI: Rigoletto: Rede wir sind allein...Wein weine. La traviata:
Fraulein Valery...Ach ich verstehe...Saget der Jungfrau...Ich
sterbe. Hermine Bosetti, s; Various accompaniments.
      +NR 7-75 p9

KGC 2
2619 BIZET: Les pêcheurs de perles: De mon amie. DONIZETTI: Il Duca
d'Alba: Angelo casto e bel. GOMES: Guarany: Sento una forza.
GOLDMARK: Regina di Saba: Magiche note. HALEVY: La Juive: Rachel
quand du Seigneur. LEONCAVALLO: La bohème: Testa adorata.
MASSENET: Le Cid: O Souverain. Manon: On l'appelle Manon.
MEYERBEER: L'Africana: Deh ch'io ritorno. PUCCINI: La bohème:
Vecchia zimarra senti. SAINT-SAENS: Samson et Dalila: Je viens
celebrer. VERDI: Don Carlo: Dio che nell'alma. I Lombardi:
Qual voluttà trascorrere. Macbeth: Ah la paterna mano. Enrico
Caruso, t; Various accompaniments.
      +NR 7-75 p10

KGM 2
2620 FLOTOW: Martha: Ja was nun. MEYERBEER: L'Africana: Schlummerarie.
Le prophète: Ach mein Sohn Gebt O Gebt. THOMAS: Mignon: Styri-
enne, Schwalbenduett. VERDI: Il trovatore: Gern will ich
schlieben; Nicht darfst du von meiner Seite; Lodernde Flammen.
WAGNER: Götterdämmerung: Erzahlung der Waltraute. Lohengrin:
Entweihte Götter. Das Rheingold: Weiche Wotan. Die Walküre:
Walkurenruf; Nicht weise bin ich. Margarethe Matzenauer, con;
Various accompaniments.
      +NR 7-75 p9

KGS 3
2621 CHOPIN: Zyczenie. FISHER: I heard a cry. HANDEL: Alessandro:
Lusinghe piu care. LA FORGE: To a messenger. MANNEY: Consecration. MASSENET: Ouvre tes yeux bleus (versions). PADER-
EWSKI: Moja piesczotka. RUBINSTEIN: Now shines the dew; Spring
song. STEBBINS: If the apples bloom; The longest day in June;
Prelly rally Oliver. STRAUSS, J. II: Frühlingstimmen, op. 410.
Geschichten aus dem Wienerwald, op. 325. VERDI: Ernani: Ernani
involami. La traviata: Ah fors'e liu (2 versions); Sempre
libera. Marcella Sembrich, s; Various accompaniments.
      +NR 7-75 p9   *ON 10-75 p46

KGP 4
2622 BEETHOVEN: Fidelio, op. 72: Leonorenari. BIZET: Carmen; Ich sprach
dass ich furchtlos mich fühle. FLOTOW: Martha: Letzte Rose.
MASCAGNI: Cavalleria rusticana: Als euer Sohn. OFFENBACH: Les
contes de Hoffmann: Barcarolle. PUCCINI: La bohème: Man nennt
mich jetzt nur Mimi. Tosca: Nur der Schonheit (2). STRAUSS, R.:

Die Agyptische Helena: Bei jener Nacht, Zweite Nacht, Zauber-
nacht. THOMAS: Mignon: Kennst du das Land. VERDI: Otello:
Ave Maria. Rose Pauli-Dreesen, Hildegard Ranczak, s; Various
accompaniments.
+NR 7-75 p10

DRACO

DR 1333
2623 COUPERIN: Livres de clavecin, Bk IV: Ordre, no. 26. DEBUSSY:
Preludes. RAVEL: Tombeau de Couperin: Rigaudon. Sonatine.
SCRIABIN: Etude, op. 2, no. 1, C sharp minor. Prelude, op. 11,
no. 14. LISZT: Legend, G 175: St. Francis of Assisi. Iren
Marik, pno.
+CL 9-75 p16

DUKE UNIVERSITY PRESS

DWR 7306 AX
2624 The art song in America, vol. 2. CUMMING: Go lovely rose; The
little black boy; Memory, hither come. DUKE: I carry your
heart; In just spring; The mountains are dancing. EARLS: Arise,
my love; Entreat me not to leave you. PERSICHETTI: The death
of a soldier; The grass; Of the surface of things; The snow man;
Thou child so wise. ROREM: A Christmas carol; For Susan; Clouds;
Guilt; What sparks and wiry cries. TRIMBLE: Love seeketh not
itself to please; Tell me where is fancy bred. John Kennedy
Hanks, t; Ruth Friedberg-Erickson, pno.
-HF 8-74 p111   +-St 6-75 p107

EMBER

GVC 37 (Reissue from Cetra LPC 55006)
2625 GIORDANO: Andrea Chénier: Nemico della patria. MASSENET: Hérodiade:
Vision fuggitiva. MEYERBEER: L'Africana: Ballata; Adamastor,
Re dell'acque profonde. MOZART: Don Giovanni, K 527: Deh, vieni
alla finestra. Le nozze di Figaro, K 492: Aprite un po' quegli
occhi. PUCCINI: La fanciulla del West: Minnie dalla mia casa.
ROSSINI: Guglielmo Tell: Resta immobile. VERDI: Falstaff:
L'onore. Otello: Credo in un Dio crudel. Rigoletto: Pari siamo.
Giuseppe Taddei, bar; Italian Radio and Television Symphony
Orchestra; Arturo Basile, Angelo Questa.
+Gr 12-74 p1212   +-RR 1-75 p24
GVC 40 (Reissue from Decca SDDR 167)
2626 GLAZUNOV: Morceaux, op. 49, no. 3: Gavotte. MIASKOVSKY: Grillen,
op. 25, nos. 1, 6. MUSSORGSKY: Pictures at an exhibition: Bydlo
and ballet (orch. Ravel). PROKOFIEV: Episodes, op. 12, nos.
1-3, 7, 10. Love for three oranges, op. 33: Intermezzo. Sar-
casms, op. 17, nos. 1, 2. Tale of the old grandmother, op. 31,
no. 3. Toccata, op. 11. RIMSKY-KORSAKOV: Scheherazade, op. 35:
Fantasy. SCRIABIN: Prelude, op. 45, no. 3, E flat major. Winged poem (Poeme aile), op. 51, no. 3. Sergei Prokofiev, pno.
+-Gr 7-75 p223   +RR 6-75 p71
+-HFN 6-75 p105

GVC 41 (Reissues from HMV 7-52110-2, 4; 2-054086, 2-054109, 2-054105,
       2-054114, various film soundtracks)
2627 BOITO: Mefistofele: Dai campi, dai prati; Se tu mi doni; Lontano,
     Lontano; Giunto sul passo estremo. BIZET: Les pêcheurs de
     perles: Au fond du temple saint. CAPUA: O sole mio; Sweet
     dream of love. GOUNOD: Faust: Love duet. GIORDANO: Caro mio
     ben. PONCHIELLI: La gioconda: Cielo e mar. PUCCINI: Tosca:
     Recondita armonia. VERDI: Aida: O terra, addio. Il trovatore:
     Deserto sulla terra. WAGNER: Die Walküre: Winterstürme. Beni-
     amino Gigli, t; Orchestra.
             +-Gr 7-75 p241   ++RR 5-75 p23
             +-HFN 6-75 p97
GVC 42 (Reissue from Columbia 7157, 7210, 7161, 7261, A 5654, 7153, 7144,
       7255, 7138, 7151)
2628 BACH: Suite, orchestra, S 1068, D major: Air on the G string.
     BOCCHERINI: Sonata, violoncello, no. 6, A major: Allegro.
     CAMPAGNOLI: Romanza. CHOPIN: Nocturne, op. 9, no. 2, E flat
     major (arr. Popper). GOLTERMANN: Concerto, violoncello, op. 14,
     A minor: Cantilena. MOZART: Quintet, clarinet, K 581, A major:
     Larghetto. POPPER: Mazurka, op. 11, no. 3, G major. SCHUMANN:
     Abendlied, op. 85, no. 12. TARTINI: Concerto, violoncello, D
     minor: Adagio. WAGNER: Tannhäuser: O star of eve. Pablo Casals,
     vlc; Various accompaniments.
             +Gr 10-75 p652   -RR 9-75 p50
             +-HFN 8-75 p85
GVC 46 (Reissues)
2629 CURTIS: Carme. DVORAK: Indian lament. GODARD: Jocelyn: Berceuse.
     KREISLER: Liebesfreud. Tambourin chinois. Caprice viennois.
     MASSENET: Thais: Meditation. MATTULATH (Böhm): Calm as the
     night. RACHMANINOFF: When night descends. SMETANA: Aus der
     Heimat. John McCormack, t; Fritz Kreisler, vln; Various accom-
     paniments.
             +-Gr 8-75 p363   +RR 7-75 p48
             +-HFN 8-75 p85
ECL 9040
2630 BACH: Suite, orchestra, S 1068, D major: Air (arr. Rampal). KREBS:
     Suite, flute and harpsichord, G major. MARCELLO, B.: Sonata,
     flute and harpsichord, B major. PEPUSCH: Tio sonata, flute and
     harpsichord, G major. TELEMANN: Sonata, 2 flutes, E major.
     Jean-Pierre Rampal, Mario Duschenes, flt; Kenneth Gilbert, hpd;
     Paris Festival Strings.
             +Gr 12-75 p1054   ++RR 8-75 p58
             /HFN 11-75 p154

                                ERA
1001 (Reissue from Mercury ME MG 50076)
2631 Americana for solo winds and string orchestra. BARLOW: The winter's
     past, rhapsody for oboe. COPLAND: Quiet city. HANSON: Pastor-
     ale, oboe, strings and harp, op. 35. Serenade, flute, strings
     and harp. KELLER: Serenade, clarinet and strings. KENNAN:
     Night soliloquy, flute and orchestra. ROGERS: Soliloquy, flute
     and strings. Eastman-Rochester Symphony Orchestra; Howard
     Hanson.
             -Audio 9-75 p69   +NR 10-74 p5

STU 70847
2632 BENDUSI: Cortesa padana e fusta. BERTOLDO: Canzon fancese. Petit
        fleur. CANOVA: Fantasia, 2 lutes (Matelart). GABRIELI, G.:
        Canzona a 5, a 6, a 7. GALILEI: Capriccio a due voci. Contra-
        punto, 2 lutes. GASTOLDI: Capriccio a due voci (2). Intradas
        a 5 (2). MAINERIO: Primo libro de balli: Suite. MARENZIO:
        Tirsir morir volea. OROLOGIO: Intradas a 5 (2). RUFFO: Dormen-
        do un giorno: Capriccio. La gamba in basso e soprano. ANON.:
        Le force d'Hercole e tripla. Zurich Ricercare Ensemble; Michel
        Piguet.
            +-Gr 12-75 p1090   +RR 12-75 p73
STU 70871
2633 ALBINONI: Concerto, trumpet, B flat major. HANDEL: Concerto,
        trumpet, G minor. PURCELL: Sonatas, trumpet, nos. 1 and 2,
        D major. TARTINI: Concerto, trumpet, D major. TELEMANN: Con-
        certo, trumpet, D major. Maurice André, tpt; AMF; Neville
        Marriner.
            +Gr 11-74 p915   ++RR 11-74 p61
            +HFN 12-75 p148   ++RR 12-75 p67
STU 70880
2634 BELLINI: Norma: Overture. MANCINELLI: Cleopatra: Overture. PONCH-
        IELLI: I promessi sposi: Overture. ROSSINI: Semiramide: Over-
        ture. SPONTINI: La Vestale: Overture. VERDI: Giovanna d'Arco:
        Overture. ORTF: Claudio Scimone.
            +-RR 3-75 p34

SDBR 3380
2635 NARVAEZ: Diferencias sobre "Guárdame las vacas". PIPO: Canción
        y danza. SANZ: Suite española. SOR: Minuetto en re mayor y
        variaciones sobre un tema de la "Flauta encantada" de Mozart.
        TARREGA: Sueño. TURINA: Garrotín y soleares (Homenaje a
        Tarrega). VILLA-LOBOS: Preludio estudio chôros. Narciso Yepes,
        gtr.
            ++NR 12-75 p14

GM 2022
2636 BLOW: Behold, O God our defender. BYRD: Alleluia, ascendit Deus.
        Ave verum corpus. GOW: Ave maris stella. FORBES: Gracious
        spirit, Holy Ghost. HEWITT-JONES: O clap your hands together.
        PALESTRINA: Missa Sine Nomine: Kyrie, Sanctus. RIMMER: Singe we
        merrily. RUTTER: Praise ye the Lord. SEARLE: Toccata alla
        passacaglia. VAUGHAN WILLIAMS: Preludes on Welsh hymn tunes,
        no. 2: Rhosymedre, O taste and see. VICTORIA: Estote fortes in
        bello. WEELKES: All people clap your hands. ANON.: Ecce novum
        gaudium (arr. Elliott). Ding dong merrily on high (French).
        Patapan (Burgundian). Edward Garden, Richard Wardell, org;
        Glasgow University Chapel Choir; Edward Garden.
            +HFN 6-75 p105   /RR 6-75 p76

GOLDEN CREST

CRS 4132
2637 Rags and tangos. BARBER: Hesitation tango. CASTRO: Tangos.
     MILHAUD: Tango des Fratellini. Caramel-Mou. Rag-Caprices (3).
     STRAVINSKY: Piano rag music. Ragtime. Tango. THOMSON: Parallel
     chords (tango). Ragtime bass. Grant Johannesen, pno.
          +St 2-75 p115

GRANGE

SGR 1124
2638 CORELLI, A.: Concerto grosso, op. 6, no. 8, C minor. FERRABOSCO:
     Pavane, no. 4. FARINA: Pavane. VIVALDI: Sinfonia, G major.
     WIDMANN: Canzona, galliard, intrada. Cantilena Chamber Players;
     Adrian Shepherd.
          ++Gr 11-75 p840   +RR 12-75 p48
          +HFN 12-75 p151

HARMONIA MUNDI

HMU 133
2639 Russian and Bulgarian Orthodox chants. DEGTIAREV: Preslavnia dnes.
     MANOLOV: Verouiou. MAREV: Otche nach. SAPOJNIKOV: Tebe poem.
     STROKIN: Nini otpuchtalechi. VEDEL: Na rekah Vavilonskih.
     Nikolai Ghiuselev, bs; Men's Chorus; Krustiu Marev.
          +Gr 12-75 p1093

HMD 204
2640 CORNYSHE: Ah, Robin; Hoyda jolly Rutterkin. GENTIAN: Je suis
     Robert. GIBBONS: The cries of London. JOHNSON: Care-charming
     sleep. MONTEVERDI: Baci soavi e cari; Lamento della Ninfa.
     SERMISY: Tant que vivray. VAUTOR: Mother, I will have a hus-
     band. WEELKES: Cease sorrows now. WILBYE: Lady, when I behold
     the roses. Deller Consort, Bulgarian Quartet; René Saorgin,
     hpd, org; Raphael Perulli, vla da gamba.
          +—Gr 11-75 p888

HMU 334
2641 d'ANGLEBERT: Chaconne, D major. Le tombeau de M. de Chambonnieres.
     CHAMBONNIERES: Chaconne, F major. Rondeau. COUPERLIN, L.: La
     Piémontaise, A minor. Passacaille, C major. DUMONT: Pavane,
     D minor. RAMEAU: Suite, E minor. L'Enharmonique, G minor.
     Kenneth Gilbert, hpd.
          ++RR 11-75 p76

HM 593
2642 English madrigals and folksongs. CAVENDISH: Sly thief, if so you
     will believe. FARMER: A little pretty bonny lass. MORLEY:
     My bonny lass she smileth. PILKINGTON: Sweet Phillida. TOMKINS:
     See, see, the shepherd's queen; Weep no more, thou sorry boy.
     VAUTOR: Shepherds and nymphs. WEELKES: Hark, all ye lovely
     saints above; Say, dear, when will your frowning leave. ANON.:
     The cuckoo; Cold blows the wind (arr. Morris); The jolly carter;
     O' 'twas on a Monday morning (arr. Holst); O waly, waly (arr.
     Cashmore); The sailor and young Nancy (arr. Moeran); The sheep
     sharing (arr. Sharp); The turtle-dove (arr. Morris). Deller
     Consort.
          /Gr 11-75 p887

Tape (c) TC MCS 13
2643 BACH: Chorale prelude, In dulci jubilo, S 729. BYRD: Haec dies.
     CHARPENTIER: Messe de minuit: Kyrie. FAURE: Requiem, op. 48:
     Agnus Dei. HANDEL: Dixit Dominus: Judicabit in nationibus.
     Messiah: For unto us a child is born. HAYDN: Mass, no. 7, C
     major: Credo, excerpts. PARRY: O how amiable. POSTON: Jesus
     Christ the apple tree. SWEELINCK: Hodie Christus natus est.
     VAUGHAN WILLIAMS: Mystical songs: Let all the world. ANON.:
     On Christmas night (arr. Vaughan Williams). Cambridge, King's
     College Choir; David Willcocks; Various Orchestras.
          +Gr 7-75 p254 tape   +RR 10-75 p98 tape
SEOM 14 (Reissues from HMV SLS 834, ASD 2754, 2784, 2800)
2644 ENESCO: Rumanian rhapsody, op. 11, no. 1, A major. GERSHWIN: An
     American in Paris. An American in Paris, rehearsal excerpt.
     PROKOFIEV: Alexander Nevsky, op. 78: Alexander's entry into
     Pskov. TCHAIKOVSKY: The nutcracker, op. 71: Waltz of the flow-
     ers; Pas de deux. LSO and Chorus; André Previn.
          +Gr 8-73 p379   +RR 10-73 p81
Tape (c) TC MCS 14
2645 BACH: St. Matthew Passion, S 244: O Haupt, voll Blut und Wunden.
     BERLIOZ: L'Enfance du Christ, op. 25: Shepherds chorus. FAURE:
     Requiem, op. 48: Sanctus. ELGAR: The dream of Gerontius, op.
     38: Praise to the holiest. GOUNOD: Messe solennelle a St.
     Cecile: Domine salvum. HANDEL: Messiah: For unto us a child
     is born. MENDELSSOHN: Elijah, op. 70: And then shall your light
     break forth. MOZART: Mass: Dies Irae. POULENC: Gloria. VAUGHAN
     WILLIAMS: Mass, G minor: Kyrie. VERDI: Requiem: Sanctus. Var-
     ious artists.
          +Gr 8-75 p376 tape   +RR 10-75 p96 tape
          +HFN 10-75 p155 tape
SEOM 19 (Reissues)
2646 BACH: Suite, solo violoncello, no. 3, S 1009, C major: Bourrées,
     nos. 1, 2. BEETHOVEN: Sonata, violoncello, no. 3, op. 69, A
     major: 3rd movement. CHOPIN: Sonata, violoncello, op. 65, G
     minor: 3rd movement. ELGAR: Concerto, violoncello, op. 85, E
     minor: 3rd movement. FAURE: Après un rêve, op. 7, no. 1. GRIEG:
     Holberg suite, op. 40: Prelude. PAGANINI: Variations on a theme
     by Rossini, D minor. RACHMANINOFF: Sonata, violoncello, op.
     19, G minor: 2d movement. STRAUSS, R.: Don Quixote, op. 35:
     Finale. TORTELIER: Pishnetto. WALTON: Concerto, violoncello:
     2d movement. Paul Tortelier, vlc; Eric Heidsieck, Aldo Cicco-
     lini, Shuku Iwasaki, pno; Max Rostal, vla; Maud Martin-Tortelier,
     vlc; ECO, Northern Sinfonia Orchestra, LPO, Bournemouth Symphony
     Orchestra, Dresden Staatskapelle; Paul Tortelier, Adrian Boult,
     Paavo Berglund, Rudolf Kempe.
          +Gr 4-75 p1819   ++RR 4-75 p42
          +HFN 6-75 p103
SEOM 20
2647 BIZET (Shchedrin): Carmen: Bolero; Toreador. BORTNIANSKY: Cherubim
     hymn, no. 7. GLAZUNOV: Raymonda, op. 57: Valse fantastique.
     GLIERE: The red poppy, op. 70: Russian sailors' dance. PROKOF-
     IEV: March, op. 99, B flat major. RACHMANINOFF: Aleko: Men's
     dance. RIMSKY-KORSAKOV: Symphony, no. 2, op. 9. SHOSTAKOVITCH:
     Quartet, strings, no. 3, op. 73: Moderato con moto. Symphony,
     no. 5, op. 47, D minor: Allegretto. TCHAIKOVSKY: Quartet,

strings, no. 1, op. 11, D major: Andante cantabile. The maid
of Orleans: Hymn. Irina Arkhipova, ms; USSR Symphony Orchestra,
Borodin Quartet, Moscow Radio Symphony Orchestra and Chorus,
Bolshoi Theatre Orchestra, USSR Ministry of Defense Symphonic
Band; USSR Chorus; Nicolai Sergeyev, Alexander Yurlov, Yevgeny
Svetlanov, Gennady Rozhdestvensky, Konstantin Ivanov, Yuri Fayer,
Maxim Shostakovitch.

    +HFN 10-75 p153   +RR 9-75 p46

RLS 714 (2) (Reissues
2648 ARNE: O ravishing delight; Where the bee sucks. BRAHMS: Sister
    dear. CADMAN: At dawning. GRIEG: To a waterlily; A dream.
    HANDEL: Joshua: O had I Jubal's lyre. Messiah: Rejoice greatly,
    O daughter of Zion; Come unto Him; How beautiful are the feet;
    I know that my Redeemer liveth; If God be for us. Rodelinda:
    Art thou troubled. Samson: Let the bright seraphim. HAYDN:
    The creation: On mighty pens. MENDELSSOHN: Elijah: What have
    I to do with thee; Hear ye Israel. PUCCINI: Madama Butterfly:
    Give me your darling hands. PURCELL: Stript of their green.
    SCHUBERT: To music, D 547; The brook. SCHUMANN: The snowdrop;
    Ladybird. SCOTT: Blackbird's songs. TRAD.: Annie Laurie; Have
    you seen but a whyte lily grow; I will walk with my love.
    Isobel Baillie, s; Various accompaniments.

    +Gr 5-75 p2009   +RR 5-75 p60

RLS 717 (Reissues from Columbia LX 532/7, 712/3, 899/903, 918, 909, 861,
    877/8, 897/8, 860/1, 868, 866, 898, CLX 2189/90, 2197/8,
    2187/8, 2165/6, 2167/8, WAX 6050/1, DC 266, CAX 8184/7,
    8717/26, 8733/6, 8737/9, 8742)
2649 BEETHOVEN: Leonore overture, no. 2, op. 72. Die Ruinen von Athens,
    op. 113: Overture. Symphony, no. 3, op. 55, E flat major.
    BERLIOZ: Les Troyens: Trojan march. BRAHMS: Symphony, no. 2,
    op. 73, D major. HANDEL: Alcina: Dream music (arr. Whittaker).
    LISZT: Les préludes, G 97. Mephisto waltz. STRAUSS, J. II:
    Voices of spring, op. 410. Wein, Weib und Gesange, op. 333.
    WAGNER: Rienzi: Overture. Tannhäuser: Prelude, Act 3. Tristan
    und Isolde: Prelude, Act 3. VPO, LSO, LPO, OSCCP, British
    Symphony Orchestra; Felix Weingartner.

    +Gr 12-75 p1061   +RR 12-75 p37

SLA 870 (3) (Reissues from Columbia 33 CXL 1406, 33 CS 1617, SAX 2433)
2650 Hoffnung's music festivals, November 13th, 1956; Interplanetary
    music festival, November 21-22, 1958; Astronautical music fes-
    tival, November 28, 1961. ARNOLD: A grand grand overture.
    United Nations, excerpts. BAINES: Fanfare. Hoffnung Festival
    overture, excerpts. Introductory music. BEETHOVEN: Leonore
    overture, no. 3, op. 72. CHAGRIN: Ballad of County Down, D
    major. Concerto, conductor and orchestra: A movement. CHOPIN:
    Mazurka, op. 62, no. 2, A minor. FRICKER: Waltz, restricted
    orchestra. HAYDN: Symphony, no. 94, G major: Andante (arr.
    Swann). HOROVITZ: Horrortorio. Metamorphosis on a bedtime
    theme. JACOB: Variations on "Annie Laurie". LEONARD: Mobile,
    7 orchestras. MANN-REIZENSTEIN-WETHERELL: Let's fake an opera
    or "The tales of Hoffnung". MOZART, L.: Concerto, hose-pipe
    and strings: 3hd movement. POSTON: Sugar plums. REIZENSTEIN:
    Concerto popolare. SEARLE: The barber of Darmstadt: Duet. Loch-
    invar, speakers and percussion. Punkt Contrapunkt. SEIBER: The
    famous Tay whale. WALTON: Belshazzar's feast, excerpts. Various
    artists.

    +Gr 3-75 p1668

HQS 1337 (Reissues from ASD 2551)
2651 BLAKE: Variations, piano. GOEHR: Pieces, op. 18 (3). HEADINGTON:
Toccata. RAWSTHORNE: Ballade. SHERLAW JOHNSON: Sonata, piano,
no. 2. John Ogdon, pno.
+Gr 4-75 p1843   +RR 4-75 p53
+HFN 6-75 p99
HQS 1350
2652 Chichester Cathedral, 900 years. BAIRSTOW: Let all mortal flesh
keep silence. DAVIES: Ave Maria. HOLST: Festival chorus,
op. 36, no. 2: Turn back, O man. HOWELLS: Magnificat. IRELAND:
Greater love hath no man. LEIGHTON: Give me the wings of faith.
SHAW: Anglican folk mass: Creed. STANFORD: Beati quorum via,
op. 51, no. 3. VAUGHAN WILLIAMS: O taste and see. WALTON:
Set me as a seal upon thine heart. WOOD: Short communion serv-
ice in the Phrygianmode: Sanctus and benedictus. Chichester
Cathedral Choir; John Birch, org; Richard Seal.
+Gr 9-75 p506   +RR 8-75 p63
HQS 1353
2653 BRAHMS: Rhapsodie, op. 79, no. 2, G minor. CHOPIN: Nocturne,
op. 9, no. 2, E flat major. Polonaise, op. 40, no. 1, C major.
Waltz, op. 64, no. 2, C sharp minor. LISZT: Etude de concert,
no. 3, G 144, D flat major: Un sospiro. Hungarian rhapsody,
no. 11, G 244, A minor. SCHUBERT: Impromptu, op. 90, no. 4,
D 899, A flat major. Moment musicaux, op. 94, D 780, F major.
SCHUMANN: Arabesque, op. 18, C major. Romance, op. 28, no. 2,
F sharp major. Daniel Adni, pno.
+-Gr 11-75 p869   +RR 11-75 p74
+-HFN 11-75 p165
ASD 2938. Tape (c) TC ASD 2938
2654 Works for trumpet and orchestra. FRANCESCHINI: Sonata, 2 trumpets
and strings, D major. HAYDN: Concerto, trumpet, E flat major.
TELEMANN: Concerto, trumpet, D major. TORELLI: Suite, no. 7,
a cinque, D major. VIVIANI: Sonata, trumpet, no. 1, C major.
Sonata, trumpet, no. 2, D major. John Wilbraham, Michael Laird,
tpt; Christopher Hogwood, hpd; Kenneth Heath, vlc; AMF; Neville
Marriner.
+Gr 4-74 p1862   +-HFN 10-75 p155 tape
+Gr 10-75 p721 tape   +RR 3-74 p39
+HFN 3-74 p122   +RR 11-75 p92 tape
ASD 3008. (Q) Q4 ASD 3008 Tape (c) TC ASD 3008
2655 French orchestral works. CHABRIER: España: Rapsodie. DEBUSSY:
Prélude à l'après-midi d'un faune. DUKAS: L'Apprenti sorcier.
RAVEL: Bolero. SAINT-SAENS: Danse macabre, op. 40. Birmingham
City Symphony Orchestra; Louis Frémaux.
+Gr 9-74 p522   +-RR 9-74 p59
+-Gr 6-75 p55 Quad   +RR 2-75 p76 tape
+-Gr 2-75 p1561 tape
ASD 3017. Tape (c) TC ASD 3017 (also Angel S 37044)
2656 ALBINONI: Adagio, organ and strings, G minor. BACH: Christmas ora-
torio, S 248: Sinfonia. Suite, orchestra, S 1068, D major: Air.
BEETHOVEN: Contradances (12). HANDEL: Berenice: Minuet. Mes-
siah: Pastoral symphony. MENDELSSOHN: Octet, op. 20, E flat
major: Scherzo. MOZART: German dance, no. 3, K 605. March, K
335, D major. PACHELBEL: Canon a 3 on a ground, D major. AMF;
Neville Marriner.
++Gr 8-75 p376 tape   ++RR 10-74 p33

```
        +HFN 10-75 p155 tape        +RR 11-75 p91 tape
        ++NR 1-75 p4
```

Melodiya ASD 3102
2657 Russian choral works of the seventeenth and eighteenth centuries.
     BEREZOVSKY: Do not reject me in my old age. BORTNYANSKY: Cher-
     ubim hymn, no. 7. I will lift up my eyes to the hills. DILET-
     ZKY: Glorify the name of the Lord. KALASHNIKOV: Concerto, 12
     voice choir: Cherubic hymn. KRESTYANIN: Befittingly. VEDEL:
     How long, O Lord, how long. USSR Russian Chorus; Alexander
     Yurlov.
          +Gr 9-75 p506   +RR 8-75 p62
          +HFN 8-75 p83

ASD 3116
2658 ALEXANDROV, A.: Patriotic war; Soviet army song. ALEXANDROV, B.:
     The guards greeted the spring in Berlin. BLANTER: Rostov town.
     KATZ: The Briansk forest. NOVIKOV: Vassya-Vassilyok. SOLOVYEV
     (Sedoi): Ballad about a soldier; Evening out in the bay; Night-
     ingales. Soviet Army Chorus and Band; Boris Alexandrov.
          +RR 9-75 p67

(Q) ASD 3131. Tape (c) TC ASD 3131 (ct) 8XASD 3131
2659 ALBINONI (Giozotto): Adagio, organ and string, G minor. DUKAS:
     L'Apprenti sorcier. DVORAK: Slavonic dance, op. 72, no. 1, B
     major. HUMPERDINCK: Hänsel und Gretel: Overture. PREVIN: André
     Previn's music night: Signature tune. RAVEL: La valse. WALTON:
     Orb and sceptre. LSO; André Previn.
          +Gr 9-75 p479   +-RR 10-75 p61
          ++HFN 10-75 p135

SLS 5003 (5) (Reissues from Columbia SAX 2458, 2488, 2514, 2537, 2554,
          5276)
2660 BERLIOZ: Symphonie fantastique, op. 14. DVORAK: Symphony, no. 9,
     op. 95, E minor. FRANCK: Symphony, D minor. MOZART: Symphony,
     no. 40, K 550, G minor. SCHUBERT: Symphony, no. 8, D 759, B
     minor. TCHAIKOVSKY: Symphony, no. 6, op. 74, B minor. PhO,
     NPhO; Otto Klemperer.
          ++Gr 6-75 p55   ++RR 6-75 p44
          ++HFN 9-75 p109

SLS 5004 (5) (Reissues from Columbia SAX 2315, 2411, 33 CX 1268, 33 C
          1036, 33 C 1390, 33 CX 1660)
2661 BEETHOVEN: Concerto, violin, op. 61, D major. BRAHMS: Concerto,
     violin, op. 77, D major. BRUCH: Concerto, violin, no. 1, op.
     26, G minor. PROKOFIEV: Concerto, violin, no. 1, op. 19, D
     major. SIBELIUS: Concerto, violin, op. 47, D minor. TANEIEV:
     Suite, op. 28. David Oistrakh, vln; French National Radio Orch-
     estra, LSO, Stockholm Festival Orchestra, PhO; André Cluytens,
     Otto Klempere, Lovro von Mataćić, Sixten Ehrling, Nicolai Malko,
     Alceo Galliera.
          ++Gr 10-75 p639

SLS 5012 (3)
2662 The incomparable Victoria de los Angeles, songs and operatic arias.
     BARBIERI: Canción de paloma. BIZET: Carmen: Habañera. DALZA:
     Enfermo estaba Antióco. ESPLA: Prego. FALLA: Spanish popular
     songs (7). La vida breve: Vivan los que rien. FUENLLANA: De
     los álamos vengo. FUSTE-VILA: Háblamo de amores. GOUNOD: Faust:
     O Dieu, que de bijoux. GRANADOS: Goyescos: La maja y el ruise-
     ñor. Tonadillas: La maja dolorosa. GURIDI: Jota. LASERNA:
     Jilguerillo con pico de oro; Las maja de Paris. MASCAGNI:

Cavalleria rusticana: Voi lo sapete. MASSENET: Manon: Je ne
suis que faiblesse...Adieu, notre petite table. MOMPOU: El
combat del Somni: Damunt de tu nomes. MONTSALVATGE: Canciones
negras: Canción de cuna para dormir. ORTEGA: Pues aun me
tienes, Miguel. PUCCINI: La bohème: Donde lieta usci. Madama
Butterfly: Un bel di vedremo. RODRIGO: De los álamos vengo,
madre. ROSSINI: Il barbiere di Siviglia: Una voce poco fa.
TOLDRA: Maig; Anacreontica. TURINA: Farruca jota. La giralda.
VERDI: Otello: Era più calmo...Piangea cantando...Ave Maria
piena. VILLA-LOBOS: Bachianas brasileiras, no. 5: Aria. VIVES:
El retrato de Isabela. WAGNER: Lohengrin: Einsam in trüben
Tagen. Tannhäuser: Dich teure Halle. ANON.: Canción de cuna;
El rossinyol; Adiós neu homiño; Parado de Valldemosa; Mariam
matrem; Una hija tiene el Rey. Pastorcico noh aduerma; Estava
la mora en su bel estar; Ay, que non hay. Victoria de los
Angeles, s; Various accompaniments.
        ++Gr 8-75 p356   +RR 8-75 p861
        +-HFN 9-75 p109

SLS 5017 (4)
2663 KOMZAK: An der schönen grunen Narenta, op. 227. Echtes Wiener
Blut, op. 189. Erzherzog-Albrecht-Marsch, op. 136. Fidels
Wien, op. 190. LANNER: Abendsterne, op. 180. Jagd, op. 82.
MILLOCKER: Carlotta. Traumwälzer. STRAUSS, E.: Ohne bremse,
op. 238. STRAUSS, J. II: Blindekuh overture. Eljen a Magyar,
op. 332. Im Krapfenwald, op. 336. Immer Heiterer, op. 235.
Kaiserwälzer, op. 437. Künstlerleben, op. 316. Neu-Wien, op.
342. Neue Pizzicato Polka, op. 449. A night in Venice overture.
Wo die Zitronen blühn, op. 364. STRAUSS, Josef: Allerlei, op.
219. Künstlergruss, op. 274. Perlen der Liebe, op. 39. SUPPE:
Morning, noon and night in Vienna: Overture. Poet and peasant
overture. ZIEHRER: Faschingskinder, op. 382. Fidels Wien.
Hoch und Nieder, op. 372. Die Landstreicher overture. Das
liegt bei uns im Blut, op. 374. Loslassen, op. 386. Schönfeld,
op. 422. Weaner Madeln, op. 388. Johann Strauss Orchestra;
Willi Boskovsky.
        ++Gr 6-75 p98   +RR 6-75 p54
        ++HFN 6-75 p107

SLS 5019 (5)
2664 BERLIOZ: Le carnaval romain, op. 9. La damnation de Faust, op. 24:
Hungarian march. Les Troyens: Royal hunt and storm. BIZET:
L'Arlésienne: Suites, nos. 1, 2; Intermezzo; Farandole. BORO-
DIN: Prince Igor: Polovtsian dances, Act 2. CHABRIER: Marche
joyeuse. GRANADOS: Goyescas: Intermezzo. LISZT: Hungarian
rhapsody, no. 2, G 244, C sharp minor (orch. Muller-Berghaus).
Les préludes, G 97. LEONCAVALLO: I Pagliacci: Intermezzo.
MASCAGNI: L'Amico Fritz: Intermezzo, Act 3. MUSSORGSKY: Pictures at an exhibition (orch. Ravel). OFFENBACH: Gaité parisiénne,
excerpts (orch. Rosenthal). PONCHIELLI: La gioconda: Dance of
the hours, Act 3. PUCCINI: Manon Lescaut: Intermezzo, Act 3.
RESPIGHI: The pines of Rome. ROSSINI: Il barbiere di Siviglia:
Overture. La gazza ladra: Overture. SCHMIDT: Notre Dame: Intermezzo. SIBELIUS: Finlandia, op. 26. Kuolema, op. 44: Valse
triste. VERDI: La traviata: Prelude, Act 3. WEBER: Invitation
to the dance, op. 63 (orch. Berlioz). WEINBERGER: Schwanda the
bagpiper: Polka. PhO; Herbert von Karajan.
        +Gr 12-75 p1097   ++RR 12-75 p60

SLS 5022 (2)
2665 The art of the recorder. ARNE: As you like it: Under the green-
        wood tree. BACH: Cantata, no. 208, Schafe können sicher weiden.
        Cantata, no. 106: Sonatina. Magnificat, S 243, D major: Esur-
        ientes. BARBIREAU: Een vrolic Wesen. BASTON: Concerto, D major.
        BRITTEN: Scherzo. BUTTERLY: The white throated warbler. BYRD:
        The leaves be green. COUPERIN, F.: Muséte de choisi. Muséte
        de taverni. DICKINSON: Recorder music. HANDEL: Acis and Gala-
        tea: O ruddier than cherry. HEURTEUR: Troys jeunes bourgeoises.
        HINDEMITH: Trio, soprano and 2 alto recorders. HOLBORNE: The
        choice. Muylinda. Pavan and galliard. Sic semper soleo.
        JACOTIN: Voyant souffrir. PURCELL: Three parts upon a ground:
        Fantasia. SCHMELZER: Sonata à 7 flauti. SERMISY: Allez sous-
        pirs. Amour me voyant. VIVALDI: Concerto, P 77, A minor.
        ANON.: English dance. Saltarello. London, Early Music Consort;
        David Munrow.
                +Gr 9-75 p479   +RR 9-75 p51
                +HFN 9-75 p91
SLS 5033 (4)
2666 FRANCK: Symphonic variations, piano and orchestra. GRIEG: Concerto,
        piano, op. 16, A minor. LISZT: Hungarian fantasia, G 123.
        LITOLFF: Concerto symphonique, no. 4, op. 102, D minor: Scherzo.
        MENDELSSOHN: Concerto, piano, no. 1, op. 25, G minor. Rondo
        brillant, op. 29, E flat major. RACHMANINOFF: Concerto, piano,
        no. 2, op. 18, C minor. SAINT-SAENS: Le carnaval des animaux.
        SCHUMANN: Concerto, piano, op. 54, A minor. TCHAIKOVSKY: Con-
        certo, piano, no. 1, op. 23, B flat minor. John Ogdon, Brenda
        Lucas, pno; PhO, NPhO, LSO, Birmingham City Symphony Orchestra;
        John Barbirolli, John Pritchard, Paavo Berglund, Aldo Ceccato,
        Louis Frémaux.
                +-RR 12-75 p68
EMD 5521
2667 BENNETT: The house of sleepe. DICKINSON: Winter afternoons.
        PATTERSON: Time piece. PENDERECKI: Ecloga VIII. WILLIAMSON:
        The musicians of Bremen. Rodney Slatford, double bs; King's
        Singers.
                +Gr 6-75 p80   +RR 5-75 p68
                ++HFN 8-75 p74
HLM 7065 (Reissue from HQM 1199)
2668 BACH: Prelude and fugue, S 546, C minor. Toccata, adagio and fugue,
        S 564, C major: Toccata. Fantasia and fugue, S 537, C minor.
        ELGAR: Sonata, organ, op. 28, G major: 1st movement. MacDOWELL:
        Sea pieces, op. 55, no. 1. WAGNER: Die Walküre: Ride of the
        Valkyries. Lohengrin: Bridal march. Tannhäuser: Grand march.
        WESLEY: Holsworthy church bells. WIDOR: Symphony, organ, no.
        5, op. 42, no. 1, F minor: Variations. Symphony, organ, no. 6,
        op. 42, no. 2, G minor: Allegro. Marcel Dupre, G. D. Cunningham,
        W. G. Alcock, R. Goss-Custard, George Thalben-Ball, org.
                +Gr 6-75 p75   +-RR 7-75 p44
Melodiya SXL 30190
2669 BORODIN: Prince Igor: Boyars' chorus with Yaroslavna. GLINKA:
        Russlan and Ludmila: Mysterious Lel. MUSSORGSKY: Khovanschina:
        Father, come out; The fair swan swims. Boris Godunov: Chorus
        from scene at Kromy. RIMSKY-KORSAKOV: The legend of the invis-
        ible city of Kitezh and the maiden Fevronia: Fevronia's wedding
        train; Prayer. The tale of the Tsar Sultan: May you grow to be

like a mighty oak. The Tsar's bride: Scene from Act 2. TCHAI-
KOVSKY: Eugene Onegin: Beautiful maidens. Mazeppa: No, oh no,
there is no bridge. Bolshoi Theatre Orchestra and Chorus; Mark
Ermler.
    +-Gr 9-75 p520   +RR 10-75 p28
    +HFN 9-75 p103

SXLP 30193 (Reissues from Capitol P 8414, Columbia SAX 2563)
2670 GLAZUNOV: Meditation, op. 32. GOLDMARK: Concerto, violin, op. 18,
    A minor. MUSSORGSKY: The fair at Sorochinsk: Gopak (trans.
    Rachmaninoff). RACHMANINOFF: Vocalise, op. 34, no. 14. RIMSKY-
    KORSAKOV: Fantasy on Russian themes, op. 33 (arr. Kreisler).
    Nathan Milstein, vln; PhO; Harry Blech, Robert Irving.
    +Gr 10-75 p639   +RR 9-75 p36
    +-HFN 9-75 p108

HUNGAROTON

SLPX 11491/3 (3)
2671 Musical life in old Hungary, 13-18th centures. Various performers.
    +HF 12-75 p84   +RR 10-75 p86
    +NR 10-75 p14

SLPX 11548
2672 BACH: Chorale preludes: In dulci jubilo, S 729; Vom Himmel hoch
    da komm ich her, S 700. COUPERIN: Messe pour les couvents: Et
    in Terra pax. DAQUIN: Noël X. FRANCK: Pastorale, op. 19.
    GERGELY: Improvisation on Hungarian Christmas tunes. KOLOSS:
    Partita. LISZYNYAI-SZABO: Két magyar pasztorál. REGER: Pieces,
    op. 59: Gloria in Excelsis. ZIPOLI: Pastorale. Ferenc Gergely,
    org.
    +NR 4-75 p10   ++SFC 12-22-74 p24

SLPX 11549
2673 ARCADELT: Si grand'è la pieta. Il ciel che rado (trans. Bakfark).
    BAKFARK: Fantasies, nos. 1, 8-10. Gagliarda. Non dite mal
    (gagliarda). Schöner Deutscher Tanz. CRECQUILLON: Le corps
    absent. Un gay bergier (trans Bakfark). JANNEQUIN: Or vien
    ca vien (trans. Bakfark). ROGIER (Pathie): D'amours me plains
    (trans. Bakfark). SANDRIN: O combien (trans. Bakfark). ANON.:
    Czarna krowa (trans. Bakfark). Daniel Benko, lt.
    +RR 4-75 p49

SLPX 11601/2 (2)
2674 ALBRECHTSBERGER: Fugue on Ite Missa est, Alleluia. BRIXI: Prelude
    and fugue, C major. DIRUTA: Toccata in Ionian mode. EBERLIN:
    Toccata and fugue, A minor. HOFHAIMER: Recordare. HUMMEL: Pre-
    lude and fugue, C minor. LISZT (arr. Gottschalk): Introduction,
    fugue and magnificat. MUFFAT: Toccata, no. 6. Versetti (6).
    MONN: Fugue, F major. MULLER: Prelude and fugue, E flat major.
    NIGER: Preludes, G major, C major, F major. NOVOTNY: Prelude,
    G major. Prelude and fugue from the Stark Tablature Book.
    Prelude, D major. RATHGEBER: Aria pastorella, G major. Schlag-
    arie, E flat major, F major, G major, D minor. Aria pastorella,
    C major. RIGLER: Fugue, A major. ROMANINI: Toccata in Mixoly-
    dian mode. STARK: Pastorale. VANHAL: Fugue, G major. ZAREWUT-
    IUS: Postlude, Benedicamus Dominicale. ZIMMERMAN: Fugue on
    Easter Alleluia. Prelude, A major. ANON.: Interlude. Gábor
    Lehotka, István Baróti, org.
    +NR 6-75 p14   +-RR 11-75 p80

SLPX 11629
2675 FARKAS: Pieces, guitar (6). HUZELLA: Dances (3). KOVATS: Three
     movements for guitar. PETRASSI: Suoni notturni. PONCE: Varia-
     tions and fugue on "Las folias". László Szendrey-Karper, gtr.
          +-NR 10-75 p12   -RR 3-75 p46
SLPX 11677
2676 PAGANINI: Caprices, op. 1, nos. 9, 21. PROKOFIEV: Romeo and Juliet,
     op. 64: 3 pieces. RAVEL: Tzigane. SAINT-SAENS: Havanaise, op.
     83. TCHAIKOVSKY: Valse scherzo, op. 34. YSAYE: Sonata, solo
     violin, no. 2. Miklós Szenthelyi, vln; Judit Szenthelyi, pno.
          +HFN 10-75 p150   -RR 8-75 p59
          ++NR 8-75 p15
SLPX 11692
2677 KALMAR: Pieces, piano (3). KOCSAR: Improvvisazioni. LANG: Inter-
     mezzi. SARY: Sounds for piano. SOPRONI: Invenzioni sul B-A-C-H.
     SZEKELY: Sonata, piano, no. 3. Adám Fellegri, pno.
          +Gr 7-75 p182   +RR 6-75 p71
          +NR 6-75 p13

                    INTERNATIONAL PIANO LIBRARY
IPL 5005/6
2678 ALBENIZ: Marcha militar. BACH: Invention, 2 part, no. 1, S 772,
     C major. Partita, B flat major: Gigue. CAMPOS-PARSI: Plena
     de concierto. CASALS: Preludio en do mayor. CHOPIN: Polonaise,
     op. 53, A flat major. CZERNY: Fantasia on Scottish airs, op.
     471. GAGNON: Sillsiana. GRANADOS: Exquise. Valse Tzigane.
     GOTTSCHALK: Variations on the Brazilian national anthem. LISZT:
     Paraphrases on DONIZETTI: Lucia di Lammermoor, G 397. VERDI:
     Rigoletto, G 434. MARSHALL: Suite catalonia: Foc-Follets.
     PADEREWSKI: Caprice. Genre Scarlatti. RAMEAU (Godowsky): Ren-
     aissance suite: Sarabande. REINECKE: Children's symphony: Slow
     movement. SCHUBERT: Ländler. TORRA: Sardana (Viladrau). Jorge
     Bolet, Ivan Davis, Gunnar Johansen, Alicia de Larrocha, Raymond
     Lewenthal, Michael May, Jesús María Sanroma, Fernando Valenti,
     Rosalyn Tureck, Bruce Hungerford, Roland Gagnon, Guiomar
     Novaes, pno; Beverly Sills, s.
          +St 3-75 p112
IPA 5007/8 (2)
2679 BEETHOVEN: Sonata, piano, no. 21, op. 53, C major. CHOPIN: Ballade,
     no. 4, op. 52, F minor. Nocturne, op. 9, no. 3, B major. Polo-
     naise, op. 26, no. 2, E flat minor. Waltz, op. 18, E flat major.
     Waltz, op. 64, no. 1, D flat major. SCHUBERT (Godowsky): Mom-
     ent musical, op. 94, D 780, F major. SCHUMANN: Kreisleriana,
     op. 16. STOKOWSKI: Caprice oriental. HOFMANN: Kaleideskop.
     Penguine. Josef Hofmann, pno.
          +St 12-75 p130

                              KENDOR
KE 22174
2680 BACH: Contrapunctus IX. BRAHMS: Chorale prelude, no. 8 (arr.).
     BUXTEHUDE: Prelude (arr.). FRACKENPOHL: Pop suite. LIEB:
     Feature suite. O'REILLY: Metropolitan quintet. SCHORGE: Way-
     ward waltz. VECCHI: Saltarello. WILDER: Brassininity. Potsdam
     Brass Quintet.
          +MJ 3-75 p33

KLAVIER

KLAVIER

KS 536

2681 Baroque brass. ADSON: Ayres for cornetts and sagbutts. BACH:
Contrapunctus IX. Contrapunctus. Fantasia. Cantata, no. 147,
Jesu, joy of man's desiring. Menuet. Sarabande. COUPERIN: La
tromba. Air tendre. GABRIELI, G.: Canzona per sonare, no. 2.
MONTEVERDI: Qui rise Tirsi; O Rosetta. MOURET: Rondeau. PUR-
CELL: Voluntary on the old 100th. SCHEIN: Die mit Tränen Säen.
WILBYE: Down in a valley. Eastern Brass Quintet.
        +Audio 12-75 p106   +NR 5-75 p8

KS 537

2682 BACH: Sonata, flute and harpsichord, G minor (trans. Macaluso).
CASTELNUOVO-TEDESCO: Sonatina, flute and guitar. DORAN: Andante,
flute and guitar. Suite, flute and guitar: Finale. KOECHLIN:
Miniatures (4) (trans. Macaluso). LECLAIR: Gigue. Musette
(trans. Macaluso). LOCATELLI: Sonata, flute and harpsichord
(trans. Macaluso). Vicenzo Macaluso, gtr; Floyd Stancliff, flt.
        +-NR 10-75 p12

LONDON

CS 6856

2683 MOZART: Le nozze di Figaro, K 492: Overture. ROSSINI: La gazza
ladra: Overture. STRAUSS, J. II: Die Fledermaus, op. 363:
Overture. WAGNER: Rienzi: Overture. WEBER: Der Freischütz:
Overture. LAPO: Zubin Mehta.
        +-HF 5-75 p94

STS 15160

2684 BACH: Chorale prelude: Wachet auf, ruft uns die Stimme, S 645 (arr.
Bantock). DEBUSSY: Suite bergamasque: Clair de lune (orch.
Mouton). ELGAR: Dream children, op. 43, nos. 1 and 2. FAURE:
Pavane. GLUCK: Orfeo ed Euridice: Dance of the blessed spirits.
MASSENET: Thais: Méditation. La vièrge: Dernier sommeil de la
vièrge. TCHAIKOVSKY: Chant sans paroles, op. 2, no. 3, F major.
Quartet, strings, no. 1, op. 11, D major: Andante cantabile.
NSL; Raymond Agoult.
        +NR 7-75 p4

STS 15198

2685 Baroque flute sonatas. BLAVET: Sonata, flute, no. 2, F major.
GAULTIER: Suite, G minor. HANDEL: Sonata, flute, op. 1, no. 5,
G major. LOEILLET: Sonata, flute, F major. TELEMANN: Sonata,
flute, F major. VINCI: Sonata, flute, D major. André Previn,
flt; Raymond Leppard, hpd; Claude Viala, vlc.
        +NR 10-74 p8   +St 2-75 p112

STS 15224

2686 ALBENIZ: Piezas características, no. 12: Torre bermeja. Suites
españolas: no. 1, Granada; no. 5, Asturias. Recuerdos de
viaje, op. 71, no. 6: Rumores de la caleta. FALLA: El sombrero
de tres picos: The miller's dance. SOR: Variations on a theme
by Mozart, op. 9. TARREGA: Recuerdos de la Alhambra. TORROBA:
Madroños. TURINA: Garrotin y soleares. Rafaga. VILLA-LOBOS:
Chôros, no. 1, E minor. Prelude, no. 1, E minor. Narciso
Yepes, gtr.
        +NR 10-75 p12

SPC 21100 (also Decca PFS 4325)

2687 ALBINONI: Concerto, oboe, op. 7, no. 6, D major: 1st movement (arr.

Wilbraham). ARBAN: Carnival in Venice (arr. Camerata). BACH:
Das Wohltemperierte Clavier, Bk II, S 870-893: Prelude, G major
(arr. Camerata). GLUCK: Orpheus and Eurydice: Dance of the
blessed spirits (arr. Camerata). HANDEL: Solomon: Arrival of
the Queen of Sheba (arr. Camerata). MOZART: Exsultate jubilate,
K 165: Alleluia (arr. Wilbraham). The magic flute, K 620: Queen
of the night's aria, Act 1 and Act 2 (arr. Wilbraham). PURCELL:
Trumpet tunes (4) (arr. Pearson). SCHUBERT: Ave Maria, D 839
(arr. Pearson). STANLEY: Trumpet tune, D major (arr. Pearson).
John Wilbraham, tpt; Leslie Pearson, org; London Festival Orchestra; Tutti Camarata.
    +-Gr 3-75 p1710   -NR 4-75 p14
     -HF 2-75 p109   -RR 3-75 p20

OS 26249
2688 Russian songs. BORODIN: For the shore of your far-off native land.
    DARGOMIZHSKY: Songs: The worm; Nocturnal breeze; The old corpor-
    al. GLINKA: The midnight review. RUBINSTEIN: Melody. TCHAIK-
    OVSKY: Don Juan's serenade, op. 38, no. 1; I bless you woods,
    op. 47, no. 5; It was in the early spring, op. 38, no. 2; Mid
    the noisy stir of a ball, op. 38, no. 3; None but the lonely
    heart, op. 6, no. 6; Not a word, oh my love, op. 6, no. 2.
    Nicolai Ghiaurov, bs; Zlatina Ghiaurov, pno.
        +HF 2-73 p100   +-Op 1-25-75 p32
        +NR 6-73 p12   +St 4-73 p126

OS 26277
2689 The art of Marilyn Horne. ARNE: Artaxerxes: Arias. BELLINI:
    Norma: Arias. DONIZETTI: Lucrezia Borgia: Arias. GLUCK: Orfeo
    ed Euridice: Arias. HANDEL: Semele: Arias. LAMPUGNANI: Meras-
    pe: Arias. MOZART: Don Giovanni, K 527: Arias. PONCHIELLA:
    La gioconda: Arias. ROSSINI: Semiramide: Arias. Marilyn Horne,
    ms; Other vocalists, various orchestras and conductors.
        +NR 1-75 p12   +St 4-75 p104

OS 26306
2690 Joan Sutherland, coloratura spectacular. BELLINI: La sonnambula:
    Arias and scenes. DONIZETTI: Linda di Chamounix: Arias and
    scenes. HANDEL: Samson: Arias and scenes. MOZART: Die Entfüh-
    rung aus dem Serail, K 384: Arias and scenes. Die Zauberflöte,
    K 620: Arias and scenes. ROSSINI: La cambiale di matrimonio:
    Arias and scenes. VERDI: Atila: Arias and scenes. I Masnadi-
    eri: Arias and scenes. Joan Sutherland, s; Other vocalists,
    various orchestras and conductors.
        +MJ 2-75 p30   +SFC 11-10-74 p27
        +NR 1-75 p12   +St 4-75 p104

OS 26315 (also Decca SXL 6585)
2691 CILEA: Adriana Lecouvreur: Ma dunque e vero. LEONCAVALLO: Manon
    Lescaut: Tu, tu amore. PONCHIELLI: La gioconda: Ma chi vien...
    Oh, la sinistra voce. VERDI: Aida: L'abborrita rivale...Gia
    i sacerdoti adunanso. ZANDONAI: Francesca da Rimini: No smar-
    agdi no...Inghirlandata di violiette. Renata Tebaldi, s; Franco
    Corelli, t; OSR; Anton Guadagno.
        +-Gr 12-73 p1242   +-ON 12-22-73 p36
         -HF 2-74 p109   -Op 5-74 p422
        +-NR 2-74 p9   +-St 3-74 p125
        +OC 12-75 p49

OS 26373 (also Decca Tape (c) KSXC 6658)
2692 Luciano Pavarotti, King of the high C's. BELLINI: I Puritani:

Arias. DONIZETTI: La fille du régiment: Arias. La favorita:
Arias. PUCCINI: La bohème: Arias. ROSSINI: William Tell: Arias
STRAUSS, R.: Der Rosenkavalier, op. 59: Arias. VERDI: Il trova-
tore: Arias. Luciano Pavarotti, t; Various orchestras; Richard
Bonynge, Edward Downes, Nicola Rescigno, Georg Solti, Herbert
von Karajan.

 +-Gr 2-75 01562 tape   +SFC 2-19-74 p30
 +RR 3-75 p74 tape   +St 4-75 p104

OS 26381
2693 Pilar Lorengar, aria recital. DVORAK: Rusalka, op. 114: Arias.
KORNGOLD: Die Tote Stadt, op. 12: Arias. MOZART: Don Giovanni,
K 527: Arias. Die Zauberflöte, K 620: Arias. PUCCINI: La
bohème: Arias. Madama Butterfly: Arias. La rondine: Arias.
VERDI: La traviata: Arias. WEBER: Der Freischütz: Arias. Pilar
Lorengar, s; Various orchestras and conductors.

 +NR 3-75 p8   +St 4-75 p104
 ++SFC 11-10-74 p27

OS 26391 (also Decca SXL 6650)
2694 BELLINI: Songs: Bella nice che d'amore; Dolente immagine di fille
mia; Ma rendi pur contento; Malinconia, ninfa gentile; Vanne
o rosa fortunata. BONONCINI: Per la gloria d'adorarvi. HANDEL:
Atalanta: Care selve. RESPIGHI: Songs: Nevicata; Nebbie; Piog-
gia. ROSSINI: La danza. SCARLATTI: Pompeo: Già il sole dal
Gange. TOSTI: Songs: La serenata; Luna d'estate; Malia; Non
t'amo più. Luciano Pavarotti, t; Bologna Teatro Comunale Orchestra; Richard Bonynge.

 +Gr 1-75 p1379   +RR 1-75 p54
 +-HF 3-75 p90   ++SFC 11-17-74 p31
 +NR 3-75 p9   +SR 2-8-75 p37
 +ON 2-8-75 p33   ++St 12-74 p152

MICHIGAN UNIVERSITY

SM 0002
2695 200 years of American marches. BAGLEY: National emblem. BELSTER-
LING: March of the steel men. BILLINGS: Chester. CHAMBERS:
The boys of the old brigade. FARRAR: Bombasto. FILLMORE: Amer-
icans we. GRAFULLA (Reeves): Washington grays. HUFFER: Black
Jack. HOLLOWAY: Wood-up quick-step. KLOHR: The billboard.
KING: Hosts of freedom. PANELLA: On the square. PHILE: Hail,
Columbia. SCHREINER: General Lee's grand march. SOUSA: El
Capitán. WELDON: Gate city. TRAD.: The rose tree. The roving
sailor. The white cockade. The world turned upside down.
Yankee doodle. Michigan School of Music Winds and Percussion;
Clifford P. Lillya.

 +St 12-75 p135

MONITOR

MFS 757
2696 Slavonic Orthodox liturgy. CHESNOKOV: Spassi, Bozhe Lyudi Tvoya
(Save, God, our people). ARKANGELSKII: Blazhen Razumevayei
(Blessed are they that understand). CHRISTOV: Velika ekteniya
(Great litany); Vo tsarstve tvoyem (In Thy kingdom); Kheruvimska
no. 2 (Hymn of the cherubim, no. 2); Tebe poyem (To Thee we
sing); Dostoyno yest (It is fitting). GRETCHANINOV: Veruyu, op.

no. 8 (Credo); Slava I Edinorodni (Glory and only begotten).
KEDROV, N., Sr.: Otche Nash (Our Father). S. Marcheva, s; N.
Peneva, ms; Christo Kamenov, t; B. Spasov, Ivan Petrov, bs;
Svetoslav Obretenov Bulgarian Choir; Georgi Robev.
      ++St 4-75 p107

MCS 2143
2697 BACH: Toccata and fugue, D minor. BEETHOVEN: Sonata, piano, no.
      14, op. 27, no. 2, C sharp minor: Presto. IBERT: Little white
      donkey. KHACHATURIAN: Gayaneh: Sabre dance. LISZT: Hungarian
      rhapsody, no. 2, G 244, C sharp minor. Etudes d'exécution
      transcendente d'après Paganini, no. 3, A flat minor, G 140.
      MESSIAEN: La nativité du Seigneur: Les anges. RAVEL: Jeux d'eau.
      RIMSKY-KORSAKOV: The tale of the Tsar Sultan: Flight of the
      bumblebee. Christian di Maccio, accordion.
      ++NR 10-75 p13

                    MUSICAL HERITAGE SOCIETY
MHS 1345 (also Odyssey Y 33520)
2698 Music for flute and harp. DAMASE, J. M.: Sonata, G major. FAURE:
      Berceuse, op. 16. IBERT: Entr'acte. KRUMPHOLZ: Sonata, F
      major. ROSSINI: Introduction and variations. ANON.: Green-
      sleeves. Jean-Pierre Rampal, flt; Lilly Laskine, hp.
      +AR 2-73 p25   ++NR 10-75 p4

MHS 1790
2699 Battle music for organ. BANCHIERI: Battaglia, Canzone Italiana
      dialogo. BULL: Coranto battle. CABANILLES: Batalla II. DIEGO
      DE CONCEICAO: Batalha de 5. FRESCOBALDI: Capriccio sopra la
      battaglia. JIMENZ: Batalla de sexto tono. KRIEGER: Schlacht.
      LOFFELHOLTZ: Die kleine Schlacht. KERLL: Feldschlacht. Franz
      Haselböck, org.
      /St 9-75 p119

MHS 1853
2700 BOISMORTIER: Sonata, bassoon, no. 5, G minor. FASCH: Sonata,
      bassoon, C major. GALLIARD: Sonata, bassoon, no. 3, F major.
      SCHUTZ: Symphoniae sacrae, Bk I, nos. 16, 17. TELEMANN: Sonata,
      viola da gamba, E minor. Robert Thomas, bassoon; Thomas Tro-
      baugh, hpd.
      +HF 7-75 p93

MHS 1951
2701 CARDILLO: Core 'ngrato. CIOFFI: Na sera 'e maggio. CAPUA: I te
      vurria vasa. CURTIS: Torna a Surrient; Voce e notte. FALVO:
      Dicitencello vuie. NARDELLA: Chiove. PENNINO: Pecchê. TAGLIA-
      FERRI: Passione; Piscatore e pusilleco. Carlo Bergonzi, t;
      Madrid Chamber Orchestra; Enrico Pessina.
      ++St 7-75 p107

MHS 1959
2702 Waltzes of Johann Strauss II--paraphrases. DOHNANYI: Treasure
      waltz. GRUNFELD: Emperor waltz; Voices of spring; Paraphrase
      on themes from Aschenbrödel. SCHNABEL: Four old Vienna waltzes.
      SCHULHOFF: Pizzicato polka. SCHULZ-ELVER: Arabesque on "The
      blue Danube". Hans Kann, pno.
      +St 6-75 p108

MHS 1976
2703 Music for voice and violin. BLACHER: Francesca da Rimini, op. 47.
      HOLST: Songs for voice and violin, op. 35 (4). HOVHANESS:

Hercules, op. 56, no. 4. VAUGHAN WILLIAMS: Along the field.
VILLA-LOBOS: Suite for voice and violin. Catherine Malfitano,
s; Joseph Malfitano, vln.
++St 6-75 p108

NEW ENGLAND CONSERVATORY

NEC 111
2704 BILLINGS: Be glad, then America. ELLINGTON: Daybreak express.
Koko. HOLYOKE: Processional march. HOPKINSON: Beneath a weep-
ing willow's shade; My generous heart disdains. IVES: March II.
March intercollegiate. Serenity. SHERWIN: Sign tonight. ANON.:
Band of music. His name so sweet. Colonel Orne's march. George
Washington's march. Psalm 21. White cockade. Wonderous love.
Yankee doodle. At the river (arr. Copland and Wilding-White).
New England Conservatory Chmaber Singers; Lorna Cooke de Varon.
+St 9-75 p119

NONESUCH

H 71141 (Reissue)
2705 BEETHOVEN: Country dances, op. 141 (12). HAYDN: Raccolta de
menuetti ballabili, nos. 1, 14. LANNER: Mitternachtswalzer, op.
8. Regata-Galopp, op. 134. MOZART: Country dances, K 609 (5).
SCHUBERT: Minuets with 6 trios, D 89 (5). Vienna State Opera
Orchestra; Paul Angerer.
+HFN 12-75 p171   +RR 12-75 p50

H 71233
2706 ALBENIZ: Recuerdos de viaje, op. 71, no. 2: Leyenda (trans.
Segovia). Piezas características, no. 12: Torre bermeja (trans.
Tarrega). FALLA: Homenaje a Debussy. GRANADOS: Tonadillas al
estilo antiguo: La maja de Goya. Danza española, op. 37, no. 5:
Andaluza. NIN-CULMELL: Variations on a theme by Milan (6).
ORBON: Preludio y tocata. RODRIGO: Zarabanda lejana. Rey de
la Torre, gtr.
/RR 3-75 p39

H 71276 (Q) HQ 1276
2707 Early American vocal music, New England anthems and Southern folk
hymns. CHAPIN: Rockbridge. BILLINGS: I am come into my garden;
I am the rose of Sharon; I charge you; An anthem for Thanks-
giving: O praise the Lord. DARE: Babylonian captivity. INGALLS:
Northfield. LAW: Bunker Hill. LEWER: Fidelia. MORGAN: Judg-
ment anthem; Amanda. READ: Newport. ROBISON: Fiducia. WHITE:
Power. ANON.: Animation; Canaan; Concert; Lonsdale; Messiah;
Pilgrim's farewell; Springhill; Triumph; Washington. The
Western Vocal Wind; Mary Lesnick, ms; Raymond Murcell, bs-bar;
Stuart Schulman, vln; Bonney McDowell, bs viol; Paul Fleischer,
pic; Allen Herman, snare drum.
+Gr 3-75 p1711   +NR 7-73 p9
+HF 3-73 p100   +RR 3-75 p14
+HF 1-74 p84 Quad   ++SFC 2-18-73 p34
+LJ 5-73 p34   ++St 4-73 p80

H 71292
2708 Music in honor of St. Thomas of Canterbury. BENET (attrib.): Jacet
granum, Sanctus. POWER: Opem nobis, Credo. ANON.: Ante thronum
regentis omnia, Sequence. Clangat tuba, martyr Thoma, Carol.

In Rama sonat gemitus, Threnody. Jacet granum, Responsory and
Prosa. Laetare Cantuaria, Carol. Novus miles sequitur, Conductus. Pastor caesus in gregis medio, Antiphon. Solemne
canticum, Sequence. Thomas gemma Cantuariae, Motet. Accademia
Monteverdiana; Trinity Boys' Choir; Denis Stevens.

    +Gr 1-75 p1380   +RR 12-74 p64
    +MQ 7-75 p496   ++St 9-74 p135
    +NR 5-74 p8

H 71298
2709 Cornet favorites. ARBAN: Fantasie and variations on "The carnival
    of Venice". HOHNE: Slavische fantasie. SIMON: Willow echoes.
    THOMSON: At the beach: Concert waltz. CLARKE: The bride of the
    waves: Polka brillante. The debutante: Caprice brillante.
    From the shores of the mighty Pacific: Rondo caprice. Sounds
    from the Hudson: Valse brillante. Gerard Schwarz, cor; William
    Bolcom, pno.

    ++Gr 3-75 p1710   ++RR 3-75 p14
    +HF 11-74 p129   ++SFC 7-21-74 p25
    +NR 7-74 p14   +St 8-74 p112

H 71301
2710 BIBER, C. H.: Sonata, trumpet, C major. Sonata, trumpet, 2 choirs.
    BIBER, H.: Sonata à 7, C major. GABRIELI: Sonata, trumpet.
    MOLTER: Symphony, 4 trumpets, C major. PEZEL: Sonatinas, nos.
    61-62, 65-66. RATHGEBER: Concerto, trumpet, op. 6, no. 15, E
    flat major. SCHEDIT: Canzon cornetto. Gerard Schwarz, tpt;
    New York Trumpet Ensemble; Kenneth Cooper, hpd and org.

    ++HF 2-75 p108   ++NR 4-75 p14
    ++HFN 6-75 p99   +RR 4-75 p22
    ++MT 9-75 p797   ++SFC 3-16-74 p25

H71302/3
2711 Spectrum: New American music. ANDERSON: Variations on a theme by
    M. B. Tolson. BABBITT: All set. JONES: Ambiance. ROCHBERG:
    Blake songs. WERNICK: Kaddish-requiem. WOLPE: Quartet, trumpet,
    tenor saxophone, percussion and piano. Phyllis Bryn-Julson, s;
    Jan DeGaetani, ms; Ramon Gilbert, cantor; Contemporary Chamber
    Ensemble; Arthur Weisberg.

    ++HF 4-75 p98   -RR 12-75 p71
    ++NR 1-75 p6   +St 6-75 p110

H 71315
2712 A Medieval Christmas: Isaiah's prophecy; The Sybil's prophecy
    (Iudicii Signum); Conductus (Adest sponsuo); Gabriel's prophecy
    (Oiet, virgines); Clausula (Domino); Hymn (Conditor alme sider-
    um); Lauda (Verbum caro factum est); Reading (On frymde waes
    work): Organum (Judea et Iherusalem): Conductus (Gedonis area):
    Reading (O moder mayde); Hymn (Ave maris stella); Sacred song
    (O Maria, Deu maire); Prosa (Ave Maria); Responsory (Gaude
    Marie); Hymn (Joseph, Liber nefe min); Conductus with refrain
    (Conguadet hodie, In dulci jubilo); Reading (Hand by hand we
    shule us take); Conductus (Lux hodie...Orientis partibus).
    Boston Camerata; Joel Cohen.

    +HFN 12-75 p157   +NR 12-75 p1

HB 73028 (2)
2713 BERIO: Sequenza. DAVIDOVSKY: Junctures. FUKUSHIMA: Pieces from
    Chu-u. LEVY, B.: Orbs with flute. REYNOLDS: Ambages. ROUSSAKIS:
    Short pieces, 2 flutes (6). TROMBLY: Kinetics III. VARESE:
    Density 21.5. WESTERGAARD: Divertimento on Discobbolic fragments.

WUORINEN: Variations, flute, I and II. Harvey Sollberger,
Sophie Sollberger, flt; Allen Blustine, clt; Jeanne Benjamin,
vln; Charles Wuorinen, pno.

    ++HF 9-75 p102   +NR 9-75 p9
    +HFN 11-75 p169   +NYT 4-27-75 pD19

ODYSSEY

Y2 32880 (2) (Recorded 1942-52)
2714 Tibute to Jennie Tourel. BELLINI: Norma: Sgombra è la sacra selva.
    BIZET: Carmen: Habañera; Seguidilla; Chanson bohème; Card scene.
    Songs: Adieu de l'hôtesse arabe. CHOPIN: Polish songs, op. 74,
    no. 1: Maiden's wish; Melancholy. DEBUSSY: Chansons de Bilitis
    (3). GINASTERA: Triste. MUSSORGSKY: Songs and dances of death.
    NIN: Paño murciano. OBRADORS: Coplas de curro dulce. RAVEL:
    Songs: Chansons madécasses; Vocalise. ROSSINI: Il barbiere di
    Siviglia: Una voce poco fa. La cenerentola: Nacqui all'affanno.
    L'Italiana in Algeri: Cruda sorte. Semiramide: Bel raggio
    lusinghier. VILLA-LOBOS: Cancão do carreiro. Jennie Tourel,
    ms; Various accompaniments.

    +HF 3-75 p68   ++NYT 9-22-74 pD32
    +NR 1-75 p12

L'OISEAU-LYRE

DSLO 3
2715 BEDFORD: You asked for it. BRITTEN: Nocturnal after John Dowland,
    op. 70. LENNON(McCartney): Yesterday. DAVIES: Lullaby for
    Ilian Rainbow. SWAYNE: Canto 1, Mr. Timothy's troubles. WALKER:
    Lorelei. Etude. Timothy Walker, gtr.

    +Gr 7-74 p236   +RR 7-74 p60
    +NR 3-75 p14

DSLO 7
2716 ALBENIZ: España, op. 165, no. 2: Tango (arr. Godowsky). CHAMINADE:
    Autrefois, op. 87, A minor. GLAZUNOV: Waltz, op. 42, no. 3,
    D major. GODOWSKY: Waltz-poem, no. 4, for the left hand. Alt
    Wien. HOFMANN: Kaleideskop, op. 40. MOSZKOWSKI: Caprice espag-
    nol, op. 37. RAMEAU: Tambourin (arr. Godowsky). RUBINSTEIN:
    Melody, op. 3, no. 1, F major. SAINT-SAENS: Le carnaval des
    animaux: The swan (arr. Godowsky). SCHUBERT: Moment musicaux,
    op. 94, D 780, F major. (arr. Godowsky). STRAUSS, J. II: Wine,
    women and song, op. 333 (arr. Godowsky). TCHAIKOVSKY: Nocturne,
    op. 19, no. 4. Shura Cherkassky, pno.

    +Gr 6-75 p75   +SFC 11-16-75 p32
    +RR 6-75 p60

12BB 203/6
2717 Musicke of sundre kindes: Renaissance secular music, 1480-1620.
    ALISON: Dolorosa pavan. ATTAINGNANT: Content desir basse danse.
    AZZAIOLO: Quando le sera; Sentomi la formicula. BARBERIIS:
    Madonna, qual certezza. BOTTEGARI: Mi stare pone Totesche.
    CAURROY: Fantasia; Prince la France te veut. CAVENDISH: Wand'
    ring in this place. COMPERE: Virgo celesti. COSTELEY: Hélas,
    hélas, que de mal. DALZA: Recercar; Suite ferrarese; Tastar
    de corde. DUNSTABLE (Anon.): O rosa bella; Hastu mir. FAYRFAX:
    I love, loved; Thatt was my woo. FONTANA: Sonata, violin.
    FORSTER: Vitrum nostrum glorosum. FESCOBALDI: Toccata.

GABRIELI, A.: Canzona françese. GABRIELI, G.: Sanctus Dominus
Deus. GESUALDO: Canzona françese; Mille volte il dir moro.
GIBBONS: Now each flowery bank. GOMBERT: Caeciliam cantate.
GUAMI: La brillantina. HECKEL: Mille regretz; Nach willen dein.
HOFHEIMER: Nach willen dein. HUME: Musick and mirth. ISAAC, A.:
Ne più bella di queste; Palle, palle; Quis dabit pacem. ISAAC,
H.: La la hö hö. JANNEQUIN: Les cris de Paris. JEUNE: Fière
cruelle. JOSQUIN DES PRES: Mille regretz. LASSUS: Cathalina,
apra finestra; Matona mia cara. MARENZIO: O voi che sospirate;
Occhi lucenti. MERULO: Canzona françese. MODENA: Ricercare
a 4. MONTEVERDI: Lamento d'Olimpia. MOUTON: La, la, la l'oysillon du bois. MUDARRA: Dulces exuyiae. NARVAEZ: Fantasia; Mille
regretz. OBRECHT: Ic draghe die mutze clutze; Mijn morken gaf;
Pater noster. ORTIZ: Dulce memoire; Recercada. PACOLONI: Pado-
ana commun; Passamezzo commun. PORTER: Thus sang Orpheus.
RONTANI: Nerinda bella. RORE: De la belle contrade. RUE: Pour
ung jamais. SANDRIN: Doulce memoire. SERMISY: Content désir;
La, je m'y plains. TERZI: Canzona françese. TIBURTINO: Ricer-
care a 3. TORRE: Adormas te, Señor; La Spagna. TROMBONCINO:
Ave Maria; Hor ch'el ciel e la terra; Ostinato vo seguire. TYE:
In nomine. VASQUEZ: Lindos ojos aveys, señora. VERDELOT: Mad-
onna, qual certezza. WILLAERT (arr. Cabezon): E qui, la dira;
Madonna qual certezza. ANON.: Belle, tenés mo; Calata; Celle
qui m'a demande; Chui dicese e non l'amare; Der Katzenföte;
Lady Wynkfyldes rownde; Las, je n'ecusse; L'e pur morto Feragú;
Mignonne allons; Pavana de la morte de la ragione; Pavana el
tedescho; Pavane Venetiana; Der rather Schwanz; Le rossignol;
Saltarello de la morte de la ragione; Se mai per maraveglia;
Shooting the guns pavan; Se mai per maraveglia; Sta notte; Suite
regina; La triquotée. ANON/AZZAIOLO: Girometa. ANON/EDWARDS:
Where griping griefs. ANON/ISAAC: Bruder Konrad. ANON/JUDEN-
KUNIG: Christ der ist erstanden. ANON/PACOLINI: La bella Fran-
ceschina. ANON/SPINACINO: Je ne fay. ANON/WYATT: Blame not my
lute. Consorte of Musicke; Anthony Rooley.
 +-Gr 12-75 p1066   +RR 12-75 p89
 ++HFN 12-75 p159

SOL 329

2718 Courtly pastimes of 16th century England. ATTAIGNANT: Tordion.
COOPER: I have been a foster. CORNYSHE: Adieu mes amours. Blow
thy horn, hunter. Fa la sol. While life or breath. GERVAISE:
Allemande. Basse danse. La volunte. HENRY VIII, King: Tho'
some saith. If love now reigned. The time of youth. MASCHERA:
Canzona seconda. Canzona quarta. RYSBYE: Who so that will him-
self apply. SUSATO: Basse danse. Le cueur est bon. Entrée du
fol. La Mourisque. Pour quoy. Saltarelle. ANON.: And I were
a maiden. Instrumental consort. Hey trolly lolly lo. Let not
us that young men be. Time to pass. John Whitworth, Derek
Harrison, c-t; Philip Langridge, t; Francis Grubb, bar; St.
George's Canzona.
 +Gr 2-73 p1540   +NR 7-74 p9
 +HF 4-75 p94   +RR 1-73 p67
 +HFN 1-73 p126

SOL 336

2719 CUTTING (Johnson): Greensleeves. DANYEL: A fancy. Rosamunde
pavan. Passymeasures galliard. DOWLAND: Master Piper's pavan.
Semper Dowland semper dolens. My Lord Willoughbie's welcome

home. JOHNSON: Rogero. The delight pavan and galliard. Dump,
no. 3. PHILIPS Philips'a pavan and galliard. ROBINSON: A toy:
Bo Peep. Fantasia. A plainsong. ANON.: Galliard fancy; A
merry moode; Jigge. Anthony Rooley, James Tyler, lt.
　　　+Gr 6-74 p85　　　　　　　　++NR 11-74 p14
　　　+HF 3-75 p89　　　　　　　　+RR 7-74 p62

SOL 337
2720 Deus. Cantores in Ecclesia: Michael Howard.
　　　++St 3-75 p109

OLD NORTH BRIDGE RECORDS

ONB 1775
2721 Fife and drum tunes: White cockade; British Grenadiers; Olde
Saybrooke; Rakes of Mallow; Grandfather's clock; La belle
Catherine; Slow Scotch; Garry Owen; Wrecker's daughter; Welcome
here again; Duke of York's march; Yankee Doodle; Sisters; Road
to Boston; Green cockade; The Harriette; Katy Hill; Rose tree;
The girl I left behind me. The Davis Blues of the Acton Minute-
man Company; The Bedford Minuteman Fifers and Drummers; Lincoln
Minute Men Fifes and Drums.
　　　+AR 2-75 p24

OLYMPIC

8136
2722 The golden age of Viennese music. KOMZAK: Bad'ner Mäd'ln, op. 252.
KALMAN: Grandioso. LEHAR: Gold und Silber, op. 79. LANNER:
Die Schönbrunner, op. 200. STOLZ: Wiener Cafe. STRAUSS, E.:
Unter der Enns. STRAUSS, J. I: Radetzky march, op. 228. STRAUSS
J. II: Die Fledermaus, op. 363; Overture. STRAUSS, Josef: Dor-
schwalben aus Osterreich, op. 164. STRAUSS, J. II/Josef: Pizzi-
cato polka. VSO; Robert Stolz.
　　　+NR 6-75 p3

OPUS ONE

22
2723 ALBRIGHT: Take that. BERTONCINI: Tune. CAGE: Amores. GARLAND:
Apple blossom. HARRISON: Fugue. MILLER: Quartet variations.
Blackearth Percussion Group.
　　　++St 8-75 p108

ORION

ORS 73114
2724 BACH: Suite, orchestra, S 1068, D major: Air. KREBS: Suite, G
minor. MARCELLO: Sonata, flute, op. 2, B minor. PEPUSCH: Trio
sonata, G minor. TELEMANN: Sonata, flute and continuo, op. 2,
E minor. Jean-Pierre Rampal, Mario Duschenes, flt; Kenneth
Gilbert, hpd; Paris Festival Strings.
　　　+Audio 9-75 p70　　　　　　　+-NR 11-74 p7

ORS 74160
2725 COPLAND: Serenade, ukulele. CRESTON: Suite, op. 18. ENGEL: Sea
shell (arr. Zimbalist). KODALY: Adagio. SARASATE: Introduction
and tarantelle, op. 43. SUK: Pieces, op. 17 (4). Diana Steiner,

vln; David Berfield, pno.
-NR 3-75 p12

ORS 74161
2726 ALAIN: Fantasie, no. 2. BUXTEHUDE: Toccata and fugue, F major.
MENDELSSOHN: Sonata, organ, no. 2. MESSIAEN: Les corps glorieux:
Joie et clarté des corps glorieux. SOWERBY: Toccata. STANLEY:
Voluntary, organ, A major. David McVey, org.
+NR 3-75 p13

ORS 75181
2727 BLOCH: Jewish life: 3 pieces. Meditation hébraique. BRUCH: Kol
Nidrei, op. 47. CASALS: Song of the birds. CHOPIN: Sonata,
violoncello, op. 65, G minor: Largo. DEBUSSY (Feuillard):
Preludes, Bk I, no. 8: La fille aux cheveux de lin. FAURE:
Après un rêve, op. 7, no. 1. Elégie, op. 24. RAVEL (Bazelaire):
Pièce en forme de habañera. Lillian Rehberg Goodman, vlc;
Harold Bogin, pno.
+HF 10-75 p90   ++NR 8-75 p15

ORS 75199
2728 FINGER: Sonata, flute, D minor. GEMINIANI: Sonata, flute, D major.
LOEILLET: Sonata, flute, recorder and harpsichord, op. 1, G
minor. MATTHESON: Sonata, flute, A minor. TELEMANN: Sonata,
flute, G major. Jean-Pierre Rampal, flt; Mario Duschenes, rec;
Kenneth Gilbert, hpd.
++NR 9-75 p9

ORYX

BRL 1
2729 Renaissance hits. HENRY VIII, King: Green groweth holly. King
Harry's pavane. Pastime with good company. PRAETORIUS: Bourrée
and Der Schützkonig. Winter is an icy guest. SIMPSON: Ballet.
Mascarada. Volta. SUSATO: Ronde-Allemande-Gaillarde. ANON.:
Carols and dances from the Middle Ages: Estampie and stantipes.
Nowell and Nowell. Tanzlied. Dances from 15th century France:
Brawls. Basses danses. Jacobean England, popular tunes: Ayre.
Courante. The frog galliard. Since first I saw your face.
Music for springtime: I chose my love. The cuckoo. Gaillarde.
La Rocque. The Quiet Consort: The first canticle. Pavane des
Dieux. Pavane Lesquercarde. Musica Antiqua Vocal and Renais-
sance Instrumental Ensemble; Michael Uridge.
-Gr 7-75 p237

ORYX 1720 (Reissue from Musica Rara MUS 13)
2730 Songs and instrumental works from the Mannheim Court. CANNABICH:
Rechenmeister amor. DANZI: Quartet, flute, op. 56, no. 2, D
minor. Songs: Ach was ist die Liebe; Dir ruh o die ich liebe;
Ich denke dein; Ich hab ein Mädchen; Ich liebe dich; In des
Lebens Maien; Oft am Rande stiller Flöten; Wann erwacht der
Knabe Wieder. FILTZ: Sonata, 2 flutes, op. 2, no. 4, D major:
Allegro and minuetto. FRANZL, F.: Andenken an Elisen. RICHTER,
F. X.: Sonata, violin and harpsichord, G major: Andante. WEBER,
B. A.: An den Mond. Renate Fried, s; Herbert Bender, t; Heinz
Mayer, pno; Mannheim Instrumental Soloists.
-Gr 7-75 p237   /HFN 10-75 p153

PANDORA

PAN 101
2731 The sound of the Italian harpsichord. BACH: Concerto, harpsichord,
     S 971, F major. FROBERGER: Suite, C major: Allemande. Fantasy,
     no. 2, E minor. PESCETTI: Allegretto, C major. Presto, C minor
     SCARLATTI, D.: Allegretto, D minor. Presto, E major. ZIPOLI:
     Largo and gavotte, B minor. Martha Goldstein, hpd.
          +AR 2-75 p22

PAN 103
2732 Baroque flute sonatas on historical instruments. BACH, C. P. E.:
     Sonatas, flute and harpsichord, Wq 134 and 125, G major, B flat
     major. BLAVET: Sonata, no. 2, B minor. HOTTETERRE: Suite, D
     major. QUANTZ: Sonata, flute, D major. TELEMANN: Sonata, no.
     3, E minor. Alexander Murray, baroque flt; Martha Goldstein,
     hpd.
          +—AR 2-75 p24

PANTON

010335
2733 MONTE: O suavitas; Stella del nostro mar a l'apparir del sol.
     OROLOGIO: Occhi miei. REGNART: Ardo si, ma non t'amo. SAYVE:
     Kyrie. SCHONDORFF: Gloria. ZANGUIS: Magnificat secundi toni.
     ZANOTTI: Tirsi morir volea. Kuhn Mixed Choir; Pavel Kuhn.
          +RR 10-75 p84

110 253
2734 ISTVAN: Refrains for string trio. JIRASEK: Serenade, flute, bass
     clarinet and tape. KLUSAK: Invention, no. 5. KUCERA: Scenario,
     flute and string trio. TAUSINGER: Canto di speranza, piano
     quartet. Petr Brock, flt; Miroslav Kučaba, bs clt; Ivo Keislich
     Jaroslav Přikryl, tape; Bohuslav Burger, vln; Pavel Janda, vla;
     Viktor Moučka, vlc; Prague Wind Quintet, Moravian Quintet,
     Members, Bohuslav Martinu Piano Quintet.
          /HFN 6-75 p99   +RR 7-75 p38

110 361
2735 Festive marches. BOHAC: March. DVORAK: Festive march. JEREMIAS:
     March. MACHA: March. NEJEDLY: Vitězství bude nase. PAUER:
     Hrdinum práce. RIDKY: March. SMETANA: March for the Shakes-
     peare festivities. ZELEZNY: March. Musici di Prague; Václav
     Smetáček.
          -HFN 9-75 p106   +RR 7-75 p35

PEARL

GEM 121/2 (2)
2736 ARDITI: Parla. BELLINI: La sonnambula: Ah, non credea mirarti; Ah,
     non giunge. FONTENAILLES: Obstination. GOLDMARK: Die Königin
     von Saba: Lockruf. GOUNOD: Serenade. HANDEL: Il Pensieroso:
     Sweet bird. KREISLER: Caprice viennois. MOZART: Il Re pastore:
     L'amerò, sarò costante. PUCCINI: La bohème: Si, mi chiamano
     Mimi; O soave fanciulla; Donde lieta usci; Addio, dolce svegliar€
     Sono andati; Finale. Madama Butterfly: Un bel di, vedremo.
     REGER: Maria Wiegenlied, op. 72, no. 52. ROSSINI: Il barbiere
     di Siviglia: Una voce poco fa. TAUBERT: Der Vogel im Walde.
     THOMAS: Mignon: Styrienne. TOSELLI: Serenade. VERDI: Ernani:
     Ernani, involami. Rigoletto: Caro nome. La traviata: Addio del

passato. Il trovatore: Vivra contende. John Amadio, flt; Vása
Príhoda, vln; Selma Kurz, Grete Forst, s; Leo Slezak, t; Hein-
rich Schlusnus, Friedrich Weidemann, bar.
+Gr 11-74 p964   +RR 3-75 p17

GEM 132
2737 BACH: Concerto, 2 violins and strings, S 1043, D minor. Suite,
orchestra, S 1068, D major: Air on the G string. MOZART: Concerto, violin, no. 4, K 218, D major. SCHUBERT: L'Abeille.
SULZER: Sarabande. TCHAIKOVSKY: Chant sans paroles. Fritz
Kreisler, Efrem Zimbalist, vln; String Quartet, piano accompaniment; LSO; Landon Ronald.
+RR 12-75 p36

SHE 511
2738 Italian song recital. CESTI: Orentea: Intorno all'idol mio.
DAVICO: O lune che fa'lume; Japanese songs (5). FASOLO: Cangia,
cangia tue voglie. LOTTI: Pur dicesti, O bocca bella. SCAR-
LATTI, A.: Pompeo: Fia il sole dal Gange. TOSTI: Consolazione.
Edward Hain de Lara, bar; Miguel Zanetti, pno.
-Gr 3-75 p1691   -RR 5-75 p67

PELCA

PSR 40589
2739 BURCKHARDI: Fantasies, 2 recorders. FRESCOBALDI: Canzona, viola
da gamba and cembalo. KERLL: Capriccio, cembalo. LOGY: Aria.
Sarabande and gigue. PEUERL: Dance. REUSNER: Suite, E minor.
SCHICKHARDT: Trio sonata, F major. WILLIAMS: Sonata in imitation
of birds. Viva Musica.
+NR 12-75 p8

PSR 40598
2740 Organ concert at Inselkirche, St. Nicolai, Helgoland. BACH: Pre-
ludes, S 925, 929, 933, 941, 936. HUMMEL: Andante, A flat major.
KRIEGER: Fantasia, D minor. LUBECK: Prelude and fugue, E major.
MENDELSSOHN: Andante with variations, D major. PACHELBEL:
Chaconne, F minor. Fantasia, G minor. Toccata, F major.
WALTHER: Warum sollt ich mich denn gramen. ZACHAU (Zachow):
Prelude and fugue, C major. Vom Himmel hoch da komm ich her.
Heinz Lohmann, org.
*NR 11-75 p15

PSRK 41013/6 (4)
2741 Organs of the Swiss countryside. BACH: Chorale prelude, Meine
Seele erhebt den Herren, S 648. Concerto, organ, no. 4, S 595,
C major. Fugue on the Magnificat, S 733. Trio sonata, no. 1,
S 525, E flat major. BEETHOVEN: Adagio, F major. BIANCIARDI:
Fantasia terza. Ricercare quinto. BOHM: Prelude, D minor.
BRAHMS: Chorale preludes: Herzlich tut mich verlangen, op. 122,
no. 4; O Welt ich muss dich lassen, op. 122, no. 11. BRUCKNER:
Prelude, D major. BRUHNS: Prelude, E minor. BRUNNER: Eingang-
spiele, Weihnacht, Passion, Ostern. CAVAZZONI: Intabulatura
d'organo: Libro secondo. Intabulatura cioe recercari canzoni:
Hymnus in festo corporis Christi. DUBEN: Partiten uber "Erstan-
den ist der heil'ge Christ". EBERLIN: Toccata e Fughe per
l'Organo: Fugue, G minor. ERBACH: Introitus sexti toni. FRESCO-
BALDI: Fiori musicali, op. 12: Canzon dopo l'Epistola. Toccata
per l'Elevazione. Il secondo libro di toccata: Toccata sesta.
HANFF: Ach Gott, vom Himmel sich darein. HUBER: In te Domine

speravi. Invention über den Choral "In dich hab ich gehoffet,
Herr". MURSCHHAUSER: Octi-Tonium novum organicum: Aria pastor-
alis variata; Variationen über das Lied "Lasst uns das Kindelein
wiegen", per imitationem cuculi. SCHEIDEMANN: Preambulum, D
minor. SCHEIDT: Tabulatura nova: Veni creator. SCHUBERT: Fugue,
D 952, E minor. SPETH: Ars magna consoni et disone: Toccata
septima oder Sibendtes musikalisches Blumen-Feld. TRABACI:
Libero secondo: Capriccio sopra la, fa, sol, la. Jean-Claude
Zehnder, Bernhard Billeter, Karl Kolly, Theodor Kaser, Hansjurg
Leutert, Hans Vollenweider, org.
            +NR 5-75 p14

                           PHILIPS

6580 066 Tape (c) 7317 034
2742 BOCCHERINI: Minuet. CLARKE: Trumpet voluntary. HANDEL: Solomon:
     Arrival of the Queen of Sheba. HAYDN: Concerto, trumpet, E
     flat major: Allegro. Quartet, strings, op. 3, no. 5, F major:
     Serenade. MENDELSSOHN: A midsummer night's dream, op. 61:
     Scherzo. MOZART: Divertimento, no. 17, K 334, D major: Minuet.
     PURCELL: Rondeau. ROSSINI: Moderato. SCHUBERT: Rosamunde,
     op. 26, D 797: Ballet music. TCHAIKOVSKY: Quartet, strings,
     no. 1, op. 11, D major: Andante cantabile. John Wilbraham, tpt;
     AMF; Neville Marriner.
            +HFN 10-72 p1905   +RR 2-75 p75 tape
            +RR 10-72 p61

6580 098
2743 BACH, J. L.: Suite, G major: Minuet, Gavotte. BACH, J. S.: Suite,
     orchestra, S 1066, C major: Minuets, nos. 1 2. Suite, orchestra,
     S 1067, B minor: Polonaise. BOCCHERINI: Quintet, strings, op.
     13, no. 5, E major: Minuet. HANDEL: Water music: Sarabande,
     Rigaudon. HAYDN: Symphony, no. 48, C major: Menuetto. Symphony,
     no. 53, D major: Menuetto. MOZART: Divertimento, no. 17, K 334,
     D major: Minuet. Symphony, no. 40, K 550, G minor: Menuetto.
     ECO, AMF: Raymond Leppard, Neville Marriner.
            ++RR 1-75 p26

6580 107
2744 BEETHOVEN: Die Ruinen von Athen, op. 113: Turkish march. BERLIOZ:
     La damnation de Faust, op. 24: Hungarian march. GLINKA: Russlan
     and Ludmilla: March of the sorcerer. GOUNOD: Funeral march
     of a marionette. GRIEG: Sigurd Jorsalfar, op. 56: Homage march.
     MENDELSSOHN: Athalie, op. 74: War march of the priests. MEYER-
     BEER: Le prophète: Coronation march, Act 4. SCHUBERT: Marche
     militarie, no. 1, op. 51, no. 1, D 733, D major (arr. Guiraud).
     WAGNER: Tannhäuser: Entry of the guests. Hague Philharmonic
     Orchestra; Willem Van Otterloo.
            /RR 6-75 p47

6599 227 (Reissue)
2745 Irish songs. DAVIS: The West's awake. FERGUSON: The lark in the
     clear air. HUGHES: The stuttering lover. KICKHAM: She lived
     beside the Anner. McCALL: Kelly, the boy from Killane; Boola-
     vogue. LOVER: The low-back'd car. MOORE: Believe me, if all
     those endearing young charms; The minstrel boy; The young May
     moon. TRAD.: An raibh tu ag an gCarrig; The bard of Armagh;
     Green bushes; The maid of the sweet boy. Frank Patterson, t;
     Orchestra; Thomas C. Kelly.

+Gr 6-73 p113   +St 8-75 p107
+MJ 2-75 p31   +STL 7-8-73 p36
+RR 6-73 p89

6747 177 (2) (Reissue)
2746 GRAINGER: Hill song, no. 2. HOLST: Hammersmith, op. 52: Prelude
and scherzo. Suite, no. 1, op. 28, E flat major. Suite, no.
2, op. 28, F major. JACOB: William Byrd suite. VAUGHAN WILLIAMS:
Folk song suite. Toccata marziale. WALTON: Crown imperial.
Eastman Wind Ensemble; Frederick Fennell.
+HFN 6-75 p97   ++RR 5-75 p30

6747 199 (2) (Reissues)
2747 BACH: Brandenburg concerto, no. 3, S 1048, G major. Concerto,
violin and strings, S 1041, A minor. BEETHOVEN: Minuet, G major.
Sonata, piano, no. 14, op. 27, no. 2, C sharp minor. BERLIOZ:
Le carnaval romain, op. 9. DVORAK: Humoresque, op. 101, no.
7, G flat major. LISZT: Paraphrase VERDI: Rigoletto, G 514.
MENDELSSOHN: Hebrides overture, op. 26. A midsummer night's
dream, op. 61: Scherzo. MOZART: Die Zauberflöte, K 620: Overture. March, K 385d, D major. NICOLAI: The merry wives of
Windsor: Overture. SCHUBERT: Rosamunde, op. 26, D 797: Ballet
music, no. 2. Various orchestras and conductors.
+Gr 11-75 p915   +HFN 11-75 p173

6747 204 (2)
2748 BEETHOVEN: Bagatelle, no. 25, G 173, A minor. Egmont overture,
op. 84. Symphony, no. 9, op. 125, D minor: 3rd movement.
BIZET: Carmen: Suite, no. 1. CHOPIN: Waltz, op. 18, E flat
major. CLARKE: Trumpet voluntary. GOUNOD: Faust: Soldiers
chorus. HANDEL: Water music: Suite, no. 3. MEYERBEER: Le
prophète: Coronation march, Act 4. MOZART: Le nozze di Figaro,
K 492: Voi che sapete. Die Zauberflöte, K 620: Overture.
RIMSKY-KORSAKOV: Russian Easter festival overture, op. 36.
SCHUBERT: Ave Maria, D 839; Die Forelle, D 550. SIBELIUS:
Finlandia, op. 26. VERDI: La traviata: Prelude, Act 1. Elly
Ameling, Elizabeth Ebert, s; Gerard Souzay bar; Dalton Baldwin,
Hans Richter-Haaser, Adam Harasiewicz, pno; John Wilbraham, tpt;
BBC Symphony Orchestra, ECO, Leipzig Radio Symphony Orchestra,
Leipzig Gewandhaus Orchestra, VSO, NPhO, Lamoureux Concerts
Orchestra, Limberg Symphony Orchestra and Chorus, COA, LSO,
AMF; Colin Davis, Edo de Waart, Robert Hanell, Franz Konwitschny,
Raymond Leppard, Igor Markevitch, Martin Koekelkoren, Eduard
van Beinum, Charles Mackerras, Neville Marriner, Christoph von
Dohnani.
+-HFN 12-75 p171   +-RR 12-75 p56

6775 006 (2)
2749 Baroque organ works from Alpine countries. AMMBERBACH: Wer das
Töchterlien haben will. BLITHEMAN: Eterne rerum conditor.
EBERLIN: Tocca sexta. Toccata and fugue tertia. FISCHER, J. C.
F.: Preludes and fugues, B minor, D major, E flat major, C minor.
FROBERGER: Ricercar, no. 1. Capriccio, no. 8. FUX: Sonata,
organ, no. 5. KERLL: Canzona, G minor. Toccata con durezza e
ligature. KREBS: Klavierübung: Praeambulum sopra "Jesu meine
Freude". MERULA: Un cromatico ovvero capriccio. MUFFAT: Fugue,
G minor. NEWMAN: Pavan. PACHELBEL: Chorale prelude, Alle
Menschen müssen sterben. Magnificat fugues, nos. 4, 5, 10, 13.
Toccata and fugue, B flat major. PASQUINI: Canzone françese,
no. 7. Ricercare no. 4. STORACE: Ballo della battaglia.

TAYLOR: Pavan and galliard. ZACHAU: Prelude and fugue, G major.
ANON.: Cathaccio, gagliarda. Lodensana, gagliarda. Pavan and
galliard. Gustav Leonhardt, org.
     ++Gr 12-74 p1187   +RR 12-74 p62
     +MT 8-75 p716

6833 105 (Reissues from 6500 008, 6500 006, 6500 014, 6500 128, 6500 515,
     6500 544, 6703 039, SAL 3720/2, 3797)
2750 BACH: Christmas oratorio, S 248: Flösst, mein Heiland. Cantata,
no. 51, Jauchzet Gott, Jauchzet Gott in allen Landen. HANDEL:
Giulio Cesare: E pur così...Piangerò la sorte mia. MENDELSSOHN:
Elijah, op. 70: Höre, Israel, höre. MOZART: Exsultate jubilate,
K 165. Le nozze di Figaro, K 492: Voi che sapete. SCHUBERT:
Heinröslein, D 257. WOLF: Auch kleine Dinge; Ich hab in Penna;
Wie lange schon war immer mein Verlangen; Nein, junger Herr;
Elfenlied; Das verlassene Mägdlein; Mausfallen, Sprüchlein.
Elly Ameling, s; Maurice André, tpt; ECO, Bavarian Radio Sym-
phony Orchestra, German Bach Soloists, Leipzig Gewandhaus Orch-
estra; Dalton Baldwin, pno; Raymond Leppard, Eugen Jochum,
Helmut Winschermann, Wolfgang Sawallish, Edo de Waart.
     +Gr 7-75 p234   +RR 7-75 p59
     +HFN 9-75 p109

6833 159
2751 ALBENIZ: Suite española, no. 5: Asturias. BACH: Suite, lute, S 995,
G minor. HANDEL: Sarabande, D major (trans. Lagoya). TARREGA:
Recuerdos de la Alhambra. TORROBA: Nocturno. VILLA-LOBOS:
Etude, guitar, no. 11. WEISS: Passacaglia (trans. Lagoya).
Alexandre Lagoya, gtr.
     +NR 12-75 p14

PLEIADES

P 150
2752 Medieval and contemporary liturgical music: Plainsong mass for
the Epiphany; Eucharistic liturgy in English. St. Meinrad
Archabbey Schola Cantorum.
     ++NR 9-75 p10

P 249
2753 13th century polyphony. University of Chicago Collegium Musicum;
University of Kentucky Collegium Musicum; St. Meinrad Archabbey
Schola Cantorum.
     ++NR 9-75 p10

P 255
2754 Late 16th century music, Part 2: The Harvard University Press
historical anthology of music. University of Chicago Collegium
Musicum; University of Kentucky Collegium; Howard M. Brown,
Wesley K. Morgan.
     ++LJ 10-75p47

POLYDOR

2460 250. Tape (c) 3170 241
2755 Music for evensong. BRIDGES (Howells): All my hope on God is
founded. DAY: Psalm 84: O how amiable. GARDINER: Evening hymn.
HOWELLS: Magnificat and nunc dimittis, G major. STANFORD:
Psalm 23: The Lord is my Shepherd. WESLEY, S. S.: Magnificat
and nunc dimittis, E major: Thou wilt keep him. Worcester

Cathedral Choir; Harry Bramma, org; Christopher Robinson.
    +Gr 11-75 p887
2460 252. Tape (c) 3170 240
2756 BACH: Toccata and fugue, S 53ĝ, D minor. BOSSI: Scherzo, G minor.
    FRANCK: Pastorale, op. 19. MENDELSSOHN: Sonata, organ,
    op. 65, no. 4, B flat major. WIDOR: Symphony, organ, no. 1,
    op. 13, no. 1, C major: Marche pontificale. Alan Wicks, org.
        +Gr 11-75 p864   +-HFN 10-75 p148
2489 519. Tape (c) 3150 367
2757 The glorious voice of Fritz Wunderlich. BACH (Gounod): Ave Maria.
    CAPUA: O sole mio. CURTIS: Vergiss mein nicht. DENZA: Funiculi
    funicula. GORDIGIANI: Santa Lucia. LEONCAVALLO: Mattinata.
    MAY: Ein Lied geht um die Welt. ROSSINI: La danza. SPOLIANSKY:
    Heute Nachte oder nie. STOLZ: Mein herz ruft immer nur nach
    dir, Oh Marta; Ob blond, ob braun ich liebe alle Frau'n. ANON.:
    Tiritomba (arr. Breuer). Fritz Wunderlich, t; Various accom-
    paniments.
        +-Gr 10-75 p703   +Gr 7-75 p254 tape

                              PYE
NSPH 6. Tape (c) ZCP 6 (r) Y8P 6
2758 The magic of Vienna. HEUBERGER: The opera ball: Im chambre separée.
    LANNER: Schönbrunner waltz. LEHAR: The Count of Luxembourg:
    Fragrance of May; Polka dance; Love and age. Paganini: Love live
    forever. MILLOCKER: Der Bettelstudent: A slap in the face; The
    alligator and the Brahmin's daughter. The Dubarry: I gave my
    heart. SCHUBERT: Die Verschworenen: I must behold her. STRAUSS,
    J. II: The gypsy baron, op. 420: The pig breeder's song.
    ZELLER: Der Vogelhändler: Rector and director; Roses in Tyrol.
    ZIEHRER: Werner Marsch. June Bronhill, s; Peter Jeffes, t;
    Eric Shilling, bar; Gaiety Orchestra; David Cairns.
        +Gr 12-75 p1103   -RR 10-75 p35
        +-HFN 10-75 p143
GH 589. Tape (c) CZGH 589 (ct) Y8GH 589
2759 BACH: St. Matthew Passion, S 244: O sacred head surrounded. Can-
    tata, no. 147, Jesu, joy of man's desiring. Toccata and fugue,
    S 565, D minor. DAVIES: Psalm 121. GOUNOD: Ave Maria. HANDEL:
    Messiah: I know that my Redeemer liveth. HUMFREY: A hymne to
    God the Father. MENDELSSOHN: Hear my prayer, op. 39, no. 1; Oh
    for the wings of a dove. MOZART: Ave verum corpus, K 618.
    WESLEY, S. S.: Blessed be the God and Father. WIDOR: Symphony,
    organ, no. 5, op. 42, no. 1, F minor: Toccata. Timothy Wilson,
    treble; Winchester Cathedral Choir; Martin Neary, Clement
    McWilliam, org; Martin Neary.
        +Gr 2-75 p1541
TPLS 13066
2760 BACH: Fantasia and fugue, S 542, G minor. DUPRE: Prelude and
    fugue, op. 7, no. 3, G minor. FRANCK: Chorale, no. 3, A minor.
    MESSIAEN: Les corps glorieux: Joie et clarté des corps glorieux.
    WALOND: Voluntary, no. 5, G major. Martin Neary, org.
        +Gr 10-75 p661   +-RR 11-75 p67
        +-HFN 11-75 p171
GSGC 14153
2761 Cantata for Venice. ALBINONI: Concerto, op. 9, no. 2, D minor:
    Adagio, organ and strings. BACH: Cantata, no. 147, Jesu, joy

of man's desiring. DURUFLE: Suite, organ, op. 5: Toccata.
GERMANI: Cantata for Venice. MAURO-COTTONE: Ninna Nanna.
SOWERBY: Pageant. Fernando Germani, org.
+-Gr 11-75 p864   +-RR 7-75 p25
/HFN 6-75 p98

GSGC 14154
2762 ALBENIZ: España, op. 165, no. 5: Capriccio catalano. BACH: English
suite, no. 3, S 808, G minor: Gavotte. CASTELNUOVO-TEDESCO:
Fandango. GANGI: Ut fabulae ferunt 1970. GOULD: Symphonietta,
no. 2: Pavane. MARELLA: Suite, no. 1, A major. ROSENMULLER:
Pavane. TARREGA: Recuerdos de la Alhambra. Pasqualino Garzia,
Carlo Carfagna, gtr.
-Gr 7-75 p213   /RR 6-75 p66
+-HFN 6-75 p101

RCA

VH 020
2763 BACH: Chorale prelude: Nun komm, der Heiden Heiland (Busoni), S
659. DEBUSSY: Children's corner suite: Serenade for the doll.
HOROWITZ: Variations on themes from Bizet's Carmen. MENDELSSOHN:
Songs without words, op. 62, no. 6; op. 67, no. 5; op. 85, no.
4. A midsummer night's dream, op. 61: Wedding march and vari-
ations (arr. Liszt, Horowitz). MUSSORGSKY: By the water (arr.
Horowitz). PROKOFIEV: Toccata. MOZART: Sonata, piano, no. 11,
K 331, A major: All turca. MOSZKOWSKI: Etude, A flat major.
Etincelles. SAINT-SAENS: Danse macabre, op. 40 (arr. Liszt,
Horowitz). SOUSA: Stars and stripes forever (arr. Horowitz).
Vladimir Horowitz, pno.
+Gr 12-75 p1085   +-RR 12-75 p81

SER 5704/6 (3). (Reissues from HMV ALP 1126-8, 1326-8, 1112-3, 1388-90,
1857, DB 6163, BLP 1055, 1053, 5025, RCA RB 6620, 16149,
16051-2, 16031-2, 16011)
2764 Operatic songs and arias Jussi Björling. ALFVEN: Skogen sover.
BEETHOVEN: Adelaide, op. 46. BIZET: Carmen: Le fleur que tu
m'avais jetée. Les pêcheurs de perles: Au fond du temple saint.
BORODIN: Prince Igor: Vladimir's recitative and cavatina. BRAHMS:
Ständchen, op. 106, no. 1. DONIZETTI: L'Elisir d'amore: Una
furtiva lagrima. FLOTOW: Martha: M'appari tutt'amor. GIORDANO:
Andrea Chénier: Come un bel di di maggio. GOUNOD: Faust: Salut
demeure. LEONCAVALLO: I Pagliacci: Vesti la giubba. MASCAGNI:
Cavalleria rusticana: O Lola; Addio alla madre. MEYERBEER:
L'Africaine: O paradiso. NORDQVIST: Till havs. PUCCINI: La
bohème: Che gelida manina; O Mimi, tu più non torni. Manon
Lescaut: Donna non vidi mai; Ah, Manon, mi tradice; Presto in
fila; Ah, non v'avvicinata; No, pazzo son; Guardate. Tosca: E
lucevan le stelle. RACHMANINOFF: In the silent night, op. 4,
no. 3. RANGSTROM: Tristan's död. SCHUBERT: Die Forelle, D 550;
Ständchen, D 889; Die böse Farbe; An die Leier, D 737. SIBELIUS:
Svarta Rosor, op. 36, no. 1. STRAUSS, R.: Zueignung, op. 10,
no. 1. TCHAIKOVSKY: Eugene Onegin, op. 24: Lenski's aria.
TOSTI: Ideale; L'alba separa dalla luce ombra. VERDI: Aida: Se
quel guerrier io fossi... Celeste Aida; Tu, Amonasro; Io son
disonorato. Don Carlo: Io l'ho perduta...Qual pallor. La
forza del destino: Solenne in quest'ora. Otello: Si, pel ciel
marmoreo giuro. Rigoletto: Quest o quella; Ella mi fu rapita...

Parmi veder la lagrime; La donna è mobile; Un di, se ben rammentomi; Bella figlia dell'amore. Il trovatore: Deserto sulla
terra; De qual tetra luce; Ah si, ben mio; Di quella pira.
WAGNER: Lohengrin: In fernem Land. Jussi Björling, t; Various
accompaniments.

    +Gr 12-74 p1219   +-RR 3-75 p16

SER 5719
2765 Swedish songs. ALFVEN: I long for you; Take my heart. ALTHEN:
Thou, blessed country. KJORLING: Evening mood. PETERSON-
BERGER: When I walk alone; Among the high fir trees. SIBELIUS:
Diamonds on the March snow; Sigh, sigh, sedges. SJOBERG: The
music. SODERMAN: King Heimer and Aslög. STENHAMMER: Sweden.
Jussi Björling, t; Stockholm Royal Opera Orchestra; Nils
Grevillius.

    +-Gr 3-75 p1692

ARL 1-0456. Tape (c) ARK 1-0456 (ct) ARS 1-0456
2766 ALBENIZ: Iberia: Evocación. Bajo la palmera, op. 212. CARULLI:
Serenade, op. 96. GIULIANI: Variazioni concertante, op. 130.
GRANADOS: Danza española, op. 37, no. 6 and no. 11. Julian
Bream, John Williams, gtr.

    +Gr 4-74 p1868   ++RR 4-74 p63
    ++MJ 2-75 p40   ++RR 12-74 p87 tape
    +-NR 11-74 p12

ARL 1-0485. Tape (c) ARK 1-0485 (ct) ARS 1-0485
2767 Favorite Spanish encores. ALBENIZ: Piezas características, no. 12:
Torre bermeja. MUDARRA: Gallarda, D major. NARVAEZ: Variations
on a Spanish folk tune. PISADOR: Pavana, E minor. SOR: Folias
de España. TARREGA: Capricho arabe. TORROBA: Nocturno. Burga-
lesa. Piezas características: Albada. Sonatina, A major:
Allegretto. Suite castellana: Fandanguillo, Arada. VALDERRA-
BANO: Soneto, D major. ANON.: El mestre. La filla del Marxant.
Andrés Segovia, gtr.

    +HF 12-74 p146 tape   ++SFC 9-8-74 p30
    +MJ 11-74 p49   ++St 1-75 p76
    +-NR 11-74 p12

ARL 1-0561
2768 ADAMS: The holy city. BIZET: Agnus Dei. BENNARD: The old rugged
cross. DREW: Bless the Lord, O my soul. HEMY: Faith of our
fathers. Living still. JANOWSKI: Avinu Malkeynu. LIDDLE: How
lovely are thy dwellings. MacDERMID: He that dwelleth in the
secret place. RODNEY: Calvary. WALKER: But for the grace of
God. WARREN: God of our Fathers. ZUNDEL: Love divine, all
loves excelling. TRAD.: Amazing grace. Let us break bread together. Sherrill Milnes, bar; Jon Spong, org.

    +Gr 5-75 p2024   -RR 5-75 p57
    +-NR 4-75 p9

ARL 1-0562
2769 ALLITSEN: The Lord is my light. CROFT: O God, our help in ages
past. DYKES: Holy, holy, holy, Lord God almighty. FRANCK:
Panis angelicus. HOLDEN: All hail the power of Jesus' name.
HOLST: The heart worships. KREMSER: We gather together. Mac-
DERMID: In my Father's house are many mansions. MILES: In the
garden. MALOTTE: The Lord's prayer. SHELLEY: The King of love
my Shepherd is. SPEAKS: In the end of the Sabbath. SULLIVAN:
The lost chord. WESLEY: The church's one foundation. Sherrill
Milnes, bar; John Spong, org.

            +—HFN 10-75 p140   +RR 9-75 p72
            +—NYT 6-8-75 pD19

ARL 1-0702
2770 Arias from great Italian operas. DONIZETTI: Linda di Chamounix:
     Ah, tardai troppo, O luce di quest'anima. Lucia di Lammermoor:
     Regnava nel silenzio, Quando rapito in estasi. LEONCAVALLO:
     I Pagliacci: Qua fiamma avea nel guardo. PUCCINI: Suor Angelica:
     Senza mamma. Tosca: Vissa d'arte. Turandot: Signore, ascolta,
     Tu che di gel sei cinta. ROSSINI: Semiramide: Bel raggio lus-
     inghier. VERDI: I Lombardi: Qual prodigio, Non fu sogno. Anna
     Moffo, s; Bavarian Radio Orchestra, Rome Opera Orchestra, RCA
     Italian Opera Orchestra; Kurt Eichorn, Tullio Serafin, Georges
     Prêtre.
            +—HF 2-75 p108   +—ON 3-8-75 p34
            +NR 11-74 p9   +—St 4-75 p107

ARL 1-0735
2771 PAGANINI: Caprices, op. 1, nos. 17, 24. PROKOFIEV: Love for three
     oranges, op. 33: March. TCHAIKOVSKY: Sérénade melancolique, op.
     26. Valse scherzo. WIENIAWSKI: Polonaise. Scherzo tarantelle.
     YSAYE: Sonata, solo violin, no. 3. Eugene Fodor, vln; Jonathan
     Feldman, pno.
            +HF 1-75 p96   ++NR 3-75 p4
            -MJ 3-75 p26   ++SFC 10-27-74 p10

ARL 1-0844. Tape (c) ARK 1-0844 (ct) ARS 1-0844
2772 Heroines from great French operas. BERLIOZ: La damnation de Faust,
     op. 24: D'amour l'ardente flamme. BIZET: Les pêcheurs de per-
     les: O Dieu Brâhma. CHARPENTIER: Louise: Depuis le jour.
     DONIZETTI: La fille du régiment: Chacun le sait, chacun le dit.
     GOUNOD: Roméo et Juliette: Je veux vivre. MASSENET: Hérodiade:
     Il est doux, il est bon. Werther: Letter scene. MEYERBEER:
     Robert le diable: Robert, toi que j'aime. THOMAS: Hamlet: Mad
     scene. Anna Moffo, s; Ambrosian Opera Chorus; NPhO; Peter Maag.
            +—NR 3-75 p7   ++St 9-75 p122

ARL 1-0851
2773 The baritone voice, Sherrill Milnes. GRETRY: Richard, Coeur de
     Lion: O Richard, O mon Roi. MASSENET: Le Roi de Lahore: Aux
     troupes du Sultan...Promesse de mon avenir. SAINT-SAENS: Henry
     VIII: Donc, le pape est hostile...Qui donc commande. VERDI:
     Un ballo in maschera: Alzati...Eri tu. Luisa Miller: Ferma
     ed ascolta...Sacra la scalta e d'un consorte. Macbeth: Perfidi
     ...Pietà, rispetto, amore...Mal per me. WEBER: Euryanthe: Wo
     berg' ich mich. Sherrill Milnes, bar; NPhO; Nello Santi.
            +Gr 11-75 p904   +—ON 12-6-75 p52
            +—HFN 11-75 p149   ++RR 12-75 p33
            +NR 12-75 p11   +—SFC 11-2-75 p28

ARL 1-0864. Tape (c) ARK 1-0864 (ct) ARS 1-0864
2774 ASENCIO: Dipso. BACH: Suite, solo violoncello, no. 1, S 1007, G
     major: 3 movements. BENDA: Sonatina, D major. Sonatina, D
     minor. PONCE: Prelude, E major. SCARLATTI: Sonatas, guitar (2).
     SOR: Andante, C minor. Minuets, C major, A major, C major.
     WEISS: Bourrée. Andrés Segovia, gtr.
            +NR 10-75 p12   ++St 10-75 p119

ARL 1-5007
2775 Cathy Berberian, songs and operatic arias. BEETHOVEN: Songs:
     Wiedersehen; Though thou so blest. CUI: Statue in Tsarskoye
     Selo. DELIBES: Les filles de Cadiz. HAHN: l'Offrande.

LEHMANN, L.: There are fairies at the bottom of our garden.
LOEWE: Tom der Reimer, op. 135. MUSSORGSKY: The song of the
flea. OFFENBACH: La périchole: Tu n'est pas beau. Que voulez-
vous faire. PARKHURST: Father's a drunkard. PURCELL: Nymphs
and shepherds. RIMSKY-KORSAKOV: Enslaved by the rose, the
nightingale. ROSSINI: Petite caprice à la Offenbach (piano
solo). SAINT-SAENS: Danse macabre, op. 40. SULLIVAN: The
Mikado: Tit willow. WECKERLIN: Maman, dites-moi. WORK: My
grandfather's clock. Cathy Berberian, s; Bruno Canino, pno.

    +-Gr 4-74 p1882   +-RR 4-74 p78
    +-HF 3-75 p70   +St 7-74 p117
    *NR 2-75 p10

LFL 1-5072
2776 GREGSON: Prelude and capriccio. HOLST: A Moorside suite. LANG-
FORD: A London scherzo. A west country fantasy. WILSON: Sinfonietta. TRAD.: Blow the wind southerly. James Watson, cor;
London City Brass; Geoffrey Brand.

    ++Gr 3-75 p1710

LRL 1-5087
2777 ARNE: Sonata, harpsichord, no. 3, G major: Gigue. BERKELEY: Nocturne. CROFT: Sarabande. Ground. CASSANOVAS: Sonata, harp,
F major. PARRY: Sonata, harp, D major. SCARLATTI, D.: Sonata,
harp, G major. SOLER: Sonata, harp, E minor. SPOHR: Fantaisie,
C minor. FOURNIER: Vers la source danse le bois. WATKINS:
Petite suite. David Watkins, hp.

    +Gr 9-75 p492   ++RR 10-75 p75
    ++HFN 11-75 p159

LRL 1-5094
2778 BACH: Suite, orchestra, S 1067, B minor: Minuet; Badinerie.
CHOPIN: Waltz, op. 64, no. 1, D flat major (arr. and orch. Ger-
hardt). DINICU: Hora staccato (arr. and orch. Gerhardt).
DOPPLER: Fantaisie pastorale hongroise, op. 26. DRIGO: Les
millions d'Arlequin: Serenade (arr. and orch. Gamley). GLUCK:
Orfeo ed Euridice: Dance of the blessed spirits. GODARD: Pieces,
op. 116: Waltz. MIYAGI: Haru no umi (arr. and orch. Gerhardt).
PAGANINI: Moto perpetuo, op. 11 (arr. and orch. Gerhardt).
RIMSKY-KORSAKOV: The tale of the Tsar Sultan: The flight of the
bumblebee (arr. and orch. Gerhardt). SAINT-SAENS: Ascanio:
Ballet music, Adagio and variation. James Galway, flt; National
Philharmonic Orchestra; Charles Gerhardt.

    ++Gr 11-75 p915   +-RR 10-75 p38
    +-HFN 12-75 p164

ARL 2-0512 (2)
2779 Jorge Bolet at Carnegie Hall. BACH (Busoni): Partita, violin, no.
2, S 1004, D minor: Chaconne. CHOPIN: Preludes, op. 28. MOSZ-
KOWSKI: La jongleuse, op. 52, no. 4. RUBINSTEIN: Etude, op. 23,
no. 2, C major. STRAUSS, J. (Tausig): Man lebt nur einmal.
Nachtfalter. STRAUSS, J. II (Schulz-Elver): Arabesques on
"The beautiful blue Danube". WAGNER (Liszt): Tannhäuser: Overture. Jorge Bolet, pno.

    /HF 1-75 p80   ++SFC 9-1-74 p20
    +MJ 9-74 p65   ++St 9-74 p85
    ++NR 8-74 p11

CRL 2-7001 (2). (also Erato)
2780 Music for trumpet and organ. ALBINONI: Sonata, trumpet and organ,
D major. Sonata, trumpet and organ, F major. BACH: Cantata,
no. 147, Jesu, joy of man's desiring. GERVAISE: Bransles (2).

Dances (4). KREBS: Chorales (6). LOEILLET: Sonata, trumpet
and organ, C major. TELEMANN: Sonata, organ, G minor. VIVALDI:
Concerto, trumpet and organ, P 75, C major. VIVIANI: Sonata,
trumpet, nos. 1 and 2. ANON.: Bransle de Bourgogne. Maurice
André, tpt, Marie-Claire Alain, Hedwig Bilgram, org.
    +MJ 10-75 p45   +SFC 1-19-75 p27
    ++NR 10-74 p6

CRL 2-7002 (2)
2781 Great trumpet concertos. ALBINONI: Concerto, trumpet, B flat major.
HAYDN: Concerto, trumpet, E flat major. HUMMEL: Concerto, trum-
pet, E flat major (rest. & ed. Willemoës). TELEMANN: Concerto,
trumpet, D major. Sonate de concert, trumpet, D major (reconstr.
Oubradous). TORELLI: Concerto, trumpet, D major. VIVALDI: Con-
certo, trumpet, B flat major. Concerto, trumpet, A flat major.
Maurice André, tpt; Bamberg Symphony Orchestra, Jean-François
Paillard Chamber Orchestra, Lamoureux Orchestra, Wiener Solisten;
Theodor Guschlbauer, Jean-François Paillard, Jean-Baptiste Mari.
    +NR 11-74 p5   ++St 3-75 p110

CRL 2-7003 (2) (also Erato)
2782 Festival of flute concertos. CIMAROSA: Concertante, 2 flutes and
orchestra, G major. HANDEL: Concerto, harp, op. 4, no. 5, F
major. MOLTER: Concerto, flute and strings, G major. MOZART:
Concerto, flute and harp, K 299, C major. PLATTI: Concerto,
flute and orchestra, G major. TARTINI: Concerto, flute and
orchestra, F major. Pean-Pierre Rampal, Clementine Scimone,
flt; Lily Laskine, hp; Jean-François Paillard Chamber Orchestra,
I Solisti Veneti; Jean-François Paillard, Claudio Scimone.
    +HF 5-75 p94   ++SFC 6-1-75 p21
    ++NR 10-74 p6

ARM 4-0942/7
2783 Vol. 1, Acoustic recordings complete: ACHRON: Hebrew dance. Hebrew
lullaby. Hebrew melody, op. 33. Stimmung. d'AMBROSIO: Serenade. BAZZINI: La ronde des lutins, op. 25. BEETHOVEN: The
ruins of Athens, op. 113: Chorus of dervishes Turkish march.
BOULANGER: Cortege. Nocturne, F major. BRAHMS: Hungarian dance,
no. 1, G minor. CHOPIN: Nocturne, D flat major. Nocturne, E
flat major. DRIGO: Valse bluette. DVORAK: Slavonic dance, no.
2. Slavonic dance, op. 72, no. 2, E minor. Slavonic dance,
op. 72, no. 8. ELGAR: La capricieuse, op. 17. GLAZUNOV: Méditation, op. 32. Raymonda, op. 57: Valse grande adagio. GODOWSKY
Waltz, D major. GOLDMARK: Concerto, violin, A minor: Andante.
GRANADOS: Danza española, op. 37, no. 5: Andaluza. HAYDN:
Quartet, strings, D major: Vivace. JUON: Berceuse. KREISLER:
Minuet. Sicilienne et Rigaudon. LALO: Symphonie espagnole,
op. 21, D minor: Andante. MENDELSSOHN: Concerto, violin, op.
64, E minor: Finale. On wings of song, op. 34, no. 2. MOSZ-
KOWSKI: Stücke, op. 45, no. 2: Guitarre. MOZART: Divertimento,
no. 17, K 334, D major: Minuet. Serenade, D major: Rondo.
PAGANINI: Caprices, op. 1, nos. 13, 20. Moto perpetuo, op. 11.
SARASATE: Carmen fantasy, op. 25. Danzas españolas, op. 21,
no. 1: Malagueña; no. 2, Habañera; op. 23, no. 2: Zapateado.
Introduction and tarantelle. Zigeunerweisen, op. 20, no. 1.
SCHUBERT: Ave Maria, D 839. SCHUMANN: Myrthen, op. 25, no. 1:
Widmung. SCOTT: The gnetle maiden. TCHAIKOVSKY: Concerto,
violin, op. 35, D major: Canzonetta. Sérénade melancolique,
op. 26. Souvenir d'un lieu cher, op. 42: Scherzo. Serenade,

strings, op. 48, C major: Valse. WIENIAWSKI: Concerto, violin,
no. 2, op. 22, D minor: Romance. Scherzo tarantelle. Vol. 2,
The first electrical recordings: ACHRON: Hebrew melody, op. 33.
ALBENIZ: Suite española, no. 3: Sevillanas. BACH: Partita,
violin, no. 3, S 1006, E minor: Minuets 1 and 2. English suite,
no. 3, S 808, G minor: Sarabande, Gavotte, Musette. CASTELNUOVO-
TEDESCO: Sea murmurs. Valse. CLERAMBAULT: Largo on the G string.
COUPERIN: Livres de clavecin, Bk IV, Order no. 17: Les petits
Moulines a vent. DEBUSSY: L'Enfant prodigue: Prelude. Prelude,
Bk I, no. 8: La fille aux cheveux de lin. La plus que lente.
DOHNANYI: Ruralia Hungarica: Gypsy andante. DRIGO: Valse
bluette. ELGAR: La capricieuse, op. 17. FALLA: Spanish popular
songs: Jota. GLAZUNOV: Concerto, violin, op. 82, A minor.
Méditation, op. 32. GODOWSKY: Alt Wien. GRIEG: Lyric piece,
op. 54, no. 6: Scherzo. Lyric piece, op. 71, no. 3: Puck.
Sonata, violin and piano, op. 45, C minor. HUMMEL: Rondo, op.
11, E flat major. KORNGOLD: Much ado about nothing, op. 11:
Holzapfel und Schlehwein. MENDELSSOHN: On wings of song, op.
34, no. 2. MILHAUD: Saudades do Brasil, no. 10: Sumaré. MOSZ-
KOWSKI: Stücke, op. 45, no. 2: Guitarre. MOZART: Concerto,
violin, no. 5, K 219, A major. PAGANINI: Caprices, op. 1, nos.
13, 20, 24. PONCE: Estrellita. RAVEL: Tzigane. RIMSKY-KORSAKOV:
The tale of the Tsar Sultan: Flight of the bumblebee. SARASATE:
Danza española op. 23, no. 2: Zapateado. SCHUBERT: Ave Maria,
D 839. Impromptu, op. 90, no. 3, D 899, G flat major. Sonatina,
violin and piano, op. 137, D major: Rondo. STRAUSS: Sonata,
violin and piano, op. 18, E flat major. Stimmungsbilder: An
einsamer Quelle. VIVALDI: Sonata, op. 2, no. 2, violin, A
major. WIENIAWSKI: Scherzo tarantelle. Vol. 3, 1935-1937:
BACH: Sonata, violin, no. 1, S 1001, G minor. Sonata, violin,
no. 3, S 1005, C major. BAZZINI: La ronde des lutins, op. 25.
BRAHMS: Sonata, violin, no. 2, op. 100, A major. DINICU: Hora
staccato. FAURE: Sonata, violin, no. 1, op. 13, A major. FALLA:
La vida breve: Danza, no. 1. GRIEG: Sonata, violin, no. 2, op.
13, G minor. POULENC: Mouvements perpetuels, no. 1. SAINT-
SAENS: Introduction and rondo capriccioso, op. 28. SCOTT:
Tallahassee suite. SZYMANOWSKI: Le Roi Roger, op. 46: Chant do
Roxane. VIEUXTEMPS: Concerto, violin, no. 4, op. 31, D minor.
WIENIAWSKI: Concerto, violin, no. 2, op. 22, D minor. Polonaise,
op. 4, D major. Vol. 4, 1937-1941: BEETHOVEN: Concerto, violin,
op. 61, D major. BRAHMS: Concerto, violin, op. 77, D major.
Concerto, violin, violoncello and orchestra, op. 102, A minor.
CHAUSSON: Concerto, violin, piano and string quartet, op. 21,
D major. PROKOFIEV: Concerto, violin, no. 2, op. 63, G minor.
SAINT-SAENS: Havanaise, op. 83. SARASATE: Zigeunerweisen, op.
20, no. 1. WALTON: Concerto, violin, B minor. Vol. 5, 1946-
1949: ARENSKY: Concerto, violin, A minor: Tempo di valse. BACH:
Concerto, 2 violins and strings, S 1043, D minor. English suite,
no. 6, S 811: Gavottes, nos. 1 and 2. BAX: Mediterranean.
BEETHOVEN: German dance, no. 6. BRUCH: Scottish fantasia, op.
46. CASTELNUOVO-TEDESCO: Sea murmurs. Tango. CHOPIN: Nocturne,
E minor. DEBUSSY: Chanson de Bilitis: La chevelure. Song:
Ariettes oubliées, no. 2: Il pleure dans mon coeur. Preludes,
Bk I, no. 8: La fille aux cheveaux de lin. ELGAR: Concerto,
violin, op. 61, B minor. FALLA: El amor brujo: Pantomime.
Spanish popular songs: Jota. HALFFTER: Danza de la gitana

(Escriche). KORNGOLD: Much ado about nothing, op. 11: Holzapfel
und Schlehwein, Garden scene. MEDTNER: Fairy tale, B flat minor.
MILHAUD: Saudades do Brasil, no. 7: Corcovado. MOZART: Divertimento, no. 17, K 334, D major: Minuet. NIN: Cantilena asturiana.
POLDOWSKI: Tango. PROKOFIEV: Gavotte. March, F minor. RACH-
MANINOFF: Daisies, op. 38, no. 3. Etude tableaux, op. 37, no.
2. Oriental sketch. RAVEL: Valses nobles et sentimentales,
nos. 6 and 7. RIMSKY-KORSAKOV: The tale of the Tsar Sultan:
Flight of the bumblebee. TANSMAN: Mouvement perpetuel. VIEUX-
TEMPS: Converto, violin, no. 5, op. 37, A minor. Vol. 6, 1950-
1955: BEETHOVEN: Romance, no. 1, op. 40, G major. Romance,
no. 2, op. 50, F major. Sonata, violin and piano, no. 9, op.
47, A major. BLOCH: Sonatas, violin, nos. 1 and 2. BRAHMS:
Sonata, violin, no. 3, op. 108, D minor. BRUCH: Concerto,
violin, no. 1, op. 26, G minor. HANDEL: Sonata, violin, op. 1,
no. 13, D major. RAVEL: Tzigane. SAINT-SAENS: Sonata, violin,
no. 1, op. 75, D minor. SCHUBERT: Sonatina, violin and piano,
no. 3, G minor. TCHAIKOVSKY: Concerto, violin, op. 35, D major.
WIENIAWSKI: Polonaise, op. 4, D major. Jascha Heifetz, vln;
With assisting artists.

+HF 8-75 p71   +NYT 6-29-75 pD17
++NR 7-75 p12   +SR 10-4-75 p51

CRL 6-0720 (6)
2784 The great violin concertos. BACH: Concerto, 2 violins, S 1043, D
minor. BEETHOVEN: Concerto, violin, op. 61, D major. BRAHMS:
Concerto, violin, op. 77, D major. BRUCH: Concerto, violin,
no. 1, op. 26, G minor. GLAZUNOV: Concerto, violin, op. 82,
A minor. MENDELSSOHN: Concerto, violin, op. 64, E minor.
MOZART: Concerto, violin, no. 5, K 219, A major. PROKOFIEV:
Concerto, violin, no. 2, op. 63, G minor. SIBELIUS: Concerto,
violin, op. 47, D minor. Jascha Heifetz, vln; Various orches-
tras and conductors.
++SFC 1-12-75 p26

REDIFFUSION

15/16
2785 The golden age of Viennese waltzes. FUCIK: Vom Donauufer (arr.
Pseničný). KOMZAK: Badn'ner Mäd'ln, op. 252 (arr. Urbanec).
LANNER: Die Schönbrunner (arr. Urbanec). STRAUSS, J. II: An
der schönen blauen Donau, op. 314. Wiener Blut, op. 354 (arr.
Urbanec). ZIEHRER: Wiener Burger, op. 419 (arr. Urbanec).
Czech Concert Wind Ensemble; Rudolf Urbanec.
/Gr 8-75 p363

ROCOCO

2035
2786 BRAHMS: Hungarian dance, no. 5, G minor (arr. Joachim). CHABRIER:
Pièces pittoresques, no. 10: Scherzo valse. DVORAK: Humoresque
(arr. Ysaye). FAURE: Berceuse, op. 16. KREISLER: Caprice
viennois, op. 2. MENDELSSOHN: Concerto, violin, op. 64, E
minor: Finale. SCHUBERT: Ave Maria, D 839. VIEUXTEMPS: Rondino
op. 32. WAGNER: Die Meistersinger von Nürnberg: Preizlied (arr.
Wilhelmj). Albumblatt, C major. WIENIAWSKI: Mazurkas, op. 19:
Obertass, Menetrier. YSAYE: Lointain passé. Eugene Ysaye, vln.
-NR 3-75 p12

2049
2787 DIEMER: Valse de concerto. CHOPIN: Valse, op. 34, no. 1, A flat
     major. KIENZL: Kahn Szene, neuer Walzer. LESCHITZKY: Gavotte.
     LISZT: Hungarian rhapsody, no. 13, G 244, A minor. Liebestraum,
     no. 3, G 541, A flat major. Paraphrase on VERDI: Rigoletto, G
     434. MENDELSSOHN: Rondo capriccioso, op. 14, E major. SAINT-
     SAENS: Marche militarie française: Africa, Valse mignonne,
     Reverie a Blidah, Suite Algérienne. SCHARWENKA: Polish dance,
     op. 31, no. 1. Camille Saint-Saens, Louis Diemer, Wilhelm
     Kienzl, Ethel Leginska, Anna Esipova, Xavier Scharwenka, pno.
          +–NR 2-75 p11

2072
2788 BACH: Suite, orchesta, S 1068, D major: Air (arr. Wilhelmj).
     BEETHOVEN: Sonata, violin and piano, no. 3, op. 12, E flat
     major. GOUNOD: Faust: Fantasie sur la valse (arr. Wieniawski).
     HANDEL: Larghetto. MOSZKOWSKI: Stücke, op. 45, no. 2: Guitarre.
     PAGANINI: Caprice, op. 1, no. 2, B minor. SCHUBERT: Ave Maria,
     D 839 (arr. Vecsey). SCHUMANN: Kinderscenen, op. 15, no. 7:
     Träumerei. SINIGAGLIA: Capriccio all'antica. TARTINI: Sonata,
     violin, no. 4: Finale. VECSEY: Foglio d'album. WIENIAWSKI:
     Souvenir de Moscou, op. 6. Ferenc von Vecsey, vln.
          -NR 3-75 p12

5365
2789 Reynaldo Hahn and his songs. BIZET: Les pêcheurs de perles: O
     nadir, Chanson d'Avril. CHABRIER: Songs: Toutes les fleurs;
     Les cigales; L'Ile heureuse. GOUNOD: Songs: Biondina bella.
     HAHN: Etudes latines: Phyllis, L'Enamourée, L'Heure exquise;
     Le chien fidèle, Lettre d'amour; Je me mets en votre mercy;
     Offrande; Venezia-Ghè pecà. LULLY: Amadis: Bois épais. PALA-
     DILHE: Psyché. Reynaldo Hahn, Arthur Endreze, Guy Ferrant, bar;
     Various accompaniments.
          +NR 7-75 p10

5368
2790 BELLINI: I Capuleti ed i Montecchi: Ecco la tomba...0 tu bell anima.
     Norma: Sgomba e la sacra selva, Deh con te il prendi...Mira O
     Norma...Si fue all'ore. CAMPANA: M'hai tradito. CORDIGIANI:
     La via del pastore. PAER: Il bacio della partenza. ROSSINI:
     L'Italiana in Algeri: Amici in ogni evento...Pensa all patria.
     Tancredi: O patria, dolce, ingrata...Di tanti palpiti. VERDI:
     Don Carlo: Canzone de veli. Giuletta Simionato, ms; Various
     accompaniments.
          +NR 9-75 p11

                              RUBINI

RS 300 (3)
2791 Annus Mirabilis, 1873: AUBER: Fra Diavolo: Agnese la Zitella (Attilio
     Salvaneschi, t). BELLINI: I puritani: Sorgea la notte (Andreas
     Perello de Segurola, bs). BRAHMS: Sapphische Ode, op. 94, no. 4
     (Feodor Chaliapin, bs). DAVID: La perle de Bresil: Charmant
     oiseau (Evangelina Florence, s). DONIZETTI: La fille du régi-
     ment: Au bruit (Louise Tiphaine, s; Jean Delvoye, bs). FAURE:
     La charité (Maurice Declery, bar). GAYNOR/BROWNIEL: The slumber
     song; Four leaf clover (Anita Rio, s). GERDALIA: Little even-
     ing; When beauty is young (Natalia Tamara, s). GRIEG: Last
     spring (Antonina Nezhdanova, s). d'HARDELOT: Say yes, Mignon

(Rosa Olitzka, con). HUMPERDINCK: Konigskinder: O du liebheil-
ige Einfalt; Wohin bist du gegangen (Friedrich Brodersen, bar).
LORTZING: Zar und Zimmermann: Van Bett's aria (Peter Lardmann,
bs). MARCHETTI: Ruy Blas: O dolce volutta (Maria Grisi, s;
Pietro Lara, t). MASCAGNI: Iris: Apri la tua finestra (Attilio
Salvaneschi, t). MASSENET: Sapho: Ce que j'appelle beau (Jane
Marignan, s). Thais: Voilà donc la terrible cité (Giovanni
Polese, t). MILLOCKER: Gasparone: Tarantella (Anna Sutter, s).
MONIUSZKO: Halka: Aria, Act 1 (Salomea Kruszelnika, s). Halka:
Jontek's dream (Henryk Drzewiecki, t). MOZART: La clemenza di
Tito, K 621: Non più di fiori (Louise Kirkby-Lunn, ms). PONS:
Laura: Serenade Napolitaine (Alice Verlet, s). PUCCINI: La
bohème: Addio senza ranco (Julia Guiraudon, s). La bohème: O
soave fanciulla (Enrico Caruso, t; Geraldine Farrar, s). Manon
Lescaut: In quella trine morbide (Marie Rappold, s). ROSSINI:
Guillaume Tell: Ah Mathilde (Leo Slezak, t; Leopold Demuth, bar).
Stabat Mater: Pro peccatis (Herbert Witherspoon, bs). RUBIN-
STEIN: Es blinkt der Tan (Alexander Heinemann, bar). SAINT-
SAENS: Samson et Dalila: Mon coeur s'ouvre à ta voix (Kathleen
Howard, ms). SCOTT-FAURE: Ame nasceri; En prière (Clara Butt,
con). SMETANA: Hubička: Trio (Maria Kubatova, s; Marie Klanova,
s; Emil Pollert, bs). STERN: Coquette (Suzanne Adams, s).
TOSTI: Ridonami la calma (Tilly Koenen, con). VERDI: I Masnad-
ieri: Come un bacio (Ivo Zaccari, t; Ernerio Constantini, bs).
Nabucco: O di qual onto (Enrico Nani, bar; Giannina Russ, s).
Il trovatore: Di quella pira (Karl Jorn, t). WAGNER: Lohengrin:
Winterstürme (Alfred von Bary, t). Das Rheingold: Alberich's
Fluch (Desidor Zador, bs-bar). Tannhauser: Dich teure Halle
(Cäcilie Rusche-Endorf, s). Tannhäuser: Dir töne Lob (Francis
Maclennan, t). Die Walküre: Du bist der Lenz (Katherina
Fleischer-Edel, s). TRAD.: Comin' thro' the rye (Lilian Blauvelt,
s).
        +Gr 1-75 p1391

                                SAGA

5356
2792 DEBUSSY: Arabesques (arr. Renie). FALLA: La vida breve: Spanish
        dance, no. 1. HANDEL: Concerto, harp, op. 4, no. 6, B flat
        major. Suite, harpsichord, no. 7: Passacaglia (arr. Vito).
        MOZART: Concerto, harp, K 545, C major (arr. Vito). PROKOFIEV:
        Prelude, op. 12, no. 7, C major. SALZEDO: Jeux d'eau. Edward
        Vito, hp; Daniel Guilet, Bernard Robbins, vln; Carleton Cooley,
        vla; Alan Shulman, vlc; Philips Sklar, double bs.
            +Gr 1-75 p1375   +-RR 1-75 p51

5374
2793 Organ music of the 16th and 17th centuries. ALBINONI: Concerto,
        B flat major (arr. Walther). BOHM: Prelude and fugue, A minor.
        HANDEL: Fugue, no. 4, B minor. LUBLINA: Dances (4). SCHEIDE-
        MANN: Preambulum and canzona, F major. SCHEIDT: Ei, du feiner
        Reiter, variations. SCHMID: Passamezzo and saltarello. TUNDER:
        In Dich hab ich gehoffet, Herr, chorale prelude. ANON.: Es ist
        des Heil uns kommen her, chorale prelude. David Sanger, org.
            +-Gr 2-75 p1520   +RR 1-75 p52
            +MT 7-75 p631

<u>5402</u>
2794 BACH: Toccata, S 912, D major. BOHM: Präludium, Fuge und Postlud-
     ium, G minor. BULL: The King's hunt. BYRD: Wolsey's wilde.
     COUPERIN, L.: Le tombeau de M. Blancrocher. FARNABY: Loath to
     depart. PHILIPS, P.: Amarilli di Julio Romano. SCARLATTI, D.:
     Sonatas, harpsichord, K 87, B minor; K 201, G major; K 370,
     E flat major; K 371, E flat major. Elizabeth de la Porte, hpd.
            +Gr 10-75 p661   +RR 8-75 p55
            +HFN 12-75 p153

<u>5411</u>
2795 BEETHOVEN: Mödlinger dances, nos. 1-4, 6, 8. HAYDN: German dances,
     nos. 4, 10-11. LANNER: The parting of the ways. Summer night's
     dream. MOZART: Ländler, K 606 (6). SCHUBERT: German dances
     with coda and 7 trios. STRAUSS, J. I: Vienna carnival. Vienna
     Volksoper Orchestra; Paul Angerer.
            +RR 12-75 p50

<u>5417</u>
2796 BISHOP: Grand march, E major. CLARKE: Prince of Denmark's march.
     HAYDN: Feldpartita, B flat major. PEZEL: Suite, C major.
     PICK (attrib): March and troop. Suite, B flat major. London
     Bach Ensemble; Trevor Sharpe.
            +-RR 12-75 p44

<u>7029</u>
2797 DONIZETTI: Lucia di Lammermoor: Alfin son tua. GOUNOD: Faust:
     Ah, je rie. Roméo et Juliette: Je veux vivre dans ce rêve.
     MASSENET: Le Cid: Pleurez, mes yeux. Don César de Bazan: Sevil-
     lana. THOMAS: Hamlet: A vox veux mes amis...Partagez vous mes
     fleurs. VERDI: Rigoletto: Caro nome; Bella figlia dell'amore.
     La traviata: Ah, fors'e lui...Sempre libera. WAGNER: Lohengrin:
     Einsam in trüben Tagen. Songs: BACH (Gounod): Ave Maria.
     BIZET: Pastorale. BLANGINI: Per valli, per boschi. Enrico
     Caruso, t; Various accompaniments.
            +Gr 2-75 p1551

<u>7206</u>
2798 DONIZETTI: Il Duca d'Alba: Angelo casto e bel. FRANCHETTI: Germania:
     Ah, vieni qui...No, non chiuder gli occhi; Studenti udite.
     GOLDMARK: Die Königin von Saba: Magische Töne. LEONCAVALLO:
     La bohème: Non qui...Testa adorata. MASSENET: Manon: En fermant
     les yeux. MEYERBEER: L'Africana: Mi batte il core...O paradiso.
     PUCCINI: Manon Lescaut: Donna non vidi mai. Tosca: Recondita
     armonia. VERDI: Aida: Se quel guerrier io fossi...Celeste Aida.
     Don Carlo: Doman dal ciel...Dio che nell'alma infondere. Otello:
     O Mostruosa colpa...Si pel ciel marmoreo. Rigoletto: Questa o
     quella. Enrico Caruso, t; Various accompaniments.
            +Gr 2-75 p1551

                          SCOTTISH RECORDS
<u>SRSS 1-2</u>
2799 History of Scottish music 1250-1625. Vol. 1: All sons of Adam.
     Credo. Ex te lux oritur. Hac in anni janua. Haec dies. Lyt-
     ill bak (Black): My heartly service. Salve, rex gloriae. Sanc-
     tus (Carver). Si quis diligit me (Peebles). Trip and go, hey.
     Woe worth the tyme. Vol. 2: Care, away go thou from me. Come
     love, let's walk. Doune in yone gardeine; Evin dead behold I
     breathe; GAlliard; Galliard (Kinloch); In thou the windoes of

myn ees; Joy to the person of my love; O lusty May; Pavan (Burnett); The Queine of Ingland's lessoune; Since that my siches;
Support your servant; What mightie motion. Clifford Hughes, t;
Saltire Singers, The King's Singers; Kenneth Elliott, hpd and
org.
    +Gr 4-75 p1859

SERAPHIM

S 60234
2800 The Seraphim guide to the instruments of the orchestra. LPO Mem-
    bers; Adrian Boult, narrator.
        +NR 9-75 p15

S 60251
2801 Songs: DEBUSSY: Ariettes oubliées; Beau soir. DUPARC: L'Invita-
    tion au voyage; Phidylé. FAURE: Les berceaux, op. 23, no. 1;
    La chanson de pêcheur, op. 4, no. 1; Mai, op. 1, no. 2. GOUNOD:
    Aimons-nous; Ou voulez-vous aller. SCHUBERT: Schwanengesang,
    D 957, excerpts (4). SCHUMANN: Lust der Sturmnacht, op. 35,
    no. 1; Mein schöner Stern; Stille Liebe, op. 35, no. 8; Stille
    Tränen, op. 35, no. 10; Widmung, op. 25, no. 1. Gerard Souzay,
    bar; Dalton Baldwin, pno.
        ++NR 12-75 p12

1750 ARCH

1752
2802 10 plus 2 equals 12 American text sound pieces. AMIRKANIAN: Heavy
    aspirations. Just. ANDERSON: Torero piece. ASHLEY: In Sara
    Mencken, Christ and Beethoven there were men and women. CAGE:
    62 mesostics re Merce Cunningham. COOLIDGE: Preface. DODGE:
    Speech songs. GIORNO: Give it to me baby. GNAZZO: The popula-
    tion explosion. GYSIN: Come to the free words. O'GALLAGHER:
    Border dissolve in audiospace. SAROYAN: Crickets.
        *NR 9-75 p15   *St 12-75 p133

S 1753
2803 DUFAY: Songs: C'Est bien raison de devoir essaucier; Je me com-
    plains piteusement; Invidia nimica; Malheureux cuer que veux
    to faire; Par droit je puis bien complaindre. GRIMACE: A l'arme
    a l'arme. LANDINI: Se la nimica mie; Adiu adiu. PERUSIO: An-
    dray soulet. VAILLANT: Par maintes foys. ANON.: Istampita
    ghaetta. Music for a While.
        +NR 11-75 p11

SOUND SUPERB

SPR 90049
2804 BACH: Brandenburg concerto, no. 3, S 1048, G major: 1st movement.
    HAYDN: Concerto, trumpet, E flat major: 3rd movement. LITOLFF:
    Concerto symphonique, no. 4, op. 102, D minor: Scherzo. MENDEL-
    SSOHN: A midsummer night's dream, op. 61: Wedding march. MOZART:
    Concerto, piano, no. 21, K 467, C major: 2d movement. Serenade,
    no. 13, K 525, G major: 3rd movement. PARRY: Jerusalem, op.
    208. STRAUSS, J. II: Unter Donner und Blitz, op. 324. TCHAI-
    KOVSKY: Symphony, no. 5, op. 64, E minor: 3rd movement. VIVALDI:
    The four seasons, op. 8: Spring. Jane Parker-Smith, org; Moura

Lympany, pno; Elgar Howarth, tpt; Peter Katin, pno; Kenneth
Sillito, vln; Virtuosi of England, St. James Ensemble, Scottish
National Orchestra, LPO; Royal Choral Society; Arthur Davison,
Steuart Bedford, Alexander Gibson, Andrew Davis, John Pritchard,
Theodor Guschlbauer.
  +Gr 2-75 p1556   +-RR 3-75 p28

STENTORIAN

SC 1685
2805 BOELLMAN: Suite gothique, op. 25. FRANCK: Pièce héroïque. HANDEL:
  Concerto, organ, no. 4, op. 4, F major: Allegro. JONGEN: Choral.
  PURVIS: Les petites cloches. VIERNE: Symphony, no. 1, op. 14,
  D minor: Andante. Keith Chapman, org.
    ++NR 12-75 p13

STROBE

SRCM 123
2806 Haslemere Festival golden jubilee, 1925-1974. BABELL: Concerto, 2
  alto recorders, harpsichord and strings, no. 6, F major: Andante.
  BACH: Trio, violin, oboe, bassoon and harpsichord, S 1040.
  BIBER: Sonata, violin and harpsichord, no. 10, G minor. BINCH-
  OIS: Je ne prise point tels Baisiers, soprano, rebec and 3 viols.
  DOWLAND: Melancholy galliard. My Lady Hunsdon's puffe, lute
  solo. FIELD: Nocturne, fortepiano, A major. GALLIARD: Suite,
  bassoon and continuo, no. 2. HEINICHEN: Concerto, 4 alto recorders, strings and continuo: Pastorell. LEATHERLAND: Pavan,
  6 viols, G minor. TELEMANN: Der Gedultige Socrates: Rodisetta's
  air. VIVALDI: Concerto, alto recorder, violin, oboe, bassoon
  and continuo, G minor. ANON (17th century): Le rossignol, 2
  lutes. Angela Beal, Elizabeth Harwood, s; Anthony Camden, ob;
  Archie Camden, John Orford, bsn; David Channon, Robert Spencer,
  lt; Elizabeth Dawson, Joan Harwell, vla; Joan Davies, fortepiano; Carl Dolmetsch, alto recorder, rebec, treble viol; Cecile
  Dolmetsch, treble viol; Jean Dolmetsch, rec, alto and treble
  viols; Marguerite Dolmetsch, vla da gamba, tenor viol, alto
  recorder; Natalie Dolmetsch, Margaret Donington, Sheila Marshall,
  vla da gamba; Heather Harrison, vlc; Jean Harvey, Eduard Melkus,
  vln; Evelyn Nallen, alto recorder; Lionel Salter, Joseph Saxby,
  hpd.
    +AR 2-75 p20

SUPRAPHON

10924
2807 BOITO: Mefistofele: Giunto sul passo estremo. DONIZETTI: L'Elisir
  d'amore: Una furtiva lagrima. GIORDANO: Andrea Chénier: Come
  un bel dì di maggio. GOUNOD: Faust: Salve, dimora casta e pura.
  FLOTOW: Marta: M'appari. HANDEL: Xerxes: Ombra mai fu. MAS-
  CAGNI: Cavalleria rusticana: Tu qui, Santuzza. MEYERBEER:
  L'Africana: O paradiso. PONCHIELLI: La gioconda: Cielo e mar.
  PUCCINI: La bohème: Che gelida manina. VERDI: Rigoletto: La
  donna è mobile. Dusolina Giannini; Beniamino Gigli, t; Orchestra.
    +-RR 3-75 p18

50919
2808 ALBENIZ: España, op. 165, no. 3: Malagueña (arr. Stutchevsky/Thaler);
     no. 2: Tango (arr. Marechal). CASSADO: Requiebros. Sonata nello
     stile antico spagnuolo. FALLA: El amor brujo: Ritual fire dance
     (arr. Piatigorsky); Pantomime (arr. Sadlo). GRANADOS: Goyescas:
     Intermezzo (arr. Cassado). Spanish dance, op. 5, no. 2 (arr.
     Piatigorsky); op. 5, no. 5 (arr. Saleski). Tonadilla. Miloš
     Sádlo, vlc; Alfréd Holeček, pno.
          +RR 4-75 p49

012 0789
2809 BELLINI: I puritani: A te, O cara. DONIZETTI: L'Elisir d'amore:
     Una furtiva lagrima. La favorita: Una vergine, un angel di
     Dio; Spirto gentil. Lucia di Lammermoor: Sulla tomba. FLOTOW:
     Martha: M'appari. GIORDANO: Andrea Chénier: Come un bel dì
     di maggio; Vicino a te. PUCCINI: La bohème: Che gelida manina.
     Manon Lescaut: Tra voi belle. VERDI: La forza del destino: O
     tu che in seno. Otello: Sì pel ciel marmoreo giuro. Aureliano
     Pertile, t; Orchestra.
          +HFN 6-75 p97   +-RR 5-75 p18

110 1429
2810 BLODEK: In the well: The rising of the moon. DVORAK: Humoresque.
     Waltz, op. 54, no. 1, A major. FIBICH: At twilight; Poem.
     FORSTER: Deborah: Polka. JANACEK: Lachian dances: The saws.
     KOVAROVIC: Mr. Brouček's excursion to the exhibition: Miners'
     polka. NEDBAL: The simple Johnny: Valse triste. NOVAK: Youth,
     op. 55, no. 21: The devil's polka. SMETANA: March of the
     students' legion. To our girls. SUK: Pieces, piano, op. 7,
     no. 1: Love song. Prague Symphony Orchestra; Vaclav Smetaček.
          +HFN 12-75 p164   +RR 7-75 p35
          +NR 11-75 p1

114 1458
2811 The golden parade of Viennese waltzes. FUCIK: Vom Donauufer.
     KOMZAK: Bäd'ner Mad'ln, op. 252. LANNER: Die Schönbrunner,
     op. 200. STRAUSS, J. II: The beautiful blue Danube, op. 314.
     Wiener Blut, op. 354. ZIEHRER: Wiener Burger, op. 419. Czecho-
     slovak Brass Orchestra; Rudolf Urbaneč.
          +-St 6-75 p108

                          SWEDISH SOCIETY

SLT 33189
2812 BACH: Suite, lute, S 996, E minor. MENDELSSOHN (arr. Blanco):
     Quartet, strings, no. 1, op. 12, E flat major: Canzonetta.
     NARVAEZ: Diferencias sobre Guárdame las vacas. PRAETORIUS (arr.
     Williams): Ballet. Volta. PUJOL VILARRUBI: El abejorro. SANZ:
     Castillian dances (4). SCARLATTI, D.: Sonata, guitar, L 23/K
     380, E major. WEISS: Fantasie. Diego Blanco, gtr.
          -RR 11-75 p81

                            TELEFUNKEN

SAWT 9589
2813 Italian recorder sonatas, vol. 2. CIMA: Sonatas, recorder, D minor,
     G minor. CORELLI: Sonata, recorder and continuo, op. 5, no. 4,
     F major. FRESCOBALDI: La bernadina. MARCELLO: Sonata, recorder
     and continuo, op. 2, no. 11, D minor. VERACINI: Sonata, recorder

and continuo, A minor. Frans Brüggen, rec; Anner Bylsma, vlc;
Gustav Leonhardt, hpd and org.

+AR 2-75 p20	-MJ 12-74 p46
+Gr 11-73 p951	++NR 4-74 p5
++HF 8-74 p107	++RR 11-73 p63
+HFN 11-73 p2329	++St 12-74 p150

SAWT 9620/1 (2)
2814 Musica iberica, 1100-1600. ALONSO: La tricotea. CORNAGO: Señora,
     qual soy venido. Gentil dama. Pues que Dios. ENCINA: Ay triste
     que vengo. Si abrá en este baldrés. Qu'es de ti. Levanta
     Pascual. MILAN: Sospirastes baldovinos. MUDARRA: Claros y
     frescos rios. Tiento. Triste estaba el Rey David. Si me
     llaman. ORTIZ: Recercada (2). URREDA: De vos i de mi. Muy
     triste será mi vida. Nunca fue pena mayor. VASQUEZ: Se me
     llaman. ANON.: Al niño. Soberana Maria. Claros y frescos rios.
     Ja nao podeis. Toda noite. Porque me nao ves, Joanna. Jançu
     janto. Quant ay lo mon consirat. Nénbressete madre. A madre.
     Benedicamus: Catholicorum concido, Ex illustri nata prosapia,
     Salva Virgo Regia, Ave gloriosa. Neil Jenkins, t; Karl-Heinz
     Klein, bar; Robert Eliscu, shawm; David Fallows, guitarra moris-
     ca, crumhorn, dulcaina; Johannes Fink, vielle, viol; James
     Tyler, cittern; Craig Wilson, lt; Early Music Quartet.

+Audio 9-75 p71	++RR 12-74 p44
+-Gr 10-74 p718	

TK 11567/1-2 (2)
2815 Historic organs of Austria. FISCHER: Prelude and fugue, D minor.
     Praeludium and fugue, C major. FROBERGER: Capriccio, C major,
     G major. Ricercar, G minor. Canzona, F major. KERLL: Canzona,
     G minor, D minor. KRIEGER: Fantasia, D minor. Praeludium and
     ricercar, A minor. Toccata, D major. KRIEGER, J. P.: Toccata
     and fugue, A minor. MUFFAT: Nova Cyclopeias harmonica. MURSCH-
     HAUSER: Praembulum fugae, finale tertii toni, A minor. PACHEL-
     BEL: Chorale partita, Alle Menschen müssen sterben. Chorale
     preludes, Ein feste Burg; Komm, Gott, Schöpfer, Heiliger Geist.
     Fugue, C major. Toccata, E minor. SPETH: Toccata quarta, E
     minor. Toccata quinta. Herbert Tachezi, org.

+Gr 12-74 p1188	/RR 1-75 p48

TK 11569/1-2 (2)
2816 BINCHOIS: Filles à marier; De plus en plus. BORLET: He, tres
     doulz roussignol. ENCINA: Fata la parte. FEVIN: Faulte d'arg-
     ent. FOGLIANO: L'amor donna. FLORENTIA: Come da lupo. GANASSI:
     Ricercar. GUYARD: M'y levay par ung matin. LANDINI: Ecco la
     primavera. Gran piant. LAURENTIUS: Mij heeft een piperken.
     LURANO: Se me grato. LUZZASCHI: O dolcezze. NOLA: Chi chi li
     chi. OBRECHT: Ic draghe de mutze clutze. RUE: Mijn hert.
     SACHS: Nachdem David. VAILLANT: Par maintes fois. VERDELOT:
     Madonna il tuo bel viso. WAELRANT: Als ic u vinde. WOLKEN-
     STEIN: Car Wunniklaich. In suria. ANON.: Estampie. Samson
     dux. Te Deum. Mij Quam eyn hope. Ich spring an diese ringe.
     Rodrigo Martinez. Dale si le das. Venid a sospirar. El
     fresco ayre. Saltarello. Tres douce regard. Mes tres doul
     roussignol. Le joli tetin. He, Robinet. Filles à marier. Il
     est de bonne heure. Je suis d'Allemagne. Studio of Early
     Music; Thomas Binkley.

     +-RR 8-75 p60

SMA 25121 T/1-3 (3)
2817 BARSANTI: Sonata, recorder, C major. BIGAGLIA: Sonata, recorder,
A minor. CIMA: Sonatas, recorder, D major and G major. CORELLI:
Sonata, recorder, F major. Variations on "La follia", recorder.
FRESCOBALDI: La bernadina. MARCELLO, B.: Sonata, recorder, D
minor. SAMMARTINI: Concerto, recorder and strings, F major.
VERACINI: Sonatas, recorder and continuo, A major, G major.
VIVALDI: Concerto, recorder and strings, A minor. Sonata, re-
corder, G minor. Frans Brüggen, rec; Anner Bylsma, vlc; Gustav
Leonhardt, hpd, org; VCM; Nikolaus Harnoncourt.
    +HFN 6-75 p100   ++RR 5-75 p52
    +NR 6-75 p9   ++SFC 5-18-75 p23
Tape (c) CX 4-41203
2818 DIEUPART: Suite, recorder and harpsichord, G major. EYCK: Pavane
lachrymae. LOEILLET: Sonata, recorder and harpsichord, C minor.
PARCHAM: Solo, recorder and harpsichord, G major. TELEMANN:
Fantasies, solo recorder, A minor and C major. Frans Brüggen,
rec; Nikolaus Harnoncourt, vla da gamba; Gustav Leonhardt, hpd.
    ++HFN 10-75 p155 tape   ++RR 10-75 p97 tape
ER 6-35257 (2)
2819 DANDRIEU: Armes, amours. BINCHOIS: Dueil angoisseus. Gloria,
laus et honor. BRASSART: O flos fragrans. CESARIS: Bonté
biauté. DUFAY: L'alta bellezza. Ave virgo. Bien veignes vous.
Bon jour, bon mois. C'est bien raison. Credo. La dolce vista.
Dona i ardente ray. Ecclesie militantis. Gloria. Helas mon
dueil. J'ay mis mon cuer. Je vous pri. Kyrie. Lamentatio
Sanctae matris ecclesiae Constantinopolitanae. Mon chier amy.
Moribus et genere Christo. Qui latuit. Sanctus. Veni creator
spiritus. DUNSTABLE: Beata mater. GRENON: La plus jolie.
HASPROIS: Puisque je voy. LANTINS, A.: Puisque je voy. LANTINS,
H.: Gloria. LOQUEVILLE: Sanctus. MORTON: La perontina. SOLAGE:
Femeux fume. ANON.: Kere dame. Musicorum decus et species.
Syntagma Musicum; Kees Otten.
    +Audio 9-75 p71   ++HFN 7-75 p80
    +Gr 10-75 p662   ++NR 6-75 p12
    +HF 10-75 p74   +RR 6-75 p78
DX 6-35262 (2)
2820 FUCIK: Der alte Brummbär. Marches: Herzegovatz; Entry of the
gladiators; Florentine. Marinarella overture. Waltzes:
Winter storms; Tales of the Danube. HEUBERGER: Der Opernball:
Overture. LEHAR: Wiener Frauen: Overture. OFFENBACH: Orphée
aux enfers: Overture. STRAUSS, J. II: Die Fledermaus, op. 363:
Overture. Waldmeister overture. SUPPE: The beautiful Galathea:
Overture. CPhO; Václav Neumann.
    +HFN 9-75 p109   +RR 6-75 p36
DX 6-35265 (Reissues from SAWT 9406, 9436)
2821 BACH: Prelude and fugue, S 546, C minor. BOHM: Chorale prelude,
Wer nur den lieben Gott lässt walten. Prelude and fugue, A
minor. BUXTEHUDE: Canzonetta, C major. Fugue, C major. Partita on Auf meinen lieben Gott. LUBECK: Prelude and fugue, E
major. REINKEN: Toccata. SCHEIDEMANN: Jesu, wolt'st uns
weisen. STEENWICK: More Palatino variations. Heyligh saligh
Bethlehem variations. WALTHER: Jesu, meine Freude, chorale
partita. ANON.: Nun komm der Heiden Heiland. Piet Kee, Albert
de Klerk, org.
    +-HFN 8-75 p89   /RR 9-75 p56

AW 6-41275
2822 Chansons de trouvères. BERNEVILLE: De moi doleros vos chant.
      BRULE: Biaus m'est estez. CAMBRAI: Retrowange novelle. DIJON:
      Chanterai por mon coraige. MEAUX: Trop est mes maris jalos.
      ANON.: Lasse, pour quoi refusai. Li joliz temps d'estey.
      Studio of Early Music; Thomas Binkley.
            +Gr 7-75 p234   +RR 6-75 p78
AJ 6-41867
2823 Songs and arias, Peter Schreier. BACH: Mass, S 232, B minor: Agnus
      Dei. St. John Passion, S 245: Es ist vollbracht. Song: Es
      kostet viel, S 459. CORNELIUS: Die Könige; Simeon. DONIZETTI:
      L'Elisir d'amore: Una furtiva lagrima. MAUERSBERGER: Dresden
      requiem: De profundis. MOZART: Così fan tutte: Der Odem der
      Liebe. Don Giovanni: Nur ihrem Frieden. Die Entführung aus
      dem Serail: Constanze, dich wieder zu sehen. ROSSINI: Il
      barbiere di Siviglia: Sieh schon die Morgenröte. SCARLATTI, A.:
      La rosaura. SCHUTZ: Was hast du verwirket. Peter Schreier,
      alto and tenor; Dresden Kreuzchor; Dresden Philharmonic Orches-
      tra, BRSO, Leipzig Radio Symphony Orchestra, Berlin Chamber
      Orchestra, Berlin Staatskapelle; Walter Olbertz, hpd; Rudolf
      Mauersberger, Helmut Koch, Kurt Masur, Alfred Schönfelder,
      Otmar Suitner.
            -Gr 11-75 p888
AW 6-41890
2824 FRESCOBALDI: Toccata cromatica per l'elevazione. Bergamasca.
      GALUPPI: Sonata, organ. MERULO: Canzon à 4. ROSSI: Toccata,
      no. 7, D minor. TRABACI: Canzona francesa cromatica VII.
      ZIPOLI: Toccata. Siegfried Hildenbrand, org.
            +Gr 11-75 p864   +RR 12-75 p84
            +HFN 12-75 p155
DLP 6-48064 (2)
2825 BEETHOVEN: Song: An die ferne Geliebte, op. 98. SCHUBERT: Songs:
      Frühlingsglaube, D 686; Ganymed, op. 19, no. 3, D 544; Liebes-
      botschaft; Lied eines Schiffers an die Dioskuren, D 360; Der
      Musensohn, D 764; Nacht und Träume, D 827; Wohin. SCHUMANN:
      Songs: Die beiden Grenadiere, op. 49, no. 1; Frühlingsfahrt,
      op. 45, no. 2; Intermezzo, op. 39, no. 2; Schöne Fremde, op.
      39, no. 6; Zum Schluss, op. 25, no. 26. STRAUSS, R.: Songs:
      Breit uber mei Haupt, op. 19, no. 2; Die Georgine, op. 10, no.
      4; Heimliche Aufforderung, op. 27, no. 3; Ich trage meine Minne,
      op. 32, no. 1; Nachtgang, op. 29, no. 3; Ständchen, op. 17, no.
      2; Traum durch die Dämmerung, op. 29, no. 1. TCHAIKOVSKY: Songs:
      Einst zum Narr'n der Weise sprach, op. 25, no. 6; Inmitten des
      Balles, op. 38, no. 3; Nur wer die Sehnsucht kennt, op. 6, no.
      6; O singe mir, Mutter, diese Weise, op. 16, no. 4; Töte mich
      aber liebe mich; Unendliche Liebe; Warum sind denn die Rosen so
      blass. Peter Anders, t; Hubert Giesen, pno; BPhO; Walter Lutze.
            +HFN 8-75 p81   -RR 10-75 p89

                         TOSCANINI SOCIETY
ATS GC 1201/6 (6)
2826 The Cantelli legacy, vol. 1. BARTOK: Music for strings, percussion
      and celesta. BEETHOVEN: Symphony, no. 1, op. 21, C major.
      Symphony, no. 5, op. 67, C minor. Plus rehearsal excerpts with
      the Philharmonic Orchestra. HAYDN: Symphony, no. 88, G major.

RAVEL: Bolero. La valse. SCHUBERT: Symphony, no. 9, D 944,
C major. SCHUMANN: Symphony, no. 4, op. 120, D minor. TCHAI-
KOVSKY: Symphony, no. 4, op. 36, F minor. WAGNER: Rienzi: Over-
ture. WEBER: Euryanthe: Overture. NBC Symphony Orchestra;
Guido Cantelli.
        +-HF 6-75 p78

TRANSATLANTIC

TRA 292. Tape (c) ZCTRA 292 (ct) Y8TRA 292
2827 Divisions on a ground: An introduction to the recorder and its
        music. FINGER: Divisions on a ground. HANDEL: Sonata, record-
        er, op. 1, no. 11, F major. LOEILLET: Sonata, recorder, op.
        3, no. 3, G minor. MATTHYSZ: Variations from "Der Gooden Fluyt
        Hemel". EYCK: Variations on "Amarilli mia bella". VIVALDI:
        Concerto, recorder, op. 44, no. 11, C major. Richard Harvey,
        rec; Andrew Parrott, hpd; Monica Hugett, Eleanor Sloan, vln;
        Trevor Jones, vla; Catherine Finnis, vlc; Adam Skeaping, violone.
        +-Gr 10-75 p652   +STL 6-8-75 p36
        ++HFN 7-75 p80

TURNABOUT

TV 34569
2828 A Renaissance Christmas. ATTEY: Sweet was the song the Virgin
        sang. BYRD: Lullaby. CLEMENS NON PAPA: Vox in Roma. ECCARD:
        Nun komm der Heiden Heiland. GUERRERO: Virgen santa. MOUTON:
        Noel. PRAETORIUS: Quem pastores laudavere; Wie schön leuchtet
        der Morgenstern. VICTORIA: O magnum mysterium. RASELIUS: Nun
        komm der heiden Heiland. RESINARIUS: Nun komm der Heiden Heil-
        and. WALTHER: Joseph, lieber Joseph mein. OBRECHT: Magnificat.
        ANON.: Marvel not, Joseph; Nova, nova; Riu, riu, chiu; Salve
        lux fidelium; Veni Redemptor gentium. Boston Camerata; Joel
        Cohen.
        +-HF 4-75 p94   +NR 12-75 p2
TV 37033. Tape (c) KTVC 37033
2829 BEETHOVEN: Sonata, piano, no. 14, op. 27, no. 2, C sharp minor.
        BRAHMS: Waltz, op. 39, no. 15, A flat major. CHOPIN: Etude,
        op. 10, no. 3, E major. Waltz, op. 70, no. 3, D flat major.
        DEBUSSY: Suite bergamasque: Clair de lune. LISZT: Liebestraum,
        no. 3, G 541, A flat major. MENDELSSOHN: Song without words,
        op. 62, no. 3: Spring song. SCHUMANN: Kinderscenen, op. 15,
        no. 7: Träumerei. TCHAIKOVSKY: Chant sans paroles, op. 2, no.
        3, F major. Walter Klien, pno.
        +-Gr 1-75 p1403   +-RR 11-72 p78
TV 37071 (Reissue from Vox STGBY 619)
2830 Renaissance brass msuic. FERRABOSCO, Alfonso II: Four note pavan.
        GIBBONS: In nomine. HOLBORNE: Galliard. O'KOEVER: Fantasie.
        SCHEIDT: Benedicamus Domino. Canzona Aechiopicam. Canzona
        bergamasca. Canzona gallicam. Galliard battaglia. Wendet
        euch um ihr Anderlein. SIMMES: Fantasie. WEELKES: As wanton
        birds. Death hath deprived me. In pride of May. O care, thou
        wilt despatch me. Sit down and sing. Eastman Brass Quintet.
        +-Gr 4-75 p1829   ++RR 2-75 p46
TV 37079
2831 DERING: Country cries. GIBBONS: Do not repine, fair sun. PEERSOn:

Sing, love is blind. RAVENSCROFT: Rustic lovers. TOMKINS:
Alman à four viols (Musica britannica, no. 33). VAUTOR: Weep,
weep mine eyes. WEELKES: The cryes of London. Since Robin Hood.
Thule, the period of cosmography. ANON.: Hey down a down a down;
Take heed of time, tune and ear. Jaye Consort of Viols; Purcell
Consort of Voices; Grayston Burgess.
+--HFN 6-75 p99   +RR 4-75 p69

UNIVERSITY OF COLORADO

Unnumbered
2832 Christmas at Colorado State University. BACH: Cantata, no. 147,
     Jesu, joy of man's desiring. BRAHMS: Low, how a rose e'er
     blooming. CHARPENTIER: Messe de minuit pour Noel: Kyrie.
     DAQUIN: The Noel, G major. DISTLER: Weinachts Geschichte: Den
     Dei hirten lobten sehre; Frohlich soll mein Herze springen; Zu
     Bethlehem geboren. GABRIELI: Canzoni septimi toni. PFAUTSCH:
     A day for dancing: What shall I bring. PACHELBEL: Vom Himmel
     hoch. PRAETORIUS: Lo, how a rose e'er blooming. ANON.: Adeste
     Fidelis. He is born. A Flemish carol (arr. Christiansen).
     Noel nouvelet (arr. Zgodav). A virgin most pure (arr. Halter).
     The Colorado State University Chamber Singers; CSU Brass Choir;
     Robert Cavara, org; Edward Anderson, Jacob Larsen.
     ++CJ 5-75 p22

UNIVERSITY OF ILLINOIS

Unnumbered
2833 BERGER: Songs of sadness and gladness; Then I commended mirth;
     Let take a cat; He that loves a rosy cheek; The seasons; My
     love is dead; Farewell. FARMER: Little pretty bonnie lass.
     VECCHI: Gioite tutti. VIADANA: Missa "L'Hora passa": Kyrie,
     Gloria, Credo, Sanctus, Benedictus, Agnus Dei. Illinois State
     University Madrigal Singers; John Farrell.
     ++CJ 5-75 p22

UNIVERSITY OF IOWA PRESS

Unnumbered
2834 BACH: Passacaglia and fugue, S 582, C minor. Toccata and fugue,
     S 565, D minor. CLERAMBAULT: Suite du deuxième ton: Plain jeu,
     Duo, Trio, Basse de cromarne, Caprice sur les grands jeux.
     ERBACH: Conzon in the Phrygian mode. KRAPF: Fantasia on a
     theme by Fescobaldi. PACHELBEL: Vom Himmel hoch. REGER: Dank-
     psalm, op. 145, no. 2. Gerhard Krapf, Delbert Disselhorst, org.
     +MJ 3-75 p26   ++NR 6-74 p12
     +HFN 3-74 p121

VANGUARD

SRV 316
2835 BINCHOIS: Adieu m'amour. BYRD: My Lord of Oxenford's mask.
     CUTTING: Galliard. HAYNE VAN GHIZEGHEM: Gentil gallans. JANNE-
     QUIN: Il était une fillette. Ma peine n'est pas grande. MONIOT
     DE PARIS: Vaduri. NEUSIEDLER: Der Judentanz. PRAETORIUS: La
     rosette. ANON.: Branle de la Reine. Chançoneta Tedesca. La
     manfredina. Pavane d'Espagne. La ronde du Jaloux. Saltarello.

La Spagna. Voulez-vous que je vous chante. Les Menestriers.
+-NR 5-75 p9   ++St 11-75 p143
++SFC 11-2-75 p28

VSD 71179
2836 Douce Dame/Music of courtly love from Medieval France and Italy.
BALOGNA: Oselleto selvaggio. FIRENZE: Apposte messe. LANDINI:
Ecco la primavera. MACHAUT: Comment qu'à moy. Douce dame jolie.
Foys porter. Je sui aussi. Rose, liz, printemps. VAILLANT:
Par maintes foys. ANON.: Istampita Isabella. Lamento di Tris-
tan and Rotta. Or sus, vous dormez trop. Waverly Consort;
Michael Jaffee.
+Gr 8-75 p356   ++NR 4-75 p9
+HF 4-75 p94   +St 3-75 p111
++HFN 7-75 p81

VSD 71207
2837 Flute music of the romantic era. BOHM: Introduction and variations
on Nel cor più, op. 4. GAUBERT: Nocturne and allegro scherzando.
GENIN: Carnival of Venice variations. GODARD: Suite de trois
morceaux, op. 116. HUMMEL: Sonata, flute, D major. Paula
Robison, flt; Samuel Sanders, pno.
++HF 11-75 p132   -SFC 8-10-75 p26
+- NR 9-75 p9   +St 10-75 p119

                              VISTA

VPS 1021
2838 BACH: Bist du bei mir (arr. Grace). BUXTEHUDE: Prelude and fugue,
G minor. CLERAMBAULT: Basse et dessus de trompette. ELGAR:
Enigma variations, op. 36: Nimrod. FRICKER: Pastorale. HANDEL:
Royal fireworks music: Minuet and allegro. HAYDN: Pieces for
mechanical clock, nos. 4, 11, 12, 23. HOWELLS: Rhapsody, no.
1, G flat major. LEMMENS: Fanfare. RIDOUT: Scherzo. STANLEY:
Voluntary, no. 7. Robert Weddle, org.
+Gr 3-75 p1687   +RR 3-75 p48

                               VOX

SVBX 5304 (3)
2839 America sings, the great sentimental age--Stephen Foster to Charles
Ives. FOSTER: We are coming Father Abraham, 300,000 more;
Willie has gone to war; Jenny June; Wilt thou be true; Katy
Bell. HANBY: Nelly Gray. HAWTHORNE: Listen to the mocking
bird. IVES: An old flame; Circus band; A Civil War memory; In
the alley; Karen; Romanzo di Central Park; A son of a gambolier.
KITTRIDGE: Tenting on the old campground. Forty-six other Amer-
ican songs. Gregg Smith Singers; New York Vocal Arts Ensemble;
Instrumental Accompaniment.
+St 3-75 p86

                      WAYNE STATE UNIVERSITY

Unnumbered
2840 BARBER: Reincarnations: I got shoes (arr. Parker-Shaw); Sometimes
I feel; The coolin (arr. Parker-Shaw). BRUCKNER: Trosterin
Musik. CARTER: Heart not so heavy as mine. FOSTER: At the
round earth's imagined corner. GESUALDO: Io tacero. HASSLER:

Agnus Dei. HEATH: Beat, beat, drums; Thy word is a lantern.
LASSUS: Tibi Laus. MONTEVERDI: Adoramus te, Cantate Domino;
Zefiro torno. PEETERS: Jubilate Deo Omnis Terra. PERSICHETTI:
Gloria. SCHULTZ: O hilf, Christe. THOMAS: Close of day. ANON.:
Ain't got time to die (arr. Duey); There is a balm in Gilead
(arr. Dawson); Everytime I feel the spirit (arr. Dawson); Ain'a
that good news (arr. Dawson). Wayne State University Men's
Glee Club, Chamber Singers; Harry Langsford.
    ++CJ 5-75 p22

WEALDEN

WS 110
2841 BACH: Chorale prelude, Ich ruf zu dir, S 639. Prelude and fugue,
F minor. HALLAM: Prelude, F major. LANGLAIS: Hymne d'actions
de graces: Te deum. RHEINBERGER: Sonata, organ, no. 6, op. 119,
E flat minor. SOWANDE: Go down, Moses. WHITLOCK: Allegretto.
Folk tune. Scherzo. Erik Pask, org.
    +-HFN 9-75 p101   +-RR 9-75 p61

WS 139
2842 BACH: Prelude and fugue, S 538, D minor. BLOW: Toccata. DU MAGE:
Livre d'orgue: Recit and basse de trompette. GABRIELI: La
spiritata. LANG: Prelude on Leoni. LEIGHTON: Fanfare. MENDEL-
SSOHN: Prelude and fugue, C minor. PARRY: Chorale prelude on
Rockingham. SWEELINCK: Variations on "Est-ce mars". VAUGHAN
WILLIAMS: Rhosymedre, chorale prelude. Malcolm McKelvey, org.
    +-HFN 12-75 p159   +-RR 12-75 p82

WS 145
2843 BACH: Sonata, organ, no. 6, S 530, G major. BOHM: Prelude and
fugue, C major. DUPRE: Prelude and fugue, op. 7, no. 1, B
major. KARG-ELERT: Vom Himmel hoch, chorale prelude. LEIGHTON:
Et resurrexit, op. 49. ORR: Preludes on a Socttish psalm tune
(3). James Parsons, org.
    +-Gr 4-75 p1843   +RR 12-74 p53

WS 160
2844 ALCOCK: Voluntary IV, C minor. BACH: Fantasia, S 570, C major.
Fantasia, S 563, B minor. COUPERIN: Mass, organ: Couplet on
Domine Deus, Rex Coelestis. STANLEY: Voluntary, organ, op. 5,
no. 6. VIERNE: Berceuse. Carillon. WESLEY, C.: Voluntary III.
WESLEY, Samuel: Air and gavotte. WESLEY, S. S.: Choral song.
Erik Pask, org.
    +-HFN 12-75 p159   +-RR 12-75 p82

WESTMINSTER

WGM 8309 (Reissues from 78 rpm originals)
2845 The young Gilels. CHOPIN: Ballade, no. 1, op. 23, G minor. LISZT:
Hungarian rhapsody, no. 9, G 244, E flat major. Etudes d'exécution transcendente d'après Paganini, no. 5, G 140, E major.
LULLY (Godowsky): Gigas. MENDELSSOHN: Songs without words, op.
38, no. 6. PROKOFIEV: The love for three oranges, op. 33: March.
RAMEAU: Le Reppel des oiseaux. SCHUMANN: Der contrabandiste
(arr. Liszt). Fantasiestücke, op. 12, no. 7: Traumeswirren.
Toccata, op. 7, C major. Emil Giles, pno.
    +-ARSC 7-75 p90   +HF 11-75 p129

A madre. cf Telefunken SAWT 9620/1.
A la fontenele. cf Argo ZRG 765.
Abide with me. cf CBS 61599.
Adeste Fidelis. cf University of Colorado, unnumbered.
Adios meu homino. cf HMV SLS 5012.
Ain'a that good news (arr. Dawson). cf Wayne State University, unnum-
    bered.
Ain't got time to die (arr. Duey). cf Wayne State University, unnum-
    bered.
Al niño. cf Telefunken SAWT 9620/1.
Alas, departing is ground of woe. cf BASF 25 22286-1.
All hail to the days. cf CRD CRD 1019.
Alle psallite cum luya. cf CRD CRD 1019.
Amazing grace. cf RCA ARL 1-0561.
And I were a maiden. cf BASF 25 22286-1.
And I were a maiden. cf L'Oiseau-Lyre SOL 329.
Angelus ad virginem. cf Argo ZRG 765.
Animation. cf Nonesuch H 71276.
Ante thronum regentis omnia, Sequence. cf Nonesuch H 71292.
As I lay. cf CRD CRD 1019.
At the rive (arr. Copland and Wilding-White). cf New England Conserva-
    tory NEC 111.
Ay, que non hay. cf HMV SLS 5012.
Balletto Polacho. cf DG Archive 2533 294.
Band of music. cf New England Conservatory NEC 111.
Die Bankelsangerlieder: Sonata. cf Crystal S 201.
The bard of Armagh. cf Philips 6599 227.
Basse danse la Magdalena. cf DG Archive 2533 184.
Basses danses. cf Oryx BRL 1.
Behold the Great Redeemer die. cf CBS 61599.
La bella Franceschina. cf L'Oiseau-Lyre 12BB 203/6.
Belle, tenés mo. cf L'Oiseau-Lyre 12BB 203/6.
Bendicamus: Catholicorum concido, Ex illustri nata prosapia, Salve
    Virgo Regia, Ave gloriosa. cf Telefunken SAWT 9620/1.
Blame not my lute. cf L'Oiseau-Lyre 12BB 203/6.
Blow the wind southerly. cf RCA LFL 1-5072.
Branle de Bourgogne. cf DG Archive 2533 184.
Branle de Bourgogne. cf RCA CRL 2-7001.
Branle de la Reine. cf Vanguard SRV 316.
Branle de l'official. cf CRD CRD 1019.
The British Grenadiers. cf Classics for Pleasure CFP 40230.
Bruder Konrad. cf L'Oiseau-Lyre 12BB 203/6.
Calata. cf L'Oiseau-Lyre 12BB 203/6.
Canaan. cf Nonesuch H 71276.
Canción de cuna. cf HMV SLS 5012.

Canciones (2). cf BIS LP 3.
Canco del Ladre. cf Amberlee ACL 501X.
Cant dels Ocells. cf CBS 61579.
Carols and dances from the Middle Ages: Estampie and stantipes. cf
      Oryx BRL 1.
Cathaccio, gagliarda. cf Philips 6775 006.
Celle qui m'a demande. cf L'Oiseau-Lyre 12BB 203/6.
Ces fascheux Sotz. cf LAYOLLE: Missa Ces fascheux Sotz.
Chançoneta Tedesca. cf Vanguard SRV 316.
The cherry tree carol. cf Argo ZRG 765.
Christ der ist erstanden. cf L'Oiseau-Lyre 12BB 203/6.
Chui dicese e non l'amare. cf L'Oiseau-Lyre 12BB 203/6.
Clangat tuba, martyr Thoma, carol. cf Nonesuch H 71292.
Claros y frescos rios. cf Telefunken SAWT 9620/1.
Cold blows the wind. cf Harmonia Mundi HM 593.
Colonel Orne's march. cf New England Conservatory NEC 111.
Come follow me. cf CBS 61599.
Come, thou glorious day of promise. cf CBS 61599.
Comin' thro' the rye. cf Rubini RS 300.
Concert. cf Nonesuch H 71276.
Coventry carol. cf Decca ECS 774.
Courante. cf Oryx BRL 1.
The cuckoo. cf Harmonia Mundi HM 593.
The cuckoo. cf Oryx BRL 1.
Czarna krowa. cf Hungaroton SLPX 11549.
Dale si le das. cf Telefunken TK 11569/1-2.
Dances from 15th century France: Brawls. cf Oryx BRL 1.
Deo gracias Anglia, Agincourt carol. cf BASF 25 22286-1.
Dindirindin. cf DG 2530 504.
Ding dong. cf DANKWORTH: Tom Sawyer's Saturday.
Ding dong merrily on high (French). cf Gemini GM 2022.
The dressed ship. cf CRD CRD 1019.
Ductia. cf CRD CRD 1019.
Ductia and danse royale. cf Argo ZRG 765.
Ecce novum gaudium (arr. Elliott). cf Gemini GM 2022.
Edi beo thu. cf CRD CRD 1019.
Eli, Eli. cf Discopaedia MB 1007.
English dance. cf HMV SLS 5022.
Es ist des Heil uns kommen her, chorale prelude. cf Saga 5374.
Estampie. cf Telefunken TK 11569/1-2.
Estampie and trotto. cf BIS LP 3.
Estava la mora en su bel estar. cf HMV SLS 5012.
Everytime I feel the spirit (arr. Dawson). cf Wayne State University,
      unnumbered.
Fandango. cf CRD CRD 1019.
Faronells ground. cf Argo ZRG 746.
Father in heaven. cf CBS 61599.
La filla del Marxant. cf RCA ARL 1-0485.
Filles à marier. cf Telefunken TK 11569/1-2.
The first Nowell. cf DANKWORTH: Tom Sawyer's Saturday.
A Flemish carol (arr. Christiansen). cf University of Colorado, un-
      numbered.
Le force d'Hercole e tripla. cf Erato STU 70847.
El fresco ayre. cf Telefunken TK 11569/1-2.
The frog galliard. cf Oryx BRL 1.
Gabriel's message. cf Argo ZRG 785.

Gaillarde. cf Oryx BRL 1.
Galliard fancy. cf L'Oiseau-Lyre SOL 336.
General salute, bugle call. cf Classics for Pleasure CFP 40230.
Gentle maiden. cf Discopaedia MB 1010.
George Washington's march. cf New England Conservatory NEC 111.
Girometa. cf L'Oiseau-Lyre 12BB 203/6.
God of our Fathers. cf CBS 61599.
God rest you merry gentlemen. cf DANKWORTH: Tom Sawyer's Saturday.
God rest you merry gentlemen. cf CRD CRD 1019.
God save the Queen (arr. Britten). cf Decca 419.
Goday, my Lord, Syr Christenmasse. cf BASF 25 22286-1.
Good King Wencelas, pavan. cf CRD CRD 1019.
Green bushes. cf Philips 6599 227.
Green garters. cf CRD CRD 1019.
Greensleeves. cf Musical Heritage MHS 1345.
Gregorian chant: Ancient Spanish chants for the Ordinary and Proper of
      the mass. Santo Domingo de Silos Abbey Monks; Dom Ismael Fernán-
      dez de la Cuesta. DG Archive 2533 163.
            +RR 1-75 p57
Gregorian chant: Les vêpres du Dimance; Les complies; Fêtes de la Sainte
      Vierge; Immaculée conception; Toussaint; Christ-Roi. L'Abbaye
      Saint-Pierre de Solesmes, Monks' Choir; Dom Joseph Gajard. French
      Decca 7548/50 (3).
            +-Gr 2-75 p1542   +-RR 1-75 p57
Gregorian chant: The baptism of Christ; The transfiguration; Jubilate
      Deo, Chants of the mass; Various chants. D'Argentan Notre-Dame
      Abbey, Nuns' Choir. French Decca 7551-2 (2).
            +-Gr 5-75 p2020   +-RR 6-75 p82
            +-HFN 6-75 p101
Gregorian chant: Responsories and monodies from the Gallicad rite.
      Deller Consort; Alfred Deller. Harmonia Mundi HMD 234.
            +HFN 12-75 p152
Gregorian chant from Hungary: Ave spes nostra, Antiphon; Veni, Redemptor
      gentium, Hymn; Christus natus est nobis, Invitatorium; Jubilamen,
      Lesson; Descendit de caelis, Responsorium; Genealogia Christi,
      Evangelium; Lux fulgebit, Introitus; Alleluja, Dominus regnavit;
      Eia, recolamus, Sequentia; Procedentim sponsum, Benedicamustropus;
      Verbum caro factum est, Responsorium; Hodie Christus natus est
      Antiphon; Dies est laetitiae, Cantio; Te Deum. Schola Hungarica
      Ensemble; Janka Szendrei, László Dobszay. Hungaroton SLPX 11477.
            +-Gr 1-75 p1392
Gregorian chants: Resurrexi, Domine, probasti me. Victimae paschali
      laudes. Pascha nostrum. Post dies octo. Christus resurgens.
      Alleluia, lapis revolutus est. Te Deum. Requiem aeternam, Introit-
      us and graduale. Absolve, Domine. Dies irae. Domine Jesu Christe.
      Lux aeterna. Libera me, Domine. In paradisum. Saint Maurice and
      Saint Maur Abbey Monks. Philips 6580 105.
            +-Gr 11-75 p893   +HFN 8-75 p91
Guide us, O thou great Jehovah. cf CBS 61599.
Hamlet: How should I your true love know. cf BBC REB 191.
Have you seen but a whyte lily grow. cf HMV RLS 714.
He is born. cf University of Colorado, unnumbered.
He, Robinet. cf Telefunken TK 11569/1-2.
Here we come a-wassailing. cf CRD CRD 1019.
Hey down a down a down. cf Turnabout TV 37079.
Hey trolly lolly lo. cf L'Oiseau-Lyre SOL 329.

Una hija tiene el Rey. cf HMV SLS 5012.
His name so sweet. cf New England Conservatory NEC 111.
Hohenfriedberger Marsch. cf DG 2721 077.
Los hombres con gran plazer. cf DG 2530 504.
How firm a foundation. cf CBS 61599.
I know that my Redeemer lives. cf CBS 61599.
I saw three ships. cf CRD CRD 1019.
I will walk with my love. cf HMV RLS 714.
Ich spring an diesem ringe. cf Telefunken TK 11569/1-2.
Il est de bonne heure. cf Telefunken TK 11569/1-2.
In Rama sonat gemitus, Threnody. cf Nonesuch H 71292.
Instrumental consort. cf L'Oiseau-Lyre SOL 329.
Interlude. cf Hungaroton SLPX 11601/2.
Introit. cf Argo ZRG 785.
Istampita ghaetta. cf DG Archive 2533 184.
Istampita ghaetta. cf 1750 Arch S 1753.
Istampita Isabella. cf Vanguard VSD 71179.
It came upon the midnight clear. cf DANKWORTH: Tom Sawyer's Saturday.
Italiana. cf DG Archive 2533 184.
Ja nao podeis. cf Telefunken SAWT 9620/1.
Jacet granum, Responsory and Prosa. cf Nonesuch H 71292.
Jacobean England, popular tunes: Ayre. cf Oryx BRL 1.
Jançu, janto. cf Telefunken SAWT 9620/1.
Je ne fay. cf L'Oiseau-Lyre 12BB 203/6.
Je suis d'Allemagne. cf Telefunken TK 11569/1-2.
Jigge. cf L'Oiseau-Lyre SOL 336.
Le joli tetin. cf Telefunken TK 11569/1-2.
The jolly carter. cf Harmonia Mundi HM 593.
Der Katzenföte. cf L'Oiseau-Lyre 12BB 203/6.
Kere dame. cf Telefunken ER 6-35257.
Kiddush. cf Columbia D3M 33448.
Kol Nidre. cf Columbia D3M 33448.
Kyrie trope: Orbis factor. cf TALLIS: Missa salve intemerata.
Lady Wynkfyldes rownde. cf L'Oiseau-Lyre 12BB 203/6.
Laetare Cantuaria, Carol. cf Nonesuch H 71292.
Lamento di Tristano: Rotta; Trotto; Istampita Ghaetta; Istampita
    Cominciamento di gioia; Saltarello; Bassa danza à 2; Bassa danza
    à 3. cf DG Archive 2533 111.
Lamento di Tristan and Rotta. cf Vanguard VSD 71179.
Lamento di Tristano: Rotta. cf DG Archive 2533 184.
Las, je n'ecusse. cf L'Oiseau-Lyre 12BB 203/6.
Lasse, pour quoi refusai. cf Telefunken AW 6-41275.
L'e pur morto Feragù. cf L'Oiseau-Lyre 12BB 203/6.
Let not us that young men be. cf L'Oiseau-Lyre SOL 329.
Let us break bread together. cf RCA ARL 1-0561.
Li joliz temps d'estey. cf Telefunken AW 6-41275.
Lodensana, gagliarda. cf Philips 6775 006.
Londonderry air (arr. Kreisler). cf Decca SPA 405.
Lonsdale. cf Nonesuch H 71276.
The Lord is my shepherd. cf CBS 61599.
Lord of all hopefulness. cf Decca ECS 774.
The maid of the sweet boy. cf Philips 6599 227.
La manfredina. cf CRD CRD 1019.
La manfredina. cf Vanguard SRV 316.
Mariam matrem. cf HMV SLS 5012.
Marsch der Finnländischen Reiterei. cf DG 2721 077.

Marvel not, Joseph. cf Turnabout TV 34569.
Medieval German plainchant and polyphony: Advent/Christmas: Ad te
    levavi; Congaudeat turba; Puer natus est; Viderunt omnes; Johannes
    postquam senuit; Hodie progeditur. Passiontide: Gloria, laus et
    honor; Dominus Jesus; Christus factus est; Hely, hely. Easter:
    Confitemini...Laudate; Unicornis captivatur; Haec dies; Ad regnum
    ...Noster cetus. Pentecost: Spiritus Domini; Factus est repente;
    Catholicorum concio. Parousia: Cum natus est Jesus; Gloria in
    excelsis. Schola Antiqua; R. John Blackley. Nonesuch H 71312.
            +-Gr 11-75 p888   +NR 10-75 p6
            +HFN 11-75 p163   ++RR 12-75 p90
Merrily on high. cf DANKWORTH: Tom Sawyer's Saturday.
A merry moode. cf L'Oiseau-Lyre SOL 336.
Mes tres doul roussignol. cf Telefunken TK 11569/1-2.
Messiah. cf Nonesuch H 71276.
El mestre. cf RCA ARL 1-0485.
A mighty fortress. cf CBS 61599.
Mignonne, allons. cf L'Oiseau-Lyre 12BB 203/6.
Mijn Quam eyn hope. cf Telefunken TK 11569/1-2.
Missa Salisburgensis (attrib. Benevoli). Plaudite Tympani, Hymnus.
    James Griffet, James Lewington, Heinrich Weber, Erwin Abel, t;
    Brian Etheridge, David Thomas, Heinz Haggenmüller, Eberhard Wieder-
    hut, bs; Tölzer Boys' Choir; Escolania de Montserrat, Collegium
    Aureum; Ireneu Segarra. BASF 2522 073-7.
            +-Gr 11-75 p869   ++RR 12-75 p86
            +HFN 11-75 p148
More holiness give me. cf CBS 61599.
The morning breaks. cf CBS 61599.
Music for springtime: I chose my love. cf Oryx BRL 1.
Music of the Greek orthodox church: Byzantine hymns of Christmas;
    Byzantine hymns of the Epiphany; The service of the Akathistos
    hymn. Society for the Dissemination of National Music Chorus;
    Simon Karas. Society for the Dissemination of National Music
    SDNM 101, 102, 107 (3).
            +Gr 11-75 p888
Musicorum decus et species. cf Telefunken ER 6-35257.
My Lady Careys dompe. cf Angel SB 3816.
The national anthem. cf DANKWORTH: Tom Sawyer's Saturday.
Nearer my God to Thee. cf CBS 61599.
Nénbressete madre. cf Telefunken SAWT 9620/1.
Noel nouvelet (arr. Zgodav). cf University of Colorado, unnumbered.
Nova, nova. cf Turnabout TV 34569.
Novus miles sequitur, Conductus. cf Nonesuch H 71292.
Now the green blade riseth (French). cf Abbey LPB 734.
Nowell and Nowell. cf Oryx BRL 1.
Nuevas te traygo, Carillo. cf DG 2530 504.
O come all ye faithful. cf DANKWORTH: Tom Sawyer's Saturday.
O God, our help in ages past. cf CBS 61599.
O' 'twas on a Monday morning. cf Harmonia Mundi HM 593.
O waly, waly. cf Harmonia Mundi HM 593.
On Christmas night (arr. Vaughan Williams). cf HMV Tape (c) TC MCS 13.
Or sus, vous dormez trop. cf Vanguard VSD 71179.
Parademarsch der 18er Husaren (arr. Gursch). cf BASF 292 2116-4.
Parado de Valldemosa. cf HMV SLS 5012.
Pariser Einzugsmarsch. cf BASF 292 2116-4.
Pastor caesus in gregis medio, Antiphon. cf Nonesuch H 71292.

Pastorcico noh aduerma. cf HMV SLS 5012.
Pavan and galliard. cf Philips 6775 006.
Pavana de la morte de la ragione. cf L'Oiseau-Lyre 12BB 203/6.
Pavana el tedescho. cf L'Oiseau-Lyre 12BB 203/6.
Pavane d'Espagne. cf Vanguard SRV 316.
Pavane des Dieux. cf Oryx BRL 1.
Pavane Lesquercarde. cf Oryx BRL 1.
Pavane Venetiana. cf L'Oiseau-Lyre 12BB 203/6.
Petersburger Marsch (arr. Gursch). cf BASF 292 2116-4.
Petersburger Marsch. cf DG 2721 077.
Pieces, lute (5). cf DG 2530 561.
Pilgrim's farewell. cf Nonesuch H 71276.
Popular Russian songs: Along Petersburg Street; Bandura; The cliff;
     Dark eyes; Farewell joy; In the dark forest; The little cudgel;
     Little night; Stenka Razin; Twelve bandits; The Volga boatmen.
     Nicolai Ghiaurov, bs; Kaval Chorus and Orchestra; Atanas Margaritov.
     Decca SXL 6659. Tape (c) KSXC 6659.
              +Gr 12-74 p1203   +RR 7-75 p71 tape
Porque me nao ves, Joanna. cf Telefunken SAWT 9620/1.
Psalm 21. cf New England Conservatory NEC 111.
Psalm 122, I was glad. cf Argo ZRG 785.
Quant ay lo mon consirat. cf Telefunken SAWT 9620/1.
The Quiet Consort: The first canticle. cf Oryx BRL 1.
An raibh tu ag an gCarrig. cf Phillips 6599 227.
Der rather Schwanz. cf L'Oiseau-Lyre 12BB 203/6.
Retreat, bugle call. cf Classics for Pleasure CFP 40230.
Riu, riu, chiu. cf Turnabout TV 34569.
Rock of ages. cf CBS 61599.
La Rocque. cf Oryx BRL 1.
Rodrigo Martinez. cf Telefunken TK 11569/1-2.
Romeo and Juliet: Where griping grief. cf BBC REB 191.
La ronde du Jaloux. cf Vanguard SRV 316.
The rose tree. cf Michigan University SM 0002.
Le rossignol. cf Argo ZRG 765.
Le rossignol. cf L'Oiseau-Lyre 12BB 203/6.
Le rossignol, 2 lutes (17th century). cf Strobe SRCM 123.
El rossinyol. cf HMV SLS 5012.
The roving sailor. cf Michigan University SM 0002.
Russian orthodox church music: Lord I have creid unto Thee; Dogmatikon
     from Vespers; Stikheron for Pentecost; Troparion to Saint Sergius;
     Stikheron for the Feast of Saint Sergius; By the waters of Babylon;
     Stikhera for Good Friday and for the Dormition; Exapostilarion
     for the Dormition; Stikhera for the Saints of Russia; Selected
     hymns of Great Lent. Zagorsk Monastic Choir, Moscow Church Choir;
     Archimadrite Matthew, Nicholas V. Mateev.
              +Gr 7-75 p234   +RR 11-75 p88
The sailor and young Nancy. cf Harmonia Mundi HM 593.
Saltarello. cf DG Archive 2533 184.
Saltarello. cf HMV SLS 5022.
Saltarello. cf Telefunken TK 11569/1-2.
Saltarello. cf Vanguard SRV 316.
Saltarello (Italian). cf CRD CRD 1019.
Saltarello de la morte de la ragione. cf L'Oiseau-Lyre 12BB 203/6.
Salve nos Domine. cf TALLIS: Missa salve intemerata.
Salve lux fidelium. cf Turnabout TV 34569.
Salve regina (with tropes). cf TALLIS: Missa salve intemerata.

Samson dux. cf Telefunken TK 11569/1-2.
Sant Marti del Canigo. cf CBS 61579.
Se mai per maraveglia. cf L'Oiseau-Lyre 12BB 203/6.
The shadows flee. cf CBS 61599.
The sheep sharing. cf Harmonia Mundi HM 593.
Shepherd's madrigal. cf Argo ZRG 805.
Shooting the guns pavan. cf L'Oiseau-Lyre 12BB 203/6.
A short measure off My Lady Wynkfylds Rownde. cf Angel SB 3816.
Simple gifts. cf Advent 5012.
Since first I saw your face. cf Oryx BRL 1.
Sir John Smith his almaine. cf DG Archive 2533 157.
Soberana Maria. cf Telefunken SAWT 9620/1.
Solemne canticum, Sequence. cf Nonesuch H 71292.
Sonata, flute, G major. cf Argo ZRG 746.
Songs (3). cf Angel S 36076.
Sonnerie aux morts. cf Classics for Pleasure CFP 40230.
La Spagna. cf Vanguard SRV 316.
Springhill. cf Nonesuch H 71276.
Sta notte. cf L'Oiseau-Lyre 12BB 203/6.
Staines Morris. cf CRD CRD 1019.
Steal away (arr. Trant). cf Decca ECS 774.
The streams of lovely Nancy. cf Argo ZRG 765.
Suite, lute. cf DG 2530 561.
Suite regina. cf L'Oiseau-Lyre 12BB 203/6.
Sumer is icumen in. cf BIS LP 3.
Synge we to this mery cumpane. cf BASF 25 22286-1.
Take heed of time, tune and ear. cf Turnabout TV 37079.
Tanzlied. cf Oryx BRL 1.
Tappster, dryngker. cf BASF 25 22286-1.
Te Deum. cf Telefunken TK 11569/1-2.
Testament d'Amelia. cf Amberlee ACL 501X.
There is a balm in Gilead (arr. Dawson). cf Wayne State University,
    unnumbered.
13th and 14th century: Benedicamus "Verbum Patris"; Laudemus virginem;
    Splendens sceptrigera; Stella splendens in monte. cf ALFONSO X,
    El Sabio: Las Cantigas de Santa Maria.
Thomas gemma Cantuariae, Motet. cf Nonesuch H 71292.
Though in the outward church. cf CBS 61599.
Time to pass. cf L'Oiseau-Lyre SOL 329.
Tiritomba. cf Columbia D3M 33448.
Tiritomba (arr. Breuer). cf Polydor 2489 519.
Tis eventide. cf CBS 61599.
Toda noite. cf Telefunken SAWT 9620/1.
Tomorrow shall be my dancing day. cf Argo ZRG 765.
Tres douce regard. cf Telefunken TK 11569/1-2.
La triquotée. cf L'Oiseau-Lyre 12BB 203/6.
Triumph. cf Nonesuch H 71276.
Tunes from The Birdy Fancyer's Delight (6). cf Argo ZRG 746.
The turtle-dove. cf Harmonia Mundi HM 593.
Twelfth night: Peg O'Ramsay. cf BBC REB 191.
Veni Redemptor gentium. cf Turnabout TV 34569.
Venid a sospirar. cf Telefunken TK 11569/1-2.
A virgin most pure (arr. Halter). cf University of Colorado, unnumbered.
Voulez-vous que je vous chante. cf Vanguard SRV 316.
Washington. cf Nonesuch H 71276.
The wind is in the north. cf Argo ZRG 785.

A winter's tale: Satyr's masque. cf BBC REB 191.
Where griping griefs. cf L'Oiseau-Lyre 12BB 203/6.
The white cockade. cf New England Conservatory NEC 111.
The white cockade. cf Michigan University SM 0002.
Wonderous love. cf New England Conservatory NEC 111.
The Worcester fagments. Academia Monteverdiana Chorus and Soloists;
     Denis Stevens. Nonesuch H 71308.
             ++HF 10-75 p90   ++MT 9-75 p801
             +HFN 8-75 p85   ++NR 8-75 p9
The world turned upside down. cf Michigan University SM 0002.
Yankee Doodle. cf Michigan University SM 0002.
Yankee Doodle. cf New England Conservatory NEC 111.
Ye simple souls who stray. cf CBS 61599.
Yehi Rotzon. cf Columbia D3M 33448.
Yir'e eineinu. cf Columbia D3 M 33448.

556, 622, 725, 733, 1555, 1753, 1809, 1811, 1822, 1824, 2073, 2105, 2109
Askenase, Stefan, piano 715, 716, 721, 762
Asper, Frank, organ 2524
Atherton, David, conductor 489, 1079, 1272, 1475, 1977
Atlantov, Vladimir, tenor 2301
Attfield, Helen, mezzo-soprano 2449
Atzmon, Moshe, conductor 1800, 2313
Auer, Leopold, violin 2609
Auger, Arleen, soprano 1219, 1394, 1586, 1621
Ausensi, Manuel, bass 1918
Austin, Michael, organ 1267
Austrian Quartet (Paul Roczek, Peter Klatt, violin; Jurgen Geise, viola;
    Wilfried Tachezi, violoncello) 910
Austrian Radio (Symphony) Chorus 1101, 1970
Austrian Radio (Symphony) Orchestra 1970
Avery, James, celesta 2557
Avery, Lawrence, tenor 1099
Ax, Emanuel, piano 757
Ayars, Anne, mezzo-soprano 1705
Ayldon, John, bass 2251, 2259
Ayo, Felix, violin 113, 2425

Baccholian Singers, London 1199
The Bach Choir 85
Bach Orchestra 85
Bachauer, Gina, piano 1804
Bâcher, Mihály, piano 249
Bachmann, Alberto, violin 2536
Bachmann, Hermann, baritone 497
Backhaus, Wilhelm, piano 1453, 1553
Bacquier, Gabriel, bass-baritone 1424, 1425, 1465, 1570
Baden Chamber Orchestra 116
Baden State Opera Chorus 2436
Baden State Opera Orchestra 2436
Badura-Skoda, Paul, conductor 264
Badura-Skoda, Paul, piano 264, 286, 1526
Bagger, Louis, harpsichord 1409
Bailey, Norman, bass-baritone 2446, 2447, 2449
Baillie, Isobel, soprano 2648
Baillie, Peter, tenor 541
Bainbridge, Elizabeth, mezzo-soprano (contralto) 436, 640, 1180, 2346,
    2358, 2369
Baker, Israel, violin 1265
Baker, Janet, mezzo-soprano (contralto) 466, 633, 640, 953, 1083, 1128,
    1193, 1377, 1390, 1572, 1968, 2031, 2038, 2344
Baker, Larry, trumpet, violoncello, percussion, piano 512
Baker, Margaret, mezzo-soprano 1478
Baker, Marilyn, soprano 639
Baldwin, Dalton, piano 1054, 1501, 2035, 2036, 2748, 2750, 2801
Baldwin, Marcia, mezzo-soprano (contralto) 500
Balinger Kantorei 156
Balsam, Artur, piano 1634, 1636
Balsar, Alfred, horn 1915
Baltimore Symphony Orchestra 25, 199
Baltsa, Agnes, contralto 298

Blegen, Judith, soprano 1139, 1389, 1396, 1708, 1782, 2546
Blickman, Loren, bassoon 2533
Bliss, Arthur, conductor 458, 2571
Blomstedt, Herbert, conductor 1698, 1914, 2468
Blumenthal, Felicja (Felicia), piano 27, 741, 751, 759, 1520, 1545,
        1905
Blustine, Allen, clarinet 513, 963, 2558, 2713
Bobesco, Lola, violin 2099, 2411
Bodenham, Peter, tenor 792
Bodes, Hannelore, soprano 2444
Bodurov, Lyubomir, tenor 1674
Boehm, Mary Louise, piano 237, 1217, 1489
Boehm, Pauline, piano 1489
Boettcher, Wolfgang, violoncello 375, 425, 429, 1630
Bogaard, Ed, saxophone 1272
Bogard, Carole, soprano 836
Bogin, Harold, piano 2727
Bognár, Margit, harp 1253
Boháčová, Marta, soprano 1378, 1688, 1712
Böhm, Karl, conductor 299, 362, 374, 376, 384, 396, 403, 601, 1163,
        1493, 1509, 1553, 1554, 1569, 1571, 1583, 1586, 1646, 1650,
        1658, 2039, 2044, 2045, 2048, 2050, 2171, 2216, 2430
Böhme, Kurt, bass 2458
Bojé, Harald, elektronium 2166
Boky, Colette, soprano 500
Bolcom, William, piano 1027, 1472, 2476, 2709
Boldt, Frina, piano 1183
Boldt, Kenwyn, piano 1183
Bolet, Jorge, piano 1339, 2679, 2779
Bolling, Claude, piano 533
Bologna Teatro Comunale Orchestra (Bologna Municipal Theater Orchestra)
        2694
Bolshoi Opera Orchestra 2301
Bolshoi Symphony Orchestra 203
Bolshoi Theatre Chorus 1671, 1675, 1772, 1775, 2302, 2669
Bolshoi Theatre Orchestra 28, 1276, 1420, 1671, 1675, 1771, 1772, 1775,
        1828, 2159, 2293, 2302, 2544, 2647, 2669
Bolshoi Theatre Trumpeters' Ensemble 1275
Bolshoi Theatre Wind Ensemble 2288
Bon, Maarten, piano 1964
Bonaldi, Clara, violin 981
Bonaventura, Anthony di, piano 813, 1961
Bonazzi, Elaine, mezzo-soprano 963, 1241
Bond, Dorothy, soprano 1705
Bonisolli, Franco, tenor 1778, 2392, 2509
Bonynge, Richard, conductor 1, 40, 437, 849, 863, 865, 1426, 1465,
        2172, 2294, 2389, 2567, 2582, 2585, 2586, 2692, 2694
Bordas, György, baritone 1346
Borden, David, synthesizers and electric piano 535
Borgonovo, Otello, baritone 21
Borisenko, Veronika, mezzo-soprano 1671
Borisova, Galina, contralto 2301
Bork, Hanneke van, soprano 1668
Borodin Quartet (Rostislav Dubinsky, Yaroslav Alexandrov, violin; Dmitri
        Shebalin, viola; Valentin Berlinsky, violoncello) 2120, 2647
Bosetti, Hermine, soprano 2618

Boskovsky Ensemble 1174
Boskovsky Quartet 11
Boskovsky, Willi, conductor 577, 1504, 1576, 1592, 1601, 2169, 2170,
        2173, 2174, 2177, 2178, 2581, 2663
Boskovsky, Willi, violin 2420
Bossert, James, organ 37
Bostok, Angela, soprano 1482
Boston Boys Choir 462
Boston Camerata 2712, 2828
Boston Experimental Electronic-Music Projects 696
Boston Musica Viva 674
Boston Pops Orchestra 2185
Boston Symphony Chamber Players 41, 2245
Boston Symphony Orchestra 210, 273, 462, 615, 847, 1447, 1843, 1847,
        1856, 1858, 1874, 1884, 2051, 2143, 2234, 2316, 2318, 2385
Bottion, Aldo, tenor 2381
Böttner, Karl-Heinz, string instruments 2707
Boukoff, Yuri, piano 1805
Boulanger, Nadia, conductor 995
Bouleyn, Kathryn, soprano 1212
Boulez, Pierre, conductor 491, 842, 1119, 1854, 1863, 1967, 1969, 2222,
        2236, 2445
Boult, Sir Adrian, conductor 458, 468, 597, 613, 633, 680, 892, 937,
        941, 950, 956, 959, 1058, 1090, 1194, 1477, 1532, 1904,
        2090, 2111, 2273, 2306, 2307, 2355, 2365, 2429, 2448, 2465,
        2646, 2800
Bour, Ernest, conductor 1299
Bourdin, Roger, baritone 1422
Bournemouth Municipal Choir 1196
Bournemouth Sinfonietta Orchestra 1135
Bournemouth Symphony Orchestra 386, 1196, 1696, 1751, 2114, 2141, 2364,
        2646
Boury, Fabienne, piano 844
Bowden, Pamela, soprano 1180
Bowen, Kenneth, tenor 638, 1967, 1985, 2497
Bowen, York, piano 542
Bowers-Broadbent, Christopher, organ 1298
Bowman, James, countertenor 638, 1484, 1485, 1794
Bozzi, Guido, conductor 1129
Bradbury, Colin, clarinet 676
Bradley, Desmond, violin 680, 849, 2534
Bradshaw, Richard, conductor 578
Braithwaite, Nicholas, conductor 767
Bramma, Henry, organ 938, 2755
Brand, Geoffrey, conductor 2776
Brandis, Thomas, violin 375, 425, 429, 1493, 1630
Bratislava Radio Orchestra 1952
Braun, Hans, bass-baritone 85, 291, 293
Braun, Helena, contralto 2439
Braun, Victor, baritone 2453
Bream, Julian, guitar 455, 1040, 1085, 2766
Bredenbeek, Hans, guitar 1964
Brehm, Alvin, bass 2533
Brejšek, Jindřich, conductor 1281
Brendel, Alfred, piano 243, 248, 339, 340, 341, 343, 551, 558, 1362,
        1534, 1539, 1755, 1759, 1982, 1991, 1992, 2019, 2021

Budapest (String) Quartet (Joseph Roisman, Alexander Schneider, violin; Boris Kroyt, viola; Mischa Schneider, violoncello) 311, 976, 1149, 2001
Budapest Symphony Orchestra 206
Bulgarian Quartet (Dimo Dimov, Alexandre Thomov, violin; Dimitre Tchilikov, viola; Dimitre Kosev, violoncello) 2640
Bulgarian Radio and Television Vocal Ensemble 1799
Bunger, Richard, piano 25
Bunke, Jerome, clarinet 482
Burge, David, piano 796
Burger, Bohuslav, violin 2734
Burgess Grayston, conductor 2831
Burmanje, Ton, guitar 1964
Burmeister, Annelies, mezzo-soprano (contralto) 186, 1433, 1598, 1600, 2426
Burmester, Willy, violin 2609
Burnett, Richard, piano 2014
Burrows, Norma, soprano 1200, 1406
Burrows, Stuart, tenor 366, 462, 1668, 2281, 2481, 2576
Busch, Adolf, violin 280
Busch, Fritz, conductor 280
Busch Quartet (Adolf Busch, Gosta Andreasson, violin; Karl Doktor, viola; Hermann Busch, violoncello) 313, 318, 581, 2003
Bussotti, Carlo, piano 626
Bustamente, Carmen, soprano 1733
Butt, Clara, contralto 2791
Byers, Alan, tenor 1782
Byers, Mervyn J., organ 15
Bylsma, Anner, violoncello 286, 1122, 2813, 2817
Byng, George, conductor 2256
Byzantine, Julian guitar 1313, 2534

Caballé, Montserrat, soprano 436, 438, 764, 1572, 1782, 1792, 1793, 2367, 2380
Cable, Margaret, mezzo-soprano (alto, contralto) 1036
Cahill, Teresa, soprano 2211
Cairns, David, conductor 2758
Calabrese, Franco, bass 1791
Calabrese, Nane, viola d'amore 2415
Calachian, Goar, mezzo-soprano 1687, 1928
Caldwell, James, viola da gamba 1410
Callas, Maria, soprano 1791
Calley, Ian, tenor 1426
Camarata, Salvador, conductor 2580
Camarata, Tutti, conductor 2687
Cambridge, King's College (Chapel) Choir 160, 522, 579, 593, 1084, 2354, 2643
Camden, Anthony, oboe 2806
Camden, Archie, bassoon 2806
Camerata Holmaie 2522
Camerata Singers 1854
Cameron, Alexander, violoncello 941
Caminada, Anita, mezzo-soprano 437
Campanella, Michele, piano 1354
Campbell, James, clarinet 2561
Campbell, Sidney, conductor 637, 2500

Campoli, Alfredo, violin 569
Canali, Anna Maria, mezzo-soprano 1787
Candela, Miguel, violin 2536
Canino, Bruno, harpsichord 116
Canino, Bruno, piano 2775
Cannetti, Linda, soprano 1789
Cantelli, Guido, conductor 655, 2385, 2826
Cantelo, April, soprano 639, 1796, 1797, 2574
Canterbury Cathedral Choir 637
Cantilena Chamber Players 306, 2638
Cantores in Ecclesia 1724, 2268, 2720
Cao, Pierre, conductor 8, 698, 1181, 1361, 1429, 1949, 2428
Capecchi, Renato, baritone 1325, 1733, 1916, 1917
Capella Cordina 868, 869, 1256, 1314, 1487, 1703
Capitol Symphony Orchestra 1074
Capon, Frédéric, piano 2469
Cappella Coloniensis 2604
Cappone, Giusto, viola 1493, 2217
Cappuccilli, Piero, baritone 437, 2367, 2372, 2380, 2391
Carbonari, Virgilio, tenor 2377
Cardiff Festival Ensemble (James Barton, violin; George Isaac, violon-
          cello; Stephen Broadbent, viola; Martin Jones, piano) 771,
          1888
Carelli, Gabor, tenor 802, 1603
Carewe, John, conductor 442
Carey, Thomas, bass-baritone 2345
Carfagna, Carlo, guitar 2762
Carlin, Mario, tenor 1784
Carlos, Walter, Moog synthesizer 2541
Carlyle, Joan, soprano 2346
Carmeli, Boris, bass 1714
Carminata, Maria Rosa, soprano 498
Caron, Michel, tenor 1706
Carracilly, Yvon, violin 1882
Carreras, José, tenor 1425, 2378
Carturan, Gabriella, soprano (mezzo-soprano) 1917, 2379
Caruso, Enrico, tenor 1789, 2619, 2791, 2797, 2798
Casa, Lisa della, soprano 1097, 1603, 2576
Casadesus, Gaby, piano 692
Casadesus, Jean, piano 837
Casadesus, Robert, piano 1547, 1551, 2531
Casals, Pablo, conductor 400, 2531
Casals, Pablo, violoncello 422, 424, 626, 2523, 2628
Casapietra, Celestina, soprano 1598, 1600
Casas, Esther, soprano 2367
Case, John Carol, baritone 937
Caselli, Dario, bass 1791, 2383, 2393
Cash, Jessica, soprano 2187
Cassilly, Richard, tenor 1969, 2344
Cassinelli, Ricardo, tenor 2378
Cassolas, Constantine, tenor 16, 1037
Castel, Nico, tenor 1782
Casula, Maria, mezzo-soprano 863
Cathedral of St. John the Divine Choir, New York City 1079
Cattenhead, Harrison, tenor 1029
Cava, Carlo, bass 1917

Grisi, Maria, soprano 2791
Grist, Reri, soprano 1571, 1586, 2372, 2373
Grobe, Donald, tenor 1405, 1594
Grodberg, Harry, organ 1275
Grofé, Ferde, conductor 1074
Groschel, Ernest, pianoforte 42
Gross, Robert, violin 803
Die grosse Garde-Regimentsmusik 2511
Grossman, Ferdinand, conductor 1596
Group for Contemporary Music, Columbia University 450, 794
Groves, Charles, conductor 386, 514, 809, 856, 2134, 2144
Grubb, Francis, baritone 2718
Gruber, Emanuel, violoncello 1479
Gruber, Perry, tenor 1785
Gruenberg, Erich, violin 2586
Grumiaux, Arthur, violin 114, 284, 352, 654, 1323, 1564, 1607, 2416
Grumiaux Trio 1605, 1619
Grumliková, Nora, violin 1417
Grümmer, Elisabeth, soprano 2458
Grünfarb, Josef, violin 1737
Guadagno, Anton, conductor 2691
Guarneri Quartet (Arnold Steinhardt, John Dalley, violin; Michael Tree,
        viola; David Soyer, violoncello) 830, 904, 905, 977, 1612,
        1613, 1615, 2010
Guarrera, Frank, baritone 1326, 1419
Gueden, Hilde, soprano 1602
Gueleva, Alexei, bass 1671
Gueneux, Georges, flute 1933
Guest, George, conductor 880, 1453, 1794, 1985
Gui, Henri, baritone 1706
Guige, Paul, bass-baritone 498
Guigui, Efrain, conductor 674
Guildford Philharmonic Choir 990
Guildford Philharmonic Orchestra 990
Guilet, Daniel, violin 2792
Guiraudon, Julia, soprano 2791
Gulbenkian Foundation Chamber Orchestra 21
Gulbenkian Foundation Choir 1790
Gulbenkian Foundation Orchestra 1790
Gulda, Friedrich, piano 1535
Gumz, Hubert, organ 1486
Gunson, Ameral, contralto 1919
Gunter, Kurt, violin 185
Gunther-Arndt Choir 2462
Guschlbauer, Theodor, conductor 1135, 2781, 2804
Gutorovich, Nicolai, tenor 2230
Gutter, Robert, conductor 1012
Györ Girls' Choir (Chorus) 206, 1175
Györ Philharmonic Orchestra 1175, 1253, 1514
Gyuzelev, Nikola, bass 1799

Haas, Monique, piano 827, 834, 843, 1758
Haas, Werner, piano 1024, 1850, 1871
Hachlicki, Kurt, tenor 1586
Haebler, Ingrid, fortepiano 50, 51, 56, 1517
Haebler, Ingrid, piano 1642, 1643, 1644

Helmerson, Frans, violoncello 175
Helps, Robert, piano 2112
Helsinki Chamber Orchestra 1840
Helsinki Philharmonic Orchestra 1701
Helsinki (Radio) Symphony Orchestra 2139
Heltay, Laszlo, conductor 634
Hemberg, Eskil, conductor 1737
Hemmings, David, soprano (treble) 639, 641
Hemsley, Thomas, bass-baritone 2345
Henck, Herbert 2166
Henzi, Hans Werner, conductor 1176, 1177
Herbert, Ralph, bass 449
Herbig, Günther, conductor 1503
Herbillon, Jacques, baritone 982, 983, 1848, 1882
Herincx, Raimund, baritone 2345, 2358
Herman, Allen, snare drum 2707
Hermanjat, Jean-Claude, flute 119
Hermann, Dagmar, mezzo-soprano (contralto) 85, 1927
Hermann, Roland, bass-baritone 1380, 1969
Herrick, Christopher, organ 1427
Herrmann, Bernard, conductor 1179, 1180, 1474, 1831, 1931, 2103, 2115,
         2575
Hersh, Paul, viola 165
Hesford, Bryan, organ 91
Hess, Dame Myra, piano 626, 1525
Hesse, Ruth, mezzo-soprano 2440
Hessenbruch, Friedhelm, bass 1903
Hibbard, William, toy piano 2557
Hibbard, William, viola 2557
Hickman, David, trumpet 36, 73
Highgate School Choir 1406
Hildenbrand, Siegfried, organ 2824
Hill, Eric, guitar 1042
Hill, Jenny, soprano (contralto) 640, 2037
Hill, Martyn, tenor 1208, 2512
Hillebrand, Nikolaus, bass 2444
Hindar Quartet (Leif Jorgensen, Trond Oyen, violin; Johs Hindar, viola;
         Levi Hindar, violoncello) 1242
Hintermeyer, Walter, violin 2420
Hirte, Klaus, bass 2444
Hlinka, Jan, viola 2205
Hlobilová, Eva, soprano 1247
Hoban, John, conductor 899
Hobson, Ann, harp 843
Hock, Bertalan, oboe 1137
Hodgson, Alfreda, contralto (mezzo-soprano) 1231, 1406, 1482
Hoelscher, Ludwig, violoncello 1071
Hofer, Pierre, violin 1882
Hoffman, Herbert, harpsichord 116
Hoffman, Stanley, violin 512
Hoffmeister, Frank, tenor 2543
Hofmann, Josef, piano 324, 2678
Hofmann, Wolfgang, conductor 686
Hogwood, Christopher, conductor 31
Hogwood, Christopher, harpsichord 31, 32, 101, 2495, 2654
Hokanson, Leonard, piano 410, 2068

Magdalen College Choir, Oxford 2160
Maggio Musicale Fiorentino Chorus 862
Maggio Musicale Fiorentino Orchestra 862
Magiera, Leone, conductor 1778, 2509, 2585
Magin, Milosz, piano 709
Magnin, Alexandre, flute 1933
Maguire, Hugh, violin 1663
Maier, Franzjosef, violin 45, 286, 1134, 1623, 2005
Maigat, Pierre-Yves le, bass-baritone 1366, 1833
Maisonneuve, Claude, oboe 70
Maixerova, Martina, piano 154
Majeski, Dnaiel, violin 1021
Majkut, Erich, tenor 300, 1602
Makelian, Vartan, bass 2288
Makhov, Vladimir, tenor 2288
Malagù, Stefania, mezzo-soprano 865, 2381, 2388, 1918
Malas, Spiro, bass 863
Malcolm, George, conductor 2399
Malcolm, George, harpsichord 49, 63
Malcuzynski, Witold, piano 720
Malfitano, Catherine, soprano 2703
Malfitano, Joseph, violin 2703
Malgoire, Jean-Claude, conductor 1092, 1366, 1368, 1833, 1835, 1836
Malko, Nicolai, conductor 2661
Malmborg, Gunilla af, soprano 20
Malta, Alexander, bass 1484, 1944
Mamlock, Theodore, violin 2562
Mancini, Caterina, soprano 2395
Mandalka, Rudolf, violoncello 2005
Mandel, Alan, piano 1234
Manen, Juan, violin 2536
Manfrini, Luigi, bass 2368
Mangin, Noël, bass 1587, 1782
Manhattan Opera Children's Chorus 500
Mann, Alfred, recorder 1081
Mann, Tor, conductor 2163
Mann, Wallace, flute 1463
Mann, Werner, bass 1970
Mannberg, Karl-Ove, violin 1258, 1737
Mannes, Leopold, piano 2523
Mannheim Instrumental Soloists 2730
Manning, Jane, soprano (mezzo-soprano) 489, 1969
Manson, John, harp 811
Mansurov, Fuat, conductor 1259
Mantovani, Dino, bass 2377
Manzone, Jacques-Francis, violin 1129
Mar, Norman del, conductor 196, 456, 945, 1010, 2291
Marcellus, Robert, clarinet 1662
Marchand, Andreas, piano 2557
Marcheva, S., soprano 2696
Marciano, Rosario, piano 1993
Marcoulescou, Yolanda, soprano 961, 1929
Marev, Krustiu, conductor 2639
Marfeuil, Jacques, baritone 1706
Margaritov, Athanas, conductor 1674
Margittay, Sándor, organ 1357

Trobaugh, Thomas, harpsichord 2700
Troyanos, Tatiana, mezzo-soprano (contralto) 1594, 2543
Trythall, Richard, piano 807
Trzcka, Ludmila, piano 1712
Ts'ong, Fou, piano 1555
Tucci, Gabrielle, soprano 1325
Tucker, Richard, tenor 1324, 1326, 1419, 2547
Tuckwell, Barry, horn 196, 328, 622, 1079, 1132, 1511, 1669, 1990,
        2168
Tulder, Louis van, tenor 158
Tüller, Niklaus, bass 138
Tullio, Louise di, flute 1297, 1720
Tullio, Virginia di, piano 1297, 1720
Tuloisela, Matti, bass-baritone 1701
Tunnell, Charles, violoncello 52, 676
Tureck, Rosalyn, piano 2679
Turetschek, Gerhard, oboe 1509
Turkovic, Milan, bassoon 790
Turner, Bruno, conductor 870, 1014, 1311, 2510
Turner, Claramae, mezzo-soprano 2385
Turnovský, Martin, conductor 265, 1764
Turpinsky, Béla, tenor 1356
Tursi, Francis, viola 1618
Tusa, Erzsebet, piano 224
Tyler, James, cittern 2814
Tyler, James, lute 2719

Uberschaef, Siegbert, viola 429
Ukena, Paul, bass-baritone 2390
Ulbrich, Markus, organ 1432
Das Ulsamer-Collegium 2595, 2598, 2601
Ulsamer, Josef, viola da gamba 534
Unger, Gerhard, tenor 292, 293, 781
Unger, Thomas, conductor 208
University of Chicago Collegium Musicum 2753, 2754
University of Illinois Chamber Players (Ensemble) 674
University of Illinois Contemporary Chamber Players (Ensemble) 1363
University of Kentucky Collegium Musicum 2753, 2754
University of Maryland Chorus 802
University of Maryland Quartet 964
University of Maryland Trio 964
University of Pennsylvania Contemporary Chamber Players 1363
Unno, Yoshio, conductor 2424
Unno, Yoshio, violin 2424
Upper, Henry, piano 485
Urban, Bela, violin 1277
Urban, Virginia, piano 1277
Urbanec, Rudolf, conductor 1011, 1280, 2785, 2811
Uridge, Michael, conductor 2729
Ursuleac, Viorica, soprano 2198, 2201, 2210, 2431
U.S.S.R. Ministry of Defense Symphonic Band 1763, 2647
U.S.S.R. Russian Chorus 2647, 2657
U.S.S.R. State Philharmonic Chorus 207
U.S.S.R. State Philharmonic Orchestra 207
U.S.S.R. Symphony Orchestra 1275, 1467, 1566, 1686, 1819, 2121, 2286,
        2647

Warfield, Sandra, mezzo-soprano 1603
Warner, W. Ring, double bass 795
Warren, Leonard, baritone 2377, 2390, 2394, 2574
Warsaw (National) Philharmonic Chorus 1729
Warsaw (National) Philharmonic Orchestra 323, 731, 1463, 1536, 1729,
          1758, 2282
Washington Camerata Chorus 2476
Washington National Symphony Orchestra 802, 1019, 2283, 2285
Washington, Paolo, bass-baritone 2377
Watanabe, Akeo, conductor 769
Watkins, David, harp 2777
Watson, James, cornet 2776
Watson, Lilian, soprano 1482
Watts, André, piano 245, 303, 1004, 1978, 2272, 2528
Watts, Helen, contralto (alto) 76, 136, 139, 937, 1100, 1103, 1111,
          1385, 1394, 1985, 2358, 2574
Waverley Consort 16, 2836
Wayne State University Men's Glee Club 2840
Weaver, James, harpsichord 779, 1410
Weaver, James, organ 779
Weaver, Thomas, violin 1486
Webb, Bailus, baritone 1212
Weber, Hans-Jorn, tenor 1586
Weber, Heinrich, tenor 1944
Weber, Ludwig, bass 2210, 2432
Webner, Tibor, piano 1359
Weddle, Robert, organ 2577, 2838
Wegmann, Hans, singer 1714
Wehofschitz, Kurt, tenor 292
Wehrung, Herrad, soprano 595
Weidemann, Friedrich, baritone 2736
Weikl, Bernd, baritone 781, 1740, 1944, 2281, 2444, 2459
Weiner, Susan, oboe 13
Weingartner, Felix, conductor 1443, 2649
Weir, Gillian, organ 684
Weisberg, Arthur, conductor 450, 690, 1254, 1471, 2474, 2589, 2711
Weisbrod, Annette, piano 1204
Weissenberg, Alexis, piano 189, 269, 549
Weissman, Dick, guitar and banjo 2537
Weldon, George, conductor 1072, 1110
Weller Quartet (Walter Weller, Alfred Staar, violin; Helmut Weiss,
          viola; Robert Scheiwein, violoncello) 1998
Weller, Walter, conductor 1773, 1820
Welles, Michael, treble 637
Welsby, Norman, baritone 2449
Wennberg, Siv, soprano 1065
Wentworth, Jean, piano 2589
Wentworth, Kenneth, piano 2589
Wentworth, Richard, bass 2390
Wenzinger, August, conductor 1105, 1107
Wenzinger, August, viola da gamba 185
Werba, Erik, piano 2033, 2058
Werner, Fritz, conductor 79
Wernick, Richard, conductor 1363
Wesstein, Frank, guitar 1964
West, Charles, clarinet 2557